Fundamentals of Many-body Physics

Wolfgang Nolting

Fundamentals of Many-body Physics

Principles and Methods

Translated by William D. Brewer

Springer

Prof. Dr. Wolfgang Nolting
Humboldt-Universität Berlin
Institut für Physik
Newstonstr. 15
12489 Berlin
Germany
wolfgang.nolting@physik.hu-berlin.de

Translator
Prof. William D. Brewer, PhD
FU Berlin
FB Physik
Inst. f. Experimentalphysik
Arnimallee 14
14195 Berlin
Germany
william.brewer@fu-berlin.de

ISBN 978-3-642-09106-3 e-ISBN 978-3-540-71931-1

DOI 10.1007/978-3-540-71931-1

Cover design: eStudio Calamar S.L.

Printed on acid-free paper

9 8 7 6 5 4 3 2 1

springer.com

Preface

The goal of the present course on "Fundamentals of Theoretical Physics" is to be a direct accompaniment to the lower-division study of physics, and it aims at providing the physical tools in the most straightforward and compact form as needed by the students in order to master theoretically more complex topics and problems in advanced studies and in research. The presentation is thus intentionally designed to be sufficiently detailed and self-contained – sometimes, admittedly, at the cost of a certain elegance – to permit individual study without reference to the secondary literature. This volume deals with the quantum theory of many-body systems. Building upon a basic knowledge of quantum mechanics and of statistical physics, modern techniques for the description of interacting many-particle systems are developed and applied to various real problems, mainly from the area of solid-state physics. A thorough revision should guarantee that the reader can access the relevant research literature without experiencing major problems in terms of the concepts and vocabulary, techniques and deductive methods found there.

The world which surrounds us consists of very many particles interacting with one another, and their description requires in principle the solution of a corresponding number of coupled quantum-mechanical equations of motion (Schrödinger equations), which, however, is possible only in exceptional cases in a mathematically strict sense. The concepts of elementary quantum mechanics and quantum statistics are therefore not directly applicable in the form in which we have thus far encountered them. They require an extension and restructuring, which is termed "many-body theory".

First of all, we have to look for possibilities for formulating real many-body problems in a mathematically correct but still manageable way. If the systems considered are composed of distinguishable particles, their description can be obtained directly from the general postulates of quantum mechanics. More interesting, however, are systems of identical particles, whose N-particle wavefunctions must fulfil quite special symmetry requirements. Working directly with the required (anti-)symmetrised wavefunctions proves to be extraordinarily tedious. A first perceptible simplification is provided in this connection by the formalism of second quantisation. It allows a quite elegant description, but of course does not provide an actual solution to the problem. The student who has been confronted in lower-division courses with problems which as a rule can be treated with mathematical rigour has to become accustomed to the idea that realistic many-body problems can practically never be treated exactly. In order to nevertheless fulfil the central function of a theoretician, i.e. the description and explanation of experiments, some concessions must be made. This includes, as a first step, the construction of a theoretical model which can be understood as a caricature of the real world, in which nonessential details are suppressed and only the essence of the problem is emphasized. Finding such a theoretical model must be considered to be a nontrivial challenge for theoreticians. Chapter 2 therefore treats the formulation and justification of important standard models of theoretical physics in detail. Their presentation is carried out consistently using the formalism of second quantisation from Chap. 1.

Unfortunately, the real situation can seldom be caricatured in such a way that the resulting model is on the one hand still sufficiently realistic, and on the other can be treated with mathematical rigour. Thus, one usually has to accept additional approximations in order to find solutions. A powerful technique in this connection has proven to be the Green's function method, with its concept of quasi-particles. The abstract theory is discussed in Chap. 3 and then applied to numerous concrete problems in Chap. 4. Diagrammatic methods of solution are worked out in Chaps. 5 and 6. They should be included nowadays within the indispensable repertoire of every theoretician. A number of exercises (together with their explicit solutions) are also included in this volume and are in particular designed to help the student to acquire a facility for working with the formalism and applying it to concrete topics. The solutions given, however, should not tempt the reader to forbear making a serious effort to solve the problems independently. At the end of each major chapter, questions are included, which can be useful to test the knowledge gained by the reader and in preparing for examinations.

This book is the result of diverse special-topics lecture courses on many-body theory which I have given at the universities of Würzburg, Münster, Osnabrück, and Berlin (Germany), Warangal (India), Valladolid (Spain), Irbid (Jordan), and Harbin (China). I am very grateful to the students of those courses for their constructive criticism. It is quite clear to me that the material in this volume with certainty no longer belongs to lower-division physics. However, I also believe that it is indispensable for making the transition to independent research as a theoretician. Since the available textbook literature on the subject of many-body theory as a rule presupposes advanced knowledge and substantial experience on the part of the reader, the present book might – hopefully – be very useful for the "beginner". I am very grateful to the Springer-Verlag for their concurring assessment as well as for their professional cooperation.

Berlin, August 2008 *Wolfgang Nolting*

Contents

Second Quantisation 1

The physical world consists of interacting many-body systems. Their exact description would require the solution of the corresponding many-body Schrödinger equations, which however is as a rule not feasible. The goal of theoretical physics therefore consists of developing concepts with whose aid a many-body problem can be approximately solved in a *physically reasonable* manner.

The formalism of *second quantisation* permits a considerable simplification in the description of many-body systems, but in the end, it involves merely a reformulation of the original Schrödinger equation, and thus does not represent a concept for its solution. The second quantisation is characterised by the introduction of so-called

creation and annihilation operators,

which render the tedious construction of N-particle wavefunctions as symmetrised or antisymmetrised products of single-particle wavefunctions unnecessary. The overall statistical properties are then included in

fundamental commutation relations

of these *creation and annihilation operators*. The interaction processes which take place in many-body systems are expressed in terms of the *creation* and *annihilation* of certain particles.

If the particles in an N-body system are **distinguishable** in terms of some physical property, then the description can be obtained directly from the general postulates of quantum mechanics. In the case of indistinguishable particles, a principle comes into play which introduces special symmetry requirements for the vectors in the Hilbert space of the N-particle systems.

If the particles are distinguishable, then they can be *enumerated* in some fashion:

$\mathcal{H}_1^{(i)}$: The Hilbert space of i-th particle.

Let $\{\widehat{\varphi}^{(i)}\}$ be a complete set of commuting observables in $\mathcal{H}_1^{(i)}$; then the (mutual) eigenstates $\left|\varphi_\alpha^{(i)}\right\rangle$ form a

basis of $\mathcal{H}_1^{(i)}$,

W. Nolting, *Fundamentals of Many-body Physics*,
DOI 10.1007/978-3-540-71931-1_1, © Springer-Verlag Berlin Heidelberg 2009

which we may assume to be orthonormalised:

$$\left\langle \varphi_\alpha^{(i)} \middle| \varphi_\beta^{(i)} \right\rangle = \delta_{\alpha\beta} \quad (\text{or } \delta(\alpha - \beta)) \,.$$

$\mathcal{H}_N$: The Hilbert space of the N-particle system

$$\mathcal{H}_N = \mathcal{H}_1^{(1)} \otimes \mathcal{H}_1^{(2)} \otimes \cdots \otimes \mathcal{H}_1^{(N)} \,.$$

As a **basis of** $\mathcal{H}_N$, we employ the direct products of the corresponding single-particle basis states:

$$\begin{aligned} |\varphi_N\rangle &= \left| \varphi_{\alpha_1}^{(1)} \varphi_{\alpha_2}^{(2)} \cdots \varphi_{\alpha_N}^{(N)} \right\rangle = \\ &= \left| \varphi_{\alpha_1}^{(1)} \right\rangle \left| \varphi_{\alpha_2}^{(2)} \right\rangle \cdots \left| \varphi_{\alpha_N}^{(N)} \right\rangle \,. \end{aligned} \tag{1.1}$$

A general N-particle state $|\psi_N\rangle$ can be expanded in terms of the $|\varphi_N\rangle$:

$$|\psi_N\rangle = \sum_{\alpha_1, \ldots, \alpha_N} C(\alpha_1, \ldots, \alpha_N) \left| \varphi_{\alpha_1}^{(1)} \varphi_{\alpha_2}^{(2)} \cdots \varphi_{\alpha_N}^{(N)} \right\rangle \,. \tag{1.2}$$

The statistical interpretation of such an N-particle state is identical with that of the single-particle states. Thus, $|C(\alpha_1, \ldots, \alpha_N)|^2$ is the probability with which a measurement of the observable $\widehat{\varphi}$ in the state $|\psi_N\rangle$ will yield the eigenvalue of $\left| \varphi_{\alpha_1}^{(1)} \cdots \varphi_{\alpha_N}^{(N)} \right\rangle$. The dynamics of the N-particle system derives from a formally unmodified Schrödinger equation:

$$\mathrm{i}\hbar \left| \dot{\psi}_N \right\rangle = \widehat{H} \left| \psi_N \right\rangle \,. \tag{1.3}$$

$\widehat{H}$ is the Hamiltonian of the N-particle system.

A quantum-mechanical treatment of many-body systems with distinguishable particles presents the same difficulties as in classical physics, simply due to its greater complexity as compared to the single-particle problem. There are, however, no additional, typically quantum-mechanical complications. This is no longer the case if we consider systems of indistinguishable particles.

1.1 Identical Particles

Definition 1.1.1: *Identical particles*

Particles which behave in exactly the same way under similar physical conditions and therefore cannot be distinguished by any objective measurement.

In classical mechanics, with well-known initial conditions, the *state* of a particle is determined for all times by Hamilton's equations of motion. The particle is always identifiable, since its orbit can be calculated. In this sense, even identical particles (with the same masses, charges, spatial extensions etc.) can be distinguished in classical mechanics.

Within the range of validity of quantum mechanics, in contrast, the fundamental

principle of indistinguishability

holds. This principle states that mutually-interacting identical particles are in principle not distinguishable. Its origin lies in the fact that as a result of the uncertainty relation, no sharply defined particle orbits exist. Instead, the particle must be treated as a *spreading* wave packet. The occupation probabilities of mutually-interacting identical particles overlap, which makes their identification impossible.

Every physical problem which requires the observation of single particles is physically meaningless for systems of identical particles! It now becomes a problem that for computational reasons, an enumeration of the particles is unavoidable. This enumeration must however be carried out in such a fashion that physically relevant statements are invariant with respect to changes in the enumeration scheme. *Physically relevant* are exclusively the measurable quantities of a physical system. These are not the *bare* operators or states, but rather the expectation values of observables or scalar products of states. They must not change if the numbering of two particles in the N-particle state is exchanged. Otherwise, there would be a measurement procedure which would distinguish between the two particles. One can therefore consider the following relation as

the defining equation for systems of identical particles:

$$\left\langle \varphi_{\alpha_1}^{(1)} \cdots \varphi_{\alpha_i}^{(i)} \cdots \varphi_{\alpha_j}^{(j)} \cdots \varphi_{\alpha_N}^{(N)} \right| \widehat{A} \left| \varphi_{\alpha_1}^{(1)} \cdots \varphi_{\alpha_i}^{(i)} \cdots \varphi_{\alpha_j}^{(j)} \cdots \varphi_{\alpha_N}^{(N)} \right\rangle \stackrel{!}{=}$$
$$\stackrel{!}{=} \left\langle \varphi_{\alpha_1}^{(1)} \cdots \varphi_{\alpha_i}^{(j)} \cdots \varphi_{\alpha_j}^{(i)} \cdots \varphi_{\alpha_N}^{(N)} \right| \widehat{A} \left| \varphi_{\alpha_1}^{(1)} \cdots \varphi_{\alpha_i}^{(j)} \cdots \varphi_{\alpha_j}^{(i)} \cdots \varphi_{\alpha_N}^{(N)} \right\rangle . \tag{1.4}$$

This holds for an arbitrary observable $\widehat{A}$ and arbitrary N-particle states. From Eq. (1.4), a whole series of characteristic properties of both the operators and of the states evolves. Equation (1.4) naturally holds for all pairs (i, j) and not only for exchange of two particles, but rather for arbitrary permutations of the particle indices. Every permutation can however be written as the product of transpositions of the type (1.4).

Definition 1.1.2: *Permutation operator* $\mathcal{P}$

$$\mathcal{P} \left| \varphi_{\alpha_1}^{(1)} \varphi_{\alpha_2}^{(2)} \cdots \varphi_{\alpha_N}^{(N)} \right\rangle = \left| \varphi_{\alpha_1}^{(i_1)} \varphi_{\alpha_2}^{(i_2)} \cdots \varphi_{\alpha_N}^{(i_N)} \right\rangle . \tag{1.5}$$

$\mathcal{P}$ is assumed here to act upon the particle indices; of course state indices α_i can also be employed. $(i_1, i_2, \cdots, i_N)$ is the permuted N-tuple $(1, 2, \ldots, N)$.

Definition 1.1.3: *Transposition operator P_{ij}*

$$P_{ij}\left|\cdots\varphi_{\alpha_i}^{(i)}\cdots\varphi_{\alpha_j}^{(j)}\cdots\right\rangle=\left|\cdots\varphi_{\alpha_i}^{(j)}\cdots\varphi_{\alpha_j}^{(i)}\cdots\right\rangle . \tag{1.6}$$

We wish to discuss some of the properties of the transposition operator. Applying P_{ij} two times to an N-particle state obviously leads back to the initial state. This means that:

$$P_{ij}^2=\mathbf{1} \iff P_{ij}=P_{ij}^{-1} . \tag{1.7}$$

Equation (1.4) can now be written in the following form:

$$\langle\varphi_N|\widehat{A}|\varphi_N\rangle\stackrel{!}{=}\langle P_{ij}\varphi_N|\widehat{A}|P_{ij}\varphi_N\rangle=\left\langle\varphi_N\middle|P_{ij}^{+}\widehat{A}P_{ij}\middle|\varphi_N\right\rangle .$$

This holds for arbitrary N-particle states of the $\mathcal{H}_N$; furthermore, also for arbitrary matrix elements of the type $\langle\varphi_N|\widehat{A}|\psi_N\rangle$, since these can be brought into the above form by the decomposition

$$\begin{aligned}\langle\varphi_N|\widehat{A}|\psi_N\rangle=\frac{1}{4}\Big\{&\langle\varphi_N+\psi_N|\widehat{A}|\varphi_N+\psi_N\rangle\\ &-\langle\varphi_N-\psi_N|\widehat{A}|\varphi_N-\psi_N\rangle\\ &+\mathrm{i}\,\langle\varphi_N-\mathrm{i}\psi_N|\widehat{A}|\varphi_N-\mathrm{i}\psi_N\rangle\\ &-\mathrm{i}\langle\varphi_N+\mathrm{i}\psi_N|\widehat{A}|\varphi_N+\mathrm{i}\psi_N\rangle\Big\} .\end{aligned}$$

This leads us to the **operator identity:**

$$\widehat{A}=P_{ij}^{+}\widehat{A}P_{ij} \quad \forall(i,j) . \tag{1.8}$$

A necessary and nearly trivial precondition for the observables of a system of identical particles is therefore that they depend explicitly on the coordinates of **all** N particles.

If we choose in (1.8) in particular $\widehat{A}=\mathbf{1}$, it follows that:

$$\mathbf{1}=P_{ij}^{+}P_{ij}\Rightarrow P_{ij}=P_{ij}^{+}P_{ij}^2=P_{ij}^{+} .$$

The transposition operator P_{ij} is thus Hermitian and unitary in the space $\mathcal{H}_N$ of identical particles:

$$P_{ij}=P_{ij}^{+}=P_{ij}^{-1} . \tag{1.9}$$

From (1.8), it also follows that:

$$P_{ij}\widehat{A} = P_{ij}P_{ij}^{+}\widehat{A}P_{ij} = \widehat{A}P_{ij} \; .$$

All the observables of the N-particle system commute with P_{ij}:

$$\left[P_{ij}, \widehat{A}\right]_{-} = P_{ij}\widehat{A} - \widehat{A}P_{ij} \equiv 0 \; . \tag{1.10}$$

This is in particular true of the Hamiltonian $\widehat{H}$ of the system:

$$\left[P_{ij}, \widehat{H}\right]_{-} = 0 \; . \tag{1.11}$$

According to the principle of the indistinguishability of identical particles, the N-particle state $|\varphi_N\rangle$ can be changed only in terms of a non-essential phase factor through the action of P_{ij}; in particular, $|\varphi_N\rangle$ must be an eigenstate of P_{ij}:

$$P_{ij}\left|\cdots \varphi_{\alpha_i}^{(i)} \cdots \varphi_{\alpha_j}^{(j)} \cdots\right\rangle = \left|\cdots \varphi_{\alpha_i}^{(j)} \cdots \varphi_{\alpha_j}^{(i)} \cdots\right\rangle \stackrel{!}{=}$$
$$\stackrel{!}{=} \lambda \left|\cdots \varphi_{\alpha_i}^{(i)} \cdots \varphi_{\alpha_j}^{(j)} \cdots\right\rangle . \tag{1.12}$$

Owing to (1.7), only the real eigenvalues

$$\lambda = \pm 1 \tag{1.13}$$

need be considered, which are independent of the particular pair (i, j). This means that:

> the states of a system of identical particles are either **symmetric** or **antisymmetric** under exchange of a pair of particles!

$\mathcal{H}_N^{(+)}$: the Hilbert space of the symmetric states $\left|\psi_N^{(+)}\right\rangle$:

$$P_{ij}\left|\psi_N^{(+)}\right\rangle = \left|\psi_N^{(+)}\right\rangle \quad \forall (i, j) \; . \tag{1.14}$$

$\mathcal{H}_N^{(-)}$: the Hilbert space of the antisymmetric states $\left|\psi_N^{(-)}\right\rangle$:

$$P_{ij}\left|\psi_N^{(-)}\right\rangle = -\left|\psi_N^{(-)}\right\rangle \quad \forall (i, j) \; . \tag{1.15}$$

For the time evolution operator

$$U(t, t_0) = \exp\left(-\frac{i}{\hbar}H\,(t - t_0)\right), \qquad H \neq H(t) \, , \tag{1.16}$$

we find as a result of (1.11):

$$\left[P_{ij}, U\right]_{-} = 0 \,. \tag{1.17}$$

The states of a system of N identical particles thus retain their symmetry character for all times.

How can such (anti-)symmetrised N-particle states be constructed? A non-symmetrised N-particle state of the type (1.1) can serve as our starting point. The following symmetrisation operator is then applied to it:

$$\widehat{S}_{\varepsilon} = \sum_{\mathcal{P}} \varepsilon^{p} \mathcal{P} \,, \tag{1.18}$$

$\varepsilon = \pm$; p is the number of transpositions which construct $\mathcal{P}$. The sum runs over all the possible permutation operators $\mathcal{P}$ for the N-tuple $(1, 2, \ldots, N)$. If a $\mathcal{P}$ in the sum is multiplied by a transposition P_{ij}, then naturally a different permutation $\mathcal{P}'$, which also occurs in the sum, is obtained, with $p' = p \pm 1$. The following rearrangement is therefore plausible:

$$P_{ij} \widehat{S}_{\varepsilon} = \sum_{\mathcal{P}} \varepsilon^{p} P_{ij} \mathcal{P} = \sum_{\mathcal{P}} \varepsilon^{p} \mathcal{P}' = \varepsilon \sum_{\mathcal{P}'} \varepsilon^{p'} \mathcal{P}' \,.$$

This means that:

$$P_{ij} \widehat{S}_{\varepsilon} = \varepsilon \widehat{S}_{\varepsilon} \,. \tag{1.19}$$

The prescription

$$\left|\psi_N^{(\varepsilon)}\right\rangle = \widehat{S}_{\varepsilon} \left|\psi_{\alpha_1}^{(1)} \psi_{\alpha_2}^{(2)} \cdots \psi_{\alpha_N}^{(N)}\right\rangle \tag{1.20}$$

thus leads to a symmetrised (ε=+) or to an antisymmetrised (ε=−) N-particle state, for which (1.14) or (1.15) holds.

For a generalised permutation $\mathcal{P}$, it then clearly holds that:

$$\mathcal{P} \widehat{S}_{\varepsilon} = \varepsilon^{p} \widehat{S}_{\varepsilon} \Longleftrightarrow \mathcal{P} \left|\psi_N^{(\varepsilon)}\right\rangle = \varepsilon^{p} \left|\psi_N^{(\varepsilon)}\right\rangle \,. \tag{1.21}$$

Thus far, we have shown that the N-particle states of identical particles can be only of the type $\left|\psi_N^{(\pm)}\right\rangle$ and that they retain their particular symmetry character for all times. This can be formulated in a somewhat more precise way:

> The states of a system of N identical particles either all belong to $\mathcal{H}_N^{(+)}$, or else they all belong to $\mathcal{H}_N^{(-)}$.

We can make this plausible as follows: If $\left|\varphi_N^{(\varepsilon)}\right\rangle$ and $\left|\psi_N^{(\varepsilon')}\right\rangle$ are two possible states of the N-particle system, then it should be possible through a suitable operation, i. e. by applying a certain operator $\hat{x}$ (or a set of operators) to transform the one state into the other and *vice versa*. Formally, this means that the scalar product is

$$\left\langle \varphi_N^{(\varepsilon)} \,|\, \hat{x} \,|\, \psi_N^{(\varepsilon')} \right\rangle \neq 0 \, .$$

Then it further follows that:

$$\begin{aligned} \varepsilon \left\langle \varphi_N^{(\varepsilon)} \,|\, \hat{x} \,|\, \psi_N^{(\varepsilon')} \right\rangle &= \left\langle P_{ij}\varphi_N^{(\varepsilon)} \,|\, \hat{x} \,|\, \psi_N^{(\varepsilon')} \right\rangle = \left\langle \varphi_N^{(\varepsilon)} \,|\, P_{ij}^{+} \hat{x} \,|\, \psi_N^{(\varepsilon')} \right\rangle = \\ &= \left\langle \varphi_N^{(\varepsilon)} \,|\, P_{ij}\hat{x} \,|\, \psi_N^{(\varepsilon')} \right\rangle = \left\langle \varphi_N^{(\varepsilon)} \,|\, \hat{x} P_{ij} \,|\, \psi_N^{(\varepsilon')} \right\rangle = \\ &= \varepsilon' \left\langle \varphi_N^{(\varepsilon)} \,|\, \hat{x} \,|\, \psi_N^{(\varepsilon')} \right\rangle . \end{aligned}$$

Thus, the conjecture $\varepsilon = \varepsilon'$ must hold.

Which space, $\mathcal{H}_N^{(+)}$ or $\mathcal{H}_N^{(-)}$, is appropriate for which type of particle is determined by relativistic quantum field theory. Here, we assume without proof the validity of the

spin-statistics relation.

$\mathcal{H}_N^{(+)}$: The space of the symmetric states of N identical particles of

integer spin.

These particles are called **Bosons.**

Examples π mesons ($S = 0$), Photons ($S = 1$), Phonons ($S = 0$), Magnons ($S = 1$), α Particles, ^{4}He,...

$\mathcal{H}_N^{(-)}$: The space of the antisymmetric states of N identical particles of

half-integer spin.

These particles are called **Fermions.**

Examples Electrons, positrons, protons, neutrons, ^{3}He,...

1.2 The "Continuous" Fock Representation

In this section, we wish to introduce the creation and annihilation operators which are typical of the second quantisation. First, some preliminary remarks are in order.

Our first problem consists of constructing a basis for the space $\mathcal{H}_N^{(\varepsilon)}$ making use of appropriate single-particle states $|\varphi_\alpha\rangle$ In the process, we must distinguish the cases in which the associated single-particle observable $\hat{\varphi}$ has a discrete or a continuous spectrum. We first discuss in this section the case of a **continuous** single-particle spectrum. We thus presuppose:

$\widehat{\varphi}$: a single-particle observable with a continuous spectrum

$$\widehat{\varphi}\,|\varphi_\alpha\rangle = \varphi_\alpha\,|\varphi_\alpha\rangle \;, \tag{1.22}$$

$$\left\langle\varphi_\alpha\middle|\varphi_\beta\right\rangle = \delta\left(\varphi_\alpha - \varphi_\beta\right) \equiv \delta(\alpha-\beta)\,. \tag{1.23}$$

The eigenstates are presumed to form a basis of $\mathcal{H}_1$:

$$\int \mathrm{d}\varphi_\alpha\;|\varphi_\alpha\rangle\,\langle\varphi_\alpha\rangle = \mathbf{1} \quad \text{in } \mathcal{H}_1\,. \tag{1.24}$$

A non-symmetrised N-particle state is found as in in (1.1) simply in the form of a product state:

$$\left|\varphi_{\alpha_1}\cdots\varphi_{\alpha_N}\right) = \left|\varphi_{\alpha_1}^{(1)}\right\rangle\left|\varphi_{\alpha_2}^{(2)}\right\rangle\cdots\left|\varphi_{\alpha_N}^{(N)}\right\rangle\,. \tag{1.25}$$

The upper index refers to the particle, and the α_i's are complete sets of quantum numbers. The N-fold state indices α_i are ordered arbitrarily but in a well-defined way according to some criteria. The state symbol on the left side of (1.25) implies this **standard ordering**. Application of the operator $\widehat{S}_\varepsilon$ from (1.18) converts (1.25) into an

(anti-)symmetrised N-particle state

$$\left|\varphi_{\alpha_1}\cdots\varphi_{\alpha_N}\right\rangle^{(\varepsilon)} = \frac{1}{N!}\sum_{\mathcal{P}}\varepsilon^p\mathcal{P}\left|\varphi_{\alpha_1}\cdots\varphi_{\alpha_N}\right)\,. \tag{1.26}$$

Here, we have introduced an appropriate normalisation factor $1/N!$. When there is no danger of misinterpretation, we shall also indicate the state (1.26) simply by $\left|\varphi_N^{(\varepsilon)}\right\rangle$.

It is easy to convince oneself that in the space of $\mathcal{H}_N^{(\varepsilon)}$, every permutation operator $\mathcal{P}$ is Hermitian:

$$\left\langle \psi_N^{(\varepsilon)} \middle| \mathcal{P}^+ \middle| \varphi_N^{(\varepsilon)} \right\rangle = \left(\left\langle \varphi_N^{(\varepsilon)} \middle| \mathcal{P} \middle| \psi_N^{(\varepsilon)} \right\rangle \right)^* = \varepsilon^p \left(\left\langle \varphi_N^{(\varepsilon)} \middle| \psi_N^{(\varepsilon)} \right\rangle \right)^* =$$
$$= \varepsilon^p \left\langle \psi_N^{(\varepsilon)} \middle| \varphi_N^{(\varepsilon)} \right\rangle = \left\langle \psi_N^{(\varepsilon)} \middle| \mathcal{P} \middle| \varphi_N^{(\varepsilon)} \right\rangle .$$

It follows from this that:

$$\mathcal{P} = \mathcal{P}^+ \quad \text{within} \quad \mathcal{H}_N^{(\varepsilon)} . \tag{1.27}$$

We can derive with this a useful relation for the states (1.26). Let $\widehat{A}$ be an arbitrary observable. Then it holds that:

$$\left\langle \psi_N^{(\varepsilon)} \middle| \widehat{A} \middle| \varphi_N^{(\varepsilon)} \right\rangle = \frac{1}{N!} \sum_{\mathcal{P}} \varepsilon^p \left\langle \psi_{\alpha_1} \cdots \psi_{\alpha_N} \middle| \mathcal{P}^+ \widehat{A} \middle| \varphi_N^{(\varepsilon)} \right\rangle =$$
$$= \frac{1}{N!} \sum_{\mathcal{P}} \varepsilon^p \left\langle \psi_{\alpha_1} \cdots \psi_{\alpha_N} \middle| \widehat{A} \mathcal{P} \middle| \varphi_N^{(\varepsilon)} \right\rangle =$$
$$= \frac{1}{N!} \sum_{\mathcal{P}} \varepsilon^{2p} \left\langle \psi_{\alpha_1} \cdots \psi_{\alpha_N} \middle| \widehat{A} \middle| \varphi_N^{(\varepsilon)} \right\rangle .$$

Due to (1.10), every transposition P_{ij} commutes with $\widehat{A}$. Since $\mathcal{P}$ can be written as a product of transpositions, it follows that $\mathcal{P}$ also commutes with every *allowed* observable $\widehat{A}$. We made use of this fact together with (1.27) in the second step. Since $\varepsilon^{2p} = +1$ and the sum contains just $N!$ terms, we have:

$$\left\langle \psi_N^{(\varepsilon)} \middle| \widehat{A} \middle| \varphi_N^{(\varepsilon)} \right\rangle = \left\langle \psi_{\alpha_1} \cdots \psi_{\alpha_N} \middle| \widehat{A} \middle| \varphi_N^{(\varepsilon)} \right\rangle . \tag{1.28}$$

The bra vector on the right-hand side is thus not symmetrised. This relation holds in particular when $\widehat{A}$ is the identity:

$$^{(\varepsilon)}\left\langle \varphi_{\beta_1} \cdots \varphi_{\beta_N} \middle| \varphi_{\alpha_1} \cdots \varphi_{\alpha_N} \right\rangle^{(\varepsilon)} =$$
$$= \left\langle \varphi_{\beta_1} \cdots \varphi_{\beta_N} \middle| \varphi_{\alpha_1} \cdots \varphi_{\alpha_N} \right\rangle^{(\varepsilon)} =$$
$$= \frac{1}{N!} \sum_{\mathcal{P}_\alpha} \varepsilon^{p_\alpha} \mathcal{P}_\alpha \left\langle \varphi_{\beta_1} \cdots \varphi_{\beta_N} \middle| \varphi_{\alpha_1} \cdots \varphi_{\alpha_N} \right\rangle .$$

The index α indicates that $\mathcal{P}_\alpha$ acts only upon the quantities φ_α. Thus we have for the

scalar product of two (anti-)symmetrised N-particle states:

$$
\begin{aligned}
&{}^{(\varepsilon)}\left\langle \varphi_{\beta_1} \cdots \varphi_{\beta_N} \middle| \varphi_{\alpha_1} \cdots \varphi_{\alpha_N} \right\rangle^{(\varepsilon)} = \\
&= \frac{1}{N!} \sum_{\mathcal{P}_\alpha} \varepsilon^{p_\alpha} \mathcal{P}_\alpha \left\{ \left\langle \varphi^{(1)}_{\beta_1} \middle| \varphi^{(1)}_{\alpha_1} \right\rangle \cdots \left\langle \varphi^{(N)}_{\beta_N} \middle| \varphi^{(N)}_{\alpha_N} \right\rangle \right\} = \\
&= \frac{1}{N!} \sum_{\mathcal{P}_\alpha} \varepsilon^{p_\alpha} \mathcal{P}_\alpha \left[\delta(\beta_1 - \alpha_1) \cdots \delta(\beta_N - \alpha_N) \right] . \qquad (1.29)
\end{aligned}
$$

This is the logical generalisation of the orthonormalisation condition (1.23) for the single-particle states to the (anti-)symmetrised N-particle states.

With (1.29) one then finds:

$$
\begin{aligned}
&\int \cdots \int \mathrm{d}\beta_1 \cdots \mathrm{d}\beta_N \left| \varphi_{\beta_1} \cdots \varphi_{\beta_N} \right\rangle^{(\varepsilon)}\, {}^{(\varepsilon)}\left\langle \varphi_{\beta_1} \cdots \varphi_{\beta_N} \middle| \varphi_{\alpha_1} \cdots \varphi_{\alpha_N} \right\rangle^{(\varepsilon)} = \\
&= \frac{1}{N!} \sum_{\mathcal{P}_\alpha} \varepsilon^{p_\alpha} \mathcal{P}_\alpha \left| \varphi_{\alpha_1} \cdots \varphi_{\alpha_N} \right\rangle^{(\varepsilon)} = \frac{1}{N!} \sum_{\mathcal{P}_\alpha} \varepsilon^{2p_\alpha} \left| \varphi_{\alpha_1} \cdots \varphi_{\alpha_N} \right\rangle^{(\varepsilon)} = \\
&= \left| \varphi_{\alpha_1} \cdots \varphi_{\alpha_N} \right\rangle^{(\varepsilon)} . \qquad (1.30)
\end{aligned}
$$

Every arbitrary N-particle state $|\psi_N\rangle^{(\varepsilon)}$ represents the sum of products of N single-particle states $|\psi\rangle$. Since, by hypothesis, the $|\varphi_\alpha\rangle$ form a complete basis set in $\mathcal{H}_1$, $|\psi\rangle$ can be written as a linear combination of the $|\varphi_\alpha\rangle$. Then it is clear that $|\psi_N\rangle^{(\varepsilon)}$ can always be expanded in terms of the $\left|\varphi_{\alpha_1} \cdots\right\rangle^{(\varepsilon)}$:

$$
\begin{aligned}
|\psi_N\rangle^{(\varepsilon)} &= \widehat{S}_\varepsilon \left| \psi^{(1)} \cdots \psi^{(N)} \right\rangle = \\
&= \sum_{\alpha_1} C_{\alpha_1} \sum_{\alpha_2} C_{\alpha_2} \cdots \sum_{\alpha_N} C_{\alpha_N} \widehat{S}_\varepsilon \left| \varphi_{\alpha_1} \cdots \varphi_{\alpha_N} \right\rangle = \\
&= \sum_{\alpha_1 \cdots \alpha_N} C(\alpha_1 \cdots \alpha_N) \left| \varphi_{\alpha_1} \cdots \varphi_{\alpha_N} \right\rangle^{(\varepsilon)} . \qquad (1.31)
\end{aligned}
$$

Then from (1.30), the

completeness relation

$$
\int \cdots \int \mathrm{d}\beta_1 \cdots \mathrm{d}\beta_N \left| \varphi_{\beta_1} \cdots \varphi_{\beta_N} \right\rangle^{(\varepsilon)}\, {}^{(\varepsilon)}\left\langle \varphi_{\beta_1} \cdots \varphi_{\beta_N} \right| = \mathbf{1} \qquad (1.32)
$$

within $\mathcal{H}_N^{(\varepsilon)}$

follows. The states defined in (1.26), $\left|\varphi_{\alpha_1} \cdots \varphi_{\alpha_N}\right\rangle^{(\varepsilon)}$, thus form a complete, orthonormalised basis of $\mathcal{H}_N^{(\varepsilon)}$.

The preceding considerations make it clear how tedious it can be to work with (anti-) symmetrised N-particle states. We thus would like to construct these with the aid of special operators entirely from the so-called

$$\textbf{vacuum state } |0\rangle \ ; \quad \langle 0 \mid 0 \rangle = 1 \ . \tag{1.33}$$

The characteristic effect of this operator,

$$a^+_{\varphi_\alpha} \equiv a^+_\alpha \ ,$$

consists in linking many-particle Hilbert spaces belonging to different numbers of particles with one another:

$$a^+_\alpha : \mathcal{H}_N^{(\varepsilon)} \quad \Longrightarrow \quad \mathcal{H}_{N+1}^{(\varepsilon)} \ . \tag{1.34}$$

The operator is completely defined by its action:

$$a^+_{\alpha_1} |0\rangle = \sqrt{1} \left|\varphi_{\alpha_1}\right\rangle^{(\varepsilon)} ,$$

$$a^+_{\alpha_2} \left|\varphi_{\alpha_1}\right\rangle^{(\varepsilon)} = \sqrt{2} \left|\varphi_{\alpha_2}\varphi_{\alpha_1}\right\rangle^{(\varepsilon)}$$

$$\cdots$$

In general, it holds that:

$$a^+_\beta \underbrace{|\varphi_{\alpha_1} \cdots \varphi_{\alpha_N}\rangle^{(\varepsilon)}}_{\in \mathcal{H}_N^{(\varepsilon)}} = \sqrt{N+1} \underbrace{|\varphi_\beta \varphi_{\alpha_1} \cdots \varphi_{\alpha_N}\rangle^{(\varepsilon)}}_{\in \mathcal{H}_{N+1}^{(\varepsilon)}} \ . \tag{1.35}$$

We refer to a^+_β as a

creation operator.

In a graphic sense, it *creates* an additional particle in the single-particle state $\left|\varphi_\beta\right\rangle$. The inverse relation to (1.35) reads:

$$\left|\varphi_{\alpha_1} \cdots \varphi_{\alpha_N}\right\rangle^{(\varepsilon)} = \frac{1}{\sqrt{N!}} a^+_{\alpha_1} a^+_{\alpha_2} \cdots a^+_{\alpha_N} |0\rangle \ . \tag{1.36}$$

Here, we must be careful to observe the order of the operators. Thus, for example:

$$a^+_{\alpha_1} a^+_{\alpha_2} \left|\varphi_{\alpha_3} \cdots \varphi_{\alpha_N}\right\rangle^{(\varepsilon)} = \sqrt{N(N-1)} \left|\varphi_{\alpha_1}\varphi_{\alpha_2}\varphi_{\alpha_3} \cdots \varphi_{\alpha_N}\right\rangle^{(\varepsilon)} ,$$

$$a^+_{\alpha_2} a^+_{\alpha_1} \left|\varphi_{\alpha_3} \cdots \varphi_{\alpha_N}\right\rangle^{(\varepsilon)} = \sqrt{N(N-1)} \left|\varphi_{\alpha_2}\varphi_{\alpha_1}\varphi_{\alpha_3} \cdots \varphi_{\alpha_N}\right\rangle^{(\varepsilon)} =$$
$$= \varepsilon\sqrt{N(N-1)} \left|\varphi_{\alpha_1}\varphi_{\alpha_2}\varphi_{\alpha_3} \cdots \varphi_{\alpha_N}\right\rangle^{(\varepsilon)} .$$

Since these are basis states, we can read off the following operator identity:

$$\left[a^+_{\alpha_1}, a^+_{\alpha_2}\right]_{-\varepsilon} \equiv a^+_{\alpha_1} a^+_{\alpha_2} - \varepsilon a^+_{\alpha_2} a^+_{\alpha_1} = 0 . \tag{1.37}$$

The creation operators commute for Bosons ($\varepsilon = +$) and anticommute for Fermions ($\varepsilon = -$).

We now discuss the operator which is adjoint to a^+_α,

$$a_\alpha = \left(a^+_\alpha\right)^+ , \tag{1.38}$$

which links the Hilbert spaces $\mathcal{H}^{(\varepsilon)}_N$ and $\mathcal{H}^{(\varepsilon)}_{N-1}$ to each other:

$$a_\alpha : \mathcal{H}^{(\varepsilon)}_N \quad \Longrightarrow \quad \mathcal{H}^{(\varepsilon)}_{N-1} . \tag{1.39}$$

The term

annihilation operator

will be justified by the following considerations. Since a_α is adjoint to a^+_α, we initially have according to (1.35) or (1.36):

$$^{(\varepsilon)}\left\langle\varphi_{\alpha_1} \cdots \varphi_{\alpha_N}\right| a_\beta = \sqrt{N+1}\, ^{(\varepsilon)}\left\langle\varphi_\beta\varphi_{\alpha_1} \cdots \varphi_{\alpha_N}\right| \tag{1.40}$$

$$^{(\varepsilon)}\left\langle\varphi_{\alpha_1} \cdots \varphi_{\alpha_N}\right| = \frac{1}{\sqrt{N!}} \langle 0|\, a_{\alpha_N} \cdots a_{\alpha_2} a_{\alpha_1} . \tag{1.41}$$

The meaning of the operator a_α can be seen by computing the following matrix element:

$$^{(\varepsilon)}\left\langle \underbrace{\varphi_{\beta_2} \cdots \varphi_{\beta_N}}_{\in \mathcal{H}^{(\varepsilon)}_{N-1}} \left| a_\gamma \right| \underbrace{\varphi_{\alpha_1} \cdots \varphi_{\alpha_N}}_{\in \mathcal{H}^{(\varepsilon)}_N} \right\rangle^{(\varepsilon)} =$$

$$= \sqrt{N}\, ^{(\varepsilon)}\left\langle \varphi_\gamma\varphi_{\beta_2} \cdots \varphi_{\beta_N} \middle| \varphi_{\alpha_1} \cdots \varphi_{\alpha_N}\right\rangle^{(\varepsilon)} =$$

$$= \frac{\sqrt{N}}{N!} \sum_{\mathcal{P}_\alpha} \varepsilon^{p_\alpha} \mathcal{P}_\alpha \left(\delta(\gamma - \alpha_1)\, \delta(\beta_2 - \alpha_2)\, \delta(\beta_3 - \alpha_3) \cdots \delta(\beta_N - \alpha_N)\right) .$$

In the last step, we made use of (1.29). We re-sort the sum:

$$
\begin{aligned}
&{}^{(\varepsilon)}\left\langle \varphi_{\beta_2} \cdots \varphi_{\beta_N} \middle| a_\gamma \middle| \varphi_{\alpha_1} \cdots \varphi_{\alpha_N} \right\rangle^{(\varepsilon)} = \\
&= \frac{1}{\sqrt{N}} \frac{1}{(N-1)!} \Big\{ \delta(\gamma - \alpha_1) \sum_{\mathcal{P}} \varepsilon^{p_\alpha} \mathcal{P}_\alpha \big(\delta(\beta_2 - \alpha_2) \cdots \delta(\beta_N - \alpha_N) \big) + \\
&+ \varepsilon \delta(\gamma - \alpha_2) \sum_{\mathcal{P}_\alpha} \varepsilon^{p_\alpha} \mathcal{P}_\alpha \big(\delta(\beta_2 - \alpha_1) \delta(\beta_3 - \alpha_3) \cdots \delta(\beta_N - \alpha_N) \big) + \\
&+ \cdots + \\
&+ \varepsilon^{N-1} \delta(\gamma - \alpha_N) \sum_{\mathcal{P}_\alpha} \varepsilon^{p_\alpha} \mathcal{P}_\alpha \big(\delta(\beta_2 - \alpha_1) \delta(\beta_3 - \alpha_2) \cdots \delta(\beta_N - \alpha_{N-1}) \big) \Big\} .
\end{aligned}
$$

The sums on the right-hand side again represent scalar products, now however in $\mathcal{H}_{N-1}^{(\varepsilon)}$:

$$
\begin{aligned}
&{}^{(\varepsilon)}\left\langle \varphi_{\beta_2} \cdots \varphi_{\beta_N} \middle| a_\gamma \middle| \varphi_{\alpha_1} \cdots \varphi_{\alpha_N} \right\rangle^{(\varepsilon)} = \\
&= \frac{1}{\sqrt{N}} \Big\{ \delta(\gamma - \alpha_1)^{(\varepsilon)} \left\langle \varphi_{\beta_2} \cdots \varphi_{\beta_N} \middle| \varphi_{\alpha_2} \cdots \varphi_{\alpha_N} \right\rangle^{(\varepsilon)} + \\
&+ \varepsilon \delta(\gamma - \alpha_2)^{(\varepsilon)} \left\langle \varphi_{\beta_2} \cdots \varphi_{\beta_N} \middle| \varphi_{\alpha_1} \varphi_{\alpha_3} \cdots \varphi_{\alpha_N} \right\rangle^{(\varepsilon)} + \\
&+ \cdots + \\
&+ \varepsilon^{N-1} \delta(\gamma - \alpha_N)^{(\varepsilon)} \left\langle \varphi_{\beta_2} \cdots \varphi_{\beta_N} \middle| \varphi_{\alpha_1} \cdots \varphi_{\alpha_{N-1}} \right\rangle^{(\varepsilon)} \Big\} .
\end{aligned}
$$

Since the bra vector is an arbitrary basis vector of $\mathcal{H}_{N-1}^{(\varepsilon)}$, this relation implies that:

$$
\begin{aligned}
a_\gamma \left| \varphi_{\alpha_1} \cdots \varphi_{\alpha_N} \right\rangle^{(\varepsilon)} = &\frac{1}{\sqrt{N}} \Big\{ \delta(\gamma - \alpha_1) \left| \varphi_{\alpha_2} \cdots \varphi_{\alpha_N} \right\rangle^{(\varepsilon)} + \\
&+ \varepsilon \delta(\gamma - \alpha_2) \left| \varphi_{\alpha_1} \varphi_{\alpha_3} \cdots \varphi_{\alpha_N} \right\rangle^{(\varepsilon)} + \\
&+ \cdots + \\
&+ \varepsilon^{N-1} \delta(\gamma - \alpha_N) \left| \varphi_{\alpha_1} \cdots \varphi_{\alpha_{N-1}} \right\rangle^{(\varepsilon)} \Big\}
\end{aligned}
\tag{1.42}
$$

If the single-particle state $|\varphi_\gamma\rangle$ appears among the states $|\varphi_{\alpha_1}\rangle$ to $|\varphi_{\alpha_N}\rangle$ which construct the N-particle state $|\varphi_{\alpha_1} \cdots \varphi_{\alpha_N}\rangle^{(\varepsilon)}$, then an $(N-1)$-particle state results, in which however $|\varphi_\gamma\rangle$ is no longer present. One then says that a_γ *annihilates* a particle in the state $|\varphi_\gamma\rangle$.

If $|\varphi_\gamma\rangle$ does not occur within the symmetrised initial state, then application of a_γ causes the initial state to vanish. In particular, an important special case applies:

$$a_\gamma |0\rangle = 0 . \tag{1.43}$$

The commutation relation for the annihilation operators follows immediately from (1.37):

$$\left[a_{\alpha_1}, a_{\alpha_2}\right]_{-\varepsilon} = -\varepsilon\left(\left[a_{\alpha_1}^+, a_{\alpha_2}^+\right]_{-\varepsilon}\right)^+ .$$

Annihilation operators commute ($\varepsilon = +$; Bosons) or else they anticommute ($\varepsilon = -$; Fermions):

$$\left[a_{\alpha_1}, a_{\alpha_2}\right]_{-\varepsilon} \equiv 0 . \tag{1.44}$$

There is still a third commutation relation, i. e. the one between the creation and the annihilation operators:

$$\left[a_{\alpha_1}, a_{\alpha_2}^+\right]_{-\varepsilon} = \delta\left(\alpha_1 - \alpha_2\right) . \tag{1.45}$$

Proof Let $\left|\varphi_{\alpha_1} \cdots \varphi_{\alpha_N}\right\rangle^{(\varepsilon)}$ be an arbitrary basis state of $\mathcal{H}_N^{(\varepsilon)}$.

$$\begin{aligned} a_\beta\left(a_\gamma^+ \left|\varphi_{\alpha_1} \cdots \varphi_{\alpha_N}\right\rangle^{(\varepsilon)}\right) &= \sqrt{N+1}\, a_\beta \left|\varphi_\gamma \varphi_{\alpha_1} \cdots \varphi_{\alpha_N}\right\rangle^{(\varepsilon)} = \\ &= \delta\left(\beta - \gamma\right)\left|\varphi_{\alpha_1} \cdots \varphi_{\alpha_N}\right\rangle^{(\varepsilon)} + \\ &\quad + \varepsilon\delta\left(\beta - \alpha_1\right)\left|\varphi_\gamma \varphi_{\alpha_2} \cdots \varphi_{\alpha_N}\right\rangle^{(\varepsilon)} + \\ &\quad + \cdots + \\ &\quad + \varepsilon^N \delta\left(\beta - \alpha_N\right)\left|\varphi_\gamma \varphi_{\alpha_1} \cdots \varphi_{\alpha_{N-1}}\right\rangle^{(\varepsilon)} , \\ a_\gamma^+\left(a_\beta \left|\varphi_{\alpha_1} \cdots \varphi_{\alpha_N}\right\rangle^{(\varepsilon)}\right) &= \delta\left(\beta - \alpha_1\right)\left|\varphi_\gamma \varphi_{\alpha_2} \cdots \varphi_{\alpha_N}\right\rangle^{(\varepsilon)} + \\ &\quad + \varepsilon\delta\left(\beta - \alpha_2\right)\left|\varphi_\gamma \varphi_{\alpha_1} \varphi_{\alpha_3} \cdots \varphi_{\alpha_N}\right\rangle^{(\varepsilon)} + \\ &\quad + \cdots + \\ &\quad + \varepsilon^{N-1} \delta\left(\beta - \alpha_N\right)\left|\varphi_\gamma \varphi_{\alpha_1} \cdots \varphi_{\alpha_{N-1}}\right\rangle^{(\varepsilon)} . \end{aligned}$$

Combining these two equations, we find:

$$\left(a_\beta a_\gamma^+ - \varepsilon a_\gamma^+ a_\beta\right)\left|\varphi_{\alpha_1} \cdots \varphi_{\alpha_N}\right\rangle^{(\varepsilon)} = \delta\left(\beta - \gamma\right)\left|\varphi_{\alpha_1} \cdots \varphi_{\alpha_N}\right\rangle^{(\varepsilon)} .$$

This proves (1.45).

Thus, by using (1.36) and (1.41), we can refer all the N-particle states to the vacuum state $|0\rangle$, by repeated application of creation and annihilation operators. The effect of the *annihilator* on $|0\rangle$ is trivial (1.43). Using the commutation relations (1.37), (1.44) and (1.45), we can change the order of the operators in any desired manner.

However, the introduction of the creation and annihilation operators is advantageous only if we are able to describe the N-particle observables within the same formalism.

Using the completeness relation (1.32) for an arbitrary observable $\widehat{A}$, we initially find:

$$\widehat{A} = \mathbf{1} \cdot \widehat{A} \cdot \mathbf{1} = \\ = \int \cdots \int \mathrm{d}\alpha_1 \cdots \mathrm{d}\alpha_N \mathrm{d}\beta_1 \cdots \mathrm{d}\beta_N \left|\varphi_{\alpha_1} \cdots\right\rangle^{(\varepsilon)} \cdot \\ \cdot {}^{(\varepsilon)}\left\langle\varphi_{\alpha_1} \cdots \right| \widehat{A} \left|\varphi_{\beta_1} \cdots\right\rangle^{(\varepsilon)} {}^{(\varepsilon)}\left\langle\varphi_{\beta_1} \cdots\right| . \tag{1.46}$$

We now insert (1.36) and (1.41):

$$\widehat{A} = \frac{1}{N!} \int \cdots \int \mathrm{d}\alpha_1 \cdots \mathrm{d}\alpha_N \mathrm{d}\beta_1 \cdots \mathrm{d}\beta_N\, a^+_{\alpha_1} \cdots a^+_{\alpha_N} |0\rangle \cdot \\ \cdot {}^{(\varepsilon)}\left\langle\varphi_{\alpha_1} \cdots \right| \widehat{A} \left|\varphi_{\beta_1} \cdots\right\rangle^{(\varepsilon)} \langle 0|\, a_{\beta_N} \cdots a_{\beta_1} . \tag{1.47}$$

As a rule, $\widehat{A}$ will contain single-particle and two-particle parts:

$$\widehat{A} = \sum_{i=1}^{n} \widehat{A}_1^{(i)} + \frac{1}{2} \sum_{i,j}^{i \neq j} \widehat{A}_2^{(i,j)} . \tag{1.48}$$

We first discuss the single-particle part, for which in (1.47) the following matrix element is required:

$$^{(\varepsilon)}\left\langle\varphi_{\alpha_1} \cdots \right| \sum_{i=1}^{n} \widehat{A}_1^{(i)} \left|\varphi_{\beta_1} \cdots\right\rangle^{(\varepsilon)} = \\ = \frac{1}{N!} \sum_{\mathcal{P}_\beta} \varepsilon^{p_\beta} \mathcal{P}_\beta \left[\left\langle\varphi^{(1)}_{\alpha_1}\right| \widehat{A}_1^{(1)} \left|\varphi^{(1)}_{\beta_1}\right\rangle \left\langle\varphi^{(2)}_{\alpha_2} \middle| \varphi^{(2)}_{\beta_2}\right\rangle \cdots \left\langle\varphi^{(N)}_{\alpha_N} \middle| \varphi^{(N)}_{\beta_N}\right\rangle + \right. \\ + \cdots + \\ \left. + \left\langle\varphi^{(1)}_{\alpha_1} \middle| \varphi^{(1)}_{\beta_1}\right\rangle \cdots \left\langle\varphi^{(N)}_{\alpha_N}\right| \widehat{A}_1^{(N)} \left|\varphi^{(N)}_{\beta_N}\right\rangle \right] . \tag{1.49}$$

Here, we have already made use of (1.28). It can readily be seen that each term of the sum over the permutations gives exactly the same contribution after inserting (1.49) into (1.47). Every permuted arrangement of the $\left|\varphi^{(i)}_{\beta_i}\right\rangle$ can namely be reduced to the standard arrangement by:

1. renaming the integration variables β_i and
2. then exchanging the corresponding annihilation operators.

The exchange in Part 2. yields a factor ε^{p_β}, owing to (1.44). Overall, this gives for each permutation a coefficient $\varepsilon^{2p_\beta} = +1$.

In a similar fashion, one can show that each summand within the square brackets in (1.49) also gives the same contribution to (1.47). This is achieved by:

1. exchanging corresponding integration variables $\left(\alpha_j \Longleftrightarrow \alpha_i,\ \beta_j \Longleftrightarrow \beta_i\right)$ and
2. then regrouping of equal numbers of creation and annihilation operators.

Part 2. gives in each case a factor $\left(\varepsilon^2\right)^{n_j} = +1$. We thus obtain an intermediate result which is already greatly simplified:

$$\begin{aligned}
&\sum_{i=1}^{n} \widehat{A}_1^{(i)} = \\
&= \frac{N}{N!} \int \cdots \int \mathrm{d}\alpha_1 \cdots \mathrm{d}\beta_N \, a_{\alpha_1}^+ \cdots a_{\alpha_N}^+ |0\rangle \cdot \\
&\quad \cdot \left\{ \left\langle \varphi_{\alpha_1}^{(1)} \middle| \widehat{A}_1^{(1)} \middle| \varphi_{\beta_1}^{(1)} \right\rangle \delta(\alpha_2 - \beta_2) \cdots \delta(\alpha_N - \beta_N) \right\} \cdot \\
&\quad \cdot \langle 0 | \, a_{\beta_N} \cdots a_{\beta_1} = \\
&= \iint \mathrm{d}\alpha_1 \mathrm{d}\beta_1 \left\langle \varphi_{\alpha_1}^{(1)} \middle| \widehat{A}_1^{(1)} \middle| \varphi_{\beta_1}^{(1)} \right\rangle a_{\alpha_1}^+ \cdot \\
&\quad \cdot \left\{ \frac{1}{(N-1)!} \int \cdots \int \mathrm{d}\alpha_2 \cdots \mathrm{d}\alpha_N \, a_{\alpha_2}^+ \cdots a_{\alpha_N}^+ |0\rangle \langle 0 | \, a_{\alpha_N} \cdots a_{\alpha_2} \right\} a_{\beta_1} \,. \qquad (1.50)
\end{aligned}$$

As one can read off (1.32), the curly brackets contain the identity **1** of $\mathcal{H}_{N-1}^{(\varepsilon)}$. As the result, we then have:

$$\sum_{i=1}^{n} \widehat{A}_1^{(i)} \equiv \iint \mathrm{d}\alpha \mathrm{d}\beta \left\langle \varphi_\alpha \middle| \widehat{A}_1 \middle| \varphi_\beta \right\rangle a_\alpha^+ a_\beta \,. \qquad (1.51)$$

On the right-hand side, the particle number N no longer appears explicitly. It is of course contained implicitly in the identity, which from (1.50) should in fact be imagined to occur between a_α^+ and a_β.

In a completely analogous way, we now treat the two-particle part of the observables $\widehat{A}$:

$$\begin{aligned}
&\frac{1}{2} \sum_{i,j}^{i \neq j} \widehat{A}_2^{(i,j)} = \\
&= \frac{1}{2N!} \int \cdots \int \mathrm{d}\alpha_1 \cdots \mathrm{d}\beta_N \, a_{\alpha_1}^+ \cdots a_{\alpha_N}^+ |0\rangle \cdot \\
&\quad \cdot \left\{ \frac{1}{N!} \sum_{\mathcal{P}_\beta} \varepsilon^{p_\beta} \mathcal{P}_\beta \left[\left\langle \varphi_{\alpha_1}^{(1)} \right| \left\langle \varphi_{\alpha_2}^{(2)} \right| \widehat{A}_2^{(1,2)} \left| \varphi_{\beta_1}^{(1)} \right\rangle \left| \varphi_{\beta_2}^{(2)} \right\rangle \cdot \right. \right.
\end{aligned}$$

$$\cdot \left\langle \varphi_{\alpha_3}^{(3)} \middle| \varphi_{\beta_3}^{(3)} \right\rangle \cdots \left\langle \varphi_{\alpha_N}^{(N)} \middle| \varphi_{\beta_N}^{(N)} \right\rangle +$$

$$+ \left\langle \varphi_{\alpha_1}^{(1)} \right| \left\langle \varphi_{\alpha_3}^{(3)} \right| \widehat{A}_2^{(1,3)} \left| \varphi_{\beta_1}^{(1)} \right\rangle \left| \varphi_{\beta_3}^{(3)} \right\rangle \left\langle \varphi_{\alpha_2}^{(2)} \middle| \varphi_{\beta_2}^{(2)} \right\rangle \cdot$$

$$\cdot \left\langle \varphi_{\alpha_4}^{(4)} \middle| \varphi_{\beta_4}^{(4)} \right\rangle \cdots \left\langle \varphi_{\alpha_N}^{(N)} \middle| \varphi_{\beta_N}^{(N)} \right\rangle + \cdots \Bigg] \Bigg\} \langle 0 | \, a_{\beta_N} \cdots a_{\beta_1} \, . \tag{1.52}$$

Precisely the same argumentation can be applied here as was used above for the single-particle portion, in order to show that all the $N!$ permutations $\mathcal{P}_\beta$ contribute to the multiple integral in a similar manner, and furthermore, that all $N(N-1)$ summands in the square brackets are equivalent. This means that:

$$\frac{1}{2} \sum_{i,j}^{i \neq j} \widehat{A}_2^{(i,j)} = \frac{1}{2} \int \cdots \int \mathrm{d}\alpha_1 \mathrm{d}\alpha_2 \mathrm{d}\beta_1 \mathrm{d}\beta_2 \left\langle \varphi_{\alpha_1} \varphi_{\alpha_2} \middle| \widehat{A}_2^{(1,2)} \middle| \varphi_{\beta_1} \varphi_{\beta_2} \right\rangle \cdot$$

$$\cdot \, a_{\alpha_1}^+ a_{\alpha_2}^+ \left\{ \frac{1}{(N-2)!} \int \cdots \int \mathrm{d}\alpha_3 \cdots \mathrm{d}\alpha_N \cdot \right.$$

$$\left. \cdot \, a_{\alpha_3}^+ \cdots a_{\alpha_N}^+ |0\rangle \langle 0| \, a_{\alpha_N} \cdots a_{\alpha_3} \right\} a_{\beta_2} a_{\beta_1} \, . \tag{1.53}$$

The curly brackets now contain the identity $\mathbf{1}$ of $\mathcal{H}_{N-2}^{(\varepsilon)}$. With this, we find:

$$\frac{1}{2} \sum_{i,j}^{i \neq j} \widehat{A}_2^{(i,j)} = \frac{1}{2} \int \cdots \int \mathrm{d}\alpha \, \mathrm{d}\beta \, \mathrm{d}\gamma \, \mathrm{d}\delta \, \left\langle \varphi_\alpha \varphi_\beta \middle| \widehat{A}_2 \middle| \varphi_\gamma \varphi_\delta \right\rangle a_\alpha^+ a_\beta^+ a_\delta a_\gamma \, . \tag{1.54}$$

The matrix element can be constructed with non-symmetrised states,

$$\left\langle \varphi_\alpha \varphi_\beta \middle| \widehat{A}_2 \middle| \varphi_\gamma \varphi_\delta \right\rangle = \left\langle \varphi_\alpha^{(1)} \right| \left\langle \varphi_\beta^{(2)} \middle| \widehat{A}_2^{(1,2)} \middle| \varphi_\gamma^{(1)} \right\rangle \left| \varphi_\delta^{(2)} \right\rangle ,$$

but also with symmetrised two-particle states:

$$\left| \varphi_\gamma \varphi_\delta \right\rangle^{(\varepsilon)} = \frac{1}{2!} \left(\left| \varphi_\gamma^{(1)} \right\rangle \left| \varphi_\delta^{(2)} \right\rangle + \varepsilon \left| \varphi_\gamma^{(2)} \right\rangle \left| \varphi_\delta^{(1)} \right\rangle \right) .$$

One can again readily convince oneself that in

$$^{(\varepsilon)} \left\langle \varphi_\alpha \varphi_\beta \middle| \widehat{A}_2 \middle| \varphi_\gamma \varphi_\delta \right\rangle^{(\varepsilon)} = \frac{1}{4} \left\{ \left\langle \varphi_\alpha^{(1)} \right| \left\langle \varphi_\beta^{(2)} \middle| \widehat{A}_2^{(1,2)} \middle| \varphi_\gamma^{(1)} \right\rangle \left| \varphi_\delta^{(2)} \right\rangle + \right.$$

$$+ \, \varepsilon \left\langle \varphi_\alpha^{(1)} \right| \left\langle \varphi_\beta^{(2)} \middle| \widehat{A}_2^{(1,2)} \middle| \varphi_\gamma^{(2)} \right\rangle \left| \varphi_\delta^{(1)} \right\rangle +$$

$$+ \, \varepsilon \left\langle \varphi_\alpha^{(2)} \right| \left\langle \varphi_\beta^{(1)} \middle| \widehat{A}_2^{(1,2)} \middle| \varphi_\gamma^{(1)} \right\rangle \left| \varphi_\delta^{(2)} \right\rangle +$$

$$\left. + \, \varepsilon^2 \left\langle \varphi_\alpha^{(2)} \right| \left\langle \varphi_\beta^{(1)} \middle| \widehat{A}_2^{(1,2)} \middle| \varphi_\gamma^{(2)} \right\rangle \left| \varphi_\delta^{(1)} \right\rangle \right\}$$

every summand provides the same contribution to (1.54), so that the normalisation factor guarantees that the symmetrised matrix element in (1.54) is equivalent to the non-symmetrised one. One can thus make the choice on the basis of convenience.

Let us summarise briefly what we have achieved thus far. Through (1.36) and (1.41), we were able to replace the tedious construction of (anti-)symmetrised products of single-particle wavefunctions for the N-particle wavefunctions by sequentially applying creation operators to the vacuum state $|0\rangle$. Their application is simple. The symmetry behaviour of the wavefunctions is reproduced by the three fundamental commutation relations (1.37), (1.44) and (1.45). The N-particle observables also can be expressed in terms of the creation and annihilation operators, (1.51) and (1.54), whereby the remaining matrix elements can be computed straightforwardly. We will give some examples of the application of this procedure in Chap. 2.

We now introduce two important special operators:

The occupation-density operator

$$\hat{n}_\alpha = a_\alpha^+ a_\alpha \, . \tag{1.55}$$

The action of this operator is found by considering (1.35) and (1.42):

$$\begin{aligned}
\hat{n}_\alpha \left|\varphi_{\alpha_1} \cdots \varphi_{\alpha_N}\right\rangle^{(\varepsilon)} &= \delta(\alpha - \alpha_1) \left|\varphi_\alpha \varphi_{\alpha_2} \cdots \varphi_{\alpha_N}\right\rangle^{(\varepsilon)} + \\
&+ \varepsilon \delta(\alpha - \alpha_2) \left|\varphi_\alpha \varphi_{\alpha_1} \varphi_{\alpha_3} \cdots \varphi_{\alpha_N}\right\rangle^{(\varepsilon)} + \\
&+ \cdots + \\
&+ \varepsilon^{N-1} \delta(\alpha - \alpha_N) \left|\varphi_\alpha \varphi_{\alpha_1} \cdots \varphi_{\alpha_{N-1}}\right\rangle^{(\varepsilon)} = \\
&= \delta(\alpha - \alpha_1) \left|\varphi_\alpha \varphi_{\alpha_2} \cdots \varphi_{\alpha_N}\right\rangle^{(\varepsilon)} + \\
&+ \varepsilon \delta(\alpha - \alpha_2)\, \varepsilon \left|\varphi_{\alpha_1} \varphi_\alpha \varphi_{\alpha_3} \cdots \varphi_{\alpha_N}\right\rangle^{(\varepsilon)} + \\
&+ \cdots + \\
&+ \varepsilon^{N-1} \delta(\alpha - \alpha_N)\, \varepsilon^{N-1} \left|\varphi_{\alpha_1} \cdots \varphi_{\alpha_{N-1}} \varphi_\alpha\right\rangle^{(\varepsilon)} \, .
\end{aligned}$$

The basis states of $\mathcal{H}_N^{(\varepsilon)}$ are thus apparently eigenstates of the occupation-density operator:

$$\hat{n}_\alpha \left|\varphi_{\alpha_1} \cdots \varphi_{\alpha_N}\right\rangle^{(\varepsilon)} = \left\{ \sum_{i=1}^{n} \delta(\alpha - \alpha_i) \right\} \left|\varphi_{\alpha_1} \cdots \varphi_{\alpha_N}\right\rangle^{(\varepsilon)} \, . \tag{1.56}$$

The microscopic occupation density is contained in the curly brackets.

The particle-number operator

$$\widehat{N} = \int d\alpha\, \hat{n}_\alpha = \int d\alpha\, a_\alpha^+ a_\alpha \,. \tag{1.57}$$

It follows immediately from (1.56) that the basis states of $\mathcal{H}_N^{(\varepsilon)}$ are also eigenstates of $\widehat{N}$, whereby in every case the eigenvalue is the total particle number N.

$$\begin{aligned}\widehat{N}\left|\varphi_{\alpha_1}\cdots\varphi_{\alpha_N}\right\rangle^{(\varepsilon)} &= \int d\alpha \sum_{i=1}^{N} \delta\left(\alpha-\alpha_i\right)\left|\varphi_{\alpha_1}\cdots\varphi_{\alpha_N}\right\rangle^{(\varepsilon)} = \\ &= N\left|\varphi_{\alpha_1}\cdots\varphi_{\alpha_N}\right\rangle^{(\varepsilon)} \,.\end{aligned} \tag{1.58}$$

Making use of the fundamental commutation relations for the creation and annihilation operators, we compute the following commutator:

$$\begin{aligned}\left[\hat{n}_\alpha, a_\beta^+\right]_- &= \hat{n}_\alpha a_\beta^+ - a_\beta^+ \hat{n}_\alpha = \\ &= a_\alpha^+ a_\alpha a_\beta^+ - a_\beta^+ \hat{n}_\alpha = \\ &= a_\alpha^+ (\delta(\alpha-\beta) + \varepsilon a_\beta^+ a_\alpha) - a_\beta^+ \hat{n}_\alpha = \\ &= a_\alpha^+ \delta(\alpha-\beta) + \varepsilon^2 a_\beta^+ a_\alpha^+ a_\alpha - a_\beta^+ \hat{n}_\alpha \,.\end{aligned}$$

The last two terms just cancel:

$$\left[\hat{n}_\alpha, a_\beta^+\right]_- = a_\alpha^+ \delta(\alpha-\beta) \,. \tag{1.59}$$

In an analogous manner, one shows that:

$$\left[\hat{n}_\alpha, a_\beta\right]_- = -a_\alpha \delta(\alpha-\beta) \,. \tag{1.60}$$

With (1.57), the analogous relations for the particle number operator are obtained:

$$[\widehat{N}, a_\alpha^+]_- = a_\alpha^+ \,; \quad [\widehat{N}, a_\alpha]_- = -a_\alpha \,. \tag{1.61}$$

This can also be written as follows:

$$\widehat{N} a_\alpha^+ = a_\alpha^+ \left(\widehat{N} + \mathbf{1}\right) \,; \quad \widehat{N} a_\alpha = a_\alpha \left(\widehat{N} - \mathbf{1}\right) \,. \tag{1.62}$$

If we apply this combination of operators to a basis state,

$$\widehat{N}\left(a_\alpha^+\left|\varphi_{\alpha_1}\cdots\varphi_{\alpha_N}\right\rangle^{(\varepsilon)}\right)=(N+1)\left(a_\alpha^+\left|\varphi_{\alpha_1}\cdots\varphi_{\alpha_N}\right\rangle^{(\varepsilon)}\right),$$

$$\widehat{N}\left(a_\alpha\left|\varphi_{\alpha_1}\cdots\varphi_{\alpha_N}\right\rangle^{(\varepsilon)}\right)=(N-1)\left(a_\alpha\left|\varphi_{\alpha_1}\cdots\varphi_{\alpha_N}\right\rangle^{(\varepsilon)}\right),$$

then we can again recognise that the terms *creation operator* for a_α^+ and *annihilation operator* for a_α are clearly appropriate.

We have made the assumption in this section that the single-particle observable $\widehat{\varphi}$, from whose eigenstates we constructed the N-particle basis of the Hilbert space $\mathcal{H}_N^{(\varepsilon)}$, possesses a continuous spectrum. A prominent example of this class of observables is the

position operator $\hat{\boldsymbol{r}}$.

The associated creation and annihilation operators are called

field operators $\widehat{\psi}(\boldsymbol{r})$, $\widehat{\psi}^+(\boldsymbol{r})$.

All of the relations derived above naturally hold for these operators, however with a special notation:

$$\widehat{\psi}^+(\boldsymbol{r})\left|\boldsymbol{r}_1\cdots\boldsymbol{r}_N\right\rangle^{(\varepsilon)}=\sqrt{N+1}\left|\boldsymbol{r}\boldsymbol{r}_1\cdots\boldsymbol{r}_N\right\rangle^{(\varepsilon)}, \tag{1.63}$$

$$\left|\boldsymbol{r}_1\boldsymbol{r}_2\cdots\boldsymbol{r}_N\right\rangle^{(\varepsilon)}=\frac{1}{\sqrt{N!}}\widehat{\psi}^+(\boldsymbol{r}_1)\cdots\widehat{\psi}^+(\boldsymbol{r}_N)\left|0\right\rangle. \tag{1.64}$$

The commutation relations of the field operators follow immediately from (1.37), (1.44) and (1.45):

$$\left[\widehat{\psi}^+(\boldsymbol{r}),\widehat{\psi}^+(\boldsymbol{r}')\right]_{-\varepsilon}=\left[\widehat{\psi}(\boldsymbol{r}),\widehat{\psi}(\boldsymbol{r}')\right]_{-\varepsilon}=0,$$

$$\left[\widehat{\psi}(\boldsymbol{r}),\widehat{\psi}^+(\boldsymbol{r}')\right]_{-\varepsilon}=\delta(\boldsymbol{r}-\boldsymbol{r}'). \tag{1.65}$$

Their relationship with general creation and annihilation operators a_α, a_α^+ is important. The completeness relation yields:

$$\left|\varphi_\alpha\right\rangle=\int d^3r\,\left|\boldsymbol{r}\right\rangle\left\langle\boldsymbol{r}\middle|\varphi_\alpha\right\rangle=\int d^3r\,\varphi_\alpha(\boldsymbol{r})\left|\boldsymbol{r}\right\rangle.$$

It thus follows owing to $\left|\varphi_\alpha\right\rangle=a_\alpha^+\left|0\right\rangle$ and $\left|\boldsymbol{r}\right\rangle=\widehat{\psi}^+(\boldsymbol{r})\left|0\right\rangle$ that:

$$a_\alpha^+=\int d^3r\,\varphi_\alpha(\boldsymbol{r})\widehat{\psi}^+(\boldsymbol{r}), \tag{1.66}$$

$$a_\alpha=\int d^3r\,\varphi_\alpha^*(\boldsymbol{r})\widehat{\psi}(\boldsymbol{r}). \tag{1.67}$$

Note that $\widehat{\psi}(\boldsymbol{r})$, $\widehat{\psi}^+(\boldsymbol{r})$ are operators, whilst $\varphi_\alpha(\boldsymbol{r})$ is the scalar wavefunction belonging to the state $|\varphi_\alpha\rangle$. The inverses of (1.66) and (1.67) follow from

$$|\boldsymbol{r}\rangle = \int \mathrm{d}\alpha \; |\varphi_\alpha\rangle \langle \varphi_\alpha | \boldsymbol{r}\rangle$$

with the same considerations as above:

$$\widehat{\psi}^+(\boldsymbol{r}) = \int \mathrm{d}\alpha \, \varphi^*_\alpha(\boldsymbol{r}) a^+_\alpha \,, \tag{1.68}$$

$$\widehat{\psi}(\boldsymbol{r}) = \int \mathrm{d}\alpha \, \varphi_\alpha(\boldsymbol{r}) a_\alpha \,. \tag{1.69}$$

1.3 The "Discrete" Fock Representation

We again assume that the basis of the Hilbert space $\mathcal{H}_N^{(\varepsilon)}$ of a system of N identical particles is constructed from the eigenstates of a single-particle observable $\widehat{\varphi}$, whereby now however $\widehat{\varphi}$ is taken to have a *discrete* spectrum:

$$\widehat{\varphi}\,|\varphi_\alpha\rangle = \varphi_\alpha\,|\varphi_\alpha\rangle \,, \tag{1.70}$$

$$\langle \varphi_\alpha | \varphi_\beta \rangle = \delta_{\alpha\beta} \,, \tag{1.71}$$

$$\sum_\alpha |\varphi_\alpha\rangle \langle \varphi_\alpha| = \mathbf{1} \quad \text{in } \mathcal{H}_1 \,. \tag{1.72}$$

In principle, we can make use of the same considerations as in Sect. 1.2, and can therefore proceed somewhat more quickly.

Our starting point is a non-symmetrised N-particle state of the form (1.25):

$$\left|\varphi_{\alpha_1} \cdots \varphi_{\alpha_N}\right\rangle = \left|\varphi^{(1)}_{\alpha_1}\right\rangle \cdots \left|\varphi^{(N)}_{\alpha_N}\right\rangle \,. \tag{1.73}$$

The state indices $\alpha_1, \ldots \alpha_N$ are taken here again to be given in an arbitrary, but fixed **standard ordering**. We now apply the operator $\widehat{S}_\varepsilon$ from (1.18) to this state and obtain an

(anti-)symmetrised N-particle state

$$\left|\varphi_{\alpha_1} \cdots \varphi_{\alpha_N}\right\rangle^{(\varepsilon)} = C_\varepsilon \sum_{\mathcal{P}} \varepsilon^p \mathcal{P} \left|\varphi_{\alpha_1} \cdots \varphi_{\alpha_N}\right\rangle \,, \tag{1.74}$$

which differs formally from (1.26) only through a normalisation constant C_ε, which is still to be determined. One can see that for Fermions ($\varepsilon = -$), the antisymmetrised state may also be written in the form of a determinant:

$$\left|\varphi_{\alpha_1} \cdots \varphi_{\alpha_N}\right\rangle^{(-)} = C_- \begin{vmatrix} \left|\varphi_{\alpha_1}^{(1)}\right\rangle & \left|\varphi_{\alpha_1}^{(2)}\right\rangle & \cdots & \left|\varphi_{\alpha_1}^{(N)}\right\rangle \\ \left|\varphi_{\alpha_2}^{(1)}\right\rangle & \left|\varphi_{\alpha_2}^{(2)}\right\rangle & \cdots & \left|\varphi_{\alpha_2}^{(N)}\right\rangle \\ \vdots & \vdots & \vdots & \vdots \\ \left|\varphi_{\alpha_N}^{(1)}\right\rangle & \left|\varphi_{\alpha_N}^{(2)}\right\rangle & \cdots & \left|\varphi_{\alpha_N}^{(N)}\right\rangle \end{vmatrix}, \tag{1.75}$$

the **Slater determinant.**

If two sets of quantum numbers are the same in the N-particle state ($\alpha_i = \alpha_j$), then this means that two rows in the determinant would be the same. The determinant would then have the value zero. The probability of finding two Fermions in the same single-particle state is thus zero. This is equivalent to the statement made by the **Pauli principle**, which of course holds not only for the case discussed here of a *discrete* spectrum. Naturally, one can also write (1.26) for $\varepsilon = -$ as a Slater determinant.

As the next step, we want to determine the normalisation constant C_ε and introduce to this end the

occupation numbers n_i.

These numbers reflect the frequency with which a particular single-particle state $\left|\varphi_{\alpha_i}\right\rangle$ occurs within the N-particle state $\left|\varphi_{\alpha_1}\cdots\right\rangle^{(\varepsilon)}$, or, more intuitively, the number of identical particles in the state $\left|\varphi_{\alpha_i}\right\rangle$:

$$\begin{aligned} &\sum_i n_i = N, \\ &n_i = 0, 1 \qquad \textbf{Fermions}, \\ &n_i = 0, 1, 2, \ldots \quad \textbf{Bosons}. \end{aligned} \tag{1.76}$$

Let C_ε be real and chosen in such a way that the N-particle state $\left|\varphi_{\alpha_1} \cdots \varphi_{\alpha_N}\right\rangle^{(\varepsilon)}$ is normalised to 1. It then follows that:

$$\begin{aligned} 1 \stackrel{!}{=} \left\langle \varphi_N^{(\varepsilon)} \middle| \varphi_N^{(\varepsilon)} \right\rangle &= C_\varepsilon \sum_{\mathcal{P}} \varepsilon^p \left\langle \varphi_{\alpha_1} \cdots \varphi_{\alpha_N} \middle| \mathcal{P}^+ \middle| \varphi_N^{(\varepsilon)} \right\rangle \stackrel{(\mathcal{P}^+ = \mathcal{P})}{=} \\ &\stackrel{(\mathcal{P}^+ = \mathcal{P})}{=} C_\varepsilon \sum_{\mathcal{P}} \varepsilon^{2p} \left\langle \varphi_{\alpha_1} \cdots \middle| \varphi_N^{(\varepsilon)} \right\rangle = \\ &= N!\, C_\varepsilon \left\langle \varphi_{\alpha_1} \cdots \varphi_{\alpha_N} \middle| \varphi_N^{(\varepsilon)} \right\rangle . \end{aligned}$$

This yields:

$$\left(N!\,C_\varepsilon^2\right)^{-1} = \sum_{\mathcal{P}} \varepsilon^p \left\langle\varphi_{\alpha_1}^{(1)}\right|\left\langle\varphi_{\alpha_2}^{(2)}\right|\cdots\left\langle\varphi_{\alpha_N}^{(N)}\right|\left(\mathcal{P}\left|\varphi_{\alpha_1}^{(1)}\right\rangle\cdots\left|\varphi_{\alpha_N}^{(N)}\right\rangle\right). \tag{1.77}$$

In the case of Fermions, each state occurs once and only once, i. e. all N single-particle states are pairwise distinct. The right-hand side is thus only nonzero when $\mathcal{P}$ is the identity, and then, due to $\varepsilon^0 = +1$ and to (1.71), it is equal to 1.

$$C_- = \frac{1}{\sqrt{N!}}\,. \tag{1.78}$$

For Bosons ($\varepsilon = +$), all the permutations are allowed which simply exchange the particles in the n_i equivalent single-particle states $|\varphi_{\alpha_i}\rangle$. Clearly, there are

$$n_1!\,n_2!\,\cdots\,n_i!\,\cdots$$

such permutations, each of which contributes a summand with the value $+1$ to (1.77). This leads to:

$$C_+ = \left(N!\prod_i n_i!\right)^{-1/2}. \tag{1.79}$$

Formally, this expression is valid also for Fermions, due to $0! = 1! = 1$.

We can see that an (anti-)symmetrised N-particle state can be uniquely characterised by giving its occupation numbers. This leads to an alternate representation, which is called the

occupation-number representation:

$$\begin{aligned}\left|N;\,n_1 n_2\cdots n_i\cdots n_j\cdots\right\rangle^{(\varepsilon)} &\equiv \left|\varphi_{\alpha_1}\cdots\varphi_{\alpha_N}\right\rangle^{(\varepsilon)} = \\ &= C_\varepsilon\sum_{\mathcal{P}}\varepsilon^p\mathcal{P}\Big\{\underbrace{\left|\varphi_{\alpha_1}^{(1)}\right\rangle\left|\varphi_{\alpha_1}^{(2)}\right\rangle\cdots}_{n_1}\cdots\underbrace{\left|\varphi_{\alpha_i}^{(p)}\right\rangle\left|\varphi_{\alpha_i}^{(p+1)}\right\rangle\cdots}_{n_i}\Big\}.\end{aligned} \tag{1.80}$$

In the symbol for the state, **all** occupation numbers are given; the unoccupied single-particle states are then denoted by $n_i = 0$. Two states are clearly identical if and only if they are the same in terms of all the occupation numbers. The

orthonormalisation

$$^{(\varepsilon)}\left\langle N;\cdots n_i\cdots\right|\left.\overline{N};\cdots\bar{n}_i\cdots\right\rangle^{(\varepsilon)} = \delta_{N\overline{N}}\prod_i\delta_{n_i\bar{n}_i} \tag{1.81}$$

follows immediately from the single-particle states. This holds in the same way for the

completeness

$$\sum_{n_1}\sum_{n_2}\cdots\sum_{n_i}\cdots\,|N;\cdots n_i\cdots\rangle^{(\varepsilon)(\varepsilon)}\langle N;\cdots n_i\cdots| = \mathbf{1} \tag{1.82}$$

of the so-called **Fock states**. The sum runs over all the allowed occupation numbers with the condition $\sum_i n_i = N$.

The creation and annihilation operators, which we shall now discuss, are defined up to their normalisation factors as in Sect. 1.2:

The creation operator: $a^+_{\alpha_r} \equiv a^+_r$

$$\begin{aligned} a_r^+\,|N;\cdots n_r\cdots\rangle^{(\varepsilon)} &= \\ &= a_r^+\,\left|\varphi_{\alpha_1}\cdots\varphi_{\alpha_N}\right\rangle^{(\varepsilon)} \equiv \\ &\equiv \sqrt{n_r+1}\,\Big|\varphi_{\alpha_r}\underbrace{\varphi_{\alpha_1}\varphi_{\alpha_1}\cdots}_{n_1}\cdots\underbrace{\varphi_{\alpha_r}\varphi_{\alpha_r}\cdots}_{n_r}\cdots\Big\rangle^{(\varepsilon)} = \\ &= \varepsilon^{N_r}\sqrt{n_r+1}\,\Big|\underbrace{\varphi_{\alpha_1}\varphi_{\alpha_1}\cdots}_{n_1}\cdots\underbrace{\varphi_{\alpha_r}\varphi_{\alpha_r}\cdots}_{n_r+1}\cdots\Big\rangle^{(\varepsilon)} \end{aligned} \tag{1.83}$$

Here,

$$N_r = \sum_{i=1}^{r-1} n_i \tag{1.84}$$

is assumed to hold. The creation operator thus acts as follows:

Bosons:

$$a_r^+\,|N;\cdots n_r\cdots\rangle^{(+)} = \sqrt{n_r+1}\,|N+1;\cdots n_r+1\cdots\rangle^{(+)}\;,$$

Fermions:

$$a_r^+\,|N;\cdots n_r\cdots\rangle^{(-)} = (-1)^{N_r}\delta_{n_r,0}\,|N+1;\cdots n_r+1\cdots\rangle^{(-)}\;. \tag{1.85}$$

Every N-particle Fock state can be **created** by repeated application of the creation operators from the vacuum state:

$$|N; n_1 \cdots n_i \cdots\rangle^{(\varepsilon)} = \prod_{p=1\cdots}^{\sum n_p = N} \frac{(a_p^+)^{n_p}}{\sqrt{n_p!}} \varepsilon^{N_p} |0\rangle \, . \tag{1.86}$$

The annihilation operator: $a_r \equiv (a_r^+)^+$

is again defined as the adjoint of the creation operator. Its action can be read off the following general matrix element:

$$\begin{aligned}
&{}^{(\varepsilon)}\big\langle N; \cdots n_r \cdots \big| a_r \big| \overline{N}; \cdots \bar{n}_r \cdots \big\rangle^{(\varepsilon)} = \\
&= \varepsilon^{N_r} \sqrt{n_r + 1}\, {}^{(\varepsilon)}\big\langle N+1; \cdots n_r + 1 \cdots \big| \overline{N}; \cdots \bar{n}_r \cdots \big\rangle^{(\varepsilon)} = \\
&= \varepsilon^{N_r} \sqrt{n_r + 1}\, \delta_{N+1, \overline{N}} \left(\delta_{n_1 \bar{n}_1} \cdots \delta_{n_r + 1, \bar{n}_r} \cdots \right) = \\
&= \varepsilon^{\bar{N}_r} \sqrt{\bar{n}_r}\, \delta_{N, \overline{N} - 1} \left(\delta_{n_1 \bar{n}_1} \cdots \delta_{n_r, \bar{n}_r - 1} \cdots \right) = \\
&= \varepsilon^{\overline{N}_r} \sqrt{\bar{n}_r}\, {}^{(\varepsilon)}\big\langle N; n_1 \cdots n_r \cdots \big| \overline{N} - 1; \bar{n}_1 \cdots \bar{n}_r - 1 \cdots \big\rangle^{(\varepsilon)} \, .
\end{aligned}$$

This holds for arbitrary basis states, so that clearly it must follow that:

$$a_r \big| \overline{N}; \cdots \bar{n}_r \cdots \big\rangle^{(\varepsilon)} = \varepsilon^{\overline{N}_r} \sqrt{\bar{n}_r} \big| \overline{N} - 1; \bar{n}_1 \cdots \bar{n}_r - 1 \cdots \big\rangle^{(\varepsilon)} \, .$$

For Fermions, we still have to take into account the limitation on the occupation numbers:

Bosons:

$$a_r |N; \cdots n_r \cdots\rangle^{(+)} = \sqrt{n_r}\, |N - 1; \cdots n_r - 1 \cdots\rangle^{(+)} ,$$

Fermions:

$$a_r |N; \cdots n_r \cdots\rangle^{(-)} = \delta_{n_r, 1} (-1)^{N_r} |N - 1; \cdots n_r - 1 \cdots\rangle^{(-)} . \tag{1.87}$$

To derive the fundamental commutation relations, we start from our definition equations (1.85) and (1.87). One can directly read off the following relations:

1. Bosons ($r \neq p$):

$$\begin{aligned}
&a_r^+ a_p^+ \big| \cdots n_r \cdots n_p \cdots \big\rangle^{(+)} = \\
&\quad = \sqrt{n_r + 1} \sqrt{n_p + 1} \big| \cdots n_r + 1 \cdots n_p + 1 \cdots \big\rangle^{(+)} = \\
&\quad = a_p^+ a_r^+ \big| \cdots n_r \cdots n_p \cdots \big\rangle^{(+)} ,
\end{aligned} \tag{1.88}$$

$$
\begin{aligned}
a_r a_p \left|\cdots n_r \cdots n_p \cdots\right\rangle^{(+)} &= \\
&= \sqrt{n_r}\sqrt{n_p}\left|\cdots n_r - 1 \cdots n_p - 1 \cdots\right\rangle^{(+)} = \\
&= a_p a_r \left|\cdots n_r \cdots n_p \cdots\right\rangle^{(+)} , \\
a_r^+ a_p \left|\cdots n_r \cdots n_p \cdots\right\rangle^{(+)} &= \qquad (1.89) \\
&= \sqrt{n_p}\sqrt{n_r + 1}\left|\cdots n_r + 1 \cdots n_p - 1 \cdots\right\rangle^{(+)} = \\
&= a_p a_r^+ \left|\cdots n_r \cdots n_p \cdots\right\rangle^{(+)} . \qquad (1.90) \\
a_r^+ a_r \left|\cdots n_r \cdots\right\rangle^{(+)} &= \\
&= \sqrt{n_r} a_r^+ \left|\cdots n_r - 1 \cdots\right\rangle^{(+)} = \\
&= n_r \left|\cdots n_r \cdots\right\rangle^{(+)} , \qquad (1.91) \\
a_r a_r^+ \left|\cdots n_r \cdots\right\rangle^{(+)} &= \\
&= \sqrt{n_r + 1}\, a_r \left|\cdots n_r + 1 \cdots\right\rangle^{(+)} = \\
&= (n_r + 1)\left|\cdots n_r \cdots\right\rangle^{(+)} . \qquad (1.92)
\end{aligned}
$$

2. Fermions $(r < p)$:

$$
\begin{aligned}
a_r^+ a_p^+ \left|\cdots n_r \cdots n_p \cdots\right\rangle^{(-)} &= \\
&= (-1)^{N_p}(-1)^{N_r}\delta_{n_r,0}\delta_{n_p,0}\left|\cdots n_r + 1 \cdots n_p + 1 \cdots\right\rangle^{(-)} , \\
a_p^+ a_r^+ \left|\cdots n_r \cdots n_p \cdots\right\rangle^{(-)} &= \\
&= (-1)^{N_r}(-1)^{N_p+1}\delta_{n_r,0}\,\delta_{n_p,0}\left|\cdots n_r + 1 \cdots n_p + 1 \cdots\right\rangle^{(-)} = \\
&= -a_r^+ a_p^+ \left|\cdots n_r \cdots n_p \cdots\right\rangle^{(-)} , \qquad (1.93) \\
a_r^+ a_r \left|\cdots n_r \cdots\right\rangle^{(-)} &= \\
&= (-1)^{2N_r}\delta_{n_r,1}\left|\cdots n_r \cdots\right\rangle^{(-)} = \delta_{n_r,1}\left|\cdots n_r \cdots\right\rangle^{(-)} , \qquad (1.94) \\
a_r a_r^+ \left|\cdots n_r \cdots\right\rangle^{(-)} &= \\
&= (-1)^{2N_r}\delta_{n_r,0}\left|\cdots n_r \cdots\right\rangle^{(-)} = \delta_{n_r,0}\left|\cdots n_r \cdots\right\rangle^{(-)} ,
\end{aligned}
$$

$$a_r^+ a_p \left|\cdots n_r \cdots n_p \cdots\right\rangle^{(-)} =$$
$$= (-1)^{N_p}(-1)^{N_r}\delta_{n_p,1}\,\delta_{n_r,0}\left|\cdots n_r+1\cdots n_p-1\cdots\right\rangle^{(-)}$$
$$a_p a_r^+ \left|\cdots n_r \cdots n_p \cdots\right\rangle^{(-)} = \tag{1.95}$$
$$= (-1)^{N_r}(-1)^{N_p+1}\delta_{n_r,0}\,\delta_{n_p,1}\left|\cdots n_r+1\cdots n_p-1\cdots\right\rangle^{(-)} =$$
$$= -a_r^+ a_p \left|\cdots n_r \cdots n_p \cdots\right\rangle^{(-)} . \tag{1.96}$$

Since all of these relations hold for arbitrary basis states, the following operator identities can be directly obtained:

$$[a_r, a_s]_{-\varepsilon} = 0 , \tag{1.97}$$
$$\left[a_r^+, a_s^+\right]_{-\varepsilon} = 0 , \tag{1.98}$$
$$\left[a_r, a_s^+\right]_{-\varepsilon} = \delta_{rs} . \tag{1.99}$$

These are the fundamental commutation relations which are analogous to (1.37), (1.44) and (1.45) for the creation and annihilation operators in the *discrete* Fock representation.

In order to represent an arbitrary operator $\widehat{A}$, which as in (1.48) consists of single-particle and two-particle parts, within the formalism of the second quantisation in terms of creation and annihilation operators, we make use of exactly the same considerations as in the case of a continuous spectrum:

$$\widehat{A} = \sum_{p,r} \langle \varphi_{\alpha_p} | \widehat{A}_1 | \varphi_{\alpha_r} \rangle a_p^+ a_r + + \frac{1}{2} \sum_{\substack{p,r,\\ s,t}} \left\langle \varphi_{\alpha_p}^{(1)} \varphi_{\alpha_r}^{(2)} | \widehat{A}_2 | \varphi_{\alpha_t}^{(1)} \varphi_{\alpha_s}^{(2)} \right\rangle a_p^+ a_r^+ a_s a_t . \tag{1.100}$$

The only difference from the continuous case consists of the fact that here, the two-particle matrix element must be formed in every case with non-symmetrised two-particle states. In (1.54), we could also use the (anti-)symmetrised states. The reason for this lies exclusively in the different normalisations.

The analogy to the occupation-density operator (1.55) is, in the *discrete* case, the

occupation-number operator

$$\hat{n}_r = a_r^+ a_r . \tag{1.101}$$

One can see from (1.90) and (1.94) that the Fock states are eigenstates of $\hat{n}_r$:

$$\hat{n}_r \, |N; \cdots n_r \cdots\rangle^{(\varepsilon)} = n_r \, |N; \cdots n_r \cdots\rangle^{(\varepsilon)} \; . \tag{1.102}$$

$\hat{n}_r$ thus *asks the question*: How many particles occupy the r-th single-particle state:

Particle number operator

$$\widehat{N} = \sum_r \hat{n}_r \; . \tag{1.103}$$

Its eigenstates are the Fock states with the total particle number N as eigenvalue:

$$\begin{aligned}\widehat{N} \, |N; \cdots n_r \cdots\rangle^{(\varepsilon)} &= \left(\sum_r n_r\right) |N; \cdots n_r \cdots\rangle^{(\varepsilon)} = \\ &= N \, |N; \cdots n_r \cdots\rangle^{(\varepsilon)} \; . \end{aligned} \tag{1.104}$$

The derivation of the following useful commutation relations, which hold equally for Bosons and for Fermions, can be carried out using (1.97), (1.98) and (1.99) and is recommended as an exercise:

$$\begin{aligned} \left[\hat{n}_r, a_p^+\right]_- &= \delta_{rp} a_p^+ \; ; \qquad & \left[\hat{n}_r, a_p\right]_- &= -\delta_{rp} a_p \; , \\ \left[\widehat{N}, a_p^+\right]_- &= a_p^+ \; ; \qquad & \left[\widehat{N}, a_p\right]_- &= -a_p \; . \end{aligned} \tag{1.105}$$

1.4 Exercises

Exercise 1.4.1 Two identical particles are moving in a one-dimensional potential well with infinitely high walls:

$$V(x) = \begin{cases} 0 & for\ 0 \le x \le a, \\ \infty & for\ x < 0\ and\ x > a. \end{cases}$$

Compute their energy eigenfunctions and the energy eigenvalues of the two-particle system, in the case that a) the particles are Bosons, and b) the particles are Fermions. What is the ground-state energy in the case of $N \gg 1$ Bosons or Fermions?

Exercise 1.4.2 Consider a system of two spin 1/2 particles. The common eigenstates

$$\left|S_i, m_S^{(i)}\right\rangle \; ; \quad S_i = \frac{1}{2} \; ; \quad m_S^{(i)} = \pm\frac{1}{2} \; ; \quad i = 1, 2$$

of the spin operators $\boldsymbol{S}_i^2$, S_i^z,

$$\boldsymbol{S}_i^2 \left|\frac{1}{2}, m_S^{(i)}\right\rangle = \frac{3}{4}\hbar^2 \left|\frac{1}{2}, m_S^{(i)}\right\rangle \; ; \quad S_i^z \left|\frac{1}{2}, m_S^{(i)}\right\rangle = \hbar m_S^{(i)} \left|\frac{1}{2}, m_S^{(i)}\right\rangle ,$$

form a complete single-particle basis. For the non-symmetrised two-particle states,

$$\left|m_{S_1}^{(1)}, m_{S_2}^{(2)}\right\rangle = \left|\frac{1}{2}, m_{S_1}^{(1)}\right\rangle \left|\frac{1}{2}, m_{S_2}^{(2)}\right\rangle ,$$

let the permutation (transposition) operator P_{12} be defined as usual:

$$P_{12} \left|m_{S_1}^{(1)}, m_{S_2}^{(2)}\right\rangle = \left|m_{S_1}^{(2)}, m_{S_2}^{(1)}\right\rangle .$$

Prove the following statements:

1. The common eigenstates $|S, M_S\rangle_t$ of the operators

$$\boldsymbol{S}_1^2, \boldsymbol{S}_2^2, \boldsymbol{S}^2 = (\boldsymbol{S}_1 + \boldsymbol{S}_2)^2 \; , \quad S^z = S_1^z + S_2^z \; ,$$

$$|0, 0\rangle_t = 2^{-1/2}\Big(\left|(1/2)^{(1)}, (-1/2)^{(2)}\right\rangle - \left|(1/2)^{(2)}, (-1/2)^{(1)}\right\rangle\Big) ,$$

$$|1, 0\rangle_t = 2^{-1/2}\Big(\left|(1/2)^{(1)}, (-1/2)^{(2)}\right\rangle + \left|(1/2)^{(2)}, (-1/2)^{(1)}\right\rangle\Big) ,$$

$$|1, \pm 1_t = \left|(\pm 1/2)^{(1)}, (\pm 1/2)^{(2)}\right\rangle$$

are eigenstates of P_{12}.

2. In $\mathcal{H}_2^{(\varepsilon)}$, the following relations hold:

$$P_{12} \boldsymbol{S}_1 P_{12} = \boldsymbol{S}_2 \; ; \quad P_{12} \boldsymbol{S}_2 P_{12} = \boldsymbol{S}_1 \, .$$

3. The representation

$$P_{12} = \frac{1}{2}\left(\mathbb{1} + \frac{4}{\hbar^2} \boldsymbol{S}_1 \cdot \boldsymbol{S}_2\right) .$$

applies.

Exercise 1.4.3 Let the normalised vacuum state $|0\rangle$ ($\langle 0 \mid 0\rangle = 1$) and $|\varphi_\alpha\rangle$ be an eigenstate belonging to an observable $\widehat{\Phi}$ with a continuous spectrum:

$$\left\langle \varphi_\alpha \middle| \varphi_\beta \right\rangle = \delta(\alpha - \beta)\ .$$

a_α^+ and a_α are creation and annihilation operators for a particle in a single-particle state $|\Phi_\alpha\rangle$. Using the commutation relations for a_α^+, a_α, derive the following expressions:

$$\langle 0 \mid a_{\beta_N} \cdots a_{\beta_1} a_{\alpha_1}^+ \cdots a_{\alpha_N}^+ \mid 0\rangle = \sum_{\mathcal{P}_\alpha} \varepsilon^{p_\alpha} \mathcal{P}_\alpha \left(\delta\left(\beta_1 - \alpha_1\right) \cdots \delta\left(\beta_N - \alpha_N\right)\right)\ .$$

$\mathcal{P}_\alpha$ is the permutation operator which acts on the state indices α_i. ε is $+1$ for Bosons and -1 for Fermions.

Exercise 1.4.4 Consider a system of N identical (spinless) particles with a pair interaction which depends only on their distance

$$V_{ij} = V\left(\left|\boldsymbol{r}_i - \boldsymbol{r}_j\right|\right)\ .$$

Show that the Hamiltonian

$$H = \sum_{i=1}^{n} \frac{p_i^2}{2m} + \frac{1}{2} \sum_{i,j}^{i \neq j} V_{ij}$$

can be written as follows in the continuous $\boldsymbol{k}$ representation (plane waves!):

$$H = \int \mathrm{d}^3 k \left(\frac{\hbar^2 k^2}{2m}\right) a_{\boldsymbol{k}}^+ a_{\boldsymbol{k}} + \frac{1}{2} \iiint \mathrm{d}^3 k\, \mathrm{d}^3 p\, \mathrm{d}^3 q\, V(\boldsymbol{q}) a_{\boldsymbol{k}+\boldsymbol{q}}^+ a_{\boldsymbol{p}-\boldsymbol{q}}^+ a_{\boldsymbol{p}} a_{\boldsymbol{k}}\ .$$

Here,

$$V(\boldsymbol{q}) = (2\pi)^{-3} \int \mathrm{d}^3 r\, V(\boldsymbol{r}) \mathrm{e}^{\mathrm{i}\boldsymbol{q}\cdot\boldsymbol{r}} = V(-\boldsymbol{q})$$

is the Fourier transform of the interaction potential. You can use the following form of the δ function:

$$\delta\left(\boldsymbol{k} - \boldsymbol{k}'\right) = (2\pi)^{-3} \int \mathrm{d}^3 r\, \mathrm{e}^{-\mathrm{i}(\boldsymbol{k}-\boldsymbol{k}')\cdot\boldsymbol{r}}\ .$$

Exercise 1.4.5 Show that the particle number operator

$$\widehat{N} = \int d^3k \, a_k^+ a_k$$

commutes with the Hamiltonian from Ex. 1.4.4!

Exercise 1.4.6 For a system of N identical (spinless) particles with a pair interaction which depends only on their distance $V(|\boldsymbol{r} - \boldsymbol{r}'|)$, the Hamiltonian H can be expressed in the second quantisation in terms of the field operators:

$$H = \int d^3r \, \widehat{\psi}^+(\boldsymbol{r}) \left\{ -\frac{\hbar^2}{2m} \Delta_r \right\} \widehat{\psi}(\boldsymbol{r}) +$$

$$+ \frac{1}{2} \iint d^3r \, d^3r' \, \widehat{\psi}^+(\boldsymbol{r}) \widehat{\psi}^+(\boldsymbol{r}') \, V(|\boldsymbol{r} - \boldsymbol{r}'|) \, \widehat{\psi}(\boldsymbol{r}') \, \widehat{\psi}(\boldsymbol{r}) \, .$$

Demonstrate the equivalence of this description to the $\boldsymbol{k}$ representation for H which was derived in Ex. 1.4.4 by making use of plane waves as single-particle wavefunctions.

Exercise 1.4.7 Let $a_{\varphi_\alpha} = a_\alpha$ and $a_{\varphi_\alpha}^+ = a_\alpha^+$ be annihilation and creation operators for single-particle states $|\varphi_\alpha\rangle$ of an observable $\widehat{\Phi}$ with a discrete spectrum. Compute the following commutators using the fundamental commutation relations for Bosons and for Fermions:

1. $\left[\hat{n}_\alpha, a_\beta^+\right]_-$;
2. $\left[\hat{n}_\alpha, a_\beta\right]_-$;
3. $\left[\widehat{N}, a_\alpha^+\right]_-$;
4. $\left[\widehat{N}, a_\alpha\right]_-$.

Exercise 1.4.8 Show that with the assumptions made in Ex. 1.4.7 for Fermions, the following relations are valid:

1. $(a_\alpha)^2 = 0 \, ; \quad \left(a_\alpha^+\right)^2 = 0 \, ,$
2. $(\hat{n}_\alpha)^2 = \hat{n}_\alpha \, ,$
3. $a_\alpha \hat{n}_\alpha = a_\alpha \, ; \; a_\alpha^+ \hat{n}_\alpha = 0 \, ,$
4. $\hat{n}_\alpha a_\alpha = 0 \, ; \; \hat{n}_\alpha a_\alpha^+ = a_\alpha^+ \, .$

Exercise 1.4.9 Consider a system of non-interacting, identical Bosons or Fermions:

$$H = \sum_{i=1}^{N} H_1^{(i)} .$$

The single-particle operator $H_1^{(i)}$ is supposed to have a discrete, non-degenerate spectrum:

$$H_1^{(i)} \left|\varphi_r^{(i)}\right\rangle = \varepsilon_r \left|\varphi_r^{(i)}\right\rangle ; \quad \left\langle \varphi_r^{(i)} \middle| \varphi_S^{(i)} \right\rangle = \delta_{rs} .$$

The $\left|\varphi_r^{(i)}\right\rangle$ are used to construct the Fock states $|N; n_1, n_2, \ldots\rangle^{(\varepsilon)}$. The general state of the system is described by the non-normalised density matrix ρ, for which in the grand canonical ensemble (variable particle number!), the following holds:

$$\rho = \exp\left[-\beta \left(H - \mu \widehat{N}\right)\right] .$$

1. What is the Hamiltonian in second quantisation?
2. Verify that for the grand canonical partition function, the following relation holds:

$$\Xi(T, V, \mu) = \mathrm{Tr}\rho = \begin{cases} \prod_i \left\{1 - \exp\left[-\beta \left(\varepsilon_i - \mu\right)\right]\right\}^{-1} & \textit{Bosons}, \\ \prod_i \left\{1 + \exp\left[-\beta \left(\varepsilon_i - \mu\right)\right]\right\} & \textit{Fermions}. \end{cases}$$

3. Compute the expectation value of the particle number:

$$\left\langle \widehat{N} \right\rangle = \frac{1}{\Xi} Tr\left(\rho \widehat{N}\right) .$$

4. Compute the internal energy:

$$U = \langle H \rangle = \frac{1}{\Xi} Tr(\rho H) .$$

5. Compute the mean occupation number of the i-th single-particle state,

$$\langle \hat{n}_i \rangle = \frac{1}{\Xi} Tr\left(\rho a_i^+ a_i\right)$$

and show that the following relation holds:

$$U = \sum_i \varepsilon_i \langle \hat{n}_i \rangle ; \quad \left\langle \widehat{N} \right\rangle = \sum_i \langle \hat{n}_i \rangle .$$

Exercise 1.4.10 Consider a system of electrons which stem from two different energy levels, ε_1 and ε_2. They are described by the following Hamiltonian:

$$H = \sum_{\sigma} \left[\varepsilon_1 a^+_{1\sigma} a_{1\sigma} + \varepsilon_2 a^+_{2\sigma} a_{2\sigma} + V \left(a^+_{1\sigma} a_{2\sigma} + a^+_{2\sigma} a_{1\sigma} \right) \right] \quad (\sigma = \uparrow \text{ or } \downarrow) .$$

1. Show that H commutes with the particle number operator

$$\widehat{N} = \sum_{\sigma} \left(a^+_{1\sigma} a_{1\sigma} + a^+_{2\sigma} a_{2\sigma} \right) .$$

2. Develop a general procedure for computing the energy eigenvalues for arbitrary total electron numbers N $(N = 0, 1, 2, 3, 4)$, making use of the Fock states

$$|N; F\rangle = \left| N; n_{1\uparrow} n_{1\downarrow}; n_{2\uparrow} n_{2\downarrow} \right\rangle^{(-)} .$$

3. Calculate the energy eigenvalues for $N = 0$ and $N = 1$.
4. Show that of the six possible Fock states for $N = 2$, two are already eigenstates of H. Solve the remaining 4×4 secular determinant.
5. Find the energy eigenvalues for $N = 3$ and $N = 4$.

1.5 Self-Examination Questions

For Section 1.1

1. What is meant by *identical* particles?
2. Why are even identical particles distinguishable in classical physics?
3. What does the principle of indistinguishability state?
4. Justify for an arbitrary observable $\widehat{A}$ of a system of identical particles the operator identity $\widehat{A} = P_{ij}^{+}\widehat{A}P_{ij}$, where P_{ij} is the transposition operator.
5. How does one construct (anti-)symmetrised N-particle states?
6. Can the symmetry character of a state of N identical particles change with time?
7. Justify why *all* the states of a system of N identical particles have the same symmetry character.
8. Formulate the relation between spin and statistics.
9. What are Bosons, and what are Fermions? Name some examples.

For Section 1.2

1. Why is every permutation operator $\mathcal{P}$ in the space $\mathcal{H}_N^{(\pm)}$ of a system of N identical particles Hermitian?
2. What is the scalar product of two (anti-)symmetrised N-particle states, which are constructed of single-particle states $|\varphi_\alpha\rangle$ with a continuous spectrum?
3. Formulate the completeness relation for states as in 1.5.
4. How can an (anti-)symmetrised N-particle state $\left|\varphi_{\alpha_1} \dots \varphi_{\alpha_N}\right\rangle^{(\pm)}$ be constructed from the vacuum state $|0\rangle$ with the aid of creation operators?
5. How does the annihilation operator a_γ act on the N-particle state $\left|\varphi_{\alpha_1} \dots \varphi_{\alpha_N}\right\rangle^{(\pm)}$?
6. How do a_α and a_α^+ act on the vacuum state $|0\rangle$?
7. Explain the concepts *creation operator* and *annihilation operator*.
8. Formulate the three fundamental commutation relations.
9. Express a general single-particle operator in terms of creation and annihilation operators.
10. Are there any restrictions on the single-particle basis $\{|\varphi_\alpha\rangle\}$ from which the (anti-) symmetrised N-particle basis states of $\mathcal{H}_N^{(\varepsilon)}$ are constructed? What aspects could influence your choice of states?
11. How are the occupation-density and particle number operators defined? What form do their eigenstates take?
12. What is meant by field operators?
13. What relation exists between field operators and the general creation and annihilation operator a_α, a_α^+?

For Section 1.3

1. What does the Slater determinant describe?
2. What is the relation between the Slater determinant and the Pauli principle?
3. What is meant by the occupation number n_i?
4. How does one formulate an N-particle state in the occupation representation?
5. Formulate the orthonormalisation and completeness relations for Fock states.
6. Describe the action of creation and annihilation operators on N-particle states in the occupation representation.
7. How can an N-particle Fock state be created from the vacuum state $|0\rangle$?
8. What are the fundamental commutation relations in the *discrete* case?
9. Show that the Fock states are eigenstates of the occupation number and the particle number operators.

Many-Body Model Systems

2

In this section, we introduce some model systems which are frequently treated and with which we shall later demonstrate and test the elements of the abstract theory. In formulating the model Hamiltonians, we can practice the transformation from the first to the second quantisation. The examples chosen are all taken from the field of theoretical solid-state physics and will be preceded by some introductory remarks.

A solid is certainly a many-body system,

$$\text{Solid} = \sum_{i=1}^{N} (\text{particles})_i \ ,$$

composed of atoms or molecules which interact with one another. Each *particle* consists of one or more positively-charged atomic nuclei and a negatively-charged electron cloud. One distinguishes between **core electrons** and **valence electrons**. The core electrons are strongly bound and are localised in the immediate neighbourhood of the nuclei. They as a rule occupy closed electronic shells – exceptions are e.g. the $4f$ electrons of the rare earths – and thus have hardly any influence on the characteristic properties of the solid. This is in contrast to the valence electrons, which occupy non-closed shells and are responsible for the bonding to form a solid. Of course, this separation into core and valence electrons is not always clear cut. It already represents a certain approximation. A **lattice ion** refers in this sense to the ensemble of the atomic nucleus plus the core electrons. This leads to the following **model**:

Solid:
an interacting system of particles consisting of lattice ions and valence electrons.

How is the corresponding Hamiltonian constructed?

$$H = H_e + H_i + H_{ei} \ . \tag{2.1}$$

W. Nolting, *Fundamentals of Many-body Physics*,
DOI 10.1007/978-3-540-71931-1_2,

The **subsystem of the electrons** is described by the operator H_e:

$$H_{\mathrm{e}} = \sum_{i=1}^{N_{\mathrm{e}}} \frac{p_{\mathrm{i}}^2}{2m} + \frac{1}{2}\frac{1}{4\pi\varepsilon_0} \sum_{i,j}^{i\neq j} \frac{e^2}{|\boldsymbol{r}_i - \boldsymbol{r}_j|} \equiv H_{\mathrm{e,kin.}} + H_{\mathrm{ee}} \,. \tag{2.2}$$

N_e is the number of valence electrons. The first term represents their kinetic energy, the second term is their Coulomb interaction. $\boldsymbol{r}_i$, $\boldsymbol{r}_j$ are the position vectors of the electrons.

The **subsystem of the ions** is defined by the operator H_i:

$$H_{\mathrm{i}} = \sum_{\alpha=1}^{N_{\mathrm{i}}} \frac{p_\alpha^2}{2M_\alpha} + \frac{1}{2} \sum_{\alpha,\beta}^{\alpha\neq\beta} V_{\mathrm{i}}(\boldsymbol{R}_\alpha - \boldsymbol{R}_\beta) \equiv H_{\mathrm{i,kin}} + H_{\mathrm{ii}} \,. \tag{2.3}$$

The ion-ion interaction need not be precisely specified at this point. It is in every case a pairwise interaction. It is partially responsible for the fact that the equilibrium positions of the ions, $\boldsymbol{R}_\alpha^{(0)}$, define a strictly periodic crystal lattice. The ions exhibit oscillations around these equilibrium positions; the oscillation energy is quantised. The elementary quantum is called a **phonon**. It is therefore expedient to separate H_{ii} further into

$$H_{\mathrm{ii}} = H_{\mathrm{ii}}^{(0)} + H_{\mathrm{p}} \,. \tag{2.4}$$

$H_{ii}^{(0)}$ determines for example the bonding in the solid, and H_p the lattice dynamics.

The **interaction of the two subsystems** is finally given by

$$H_{\mathrm{ei}} = \sum_{i=1}^{N_{\mathrm{e}}} \sum_{\alpha=1}^{N_{\mathrm{i}}} V_{\mathrm{ei}}(\boldsymbol{r}_i - \boldsymbol{R}_\alpha) \,, \tag{2.5}$$

where here also, a further separation is expedient:

$$H_{\mathrm{ei}} = H_{\mathrm{ei}}^{(0)} + H_{\mathrm{e-p}} \,. \tag{2.6}$$

$H_{ei}^{(0)}$ refers to the interaction of the electrons with the ions in their equilibrium positions. H_{e-p} is the electron-phonon interaction.

An exact solution for the overall system (2.1) would appear to be impossible. An approximation can be formulated in the following three steps:

1. Electronic motions, e.g. in a rigid ionic lattice: $H_e + H_{ei}^{(0)}$.
2. Ionic motions, e.g. in a homogeneous electron gas H_p.
3. Coupling, e.g. the perturbation-theoretical treatment of H_{e-p}.

Following this concept, in the following section, we discuss the electronic subsystem.

2.1 Crystal Electrons

2.1.1 Non-interacting Bloch Electrons

We first consider electrons in a rigid ionic lattice, which do not interact with each other, but rather only with the periodic lattice potential, i.e. we are looking for the solutions corresponding to the eigenstates of the following Hamiltonian:

$$H_0 = H_{\mathrm{e,kin}} + H_{\mathrm{ei}}^{(0)} . \tag{2.7}$$

The so-called **lattice potential** is defined by the ions which are fixed in their equilibrium positions

$$\widehat{V}(\boldsymbol{r}_i) = \sum_{\alpha=1}^{N_\mathrm{i}} V_{\mathrm{ei}}\left(\boldsymbol{r}_i - \boldsymbol{R}_\alpha^{(0)}\right) . \tag{2.8}$$

More precisely, we have for the positions of the ions $\boldsymbol{R}_\alpha^{(0)}$:

$$\begin{aligned} \boldsymbol{R}_\alpha^{(0)} &\Rightarrow \boldsymbol{R}_s^{\boldsymbol{n}} = \boldsymbol{R}^{\boldsymbol{n}} + \boldsymbol{R}_s , \\ \boldsymbol{n} &= (n_1, n_2, n_3) ; \quad n_\mathrm{i} \in \boldsymbol{Z} . \end{aligned} \tag{2.9}$$

Here, $\boldsymbol{R}^{\boldsymbol{n}}$ defines the Bravais lattice:

$$\boldsymbol{R}^{\boldsymbol{n}} = \sum_{i=1}^{3} n_\mathrm{i} \boldsymbol{a}_\mathrm{i} . \tag{2.10}$$

$\boldsymbol{a}_1$, $\boldsymbol{a}_2$, $\boldsymbol{a}_3$ are the primitive translations, and $\boldsymbol{R}_s$ are the position vectors of the basis atoms. The periodicity mentioned above refers to the Bravais lattice:

$$\widehat{V}(\boldsymbol{r}_i + \boldsymbol{R}^{\boldsymbol{n}}) \stackrel{!}{=} \widehat{V}(\boldsymbol{r}_i) . \tag{2.11}$$

$\widehat{V}(\boldsymbol{r}_i) = \widehat{V}(\hat{\boldsymbol{r}}_\mathrm{i})$ is a single-particle operator, and this can be inserted into:

$$H_{\mathrm{ei}}^{(0)} = \sum_{i=1}^{N_\mathrm{e}} \widehat{V}(\hat{\boldsymbol{r}}_\mathrm{i}) . \tag{2.12}$$

We thus have to solve the following eigenvalue equation:

$$h_0 \psi_{\boldsymbol{k}}(\boldsymbol{r}) = \varepsilon(\boldsymbol{k}) \psi_{\boldsymbol{k}}(\boldsymbol{r}) . \tag{2.13}$$

We refer to $\psi_k(r)$ as a **Bloch function** and $\varepsilon(k)$ as the corresponding **Bloch energy**. k is a wave vector within the first Brillouin zone. h_0 refers to the operator

$$h_0 = \frac{p^2}{2m} + \widehat{V}(\hat{r}) . \tag{2.14}$$

The solution of (2.13) for realistic lattices is a non-trivial problem. Using the periodicity (2.11) of the lattice potential, one can derive the fundamental **Bloch's Theorem**:

$$\psi_k\left(r + R^n\right) = \mathrm{e}^{\mathrm{i}k \cdot R^n} \psi_k(r) . \tag{2.15}$$

Employing the usual *ansatz*

$$\psi_k(r) = u_k(r)\mathrm{e}^{\mathrm{i}k \cdot r} , \tag{2.16}$$

the amplitude function must have the periodicity of the lattice:

$$u_k\left(r + R^n\right) = u_k(r) . \tag{2.17}$$

The Bloch functions $\psi_k(r)$ form a complete, orthonormalised system:

$$\int \mathrm{d}^3r\, \psi_k^*(r)\psi_{k'}(r) = \delta_{k,k'} , \tag{2.18}$$

$$\sum_k^{1.\mathrm{BZ}} \psi_k^*(r)\psi_k\left(r'\right) = \delta\left(r - r'\right) . \tag{2.19}$$

The sum runs over all the wave vectors k in the first Brillouin zone. Owing to the periodic boundary conditions, these are discrete. Since h_0 contains no spin parts, its eigenfunctions can be factored into a spin and a configuration-space function:

$$\begin{aligned} |k\sigma\rangle \quad &\Longleftrightarrow \quad \text{Bloch state} , \\ \langle r \mid k\sigma\rangle &= \psi_{k\sigma}(r) = \psi_k(r)\chi_\sigma , \\ \chi_\uparrow = \begin{pmatrix}1\\0\end{pmatrix}&; \quad \chi_\downarrow = \begin{pmatrix}0\\1\end{pmatrix} . \end{aligned} \tag{2.20}$$

If we consider electrons from different energy bands, the Bloch function is also characterised by a band index n. We limit ourselves here, however, to electrons within a single band.

We define:

$a_{k\sigma}^+ \quad (a_{k\sigma})$: **creation (annihilation) operator for a Bloch electron.**

Since H_0 is a single-particle operator, it follows from (1.100) that:

$$H_0 = \sum_{\substack{k\sigma \\ k'\sigma'}} \langle k\sigma \mid h_0 \mid k'\sigma' \rangle \, a^+_{k\sigma} \, a_{k'\sigma'} \, .$$

The matrix elements can be computed in a straightforward manner:

$$\langle k\sigma \mid h_0 \mid k'\sigma' \rangle = \varepsilon(k') \, \langle k\sigma \mid k'\sigma' \rangle = \varepsilon(k) \, \delta_{kk'} \, \delta_{\sigma\sigma'} \, , \tag{2.21}$$

since $|k\sigma\rangle$ is an eigenstate of h_0. It then follows that:

$$H_0 = \sum_{k\sigma} \varepsilon(k) a^+_{k\sigma} \, a_{k\sigma} = \sum_{k\sigma} \varepsilon(k) n_{k\sigma} \, . \tag{2.22}$$

The *Bloch operators* $a_{k\sigma}$, $a^+_{k\sigma}$ of course fulfil the fundamental commutation relations:

$$[a_{k\sigma}, \, a_{k'\sigma'}]_+ = \left[a^+_{k\sigma}, \, a^+_{k'\sigma'}\right]_+ = 0 \, , \tag{2.23}$$

$$\left[a_{k\sigma}, \, a^+_{k'\sigma'}\right]_+ = \delta_{kk'} \, \delta_{\sigma\sigma'} \, . \tag{2.24}$$

If we neglect the crystalline structure of the solid and consider the ionic lattice merely as a positively-charged background for the electronic system, ($\widehat{V}(r) = \text{const}$), then the Bloch functions become plane waves,

$$\psi_k(r) \underset{[\widehat{V}=\text{const}]}{\Rightarrow} \frac{1}{\sqrt{V}} \mathrm{e}^{\mathrm{i} k \cdot r} \, , \tag{2.25}$$

and the Bloch energies, due to $p^2/2m = -(\hbar^2/2m)\Delta$, are:

$$\varepsilon(k) \underset{[\widehat{V}=\text{const}]}{\Rightarrow} \frac{\hbar^2 k^2}{2m} \, . \tag{2.26}$$

(V is the volume of the solid. It is important to distinguish between V and the lattice potential $\widehat{V}$!) We will discuss two other representations of H_0 which are important for applications, e.g. the

field operators

$$\widehat{\psi}^+_\sigma(r), \quad \widehat{\psi}_\sigma(r) \, ,$$

which are to be understood as in (1.63) through (1.69), with the addition that we now also take the spin of the electron into account. The generalisation of the formulas given in Chap. 1 is evident. Thus, for example:

$$\left[\widehat{\psi}_\sigma(\boldsymbol{r}), \widehat{\psi}^+_{\sigma'}(\boldsymbol{r}')\right]_+ = \delta\left(\boldsymbol{r}-\boldsymbol{r}'\right)\delta_{\sigma\sigma'} . \tag{2.27}$$

From this it follows for H_0:

$$\begin{aligned} H_0 &= \sum_{\sigma,\sigma'} \iint d^3r\, d^3r' \left\langle \boldsymbol{r}\sigma \mid h_0 \mid \boldsymbol{r}'\sigma' \right\rangle \widehat{\psi}^+_\sigma(\boldsymbol{r}) \widehat{\psi}_{\sigma'}(\boldsymbol{r}') = \\ &= \sum_{\sigma,\sigma'} \iint d^3r\, d^3r'\, \delta_{\sigma\sigma'} \left(-\frac{\hbar^2}{2m}\Delta_{r'} + \widehat{V}(\boldsymbol{r}')\right) \delta\left(\boldsymbol{r}-\boldsymbol{r}'\right) \widehat{\psi}^+_\sigma(\boldsymbol{r}) \widehat{\psi}_{\sigma'}(\boldsymbol{r}') = \\ &= \sum_\sigma \int d^3r\, \widehat{\psi}^+_\sigma(\boldsymbol{r}) \left(-\frac{\hbar^2}{2m}\Delta_r + \widehat{V}(\boldsymbol{r})\right) \widehat{\psi}_\sigma(\boldsymbol{r}) . \end{aligned} \tag{2.28}$$

An additional, frequently-used particular configuration representation makes use of

Wannier functions

$$\omega_\sigma\left(\boldsymbol{r}-\boldsymbol{R}_\mathrm{i}\right) = \frac{1}{\sqrt{N_\mathrm{i}}} \sum_{\boldsymbol{k}}^{1.\mathrm{BZ}} e^{-i\boldsymbol{k}\cdot\boldsymbol{R}_\mathrm{i}} \psi_{\boldsymbol{k}\sigma}(\boldsymbol{r}) . \tag{2.29}$$

A typical feature of these functions is their relatively strong concentration around each lattice position $\boldsymbol{R}_\mathrm{i}$ (Fig. 2.1). With (2.18) as well as

$$\frac{1}{N_\mathrm{i}} \sum_{\boldsymbol{k}}^{1.\mathrm{BZ}} e^{i\boldsymbol{k}\cdot(\boldsymbol{R}_\mathrm{i}-\boldsymbol{R}_j)} = \delta_{ij} , \tag{2.30}$$

one can readily prove the orthogonality relation:

$$\int d^3r\, \omega^*_\sigma\left(\boldsymbol{r}-\boldsymbol{R}_\mathrm{i}\right) \omega_{\sigma'}\left(\boldsymbol{r}-\boldsymbol{R}_j\right) = \delta_{\sigma\sigma'}\delta_{ij} . \tag{2.31}$$

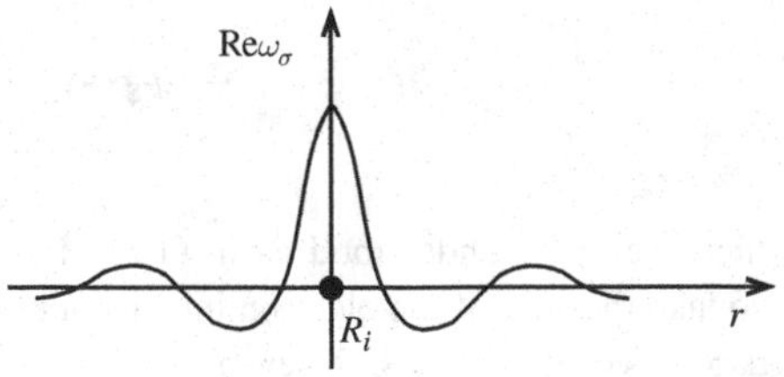

Fig. 2.1 The qualitative position dependence of the real part of a Wannier function

Using the notations

$$
\begin{aligned}
|i\sigma\rangle &\iff \text{Wannier state,} \\
\langle \boldsymbol{r} \mid i\sigma\rangle &= \omega_\sigma(\boldsymbol{r}-\boldsymbol{R}_\mathrm{i}), \\
a_{i\sigma}^{+} \quad (a_{i\sigma}) &: \text{creation (annihilation) operator for an electron} \\
&\quad \text{in a Wannier state at the lattice site } \boldsymbol{R}_\mathrm{i},
\end{aligned}
\tag{2.32}
$$

in second quantisation, H_0 is given by

$$
H_0 = \sum_{ij\sigma} T_{ij}\, a_{i\sigma}^{+}\, a_{j\sigma}\,, \tag{2.33}
$$

and describes in an intuitively clear manner the *hopping* of an electron with spin σ from the lattice site $\boldsymbol{R}_j$ – where it is annihilated – to the lattice site $\boldsymbol{R}_\mathrm{i}$, where it is created. T_{ij} is therefore also called the

"hopping" integral.

We start with:

$$
\begin{aligned}
\left\langle i\sigma \mid h_0 \mid j\sigma'\right\rangle &= \delta_{\sigma\sigma'}\langle i\sigma \mid h_0 \mid j\sigma\rangle = \\
&= \delta_{\sigma\sigma'} \sum_{\substack{\boldsymbol{k},\boldsymbol{k}'\\ \sigma_1,\sigma_2}} \langle i\sigma \mid \boldsymbol{k}\sigma_1\rangle \left\langle \boldsymbol{k}\sigma_1 \mid h_0 \mid \boldsymbol{k}'\sigma_2\right\rangle \left\langle \boldsymbol{k}'\sigma_2 \mid j\sigma\right\rangle = \\
&= \delta_{\sigma\sigma'} \sum_{\substack{\boldsymbol{k},\boldsymbol{k}'\\ \sigma_1,\sigma_2}} \varepsilon(\boldsymbol{k}')\,\langle i\sigma \mid \boldsymbol{k}\sigma_1\rangle \left\langle \boldsymbol{k}\sigma_1 \mid \boldsymbol{k}'\sigma_2\right\rangle \left\langle \boldsymbol{k}'\sigma_2 \mid j\sigma\right\rangle = \\
&= \delta_{\sigma\sigma'} \sum_{\boldsymbol{k},\sigma_1} \varepsilon(\boldsymbol{k})\,\langle i\sigma \mid \boldsymbol{k}\sigma_1\rangle\,\langle \boldsymbol{k}\sigma_1 \mid j\sigma\rangle\,.
\end{aligned}
\tag{2.34}
$$

The remaining matrix elements can then be computed as follows:

$$
\begin{aligned}
\langle i\sigma \mid \boldsymbol{k}\sigma_1\rangle &= \int \mathrm{d}^3r\,\langle i\sigma \mid \boldsymbol{r}\rangle\,\langle \boldsymbol{r} \mid \boldsymbol{k}\sigma_1\rangle = \\
&= \int \mathrm{d}^3r\,\omega_\sigma^*(\boldsymbol{r}-\boldsymbol{R}_\mathrm{i})\,\psi_{\boldsymbol{k}\sigma_1}(\boldsymbol{r}) = \\
&= \frac{1}{\sqrt{N_\mathrm{i}}}\sum_{\boldsymbol{k}'} \mathrm{e}^{\mathrm{i}\boldsymbol{k}'\cdot\boldsymbol{R}_\mathrm{i}} \int \mathrm{d}^3r\,\psi_{\boldsymbol{k}'\sigma}^*(\boldsymbol{r})\psi_{\boldsymbol{k}\sigma_1}(\boldsymbol{r}) = \\
&= \frac{1}{\sqrt{N_\mathrm{i}}}\sum_{\boldsymbol{k}'} \mathrm{e}^{\mathrm{i}\boldsymbol{k}'\cdot\boldsymbol{R}_\mathrm{i}}\,\delta_{\boldsymbol{k}\boldsymbol{k}'}\,\delta_{\sigma\sigma_1} = \delta_{\sigma\sigma_1}\frac{\mathrm{e}^{\mathrm{i}\boldsymbol{k}\cdot\boldsymbol{R}_\mathrm{i}}}{\sqrt{N_\mathrm{i}}}\,.
\end{aligned}
$$

This yields in (2.34):

$$\langle i\sigma \mid h_0 \mid j\sigma' \rangle = \delta_{\sigma\sigma'} T_{ij} \tag{2.35}$$

with

$$T_{ij} = \frac{1}{N_\mathrm{i}} \sum_{\boldsymbol{k}} \varepsilon(\boldsymbol{k}) \mathrm{e}^{\mathrm{i}\boldsymbol{k}\cdot(\boldsymbol{R}_\mathrm{i}-\boldsymbol{R}_j)} . \tag{2.36}$$

The inverse relation is given by:

$$\varepsilon(\boldsymbol{k}) = \frac{1}{N_\mathrm{i}} \sum_{i,j} T_{ij} \mathrm{e}^{-\mathrm{i}\boldsymbol{k}\cdot(\boldsymbol{R}_\mathrm{i}-\boldsymbol{R}_j)} , \tag{2.37}$$

as can be verified by substituting in (2.36) and employing (2.30).

The relation between the Bloch and the Wannier operators can be found in the same way as shown in (1.66) for the example of the field operators:

$$a_{i\sigma} = \frac{1}{\sqrt{N_\mathrm{i}}} \sum_{\boldsymbol{k}}^{1.\,\mathrm{BZ}} \mathrm{e}^{\mathrm{i}\boldsymbol{k}\cdot\boldsymbol{R}_\mathrm{i}} a_{\boldsymbol{k}\sigma} , \tag{2.38}$$

$$a_{\boldsymbol{k}\sigma} = \frac{1}{\sqrt{N_\mathrm{i}}} \sum_{i=1}^{N_\mathrm{i}} \mathrm{e}^{-\mathrm{i}\boldsymbol{k}\cdot\boldsymbol{R}_\mathrm{i}} a_{i\sigma} . \tag{2.39}$$

From the commutation relations for the Bloch operators (2.23) and (2.24), the commutation relations for the Wannier operators then follow immediately:

$$\left[a_{i\sigma}, a_{j\sigma'}\right]_+ = \left[a_{i\sigma}^+, a_{j\sigma'}^+\right]_+ = 0, \tag{2.40}$$

$$\left[a_{i\sigma}, a_{j\sigma'}^+\right]_+ = \delta_{ij}\, \delta_{\sigma\sigma'} . \tag{2.41}$$

2.1.2 The Jellium Model

This model is adequate for the description of simple metals and is based on the following assumptions:

1. N_e electrons within the volume $V = L^3$ interact with each other via the Coulomb interaction

$$H_{\mathrm{ee}} = \frac{e^2}{8\pi\varepsilon_0} \sum_{i,j}^{i\neq j} \frac{1}{|\boldsymbol{r}_i - \boldsymbol{r}_j|} . \tag{2.42}$$

2. The ions are singly positively charged:

$$N_e = N_i = N \,. \tag{2.43}$$

3. The ions form a *homogeneously distributed* background and thus guarantee (a) charge neutrality, (b) a constant lattice potential.
   The Bloch functions then become plane waves:

$$\psi_{k\sigma}(\boldsymbol{r}) \;\Rightarrow\; \frac{1}{\sqrt{V}} \mathrm{e}^{\mathrm{i}\boldsymbol{k}\cdot\boldsymbol{r}} \chi_\sigma \,. \tag{2.44}$$

4. Periodic boundary conditions for V give rise to discrete wave numbers:

$$\boldsymbol{k} = \frac{2\pi}{L}\left(n_x, n_y, n_z\right), \qquad n_{x,y,z} \in \boldsymbol{Z} \,. \tag{2.45}$$

How is the Hamiltonian for the model corresponding to these assumptions formulated in first quantisation? It should contain three terms:

$$H = H_e + H_+ + H_{e+} \,. \tag{2.46}$$

H_e is to be interpreted as in (2.2) and is the pivotal term. H_+ describes the homogeneously distributed ionic charges, where *homogeneously distributed* is taken to imply that the ion density $n(\boldsymbol{r})$ is position-independent:

$$n(\boldsymbol{r}) \;\Rightarrow\; \frac{N}{V} \,. \tag{2.47}$$

Then we have for H_+:

$$H_+ = \frac{e^2}{8\pi\varepsilon_0} \iint \mathrm{d}^3r\, \mathrm{d}^3r' \frac{n(\boldsymbol{r})\cdot n(\boldsymbol{r}')}{|\boldsymbol{r}-\boldsymbol{r}'|} \mathrm{e}^{-\alpha|\boldsymbol{r}-\boldsymbol{r}'|} \,. \tag{2.48}$$

Due to the 4th assumption, we must discuss our results in the thermodynamic limit, i.e. for $N \to \infty$, $V \to \infty$, $N/V \to$ const. Owing to the long range of the Coulomb forces, the integrals then diverge. For this reason, a convergence factor $\exp(-\alpha|\boldsymbol{r}-\boldsymbol{r}'|)$ with $\alpha > 0$ is introduced. After evaluating the integrals, the limit $\alpha \to 0$ is taken.

Because of (2.47), we require the following integral in (2.48):

$$\iint \mathrm{d}^3r\, \mathrm{d}^3r' \frac{\mathrm{e}^{-\alpha|\boldsymbol{r}-\boldsymbol{r}'|}}{|\boldsymbol{r}-\boldsymbol{r}'|} = V \int_V \mathrm{d}^3r \frac{\mathrm{e}^{-\alpha r}}{r} \;\xrightarrow[V\to\infty]{}\; \frac{4\pi V}{\alpha^2} \,.$$

We then obtain:

$$H_+ = \frac{e^2}{8\pi\varepsilon_0} \frac{\widehat{N}^2}{V} \frac{4\pi}{\alpha^2} \,. \tag{2.49}$$

H_+ indeed diverges for $\alpha \to 0$, but it is compensated by other terms which are yet to be discussed. H_{e+} in (2.46) describes the interactions of the electrons with the homogeneous background of ions:

$$H_{e+} = -\frac{e^2}{4\pi\varepsilon_0}\sum_{i=1}^{N}\int d^3r\,\frac{n(\boldsymbol{r})}{|\boldsymbol{r}-\boldsymbol{r}_i|}e^{-\alpha|\boldsymbol{r}-\boldsymbol{r}_i|}\,. \tag{2.50}$$

With the same considerations as used for H_+, we find:

$$\begin{aligned} H_{e+} &= -\frac{e^2}{4\pi\varepsilon_0}\frac{N}{V}\sum_{i=1}^{N}\int d^3r\,\frac{e^{-\alpha|\boldsymbol{r}-\boldsymbol{r}_i|}}{|\boldsymbol{r}-\boldsymbol{r}_i|} = \\ &= -\frac{e^2}{4\pi\varepsilon_0}\frac{N}{V}\sum_{i=1}^{N}\frac{4\pi}{\alpha^2}\,. \end{aligned}$$

We now replace the classical particle number N by the particle-number operator $\widehat{N}$; this yields:

$$H_{e+} = -\frac{e^2}{4\pi\varepsilon_0}\frac{\widehat{N}^2}{V}\frac{4\pi}{\alpha^2}\,. \tag{2.51}$$

All together, this gives for our model:

$$H = H_e - \frac{1}{2}\frac{e^2}{4\pi\varepsilon_0}\frac{\widehat{N}^2}{V}\frac{4\pi}{\alpha^2}\,. \tag{2.52}$$

This still looks critical for $\alpha \to 0$, but as we shall see, H_e contains an exactly corresponding term, which just cancels with the second term in (2.52). H_e is in fact the decisive operator, and according to (2.2), it is composed of the kinetic energy H_0 (2.7) and the Coulomb interaction H_{ee} (2.42). H_0 was already transformed to second quantisation in the previous section. H_{ee} is a typical two-particle operator, for which, according to (1.100), we find in the Bloch representation:

$$H_{ee} = \frac{1}{2}\sum_{\substack{\boldsymbol{k}_1\cdots\boldsymbol{k}_4\\ \sigma_1\cdots\sigma_4}} v(\boldsymbol{k}_1\sigma_1,\ldots,\boldsymbol{k}_4\sigma_4)\, a^+_{\boldsymbol{k}_1\sigma_1}a^+_{\boldsymbol{k}_2\sigma_2}a_{\boldsymbol{k}_4\sigma_4}a_{\boldsymbol{k}_3\sigma_3}\,. \tag{2.53}$$

The matrix element

$$\begin{aligned} &v(\boldsymbol{k}_1\sigma_1,\ldots,\boldsymbol{k}_4\sigma_4) = \\ &= \frac{e^2}{4\pi\varepsilon_0}\left\langle (\boldsymbol{k}_1\sigma_1)^{(1)}(\boldsymbol{k}_2\sigma_2)^{(2)}\left|\frac{1}{|\hat{\boldsymbol{r}}^{(1)}-\hat{\boldsymbol{r}}'^{(2)}|}\right|(\boldsymbol{k}_3\sigma_3)^{(1)}(\boldsymbol{k}_4\sigma_4)^{(2)}\right\rangle \end{aligned}$$

is with certainty nonzero only for

$$\sigma_1 = \sigma_3 \quad \text{and} \quad \sigma_2 = \sigma_4 \ ,$$

since the operator itself is spin-independent:

$$\begin{aligned} v(\boldsymbol{k}_1\sigma_1, \dots, \boldsymbol{k}_4\sigma_4) &= \frac{e^2}{4\pi\varepsilon_0} \iint \mathrm{d}^3r_1\,\mathrm{d}^3r_2 \left\langle \boldsymbol{k}_1^{(1)}\boldsymbol{k}_2^{(2)} \right| \frac{1}{|\hat{\boldsymbol{r}}^{(1)} - \hat{\boldsymbol{r}}'^{(2)}|} \cdot \\ &\cdot \left| \boldsymbol{r}_1^{(1)}\boldsymbol{r}_2^{(2)} \right\rangle \left\langle \boldsymbol{r}_1^{(1)}\boldsymbol{r}_2^{(2)} \middle| \boldsymbol{k}_3^{(1)}\boldsymbol{k}_4^{(2)} \right\rangle \delta_{\sigma_1\sigma_3}\,\delta_{\sigma_2\sigma_4} = \\ &= \frac{e^2}{4\pi\varepsilon_0} \iint \mathrm{d}^3r_1\,\mathrm{d}^3r_2 \frac{1}{|\boldsymbol{r}_1 - \boldsymbol{r}_2|} \left\langle \boldsymbol{k}_1^{(1)}\boldsymbol{k}_2^{(2)} \middle| \boldsymbol{r}_1^{(1)}\boldsymbol{r}_2^{(2)} \right\rangle \cdot \\ &\cdot \left\langle \boldsymbol{r}_1^{(1)}\boldsymbol{r}_2^{(2)} \middle| \boldsymbol{k}_3^{(1)}\boldsymbol{k}_4^{(2)} \right\rangle \delta_{\sigma_1\sigma_3}\,\delta_{\sigma_2\sigma_4} = \\ &= \frac{e^2}{4\pi\varepsilon_0} \iint \mathrm{d}^3r_1\,\mathrm{d}^3r_2 \frac{1}{|\boldsymbol{r}_1 - \boldsymbol{r}_2|} \psi^*_{\boldsymbol{k}_1}(\boldsymbol{r}_1)\,\psi^*_{\boldsymbol{k}_2}(\boldsymbol{r}_2) \cdot \\ &\cdot \psi_{\boldsymbol{k}_3}(\boldsymbol{r}_1)\,\psi_{\boldsymbol{k}_4}(\boldsymbol{r}_2)\,\delta_{\sigma_1\sigma_3}\,\delta_{\sigma_2\sigma_4} \ . \end{aligned}$$

Making use of Bloch's theorem (2.15), we can furthermore show that in addition,

$$\boldsymbol{k}_1 + \boldsymbol{k}_2 = \boldsymbol{k}_3 + \boldsymbol{k}_4$$

must hold. We then have:

$$\begin{aligned} v(\boldsymbol{k}_1\sigma_1, \dots, \boldsymbol{k}_4\sigma_4) &= \delta_{\sigma_1\sigma_3}\,\delta_{\sigma_2\sigma_4}\,\delta_{\boldsymbol{k}_1+\boldsymbol{k}_2,\,\boldsymbol{k}_3+\boldsymbol{k}_4}\, v(\boldsymbol{k}_1, \dots \boldsymbol{k}_4) \ , \\ v(\boldsymbol{k}_1, \dots, \boldsymbol{k}_4) &= \frac{e^2}{4\pi\varepsilon_0} \iint \mathrm{d}^3r_1\,\mathrm{d}^3r_2\, \psi^*_{\boldsymbol{k}_1}(\boldsymbol{r}_1)\,\psi^*_{\boldsymbol{k}_2}(\boldsymbol{r}_2) \cdot \\ &\cdot \frac{1}{|\boldsymbol{r}_1 - \boldsymbol{r}_2|} \psi_{\boldsymbol{k}_3}(\boldsymbol{r}_1)\,\psi_{\boldsymbol{k}_4}(\boldsymbol{r}_2) \ . \end{aligned} \tag{2.54}$$

For the Coulomb interaction H_{ee}, we thus obtain the following expression:

$$H_{\mathrm{ee}} = \frac{1}{2} \sum_{\substack{\boldsymbol{k}_1, \dots, \boldsymbol{k}_4 \\ \sigma, \sigma'}} v(\boldsymbol{k}_1, \dots, \boldsymbol{k}_4)\, \delta_{\boldsymbol{k}_1+\boldsymbol{k}_2,\,\boldsymbol{k}_3+\boldsymbol{k}_4}\, a^+_{\boldsymbol{k}_1\sigma}\, a^+_{\boldsymbol{k}_2\sigma'}\, a_{\boldsymbol{k}_4\sigma'}\, a_{\boldsymbol{k}_3\sigma} \ . \tag{2.55}$$

In the jellium model, the $\psi_{\boldsymbol{k}}(\boldsymbol{r})$ are plane waves, so that we still must calculate:

$$\begin{aligned} v_\alpha(\boldsymbol{k}_1, \dots, \boldsymbol{k}_4) &= \\ &= \frac{e^2}{4\pi\varepsilon_0} \frac{1}{V^2} \iint \mathrm{d}^3r_1\,\mathrm{d}^3r_2 \frac{\mathrm{e}^{-\mathrm{i}(\boldsymbol{k}_1-\boldsymbol{k}_3)\cdot\boldsymbol{r}_1}\,\mathrm{e}^{-\mathrm{i}(\boldsymbol{k}_2-\boldsymbol{k}_4)\cdot\boldsymbol{r}_2}}{|\boldsymbol{r}_1 - \boldsymbol{r}_2|} \mathrm{e}^{-\alpha|\boldsymbol{r}_1-\boldsymbol{r}_2|} \ . \end{aligned} \tag{2.56}$$

We set

$$
\begin{aligned}
& r = r_1 - r_2; \qquad R = \frac{1}{2}(r_1 + r_2) \\
\Longleftrightarrow \quad & r_1 = \frac{1}{2}r + R; \qquad r_2 = -\frac{1}{2}r + R\,.
\end{aligned}
\tag{2.57}
$$

and must then solve:

$$
\begin{aligned}
v_\alpha(\boldsymbol{k}_1, \ldots, \boldsymbol{k}_4) &= \frac{e^2}{4\pi\varepsilon_0}\frac{1}{V}\int \mathrm{d}^3R\,\mathrm{e}^{-\mathrm{i}(\boldsymbol{k}_1 - \boldsymbol{k}_3 + \boldsymbol{k}_2 - \boldsymbol{k}_4)\cdot\boldsymbol{R}}\,. \\
&\cdot \frac{1}{V}\int \mathrm{d}^3r\,\frac{1}{r}\,\mathrm{e}^{-\alpha r}\,\mathrm{e}^{-(\mathrm{i}/2)(\boldsymbol{k}_1 - \boldsymbol{k}_3 - \boldsymbol{k}_2 + \boldsymbol{k}_4)\cdot\boldsymbol{r}} = \\
&= \frac{e^2}{4\pi\varepsilon_0}\,\delta_{\boldsymbol{k}_1+\boldsymbol{k}_2,\,\boldsymbol{k}_3+\boldsymbol{k}_4}\,\frac{1}{V}\int \mathrm{d}^3r\,\frac{\mathrm{e}^{-\mathrm{i}(\boldsymbol{k}_1-\boldsymbol{k}_3)\cdot\boldsymbol{r}}\,\mathrm{e}^{-\alpha r}}{r}\,.
\end{aligned}
$$

Using

$$
\int \mathrm{d}^3r\,\frac{\mathrm{e}^{-\mathrm{i}\boldsymbol{q}\cdot\boldsymbol{r}}}{r}\,\mathrm{e}^{-\alpha r} = \frac{4\pi}{q^2 + \alpha^2}\,,
\tag{2.58}
$$

we finally obtain:

$$
v_\alpha(\boldsymbol{k}_1, \ldots, \boldsymbol{k}_4) = \frac{e^2}{\varepsilon_0 V\left[(\boldsymbol{k}_1 - \boldsymbol{k}_3)^2 + \alpha^2\right]}\,\delta_{\boldsymbol{k}_1-\boldsymbol{k}_3,\,\boldsymbol{k}_4-\boldsymbol{k}_2}\,.
\tag{2.59}
$$

We insert this into (2.55):

$$
H_{\mathrm{ee}}^{(\alpha)} = \frac{1}{2}\sum_{\substack{\boldsymbol{k},\,\boldsymbol{p},\,\boldsymbol{q}\\ \sigma,\sigma'}} v_\alpha(\boldsymbol{q})\,a^+_{\boldsymbol{k}+\boldsymbol{q}\sigma}\,a^+_{\boldsymbol{p}-\boldsymbol{q}\sigma'}\,a_{\boldsymbol{p}\sigma'}\,a_{\boldsymbol{k}\sigma}\,,
\tag{2.60}
$$

$$
v_\alpha(\boldsymbol{q}) = \frac{e^2}{\varepsilon_0 V\left(q^2 + \alpha^2\right)}\,.
\tag{2.61}
$$

We consider now the $q = 0$ term of the Coulomb interaction:

$$
\begin{aligned}
&\frac{1}{2}\frac{e^2}{\varepsilon_0 V\alpha^2}\sum_{\substack{\boldsymbol{k},\,\boldsymbol{p}\\ \sigma,\sigma'}} a^+_{\boldsymbol{k}\sigma}\,a^+_{\boldsymbol{p}\sigma'}\,a_{\boldsymbol{p}\sigma'}\,a_{\boldsymbol{k}\sigma} = \\
&= \frac{1}{2}\frac{e^2}{\varepsilon_0 V\alpha^2}\sum_{\substack{\boldsymbol{k},\,\boldsymbol{p}\\ \sigma,\sigma'}}\left(-\delta_{\sigma\sigma'}\,\delta_{\boldsymbol{k}\boldsymbol{p}}\,n_{\boldsymbol{k}\sigma} + n_{\boldsymbol{p}\sigma'}\,n_{\boldsymbol{k}\sigma}\right) = \\
&= \frac{e^2}{2\varepsilon_0 V\alpha^2}\left[-\widehat{N} + \left(\widehat{N}\right)^2\right].
\end{aligned}
\tag{2.62}
$$

We can see that the second term in (2.62) just compensates the second term in (2.52), i.e. the contributions from H_+ and H_{e+} just cancel. The first term in (2.62) leads to an energy per particle which vanishes in the thermodynamic limit,

$$-\frac{e^2}{2\varepsilon_0 V\alpha^2} \xrightarrow[N\to\infty;\,V\to\infty]{} 0\,,$$

and therefore can be left off from the beginning. If we now finally take the limit $\alpha \to 0$, we find for the

Hamiltonian of the jellium model:

$$H = \sum_{k\sigma} \varepsilon_0(\boldsymbol{k})\, a^+_{k\sigma}\, a_{k\sigma} + \frac{1}{2} \sum_{\substack{k,\,p,\,q\\ \sigma,\,\sigma'}}^{q\neq 0} v_0(\boldsymbol{q})\, a^+_{k+q\sigma}\, a^+_{p-q\sigma'}\, a_{p\sigma'}\, a_{k\sigma}\,. \tag{2.63}$$

From (2.26), we have

$$\varepsilon_0(\boldsymbol{k}) = \frac{\hbar^2 k^2}{2m} \tag{2.64}$$

as the matrix element of the kinetic energy, and

$$v_0(\boldsymbol{q}) = \frac{1}{V}\frac{e^2}{\varepsilon_0 q^2} \tag{2.65}$$

as that of the Coulomb interaction.

In addition, we would like to derive a useful alternative representation of H, making use of the

electron density operator:

$$\hat{\rho}(\boldsymbol{r}) = \sum_{i=1}^{N} \delta\left(\boldsymbol{r} - \hat{\boldsymbol{r}}_{\mathrm{i}}\right). \tag{2.66}$$

This is a single-particle operator. The site of the electron $\hat{\boldsymbol{r}}_{\mathrm{i}}$ is an operator here, whilst the variable $\boldsymbol{r}$ is naturally not. From (1.100), we find for $\hat{\rho}$ in the second-quantisation formalism using the Bloch representation:

$$\hat{\rho}(\boldsymbol{r}) = \sum_{\substack{k,\,k'\\ \sigma,\,\sigma'}} \langle \boldsymbol{k}\sigma|\,\delta\left(\boldsymbol{r} - \hat{\boldsymbol{r}}'\right)|\boldsymbol{k}'\sigma'\rangle\, a^+_{k\sigma}\, a_{k'\sigma'}\,. \tag{2.67}$$

For the matrix element, we need to calculate the following:

$$\begin{aligned}\langle k\sigma|\,\delta\left(r-\hat{r}'\right)|k'\sigma'\rangle &= \sum_{\sigma''}\int d^3r''\,\langle k\sigma|\,\delta\left(r-\hat{r}'\right)|r''\sigma''\rangle\langle r''\sigma''\mid k'\sigma'\rangle = \\ &= \sum_{\sigma''}\int d^3r''\,\delta\left(r-r''\right)\langle k\sigma\mid r''\sigma''\rangle\langle r''\sigma''\mid k'\sigma'\rangle = \\ &= \sum_{\sigma''}\delta_{\sigma\sigma''}\,\delta_{\sigma''\sigma'}\,\langle k\sigma\mid r\sigma\rangle\langle r\sigma\mid k'\sigma\rangle = \\ &= \delta_{\sigma\sigma'}\psi_k^*(r)\psi_{k'}(r)\,.\end{aligned}$$

If we confine ourselves to plane waves, as in the jellium model, then we have

$$\langle k\sigma|\,\delta\left(r-\hat{r}'\right)|k'\sigma'\rangle = \delta_{\sigma\sigma'}\frac{1}{V}\mathrm{e}^{\mathrm{i}(k'-k)\cdot r}\,. \tag{2.68}$$

In terms of (2.67), this means:

$$\hat{\rho}(r) = \frac{1}{V}\sum_{k,\,q,\,\sigma} a_{k\sigma}^{+}\,a_{k+q\sigma}\mathrm{e}^{\mathrm{i}q\cdot r}\,. \tag{2.69}$$

For the Fourier component of the electron-density operator, we thus find:

$$\hat{\rho}_q = \sum_{k\sigma} a_{k\sigma}^{+}\,a_{k+q\sigma}\,. \tag{2.70}$$

One can read off, among other things:

$$\hat{\rho}_q^{+} = \hat{\rho}_{-q};\quad \hat{\rho}_{q=0} = \widehat{N}\,. \tag{2.71}$$

With this result, we can express the Hamiltonian of the jellium model in terms of density operators. The kinetic energy remains unchanged:

$$\begin{aligned}H_{\mathrm{ee}} &= \frac{1}{2}\sum_{\substack{k,\,p,\,q\\ \sigma,\,\sigma'}}^{q\neq 0} v_0(q)\,a_{k+q\sigma}^{+}\,a_{p-q\sigma'}^{+}\,a_{p\sigma'}\,a_{k\sigma} = \\ &= \frac{1}{2}\sum_{\substack{k,\,pq\\ \sigma,\,\sigma'}}^{q\neq 0} v_0(q)\,a_{k+q\sigma}^{+}\left\{-\delta_{\sigma\sigma'}\,\delta_{k,\,p-q} + a_{k\sigma}\,a_{p-q\sigma'}^{+}\right\}a_{p\sigma'} = \\ &= -\frac{1}{2}\sum_{q,\,p,\,\sigma}^{q\neq 0} v_0(q)\,a_{p\sigma}^{+}\,a_{p\sigma} + \frac{1}{2}\sum_{q}^{q\neq 0} v_0(q)\sum_{k\sigma} a_{k+q\sigma}^{+}\,a_{k\sigma}\,\cdot \\ &\quad\cdot\sum_{p,\,\sigma'} a_{p-q\sigma'}^{+}\,a_{p\sigma}\,.\end{aligned}$$

Thus, all together, the Hamiltonian of the jellium model becomes:

$$H = \sum_{k\sigma} \varepsilon_0(\boldsymbol{k})\, a^+_{k\sigma}\, a_{k\sigma} + \frac{1}{2} \sum_{q}^{q\neq 0} v_0(\boldsymbol{q}) \left\{ \hat{\rho}_q \hat{\rho}_{-q} - \widehat{N} \right\} . \tag{2.72}$$

In order to obtain a certain insight into the *physics* of the model, we now investigate the ground-state energy of the jellium model. To this end, we make use of first-order perturbation theory, which according to the variational principle will in any case give us an upper limit for the ground-state energy. We consider the Coulomb interaction H_{ee} as a *perturbation*; the *unperturbed* system is thus given by

$$H_0 = \sum_{k\sigma} \varepsilon_0 \boldsymbol{k}\, a^+_{k\sigma}\, a_{k\sigma} \tag{2.73}$$

(**Sommerfeld model**). It can be solved exactly. In the

"unperturbed" ground state $|E_0\rangle$,

the N electrons occupy all the states with energies which are not greater than a limiting energy ε_F, which is referred to as the **Fermi energy**:

$$\varepsilon_0(\boldsymbol{k}) = \frac{\hbar^2 k^2}{2m} \leq \varepsilon_F = \frac{\hbar^2 k_F^2}{2m} . \tag{2.74}$$

k_F is the **Fermi wavevector**, which can readily be computed as follows: owing to the isotropic energy dispersion

$$\varepsilon_0(\boldsymbol{k}) = \varepsilon_0(k) , \tag{2.75}$$

the electrons occupy all the states in $\boldsymbol{k}$ space within a sphere of radius k_F. Since the $\boldsymbol{k}$-points are discrete in k space due to the periodic boundary conditions (cf. (2.45)), each $\boldsymbol{k}$-point occupies an available

$$\textit{grid volume} \quad \Delta k = \frac{(2\pi)^3}{L^3} = \frac{(2\pi)^3}{V} . \tag{2.76}$$

If we now take the spin degeneracy into account, we find the following relation between the electron number N and the Fermi wavevector k_F:

$$N = 2\frac{1}{\Delta k}\left(\frac{4\pi}{3}k_F^3\right) = \frac{V}{3\pi^2}k_F^3 .$$

This means that:

$$k_F = \left(3\pi^2 \frac{N}{V}\right)^{1/3} , \tag{2.77}$$

$$\varepsilon_\mathrm{F} = \frac{\hbar^2}{2m}\left(3\pi^2\frac{N}{V}\right)^{2/3} . \tag{2.78}$$

We can readily compute the mean energy per particle $\bar{\varepsilon}$, finding:

$$\bar{\varepsilon} = \frac{2}{N}\left(\int\limits_{k\le k_\mathrm{F}} \mathrm{d}^3k\frac{\hbar^2k^2}{2m}\right)\frac{1}{\Delta k} = \frac{3}{5}\varepsilon_\mathrm{F} . \tag{2.79}$$

We thus have obtained the ground-state energy:

$$E_0 = N\bar{\varepsilon} = \frac{3}{5}N\varepsilon_\mathrm{F} . \tag{2.80}$$

We introduce some standard abbreviations:

$$n_\mathrm{e} = \frac{N}{V}: \quad \text{mean electron density,} \tag{2.81}$$

$$v_\mathrm{e} = \frac{1}{n_\mathrm{e}}: \quad \text{mean volume per electron.} \tag{2.82}$$

v_e determines via

$$v_\mathrm{e} = \frac{4\pi}{3}(a_\mathrm{B}r_s)^3 \tag{2.83}$$

the dimensionless **density parameter** r_s, where

$$a_\mathrm{B} = \frac{4\pi\,\varepsilon_0\hbar^2}{me^2} = 0.529\,\text{Å} \tag{2.84}$$

is the **Bohr radius**. If we introduce an energy parameter in a similar fashion,

$$1\ \mathrm{ryd} = \frac{1}{4\pi\,\varepsilon_0}\frac{e^2}{2a_\mathrm{B}} = 13.605\,\mathrm{eV} , \tag{2.85}$$

then for the Fermi energy ε_F, we find:

$$\varepsilon_\mathrm{F} = \frac{\alpha^2}{r_s^2}[\mathrm{ryd}]; \quad \alpha = \left(\frac{9\pi}{4}\right)^{1/3} . \tag{2.86}$$

Then the *unperturbed* ground-state energy is given by:

$$E_0 = N\frac{2.21}{r_s^2}[\mathrm{ryd}] . \tag{2.87}$$

We now switch on the *perturbation* H_{ee} and compute the energy correction to first order:

$$\varepsilon^{(1)} = \frac{1}{2N} \sum_{\substack{k,p,q \\ \sigma,\sigma'}}^{q \neq 0} v_0(q) \left\langle E_0 \left| a^+_{k+q\sigma} \, a^+_{p-q\sigma'} \, a_{p\sigma'} \, a_{k\sigma} \right| E_0 \right\rangle . \tag{2.88}$$

Only those terms contribute for which the annihilation operator acts on states **within** the Fermi sphere, and the creation operator subsequently fills the resulting holes within the Fermi sphere:

1) Direct Term:

$$k = k + q; \quad p = p - q \iff q = 0 . \tag{2.89}$$

According to our preliminary considerations, terms of this type however do not occur in the sum!

2) Exchange Term:

$$\sigma = \sigma'; \quad k + q = p; \quad p - q = k . \tag{2.90}$$

This is a typically quantum-mechanical term, which is not classically understandable. It results from the antisymmetrisation principle for the N-particle states:

$$\begin{aligned} \varepsilon^{(1)} &= \frac{1}{2N} \sum_{k,q,\sigma}^{q \neq 0} v_0(q) \left\langle E_0 \left| a^+_{k+q\sigma} \, a^+_{k\sigma} \, a_{k+q\sigma} \, a_{k\sigma} \right| E_0 \right\rangle = \\ &= -\frac{1}{2N} \sum_{k,q,\sigma}^{q \neq 0} v_0(q) \left\langle E_0 \left| \hat{n}_{k+q\sigma} \hat{n}_{k\sigma} \right| E_0 \right\rangle . \end{aligned} \tag{2.91}$$

Since in the *unperturbed* ground state $|E_0\rangle$, all the states within the Fermi sphere are occupied and all those outside it are unoccupied, it follows that:

$$\varepsilon^{(1)} = -\frac{1}{2N} \sum_{k,q,\sigma}^{q \neq 0} v_0(q) \, \Theta \left(k_F - |k + q| \right) \Theta \left(k_F - k \right) . \tag{2.92}$$

In the thermodynamic limit, we can replace the sums by integrals:

$$\sum_k \Rightarrow \frac{1}{\Delta k} \int d^3k = \frac{V}{(2\pi)^3} \int d^3k .$$

After carrying out the summation over spins, we still need to compute:

$$\varepsilon^{(1)} = -\frac{V}{N}\frac{e^2}{\varepsilon_0(2\pi)^6}\int \mathrm{d}^3k \int \mathrm{d}^3q \, \frac{1}{q^2}\,\Theta\left(k_\mathrm{F} - |\boldsymbol{k}+\boldsymbol{q}|\right)\Theta\left(k_\mathrm{F} - k\right) .$$

The substitution

$$\boldsymbol{k} \Rightarrow \boldsymbol{x} = \boldsymbol{k} + \frac{1}{2}\boldsymbol{q}$$

leads to

$$\varepsilon^{(1)} = -\frac{V}{N}\frac{e^2}{\varepsilon_0(2\pi)^6}\int \mathrm{d}^3q \, \frac{1}{q^2} 2S(q), \tag{2.93}$$

$$S(q) = \frac{1}{2}\int \mathrm{d}^3x \, \Theta\left(k_\mathrm{F} - \left|\boldsymbol{x}+\frac{1}{2}\boldsymbol{q}\right|\right)\Theta\left(k_\mathrm{F} - \left|\boldsymbol{x}-\frac{1}{2}\boldsymbol{q}\right|\right) . \tag{2.94}$$

For the spherical segment sketched in Fig. 2.2, we clearly need to calculate:

$$S(q) = \Theta\left(k_\mathrm{F} - \frac{q}{2}\right)\int\limits_{\frac{q/2}{k_\mathrm{F}}}^{1} \mathrm{d}\cos\vartheta \int \mathrm{d}\varphi \int\limits_{y(\vartheta)}^{k_\mathrm{F}} \mathrm{d}x\, x^2 ,$$

$$y(\vartheta) = \frac{q/2}{\cos\vartheta} .$$

The integration can be readily carried out:

$$S(q) = \frac{2\pi}{3}\Theta\left(k_\mathrm{F} - \frac{q}{2}\right)\left\{k_\mathrm{F}^3 - \frac{3}{4}qk_\mathrm{F}^2 + \frac{1}{16}q^3\right\} . \tag{2.95}$$

The remaining evaluation of (2.93) is then simple:

$$\varepsilon^{(1)} = -\frac{0.916}{r_s}[\mathrm{ryd}] .$$

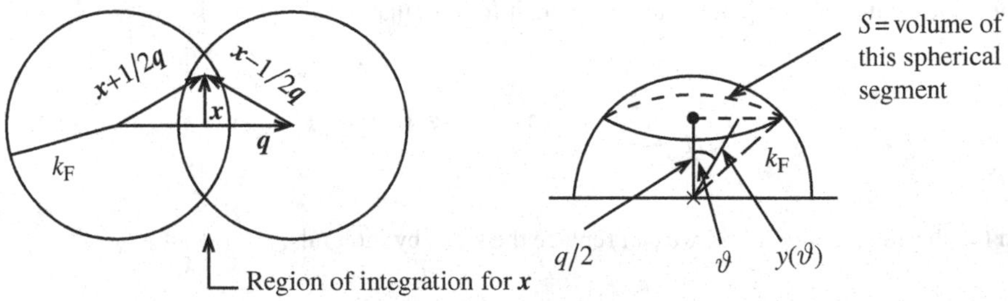

Fig. 2.2 A schematic representation of the integration region for computing the ground-state energy in the jellium model to first order in perturbation theory as in (2.93)

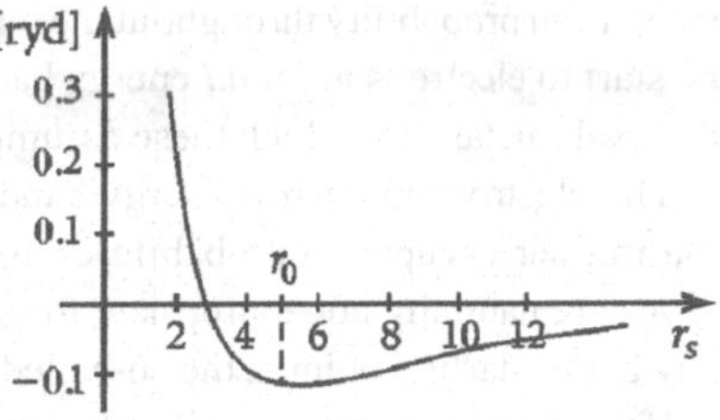

Fig. 2.3 Ground-state energy per particle in the jellium model as a function of the density parameter r_S

This yields finally for the ground-state energy per particle:

$$\frac{1}{N} E_{\min}[\text{ryd}] = \frac{2.21}{r_S^2} - \frac{0.916}{r_S} + \varepsilon_{\text{corr}} = \varepsilon \,. \tag{2.96}$$

The first term is the kinetic energy (2.87), the second represents the so-called **exchange energy**. The latter is typical of systems of identical particles and is a direct result of the principle of indistinguishability and thus for Fermions of the Pauli principle. It guarantees that electrons with parallel spins do not approach each other too closely. Every effect which keeps particles of the same charge *at a distance* leads to a reduction of their ground-state energy. This is the reason for the minus sign in (2.96). The last term is called the **correlation energy**. It gives the deviation of the perturbation-theoretical energy from the exact result and is thus naturally unknown. Modern methods of many-body theory lead to the following series (see (5.177)):

$$\varepsilon_{\text{corr}} = \frac{2}{\pi^2}(1 - \ln 2) \ln r_S - 0.094 + O\left(r_S \ln r_S\right) [\text{ryd}] \,. \tag{2.97}$$

The simple jellium model already gives useful results, e.g. $\varepsilon - \varepsilon_{\text{corr}}$ passes through a minimum at

$$r_0 = (r_S)_{\min} = 4.83,$$
$$(\varepsilon - \varepsilon_{\text{corr}})_{\min} = -0.095\,[\text{ryd}] = -1.29\,[\text{eV}] \,.$$

This indicates an optimal value of the electron density, which corresponds finally to the energetically most favourable ionic spacing, and thus explains, at least qualitatively, the phenomenon of **metallic bonding**.

2.1.3 The Hubbard Model

The decisive simplification achieved by the jellium model consists of the fact that it treats the ions in a solid merely as a positively-charged, homogeneously distributed background, i.e. the crystalline structure is completely ignored. The Bloch functions then become plane waves (2.44), so that within the framework of this model, the electrons have a constant occupation probability throughout the entire crystal. The jellium model is thus limited from the start to electrons in *broad* energy bands, i.e. for example to the conduction electrons of the alkali metals, for which these assumptions are valid to a good approximation.

The electrons in *narrow* energy bands have a relatively low mobility and distinct maxima in their occupation probabilities at the locations of the individual lattice ions. Plane waves are naturally not appropriate for the description of such band electrons. A considerably better starting point is the so-called **tight-binding approximation**.

If we assume a strong lattice potential $\widehat{V}(\boldsymbol{r})$ and a low mobility of the band electrons, then in the neighbourhood of the lattice ions, the atomic Hamiltonian

$$H_{\text{at}} = \sum_{i=1}^{N_{\text{i}}} h_{\text{at}}^{(i)} , \tag{2.98}$$

which is the sum of the Hamiltonians for the individual atoms, should yield a fairly reasonable description, that is, it should be quite similar to H_0 as in (2.7):

$$h_{\text{at}}^{(i)} \varphi_n (\boldsymbol{r} - \boldsymbol{R}_{\text{i}}) = \varepsilon_n \varphi_n (\boldsymbol{r} - \boldsymbol{R}_{\text{i}}) . \tag{2.99}$$

φ_n is an atomic wavefunction, which we can take to be known. The index n symbolises a set of quantum numbers. We are interested in the case that the functions φ_n have only a limited overlap when they are centered at different locations $\boldsymbol{R}_{\text{i}}$, $\boldsymbol{R}_j$. This results in a low tunneling probability for the electrons from atom to atom and therefore only a weak splitting of the atomic levels in the solid – i.e. a narrow energy band.

For the Hamiltonian of the non-interacting electrons (2.7),

$$H_0 = \sum_{i=1}^{N_{\text{e}}} h_0^{(i)} , \tag{2.100}$$

we use the following approach:

$$h_0 = h_{\text{at}} + V_1(\boldsymbol{r}) . \tag{2.101}$$

The correction $V_1(\boldsymbol{r})$ should thus be small in the neighbourhood of the lattice ions, but in contrast relatively large in the intermediate regions, where however the φ_n have dropped to nearly zero. From (2.13), we in fact must solve the following problem:

$$h_0 \psi_{n\boldsymbol{k}}(\boldsymbol{r}) = \varepsilon_n(\boldsymbol{k}) \psi_{n\boldsymbol{k}}(\boldsymbol{r}) . \tag{2.102}$$

The complete solution of this eigenvalue problem appears to be extremely complicated. We therefore use the following trial functions for the Bloch functions $\psi_{n\boldsymbol{k}}(\boldsymbol{r})$:

$$\psi_{n\boldsymbol{k}}(\boldsymbol{r}) = \frac{1}{\sqrt{N_{\text{i}}}} \sum_{j=1}^{N_{\text{i}}} \mathrm{e}^{\mathrm{i}\boldsymbol{k}\cdot\boldsymbol{R}_j} \varphi_n \left(\boldsymbol{r} - \boldsymbol{R}_j\right) . \tag{2.103}$$

This *ansatz* obeys the Bloch theorem (2.15), and it is practically exact near the ionic cores ($V_1(\boldsymbol{r}) \approx 0$), whilst the errors in the interatomic regions are not too great, due to the small overlap of the wavefunctions there. A comparison with (2.29) shows that we have replaced the exact Wannier functions by the atomic wavefunctions. Using (2.102), we now compute approximately the Bloch energies $\varepsilon_n(\boldsymbol{k})$. To start, the following expressions are strictly valid:

$$\int \varphi_n^*(\boldsymbol{r}) h_0 \psi_{n\boldsymbol{k}}(\boldsymbol{r}) \mathrm{d}^3 r = \varepsilon_n(\boldsymbol{k}) \int \varphi_n^*(\boldsymbol{r}) \psi_{n\boldsymbol{k}}(\boldsymbol{r}) \mathrm{d}^3 r ,$$

$$\int \varphi_n^*(\boldsymbol{r}) V_1(\boldsymbol{r}) \psi_{n\boldsymbol{k}}(\boldsymbol{r}) \mathrm{d}^3 r = (\varepsilon_n(\boldsymbol{k}) - \varepsilon_n) \int \varphi_n^*(\boldsymbol{r}) \psi_{n\boldsymbol{k}}(\boldsymbol{r}) \mathrm{d}^3 r .$$

Here, we now apply the *ansatz* (2.103). With the abbreviations

$$v_n = \int \mathrm{d}^3 r \; V_1(\boldsymbol{r}) \left| \varphi_n(\boldsymbol{r}) \right|^2 , \tag{2.104}$$

$$T_0^{(n)} = \varepsilon_n + v_n , \tag{2.105}$$

$$\alpha_n^{(j)} = \int \mathrm{d}^3 r \; \varphi_n^*(\boldsymbol{r}) \varphi_n \left(\boldsymbol{r} - \boldsymbol{R}_j \right) , \tag{2.106}$$

$$\gamma_n^{(j)} = \int \mathrm{d}^3 r \; \varphi_n^*(\boldsymbol{r}) V_1(\boldsymbol{r}) \varphi_n \left(\boldsymbol{r} - \boldsymbol{R}_j \right) \tag{2.107}$$

we obtain:

$$(\varepsilon_n(\boldsymbol{k}) - \varepsilon_n) = v_n + \frac{1}{\sqrt{N_\mathrm{i}}} \sum_j^{\boldsymbol{R}_j \neq 0} \left[\gamma_n^{(j)} - (\varepsilon_n(\boldsymbol{k}) - \varepsilon_n) \, \alpha_n^{(j)} \right] \mathrm{e}^{\mathrm{i}\boldsymbol{k} \cdot \boldsymbol{R}_j} ,$$

where we have presumed that the atomic wavefunctions are normalised. We then find for the Bloch energies:

$$\varepsilon_n(\boldsymbol{k}) = \varepsilon_n + \frac{v_n + \frac{1}{\sqrt{N_\mathrm{i}}} \sum_j^{\neq 0} \gamma_n^{(j)} \mathrm{e}^{\mathrm{i}\boldsymbol{k} \cdot \boldsymbol{R}_j}}{1 + \frac{1}{\sqrt{N_\mathrm{i}}} \sum_j^{\neq 0} \alpha_n^{(j)} \mathrm{e}^{\mathrm{i}\boldsymbol{k} \cdot \boldsymbol{R}_j}} . \tag{2.108}$$

The *overlap integrals* $\gamma_n^{(j)}$ and $\alpha_n^{(j)}$ are by assumption for $\boldsymbol{R}_j \neq 0$ only very small quantities, so that we can with confidence simplify further:

$$\varepsilon_n(\boldsymbol{k}) = T_0^{(n)} + \gamma_n^{(1)} \sum_\Delta \mathrm{e}^{\mathrm{i}\boldsymbol{k} \cdot \boldsymbol{R}_\Delta} . \tag{2.109}$$

Δ indicates the nearest neighbours to the atom at the origin of the coordinate system. The sum can as a rule be readily computed. Thus, for a simple **cubic lattice**:

$$\begin{gathered} \boldsymbol{R}_\Delta = a(\pm 1, 0, 0); \quad a(0, \pm 1, 0); \quad a(0, 0, \pm 1) , \\ \varepsilon_n^{\mathrm{s.c.}}(\boldsymbol{k}) = T_0^{(n)} + 2\gamma_n^{(1)} \left(\cos(k_x a) + \cos(k_y a) + \cos(k_z a) \right) . \end{gathered} \tag{2.110}$$

a is the lattice constant, and $T_0^{(n)}$ and $\gamma_n^{(1)}$ are parameters which must be determined experimentally. $\gamma_n^{(1)}$ is determined by the width W of the band:

$$W_n^{\text{s.c.}} = 12\left|\gamma_n^{(1)}\right| . \tag{2.111}$$

The tight-binding approximation, which led to (2.109), is strictly speaking allowed only for so-called s bands. For p-, d-, f- ... bands, a certain degree of degeneracy must be taken into account, but we shall not discuss this point further here. In the following, we limit our treatment to s bands and thus leave off the index n from here on.

The Bloch energies, (2.109) or (2.110), now clearly exhibit the influence of the crystal structure. Only for very small $|\boldsymbol{k}|$ values near the bottom of the band does the parabolic dispersion, which applies within the jellium model, hold approximately, $\varepsilon(\boldsymbol{k}) \Rightarrow \varepsilon_0(\boldsymbol{k})/\hbar^2k^2/2m$.

In second quantisation, H_0 takes the same form as in (2.33):

$$H_0 = \sum_{ij\sigma} T_{ij} a_{i\sigma}^+ a_{j\sigma} . \tag{2.112}$$

The tight-binding approximation permits electronic transitions via the hopping integral

$$T_{ij} = \frac{1}{N_{\text{i}}} \sum_{\boldsymbol{k}} \varepsilon(\boldsymbol{k}) \mathrm{e}^{\mathrm{i}\boldsymbol{k}\cdot(\boldsymbol{R}_{\text{i}}-\boldsymbol{R}_j)} \tag{2.113}$$

only between nearest-neighbour lattice positions. For the Coulomb interaction of the band electrons, (2.55) of course still applies. The transformation to real space then yields:

$$H_{\text{ee}} = \frac{1}{2} \sum_{\substack{ijkl \\ \sigma,\sigma'}} v(ij;kl)\, a_{i\sigma}^+ a_{j\sigma'}^+ a_{l\sigma'} a_{k\sigma} , \tag{2.114}$$

where the matrix element is to be computed with atomic wavefunctions:

$$v(ij;kl) = \\ = \frac{e^2}{4\pi\varepsilon_0} \iint \mathrm{d}^3r_1\, \mathrm{d}^3r_2\, \frac{\varphi^*(\boldsymbol{r}_1 - \boldsymbol{R}_{\text{i}})\, \varphi^*\left(\boldsymbol{r}_2 - \boldsymbol{R}_j\right) \varphi(\boldsymbol{r}_2 - \boldsymbol{R}_l)\, \varphi(\boldsymbol{r}_1 - \boldsymbol{R}_k)}{|\boldsymbol{r}_1 - \boldsymbol{r}_2|} . \tag{2.115}$$

Owing to the small overlap of the atomic wavefunctions which are centered on different lattice positions, the intra-atomic matrix element

$$U = v\,(ii;ii) \tag{2.116}$$

predominates. Hubbard made the suggestion that the electron-electron interaction therefore be limited to this term:

Hubbard model

$$H = \sum_{ij\sigma} T_{ij}\, a^+_{i\sigma}\, a_{j\sigma} + \frac{1}{2} U \sum_{i,\sigma} \hat{n}_{i\sigma}\, \hat{n}_{i-\sigma} \qquad (2.117)$$

(Notation: $\sigma = \uparrow\ (\downarrow) \Longleftrightarrow -\sigma = \downarrow\ (\uparrow)$). The Hubbard model must thus be the simplest model with which one can study the interplay of the kinetic energy, the Coulomb interactions, the Pauli principle and the lattice structure.

The drastic simplifications which led to (2.117) of course entail a correspondingly limited applicability of the model.

The model is used in the discussion of

1. the electronic properties of solids with narrow energy bands (e.g. transition metals),
2. band magnetism (Fe, Co, Ni, ...),
3. metal-insulator transitions ("Mott transitions"),
4. general principles of statistical mechanics,
5. high-temperature superconductivity.

In spite of its simple structure, the exact solution of the Hubbard model has thus far not been achieved. One must still resort to approximate solutions. Examples will be discussed in the following sections.

2.1.4 Exercises

Exercise 2.1.1 A solid contains $N = N'^3$ (N' even) unit cells in the volume $V = L^3$ $(L = aN')$. For the allowed wave vectors, using periodic boundary conditions, the following holds:

$$\boldsymbol{k} = \frac{2\pi}{L}\left(n_x, n_y, n_z\right); \quad n_{x,y,z} = 0, \pm 1, +2, \ldots, \pm\left(\frac{N'}{2} - 1\right), N'/2 .$$

Prove the orthogonality relation

$$\delta_{ij} = \frac{1}{N} \sum_{\boldsymbol{k}}^{1.\,\mathrm{BZ}} \exp\left[i\boldsymbol{k} \cdot (\boldsymbol{R}_i - \boldsymbol{R}_j)\right] .$$

The sum runs over all the wavenumbers within the first Brillouin zone.

Exercise 2.1.2 Based on the fundamental commutation relations for Bloch operators, $a_{k\sigma}^{+}$, $a_{k\sigma}$, derive the corresponding relations for Wannier operators $a_{i\sigma}^{+}$, $a_{j\sigma}$.

Exercise 2.1.3 In theoretical solid-state physics, one often has to deal with integrals of the type

$$I(T) = \int_{-\infty}^{+\infty} \mathrm{d}x \, g(x) f_{-}(x), \quad f_{-}(x) = \{\exp[\beta(x-\mu)] + 1\}^{-1} .$$

These deviate from their values at $T = 0$

$$I(T=0) = \int_{-\infty}^{\varepsilon_F} \mathrm{d}x \, g(x)$$

by an expression which is determined almost exclusively by the behaviour of the function $g(x)$ within the *Fermi layer* $(\mu - 2k_B T; \mu + 2k_B T)$, where μ represents the chemical potential. Power series are therefore very promising! Assume that $g(x) \to 0$ for $x \to -\infty$, and that $g(x)$ for $x \to +\infty$ diverges at most as a power of x and is regular within the *Fermi layer*.

1. Show that

$$I(T) = -\int_{-\infty}^{+\infty} \mathrm{d}x \, p(x) \frac{\partial}{\partial x} f_{-}(x)$$

holds, with

$$p(x) = \int_{-\infty}^{x} \mathrm{d}y \, g(y) .$$

2. Use a Taylor series for $p(x)$ around μ (chemical potential) for the following representation of the integral:

$$I(T) = p(\mu) + 2 \sum_{n=1}^{\infty} \left(1 - 2^{1-2n}\right) \beta^{-2n} \zeta(2n) g^{(2n-1)}(\mu) .$$

Here, $g^{(2n-1)}(\mu)$ is the $(2n-1)$-th derivative of the function $g(x)$ at the position $x = \mu$, and $\zeta(n)$ is Riemann's ζ function:

$$\zeta(n) = \sum_{p=1}^{\infty} p^{-n} = \frac{1}{(1-2^{1-n})\Gamma(n)} \int_0^{\infty} du \, \frac{u^{n-1}}{e^u + 1} \; .$$

3. Calculate explicitly the first three terms of the series for $I(T)$.

Exercise 2.1.4 The *Sommerfeld model* can explain many electronic properties of the so-called *simple metals* such as Na, K, Mg, Cu, ... to a good approximation. It is defined by the following model assumptions:

a) An ideal Fermi gas within the volume $V = L^3$.
b) Periodic boundary conditions on V.
c) A constant lattice potential $V(\boldsymbol{r}) = \text{const.}$

1. Give the eigenstate energies and the eigenfunctions.
2. Calculate the Fermi energy and the Fermi wavevector as functions of the electron density $n = N/V$.
3. How does the average energy per electron depend on the Fermi energy?
4. Determine the electronic density of states $\rho_0(E)$.
5. Make use of the dimensionless density parameter r_s from Eq. (2.83) to compute the ground-state energy E_0:

$$E_0 = N \frac{2,21}{r_s^2} [\text{ryd}] \; .$$

Exercise 2.1.5 Discuss some of the thermodynamic properties of the Sommerfeld model which was introduced in Ex. 2.1.4.

1. Calculate the temperature dependence of the mean occupation number of a single-particle level.
2. How are the total particle number N and the internal energy $U(T)$ related to the density of states $\rho_0(E)$?
3. Verify, using the Sommerfeld series from Ex. 2.1.2, that the following relation holds for the chemical potential μ:

$$\mu = \varepsilon_{\mathrm{F}} \left[1 - \frac{\pi^2}{12} \left(\frac{k_{\mathrm{B}} T}{\varepsilon_{\mathrm{F}}} \right)^2 \right] \; .$$

4. Compute to a precision of $(k_B T/\varepsilon_{\mathrm{F}})^4$ the internal energy $U(T)$ and the specific heat c_V of the itinerant metal electrons.

5. Calculate and discuss the entropy

$$S = \frac{\partial}{\partial T}\left(k_{\mathrm{B}} T \ln \Xi\right) .$$

Test the Third Law! Ξ is the grand canonical partition function.

Exercise 2.1.6

1. Transform the operator for the electron density

$$\widehat{\rho} = \sum_{i=1}^{N} \delta\left(\boldsymbol{r} - \widehat{\boldsymbol{r}}_{\mathrm{i}}\right)$$

to the second quantisation with Wannier states as the single-particle basis.
2. Derive, using the result of 1, the relation between the electron number and the electron density operator.
3. What form does the electron density operator from part 1 take in the special case of the jellium model?

Exercise 2.1.7 Represent the operator for the electron density

$$\widehat{\rho} = \sum_{i=1}^{N} \delta\left(\boldsymbol{r} - \widehat{\boldsymbol{r}}_{\mathrm{i}}\right)$$

in the formalism of second quantisation using field operators.

Exercise 2.1.8 Transform the Hamiltonian of the jellium model into second quantisation using Wannier states as a single-particle basis.

Exercise 2.1.9 Making use of the electron density operator

$$\widehat{\rho} = \sum_{i=1}^{N} \delta\left(\boldsymbol{r} - \widehat{\boldsymbol{r}}_{\mathrm{i}}\right) ,$$

one can calculate the so-called density correlation

$$G(\boldsymbol{r}, t) = \frac{1}{N} \int \mathrm{d}^3 r' \left\langle \rho\left(\boldsymbol{r}' - \boldsymbol{r}, 0\right) \rho\left(\boldsymbol{r}', t\right) \right\rangle$$

as well as the dynamic structure factor

$$S(\boldsymbol{q},\omega) = \int \mathrm{d}^3 r \int_{-\infty}^{+\infty} \mathrm{d}t\, G(\boldsymbol{r},t)\mathrm{e}^{\mathrm{i}(\boldsymbol{q}\cdot\boldsymbol{r}-\omega t)}\,.$$

The expression

$$S(\boldsymbol{q}) = \int_{-\infty}^{+\infty} \mathrm{d}\omega\, S(\boldsymbol{q},\omega)$$

is termed the static structure factor,
whilst the static pair distribution function $g(\boldsymbol{r})$ is defined by

$$G(\boldsymbol{r},0) = \delta(\boldsymbol{r}) + n g(\boldsymbol{r}) \quad (n = N/V)\,.$$

1. Show that for the density correlation,

$$G(\boldsymbol{r},t) = \frac{1}{NV}\sum_{\boldsymbol{q}} \left\langle \rho_{\boldsymbol{q}}\rho_{-\boldsymbol{q}}(t)\right\rangle \mathrm{e}^{-\mathrm{i}\boldsymbol{q}\cdot\boldsymbol{r}}$$

holds. What is the meaning of $G(\boldsymbol{r},t)$?

2. Verify the expression

$$n g(\boldsymbol{r}) = \frac{1}{N}\sum_{i,\,j}^{i\neq j} \left\langle \delta\left(\boldsymbol{r} + \boldsymbol{r}_i(0) - \boldsymbol{r}_j(0)\right)\right\rangle\,.$$

Consider an appropriate physical interpretation here, also.

3. Prove the following relations for the structure factor:

$$S(\boldsymbol{q},\omega) = \frac{1}{N}\int_{-\infty}^{+\infty} \mathrm{d}t\, \mathrm{e}^{-\mathrm{i}\omega t}\left\langle \rho_{\boldsymbol{q}}\rho_{-\boldsymbol{q}}(t)\right\rangle\,,$$

$$S(\boldsymbol{q}) = \frac{2\pi}{N}\left\langle \rho_{\boldsymbol{q}}\rho_{-\boldsymbol{q}}\right\rangle\,.$$

4. Show that at $T = 0$, the following holds:

$$S(\boldsymbol{q},\omega) = \frac{2\pi}{N}\sum_{n} \left|\left|E_n\right|\rho_{\boldsymbol{q}}^{+}\left\langle E_0\right|\right|^2 \delta\left[\omega - \frac{1}{\hbar}(E_n - E_0)\right]\,.$$

$|E_n\rangle$ are the eigenstates of the Hamiltonian, and $|E_0\rangle$ is its ground state.

Exercise 2.1.10

1. Use the general results from Ex. 2.1.9 to determine the static structure factor $S(\boldsymbol{q})$ with the exact eigenstates of the Sommerfeld model. Sketch its q dependence.
2. Compute also the static pair distribution function $g(\boldsymbol{r})$. Sketch and discuss its r dependence.

Exercise 2.1.11 Compute in the *tight-binding* approximation the Bloch energies $\varepsilon(\boldsymbol{k})$ for the body-centered cubic and for the face-centered cubic lattice structures.

Exercise 2.1.12 Show that the *tight-binding* approach for the electronic wavefunctions $\varphi_{n\boldsymbol{k}}(\boldsymbol{r})$ obeys the Bloch theorem.

2.2 Lattice Vibrations

In Sect. 2.1, the lattice ions were assumed to be motionless and only the excitations of the electronic system were investigated. Following a programme as in (2.6) we now want to discuss the subsystem of the ions in more detail; i.e. the Hamiltonian of (2.3) will now be at the centre of attention.

If energy is transferred to a single lattice ion, e.g. by a particle collision, it will be rapidly distributed over the whole lattice as a result of the strong ion-ion interactions. The local excitation will become a **collective excitation**, in which finally all the lattice sites participate. It is therefore expedient to use collective coordinates, which are still to be defined, in the mathematical description instead of ion coordinates. In this representation, the lattice vibrations can then be quantised. The corresponding quanta are called **phonons**.

2.2.1 The Harmonic Approximation

The restoring forces required for lattice vibrations are the **bonding forces**, which can have rather diverse physical origins. Qualitatively, the pair potential $V_{\mathrm{i}}(|\boldsymbol{R}_\alpha - \boldsymbol{R}_\beta|)$ however always has the same form. The potential minimum defines the equilibrium distance $R^{(0)}_{\alpha\beta}$. The so-called **harmonic approximation** consists in the end in treating the potential curve approximately as a parabola, which seems reasonable for small excursions from the equilibrium distance. We shall next discuss this point more quantitatively.

Our starting point will be a Bravais lattice with a basis containing p atoms, which we describe as in (2.9) by

$$\boldsymbol{R}_s^{\boldsymbol{m}} = \boldsymbol{R}^{\boldsymbol{m}} + \boldsymbol{R}_s \tag{2.118}$$

with $s = 1, 2, \ldots, p$ and $\boldsymbol{m} \equiv (m_1, m_2, m_3);\ m_{\mathrm{i}} \in \mathbf{Z}$,

$$\boldsymbol{R}^{\boldsymbol{m}} = \sum_{i=1}^{3} m_{\mathrm{i}} \boldsymbol{a}_{\mathrm{i}} \,. \tag{2.119}$$

Let

$\boldsymbol{x}_s^{\boldsymbol{m}}(t)$ be the momentary position of the $(\boldsymbol{m}, s)$-th atom, and
$\boldsymbol{u}_s^{\boldsymbol{m}}(t)$ be the displacement of the $(\boldsymbol{m}, s)$-th atom from equilibrium.

As a result, we find:

$$\boldsymbol{x}_s^{\boldsymbol{m}}(t) = \boldsymbol{R}_s^{\boldsymbol{m}} + \boldsymbol{u}_s^{\boldsymbol{m}}(t) \,. \tag{2.120}$$

The **kinetic energy** of the lattice ions is then given by:

$$H_{i,\mathrm{kin}} = \frac{1}{2} \sum_{\substack{\boldsymbol{m} \\ s,\, i}} M_s \left(\frac{\mathrm{d}u_{s,i}^{\boldsymbol{m}}}{\mathrm{d}t} \right)^2 , \quad i = x, y, z \,. \tag{2.121}$$

For the **potential energy**, we write:

$$H_{\mathrm{ii}} = V\left(\left\{\boldsymbol{x}_s^{\boldsymbol{m}}\right\}\right) = V\left(\left\{\boldsymbol{R}_s^{\boldsymbol{m}} + \boldsymbol{u}_s^{\boldsymbol{m}}\right\}\right) . \tag{2.122}$$

Here, the quantity

$$V_0 = V\left(\left\{\boldsymbol{R}_s^{\boldsymbol{m}}\right\}\right) \tag{2.123}$$

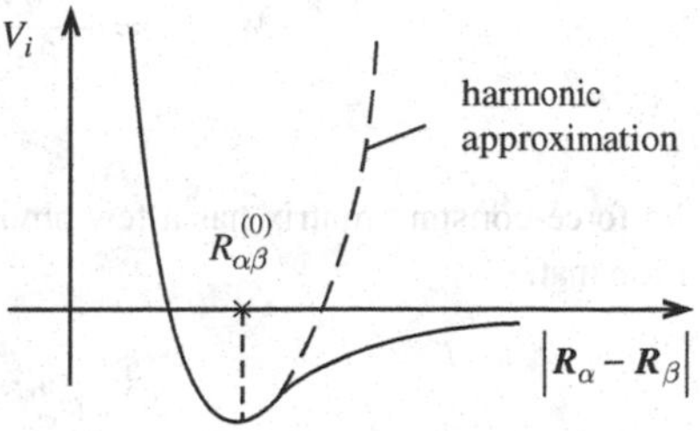

Fig. 2.4 Illustration of the harmonic approximation for the pair potential in a solid

represents the so-called **binding energy**. We expand V around the equilibrium position:

$$V\left(\{x_s^m\}\right) = V_0 + \sum_{\substack{m\\ s,i}} \varphi_{m,s,i}\, u_{s,i}^m + \\ + \frac{1}{2} \sum_{\substack{m\\ s,i}} \sum_{\substack{n\\ t,j}} \varphi_{m,s,i}^{n,t,j}\, u_{s,i}^m\, u_{t,j}^n + O\left(u^3\right) . \tag{2.124}$$

The **harmonic approximation** now consists of neglecting the remainder $O\,(u^3)$. The displacements u are as a rule less than 5% of the lattice spacing, so that the harmonic approximation is quite appropriate. Higher-order, so-called **anharmonic terms**, are therefore initially not of interest.

For the partial derivatives φ in (2.124), we find:

$$\varphi_{m,s,i} \equiv \left.\frac{\partial V}{\partial x_{s,i}^m}\right|_0 = 0 . \tag{2.125}$$

This is the definition of the equilibrium position. The second derivatives form a

matrix of the atomic force constants

$$\varphi_{m,s,i}^{n,t,j} \equiv \left.\frac{\partial^2 V}{\partial x_{t,j}^n \partial x_{s,i}^m}\right|_0 . \tag{2.126}$$

For a better understanding of this important matrix, the following statement is useful:

$-\varphi_{m,s,i}^{n,t,j} u_{t,j}^n$ is the force in the i direction, which acts on the $(\boldsymbol{m},s)$-th atom, when the $(\boldsymbol{n},t)$-th atom is displaced in the j direction by $u_{t,j}^n$, and all the other atoms remain fixed.

The harmonic approximation thus corresponds to a linear force law, as in a harmonic oscillator:

$$M_s \ddot{u}_{s,i}^m = -\frac{\partial V}{\partial u_{s,i}^m} = -\sum_{\substack{n\\ t,j}} \varphi_{m,s,i}^{n,t,j}\, u_{t,j}^n . \tag{2.127}$$

The force-constant matrix has a few obvious symmetries. It follows directly from its definition that:

$$\varphi_{m,s,i}^{n,t,j} \equiv \varphi_{n,t,j}^{m,s,i} . \tag{2.128}$$

On translating the whole solid body by $\Delta \boldsymbol{x} = (\Delta x_1, \Delta x_2, \Delta x_3)$, the forces naturally remain unchanged. It therefore follows from

$$-\sum_j \Delta x_j \sum_{n,t} \varphi_{m,s,i}^{n,t,j} = 0$$

that the relation

$$\sum_{n,t} \varphi_{m,s,i}^{n,t,j} = 0 \tag{2.129}$$

holds. Finally, the translational symmetry yields:

$$\varphi_{m,s,i}^{n,t,j} = \varphi_{s,i}^{t,j}(\boldsymbol{n} - \boldsymbol{m}) . \tag{2.130}$$

To solve (2.127), we first take a trial solution of the form:

$$u_{s,i}^{m} = \frac{\hat{u}_{s,i}^{m}}{\sqrt{M_s}} \mathrm{e}^{-\mathrm{i}\omega t} . \tag{2.131}$$

This gives the eigenvalue equation

$$\omega^2 \hat{u}_{s,i}^{m} = \sum_{\substack{n \\ t,j}} D_{m,s,i}^{n,t,j} \hat{u}_{t,j}^{n} \tag{2.132}$$

for the real and symmetric matrix

$$D = \frac{\varphi}{\sqrt{M_s M_t}} . \tag{2.133}$$

It has $3pN$ real eigenvalues $(\omega_{s,i}^{m})^2$. The eigenvalues $\omega_{s,i}^{m}$ are thus likewise real or purely imaginary. Only the real eigenvalues represent *physical* solutions. Making use of the translational symmetry (2.130), the dimensionality of the eigenvalue problem is reduced from $3pN$ to $3p$:

$$\omega^2 c_{s,i} = \sum_{t,j} K_{i,j}^{s,t} c_{t,j} . \tag{2.134}$$

Here, we have used the following definitions:

$$u_{s,i}^{m} = \frac{c_{s,i}}{\sqrt{M_s}} \exp\left[\mathrm{i}\left(\boldsymbol{q} \cdot \boldsymbol{R}_m - \omega t\right)\right] , \tag{2.135}$$

$$K_{i,j}^{s,t}(\boldsymbol{q}) = \sum_{p} \frac{\varphi_{0,s,i}^{p,t,j}}{\sqrt{M_s M_t}} \exp\left(\mathrm{i}\boldsymbol{q} \cdot \boldsymbol{R}^{p}\right) . \tag{2.136}$$

Equation (2.134) is an eigenvalue equation for the matrix K with $3p$ eigenvalues:

$$\omega = \omega_r(\boldsymbol{q}), \quad r = 1, 2, \ldots, 3p \,. \tag{2.137}$$

Crystals are anisotropic. The **dispersion branches** $\omega_r(\boldsymbol{q})$ therefore have to be determined for each direction $\boldsymbol{q}/|\boldsymbol{q}|$ as functions of $q = |\boldsymbol{q}|$. Details can be found for the standard example of a *diatomic, linear chain* in the textbook literature of solid-state physics. One finds there (Ex. 2.2.1):

$$3 \textbf{ acoustic branches} \Longleftrightarrow \omega(q = 0) = 0,$$

$$3(p-1) \quad \textbf{optical branches} \quad \Longleftrightarrow \quad \omega(q = 0) \neq 0 \,.$$

Owing to the periodic boundary conditions, the wavenumbers $\boldsymbol{q}$ are discrete. If $\boldsymbol{G}$ is an arbitrary vector in the reciprocal lattice, then because of $\exp(\mathrm{i}\boldsymbol{G} \cdot \boldsymbol{R}^m) = 1$, we have:

$$\omega_r(\boldsymbol{q} + \boldsymbol{G}) = \omega_r(\boldsymbol{q}) \,. \tag{2.138}$$

This means that one needs only consider wavenumbers $\boldsymbol{q}$ within the first Brillouin zone. Time-reversal invariance of the equations of motion finally leads to:

$$\omega_r(\boldsymbol{q}) = \omega_r(-\boldsymbol{q}) \,. \tag{2.139}$$

For each of the $3p$ ω_r values, Eq. (2.134) has a solution

$$c_{s,i} = \varepsilon_{s,i}^{(r)}(\boldsymbol{q}) \,, \tag{2.140}$$

which can be chosen so that the orthonormality relation

$$\sum_{s,i} \varepsilon_{s,i}^{(r)*}(\boldsymbol{q})\, \varepsilon_{s,i}^{(r')}(\boldsymbol{q}) = \delta_{r,r'} \tag{2.141}$$

is fulfilled. The general solution of the equation of motion (2.127) is thus finally found to be:

$$u_{s,i}^{m}(t) = \frac{1}{\sqrt{N M_s}} \sum_{r=1}^{3p} \sum_{\boldsymbol{q}}^{1.\mathrm{BZ}} Q_r(\boldsymbol{q}, t)\, \varepsilon_{s,i}^{(r)}(\boldsymbol{q})\, \mathrm{e}^{\mathrm{i}\boldsymbol{q} \cdot \boldsymbol{R}^m} \,. \tag{2.142}$$

Here, we have included the time factor $\exp(-\mathrm{i}\omega_r(\boldsymbol{q})t)$ within the coefficients $Q_r(\boldsymbol{q}, t)$. With

$$\frac{1}{N} \sum_m \exp\left(\mathrm{i}\, (\boldsymbol{q} - \boldsymbol{q}') \cdot \boldsymbol{R}^m\right) = \delta_{\boldsymbol{q},\boldsymbol{q}'} \,,$$

we find the **normal coordinates** $Q_r(\boldsymbol{q},t)$

$$Q_r(\boldsymbol{q},t) = \frac{1}{\sqrt{N}} \sum_{\substack{\boldsymbol{m} \\ s,i}} \sqrt{M_s}\, u^{\boldsymbol{m}}_{s,i}(t)\, \varepsilon^{(r)*}_{s,i}(\boldsymbol{q})\, \mathrm{e}^{-\mathrm{i}\boldsymbol{q}\cdot\boldsymbol{R}^{\boldsymbol{m}}} , \tag{2.143}$$

which obey the equation of motion of the harmonic oscillator

$$\ddot{Q}_r(\boldsymbol{q},t) + \omega_r^2(\boldsymbol{q}) Q_r(\boldsymbol{q},t) = 0 . \tag{2.144}$$

2.2.2 The Phonon Gas

The harmonic approximation of the previous sections gives the following expression for the Lagrange function $L = T - V$ of the ion system:

$$L = \frac{1}{2} \sum_{\substack{\boldsymbol{m} \\ s,i}} M_s \left(\dot{u}^{\boldsymbol{m}}_{s,i}\right)^2 - \frac{1}{2} \sum_{\substack{\boldsymbol{m},s,i \\ \boldsymbol{n},t,j}} \varphi^{\boldsymbol{n},t,j}_{\boldsymbol{m},s,i} u^{\boldsymbol{m}}_{s,i} u^{\boldsymbol{n}}_{t,j} . \tag{2.145}$$

We wish to represent L in normal coordinates. We rearrange, making use of:

$$\frac{1}{N} \sum_{\boldsymbol{m}} \exp\left[\mathrm{i}\,(\boldsymbol{q}-\boldsymbol{q}')\cdot\boldsymbol{R}^{\boldsymbol{m}}\right] = \begin{cases} 1, & \text{if } \boldsymbol{q}-\boldsymbol{q}' = 0 \text{ or } \boldsymbol{G}, \\ 0 & \text{otherwise,} \end{cases} \tag{2.146}$$

$$\left[Q_r(\boldsymbol{q},t)\,\varepsilon^{(r)}_{s,i}(\boldsymbol{q})\right]^* = Q_r(-\boldsymbol{q},t)\varepsilon^{(r)}_{s,i}(-\boldsymbol{q}) . \tag{2.147}$$

Equation (2.147) must hold, so that the displacements $u^{\boldsymbol{m}}_{s,i}$ are real. We have already used Eq. (2.146) in various contexts.

$$\begin{aligned} \frac{1}{2} \sum_{\substack{\boldsymbol{m} \\ s,i}} M_s \left(\dot{u}^{\boldsymbol{m}}_{s,i}\right)^2 &= \frac{1}{2} \sum_{\substack{\boldsymbol{m} \\ s,i}} M_s \frac{1}{N M_s} \sum_{\boldsymbol{q},\boldsymbol{q}'} \sum_{r,r'} \dot{Q}_r(\boldsymbol{q},t)\, \dot{Q}_{r'}(\boldsymbol{q}',t)\, \varepsilon^{(r)}_{s,i}(\boldsymbol{q}) \cdot \\ &\quad \cdot \varepsilon^{(r')}_{s,i}(\boldsymbol{q}')\, \mathrm{e}^{\mathrm{i}(\boldsymbol{q}+\boldsymbol{q}')\cdot\boldsymbol{R}^{\boldsymbol{m}}} = \\ &= \frac{1}{2} \sum_{\boldsymbol{q}} \sum_{r,r'} \dot{Q}_r(\boldsymbol{q},t)\, \dot{Q}_{r'}(-\boldsymbol{q},t) \sum_{s,i} \varepsilon^{(r)}_{s,i}(\boldsymbol{q})\, \varepsilon^{(r')}_{s,i}(-\boldsymbol{q}) = \\ &= \frac{1}{2} \sum_{\boldsymbol{q},r} \dot{Q}^*_r(\boldsymbol{q},t)\, \dot{Q}_r(\boldsymbol{q},t) . \end{aligned} \tag{2.148}$$

In an analogous manner, we find the potential energy:

$$
\begin{aligned}
&\frac{1}{2}\sum_{\substack{m,s,i\\n,t,j}} \varphi_{m,s,i}^{n,t,j} u_{s,i}^{m} u_{t,j}^{n} = \\
&= \frac{1}{2N}\sum_{\substack{m,s,i\\n,t,j}} \varphi_{m,s,i}^{n,t,j} \frac{1}{\sqrt{M_s M_t}} \sum_{q,q'}\sum_{r,r'} Q_r(q,t)\, Q_{r'}(q',t)\cdot \\
&\cdot \varepsilon_{s,i}^{(r)}(q)\, \varepsilon_{t,j}^{(r')}(q')\, e^{iq\cdot R^m} e^{iq'\cdot R^n} = \\
&= \frac{1}{2N}\sum_{\substack{s,i\\n,t,j}}\sum_{qq'}\sum_{r,r'} Q_r(q,t)\, Q_{r'}(q',t)\, \varepsilon_{s,i}^{(r)}(q)\, \varepsilon_{t,j}^{(r')}(q')\cdot \\
&\cdot \sum_{m} \frac{\varphi_{s,i}^{t,j}(n-m)}{\sqrt{M_s M_t}} e^{iq\cdot(R^m-R^n)} e^{i(q+q')\cdot R^n} = \\
&= \frac{1}{2}\sum_{\substack{s,i\\t,j}}\sum_{q,q'}\sum_{r,r'} Q_r(q,t)\, Q_{r'}(q',t)\, \varepsilon_{s,i}^{(r)}(q)\, \varepsilon_{t,j}^{(r')}(q')\cdot \\
&\cdot K_{i,j}^{s,t}(q)\frac{1}{N}\sum_{n} e^{i(q+q')\cdot R^n} = \\
&= \frac{1}{2}\sum_{s,i}\sum_{q}\sum_{r,r'} Q_r(q,t)\, Q_{r'}(-q,t)\, \varepsilon_{s,i}^{(r)}(q)\sum_{t,j} K_{ij}^{s,t}(q)\, \varepsilon_{t,j}^{(r')}(-q) = \\
&= \frac{1}{2}\sum_{q}\sum_{r,r'} \omega_{r'}^2(-q)\, Q_r(q,t)\, Q_{r'}(-q,t)\sum_{s,i} \varepsilon_{s,i}^{(r)}(q)\, \varepsilon_{s,i}^{(r')}(-q) = \\
&= \frac{1}{2}\sum_{q,r} \omega_r^2(q)\, Q_r(q,t)\, Q_r^*(q,t)\,.
\end{aligned} \tag{2.149}
$$

All together, we then have for the Lagrange function:

$$L = \frac{1}{2}\sum_{r,q}\left\{\dot{Q}_r^*(q,t)\,\dot{Q}_r(q,t) - \omega_r^2(q)\, Q_r^*(q,t)\, Q_r(q,t)\right\}\,. \tag{2.150}$$

The momenta which are canonically conjugate to the normal coordinates,

$$\Pi_r(q,t) = \frac{\partial L}{\partial \dot{Q}_r} = \dot{Q}_r^*(q,t)\,, \tag{2.151}$$

are required to formulate the classical Hamilton function:

$$H = \frac{1}{2}\sum_{r,q}\left\{\Pi_r^*(q,t)\,\Pi_r(q,t) + \omega_r^2(q)\, Q_r^*(q,t)\, Q_r(q,t)\right\}\,. \tag{2.152}$$

This is a notable result, since by transforming to the normal coordinates, we have shown that the Hamilton function decomposes into a sum of $3pN$ non-coupled, linear harmonic oscillators.

The next step is the **quantisation** of the classical variables. The displacements $u^{m}_{s,i}$ and the momenta $M_s \dot{u}^{m}_{s,i}$ now become operators with the fundamental commutation relations:

$$\left[u^{m}_{s,i}, u^{n}_{t,j}\right]_{-} = \left[M_s \dot{u}^{m}_{s,i}\,, M_t \dot{u}^{n}_{t,j}\right]_{-} = 0, \tag{2.153}$$

$$\left[M_s \dot{u}^{m}_{s,i}, u^{n}_{t,j}\right]_{-} = \frac{\hbar}{\mathrm{i}}\,\delta_{m,n}\,\delta_{s,t}\,\delta_{i,j}\,. \tag{2.154}$$

By substitution, we find from them the commutation relations for the normal coordinates and their canonically conjugated momenta. With (2.143) and (2.153), we immediately obtain:

$$\left[Q_r(\boldsymbol{q}), Q_{r'}(\boldsymbol{q}')\right]_{-} = \left[\Pi_r(\boldsymbol{q}), \Pi_{r'}(\boldsymbol{q}')\right]_{-} = 0\,. \tag{2.155}$$

For the third relation, we make use of (2.154):

$$\begin{aligned}
\left[\Pi_r(\boldsymbol{q}), Q_{r'}(\boldsymbol{q}')\right]_{-} &= \frac{1}{N}\sum_{\substack{m\\ s,i}}\sum_{\substack{n\\ t,j}} \sqrt{M_s M_t}\,\varepsilon^{(r)}_{s,i}(\boldsymbol{q})\,\mathrm{e}^{\mathrm{i}\boldsymbol{q}\cdot\boldsymbol{R}^m}\,.\\
&\quad\cdot \varepsilon^{(r')*}_{t,j}(\boldsymbol{q}')\,\mathrm{e}^{-\mathrm{i}\boldsymbol{q}'\cdot\boldsymbol{R}^n}\,\frac{1}{M_s}\left[M_s \dot{u}^{m}_{s,i}, u^{m}_{t,j}\right] =\\
&= \frac{\hbar}{\mathrm{i}}\frac{1}{N}\sum_{\substack{m\\ s,i}} \mathrm{e}^{\mathrm{i}(\boldsymbol{q}-\boldsymbol{q}')\cdot\boldsymbol{R}^m}\,\varepsilon^{(r)}_{s,i}(\boldsymbol{q})\,\varepsilon^{(r')*}_{s,i}(\boldsymbol{q}') =\\
&= \frac{\hbar}{\mathrm{i}}\sum_{s,i} \varepsilon^{(r)}_{s,i}(\boldsymbol{q})\,\varepsilon^{(r')*}_{s,i}(\boldsymbol{q})\,\delta_{\boldsymbol{q},\boldsymbol{q}'}\,.
\end{aligned}$$

With (2.141), it finally follows that:

$$\left[\Pi_r(\boldsymbol{q}), Q_{r'}(\boldsymbol{q}')\right]_{-} = \frac{\hbar}{\mathrm{i}}\,\delta_{\boldsymbol{q},\boldsymbol{q}'}\,\delta_{r,r'}\,. \tag{2.156}$$

We now introduce new operators $b_{\boldsymbol{q}r}$ and $b^{+}_{\boldsymbol{q}r}$:

$$Q_r(\boldsymbol{q}) = \sqrt{\frac{\hbar}{2\omega_r(\boldsymbol{q})}}\left\{b_{\boldsymbol{q}r} + b^{+}_{-\boldsymbol{q}r}\right\}, \tag{2.157}$$

$$\Pi_r(\boldsymbol{q}) = \mathrm{i}\sqrt{\frac{1}{2}\hbar\omega_r(\boldsymbol{q})}\left\{b^{+}_{\boldsymbol{q}r} - b_{-\boldsymbol{q}r}\right\}. \tag{2.158}$$

We can read off directly:

$$Q^{+}_r(-\boldsymbol{q}) = Q_r(\boldsymbol{q}); \quad \Pi^{+}_r(-\boldsymbol{q}) = \Pi_r(\boldsymbol{q})\,. \tag{2.159}$$

The inverses of (2.157) and (2.158) are given by:

$$b_{qr} = (2\hbar\omega_r(\boldsymbol{q}))^{-1/2}\left\{\omega_r(\boldsymbol{q})Q_r(\boldsymbol{q}) + \mathrm{i}\,\Pi_r^+(\boldsymbol{q})\right\}, \tag{2.160}$$

$$b_{qr}^+ = (2\hbar\omega_r(\boldsymbol{q}))^{-1/2}\left\{\omega_r(\boldsymbol{q})Q_r^+(\boldsymbol{q}) - \mathrm{i}\,\Pi_r(\boldsymbol{q})\right\}. \tag{2.161}$$

We compute the commutation relations:

$$\begin{aligned}
&\left[b_{qr}, b_{q'r'}\right]_- = \\
&= \left(4\hbar^2\omega_r(\boldsymbol{q})\omega_{r'}(\boldsymbol{q}')\right)^{-1/2}\cdot \\
&\cdot\left\{\mathrm{i}\omega_r(\boldsymbol{q})\left[Q_r(\boldsymbol{q}), \Pi_{r'}^+(\boldsymbol{q}')\right]_- + \mathrm{i}\omega_{r'}(\boldsymbol{q}')\left[\Pi_r^+(\boldsymbol{q}), Q_{r'}(\boldsymbol{q}')\right]_-\right\} = \\
&= \left(4\hbar^2\omega_r(\boldsymbol{q})\omega_{r'}(\boldsymbol{q}')\right)^{-1/2}\cdot \\
&\cdot\left\{\mathrm{i}\omega_r(\boldsymbol{q})\left(-\frac{\hbar}{\mathrm{i}}\,\delta_{rr'}\,\delta_{\boldsymbol{q},-\boldsymbol{q}'}\right) + \mathrm{i}\omega_{r'}(\boldsymbol{q}')\left(\frac{\hbar}{\mathrm{i}}\,\delta_{rr'}\,\delta_{-\boldsymbol{q},\boldsymbol{q}'}\right)\right\} = \\
&= 0, \\
&\left[b_{qr}, b_{q'r'}^+\right]_- = \\
&= \left(4\hbar^2\omega_r(\boldsymbol{q})\omega_{r'}(\boldsymbol{q}')\right)^{-1/2}\cdot \\
&\cdot\left\{-\mathrm{i}\omega_r(\boldsymbol{q})\left[Q_r(\boldsymbol{q}), \Pi_{r'}(\boldsymbol{q}')\right]_- + \mathrm{i}\omega_{r'}(\boldsymbol{q}')\left[\Pi_r^+(\boldsymbol{q}), Q_{r'}^+(\boldsymbol{q}')\right]_-\right\} = \\
&= \left(4\hbar^2\omega_r(\boldsymbol{q})\,\omega_{r'}(\boldsymbol{q}')\right)^{-1/2}\cdot \\
&\cdot\left\{-\mathrm{i}\omega_r(\boldsymbol{q})\left(-\frac{\hbar}{\mathrm{i}}\,\delta_{r,r'}\,\delta_{\boldsymbol{q}\boldsymbol{q}'}\right) + \mathrm{i}\omega_{r'}(\boldsymbol{q}')\left(\frac{\hbar}{\mathrm{i}}\,\delta_{r,r'}\,\delta_{-\boldsymbol{q},-\boldsymbol{q}'}\right)\right\} = \\
&= \delta_{rr'}\,\delta_{\boldsymbol{q}\boldsymbol{q}'}\,.
\end{aligned}$$

b_{qr} and b_{qr}^+ are thus **Bosonic operators**:

$$\left[b_{qr}, b_{q'r'}\right]_- = \left[b_{qr}^+, b_{q'r'}\right]_- = 0\,, \tag{2.162}$$

$$\left[b_{qr}, b_{q'r'}^+\right]_- = \delta_{\boldsymbol{q}\boldsymbol{q}'}\,\delta_{rr'}\,. \tag{2.163}$$

We are now in a position to quantise the Hamilton function:

$$H = \sum_{\boldsymbol{q},r}\frac{1}{2}\left\{\Pi_r^+(q)\,\Pi_r(\boldsymbol{q}) + \omega_r^2(\boldsymbol{q})\,Q_r^+(q)\,Q_r(\boldsymbol{q})\right\} =$$

$$= \frac{1}{4}\sum_{qr} \hbar\omega_r(\mathbf{q})\left\{\left(b_{qr} - b^+_{-qr}\right)\left(b^+_{qr} - b_{-qr}\right) + \left(b^+_{qr} + b_{-qr}\right)\left(b_{qr} + b^+_{-qr}\right)\right\} =$$

$$= \frac{1}{4}\sum_{qr} \hbar\omega_r(\mathbf{q})\left\{b_{qr}\, b^+_{qr} + b^+_{-qr}\, b_{-qr} + b^+_{qr}\, b_{qr} + b_{-qr}\, b^+_{-qr}\right\} =$$

$$= \frac{1}{4}\sum_{qr} \hbar\omega_r(\mathbf{q})\left\{2b^+_{qr}\, b_{qr} + 2b^+_{-qr}\, b_{-qr} + 2\right\} .$$

We can also make use of (2.139) and then obtain within the harmonic approximation the **Hamiltonian for the quantised vibrations of the ion lattice**:

$$H = \sum_{qr} \hbar\omega_r(\mathbf{q})\left\{b^+_{qr}\, b_{qr} + \frac{1}{2}\right\} . \tag{2.164}$$

We are dealing here with a system of $3pN$ non-coupled harmonic oscillators.

In Eqs. (2.157) and (2.158), we suppressed the time dependence of the normal coordinates Q_r and their canonical momenta. As set out in (2.142), it is given simply by:

$$Q_r(\mathbf{q}t) = Q_r(\mathbf{q})\, \mathrm{e}^{-\mathrm{i}\omega_r(\mathbf{q})t} . \tag{2.165}$$

This implies according to (2.157) that:

$$b_{qr}(t) = b_{qr}\, \mathrm{e}^{-\mathrm{i}\omega_r(\mathbf{q})t} . \tag{2.166}$$

We wish to show that this result agrees with

$$b_{qr}(t) = \exp\left(\frac{\mathrm{i}}{\hbar}Ht\right) b_{qr} \exp\left(-\frac{\mathrm{i}}{\hbar}Ht\right) . \tag{2.167}$$

To this end, we first prove the assertion

$$b_{qr}H^n = \{\hbar\omega_r(\mathbf{q}) + H\}^n\, b_{qr} , \tag{2.168}$$

using the method of complete induction:
$n = 1$:

$$\left[b_{qr}, H\right]_- = \sum_{q',r'} \hbar\omega_{r'}(\mathbf{q}')\left[b_{qr}\, , b^+_{q'r'}\, b_{q'r'}\right]_- = \hbar\omega_r(\mathbf{q})\, b_{qr}$$

$$\Rightarrow b_{qr}H = (\hbar\omega_r(\mathbf{q}) + H)\, b_{qr} .$$

$n \Rightarrow n + 1$:

$$b_{qr}H^{n+1} = \left(b_{qr}H^n\right)H = (\hbar\omega_r(\mathbf{q}) + H)^n\, b_{qr}H =$$

$$= (\hbar\omega_r(\mathbf{q}) + H)^{n+1}\, b_{qr} .$$

This proves the assertion in (2.168). It then follows that:

$$b_{qr} \exp\left(\frac{-\mathrm{i}}{\hbar} H t\right) = \sum_{n=0}^{\infty} \frac{(-\mathrm{i}/\hbar)^n}{n!} t^n \, b_{qr} H^n =$$

$$= \exp\left[-\frac{\mathrm{i}}{\hbar}\left(\hbar\omega_r(\boldsymbol{q}) + H\right) t\right] b_{qr} \, .$$

After insertion into (2.167), we find the result (2.166). The two relations are therefore equivalent.

The essential result of this section is (2.164). This makes it clear that the energy of the lattice vibrations is quantised. The elementary quantum $\hbar\omega_r(\boldsymbol{q})$ is interpreted as the energy of the quasi-particle **phonon**. In detail, one makes the following associations:

b^+_{qr}: Creation operator for a $(\boldsymbol{q}, r)$ phonon,

b_{qr}: Annihilation operator for a $(\boldsymbol{q}, r)$ phonon,

$\hbar\omega_r(\boldsymbol{q})$: Energy of the $(\boldsymbol{q}, r)$ phonon.

Phonons are Bosons! Each vibrational state can therefore be occupied by arbitrarily many phonons.

The harmonic approximation which underlies this section models the ion lattice as a non-interacting **phonon gas**. The terms neglected in the series expansion (2.124) for the potential V, which are of third or higher order in the displacements $u^{\boldsymbol{m}}_{s,i}$ (*anharmonicity* of the lattice), can be interpreted as a coupling, i.e. an interaction between the phonons. They are important for the description of effects such as thermal expansion, the approach to thermal equilibrium, heat conductivity, the high-temperature behaviour of c_p, c_V, etc.

2.2.3 Exercises

Exercise 2.2.1 Consider a linear chain composed of two different types of atoms (masses m_1, m_2) alternating along the chain:

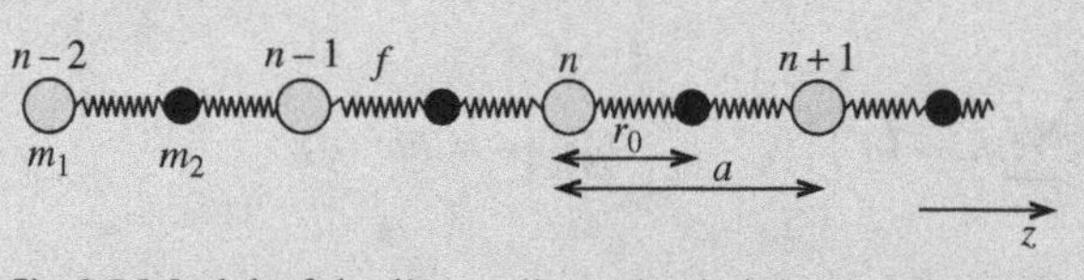

Fig. 2.5 Model of the linear diatomic chain

The interaction between the atoms can be taken to a good approximation to be limited to nearest neighbours. Within the harmonic approximation (linear force law), the coupling between neighbouring atoms can be expressed in terms of a force constant f.

1. Describe the chain as a linear Bravais lattice with a diatomic basis. Determine the primitive translations and the vectors of the (reciprocal) lattice as well as the first Brillouin zone.
2. Formulate the equation of motion for longitudinal lattice vibrations.
3. Justify and make use of the trial solution

$$u_\alpha^n = \frac{c_\alpha}{\sqrt{m_\alpha}} \exp\left[\mathrm{i}\left(q R^n - \omega t\right)\right]$$

for the displacement of the (n, α)-th atom from its equilibrium position.

4. Sketch the dispersion branches for a qualitative discussion. Investigate in particular the special cases $q = 0, \; +\pi/a, -\pi/a, 0 < q \ll \pi/a$.

Exercise 2.2.2 Compute the density of states $D(\omega)$ of the linear chain:

$$D(\omega)\mathrm{d}\omega = \text{The number of eigenfrequencies in the interval } (\omega; \omega + \mathrm{d}\omega).$$

Use appropriate periodic boundary conditions. How does $D(\omega)$ depend on the *group velocity* $v_g = \mathrm{d}\omega/\mathrm{d}q_z$? Give a qualitative sketch of $D(\omega)$!

Exercise 2.2.3 Compute the density of states $D(\omega)$ for the lattice vibrations of a three-dimensional crystal. The crystal has the primitive translations $\boldsymbol{a}_i$, $i = 1, 2, 3$, which are not necessarily orthogonal.

1. Introduce periodic boundary conditions on a parallelepiped with the edges $N_i a_i$, $i = 1, 2, 3$. Express the allowed wavenumbers in terms of the primitive translations of the reciprocal lattice.
2. Calculate the grid volume in $\boldsymbol{q}$ space, which contains one and only one wavevector.
3. Express the density of states for one dispersion branch $\omega_r(\boldsymbol{q})$ in terms of a volume integral in $\boldsymbol{q}$ space.
4. Make use of the group velocity to find an alternative representation of the density of states:

$$v_{\mathrm{g}}^{(r)} = \left|\Delta_q \omega_r(\boldsymbol{q})\right| .$$

5. What is the expression for the overall density of states?

Exercise 2.2.4 The so-called Debye model for the lattice vibrations of a pure Bravais lattice ($p = 1$, monatomic basis) makes use of the following two assumptions:

1. A linear, isotropic approximation for the acoustic branches:

$$\omega_r = \bar{v}_r q \, .$$

2. Replacement of the Brillouin zone by a sphere of the same volume.

Due to 2., there must be a limiting frequency ω_r^D (the Debye frequency). Calculate it! Derive the density of states $D_D(\omega)$ corresponding to this model.

Exercise 2.2.5

1. Calculate in the harmonic approximation the internal energy $U(T) = \langle H \rangle$ ($\langle \cdots \rangle$: thermal average) of the lattice vibrations of a three-dimensional crystal. Discuss the limiting cases of high and low temperatures (Hint: $\langle b_{qr}^+ \, b_{qr} \rangle \Rightarrow$ Bose-Einstein distribution).
2. Use the Debye model (Ex. 2.2.4) to compute the specific heat at low temperatures.

2.3 The Electron-Phonon Interaction

Having discussed in Sect. 2.1 the crystal electrons and in Sect. 2.2 the lattice ions, essentially with no mutual coupling, or at most coupled in a very simple manner via H_{e+} (2.50), we now examine the interaction between these two subsystems in more detail. Within our general model of the solid state (2.1), we will now consider the operator H_{ei}.

2.3.1 The Hamiltonian

Our starting point is the operator (2.5):

$$H_{ei} = \sum_{j=1}^{N_e} \sum_{\alpha=1}^{N_i} V_{ei} \left(\boldsymbol{r}_j - \boldsymbol{x}_\alpha \right) = H_{ei}^{(0)} + H_{e-p} \, . \tag{2.169}$$

The interaction $H_{ei}^{(0)}$ of the electrons with the rigid ion lattice was already included in our model H_0 for the crystal electrons (see (2.7)). H_{e-p} is the electron-phonon interaction *per se*.

Following the considerations of the previous section, we know that every lattice vibration is characterised by the states defined by the wavenumber $\boldsymbol{q}$ and the branch r of the dispersion spectrum $\omega_r(\boldsymbol{q})$. The electron-phonon interaction thus implies the

absorption and emission of $(\boldsymbol{q}, r)$ phonons.

The conceivable **elementary processes** can be shown graphically in a simple way (see Fig. 2.6).

All the interactions can be composed out of these four elementary processes. They should therefore be reflected in a corresponding model Hamiltonian.

We assume that in these interactions, the ion is displaced as a rigid body and is not deformed, which is of course by no means to be taken for granted. Deformations of the ions however represent *higher-order* effects. In the framework of the *harmonic approximation* for the lattice vibrations, we expand the interaction energy V_{ei} up to the first non-vanishing term. It is in this case the linear term:

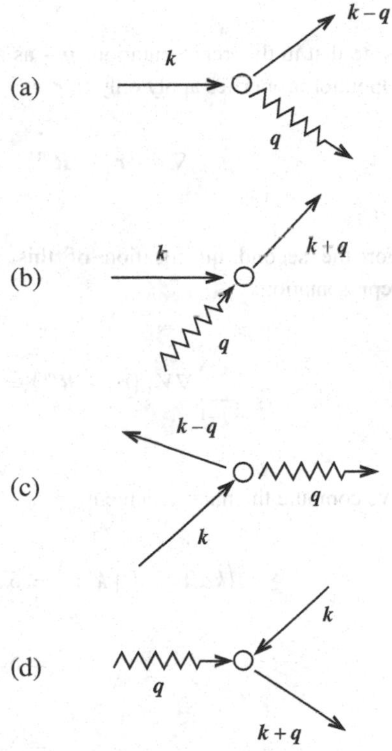

Fig. 2.6 Elementary processes of the electron-phonon interaction; *straight arrows* stand for electrons, *wavy arrows* for phonons: (**a**) Phonon emission by an electron; (**b**) Phonon absorption by an electron; (**c**) Phonon emission from electron-hole recombination; (**d**) Creation of an electron-hole pair by phonon annihilation

$$
\begin{aligned}
V_{\mathrm{ei}}\left(\boldsymbol{r}_j-\boldsymbol{x}_s^{\boldsymbol{m}}\right) &\equiv V_{\mathrm{ei}}\left(\boldsymbol{r}_j-\boldsymbol{R}_s^{\boldsymbol{m}}-\boldsymbol{u}_s^{\boldsymbol{m}}\right)= \\
&= V_{\mathrm{ei}}\left(\boldsymbol{r}_j-\boldsymbol{R}_s^{\boldsymbol{m}}\right)-\boldsymbol{u}_s^{\boldsymbol{m}}\cdot\nabla V_{\mathrm{ei}}+O\left(u^2\right) .
\end{aligned} \tag{2.170}
$$

The first term leads to $H_{\mathrm{ei}}^{(0)}$ and was already taken into account in the treatment of the crystal electrons (see Sect. 2.1) e.g. in the Bloch energies $\varepsilon(\boldsymbol{k})$. The second term contains the actual electron-phonon interaction. We assume singly-charged ions, ($N_{\mathrm{e}}=N_{\mathrm{i}}=N$), and use expression (2.142) for the displacements $\boldsymbol{u}_s^{\boldsymbol{m}}$:

$$
\begin{aligned}
H_{\mathrm{e-p}} = &-\sum_{j=1}^{N}\sum_{m,s}\sum_{r=1}^{3p}\sum_{q}^{1.\mathrm{BZ}}\frac{1}{\sqrt{NM_s}}Q_r(\boldsymbol{q})\,\mathrm{e}^{\mathrm{i}\boldsymbol{q}\cdot\boldsymbol{R}_m}\,. \\
&\cdot\boldsymbol{\varepsilon}_s^{(r)}(\boldsymbol{q})\cdot\nabla V_{\mathrm{ei}}\left(\boldsymbol{r}_j-\boldsymbol{R}_s^{\boldsymbol{m}}\right) .
\end{aligned} \tag{2.171}
$$

$Q_r(\boldsymbol{q})$ is already familiar from (2.157) in second quantisation. We still have to transform the electronic part. In ∇V_{ei}, the electronic variable $\boldsymbol{r}_j$ appears. We choose the Fourier representation for V_{ei}:

$$
V_{\mathrm{ei}}\left(\boldsymbol{r}_j-\boldsymbol{R}_s^{\boldsymbol{m}}\right)=\sum_{\boldsymbol{p}}V_{\mathrm{ei}}^{(s)}(\boldsymbol{p})\,\mathrm{e}^{\mathrm{i}\boldsymbol{p}\cdot(\boldsymbol{r}_j-\boldsymbol{R}^m)}\,. \tag{2.172}
$$

Note that in this representation, $\boldsymbol{p}$ – as a wavenumber – is a variable and not an operator. Operator properties apply only to $\boldsymbol{r}_j$.

$$
\nabla V_{\mathrm{ei}}\left(\boldsymbol{r}_j-\boldsymbol{R}_s^{\boldsymbol{m}}\right)=\mathrm{i}\sum_{\boldsymbol{p}}V_{\mathrm{ei}}^{(s)}(\boldsymbol{p})\,\boldsymbol{p}\,\mathrm{e}^{\mathrm{i}\boldsymbol{p}\cdot(\boldsymbol{r}_j-\boldsymbol{R}^m)}\,. \tag{2.173}
$$

For the second quantisation of this single-electron operator, we choose the Bloch representation:

$$
\sum_{j=1}^{N}\nabla V_{\mathrm{ei}}\left(\boldsymbol{r}_j-\boldsymbol{R}_s^{\boldsymbol{m}}\right)=\sum_{\substack{\boldsymbol{k},\boldsymbol{k}'\\ \sigma,\sigma'}}\left\langle\boldsymbol{k}\sigma\mid\nabla V_{\mathrm{ei}}\mid\boldsymbol{k}'\sigma'\right\rangle a_{\boldsymbol{k}\sigma}^{+}\,a_{\boldsymbol{k}'\sigma'}\,. \tag{2.174}
$$

We compute the matrix element:

$$
\begin{aligned}
\left\langle\boldsymbol{k}\sigma\mid\mathrm{e}^{\mathrm{i}\boldsymbol{p}\cdot\hat{\boldsymbol{r}}}\mid\boldsymbol{k}'\sigma'\right\rangle &= \delta_{\sigma\sigma'}\int\mathrm{d}^3r\left\langle\boldsymbol{k}\mid\mathrm{e}^{\mathrm{i}\boldsymbol{p}\cdot\hat{\boldsymbol{r}}}\mid\boldsymbol{r}\right\rangle\left\langle\boldsymbol{r}\mid\boldsymbol{k}'\right\rangle= \\
&= \delta_{\sigma\sigma'}\int\mathrm{d}^3r\,\mathrm{e}^{\mathrm{i}\boldsymbol{p}\cdot\boldsymbol{r}}\left\langle\boldsymbol{k}\mid\boldsymbol{r}\right\rangle\left\langle\boldsymbol{r}\mid\boldsymbol{k}'\right\rangle= \\
&= \delta_{\sigma\sigma'}\int\mathrm{d}^3r\,\mathrm{e}^{\mathrm{i}\boldsymbol{p}\cdot\boldsymbol{r}}\,\psi_{\boldsymbol{k}}^{*}(\boldsymbol{r})\psi_{\boldsymbol{k}'}(\boldsymbol{r})\,.
\end{aligned}
$$

For the Bloch functions, we use (2.16):

$$\left\langle k\sigma \mid e^{ip\cdot\hat{r}} \mid k'\sigma' \right\rangle = \delta_{\sigma\sigma'} \int d^3r\, e^{i(p-k+k')\cdot r}\, u_k^*(r)\cdot u_{k'}(r)\,. \tag{2.175}$$

The amplitude function $u_k(r)$ which reflects the periodicity of the lattice is not to be confused with the displacements u_s^m. Inserting (2.175) into (2.174), we now find the following intermediate result:

$$\begin{aligned}\sum_{j=1}^{N} \nabla V_{ei}\left(r_j - R_s^m\right) = i \sum_{\substack{k,k'\\ p,\sigma}} V_{ei}^{(s)}(p)\, p e^{-ip\cdot R^m}\, a_{k\sigma}^+ a_{k'\sigma}\,\cdot \\ \cdot \int d^3r\, e^{i(p-k+k')\cdot r} u_k^*(r) u_{k'}(r)\,.\end{aligned} \tag{2.176}$$

The product of the displacements has the periodicity of the lattice, owing to (2.17). The integral can therefore be nonzero only for $k = k' + p$. Inserting into (2.171) then yields the following result (making use of

$$\frac{1}{N}\sum_m e^{i(q-p)\cdot R^m} = \sum_K \delta_{p,\,q+K}\,, \tag{2.177}$$

where K is a vector in the reciprocal lattice):

$$\begin{aligned}H_{e\text{-}p} = -\sum_{s,r}\sum_{q,k',K,\sigma} i\sqrt{\frac{N}{M_s}}\, Q_r(q) V_{ei}^{(s)}(q+K)\,\cdot \\ \cdot \left(\varepsilon_s^{(r)}(q)\cdot(q+K)\right) a_{k'+q+K\sigma}^+\, a'_{k\sigma}\,\cdot \\ \cdot \int d^3r\, u_{k'+q+K}^*(r)\, u_{k'}(r)\,.\end{aligned}$$

We now use (2.157) for the normal coordinates $Q_r(q,\, t)$, and define as an abbreviation the

Matrix element of the electron-phonon coupling

$$\begin{aligned}T_{k,q,K}^{(s,r)} = -i\sqrt{\frac{\hbar N}{2M_s\,\omega_r(q)}}\, V_{ei}^{(s)}(q+K)\left[\varepsilon_s^{(r)}(q)\cdot(q+K)\right]\cdot \\ \cdot \int d^3r\, u_{k+q+K}^*(r)\, u_k(r)\,.\end{aligned} \tag{2.178}$$

Then the Hamiltonian for the electron-phonon interaction is given by:

$$H_{\text{e-p}} = \sum_{k\sigma} \sum_{\boldsymbol{q},\boldsymbol{K}} \sum_{s,r} T^{(s,r)}_{\boldsymbol{k},\boldsymbol{q},\boldsymbol{K}} \left(b_{\boldsymbol{q}r} + b^+_{-\boldsymbol{q}r}\right) a^+_{\boldsymbol{k}+\boldsymbol{q}+\boldsymbol{K}\sigma}\, a_{\boldsymbol{k}\sigma} \,. \tag{2.179}$$

Upon emission (creation) of a $(-\boldsymbol{q}, r)$ phonon, or upon absorption (annihilation) of a $(\boldsymbol{q}, r)$ phonon, the wavenumber $\boldsymbol{k}$ of the electron becomes $\boldsymbol{k} + \boldsymbol{q} + \boldsymbol{K}$. One therefore defines the

$\hbar(\boldsymbol{q} + \boldsymbol{K})$: **quasi-(crystal-)momentum of the phonons,**

where $\boldsymbol{q}$ originates in the first Brillouin zone, whilst $\boldsymbol{K}$ can be an arbitrary reciprocal-lattice vector. In (2.179), $\boldsymbol{K}$ is fixed by the requirement

$\boldsymbol{k} + \boldsymbol{q} + \boldsymbol{K} \in$ **the first Brillouin zone.**

We distinguish between:

$\boldsymbol{K} = \boldsymbol{0}$: **normal processes,** and

$\boldsymbol{K} \neq \boldsymbol{0}$: **umklapp processes.**

The complicated matrix element (2.178) can be greatly simplified if the following assumptions can be made:

1. A simple Bravais lattice: $p = 1 \Rightarrow \sum_s$ is omitted,
2. Only normal processes: $\boldsymbol{K} = \boldsymbol{0} \Rightarrow \sum_{\boldsymbol{K}}$ is omitted,
3. The phonons are uniquely longitudinally or transversally polarised:

$\boldsymbol{\varepsilon}^{(r)}(\boldsymbol{q}) \cdot \boldsymbol{q}$ → $\neq 0$: longitudinal,
→ $= 0$: transverse.

Under these assumptions, only the longitudinal acoustic phonons interact with the electrons. With the matrix element

$$T_{\boldsymbol{k},\boldsymbol{q}} = -\mathrm{i}\sqrt{\frac{\hbar N}{2M\omega(\boldsymbol{q})}}\, V_{\text{ei}}(\boldsymbol{q})\,[\boldsymbol{\varepsilon}(\boldsymbol{q}) \cdot \boldsymbol{q}] \int \mathrm{d}^3r\, u^*_{\boldsymbol{k}+\boldsymbol{q}}(\boldsymbol{r}) u_{\boldsymbol{k}}(\boldsymbol{r}) \,, \tag{2.180}$$

the electron-phonon interaction can be simplified to:

$$H_{\text{e-p}} = \sum_{\boldsymbol{k}\boldsymbol{q}\sigma} T_{\boldsymbol{k}\boldsymbol{q}} \left(b_{\boldsymbol{q}} + b^+_{-\boldsymbol{q}}\right) a^+_{\boldsymbol{k}+\boldsymbol{q}\sigma}\, a_{\boldsymbol{k}\sigma} \,. \tag{2.181}$$

2.3.2 The Effective Electron-Electron Interaction

The elementary processes sketched in Fig. 2.6 may be combined into additional, more complex types of coupling. In particular, phonon-induced electron-electron interactions can be described. Figure 2.7 symbolises a process in which a $(\boldsymbol{k}, \sigma)$ electron emits a $\boldsymbol{q}$ phonon, which is then absorbed by a $(\boldsymbol{k}', \sigma')$ electron. The spin of the electron is of course not involved in this process. The first electron deforms the lattice in its immediate neighbourhood, i.e. as a negatively-charged particle, it displaces the positively-charged ions slightly. *Deformation* means abstractly always absorption or emission of phonons. A second electron "*sees*" this lattice deformation and reacts to it. The result is thus an effective electron-electron interaction, which naturally has nothing to do with the usual Coulomb interaction and can therefore be either attractive or repulsive. In the case of an attractive interaction, it can lead to the formation of electron pairs (**Cooper pairs**) with an accompanying decrease in the ground-state energy. This process forms the basis for conventional superconductivity. We consider the electron-phonon interaction in the form (2.181) and neglect electron-electron as well as phonon-phonon interactions. The matrix element $T_{\boldsymbol{kq}}$ (2.180) can be computed for simplicity with plane waves, which also eliminates the $\boldsymbol{k}$-dependence $\left(u_{\boldsymbol{k}}(\mathbf{r}) \Rightarrow 1/\sqrt{V}\right)$:

$$T_{\boldsymbol{q}} = -\mathrm{i}\sqrt{\frac{\hbar N}{2M\omega(\boldsymbol{q})}}\, V_{\mathrm{ei}}(\boldsymbol{q})\, [\boldsymbol{\varepsilon}(\boldsymbol{q}) \cdot \boldsymbol{q}] \ . \tag{2.182}$$

One can see from (2.172) that

$$V_{\mathrm{ei}}^{*}(\boldsymbol{q}) = V_{\mathrm{ei}}(-\boldsymbol{q})$$

must hold. Due to (2.147), we also can assume

$$[\boldsymbol{\varepsilon}(\boldsymbol{q}) \cdot \boldsymbol{q}]^{*} = \boldsymbol{\varepsilon}(-\boldsymbol{q}) \cdot \boldsymbol{q} \ ,$$

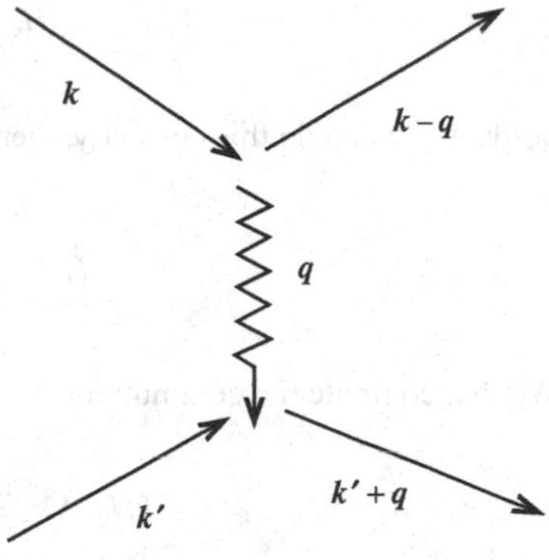

Fig. 2.7 Elementary process of the phonon-induced effective electron-electron interaction

so that

$$T_q^* = T_{-q} \tag{2.183}$$

follows. We now investigate whether the following model Hamiltonian contains terms representing an effective electron-electron interaction, as presumed:

$$H = \sum_{k\sigma} \varepsilon(k)\, a_{k\sigma}^+ a_{k\sigma} + \sum_q \hbar\omega(q)\, b_q^+ b_q + \sum_{kq\sigma} T_q \left(b_q + b_{-q}^+\right) a_{k+q\sigma}^+ a_{k\sigma} \,. \tag{2.184}$$

We carry out an appropriate **canonical transformation** and try to eliminate linear terms in $H_{\text{e-p}}$.

$$\widetilde{H} = \mathrm{e}^{-S} H \mathrm{e}^{S} = \left(1 - S + \frac{1}{2}S^2 + \cdots\right) H \left(1 + S + \frac{1}{2}S^2 + \cdots\right) =$$

$$= H + [H, S]_- + \frac{1}{2}\,[[H, S]_-, S]_- + \cdots ,$$

$$\widetilde{H} = \mathrm{e}^{-S} H \mathrm{e}^{S} = H_0 + H_{\text{e-p}} + [H_0, S]_- + \left[H_{\text{e-p}}, S\right]_- + \frac{1}{2}\,[[H_0, S]_-, S]_- + \cdots \tag{2.185}$$

We take $H_{\text{e-p}}$ to be a small perturbation. S should be of the same order of magnitude. We therefore neglect all the terms in the expansion (2.185) which are of higher than quadratic order in S or $H_{\text{e-p}}$. H_0 combines the first two terms in (2.184).

For S, we take the *ansatz*

$$S = \sum_{kq\sigma} T_q \left(x b_q + y b_{-q}^+\right) a_{k+q\sigma}^+ a_{k\sigma} \tag{2.186}$$

and fix the parameters x and y in such a way that

$$H_{\text{e-p}} + [H_0, S]_- \stackrel{!}{=} 0 \tag{2.187}$$

holds. If we can do this correctly, then the effective operator $\widetilde{H}$ is given by:

$$\widetilde{H} \approx H_0 + \frac{1}{2}\left[H_{\text{e-p}}, S\right]_- \,. \tag{2.188}$$

We first compute the commutator:

$$[H_0, S]_- = [H_{\text{e}}, S]_- + \left[H_{\text{p}}, S\right]_- \,.$$

Here,

$$[H_e, S]_- =$$
$$= \sum_{p,\sigma'}\sum_{kq\sigma} \varepsilon(\boldsymbol{p})T_q\left[a^+_{p\sigma'}\,a_{p\sigma'}\,,\,(xb_q + yb^+_{-q})\,a^+_{k+q\sigma}\,a_{k\sigma}\right]_- =$$
$$= \sum_{\substack{p,k,q\\ \sigma,\sigma'}} \varepsilon(\boldsymbol{p})T_q\,(xb_q + yb^+_{-q})\left[a^+_{p\sigma'}\,a_{p\sigma'}\,,\,a^+_{k+q\sigma}\,a_{k\sigma}\right]_- =$$
$$= \sum \varepsilon(\boldsymbol{p})T_q\,(xb_q + yb^+_{-q})\,\delta_{\sigma\sigma'}\left(\delta_{p,k+q}\,a^+_{p\sigma'}\,a_{k\sigma} - \delta_{kp}\,a^+_{k+q\sigma}\,a_{p\sigma'}\right) =$$
$$= \sum_{kq\sigma} T_q\,(\varepsilon(\boldsymbol{k}+\boldsymbol{q}) - \varepsilon(\boldsymbol{k}))\,a^+_{k+q\sigma}\,a_{k\sigma}\,(xb_q + yb^+_{-q})\ .$$

We have repeatedly made use of the fact that the creation and annihilation operators for electrons and phonons are of course mutually commuting.

$$[H_p, S]_- = \sum_p\sum_{kq\sigma} \hbar\omega(\boldsymbol{p})T_q\left[b^+_p\,b_p,\,(xb_q + yb^+_{-q})\right]_-\,a^+_{k+q\sigma}\,a_{k\sigma} =$$
$$= \sum_p\sum_{kq\sigma} \hbar\omega(\boldsymbol{p})T_q\,(-x\delta_{qp}\,b_p + y\delta_{-qp}\,b^+_p)\,a^+_{k+q\sigma}\,a_{k\sigma} =$$
$$= \sum_{kq\sigma} T_q\hbar\omega(\boldsymbol{q})\,(-xb_q + yb^+_{-q})\,a^+_{k+q\sigma}\,a_{k\sigma}\ .$$

All together, we obtain:

$$[H_0, S]_- = \sum_{kq\sigma} T_q\Big\{x\,(\varepsilon(\boldsymbol{k}+\boldsymbol{q}) - \varepsilon(\boldsymbol{k}) - \hbar\omega(\boldsymbol{q}))\,b_q +$$
$$+\,y\,(\varepsilon(\boldsymbol{k}+\boldsymbol{q}) - \varepsilon(\boldsymbol{k}) + \hbar\omega(\boldsymbol{q}))\,b^+_{-q}\Big\}\,a^+_{k+q\sigma}\,a_{k\sigma}\ . \tag{2.189}$$

Equation (2.187) can thus be obtained using the following parameters x and y:

$$x = \{\varepsilon(\boldsymbol{k}) - \varepsilon(\boldsymbol{k}+\boldsymbol{q}) + \hbar\omega(\boldsymbol{q})\}^{-1}\ , \tag{2.190}$$

$$y = \{\varepsilon(\boldsymbol{k}) - \varepsilon(\boldsymbol{k}+\boldsymbol{q}) - \hbar\omega(\boldsymbol{q})\}^{-1}\ . \tag{2.191}$$

In the last step, we have inserted the expression for S thus obtained into (2.188). The essential task is the computation of the following commutator:

$$\left[\left(b_{q'} + b^+_{-q'}\right)a^+_{k'+q'\sigma'}\,a_{k'\sigma'},\,(xb_q + yb^+_{-q})\,a^+_{k+q\sigma}\,a_{k\sigma}\right]_- =$$
$$= \left(b_{q'} + b^+_{-q'}\right)(xb_q + yb^+_{-q})\left[a^+_{k'+q'\sigma'}\,a_{k'\sigma'},\,a^+_{k+q\sigma}\,a_{k\sigma}\right]_- +$$
$$+\left[\left(b_{q'} + b^+_{-q'}\right),\,(xb_q + yb^+_{-q})\right]_-\,a^+_{k'+q'\sigma'}\,a_{k'\sigma'}\,a^+_{k+q\sigma}\,a_{k\sigma}\ .$$

Only the last term leads to an effective electron-electron interaction. We thus concentrate exclusively on this term:

$$\left[\left(b_{q'} + b^+_{-q'}\right), \left(x b_q + y b^+_{-q}\right)\right]_- = x\left[b^+_{-q'}, b_q\right]_- + y\left[b_{q'}, b^+_{-q}\right]_- = \\ = -x\delta_{q',-q} + y\delta_{q',-q} \,. \tag{2.192}$$

This yields the following contribution to $\widetilde{H}$:

$$\widetilde{H}_{\text{eff}} = \frac{1}{2}\sum_{\substack{kq\sigma \\ k'q'\sigma'}} T_{q'} T_q (y - x)\,\delta_{q',-q}\, a^+_{k'+q'\sigma'}\, a_{k'\sigma'}\, a^+_{k+q\sigma}\, a_{k\sigma} =$$

$$= \frac{1}{2}\sum_{\substack{kq\sigma \\ k'\sigma'}} T_{-q} T_q (y - x)\left(a^+_{k+q\sigma}\, a^+_{k'-q\sigma'}\, a_{k'\sigma'}\, a_{k\sigma} + \delta_{k',k+q}\hat{n}_{k\sigma}\right).$$

The final term is uninteresting in this context. However, we can see that the electron-phonon interaction brings about a term of the following form:

$$\widetilde{H}_{\text{ee}} = \sum_{kpq\sigma,\sigma'} \left|T_q\right|^2 \frac{\hbar\omega(\boldsymbol{q})}{(\varepsilon(\boldsymbol{k}+\boldsymbol{q}) - \varepsilon(\boldsymbol{k}))^2 - (\hbar\omega(\boldsymbol{q}))^2}\, a^+_{k+q\sigma}\, a^+_{p-q\sigma'}\, a_{p\sigma'}\, a_{k\sigma} \,. \tag{2.193}$$

This interaction is

repulsive, when $(\varepsilon(\boldsymbol{k}+\boldsymbol{q}) - \varepsilon(\boldsymbol{k}))^2 > (\hbar\omega(\boldsymbol{q}))^2$,

attractive, when $(\varepsilon(\boldsymbol{k}+\boldsymbol{q}) - \varepsilon(\boldsymbol{k}))^2 < (\hbar\omega(\boldsymbol{q}))^2$.

The latter effect explains the stability of Cooper pairs, and thus forms the basis for our understanding of superconductivity.

2.3.3 Exercises

Exercise 2.3.1 The initial idea of the BCS theory of superconductivity is the correlation of conduction electrons through virtual phonon exchange into so-called **Cooper pairs**, each consisting of two electrons with oppositely-directed wavevectors and spins,

$$(\boldsymbol{k}\uparrow, -\boldsymbol{k}\downarrow)\,,$$

which form a bound state. Define suitable creation and annihilation operators for the Cooper pairs! Compute the associated fundamental commutation relations! Are Cooper pairs Bosons?

Exercise 2.3.2 The normal electron-phonon interaction generates an effective electron-electron interaction induced by phonon exchange, which under certain circumstances can also be attractive (Sect. 2.3.2). Consider the following model:

a) N interaction-free electrons in states $k \leq k_F$, all states with $k > k_F$ unpopulated $\Longleftrightarrow$ a filled Fermi sphere $|\text{FS}\rangle$.

b) Two additional electrons with oppositely-directed wavevectors and spins (Cooper pair, see Ex. 2.3.1) interact according to

$$V_k(q) = \begin{cases} -V\,, & \text{if} \quad |\varepsilon(k+q) - \varepsilon(k)| \leq \hbar\omega_{\text{D}}\,, \\ 0\,, & \text{otherwise} \end{cases}$$

(ω_{D}: Debye frequency).

1. Formulate the model Hamiltonian.
2. Justify the *ansatz*

$$|\psi\rangle = \sum_{k,\sigma} \alpha_\sigma(k)\, a^+_{k\sigma}\, a^+_{-k-\sigma}\, |\text{FS}\rangle$$

for the Cooper-pair state and show that

$$\alpha_\sigma(k) = -\alpha_{-\sigma}(-k)$$

must hold.

3. Verify that due to the normalisation of $|\psi\rangle$ and $|\text{FS}\rangle$ the following relation must hold:

$$\sum_{k,\sigma}^{k>k_{\text{F}}} |\alpha_\sigma(k)|^2 = 1\,.$$

Exercise 2.3.3 Consider again the *Cooper model* defined in Ex. 2.3.2 with the *ansatz* $|\psi\rangle$ for the *Cooper-pair state*:

1. Show that for the expectation value of the kinetic energy in the state $|\psi\rangle$, the following holds:

$$\langle\psi \mid T \mid \psi\rangle = 2 \sum_{k,\sigma}^{k>k_{\text{F}}} \varepsilon(k)\, |\alpha_\sigma(k)|^2 + 2 \sum_{k}^{k<k_{\text{F}}} \varepsilon(k)\,.$$

2. Show that for the expectation value of the potential energy in the state $|\psi\rangle$, the following holds:

$$\langle\psi \mid V \mid \psi\rangle = 2 \sum_{k,q,\sigma}^{k,\,|k+q|>k_{\text{F}}} V_k(q)\, \alpha^*_\sigma(k+q)\, \alpha_\sigma(k)\,.$$

Exercise 2.3.4 Consider still further the *Cooper model* defined in Ex. 2.3.2 with the *ansatz* $|\psi\rangle$ for the *Cooper-pair state*:

1. Determine the *optimum* expansion coefficients $\alpha_\sigma(\boldsymbol{k})$ by minimising the energy calculated in Ex. 2.3.3, $E = \langle\psi|H|\psi\rangle$. Note the side condition from Ex. 2.3.2, 3, which follows from the normalisation of $|\psi\rangle$.
2. Show that the energy of the Cooper pair is less than the energy of two non-interacting electrons at the Fermi edge. What conclusions can you draw from this?

   Hint: summations over $\boldsymbol{k}$ can often be advantageously converted into simpler integrals over energy by making use of the *free* Bloch density of states:

$$\rho_0(\varepsilon) = \frac{1}{N}\sum_{\boldsymbol{k}} \delta\left(\varepsilon - \varepsilon(\boldsymbol{k})\right)!$$

Exercise 2.3.5 On the BCS theory of superconductivity (Phys. Rev. **108**, 1175 (1957)): The BCS model suppresses from the beginning all those interactions which give the same contributions in the normal and the superconducting phase. It considers only the attractive part of the phonon-induced electron-electron interaction. As test states for a variational calculation of the BCS ground-state energy ($\Longleftrightarrow$ difference between the ground-state energies in the normal and the superconducting phases), products of Cooper-pair states are used, since according to Ex. 2.3.4, the latter lead to an energy decrease:

$$|\mathrm{BCS}\rangle = \left[\prod_{\boldsymbol{k}} \left(u_{\boldsymbol{k}} + v_{\boldsymbol{k}}\, b_{\boldsymbol{k}}^{+}\right)\right]|0\rangle\,, \qquad |0\rangle: \quad \text{particle vacuum,}$$

$b_{\boldsymbol{k}}^{+} = a_{\boldsymbol{k}\uparrow}^{+}\, a_{-\boldsymbol{k}\downarrow}^{+}$: *Cooper-pair creation operator* (see Ex. 2.3.1). The coefficients $u_{\boldsymbol{k}}$ and $v_{\boldsymbol{k}}$ can be taken to be real.

1. Show that due to the normalisation of the state $|\mathrm{BCS}\rangle$,

$$u_{\boldsymbol{k}}^2 + v_{\boldsymbol{k}}^2 = 1$$

   must hold.
2. Calculate the following expectation values:

$$\langle \mathrm{BCS} \mid b_{\boldsymbol{k}}^{+}\, b_{\boldsymbol{k}} \mid \mathrm{BCS}\rangle; \qquad \langle \mathrm{BCS} \mid b_{\boldsymbol{k}}^{+}\, b_{\boldsymbol{k}} b_{\boldsymbol{p}}^{+}\, b_{\boldsymbol{p}} \mid \mathrm{BCS}\rangle;$$

$$\langle \mathrm{BCS} \mid b_{\boldsymbol{k}}^{+}\, b_{\boldsymbol{k}}\left(1 - b_{\boldsymbol{p}}^{+}\, b_{\boldsymbol{p}}\right) \mid \mathrm{BCS}\rangle; \quad \langle \mathrm{BCS} \mid b_{\boldsymbol{p}}^{+}\, b_{\boldsymbol{k}} \mid \mathrm{BCS}\rangle.$$

Exercise 2.3.6 On the BCS theory of superconductivity (Phys. Rev. **108**, 1175 (1957)): The BCS model of superconductivity limits itself, as explained in Ex. 2.3.5, to treating the attractive contribution to the phonon-induced electron-electron interaction (see Ex. 2.3.2). Using the variational expression |BCS⟩ from Ex. 2.3.5, an upper limit to the ground-state energy can be calculated.

1. Justify the model Hamiltonian:

$$H_{\mathrm{BCS}} = \sum_{k,\sigma} t(k)\, a^+_{k\sigma}\, a_{k\sigma} - V \sum_{k,\,p}^{k\neq p} b^+_p\, b_k ;$$

$$t(k) = \varepsilon(k) - \mu \, .$$

2. Calculate:

$$E = \langle \mathrm{BCS} \mid H_{\mathrm{BCS}} \mid \mathrm{BCS} \rangle \, .$$

3. Show that for the *gap parameter*

$$\Delta_k = V \sum_p^{\neq k} u_p v_p \, ,$$

the minimum condition for $E = E(\{v_k\})$ leads to the result:

$$\Delta_k = \frac{V}{2} \sum_p^{\neq k} \Delta_p \left(t^2(p) + \Delta_p^2\right)^{-1/2} \, .$$

4. Express v_k^2, u_k^2, $E_0 = (E(\{v_k\}))_{\min}$ in terms of Δ_k and $t(k)$.

Exercise 2.3.7 In order to derive the effective electron-electron interaction $\widetilde{H}$ from the actual electron-phonon interaction H, a canonical transformation (2.185)

$$\widetilde{H} = \mathrm{e}^{-S} H \mathrm{e}^{S} \, ,$$

is carried out. Why must $S^+ = -S$ be required? Is this requirement fulfilled by the solutions (2.186), (2.190), (2.191)?

2.4 Spin Waves

The concepts of many-body theory have a particularly rich field of application in the area of magnetism. For this in fact rather old phenomenon, there is thus far no complete theory. Model concepts are necessary, and they are adapted to particular manifestations of magnetism. We develop the most important of these in this section.

2.4.1 Classification of Magnetic Solids

Using the magnetic susceptibility

$$\chi = \left(\frac{\partial M}{\partial H}\right)_T \qquad (M : \textbf{magnetisation}) , \tag{2.194}$$

the various magnetic phenomena can be divided roughly into three classes:

diamagnetism, paramagnetism, and "collective" magnetism.

In the case of

1) Diamagnetism
In diamagnetism, we are dealing basically with a purely inductive effect. The applied magnetic field H induces magnetic dipoles which are, according to Lenz's rule, opposed to the field which induces them. A negative susceptibility is thus typical of diamagnets:

$$\chi^{\text{dia}} < 0; \quad \chi^{\text{dia}}(T, H) \approx \text{const} . \tag{2.195}$$

Diamagnetism is naturally a property of **all** materials. One therefore refers to a diamagnet only when there is no *additional* paramagnetism or *collective* magnetism present which would overcompensate the relatively weak diamagnetism.

The decisive precondition for

2) Paramagnetism
In paramagnetism is the existence of **permanent** magnetic moments, which can be oriented by the applied field H in competition with the thermal motion of the elementary magnets. It is thus typified by:

$$\chi^{\text{para}} > 0; \quad \chi^{\text{para}}(T, H) \overset{\text{i.g.}}{=} \chi^{\text{para}}(T) . \tag{2.196}$$

The permanent moments can be

2a) localised moments
which result from electron shells which are only partly filled. If these are sufficiently well shielded from **environmental influences** by outer, filled shells, then the electrons of the

unfilled shell will not contribute to an electric current in the solid, but rather will remain localised in the region of their *mother ion*. Prominent examples are the $4f$ electrons of the **rare earths**. An incompletely filled electronic shell has as a rule a resultant magnetic moment. Without an applied magnetic field, the moments are statistically distributed over all directions, so that the solid as a whole has no net moment. In an applied field, the moments become oriented, and their magnetic susceptibility follows the so-called **Curie law** at temperatures which are not too low:

$$\chi^{\text{para}}(T) \approx \frac{C}{T} \qquad (C = \text{const}) . \tag{2.197}$$

Such a system is called a **Langevin paramagnet**.
The permanent magnetic moments of a paramagnet can however also be the

2b) itinerant moments
of quasi-free conduction electrons, of which each carries a moment of one Bohr magneton ($1\mu_B$). In this case, one refers to **Pauli paramagnetism**, whose susceptibility is to first order temperature **in**dependent as a result of the Pauli principle.

Dia- and paramagnetism can be regarded as essentially understood. They are more or less properties of individual atoms, and thus not typical many-body phenomena. Here, we are interested only in

3) "Collective" Magnetism
"Collective" magnetism results from a characteristic interaction which is understandable only in terms of quantum mechanics, the *exchange interaction* between permanent magnetic dipole moments. These permanent moments can again be

localised (Gd, EuO, $Rb_2 MnCl_4$)

or else they can be

itinerant (Fe, Co, Ni) .

The exchange interaction leads to a

critical temperature T* ,

below which the moments order **spontaneously**, i.e. without an applied magnetic field. Above T^*, they behave as in a *normal* paramagnet. The susceptibility for $T < T^*$ is in general a complicated function of the applied field and the temperature, which in addition depends on the previous treatment (history) of the sample:

$$\chi^{KM} = \chi^{KM}(T, H, \text{history}) \quad (T \le T^*) . \tag{2.198}$$

Collective magnetism can be divided into three major subclasses:

3a) Ferromagnetism
In this case, the critical temperature is referred to as

$$T^* = T_\mathrm{C}: \quad \textbf{Curie temperature.}$$

At $T = 0$, all the moments are oriented parallel to one another (*ferromagnetic saturation*). This ordering decreases with increasing temperature. In the range $0 < T < T_\mathrm{C}$, however, a preferred axis persists, i.e. a **spontaneous magnetisation** of the sample is still present; it then vanishes at T_C. Above T_C, the system is paramagnetic with a characteristic high-temperature behaviour of its susceptibility, which is called the **Curie-Weiss law**:

$$\chi(T) = \frac{C}{T - T_\mathrm{C}} \quad (T \gg T_\mathrm{C}) \; . \tag{2.199}$$

3b) Ferrimagnetism
The lattice in this case is composed of *two* ferromagnetic sublattices A and B with differing spontaneous magnetisations:

$$\boldsymbol{M}_A \neq \boldsymbol{M}_\mathrm{B}: \quad \boldsymbol{M}_A + \boldsymbol{M}_\mathrm{B} = \boldsymbol{M} \neq 0 \quad \text{for } T < T_\mathrm{C}. \tag{2.200}$$

3c) Antiferromagnetism
This is a special case of ferrimagnetism. Below a critical temperature, which in this case is termed

$$T^* = T_\mathrm{N}: \quad \text{the } \textbf{Néel temperature} \; ,$$

the two sublattices order ferromagnetically with opposite but equal spontaneous magnetisations:

$$T < T_\mathrm{N}: \quad |\boldsymbol{M}_A| = |\boldsymbol{M}_\mathrm{B}| \neq 0; \quad \boldsymbol{M} = \boldsymbol{M}_A + \boldsymbol{M}_\mathrm{B} \equiv 0 \; . \tag{2.201}$$

Above T_N, the system is normally paramagnetic, with a linear high-temperature behaviour of the inverse susceptibility, as in a ferromagnet:

$$\chi(T) = \frac{C}{T - \Theta} \quad (T \gg T_\mathrm{N}) \; . \tag{2.202}$$

Θ is called the **paramagnetic Curie temperature**. As a rule, it is negative.

2.4.2 Model Concepts

Models are indispensable owing to the lack of a complete theory of magnetism; they relate specifically to particular magnetic phenomena. Here, we refer exclusively to collective magnetism. The collective magnetism of insulators and of metals must be treated separately.

1) Insulators
Magnetism is produced by localised magnetic moments which are due to incompletely filled electronic shells ($3d$-, $4d$-, $4f$- or $5f$-) in the atoms.
Examples:

Ferromagnets:	$CrBr_3$, K_2CuF_4, EuO, EuS, $CdCr_2Se_4$, Rb_2CrCl_4, ...
Antiferromagnets:	MnO, EuTe, NiO, $RbMnF_3$, Rb_2MnCl_4, ...
Ferrimagnets:	$MO \cdot Fe_2O_3$ (M = divalent metal ion such as Fe, Ni, Cd, Mg, Mn, ...)

These substances are described quite realistically by the so-called

Heisenberg model

$$H = -\sum_{i,j} J_{ij}\, S_i \cdot S_j \,. \tag{2.203}$$

Each localised magnetic moment is associated with an angular momentum $\boldsymbol{J}_i$ (magneto-mechanical parallelism):

$$\boldsymbol{m}_i = \mu_B (\boldsymbol{L}_i + 2\boldsymbol{S}_i) \equiv \mu_B g_J \cdot \boldsymbol{J}_i \,. \tag{2.204}$$

$\boldsymbol{L}_i$ is here the orbital contribution, $\boldsymbol{S}_i$ the spin contribution, and g_J is the Landé g-factor. Due to

$$\boldsymbol{S}_i = (g_J - 1)\, \boldsymbol{J}_i \,, \tag{2.205}$$

the exchange interaction between the moments can be formulated as an interaction between their associated spins. The index i refers to the lattice site. The coupling constants J_{ij} are called **exchange integrals**.

The Heisenberg Hamiltonian (2.203) is to be understood as an *effective* operator. The spin-spin interaction ($\boldsymbol{S}_i \cdot \boldsymbol{S}_j$), applied to corresponding spin states, *simulates* the contribution of the exchange matrix elements of the Coulomb interaction (cf. (2.90)), which is presumed to be at the origin of the spontaneous magnetisation.

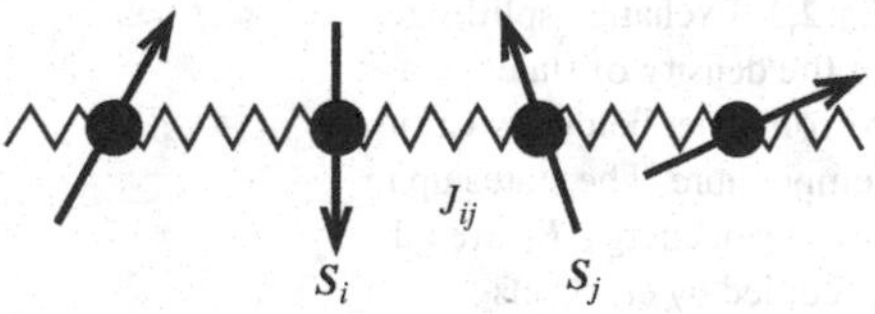

Fig. 2.8 Model of a ferromagnet with localised magnetic moments. J_{ij} are the exchange integrals

Although the Heisenberg model works well for the magnetic insulators, it is practically useless for the description of magnetic metals.

2) Metals
It is expedient to subdivide this topic into those magnetic metals in which the magnetism and the electrical conductivity are due to the same group of electrons, and those in which these properties can be ascribed to different groups of electrons. In the former case, one refers to

2a) Band magnetism
Prominent representatives of this class are Fe, Co and Ni. A quantum-mechanical exchange interaction causes a spin-dependent band shift below $T < T_C$. Since the two spin subbands are each filled with electrons up to the common Fermi energy E_F, it follows that

$$N_\uparrow > N_\downarrow \quad (T < T_C) \ ,$$

and thus a spontaneous magnetic moment is observed. It is found that band magnetism is possible especially with narrow energy bands, and it is therefore thought that the phenomenon can be explained by the **Hubbard model** which was discussed in Sect. 2.1.

2b) "Localised" magnetism
The prototype of this class is the $4f$ metal Gd. Its magnetism is carried by localised $4f$ moments, which can be described realistically by the Heisenberg model (2.203). The electric current in Gd is carried by quasi-free, mobile conduction electrons, which can be understood with the aid of e.g. the jellium model (Sect. 2.1.2), or also with the Hubbard model (Sect. 2.1.3). Interesting phenomena result from an interaction between the localised $4f$ moments and the itinerant conduction electrons. It can for example lead to an effective coupling of the $4f$ moments and thus can amplify the collective magnetism. It can however also contribute to the electrical resistance via scattering of the conduction electrons from the local moments. An appropriate model is the so-called

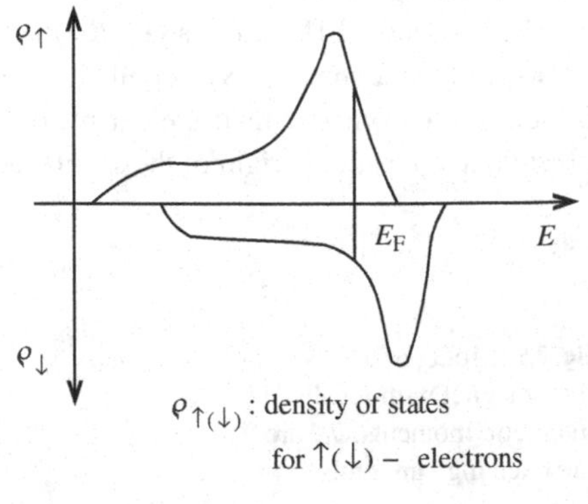

Fig. 2.9 Exchange splitting of the density of states of a ferromagnet below its Curie temperature. The states up to the Fermi energy E_F are all occupied by electrons

s-f (s-d) model

$$H = H\,(\text{Hubbard, jellium}) + H\,(\text{Heisenberg}) - g \sum_{\mathrm{i}} \boldsymbol{\sigma}_{\mathrm{i}} \cdot \boldsymbol{S}_{\mathrm{i}}\,. \tag{2.206}$$

σ is the spin operator for the conduction electrons at the site $\boldsymbol{R}_{\mathrm{i}}$, and g is a corresponding coupling constant.

2.4.3 Magnons

There are interesting analogies between the lattice vibrations treated in Sect. 2.2 and the elementary excitations in a ferromagnet. The oscillations of the lattice ions about their equilibrium positions can be decomposed into normal modes with quantised amplitudes. The unit of quantisation is called the **phonon**. The oscillations in a ferromagnet corresponding to the normal modes are called

spin waves,

following Bloch, and their unit of quantisation is the

magnon.

We want to analyse these excitations within the framework of the Heisenberg model (2.203) in more detail. With the usual conventions

$$J_{ij} = J_{ji}; \quad J_{\mathrm{ii}} = 0; \quad J_0 = \sum_{\mathrm{i}} J_{\mathrm{i}j} = \sum_{j} J_{\mathrm{i}j} \tag{2.207}$$

and the well-known spin operators

$$\boldsymbol{S}_j = \left(S_j^x, S_j^y, S_j^z\right), \tag{2.208}$$

$$S_j^{\pm} = S_j^x \pm \mathrm{i}\, S_j^y\,, \tag{2.209}$$

$$S_j^x = \frac{1}{2}\left(S_j^+ + S_j^-\right); \quad S_j^y = \frac{1}{2\mathrm{i}}\left(S_j^+ - S_j^-\right), \tag{2.210}$$

we can decompose the scalar product in the Heisenberg Hamiltonian into its components:

$$\begin{aligned} \boldsymbol{S}_{\mathrm{i}} \cdot \boldsymbol{S}_j &= \frac{1}{2}\left(S_{\mathrm{i}}^+ S_j^- + S_{\mathrm{i}}^- S_j^+\right) + S_{\mathrm{i}}^z S_j^z \\ \Rightarrow\ H &= -\sum_{\mathrm{i},j} J_{\mathrm{i}j}\left(S_{\mathrm{i}}^+ S_j^- + S_{\mathrm{i}}^z S_j^z\right) - \frac{1}{\hbar} g_J \mu_{\mathrm{B}} B_0 \sum_{\mathrm{i}} S_{\mathrm{i}}^z\,. \end{aligned} \tag{2.211}$$

Compared to (2.203), we have added to the Hamiltonian a Zeeman term, in order to take account of the interaction of the local moments with the applied magnetic field $B_0 = \mu_0 H$.

It is often expedient to make use of the spin operators in $\boldsymbol{k}$ space:

$$S^\alpha(\boldsymbol{k}) = \sum_{\mathrm{i}} \mathrm{e}^{-\mathrm{i}\boldsymbol{k}\cdot\boldsymbol{R}_\mathrm{i}} S_\mathrm{i}^\alpha \,, \tag{2.212}$$

$$S_\mathrm{i}^\alpha = \frac{1}{N}\sum_{\boldsymbol{k}} \mathrm{e}^{\mathrm{i}\boldsymbol{k}\cdot\boldsymbol{R}_\mathrm{i}} S^\alpha(\boldsymbol{k}) \,, \tag{2.213}$$

with $(\alpha = x, y, z, +, -)$.

From the commutation relations in real space,

$$\left[S_\mathrm{i}^x, S_j^y\right]_- = \mathrm{i}\hbar\, \delta_{ij} S_\mathrm{i}^z \quad \text{and cyclic permutations} \,, \tag{2.214}$$

$$\left[S_\mathrm{i}^z, S_j^\pm\right]_- = \pm\hbar\, \delta_{ij} S_\mathrm{i}^\pm \,, \tag{2.215}$$

$$\left[S_\mathrm{i}^+, S_j^-\right]_- = 2\hbar\, \delta_{ij} S_\mathrm{i}^z \,, \tag{2.216}$$

the commutation relations in $\boldsymbol{k}$ space follow immediately:

$$\left[S^+(\boldsymbol{k}_1), S^-(\boldsymbol{k}_2)\right]_- = 2\hbar S^z(\boldsymbol{k}_1 + \boldsymbol{k}_2) \,, \tag{2.217}$$

$$\left[S^z(\boldsymbol{k}_1), S^\pm(\boldsymbol{k}_2)\right]_- = \pm\hbar S^\pm(\boldsymbol{k}_1 + \boldsymbol{k}_2) \,, \tag{2.218}$$

$$\left(S^+(\boldsymbol{k})\right)^+ = S^-(-\boldsymbol{k}) \,. \tag{2.219}$$

With the wavenumber-dependent exchange integrals,

$$J(\boldsymbol{k}) = \frac{1}{N}\sum_{i,j} J_{ij}\, \mathrm{e}^{\mathrm{i}\boldsymbol{k}\cdot(\boldsymbol{R}_\mathrm{i}-\boldsymbol{R}_j)} \,, \tag{2.220}$$

we can then rewrite the Hamiltonian (2.211) in terms of wavenumbers:

$$H = -\frac{1}{N}\sum_{\boldsymbol{k}} J(\boldsymbol{k})\left\{S^+(\boldsymbol{k})S^-(-\boldsymbol{k}) + S^z(\boldsymbol{k})S^z(-\boldsymbol{k})\right\} - \\ -\frac{1}{\hbar} g_J \mu_\mathrm{B} B_0 S^z(\boldsymbol{0}) \,. \tag{2.221}$$

The ground state $|S\rangle$ of a Heisenberg ferromagnet corresponds to an overall parallel orientation of all the spins. We first compute its energy eigenvalue. The effect of the spin operators on $|S\rangle$ is immediately clear:

$$S_\mathrm{i}^z\,|S\rangle = \hbar S\,|S\rangle \;\Rightarrow\; S^z(\boldsymbol{k})\,|S\rangle = \hbar N S\,|S\rangle\; \delta_{\boldsymbol{k},0} \,, \tag{2.222}$$

$$S_\mathrm{i}^+\,|S\rangle = 0 \;\Rightarrow\; S^+(\boldsymbol{k})\,|S\rangle = 0 \,. \tag{2.223}$$

It thus follows that:

$$\begin{aligned}
&-\frac{1}{N}\sum_{k} J(k) S^{+}(k) S^{-}(-k)\,|S\rangle = \\
&= -\frac{1}{N}\sum_{k} J(k)\Big[S^{-}(-k)S^{+}(k) + 2\hbar S^{z}(0)\Big]\,|S\rangle = \\
&= -2N\hbar^{2} S J_{ii}\,|S\rangle = 0\,, \\
&-\frac{1}{N}\sum_{k} J(k) S^{z}(k) S^{z}(-k)\,|S\rangle = \\
&= -\hbar N S \frac{1}{N} J(0) S^{z}(0)\,|S\rangle = -N J_0 \hbar^{2} S^{2}\,|S\rangle\;.
\end{aligned}$$

This yields the **ground state energy** E_0 of the Heisenberg ferromagnet:

$$\begin{aligned}
H\,|S\rangle &= E_0\,|S\rangle\,, \\
E_0 &= -N J_0 \hbar^{2} S^{2} - N g_J \mu_B B_0 S\;.
\end{aligned} \tag{2.224}$$

We now show that the state

$$S^{-}(k)\,|S\rangle$$

is likewise an eigenstate of H. To do so, we calculate the following commutator:

$$\begin{aligned}
&\big[H, S^{-}(k)\big]_{-} \\
&= -\frac{1}{N}\sum_{p} J(p)\Big\{\big[S^{+}(p), S^{-}(k)\big]_{-} S^{-}(-p) + \\
&\quad + S^{z}(p)\big[S^{z}(-p), S^{-}(k)\big]_{-} + \big[S^{z}(p), S^{-}(k)\big]_{-}\, S^{z}(-p)\Big\} - \\
&\quad - \frac{1}{\hbar} g_J \mu_B B_0 \big[S^{z}(0), S^{-}(k)\big] = \\
&= -\frac{1}{N}\sum_{p} J(p)\Big\{2\hbar S^{z}(k+p) S^{-}(-p) - \hbar S^{z}(p) S^{-}(k-p) - \\
&\quad - \hbar S^{-}(k+p) S^{z}(-p)\Big\} + g_J \mu_B B_0 S^{-}(k) = \\
&= g_J \mu_B B_0 S^{-}(k) - \frac{1}{N}\sum_{p} J(p)\Big\{-2\hbar^{2} S^{-}(k) + \\
&\quad + 2\hbar S^{-}(-p) S^{z}(k+p) + \hbar^{2} S^{-}(k) - \hbar S^{-}(k-p) S^{z}(p) - \\
&\quad - \hbar S^{-}(k+p) S^{z}(-p)\Big\}\;.
\end{aligned}$$

Due to

$$J_{\text{ii}} = \frac{1}{N}\sum_{p} J(p) = 0 \tag{2.225}$$

we finally find:

$$\begin{aligned}\left[H, S^-(k)\right]_- &= g_J\mu_B B_0 S^-(k) - \frac{\hbar}{N}\sum_{p} J(p)\Big\{2S^-(-p)S^z(k+p) - \\ &\quad - S^-(k-p)S^z(p) - S^-(k+p)S^z(-p)\Big\}\,.\end{aligned} \tag{2.226}$$

The application of this commutator to the ground state $|S\rangle$ yields:

$$\left[H, S^-(k)\right]_- |S\rangle = \hbar\omega(k)\left(S^-(k)\,|S\rangle\right)\,, \tag{2.227}$$

$$\hbar\omega(k) = g_J\mu_B B_0 + 2S\hbar^2\left(J_0 - J(k)\right)\,. \tag{2.228}$$

Here, we have also made use of $J(k) = J(-k)$. Our assertion that $S^-(k)|S\rangle$ is an eigenstate of H can now be readily demonstrated:

$$\begin{aligned}H\left(S^-(k)\,|S\rangle\right) &= S^-(k)H\,|S\rangle + \left[H, S^-(k)\right]_-\,|S\rangle = \\ &= E(k)\left(S^-(k)\,|S\rangle\right)\,,\end{aligned} \tag{2.229}$$

$$E(k) = E_0 + \hbar\omega(k)\,. \tag{2.230}$$

If we presume the ground state $|S\rangle$ to be normalised, then it follows that:

$$\langle S|\,S^+(-k)S^-(k)\,|S\rangle = \langle S|\left(2\hbar S^z(\mathbf{0}) + S^-(k)S^+(-k)\right)|S\rangle = 2\hbar^2 NS.$$

We thus have the following important final result: The

normalised single-magnon state

$$|k\rangle = \frac{1}{\hbar\sqrt{2SN}}S^-(k)\,|S\rangle \tag{2.231}$$

is an eigenstate belonging to the the energy

$$E(k) = E_0 + \hbar\omega(k)\,.$$

This corresponds to the **excitation energy**

$$\hbar\omega(k) = g_J\mu_B B_0 + 2S\hbar^2\left(J_0 - J(k)\right) \tag{2.232}$$

which is ascribed to the quasi-particle **magnon**. The magnetic field term $g_J \mu_B B_0$ contains more information. One can see from it that the magnetic moment of the sample in the state $|\boldsymbol{k}\rangle$ has been modified relative to the ground state $|S\rangle$ only by a term $g_J \mu_B$. The magnon thus has a spin of $S = 1$:

magnons are Bosons!

Another interesting result can be found from the expectation value of the local spin operator S_i^z in the single-magnon state $|\boldsymbol{k}\rangle$:

$$
\begin{aligned}
&\langle \boldsymbol{k} \mid S_i^z \mid \boldsymbol{k}\rangle = \\
&= \frac{1}{2SN\hbar^2} \langle S \mid S^+(-\boldsymbol{k}) S_i^z S^-(\boldsymbol{k}) \mid S\rangle = \\
&= \frac{1}{2SN^2\hbar^2} \sum_{\boldsymbol{q}} \mathrm{e}^{\mathrm{i}\boldsymbol{q}\cdot\boldsymbol{R}_i} \langle S \mid S^+(-\boldsymbol{k}) S^z(\boldsymbol{q}) S^-(\boldsymbol{k}) \mid S\rangle = \\
&= \frac{1}{2SN^2\hbar^2} \sum_{\boldsymbol{q}} \mathrm{e}^{\mathrm{i}\boldsymbol{q}\cdot\boldsymbol{R}_i} \langle S \mid S^+(-\boldsymbol{k}) \left(-\hbar S^-(\boldsymbol{k}+\boldsymbol{q}) + S^-(\boldsymbol{k}) S^z(\boldsymbol{q})\right) \mid S\rangle = \\
&= \frac{1}{2SN^2\hbar^2} \sum_{\boldsymbol{q}} \mathrm{e}^{\mathrm{i}\boldsymbol{q}\cdot\boldsymbol{R}_i} \left\{-2\hbar^2 \langle S \mid S^z(\boldsymbol{q}) \mid S\rangle + \hbar N S \delta_{\boldsymbol{q},\boldsymbol{0}} 2\hbar \langle S \mid S^z(\boldsymbol{0}) \mid S\rangle\right\} = \\
&= \frac{1}{2SN^2\hbar^2} \left\{-2\hbar^2 N\hbar S + 2\hbar^2 N S N \hbar S\right\} = \\
&= \hbar S - \frac{\hbar}{N} \,.
\end{aligned}
$$

We thus have the notable result

$$
\langle \boldsymbol{k} \mid S_i^z \mid \boldsymbol{k}\rangle = \hbar \left(S - \frac{1}{N}\right) \quad \forall i, \boldsymbol{k} \,. \tag{2.233}
$$

The right-hand side is not dependent on i and $\boldsymbol{k}$. That means that the spin deviation $1\hbar$ in the single-magnon state $|\boldsymbol{k}\rangle$ is uniformly distributed over all the lattice sites $\boldsymbol{R}_i$. As compared to the completely ordered ground state $|S\rangle$, with

$$
\langle S \mid S_i^z \mid S\rangle = \hbar S \quad \forall i \,, \tag{2.234}
$$

we find a deviation of the local spin per lattice site of $\hbar/N$. This leads immediately to the concept of a **spin wave**, which implies just this **collective** excitation $|\boldsymbol{k}\rangle$. Every existing spin wave thus implies for the entire lattice a spin deviation of exactly one unit of angular momentum. The spin wave is characterised by its wavevector $\boldsymbol{k}$, which can be visualised in a semiclassical vector model as follows: The local spin $\boldsymbol{S}_i$ precesses about the z-axis with an axial angle which has just the right value so that the projection of the spins of length $\hbar S$ onto the z-axis has the value $\hbar(S - 1/N)$. The precessing spins have a fixed, constant phase shift from lattice site to lattice site corresponding to $k = 2\pi \,/\, \lambda$. They thus clearly define a wave.

2.4.4 The Spin-Wave Approximation

The Heisenberg model (2.211) is not exactly solvable for the general case. In order to arrive at an approximate solution, it is often expedient to transform the somewhat *unwieldy* spin operators to creation and annihilation operators in the second quantisation:

Holstein-Primakoff transformation:

$$S_i^+ = \hbar\,\sqrt{2S}\,\varphi(n_i)\,a_i\,, \tag{2.235}$$

$$S_i^- = \hbar\,\sqrt{2S}\,a_i^+\varphi(n_i)\,, \tag{2.236}$$

$$S_i^z = \hbar\,(S - n_i)\,. \tag{2.237}$$

Here, the following abbreviations were used:

$$n_i = a_i^+\,a_i; \quad \varphi\,(n_i) = \sqrt{1 - \frac{n_i}{2S}}\,. \tag{2.238}$$

By insertion, one can verify that the commutation relations for the spin operators (2.214), (2.215), and (2.216) are fulfilled if and only if the creation and annihilation operators a_i^+, a_i are **Bosonic operators**:

$$\begin{aligned} \left[a_i,\, a_j\right]_- = \left[a_i^+,\, a_j^+\right]_- &= 0, \\ \left[a_i,\, a_j^+\right]_- &= \delta_{ij}\,. \end{aligned} \tag{2.239}$$

The corresponding Fourier transforms

$$a_q = \frac{1}{\sqrt{N}}\sum_i e^{-i q \cdot R_i}\,a_i; \quad a_q^+ = \frac{1}{\sqrt{N}}\sum_i e^{i q \cdot R_i}\,a_i^+ \tag{2.240}$$

can be interpreted as magnon annihilation or creation operators. The model Hamiltonian (2.211) then takes on the following form as a result of the transformation:

$$H = E_0 + 2S\hbar^2 J_0 \sum_i n_i - 2S\hbar^2 \sum_{i,j} J_{ij}\varphi\,(n_i)\,a_i\,a_j^+\varphi(n_j) - \hbar^2 \sum_{i,j} J_{ij}\,n_i\,n_j\,. \tag{2.241}$$

Here, E_0 is the ground-state energy (2.224). A disadvantage of the Holstein-Primakoff transformation is obvious: working explicitly with H required us to carry out an expansion of the square root in $\varphi(n_i)$:

$$\varphi(n_i) = 1 - \frac{n_i}{4S} - \frac{n_i^2}{32S^2} - \cdots . \tag{2.242}$$

This means that H in principle consists of infinitely many terms. The transformation is thus only reasonable when there is a physical justification for terminating the infinite series. Since n_i can be interpreted as the operator for the magnon number at the site $\boldsymbol{R}_i$, but at low temperatures only a few magnons are excited, in such a case one can limit n_i to only its lowest powers. The simplest approximation in this sense is the so-called **spin-wave approximation**:

$$H^{\text{SW}} = E_0 + 2S\hbar^2 \sum_{i,j} \left(J_0\, \delta_{ij} - J_{ij}\right) a_i^+ a_j . \tag{2.243}$$

After the transformation to wavenumbers, H^{SW} is diagonal

$$H^{\text{SW}} = E_0 + \sum_{\boldsymbol{k}} \hbar\omega(\boldsymbol{k})\, a_{\boldsymbol{k}}^+ a_{\boldsymbol{k}} \tag{2.244}$$

with $\hbar\omega(\boldsymbol{k})$ as in (2.232). In this low-temperature approximation, the ferromagnet is thus described as a *gas* of non-interacting magnons. According to the rules of statistical mechanics, the mean magnon number $\langle n_{\boldsymbol{k}}\rangle$ at $T > 0$ is then given by the Bose-Einstein distribution function:

$$\langle n_{\boldsymbol{k}}\rangle = \frac{1}{\exp\left(\beta\hbar\omega(\boldsymbol{k})\right) - 1} . \tag{2.245}$$

Then we find for the magnetisation of the ferromagnet:

$$M(T, H) = g_J \mu_{\text{B}} \frac{N}{V} \left(S - \frac{1}{N} \sum_{\boldsymbol{k}} \langle n_{\boldsymbol{k}}\rangle\right) . \tag{2.246}$$

At low temperatures, this result is experimentally confirmed to high precision.

2.4.5 Exercises

Exercise 2.4.1 Derive the corresponding relations, using the commutation relations of the spin operators in real space, for the wavenumber-dependent spin operators (i.e. in reciprocal space):

$$S^\alpha(\boldsymbol{k}) = \sum_i \mathrm{e}^{-i\boldsymbol{k}\cdot\boldsymbol{R}_i} S_i^\alpha .$$

Exercise 2.4.2 Reformulate the Heisenberg-model Hamiltonian,

$$H = -\sum_{i,j} J_{ij}\left(S_i^+ S_j^- + S_i^z S_j^z\right) - g_J \frac{\mu_B}{\hbar} B_0 \sum_i S_i^z ,$$

making use of the $\boldsymbol{k}$-space spin operators from Ex. 2.4.1.

Exercise 2.4.3 Carry out the Holstein-Primakoff transformation on the Heisenberg model Hamiltonian from (Ex. 2.4.2).

Exercise 2.4.4 In the *spin-wave approximation*, the spontaneous magnetisation of a Heisenberg ferromagnet at low temperatures is given by:

$$\frac{M_0 - M_S(T)}{M_0} = \frac{1}{NS} \sum_{\boldsymbol{q}} \frac{1}{\exp[\beta\hbar\omega(\boldsymbol{q})] - 1} . \qquad \text{(s. (2.246))}$$

$M_0 = g_J \mu_B S \frac{N}{V}$ is the saturation magnetisation and

$$\hbar\omega(\boldsymbol{q}) = 2S\hbar^2 \left(J_0 - J(\boldsymbol{q})\right)$$

is the magnon energy. Prove Bloch's $T^{3/2}$ law:

$$\frac{M_0 - M_S(T)}{M_0} \sim T^{3/2} .$$

Hints:

a) Transform the summation over $\boldsymbol{q}$ into an integral.

b) Keep in mind that at low temperatures, it suffices to use the magnon energies in the form which is valid for small q-values:

$$\hbar\omega(\boldsymbol{q}) = \frac{D}{2S\hbar^2} q^2 ,$$

and that it is allowed to extend the integration over q to the entire q-space rather than limiting it to the first Brillouin zone.

Exercise 2.4.5 Let the following be given:

$$H: \quad \text{Hamiltonian with } H\,|n\rangle = E_n\,|n\rangle; \quad W_n = \frac{\exp(-\beta E_n)}{\mathrm{Tr}[\exp(-\beta H)]}\,,$$

A, B, C: arbitrary operators.

1. Show that

$$(A, B) = \sum_{n,m}^{E_n \neq E_m} \langle n \mid A^+ \mid m\rangle \langle m \mid B \mid n\rangle \frac{W_m - W_n}{E_n - E_m}$$

represents a (semidefinite) scalar product.

2. Show that with $B = [C^+, H]_-$, the following relations hold:

$$(A, B) = \left\langle [C^+, A^+]_- \right\rangle; \quad (B, B) = \left\langle [C^+, [H, C]_-]_- \right\rangle \geq 0\,,$$

$$(A, A) \leq \frac{1}{2}\beta \left\langle [A, A^+]_+ \right\rangle\,.$$

3. Prove the Bogoliubov inequality using 2.:

$$\frac{\beta}{2}\left\langle [A, A^+]_+ \right\rangle \left\langle [[C, H]_-, C^+]_- \right\rangle \geq |\langle [C, A]_- \rangle|^2\,.$$

Exercise 2.4.6

1. Show that for the scalar product defined in Ex. 2.4.5, $(H, H) = 0$ holds when H is the Hamiltonian of the system.
2. Let C be an operator which commutes with the Hamiltonian H. Show that for C, the Bogoliubov relation from Ex. 2.4.5 can be taken as an equation.

Exercise 2.4.7 Discuss the isotropic Heisenberg model:

$$H = -\sum_{i,j} J_{ij} \boldsymbol{S}_\mathrm{i} \cdot \boldsymbol{S}_j - bB_0 \sum_\mathrm{i} S_\mathrm{i}^z \exp(-\mathrm{i}\,\boldsymbol{K} \cdot \boldsymbol{R}_\mathrm{i}); \quad b = \frac{g_J \mu_\mathrm{B}}{\hbar}\,.$$

The wavevector $\boldsymbol{K}$ is a help in distinguishing different magnetic configurations. Thus, $\boldsymbol{K} = 0$ leads to ferromagnetism. We assume that

$$Q = \frac{1}{N} \sum_{i,j} \left|\boldsymbol{R}_\mathrm{i} - \boldsymbol{R}_j\right|^2 \left|J_{ij}\right| < \infty\,,$$

which is not a major limitation of generality. For the magnetisation, we then have:

$$M(T, B_0) = b\frac{1}{N}\sum_{i} \exp(\mathrm{i}\,\boldsymbol{K}\cdot\boldsymbol{R}_i)\langle S_i^z\rangle\ .$$

In the case of an antiferromagnet, ($\boldsymbol{K} = (1/2)\boldsymbol{Q}$, $\boldsymbol{Q}$: the smallest reciprocal lattice vector), M represents the **sublattice** magnetisation.

1. Choose

$$A = S^-(-\boldsymbol{k}-\boldsymbol{K});\quad C = S^+(\boldsymbol{k})$$

and then prove that

a) $\langle[C, A]_-\rangle = \frac{2\hbar N}{b} M(T, B_0)$,

b) $\sum_{\boldsymbol{k}}\langle[A, A^+]_+\rangle \le 2\hbar^2 N S(S+1)$,

c) $\langle[[C, H]_-, C^+]_-\rangle \le 4N\hbar^2(|B_0 M| + \hbar^2 k^2 Q S(S+1))$.

2. Prove the **Mermin-Wagner theorem** (Phys. Rev. Lett. **17**, 1133 (1966)), using the Bogoliubov inequality, (Ex. 2.4.5): **In the** $d = 1$- **and** $d = 2$-**dimensional, isotropic Heisenberg model, there can be no spontaneous magnetisation for** ($T \neq 0$).

a) Show that the following holds in this connection:

$$S(S+1) \ge \frac{M^2 v_d \Omega_d}{\beta\hbar^2 b^2 (2\pi)^d}\int_0^{k_0} dk\, \frac{k^{d-1}}{|BM| + \hbar^2 k^2 Q S(S+1)}\ .$$

Here, k_0 is the radius of a sphere which lies completely within the Brillouin zone, Ω_d is the surface area of the d-dimensional unit sphere ($\Omega_1 = 1$, $\Omega_2 = 2\pi$, $\Omega_3 = 4\pi$), and $v_d = V_d\ /\ N_d$ is the specific volume of the d-dimensional system in the thermodynamic limit.

b) Verify for the **spontaneous magnetisation** that:

$$M_S(T) = \lim_{B_0\to 0} M(T, B_0) = 0 \quad \text{for } T \neq 0 \text{ and } d = 1 \text{ and } 2\ .$$

2.5 Self-Examination Questions

For Section 2.1

1. Which eigenvalue equation leads to the Bloch functions and the Bloch energies?
2. What is stated by Bloch's Theorem?
3. What are the orthogonality and completeness relations for Bloch functions?
4. Give the Hamiltonian H_0 for non-interacting crystal electrons in second quantistion for the Bloch representation, for the real-space representation with field operators, and for the Wannier representation.
5. What are the commutation relations for Bloch operators $a_{k\sigma}^{+}$, $a_{k\sigma}$ and for Wannier operators $a_{i\sigma}^{+}$, $a_{i\sigma}$?
6. When does a Bloch function become a plane wave?
7. What is meant by a *hopping integral*?
8. What relationship exists between Bloch and Wannier operators?
9. Which assumptions define the jellium model?
10. Justify the necessity of a *convergence-producing* factor in the Coulomb integrals of the jellium model.
11 What is the Hamiltonian of the jellium model? What is the effect of the *homogeneously distributed* positive ion charges?
12 How is the operator for the electron density written in the formalism of second quantisation if plane waves are used as a single-particle basis?
13 What relationship exists between the electron density operator and the particle number operator?
14 Formulate the Hamiltonian of the jellium model using the electron density operator.
15 Define the concepts of Fermi energy and Fermi wavevector.
16 What is meant by the *direct term* and the *exchange term* in the Coulomb interaction of the jellium model?
17 Give the two leading terms in the expansion of the ground-state energy of the jellium model in terms of the dimensionless density parameter r_s, and interpret them.
18 What is meant by *correlation energy*?
19 Why is the jellium model not useful for the description of electrons in narrow energy bands?
20 Describe the so-called *tight-binding approximation*.
21 What are the decisive simplifications which finally lead to the Hubbard model?
22 What is the Hamiltonian of the Hubbard model?
23 Which physical parameters mainly influence the statements of the Hubbard model?
24 Name some of the important areas of application of the Hubbard model.

For Section 2.2

1. Why is it reasonable in the description of lattice vibrations to use collective coordinates instead of the ion coordinates?
2. How can the *harmonic approximation* be justified?
3. How is the *matrix of the atomic force constants* defined? What is the meaning of its elements?
4. Name some of the obvious symmetries of the force-constant matrix.
5. Justify the terms *acoustic* and *optical dispersion branch*.
6. What equation of motion is obeyed by the so-called *normal coordinates*? How are they related to the real displacements of the ions?
7. How is the Lagrangian function of the ion system written in terms of the normal coordinates?
8. What are the momenta which are canonically conjugate to the normal coordinates?
9. Give the classical Hamilton function of the ion system. Interpret it.
10. State the commutation relations for the normal coordinates and for the momenta which are canonically conjugate to them.
11. How are the creation and annihilation operators b_{qr}^{+}, b_{qr} related to the normal coordinates and their canonically conjugated momenta?
12. Why are b_{qr} and b_{qr}^{+} Bosonic operators?
13. Give the Hamiltonian for the ion system in the harmonic approximation in terms of the creation and annihilation operators b_{qr} and b_{qr}^{+}.
14. What is a *phonon*?

For Section 2.3

1. Describe the elementary processes which lead to an electron-phonon interaction.
2. Which approximation for the electron-phonon interaction corresponds to the harmonic approximation for the lattice vibrations?
3. Which operator combination defines the electron-phonon interaction within the formalism of second quantisation?
4. What is meant by normal and umklapp processes?
5. Describe how the elementary processes of the electron-phonon interaction can be combined.
6. Which method of theoretical physics allows us to recognise that the electron-phonon interaction contains terms describing an effective phonon-induced electron-electron interaction?
7. Can this effective electron-electron interaction also be attractive?

For Section 2.4

1. Which physical quantity would appear to be particularly suited for the classification of magnetic solids?
2. Why is diamagnetism a property of **all** materials?
3. What is the decisive precondition for the occurrence of paramagnetism and collective magnetism?
4. What distinguishes Langevin paramagnetism from Pauli paramagnetism?
5. Comment on the Curie law.
6. Into which three major subclasses can collective magnetism be subdivided?
7. What is the Hamiltonian of the Heisenberg model? For which class of magnetic substances is the model suited?
8. When does one speak of *band magnetism*?
9. Which magnetic materials are described by the s-f (or s-d) model?
10. Sketch the derivation of the so-called *single-magnon state*

$$|\boldsymbol{k}\rangle = \left(\hbar^2 2SN\right)^{-1/2} S^-(\boldsymbol{k})\,|S\rangle \quad (|S\rangle \Longleftrightarrow \text{ferromagnetic saturation})$$

as an eigenstate of the Heisenberg Hamiltonian.

11 What is the spin of magnons?

12 What is the expectation value of the local spin operator S_i^z in the single-magnon state $|\boldsymbol{k}\rangle$? Interpret the result.

13 Explain the concept of a *spin wave*.

14 Formulate the Holstein-Primakoff transformation of the spin operators.

15 What is meant by the *spin-wave approximation*? Under which conditions is it justified?

Green's Functions 3

The goal of theoretical physics consists in developing methods for the calculation of measurable physical quantities. **Measurable physical quantities** are:

1. the eigenvalues of observables,
2. the expectation values of observables $\langle \widehat{A}(t) \rangle, \langle \widehat{B}(t') \rangle, \ldots,$
3. the correlation functions between observables $\langle \widehat{A}(t) \cdot \widehat{B}(t') \rangle \ldots$

Within the framework of statistical mechanics, calculations of measurable quantities of category 2. or 3. are possible only when the partition function of the physical system under consideration is known. This presupposes, on the other hand, a knowledge of the eigenvalues and the eigenstates of the Hamiltonian, which is as a rule not the case for realistic many-body problems. The **Green's-function method** allows a determination, in general necessarily approximate, of the expectation values and correlation functions **without** an explicit knowledge of the partition function. The corresponding methods will be discussed in this chapter and in the following ones. To this end, we require some preliminary information.

3.1 Preliminary Considerations

3.1.1 Representations

For the description of the time dependence of physical systems, we use one of the three equivalent *representations*, depending on which is most expedient:

Schrödinger, Heisenberg, Dirac representation.

We shall begin with the representation which is used almost exclusively in **Quantum Mechanics.**

W. Nolting, *Fundamentals of Many-body Physics*,
DOI 10.1007/978-3-540-71931-1_3, © Springer-Verlag Berlin Heidelberg 2009

1) The Schrödinger representation (state representation)
In this representation, the time dependence is carried by the states, whilst the operators are independent of time, unless they have an *explicit* time dependence, e.g. due to switching-on and -off processes. We adopt the

equations of motion

from elementary quantum mechanics
a) for pure states:

$$\mathrm{i}\hbar \left| \dot{\psi}_s(t) \right\rangle = H \left| \psi_s(t) \right\rangle , \tag{3.1}$$

b) for mixed states:

$$\dot{\rho}_S = \frac{\mathrm{i}}{\hbar} \left[\rho_S, H \right]_- . \tag{3.2}$$

Here, ρ_S is the **density matrix** with its well-known properties:

$$\rho_S = \sum_m p_m \left| \psi_m \right\rangle \left\langle \psi_m \right| \tag{3.3}$$

where p_m is the probability that the system is to be found in the state $\left| \psi_m \right\rangle$,

$$\left\langle \widehat{A} \right\rangle = \mathrm{Tr} \left(\rho_s \widehat{A} \right) , \tag{3.4}$$

$$\mathrm{Tr} \rho_s = 1 , \tag{3.5}$$

$$\mathrm{Tr} \rho_s^2 = \begin{cases} 1 : & \text{pure state,} \\ < 1 : & \text{mixed state.} \end{cases} \tag{3.6}$$

For the following, the

time-evolution operator $U_S(t, t_0)$,

is important; it is defined by

$$\left| \psi_S(t) \right\rangle = U_S(t, t_0) \left| \psi_S(t_0) \right\rangle . \tag{3.7}$$

Essential properties of this operator are:

$$1. \quad U_S^+(t, t_0) = U_S^{-1}(t, t_0) , \tag{3.8}$$

$$2. \quad U_S(t_0, t_0) = \mathbf{1} \tag{3.9}$$

$$3. \quad U_S(t, t_0) = U_S\left(t, t'\right) U_S\left(t', t_0\right) . \tag{3.10}$$

If we use (3.7) in (3.1), then we obtain an equivalent equation of motion for the time-evolution operator:

$$\mathrm{i}\hbar \dot{U}_S(t, t_0) = H_t U_S(t, t_0) . \tag{3.11}$$

The index t of the Hamiltonian indicates a possible explicit time dependence. (3.11) can be formally integrated by taking (3.9) into account:

$$U_S(t, t_0) = \mathbf{1} - \frac{\mathrm{i}}{\hbar} \int_{t_0}^{t} \mathrm{d}t_1 \, H_{t_1} U_S(t_1, t_0) \; . \tag{3.12}$$

After some iterations, we obtain

von Neumann's series

$$U_S(t, t_0) = \mathbf{1} + \sum_{n=1}^{\infty} U_S^{(n)}(t, t_0) \; , \tag{3.13}$$

$$U_S^{(n)}(t, t_0) = \left(-\frac{\mathrm{i}}{\hbar}\right)^n \int_{t_0}^{t} \mathrm{d}t_1 \int_{t_0}^{t_1} \mathrm{d}t_2 \cdots \int_{t_0}^{t_{n-1}} \mathrm{d}t_n \, H_{t_1} H_{t_2} \cdots H_{t_n} \tag{3.14}$$

$$\left(t \geq t_1 \geq t_2 \geq \cdots \geq t_n \geq t_0\right) .$$

The time ordering must be strictly observed, since the operators H_{t_i} at different times do not necessarily commute.

For further rearrangements, we introduce a special operator:

Dyson's time-ordering operator

$$T_D\left(A(t_1) B(t_2)\right) = \begin{cases} A(t_1) B(t_2) & \textit{for} \quad t_1 > t_2 \, , \\ B(t_2) A(t_1) & \textit{for} \quad t_2 > t_1 \, . \end{cases} \tag{3.15}$$

The generalisation to more than two operators is obvious. The following relations can be seen from Fig. 3.1:

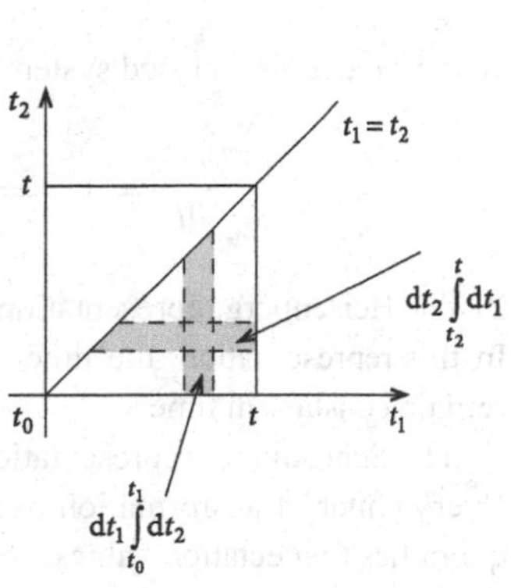

Fig. 3.1 Illustration of the rearrangement of the time-ordering operator from (3.14) as in (3.17)

$$\int_{t_0}^{t} dt_1 \int_{t_0}^{t_1} dt_2\, H_{t_1} H_{t_2} = \int_{t_0}^{t} dt_2 \int_{t_2}^{t} dt_1\, H_{t_1} H_{t_2} \,.$$

On the right-hand side of the equation, we interchange t_1 and t_2:

$$\int_{t_0}^{t} dt_1 \int_{t_0}^{t_1} dt_2\, H_{t_1} H_{t_2} = \int_{t_0}^{t} dt_1 \int_{t_1}^{t} dt_2\, H_{t_2} H_{t_1} \,.$$

When we combine the last two relations, this yields:

$$\begin{aligned}
&\int_{t_0}^{t} dt_1 \int_{t_0}^{t_1} dt_2\, H_{t_1} H_{t_2} = \\
&= \frac{1}{2} \int_{t_0}^{t} dt_1 \int_{t_0}^{t} dt_2 \left(H_{t_1} H_{t_2} \Theta(t_1 - t_2) + H_{t_2} H_{t_1} \Theta(t_2 - t_1) \right) = \qquad (3.16)\\
&= \frac{1}{2!} \iint_{t_0}^{t} dt_1 dt_2\, T_{\mathrm{D}} \left(H_{t_1} H_{t_2} \right) .
\end{aligned}$$

This result can be generalised to n terms, so that from (3.14), we now obtain:

$$U_{\mathrm{S}}^{(n)}(t, t_0) = \frac{1}{n!} \left(-\frac{\mathrm{i}}{\hbar} \right)^n \int_{t_0}^{t} \cdots \int_{t_0}^{t} dt_1 \cdots dt_n\, T_{\mathrm{D}} \left(H_{t_1} H_{t_2} \cdots H_{t_n} \right) . \qquad (3.17)$$

Thus, the time-evolution operator can be represented compactly in the following form:

$$U_{\mathrm{S}}(t, t_0) = T_{\mathrm{D}} \exp \left(-\frac{\mathrm{i}}{\hbar} \int_{t_0}^{t} dt'\, H_{t'} \right) . \qquad (3.18)$$

A special case is a closed system:

$$\frac{\partial H}{\partial t} = 0 \implies U_{\mathrm{S}}(t, t_0) = \exp\left(-\frac{\mathrm{i}}{\hbar} H (t - t_0) \right) . \qquad (3.19)$$

2) The Heisenberg representation (operator representation)
In this representation, the time dependence is carried by the operators, whilst the states remain constant in time.

The Schrödinger representation discussed in 1) is of course by no means compulsory. Every unitary transformation of the operators and the states which leaves the measurable quantities (expectation values, scalar products) invariant is naturally allowed.

For the states in the Heisenberg representation, we assume:

$$\left|\psi_{\mathrm{H}}(t)\right\rangle \equiv \left|\psi_{\mathrm{H}}\right\rangle \stackrel{!}{=} \left|\psi_{\mathrm{S}}\left(t_0\right)\right\rangle . \tag{3.20}$$

Here, t_0 is an arbitrary but fixed time, e.g. $t_0 = 0$. With (3.7), (3.9) and (3.10), it follows that:

$$\left|\psi_{\mathrm{H}}\right\rangle = U_{\mathrm{S}}^{-1}\left(t, t_0\right)\left|\psi_{\mathrm{S}}(t)\right\rangle = U_{\mathrm{S}}\left(t_0, t\right)\left|\psi_{\mathrm{S}}(t)\right\rangle . \tag{3.21}$$

Due to

$$\left\langle\psi_{\mathrm{H}}\right| A_{\mathrm{H}}(t)\left|\psi_{\mathrm{H}}\right\rangle \stackrel{!}{=} \left\langle\psi_{\mathrm{S}}(t)\right| A_{\mathrm{S}}\left|\psi_{\mathrm{S}}(t)\right\rangle \tag{3.22}$$

we then find for the observable A in the Heisenberg representation:

$$A_{\mathrm{H}}(t) = U_{\mathrm{S}}^{-1}\left(t, t_0\right) A_{\mathrm{S}} U_{\mathrm{S}}\left(t, t_0\right) . \tag{3.23}$$

If H is not explicitly time-dependent, this relation can be simplified to

$$A_{\mathrm{H}}(t) = \exp\left(\frac{\mathrm{i}}{\hbar} H\left(t - t_0\right)\right) A_{\mathrm{S}} \exp\left(-\frac{\mathrm{i}}{\hbar} H\left(t - t_0\right)\right) \quad \left(\frac{\partial H}{\partial t} = 0\right) . \tag{3.24}$$

In particular, we then see that:

$$H_{\mathrm{H}}(t) = H_{\mathrm{H}} = H_{\mathrm{S}} = H . \tag{3.25}$$

We now derive the equation of motion of the Heisenberg operators:

$$\begin{aligned}
\frac{\mathrm{d}}{\mathrm{d}t} A_{\mathrm{H}}(t) = {} & \dot{U}_{\mathrm{S}}^{+}\left(t, t_0\right) A_{\mathrm{S}} U_{\mathrm{S}}\left(t, t_0\right) + U_{\mathrm{S}}^{+}\left(t, t_0\right) \frac{\partial A_{\mathrm{S}}}{\partial t} U_{\mathrm{S}}\left(t, t_0\right) + \\
& + U_{\mathrm{S}}^{+}\left(t, t_0\right) A_{\mathrm{S}} \dot{U}_{\mathrm{S}}\left(t, t_0\right) = \\
= {} & -\frac{1}{\mathrm{i}\hbar} U_{\mathrm{S}}^{+} H A_{\mathrm{S}} U_{\mathrm{S}} + \frac{1}{\mathrm{i}\hbar} U_{\mathrm{S}}^{+} A_{\mathrm{S}} H U_{\mathrm{S}} + U_{\mathrm{S}}^{+} \frac{\partial A_{\mathrm{S}}}{\partial t} U_{\mathrm{S}} = \\
= {} & \frac{\mathrm{i}}{\hbar} U_{\mathrm{S}}^{+} \left[H, A_{\mathrm{S}}\right]_{-} U_{\mathrm{S}} + U_{\mathrm{S}}^{+} \frac{\partial A_{\mathrm{S}}}{\partial t} U_{\mathrm{S}} .
\end{aligned}$$

We define

$$\frac{\partial A_{\mathrm{H}}}{\partial t} = U_{\mathrm{S}}^{-1}\left(t, t_0\right) \frac{\partial A_{\mathrm{S}}}{\partial t} U_{\mathrm{S}}\left(t, t_0\right) \tag{3.26}$$

and then find for the equation of motion:

$$\mathrm{i}\hbar \frac{\mathrm{d}}{\mathrm{d}t} A_{\mathrm{H}}(t) = \left[A_{\mathrm{H}}, H_{\mathrm{H}}\right]_{-}(t) + \mathrm{i}\hbar \frac{\partial A_{\mathrm{H}}}{\partial t} . \tag{3.27}$$

An intermediate role between that of the Schrödinger and the Heisenberg representation is played by

3) the Dirac representation (interaction representation)
Here, the time dependence is distributed between the states **and** the operators. The starting point is the usual situation,

$$H = H_0 + V_t \,, \tag{3.28}$$

in which the Hamiltonian is composed of a part H_0 for the *free* system and a possibly explicitly time dependent interaction V_t. Then the following *ansatz* is taken conventionally:

$$\left|\psi_D(t_0)\right\rangle = \left|\psi_S(t_0)\right\rangle = \left|\psi_H\right\rangle \,, \tag{3.29}$$

$$\left|\psi_D(t)\right\rangle = U_D\left(t, t'\right)\left|\psi_D\left(t'\right)\right\rangle \,, \tag{3.30}$$

$$\left|\psi_D(t)\right\rangle = U_0^{-1}\left(t, t_0\right)\left|\psi_S(t)\right\rangle \,. \tag{3.31}$$

Here,

$$U_0\left(t, t'\right) = \exp\left[-\frac{i}{\hbar} H_0\left(t - t'\right)\right] \tag{3.32}$$

is the time-evolution operator of the *free* system. We find from this that in the absence of interactions, the Dirac and the Heisenberg representations are identical.

As a result of (3.29) through (3.31), the following rearrangement holds:

$$\begin{aligned} \left|\psi_D(t)\right\rangle &= U_0^{-1}\left(t, t_0\right)\left|\psi_S(t)\right\rangle = U_0^{-1}\left(t, t_0\right) U_S\left(t, t'\right)\left|\psi_S\left(t'\right)\right\rangle = \\ &= U_0^{-1}\left(t, t_0\right) U_S\left(t, t'\right) U_0\left(t', t_0\right)\left|\psi_D\left(t'\right)\right\rangle \stackrel{!}{=} U_D\left(t, t'\right)\left|\psi_D\left(t'\right)\right\rangle \,. \end{aligned}$$

We have thus found the expression which relates the Dirac to the Schrödinger time-evolution operator:

$$U_D\left(t, t'\right) = U_0^{-1}\left(t, t_0\right) U_S\left(t, t'\right) U_0\left(t', t_0\right) \,. \tag{3.33}$$

We can see that for $V_t \equiv 0$, i.e. $U_S = U_0$, $U_D\left(t, t'\right) \equiv 1$ holds. Dirac states are then time independent. We require

$$\left\langle\psi_D(t)\right| A_D(t)\left|\psi_D(t)\right\rangle \stackrel{!}{=} \left\langle\psi_S(t)\right| A_S\left|\psi_S(t)\right\rangle$$

for an arbitrary operator A. Making use of (3.31) and (3.32), this yields:

$$A_D(t) = \exp\left(\frac{i}{\hbar} H_0\left(t - t_0\right)\right) A_S \exp\left(-\frac{i}{\hbar} H_0\left(t - t_0\right)\right) \,. \tag{3.34}$$

The dynamics of the operators in the Dirac representation are thus determined by H_0. This can in particular be seen from the equation of motion which can be derived directly from (3.34):

$$i\hbar \frac{d}{dt} A_D(t) = [A_D, H_0]_- + i\hbar \frac{\partial A_D}{\partial t} . \tag{3.35}$$

Analogously to (3.26), here we have used the definitions:

$$\frac{\partial A_D}{\partial t} = U_0^{-1}(t, t_0) \frac{\partial A_S}{\partial t} U_0(t, t_0) . \tag{3.36}$$

For the time dependence of the states, according to (3.31) we find:

$$\begin{aligned} |\dot{\psi}_D(t)\rangle &= \dot{U}_0^+(t, t_0) |\psi_S(t)\rangle + U_0^+(t, t_0) |\dot{\psi}_S(t)\rangle = \\ &= \frac{i}{\hbar} \left(U_0^+(t, t_0) H_0 - U_0^+(t, t_0) H \right) |\psi_S(t)\rangle = \\ &= \frac{i}{\hbar} U_0^+(t, t_0)(-V_t) U_0(t, t_0) |\psi_D(t)\rangle . \end{aligned}$$

It thus follows that:

$$i\hbar |\dot{\psi}_D(t)\rangle = V_t^D(t) |\psi_D(t)\rangle . \tag{3.37}$$

The dynamics of the states are thus determined by the interaction V_t. We distinguish the two time dependencies in $V_t^D(t)$! Analogously to (3.37), the equation of motion of the density matrix can be derived:

$$\dot{\rho}_D(t) = \frac{i}{\hbar} \left[\rho_D, V_t^D \right] (t) . \tag{3.38}$$

Inserting (3.30) into (3.37), we find with

$$i\hbar \frac{d}{dt} U_D(t, t') = V_t^D(t) U_D(t, t') \tag{3.39}$$

an equation of motion for the time-evolution operator, which is formally identical to (3.11). The same logical sequence as employed following Eq. (3.13) then leads to an important relation:

$$U_D(t, t') = T_D \exp\left(-\frac{i}{\hbar} \int_{t'}^{t} dt'' \, V_{t''}^D(t'') \right) , \tag{3.40}$$

which represents the starting point for the diagram techniques which we shall discuss later. Note that $U_D(t, t')$, in contrast to $U_S(t, t')$, cannot be further simplified even when there is no explicit time dependence, since then simply the replacement $V_{t''}^D(t'') \to V^D(t'')$ is to be made. A time dependence thus remains.

3.1.2 Linear-Response Theory

We want to introduce the Green's functions in connection with a concrete physical problem:

How does a physical system react to an external perturbation?

Problems of this type are characterised by so-called

response functions,

among which in particular are

1. the electrical conductivity,
2. the magnetic susceptibility, and
3. the dielectric function.

It is found that these quantities are described by **retarded Green's functions**. To show this, we introduce the *linear-response* theory, an important tool of theoretical physics.

We describe the system under consideration by its Hamiltonian:

$$H = H_0 + V_t \; . \tag{3.41}$$

Here, V_t has a somewhat different meaning than in (3.28). It describes the interaction of the system with an applied field (the *perturbation*). H_0 describes the system of interacting particles when the field is switched off. Due to the interactions between the particles, even the eigenvalue problem belonging to H_0 usually cannot be solved exactly.

The scalar field F_t is assumed to couple to an observable $\widehat{B}$ of the system:

$$V_t = \widehat{B} F_t \; . \tag{3.42}$$

Note that $\widehat{B}$ is an operator and F_t is a c-number. Let $\widehat{A}$ be a not explicitly time-dependent observable, whose thermodynamic expectation value $\langle \widehat{A} \rangle$ can be interpreted as a measurable quantity. We wish to investigate how $\langle \widehat{A} \rangle$ reacts to the *perturbation* V_t.

Without the applied field, we have

$$\langle \widehat{A} \rangle_0 = \mathrm{Tr}\left(\rho_0 \widehat{A}\right) \; , \tag{3.43}$$

where ρ_0 is the density matrix of the field-free system:

$$\rho_0 = \frac{\exp(-\beta \mathcal{H}_0)}{\mathrm{Tr}\left[\exp(-\beta \mathcal{H}_0)\right]} \; . \tag{3.44}$$

We average over the grand canonical ensemble:

$$\mathcal{H}_0 = H_0 - \mu \widehat{N} \; . \tag{3.45}$$

μ is the chemical potential. If we now switch on the field F_t, the density matrix will be correspondingly modified:

$$\rho_0 \longrightarrow \rho_t \,. \tag{3.46}$$

This modification then affects the expectation value of $\widehat{A}$:

$$\left\langle \widehat{A} \right\rangle_t = \mathrm{Tr}\left(\rho_t \widehat{A}\right) \,. \tag{3.47}$$

We have initially used the Schrödinger representation here, but we leave off the index S. The equation of motion of the density matrix is found from (3.2):

$$\mathrm{i}\hbar\dot{\rho}_t = \left[\mathcal{H}_0, \rho_t\right]_- + \left[V_t, \rho_t\right]_- \,. \tag{3.48}$$

We assume that the field is switched on at some particular time, and we can therefore use the following as the boundary condition for the differential equation of first order (3.48):

$$\lim_{t\to-\infty} \rho_t = \rho_0 \,. \tag{3.49}$$

We now (temporarily) change to the Dirac representation, in which we find with $t_0 = 0$ from (3.34):

$$\rho_t^{\mathrm{D}}(t) = \exp\left(\frac{\mathrm{i}}{\hbar}\mathcal{H}_0 t\right) \rho_t \exp\left(-\frac{\mathrm{i}}{\hbar}\mathcal{H}_0 t\right) \,. \tag{3.50}$$

The equation of motion (3.38) leads, with the boundary condition (3.49),

$$\lim_{t\to-\infty} \rho_t^{\mathrm{D}}(t) = \rho_0 \,, \tag{3.51}$$

to the result:

$$\rho_t^{\mathrm{D}}(t) = \rho_0 - \frac{\mathrm{i}}{\hbar}\int_{-\infty}^{t} \mathrm{d}t' \left[V_{t'}^{\mathrm{D}}(t'), \rho_{t'}^{\mathrm{D}}(t')\right]_- \,. \tag{3.52}$$

This equation can be solved to arbitrary precision by iterating:

$$\rho_t^{\mathrm{D}}(t) = \rho_0 + \sum_{n=1}^{\infty} \rho_t^{D(n)}(t) \,, \tag{3.53}$$

$$\begin{aligned} \rho_t^{D(n)}(t) = \left(-\frac{\mathrm{i}}{\hbar}\right)^n \int_{-\infty}^{t} \mathrm{d}t_1 \int_{-\infty}^{t_1} \mathrm{d}t_2 \cdots \int_{-\infty}^{t_{n-1}} \mathrm{d}t_n \,\cdot \\ \cdot \left[V_{t_1}^{\mathrm{D}}(t_1), \left[V_{t_2}^{\mathrm{D}}(t_2), \left[\ldots, \left[V_{t_n}^{\mathrm{D}}(t_n), \rho_0\right]_- \ldots\right]_-\right]_-\right]_- \,. \end{aligned} \tag{3.54}$$

This formula is indeed exact, but as a rule not applicable, since the the infinite series cannot be computed. We therefore assume that the external perturbations are sufficiently small that we can limit ourselves to *linear terms* in the perturbation V:

Linear response

$$\rho_t \approx \rho_0 - \frac{\mathrm{i}}{\hbar} \int\limits_{-\infty}^{t} \mathrm{d}t' \exp\left(-\frac{\mathrm{i}}{\hbar}\mathcal{H}_0 t\right) \left[V_{t'}^{\mathrm{D}}\left(t'\right), \rho_0\right]_{-} \exp\left(\frac{\mathrm{i}}{\hbar}\mathcal{H}_0 t\right) . \quad (3.55)$$

In this expression, we have already transformed the density matrix back to the Schrödinger representation. We can now insert this expression into (3.47) in order to compute the *perturbed* expectation value:

$$\begin{aligned}\left\langle \widehat{A}\right\rangle_t &= \left\langle \widehat{A}\right\rangle_0 - \frac{\mathrm{i}}{\hbar} \int\limits_{-\infty}^{t} \mathrm{d}t' \,\mathrm{Tr}\left\{\exp\left(-\frac{\mathrm{i}}{\hbar}\mathcal{H}_0 t\right)\left[V_{t'}^{\mathrm{D}}\left(t'\right), \rho_0\right]_{-} \exp\left(\frac{\mathrm{i}}{\hbar}\mathcal{H}_0 t\right)\widehat{A}\right\} = \\ &= \left\langle \widehat{A}\right\rangle_0 - \frac{\mathrm{i}}{\hbar} \int\limits_{-\infty}^{t} \mathrm{d}t' \, F_{t'} \mathrm{Tr}\left\{\left[\widehat{B}^{\mathrm{D}}\left(t'\right), \rho_0\right]_{-} \widehat{A}^{\mathrm{D}}(t)\right\} = \\ &= \left\langle \widehat{A}\right\rangle_0 - \frac{\mathrm{i}}{\hbar} \int\limits_{-\infty}^{t} \mathrm{d}t' \, F_{t'} \mathrm{Tr}\left\{\rho_0 \left[\widehat{A}^{\mathrm{D}}(t), \widehat{B}^{\mathrm{D}}\left(t'\right)\right]_{-}\right\} .\end{aligned}$$

We were able to make use of the cyclic invariance of the trace several times here. We thus now know the reaction of the system to the external perturbation, as reflected in the observable $\widehat{A}$:

$$\Delta A_t = \left\langle \widehat{A}\right\rangle_t - \left\langle \widehat{A}\right\rangle_0 = -\frac{\mathrm{i}}{\hbar} \int\limits_{-\infty}^{t} \mathrm{d}t' \, F_{t'} \left\langle \left[\widehat{A}^{\mathrm{D}}(t), \widehat{B}^{\mathrm{D}}\left(t'\right)\right]_{-}\right\rangle_0 . \quad (3.56)$$

Note that the reaction of the system is determined by an expectation value of the unperturbed system. The Dirac representation of the operators $\widehat{A}^{\mathrm{D}}(t)$, $\widehat{B}^{\mathrm{D}}(t')$ corresponds to the Heisenberg representation when the field is off. We define the

double-time retarded Green's function

$$G_{AB}^{\mathrm{ret}}\left(t, t'\right) = \langle\langle A(t); B\left(t'\right)\rangle\rangle = -\mathrm{i}\,\Theta\left(t - t'\right)\left\langle \left[A(t), B\left(t'\right)\right]_{-}\right\rangle_0 . \quad (3.57)$$

The operators here are always taken to be in the Heisenberg representation of the field-free system. We leave off the corresponding index.

The retarded Green's function G_{AB}^{ret} thus describes the reaction of the system, as it manifests itself in the observable $\widehat{A}$ when the perturbation acts on the observable $\widehat{B}$:

$$\Delta A_t = \frac{1}{\hbar} \int\limits_{-\infty}^{+\infty} \mathrm{d}t' \, F_{t'} G_{AB}^{\mathrm{ret}}\left(t, t'\right) . \quad (3.58)$$

Using the Fourier transform $F(E)$ of the perturbation,

$$F_t = \frac{1}{2\pi\hbar}\int\limits_{-\infty}^{+\infty} dE\,\exp\left[-\frac{i}{\hbar}\left(E + i0^+\right)t\right]F(E)\,, \tag{3.59}$$

and in anticipation of a later result that the Green's function itself depends only on the time difference $t - t'$ when the Hamiltonian is not explicitly time dependent, we can write (3.58) also in the following form:

Kubo formula

$$\Delta A_t = \frac{1}{2\pi\hbar^2}\int\limits_{-\infty}^{+\infty} dE\,F(E)G^{\text{ret}}_{AB}(E + i0^+)\exp\left[-\frac{i}{\hbar}\left(E + i0^+\right)t\right]. \tag{3.60}$$

The term $i0^+$ in the exponent guarantees the fulfillment of the boundary condition (3.49). The field F_t is, as one says, thus switched on *adiabatically*. In the following three sections, we discuss some examples of applications of the important Kubo formula.

3.1.3 The Magnetic Susceptibility

The *perturbation* is caused by a spatially homogeneous, temporally oscillating magnetic field:

$$\boldsymbol{B}_t = \frac{1}{2\pi\hbar}\int\limits_{-\infty}^{+\infty} dE\,\exp\left[-\frac{i}{\hbar}\left(E + i0^+\right)t\right]\boldsymbol{B}(E)\,. \tag{3.61}$$

The field couples to the magnetic moment of the system:

$$\boldsymbol{m} = \sum_i \boldsymbol{m}_i = \frac{g_J\mu_B}{\hbar}\sum_i \boldsymbol{S}_i\,. \tag{3.62}$$

This produces the following *perturbation term* in the Hamiltonian:

$$\begin{aligned} V_t &= -\boldsymbol{m}\cdot\boldsymbol{B}_t = \\ &= -\frac{1}{2\pi\hbar}\sum_\alpha^{(x,y,z)}\int\limits_{-\infty}^{+\infty} dE\,\exp\left[-\frac{i}{\hbar}\left(E + i0^+\right)t\right]m^\alpha B^\alpha(E)\,. \end{aligned} \tag{3.63}$$

Of particular interest is of course the reaction of the magnetisation to the switched-on field. As a result of

$$\boldsymbol{M} = \frac{1}{V}\langle \boldsymbol{m}\rangle = \frac{g_J\mu_B}{\hbar V}\sum_i \langle \boldsymbol{S}_i\rangle \; , \tag{3.64}$$

in the Kubo formula (3.60) or (3.58), we choose both operators $\widehat{A}$ and $\widehat{B}$ to correspond to the magnetic-moment operator $\boldsymbol{m}$. From (3.58), we then obtain:

$$M_t^\beta - M_0^\beta = -\frac{1}{V\hbar}\sum_\alpha \int_{-\infty}^{+\infty} \mathrm{d}t' \, B_{t'}^\alpha \langle\langle m^\beta(t); m^\alpha(t')\rangle\rangle \; . \tag{3.65}$$

The field-free magnetisation M_0^β is of course nonvanishing only in the case of a ferromagnet. (3.65) defines the

magnetic-susceptibility tensor

$$\chi_{ij}^{\beta\alpha}(t,t') = -\frac{\mu_0}{V\hbar}\frac{g_J^2\mu_B^2}{\hbar^2}\langle\langle S_i^\beta(t); S_j^\alpha(t')\rangle\rangle \tag{3.66}$$

as a retarded Green's function. We then have

$$\Delta M_t^\beta = \frac{1}{\mu_0}\sum_{i,j}\sum_\alpha \int_{-\infty}^{+\infty} \mathrm{d}t' \, \chi_{ij}^{\beta\alpha}(t,t')\, B_{t'}^\alpha \; , \tag{3.67}$$

or, in the energy representation:

$$\Delta M_t^\beta = \frac{1}{2\pi\hbar\mu_0}\sum_{i,j}\sum_\alpha \int_{-\infty}^{+\infty} \mathrm{d}E \, \exp\left[-\frac{\mathrm{i}}{\hbar}\left(E + \mathrm{i}0^+\right)t\right]\chi_{ij}^{\beta\alpha}(E) B^\alpha(E) \; . \tag{3.68}$$

In applying (3.62), we assumed implicitly that the physical system under consideration contains permanent local moments (cf. (2.204)). In such a situation, two special types of susceptibilities are of particular interest:

1) The longitudinal susceptibility

$$\chi_{ij}^{zz}(E) = -\frac{\mu_0}{V\hbar}\frac{g_J^2\mu_B^2}{\hbar^2}\langle\langle S_i^z; S_j^z\rangle\rangle_E \; . \tag{3.69}$$

The index E denotes the energy-dependent Fourier transform of the retarded Green's function.

From χ_{ij}^{zz}, one can derive important statements about the stability of magnetic order. To this end, we compute the spatial Fourier transform

$$\chi_q^{zz}(E) = \frac{1}{N}\sum_{i,j}\chi_{ij}^{zz}(E)\mathrm{e}^{\mathrm{i}\boldsymbol{q}\cdot(\boldsymbol{R}_i-\boldsymbol{R}_j)} \tag{3.70}$$

for the paramagnetic phase. Given the singularities of this response function, an infinitesimal field suffices to produce a finite magnetisation in the sample, i.e. to bring about a *spontaneous* ordering of the magnetic moments. One therefore investigates under which conditions

$$\left\{\lim_{(q,E)\to 0}\chi_q^{zz}(E)\right\}^{-1} = 0 \tag{3.71}$$

holds, and reads off from this condition the characteristics of the phase transition for para- $\Longleftrightarrow$ ferromagnetism.

The

2) transverse susceptibility

$$\chi_{ij}^{+-}(E) = -\frac{\mu_0}{V\hbar}\frac{g_J^2\mu_B^2}{\hbar^2}\langle\langle S_i^+;S_j^-\rangle\rangle_E \tag{3.72}$$

also contains considerable information. Its poles are identical with the spin-wave energies (*magnons*):

$$\left\{\chi_q^{+-}(E)\right\}^{-1} = 0 \Longleftrightarrow E = \hbar\omega(\boldsymbol{q})\,. \tag{3.73}$$

These examples show that the *linear-response* theory not only represents an approximate method for weak external perturbations, but it also allows us to make statements about the unperturbed system.

3.1.4 The Electrical Conductivity

We next take the *perturbation* to be a spatially homogeneous, temporally oscillating electric field:

$$\boldsymbol{F}_t = \frac{1}{2\pi\hbar}\int\limits_{-\infty}^{+\infty}\mathrm{d}E\,\exp\left[-\frac{\mathrm{i}}{\hbar}\left(E+\mathrm{i}0^+\right)t\right]\boldsymbol{F}(E)\,. \tag{3.74}$$

We choose the symbol $\boldsymbol{F}$ for this field instead of the more usual $\boldsymbol{E}$ in order to avoid confusion with the energy E.

The electric field couples to the operator of the electric dipole moment $\boldsymbol{P}$:

$$\boldsymbol{P} = \int \mathrm{d}^3 r\, \boldsymbol{r}\rho(\boldsymbol{r}) . \tag{3.75}$$

We consider N point charges q_i at the positions $\hat{\boldsymbol{r}}_i(t)$. Then the charge density is given by

$$\rho(\boldsymbol{r}) = \sum_{i=1}^{N} q_i \delta\left(\boldsymbol{r} - \hat{\boldsymbol{r}}_i\right) , \tag{3.76}$$

and thus we have for the dipole-moment operator:

$$\boldsymbol{P} = \sum_{i=1}^{N} q_i \hat{\boldsymbol{r}}_i . \tag{3.77}$$

The electric field causes an additional term to appear in the Hamiltonian:

$$\begin{aligned} V_t &= -\boldsymbol{P} \cdot \boldsymbol{F}_t = \\ &= -\frac{1}{2\pi\hbar} \sum_{\alpha}^{(x,y,z)} \int_{-\infty}^{+\infty} \mathrm{d}E \, \exp\left[-\frac{\mathrm{i}}{\hbar}\left(E + \mathrm{i}0^+\right)t\right] P^\alpha F^\alpha(E) . \end{aligned} \tag{3.78}$$

One is of course interested in particular in the reaction of the current density to the field. The expectation value of the current-density operator,

$$\boldsymbol{j} = \frac{1}{V} \sum_{i=1}^{N} q_i \dot{\hat{\boldsymbol{r}}}_i = \frac{1}{V} \dot{\boldsymbol{P}} , \tag{3.79}$$

is certainly zero in the absence of an applied field:

$$\langle \boldsymbol{j} \rangle_0 = 0 . \tag{3.80}$$

After switching on the field, owing to (3.58), we find:

$$\left\langle j^\beta \right\rangle_t = -\frac{1}{\hbar} \sum_{\alpha} \int_{-\infty}^{+\infty} \mathrm{d}t' \, F_{t'}^\alpha \left\langle\!\left\langle j^\beta(t); P^\alpha\left(t'\right) \right\rangle\!\right\rangle . \tag{3.81}$$

In the energy representation, this gives:

$$\left\langle j^\beta \right\rangle_t = \frac{1}{2\pi\hbar} \sum_{\alpha} \int_{-\infty}^{+\infty} \mathrm{d}E \, \exp\left[-\frac{\mathrm{i}}{\hbar}\left(E + \mathrm{i}0^+\right)t\right] \sigma^{\beta\alpha}(E) F^\alpha(E) . \tag{3.82}$$

This relation (*Ohm's law*) defines the

electrical conductivity tensor

$$\sigma^{\beta\alpha}(E) \equiv -\frac{1}{\hbar}\langle\langle j^\beta; P^\alpha \rangle\rangle_E\,, \tag{3.83}$$

whose components are represented by retarded Green's functions. This expression still requires some rearrangement. For this, we make use of the temporal homogeneity of the Green's functions, which we have already used and which will be proved later:

$$\begin{aligned}
\sigma^{\beta\alpha}(E) &= -\frac{1}{\hbar}\int\limits_{-\infty}^{+\infty} \mathrm{d}t\, \langle\langle j^\beta(0); P^\alpha(-t)\rangle\rangle \exp\left[\frac{\mathrm{i}}{\hbar}\left(E + \mathrm{i}0^+\right)t\right] = \\
&= \frac{\mathrm{i}}{\hbar}\int\limits_{0}^{\infty} \mathrm{d}t \left\langle\left[j^\beta, P^\alpha(-t)\right]_-\right\rangle \exp\left[\frac{\mathrm{i}}{\hbar}\left(E + \mathrm{i}0^+\right)t\right] = \\
&= \frac{\left\langle\left[j^\beta, P^\alpha(-t)\right]_-\right\rangle}{E + \mathrm{i}0^+}\exp\left[\frac{\mathrm{i}}{\hbar}\left(E + \mathrm{i}0^+\right)t\right]\Bigg|_0^\infty - \\
&\quad - \int\limits_0^\infty \mathrm{d}t\, \frac{\exp\left[(\mathrm{i}/\hbar)\left(E + \mathrm{i}0^+\right)t\right]}{E + \mathrm{i}0^+}\frac{\mathrm{d}}{\mathrm{d}t}\left\langle\left[j^\beta, P^\alpha(-t)\right]_-\right\rangle = \\
&= -\frac{\left\langle\left[j^\beta, P^\alpha\right]_-\right\rangle}{E + \mathrm{i}0^+} + \int\limits_0^\infty \mathrm{d}t \left\langle\left[j^\beta, \dot{P}^\alpha(-t)\right]_-\right\rangle \frac{\exp\left[(\mathrm{i}/\hbar)\left(E + \mathrm{i}0^+\right)t\right]}{E + \mathrm{i}0^+} = \\
&= -\frac{\left\langle\left[j^\beta, P^\alpha\right]_-\right\rangle}{E + \mathrm{i}0^+} + \mathrm{i}V\frac{\langle\langle j^\beta; j^\alpha\rangle\rangle_E}{E + \mathrm{i}0^+}\,.
\end{aligned} \tag{3.84}$$

The first term can be readily evaluated:

$$\left[j^\beta, P^\alpha\right]_- = \frac{1}{V}\sum_{i,j} q_i q_j \left[\dot{\hat{r}}_i^\beta, \hat{r}_j^\alpha\right]_- = \frac{1}{V}\sum_{i,j} q_i q_j \frac{\hbar}{\mathrm{i}}\frac{\delta_{ij}\delta_{\alpha\beta}}{m_i}\,. \tag{3.85}$$

We assume identical charge carriers,

$$q_i = q\,; \quad m_i = m \quad \forall i\,,$$

and then, inserting (3.85) into (3.84), we find:

$$\sigma^{\beta\alpha}(E) = \mathrm{i}\hbar\frac{(N/V)q^2}{m\left(E + \mathrm{i}0^+\right)}\delta_{\alpha\beta} + \mathrm{i}V\frac{\langle\langle j^\beta; j^\alpha\rangle\rangle_E}{E + \mathrm{i}0^+}\,. \tag{3.86}$$

The first term represents the conductivity of a system of non-interacting electrons, as is known from the classical Drude theory. The influence of the particle interactions is thus brought into play exclusively by the retarded current-current Green's function.

3.1.5 The Dielectric Function

If an external charge density $\rho_{\text{ext}}(\boldsymbol{r}, t)$ is added to a metal, it will give rise to a change in the density of the quasi-free conduction electrons within the system, producing screening of the *perturbation charges*. This screening effect is described by the dielectric function $\varepsilon(\boldsymbol{q}, E)$, which is therefore a measure of the *response* of the system to the external *perturbation* $\rho_{\text{ext}}(\boldsymbol{r}, t)$. It is a further example of a *response function* and can likewise be expressed in terms of a retarded Green's function. We shall demonstrate this in the present section, first preparing the problem using a **classical** treatment.

For the external charge density, we take:

$$\rho_{\text{ext}}(\boldsymbol{r}, t) = \frac{1}{2\pi\hbar V} \int\limits_{-\infty}^{+\infty} \mathrm{d}E \sum_{\boldsymbol{q}} \rho_{\text{ext}}(\boldsymbol{q}, E) \mathrm{e}^{\mathrm{i}\boldsymbol{q}\cdot\boldsymbol{r}} \exp\left[-\frac{\mathrm{i}}{\hbar}\left(E + \mathrm{i}0^+\right) t\right]. \tag{3.87}$$

Acting between ρ_{ext} and the charge density of the conduction electrons,

$$-e\rho(\boldsymbol{r}) = -\frac{e}{V} \sum_{\boldsymbol{q}} \rho_{\boldsymbol{q}} \mathrm{e}^{\mathrm{i}\boldsymbol{q}\cdot\boldsymbol{r}}, \tag{3.88}$$

is an interaction energy

$$V_t = \frac{-e}{4\pi\varepsilon_0} \iint \mathrm{d}^3r\, \mathrm{d}^3r' \frac{\rho(\boldsymbol{r})\rho_{\text{ext}}(\boldsymbol{r}', t)}{|\boldsymbol{r} - \boldsymbol{r}'|}. \tag{3.89}$$

As in (2.58), one can show that

$$\iint \mathrm{d}^3r \mathrm{d}^3r' \frac{\exp\left[\mathrm{i}\left(\boldsymbol{q}\cdot\boldsymbol{r} + \boldsymbol{q}'\cdot\boldsymbol{r}'\right)\right]}{|\boldsymbol{r} - \boldsymbol{r}'|} = \delta_{\boldsymbol{q}, -\boldsymbol{q}'} \frac{4\pi V}{q^2}$$

holds. Then, using the definition

$$\bar{v}(\boldsymbol{q}) = \frac{v_0(\boldsymbol{q})}{-e} = \frac{1}{V} \frac{-e}{\varepsilon_0 q^2}, \tag{3.90}$$

along with (3.87) and (3.88) inserted into (3.89), we obtain for V_t:

$$V_t = \frac{1}{2\pi\hbar} \int\limits_{-\infty}^{+\infty} \mathrm{d}E \exp\left[-\frac{\mathrm{i}}{\hbar}\left(E + \mathrm{i}0^+\right) t\right] \sum_{\boldsymbol{q}} \bar{v}(\boldsymbol{q}) \rho_{-\boldsymbol{q}} \rho_{\text{ext}}(\boldsymbol{q}, E). \tag{3.91}$$

We use the jellium model for the metal, i.e. we assume that in equilibrium, the electronic and the ionic charge densities just compensate each other. Furthermore, the perturbation charge is assumed to polarise only the more mobile electronic system, whilst the ionic charges remain homogeneously distributed. The overall charge density is then the sum of ρ_{ext} and the induced charge density ρ_{ind} which it produces in the electron gas:

$$\rho_{\text{tot}}(\boldsymbol{r}, t) = \rho_{\text{ext}}(\boldsymbol{r}, t) + \rho_{\text{ind}}(\boldsymbol{r}, t) . \tag{3.92}$$

From the Maxwell equations

$$\mathrm{i}\boldsymbol{q} \cdot \boldsymbol{D}(\boldsymbol{q}, E) = \rho_{\text{ext}}(\boldsymbol{q}, E) , \tag{3.93}$$

$$\mathrm{i}\boldsymbol{q} \cdot \boldsymbol{F}(\boldsymbol{q}, E) = \frac{1}{\varepsilon_0} \left(\rho_{\text{ext}}(\boldsymbol{q}, E) + \rho_{\text{ind}}(\boldsymbol{q}, E)\right) \tag{3.94}$$

and the electric elasticity of matter

$$\boldsymbol{D}(\boldsymbol{q}, E) = \varepsilon_0 \varepsilon(\boldsymbol{q}, E) \boldsymbol{F}(\boldsymbol{q}, E) , \tag{3.95}$$

we find for the induced charge density:

$$\rho_{\text{ind}}(\boldsymbol{q}, E) = \left[\frac{1}{\varepsilon(\boldsymbol{q}, E)} - 1\right] \rho_{\text{ext}}(\boldsymbol{q}, E) . \tag{3.96}$$

We next transform our thus-far classical considerations to a quantum-mechanical representation. For the electron density ρ_{-q} in (3.91), we obtain the **density operator**, which was formulated in second quantisation in (2.70):

$$\rho_q = \sum_{k\sigma} a^+_{k\sigma} a_{k+q\sigma} ; \quad \rho_{-q} = \rho^+_q . \tag{3.97}$$

The interaction energy defined in (3.91) then likewise becomes an operator:

$$V_t = \sum_q \rho^+_q \widetilde{F}_t(\boldsymbol{q}) . \tag{3.98}$$

The *perturbation field* $\widetilde{F}_t(\boldsymbol{q})$,

$$\widetilde{F}_t(\boldsymbol{q}) = \frac{\bar{v}(\boldsymbol{q})}{2\pi\hbar} \int_{-\infty}^{+\infty} \mathrm{d}E \, \exp\left[-\frac{\mathrm{i}}{\hbar}\left(E + \mathrm{i}0^+\right) t\right] \rho_{\text{ext}}(\boldsymbol{q}, E) , \tag{3.99}$$

in contrast, remains a scalar c-number. We are interested in how the expectation value of the induced charge density (operator!),

$$\langle \rho_{\text{ind}}(\boldsymbol{q}, t) \rangle = -e \left(\langle \rho_q \rangle_t - \langle \rho_q \rangle_0 \right) , \tag{3.100}$$

reacts to the *perturbation field*. The corresponding information is contained in the Kubo formula (3.60):

$$\Delta(\rho_q)_t = \frac{1}{\hbar}\sum_{q'}\int_{-\infty}^{+\infty} dt'\, \widetilde{F}_{t'}(q')\,\langle\langle \rho_q(t); \rho^+_{q'}(t')\rangle\rangle \,. \tag{3.101}$$

The translational symmetry of the unperturbed system guarantees that the retarded Green's function is nonzero only for $q = q'$. This means that

$$\langle\rho_{\mathrm{ind}}(q,t)\rangle = \frac{-e}{\hbar}\int_{-\infty}^{+\infty} dt'\, \widetilde{F}_{t'}(q)\langle\langle \rho_q(t); \rho^+_q(t')\rangle\rangle \,, \tag{3.102}$$

or, after Fourier transformation:

$$\langle\rho_{\mathrm{ind}}(q,E)\rangle = \frac{-e\bar{v}(q)}{\hbar}\rho_{\mathrm{ext}}(q,E)\langle\langle \rho_q; \rho^+_q\rangle\rangle_E \,. \tag{3.103}$$

If we now compare this result with the classical expression (3.96), then we find that the dielectric function is also determined by a retarded Green's function:

$$\frac{1}{\varepsilon(q,E)} = 1 + \frac{1}{\hbar}v_0(q)\langle\langle \rho_q; \rho^+_q\rangle\rangle_E \,. \tag{3.104}$$

If $\varepsilon(q,E)$ is very large, then it follows from (3.103) and (3.104) that $\langle\rho_{\mathrm{ind}}\rangle \simeq -\rho_{\mathrm{ext}}$. The screening of the perturbation charges by the induced charges in the electron gas is thus practically complete. The other limiting case $\varepsilon(q,E) \to 0$ corresponds to a singularity of the Green's function $\langle\langle \rho_q; \rho^+_q\rangle\rangle_E$. According to (3.96), arbitrarily small perturbation charges then suffice to provoke finite density fluctuations in the system of conduction electrons. The poles of $\langle\langle \rho_q; \rho^+_q\rangle\rangle_E$ therefore correspond to certain *proper frequencies* (*resonances*) of the system. These collective excitations of the electronic system are associated with the quasi-particles called **plasmons**, in the same sense as we associated spin waves with the quasi-particles called *magnons*, thus defining the latter, in Sect. 2.4.3.

3.1.6 Spectroscopies, Spectral Density

An additional important motivation for dealing with Green's functions is their close connection to the

elementary excitations

of the system, which can be observed directly by means of suitable spectroscopies. Certain Green's functions thus provide an immediate access to experimental results. This is true in an even more direct form of another fundamental function, which however is very closely related to the Green's functions, namely the so-called

spectral density.

Figure 3.2 shows in schematic form which elementary processes can be used to determine the electronic structure with four well-known spectroscopic methods. *Photoemission (PES)* and *inverse photoemission (IPE)* are so-called single-particle spectroscopies, since the system (the solid) contains one particle more (or less) after the excitation process than before the excitation. In *photoemission*, the energy $\hbar\omega$ of a photon is absorbed by an electron from a (partially) occupied energy band. The increase in energy can permit the electron to leave the solid. An analysis of the kinetic energies of these photoelectrons then permits conclusions to be drawn about the energies of the occupied states of the energy bands involved. The transition operator $Z_{-1} = a_\alpha$ then corresponds to the annihilation operator a_α, if the electron was in the single-particle state $|\alpha\rangle$ before the excitation process. In *inverse photoemission*, essentially the reverse process occurs. An electron

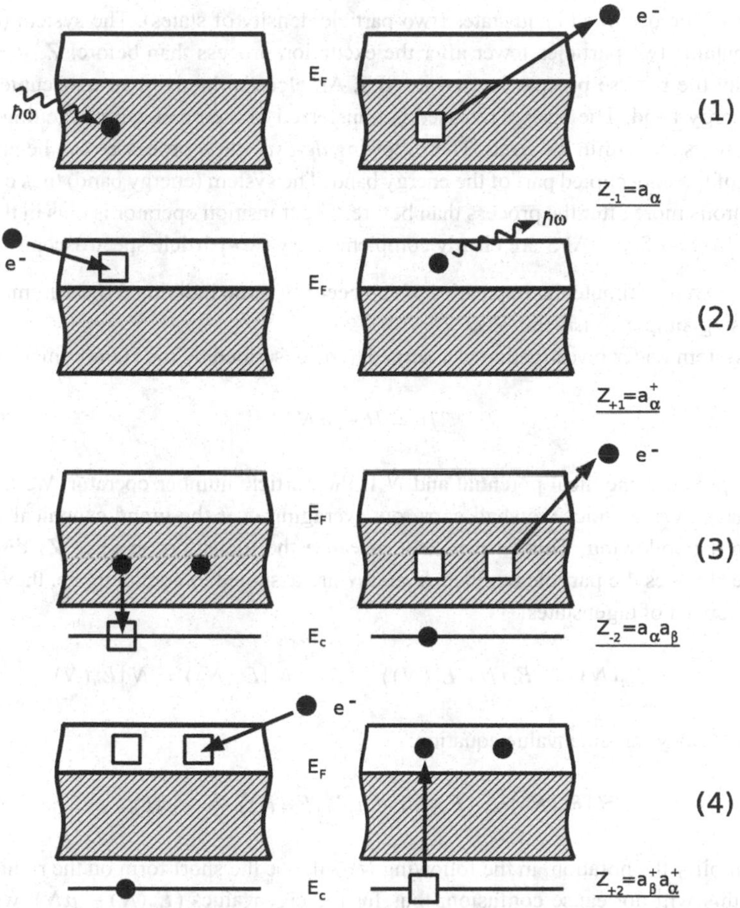

Fig. 3.2 Elementary processes relevant to four different spectroscopies: 1. Photoemission (PES), 2. Inverse Photoemission (IPE), 3. Auger Electron Spectroscopy (AES), 4. Appearance-Potential Spectroscopy (APS). Z_j is the transition operator, where j means the change in the electron numbers due to the respective excitation

is injected into the solid and lands there in a hitherto unoccupied state $|\beta\rangle$ of the partially filled energy band. The energy which is released is emitted in the form of a photon $\hbar\omega$, and this photon is detected and analysed. The system now contains one electron more than before the process. This corresponds to the transition operator $Z_{+1} = a_\beta^\dagger$. PES and IPE are to a certain extent complementary spectroscopies. The former permits statements about occupied states of the energy band to be made, the latter permits statements about the unoccupied states.

Auger electron spectroscopy (AES) and *appearance potential spectroscopy (APS)* are two-particle spectroscopies. The initial situation for AES is characterized by the existence of a hole in a deep-lying core state of an atom in the solid. An electron from the partially-filled energy band makes a transition into this core state and transfers the energy released to another electron from the same energy band; this electron can then leave the solid. The analysis of the kinetic energies of the emitted electrons yields information about the energy structure of the occupied band states (two-particle density of states). The system (energy band) contains two particles fewer after the excitation process than before: $Z_{-2} = a_\alpha a_\beta$. Practically the reverse process is used in APS. An electron lands in an unoccupied state of the energy band. The energy released is transferred to a core electron, exciting it into another free state within the band. The following *de-excitation* processes can be analysed in terms of the unoccupied part of the energy band. The system (energy band) thus contains two electrons more after the process than before. The transition operator is thus in this case $Z_{+2} = a_\beta^\dagger a_\alpha^\dagger$. AES and APS are clearly complementary two-particle spectroscopies.

We now wish to estimate the intensities which occur in these individual measurement processes using simple considerations.

- The system under investigation is assumed to be described by the Hamiltonian

$$\mathcal{H} = H - \mu\widehat{N} \,. \tag{3.105}$$

Here, μ is the chemical potential and $\widehat{N}$ is the particle number operator. We use here $\mathcal{H}$ instead of H, since we shall carry out averaging over the grand canonical ensemble in the following. This is expedient, because the transition operator Z_j discussed above changes the particle number. H and $\widehat{N}$ are assumed to commute, i.e. they have a common set of eigenstates:

$$H\,|E_n(N)\rangle = E_n(N)\,|E_n(N)\rangle \quad ; \quad \widehat{N}\,|E_n(N)\rangle = N\,|E_n(N)\rangle$$

Then $\mathcal{H}$ obeys an eigenvalue equation:

$$\mathcal{H}\,|E_n(N)\rangle = (E_n(N) - \mu N)\,|E_n(N)\rangle \rightarrow E_n\,|E_n\rangle \,. \tag{3.106}$$

To simplify the notation, in the following we will use the short form on the right whenever this will not cause confusion; thus for the eigenvalues $(E_n(N) - \mu N)$, we write briefly E_n, and for the eigenstates $|E_n(N)\rangle$, we write simply $|E_n\rangle$. The real dependence of the states or energy eigenvalues on the particle number must however always be kept in mind.

- With the probability

$$\frac{1}{\Xi}\exp(-\beta E_n)$$

the system at the temperature T can be found in an eigenstate $|E_n\rangle$ of the Hamiltonian $\mathcal{H}$. Ξ is the grand canonical partition function:

$$\Xi = Tr\,(\exp(-\beta\mathcal{H}) \tag{3.107}$$

- The transition operator Z_r causes a transition between the states $|E_n\rangle$ and $|E_m\rangle$ with the probability:

$$|\langle E_m|Z_r|E_n\rangle|^2 \qquad r = \pm 1, \pm 2$$

- The **intensity** of the elementary processes to be measured corresponds to the total number of transitions with excitation energies between E and $E + dE$:

$$I_r(E) = \frac{1}{\Xi}\sum_{m,n} e^{-\beta E_n}\,|\langle E_m|Z_r|E_n\rangle|^2\,\delta(E - (E_m - E_n))\,. \tag{3.108}$$

If the excitation energies $(E_m - E_n)$ are sufficiently closely spaced, which for example always holds for a solid, then $I_r(E)$ will be a continuous function of the energy E.
- At this point, we neglect several additional effects which can be quite important for a quantitative analysis of the corresponding experiment, but are not decisive for the actual process of interest here. This applies e.g. in PES and AES to the fact that the "photoelectron" which leaves the solid will still exhibit a coupling to the rest of the system (*the "sudden approximation"*). Furthermore, matrix elements for the transition from a band level to the vacuum state are not included here. The *"bare line shape"* of the spectroscopies listed above should however be correctly described by (3.108).

Note that for the transition operator,

$$Z_r = Z^{\dagger}_{-r} \tag{3.109}$$

holds, i.e. complementary spectroscopies will be related to one another in a certain fashion. We now proceed to investigate this point in more detail.

$$I_r(E) = \frac{1}{\Xi}\sum_{m,n} e^{\beta E}e^{-\beta E_m}\,|\langle E_m|Z_r|E_n\rangle|^2\,\delta(E - (E_m - E_n))$$

$$= \frac{1}{\Xi}\sum_{n,m} e^{\beta E}e^{-\beta E_n}\,|\langle E_n|Z_r|E_m\rangle|^2\,\delta(E - (E_n - E_m))$$

$$= \frac{e^{\beta E}}{\Xi}\sum_{n,m} e^{-\beta E_n}\,|\langle E_m|Z_{-r}|E_n\rangle|^2\,\delta((-E) - (E_m - E_n))$$

In the second step, only the summation indices n, m were exchanged; the last step then uses (3.109). We thus have derived a **symmetry relation** for *complementary* spectroscopies:

$$I_r(E) = e^{\beta E}\, I_{-r}(-E) \tag{3.110}$$

We now define the **spectral density**, which is important for the considerations that follow:

$$\frac{1}{\hbar} S_r^{(\pm)}(E) = I_{-r}(E) \mp I_r(-E) = \left(e^{\beta E} \mp 1\right) I_r(-E)\,. \tag{3.111}$$

The freedom of choice of the signs will be interpreted later. We can see from (3.110) and (3.111) that intensities of *complementary* spectroscopies are determined in a simple way by one and the same spectral density.

$$\hbar\, I_r(E) = \frac{1}{e^{-\beta E} \mp 1}\, S_r^{(\pm)}(-E) \tag{3.112}$$

$$\hbar\, I_{-r}(E) = \frac{e^{\beta E}}{e^{\beta E} \mp 1}\, S_r^{(\pm)}(E) \tag{3.113}$$

The spectral density which we have introduced is thus very closely related to the intensities of the various spectroscopies. We therefore wish to investigate it further, by carrying out a Fourier transformation into the time representation:

$$
\begin{aligned}
&\frac{1}{2\pi\,\hbar}\int_{-\infty}^{+\infty} dE e^{-\frac{i}{\hbar}E(t-t')} I_{-r}(E) \\
&= \frac{1}{2\pi\,\hbar}\frac{1}{\Xi}\sum_{m,n} e^{-\beta E_n}\, e^{-\frac{i}{\hbar}(E_m - E_n)(t-t')} \langle E_m | Z_{-r} | E_n\rangle \, \langle E_n | Z_{-r}^{\dagger} | E_m\rangle \\
&= \frac{1}{2\pi\,\hbar}\frac{1}{\Xi}\sum_{m,n} e^{-\beta E_n} \langle E_m | e^{\frac{i}{\hbar}\mathcal{H}t'}\, Z_{-r} e^{-\frac{i}{\hbar}\mathcal{H}t'} | E_n\rangle \, \langle E_n | e^{\frac{i}{\hbar}\mathcal{H}t}\, Z_r e^{-\frac{i}{\hbar}\mathcal{H}t} | E_m\rangle \\
&= \frac{1}{2\pi\,\hbar}\frac{1}{\Xi}\sum_{m,n} e^{-\beta E_n} \langle E_n | Z_r(t) | E_m\rangle \, \langle E_m | Z_r^{\dagger}(t') | E_n\rangle \\
&= \frac{1}{2\pi\,\hbar}\frac{1}{\Xi}\sum_{n} e^{-\beta E_n} \langle E_n | Z_r(t) Z_r^{\dagger}(t') | E_n\rangle \\
&= \frac{1}{2\pi\,\hbar} \langle Z_r(t)\, Z_r^{\dagger}(t')\rangle\,.
\end{aligned}
$$

In complete analogy, one finds

$$\frac{1}{2\pi\hbar}\int_{-\infty}^{+\infty} dE e^{-\frac{i}{\hbar}E(t-t')} I_r(-E) = \frac{1}{2\pi\hbar}\langle Z_r^\dagger(t')\, Z_r(t)\rangle \, .$$

With (3.111), this implies for the double-time spectral density that:

$$\begin{aligned} S_r^{(\varepsilon)}(t,t') &= \frac{1}{2\pi\hbar}\int_{-\infty}^{+\infty} dE e^{-\frac{i}{\hbar}E(t-t')}\, S_r^{(\varepsilon)}(E) \\ &= \frac{1}{2\pi}\langle [Z_r(t),\, Z_r^\dagger(t')]_{-\varepsilon}\rangle \, . \end{aligned} \tag{3.114}$$

Here, $\eta = \pm$ is merely an initially arbitrary sign factor; $[\cdots,\cdots]_{-\eta}$ is either the commutator or the anticommutator:

$$\left[Z_r(t),\, Z_r^\dagger(t')\right]_{-\varepsilon} = Z_r(t)\, Z_r^\dagger(t') - \varepsilon\, Z_r^\dagger(t')\, Z_r(t) \, . \tag{3.115}$$

We were thus able to show that the spectral density in (3.114) is of central importance for the intensities of the spectroscopies. One can furthermore verify that the generalisation of the spectral density to arbitrary operators $\widehat{A}$ and $\widehat{B}$ is closely connected to the retarded Green's function introduced in (3.57). This holds also for the other types of Green's functions which will be defined later. The **spectral density**

$$S_{AB}^{(\eta)}(t,t') = \frac{1}{2\pi}\langle [\widehat{A}(t),\, \widehat{B}(t')]_{-\varepsilon}\rangle \tag{3.116}$$

is of the same fundamental importance for many-body theory as are the Green's functions.

3.1.7 Exercises

Exercise 3.1.1 For the non-interacting electron gas (H_e) and for the non-interacting phonon gas (H_p),

$$H_e = \sum_{k,\sigma}\varepsilon(\boldsymbol{k}) a_{k\sigma}^+\, a_{k\sigma}\, ; \quad H_p = \sum_{q,r}\hbar\omega_r(\boldsymbol{q})\left(b_{qr}^+\, b_{qr} + \frac{1}{2}\right) ,$$

compute the time dependence of the annihilation operators $a_{k\sigma}(t)$, $b_{qr}(t)$ in the Heisenberg representation.

Exercise 3.1.2 Let A and B be linear operators with $A \neq A(\lambda)$ and $B \neq B(\lambda)$, $\lambda \in \mathbf{R}$.

1. Write

$$e^{\lambda A} B e^{-\lambda A} = \sum_{n=0}^{\infty} \alpha_n \lambda^n \quad (\alpha_n \text{are operators!})$$

and compute the coefficients α_n.

2. Show that from

$$[A, [A, B]_-]_- = 0 ,$$

it follows that:

$$e^{\lambda A} B e^{-\lambda A} = B + \lambda [A, B]_- .$$

3. Use the results from 1. and 2. to derive the differential equation

$$\frac{d}{d\lambda}\left(e^{\lambda A} e^{\lambda B}\right) = (A + B + \lambda [A, B]_-)\left(e^{\lambda A} e^{\lambda B}\right)$$

for $[A, [A, B]_-]_- = [B, [A, B]_-]_- = 0$.

4. Prove using 3. the following relation:

$$e^{A} e^{B} = e^{A + B + \frac{1}{2}[A,B]_-}, \qquad \text{when } [A, [A, B]_-]_- = [B, [A, B]_-]_- = 0.$$

Exercise 3.1.3 Let $A(t)$ be an arbitrary operator in the Heisenberg representation and ρ the statistical operator:

$$\rho = \frac{e^{-\beta \mathcal{H}}}{Tr\left(e^{-\beta \mathcal{H}}\right)} .$$

Prove the Kubo identity:

$$\frac{i}{\hbar}[A(t), \rho]_- = \rho \int_0^{\beta} d\lambda \, \dot{A}(t - i\lambda\hbar).$$

Exercise 3.1.4 Show by using the Kubo identity (Ex. 3.1.3), that the retarded (commutator-) Green's function can be written as follows:

$$\langle\!\langle A(t); B(t') \rangle\!\rangle^{\text{ret}} = -\hbar\Theta(t-t') \int\limits_0^\beta \mathrm{d}\lambda \langle \dot{B}(t' - \mathrm{i}\lambda\hbar) A(t) \rangle .$$

Exercise 3.1.5 Make use of the Kubo identity (Ex. 3.1.3) to express the tensor of the electrical conductivity in terms of a current-current correlation function:

$$\sigma^{\beta\alpha}(E) = V \int\limits_0^\beta \mathrm{d}\lambda \int\limits_0^\infty \mathrm{d}t \langle j^\alpha(0) j^\beta(t + \mathrm{i}\lambda\hbar) \rangle \exp\left(\frac{\mathrm{i}}{\hbar}(E + \mathrm{i}0^+)t\right) .$$

V is the volume of the system!

Exercise 3.1.6 Compute the current-density operator $\hat{j}$ in the formalism of second quantisation using

1. the Bloch representation, and
2. the Wannier representation.

Which form does the conductivity tensor take in these cases?

Exercise 3.1.7 In the so-called tight-binding model (cf. Sect.2.1.3) for strongly bound electrons in solids, the following approximate expression holds for the matrix element:

$$p_{ij\sigma} = \int \mathrm{d}^3 r\, w_\sigma^* (\boldsymbol{r} - \boldsymbol{R}_i) \boldsymbol{r} w_\sigma (\boldsymbol{r} - \boldsymbol{R}_j) \simeq \boldsymbol{R}_i \delta_{ij} .$$

Here, $w_\sigma(\boldsymbol{r} - \boldsymbol{R}_i)$ is the Wannier function (2.29) centered on the lattice site $\boldsymbol{R}_i$.

1. How are the dipole-moment operator $\widehat{\boldsymbol{P}}$ and the current-density operator $\hat{\boldsymbol{j}}$ written in second quantisation using the Wannier representation?
2. The system of interacting electrons can be described by a Hamiltonian of the form

$$H = \sum_{i,j,\sigma} T_{ij} a_{i\sigma}^+ a_{j\sigma} + \sum_{i,j,\sigma,\sigma'} V_{ij\sigma\sigma'} n_{i\sigma} n_{j\sigma} .$$

Compute the current-density operator $\hat{\boldsymbol{j}}$. Which Green's function determines the conductivity tensor $\sigma^{\alpha\beta}(E)$?

3.2 Double-Time Green's Functions

3.2.1 Equations of Motion

For the construction of the complete Green's function formalism, the retarded functions which we have thus far introduced are not sufficient. We require two additional types of Green's functions:

Retarded Green's functions

$$\begin{aligned} G_{AB}^{\mathrm{ret}}\left(t,t'\right) &\equiv \langle\langle A(t); B\left(t'\right)\rangle\rangle^{\mathrm{ret}} = \\ &= -\mathrm{i}\,\Theta\left(t-t'\right)\left\langle\left[A(t), B\left(t'\right)\right]_{-\varepsilon}\right\rangle . \end{aligned} \tag{3.117}$$

Advanced Green's functions

$$\begin{aligned} G_{AB}^{\mathrm{adv}}\left(t,t'\right) &\equiv \langle\langle A(t); B\left(t'\right)\rangle\rangle^{\mathrm{adv}} = \\ &= +\mathrm{i}\,\Theta\left(t'-t\right)\left\langle\left[A(t), B\left(t'\right)\right]_{-\varepsilon}\right\rangle . \end{aligned} \tag{3.118}$$

Causal Green's functions

$$\begin{aligned} G_{AB}^{c}\left(t,t'\right) &\equiv \langle\langle A(t); B\left(t'\right)\rangle\rangle^{c} = \\ &= -\mathrm{i}\left\langle T_{\varepsilon}\left(A(t)B\left(t'\right)\right)\right\rangle . \end{aligned} \tag{3.119}$$

The operators which generate the Green's functions are given here in their time-dependent Heisenberg representation, i.e. from (3.24) for the case of a Hamiltonian which is not explicitly time-dependent:

$$X(t) = \exp\left(\frac{\mathrm{i}}{\hbar}\mathcal{H}t\right) X \exp\left(-\frac{\mathrm{i}}{\hbar}\mathcal{H}t\right) . \tag{3.120}$$

$\mathcal{H}$ is defined as in (3.45):

$$\mathcal{H} = H - \mu\widehat{N} . \tag{3.121}$$

The averaging is carried out over the grand canonical ensemble:

$$\langle X \rangle = \frac{1}{\Xi} \operatorname{Tr}\left(e^{-\beta\mathcal{H}} X\right) , \tag{3.122}$$

$$\Xi = \operatorname{Tr}\left(e^{-\beta\mathcal{H}}\right) . \tag{3.123}$$

Ξ is the grand-canonical partition function. $\varepsilon = \pm$ is the sign index which was introduced in Chap. 1. The value of ε in the definitions (3.117) to (3.119) is completely arbitrary. If A and B are purely Fermionic (Bosonic) operators, then the choice $\varepsilon = -(+)$ proves to be expedient, as we shall later see. It is, however, by no means imperative.

We recall that

$$\left[A(t), B\left(t'\right)\right]_{-\varepsilon} = A(t)B\left(t'\right) - \varepsilon B\left(t'\right) A(t) \tag{3.124}$$

refers to the commutator when $\varepsilon = +$, and to the anticommutator when $\varepsilon = -$.

Finally, we must define **Wick's time-ordering operator** T_ε, which sorts the operators in a product according to their time arguments:

$$T_\varepsilon\left(A(t)B\left(t'\right)\right) = \Theta\left(t - t'\right) A(t)B\left(t'\right) + \varepsilon\Theta(t' - t)B\left(t'\right) A(t) . \tag{3.125}$$

Due to ε, it is **not** identical to Dyson's time-ordering operator T_D (3.15). The step function Θ,

$$\Theta\left(t - t'\right) = \begin{cases} 1 & \text{for} \quad t > t' , \\ 0 & \text{for} \quad t < t' , \end{cases} \tag{3.126}$$

is not defined for $t = t'$. This holds also for the Green's functions.

Owing to the averaging process in the defining equations (3.117) to (3.119), the Green's functions are also temperature dependent. We shall demonstrate later how the time and temperature variables can be brought into a close interrelation (see Chap. 6).

There is another very important function in many-body theory which we wish to introduce at this point. It is the so-called **spectral density**, whose information content will prove to be identical to that of the Green's functions:

$$S_{AB}\left(t, t'\right) = \frac{1}{2\pi}\left\langle\left[A(t), B\left(t'\right)\right]_{-\varepsilon}\right\rangle . \tag{3.127}$$

We can now prove the fact, already used several times in Sect. 3.1, that for Hamiltonians which are not explicitly time-dependent, the Green's functions and the spectral density are **homogeneous in time**:

$$\frac{\partial \mathcal{H}}{\partial t} = 0 \Longrightarrow G^{\alpha}_{AB}\left(t, t'\right) = G^{\alpha}_{AB}\left(t - t'\right) \quad (\alpha = \text{ret, adv, c}) , \tag{3.128}$$

$$S_{AB}\left(t, t'\right) = S_{AB}\left(t - t'\right) . \tag{3.129}$$

The proof is evidently substantiated if we can demonstrate this homogeneity for the so-called

correlation functions

$$\left\langle A(t) B\left(t'\right)\right\rangle , \quad \left\langle B\left(t'\right) A(t)\right\rangle .$$

This can be achieved by making use of the cyclic invariance of the trace:

$$\begin{aligned}
&\mathrm{Tr}\left[\exp(-\beta\mathcal{H})\, A(t) B\left(t'\right)\right] = \\
&= \mathrm{Tr}\left[\exp(-\beta\mathcal{H}) \exp\left(\frac{\mathrm{i}}{\hbar}\mathcal{H}t\right) A \exp\left(-\frac{\mathrm{i}}{\hbar}\mathcal{H}\left(t - t'\right)\right) B \exp\left(-\frac{\mathrm{i}}{\hbar}\mathcal{H}t'\right)\right] = \\
&= \mathrm{Tr}\left[\exp(-\beta\mathcal{H}) \exp\left(\frac{\mathrm{i}}{\hbar}\mathcal{H}\left(t - t'\right)\right) A \exp\left(-\frac{\mathrm{i}}{\hbar}\mathcal{H}\left(t - t'\right)\right) B\right] = \\
&= \mathrm{Tr}\left[\exp(-\beta\mathcal{H}) A\left(t - t'\right) B(0)\right] .
\end{aligned}$$

From this it follows that:

$$\left\langle A(t) B\left(t'\right)\right\rangle = \left\langle A\left(t - t'\right) B(0)\right\rangle . \tag{3.130}$$

Analogously, one finds:

$$\left\langle B\left(t'\right) A(t)\right\rangle = \left\langle B(0) A\left(t - t'\right)\right\rangle . \tag{3.131}$$

Then (3.128) and (3.129) are proved!

For the actual computation of the Green's functions, we as a rule will require their **equations of motion**. These can be obtained directly from the general equation of motion (3.27) for Heisenberg operators. Due to

$$\frac{\mathrm{d}}{\mathrm{d}t}\Theta\left(t - t'\right) = \delta\left(t - t'\right) = -\frac{\mathrm{d}}{\mathrm{d}t'}\Theta\left(t - t'\right) ,$$

one finds for all three types of Green's functions (3.117) to (3.119) formally the same equation of motion:

$$i\hbar\frac{\partial}{\partial t}G^{\alpha}_{AB}\left(t,t'\right)=\hbar\delta\left(t-t'\right)\langle[A,B]_{-\varepsilon}\rangle+\langle\langle[A,\mathcal{H}]_{-}(t);B\left(t'\right)\rangle\rangle^{\alpha}\;. \tag{3.132}$$

The solutions for the three functions are however subject to different boundary conditions:

$$G^{\mathrm{ret}}_{AB}\left(t,t'\right)=0 \qquad \text{for } t<t', \tag{3.133}$$

$$G^{\mathrm{adv}}_{AB}\left(t,t'\right)=0 \qquad \text{for } t>t', \tag{3.134}$$

$$G^{\mathrm{c}}_{AB}\left(t,t'\right)=\begin{cases}-\mathrm{i}\left\langle A\left(t-t'\right)B(0)\right\rangle & \text{for } t>t'\,,\\ -\mathrm{i}\varepsilon\left\langle B(0)A\left(t-t'\right)\right\rangle & \text{for } t<t'\,.\end{cases} \tag{3.135}$$

On the right-hand side of (3.132), a new Green's function has appeared, since the commutator $[A,\mathcal{H}]_{-}$ is itself an operator. This is as a rule a so-called *higher-order* Green's function, i.e. one which is constructed from more operators than the original function $G^{\alpha}_{AB}(t,t')$. These higher-order Green's functions of course also obey an equation of motion of the type (3.132), in which then a further new Green's function appears on the right-hand side,

$$\begin{aligned}i\hbar\frac{\partial}{\partial t}\langle\langle[A,\mathcal{H}]_{-}(t);B\left(t'\right)\rangle\rangle^{\alpha}&=\hbar\delta\left(t-t'\right)\left\langle[[A,\mathcal{H}]_{-},B]_{-\varepsilon}\right\rangle+\\&+\langle\langle[[A,\mathcal{H}]_{-},\mathcal{H}]_{-}(t);B\left(t'\right)\rangle\rangle^{\alpha}\,,\end{aligned} \tag{3.136}$$

to which the process can again be applied, etc. This leads for non-trivial problems to an infinite **chain of equations of motion**, which must be decoupled at some point in order to obtain an approximate solution. The point of decoupling should in every case be physically justifiable.

More expedient than the time representation of the Green's functions and of the spectral density is often their energy representation:

$$\begin{aligned}G^{\alpha}_{AB}(E)&\equiv\langle\langle A;B\rangle\rangle^{\alpha}_{E}=\\&=\int\limits_{-\infty}^{+\infty}\mathrm{d}\left(t-t'\right)\,G^{\alpha}_{AB}\left(t-t'\right)\exp\left(\frac{\mathrm{i}}{\hbar}E\left(t-t'\right)\right)\,,\end{aligned} \tag{3.137}$$

$$G^{\alpha}_{AB}\left(t-t'\right)=\frac{1}{2\pi\hbar}\int\limits_{-\infty}^{+\infty}\mathrm{d}E\,G^{\alpha}_{AB}(E)\exp\left(-\frac{\mathrm{i}}{\hbar}E\left(t-t'\right)\right)\,. \tag{3.138}$$

The spectral density is transformed in a similar fashion. If we now use the Fourier representations of the δ-functions,

$$\delta\left(E-E'\right)=\frac{1}{2\pi\hbar}\int\limits_{-\infty}^{+\infty}\mathrm{d}\left(t-t'\right)\exp\left(-\frac{\mathrm{i}}{\hbar}\left(E-E'\right)\left(t-t'\right)\right), \qquad (3.139)$$

$$\delta\left(t-t'\right)=\frac{1}{2\pi\hbar}\int\limits_{-\infty}^{+\infty}\mathrm{d}E\,\exp\left(\frac{\mathrm{i}}{\hbar}E\left(t-t'\right)\right), \qquad (3.140)$$

then the equation of motion (3.132) becomes:

$$E\,\langle\langle A;B\rangle\rangle_E^{\alpha}=\hbar\,\langle[A,B]_{-\varepsilon}\rangle+\left\langle\left\langle[A,\mathcal{H}]_-;B\right\rangle\right\rangle_E^{\alpha}. \qquad (3.141)$$

We are thus now no longer dealing with a differential equation, but instead with a purely algebraic equation. However, we again have an infinite chain of such equations of motion, which must be decoupled. The different boundary conditions (3.133) to (3.135) manifest themselves in the energy representation in terms of different analytic behaviour of the Green's functions $G^{\alpha}_{AB}(E)$ in the complex E-plane. We investigate this behaviour in the following section.

3.2.2 Spectral Representations

In order to supplement the system of equations which results from (3.141) with the boundary conditions, it is important to be aware of the so-called spectral representations of the Green's function.

Let E_n and $\left|E_n\right\rangle$ be the energy eigenvalues and the eigenstates of the Hamiltonian $\mathcal{H}$ of the physical system under consideration:

$$\mathcal{H}\left|E_n\right\rangle=E_n\left|E_n\right\rangle. \qquad (3.142)$$

The states $\left|E_n\right\rangle$ are assumed to form a complete, orthonormalised system:

$$\sum_n\left|E_n\right\rangle\left\langle E_n\right|=\mathbf{1}\,;\quad\left\langle E_n\,\middle|\,E_m\right\rangle=\delta_{nm}. \qquad (3.143)$$

We first want to discuss the **correlation functions** $\langle A(t)B(t')\rangle$, $\langle B(t')A(t)\rangle$:

$$
\begin{aligned}
\Xi \langle A(t)B(t')\rangle &= \mathrm{Tr}\left\{ e^{-\beta\mathcal{H}} A(t)B(t')\right\} = \sum_n \langle E_n \mid e^{-\beta\mathcal{H}} A(t)B(t') \mid E_n\rangle = \\
&= \sum_{n,m} \langle E_n \mid A(t) \mid E_m\rangle \langle E_m \mid B(t') \mid E_n\rangle e^{-\beta E_n} = \\
&= \sum_{n,m} \langle E_n \mid A \mid E_m\rangle \langle E_m \mid B \mid E_n\rangle e^{-\beta E_n} \cdot \\
&\quad \cdot \exp\left[\frac{i}{\hbar}(E_n - E_m)(t-t')\right] = \\
&= \sum_{n,m} \langle E_n \mid B \mid E_m\rangle \langle E_m \mid A \mid E_n\rangle e^{-\beta E_n} e^{-\beta(E_m - E_n)} \cdot \\
&\quad \cdot \exp\left[-\frac{i}{\hbar}(E_n - E_m)(t-t')\right] .
\end{aligned}
\tag{3.144}
$$

In the third step, we inserted the complete set of eigenstates between the operators, rendering the time dependence of the Heisenberg operators trivial. In the final step, we exchanged the indices n and m. In a quite analogous manner, we find for the second correlation function:

$$
\begin{aligned}
\Xi \langle B(t')A(t)\rangle &= \sum_{n,m} \langle E_n \mid B \mid E_m\rangle \langle E_m \mid A \mid E_n\rangle e^{-\beta E_n} \cdot \\
&\quad \cdot \exp\left[-\frac{i}{\hbar}(E_n - E_m)(t-t')\right] .
\end{aligned}
\tag{3.145}
$$

Inserting (3.144) and (3.145) into (3.127) leads after Fourier transformation to the important

spectral representation of the spectral density

$$
\begin{aligned}
S_{AB}(E) &= \frac{\hbar}{\Xi} \sum_{n,m} \langle E_n \mid B \mid E_m\rangle \langle E_m \mid A \mid E_n\rangle e^{-\beta E_n} \cdot \\
&\quad \cdot \left(e^{\beta E} - \varepsilon\right) \delta\left[E - (E_n - E_m)\right] .
\end{aligned}
\tag{3.146}
$$

Note that the arguments of the δ-functions contain the possible excitation energies of the system.

We now wish to express the Green's functions in terms of the spectral densities. To this end, we make use of the following representation of the step function:

$$\Theta\left(t-t'\right)=\frac{\mathrm{i}}{2\pi}\int\limits_{-\infty}^{+\infty}\mathrm{d}x\,\frac{\mathrm{e}^{-\mathrm{i}x(t-t')}}{x+\mathrm{i}0^{+}}\,. \tag{3.147}$$

Its proof is readily carried out using the residual theorem (Ex. 3.2.4). With (3.147), the retarded Green's function (3.117) may be rearranged as follows:

$$\begin{aligned}
G^{\mathrm{ret}}_{AB}(E) &= \int\limits_{-\infty}^{+\infty}\mathrm{d}\left(t-t'\right)\,\exp\!\left(\frac{\mathrm{i}}{\hbar}E\left(t-t'\right)\right)\left(-\mathrm{i}\Theta\left(t-t'\right)\right)\left(2\pi\,S_{AB}\left(t-t'\right)\right) = \\
&= \int\limits_{-\infty}^{+\infty}\mathrm{d}\left(t-t'\right)\,\exp\!\left(\frac{\mathrm{i}}{\hbar}E\left(t-t'\right)\right)\left(-\mathrm{i}\Theta\left(t-t'\right)\right)\cdot \\
&\quad\cdot\frac{1}{\hbar}\int\limits_{-\infty}^{+\infty}\mathrm{d}E'\,S_{AB}(E')\exp\!\left(-\frac{\mathrm{i}}{\hbar}E'\left(t-t'\right)\right) = \\
&= \int\limits_{-\infty}^{+\infty}\mathrm{d}E'\int\limits_{-\infty}^{+\infty}\mathrm{d}x\,\frac{S_{AB}\left(E'\right)}{x+\mathrm{i}0^{+}}\,\cdot \\
&\quad\cdot\frac{1}{2\pi\,\hbar}\int\limits_{-\infty}^{+\infty}\mathrm{d}\left(t-t'\right)\,\exp\!\left[-\frac{\mathrm{i}}{\hbar}\left(\hbar x-E+E'\right)\left(t-t'\right)\right] = \\
&= \int\limits_{-\infty}^{+\infty}\mathrm{d}E'\int\limits_{-\infty}^{+\infty}\mathrm{d}x\,\frac{S_{AB}\left(E'\right)}{x+\mathrm{i}0^{+}}\,\frac{1}{\hbar}\,\delta\!\left(x-\frac{1}{\hbar}\left(E-E'\right)\right).
\end{aligned}$$

With this, we obtain the

spectral representation of the retarded Green's function

$$G^{\mathrm{ret}}_{AB}(E)=\int\limits_{-\infty}^{+\infty}\mathrm{d}E'\,\frac{S_{AB}\left(E'\right)}{E-E'+\mathrm{i}0^{+}}\,. \tag{3.148}$$

The treatment of the advanced function is completely analogous:

$$G_{AB}^{\mathrm{adv}}(E) = \int\limits_{-\infty}^{+\infty} \mathrm{d}\left(t-t'\right) \exp\left(\frac{\mathrm{i}}{\hbar} E\left(t-t'\right)\right) \mathrm{i}\Theta\left(t'-t\right) 2\pi\, S_{AB}\left(t-t'\right) =$$

$$= \int\limits_{-\infty}^{+\infty} \mathrm{d}E' \int\limits_{-\infty}^{+\infty} \mathrm{d}x \, \frac{S_{AB}\left(E'\right)}{x+\mathrm{i}0^+} \cdot$$

$$\cdot \frac{-1}{2\pi\,\hbar} \int\limits_{-\infty}^{+\infty} \mathrm{d}(t-t') \exp\left[-\frac{\mathrm{i}}{\hbar}\left(-\hbar x - E + E'\right)\left(t-t'\right)\right] =$$

$$= -\int\limits_{-\infty}^{+\infty} \mathrm{d}E' \int\limits_{-\infty}^{+\infty} \mathrm{d}x \, \frac{S_{AB}\left(E'\right)}{x+\mathrm{i}0^+} \frac{1}{\hbar} \delta\left(-x - \frac{1}{\hbar}\left(E-E'\right)\right) .$$

This yields the

spectral representation of the advanced Green's function

$$G_{AB}^{\mathrm{adv}}(E) = \int\limits_{-\infty}^{+\infty} \mathrm{d}E' \frac{S_{AB}\left(E'\right)}{E-E'-\mathrm{i}0^+} . \tag{3.149}$$

The sign of $\mathrm{i}0^+$ is the only –but still important– difference between the retarded and advanced functions, and leads to their differing analytic behaviours:

$$G_{AB}^{\mathrm{ret}}$$

can be analytically continued in the upper half-plane,

$$G_{AB}^{\mathrm{adv}}$$

in the lower half-plane! The causal Green's function, which we still have to discuss, can be analytically continued neither in the upper nor in the lower half-plane, in contrast.

If we now insert the spectral representation (3.146) of the spectral density into (3.148) or (3.149), we obtain the following notable expression:

$$G_{AB}^{\substack{\mathrm{ret}\\ \mathrm{adv}}} = \frac{\hbar}{\Xi} \sum_{n,m} \left\langle E_n \,|\, B \,|\, E_m \right\rangle \left\langle E_m \,|\, A \,|\, E_n \right\rangle \mathrm{e}^{-\beta E_n} \frac{\mathrm{e}^{\beta(E_n - E_m)} - \varepsilon}{E - (E_n - E_m) \pm \mathrm{i}0^+} . \tag{3.150}$$

In both cases, we thus have a meromorphic function with simple poles at precisely the excitation energies of the interacting systems. If we are able somehow to determine the Green's function, then we could read off exactly the energies $(E_n - E_m)$ at the singularities for which the matrix elements of the operators A and B are nonzero. Thus, by a suitable choice of A and B, one can specify that particular types of excitation energies will appear as poles.

Owing to their identical physical information content, the retarded and the advanced Green's functions are sometimes combined into a single function $G_{AB}(E)$. Specifically, one considers $G^{\rm ret}_{AB}$ and $G^{\rm adv}_{AB}$ as the two branches of a unified Green's function in the complex E-plane:

$$G_{AB}(E) = \int_{-\infty}^{+\infty} dE' \frac{S_{AB}(E')}{E-E'} = \begin{cases} G^{\rm ret}_{AB}(E)\,, & \text{when} \quad \operatorname{Im} E > 0\,, \\ G^{\rm adv}_{AB}(E)\,, & \text{when} \quad \operatorname{Im} E < 0\,. \end{cases} \tag{3.151}$$

The singularities then lie on the real axis.

In (3.148), (3.149) and (3.151), we expressed the Green's functions in terms of the spectral density. Making use of the Dirac identity

$$\frac{1}{x - x_0 \pm \mathrm{i}0^+} = \mathcal{P}\frac{1}{x-x_0} \mp \mathrm{i}\pi\,\delta\,(x-x_0)\,, \tag{3.152}$$

in which $\mathcal{P}$ denotes the Cauchy principal value, we can also readily derive the *converse*:

$$S_{AB}(E) = \frac{\mathrm{i}}{2\pi}\left[G_{AB}\left(E+\mathrm{i}0^+\right) - G_{AB}\left(E-\mathrm{i}0^+\right)\right]. \tag{3.153}$$

If one presumes the spectral density in (3.151) to be real, then it follows that:

$$S_{AB}(E) = \mp\frac{1}{\pi}\operatorname{Im} G^{\substack{\rm ret\\ \rm adv}}_{AB}(E)\,. \tag{3.154}$$

We have still not given the spectral representation of the causal Green's function. Starting from the definition (3.119)

$$G^{\rm c}_{AB}(E) = -\mathrm{i}\int_{-\infty}^{+\infty} \mathrm{d}(t-t')\,\exp\left(\frac{\mathrm{i}}{\hbar}E\left(t-t'\right)\right)\left(\Theta\left(t-t'\right)\left\langle A(t)B\left(t'\right)\right\rangle + \right.$$
$$\left. +\,\varepsilon\Theta\left(t'-t\right)\left\langle B\left(t'\right)A(t)\right\rangle\right),$$

and inserting (3.144), (3.145) and (3.147), we obtain:

$$G^{\rm c}_{AB}(E) =$$
$$= \frac{1}{\Xi}\sum_{n,m}\left\langle E_n\middle|B\middle|E_m\right\rangle\left\langle E_m\middle|A\middle|E_n\right\rangle \mathrm{e}^{-\beta E_n}\frac{1}{2\pi}\int_{-\infty}^{+\infty}\mathrm{d}t''\int_{-\infty}^{+\infty}\mathrm{d}x\,\frac{1}{x+\mathrm{i}0^+}\,\cdot$$
$$\cdot\left\{\exp[\beta\,(E_n-E_m)]\exp\left\{\frac{\mathrm{i}}{\hbar}\left[E-(E_n-E_m)-\hbar x\right]\right\}t''+\right.$$
$$\left. +\,\varepsilon\exp\left[\frac{\mathrm{i}}{\hbar}\left(E-(E_n-E_m)+\hbar x\right)t''\right]\right\} =$$

$$= \frac{\hbar}{\Xi} \sum_{n,m} \langle E_n \mid B \mid E_m \rangle \langle E_m \mid A \mid E_n \rangle \mathrm{e}^{-\beta E_n} \cdot$$

$$\cdot \int_{-\infty}^{+\infty} \mathrm{d}x \, \frac{1}{x + \mathrm{i}0^+} \Big[\mathrm{e}^{\beta(E_n - E_m)} \delta \left(E - (E_n - E_m) - \hbar x\right) +$$

$$+ \varepsilon \delta \left(E - (E_n - E_m) + \hbar x\right) \Big] \,.$$

This yields the

spectral representation of the causal Green's function

$$G^{\mathrm{c}}_{AB}(E) = \frac{\hbar}{\Xi} \sum_{n,m} \langle E_n \mid B \mid E_m \rangle \langle E_m \mid A \mid E_n \rangle \mathrm{e}^{-\beta E_n} \cdot \tag{3.155}$$

$$\cdot \left[\frac{\mathrm{e}^{\beta(E_n - E_m)}}{E - (E_n - E_m) + \mathrm{i}0^+} - \frac{\varepsilon}{E - (E_n - E_m) - \mathrm{i}0^+} \right] .$$

The causal Green's function thus has singularities both in the lower and in the upper half-plane; it cannot be analytically continued in either. For specific computations, the retarded and the advanced functions are as a rule more tractable. The diagram techniques which we will introduce later can, however, be carried out only using the causal function.

3.2.3 The Spectral Theorem

In the last section, we have seen that from the Green's functions or alternatively from the spectral density, valuable microscopic information about the physical system under consideration can be obtained. The singularities of these functions are identical with the excitation energies of the system. The Green's functions however can yield considerably more information. We now want to show that the complete macroscopic thermodynamics of the system is determined by suitably-defined Green's functions. To this end, we first derive the fundamental spectral theorem.

We start with the correlation function

$$\left\langle B\left(t'\right) A(t) \right\rangle \,,$$

whose spectral representation (3.145) is very similar to the corresponding representation (3.146) of the spectral density. The correlation can therefore be expressed in terms of the spectral density. This is directly possible with the aid of the **anticommutator spectral density** ($\varepsilon = -$). Combining (3.145) and (3.146) yields:

$$\left\langle B\left(t'\right)A(t)\right\rangle = \frac{1}{\hbar}\int\limits_{-\infty}^{+\infty} \mathrm{d}E\,\frac{S_{AB}^{(-)}(E)}{\mathrm{e}^{\beta E}+1}\exp\left(-\frac{\mathrm{i}}{\hbar}E\left(t-t'\right)\right) . \tag{3.156}$$

This fundamental relation is called the **spectral theorem**. With its aid, arbitrary correlation functions and expectation values ($t = t'$) over suitably defined spectral densities can be computed. One must however keep in mind that when using *commutator spectral densities* ($\varepsilon = +$), the above expression must be complemented with a constant D, so that the complete spectral theorem must be formulated as follows:

$$\left\langle B\left(t'\right)A(t)\right\rangle = \frac{1}{\hbar}\int\limits_{-\infty}^{+\infty} \mathrm{d}E\,\frac{S_{AB}^{(\varepsilon)}(E)}{\mathrm{e}^{\beta E}-\varepsilon}\exp\left(-\frac{\mathrm{i}}{\hbar}E\left(t-t'\right)\right) + \frac{1}{2}(1+\varepsilon)D . \tag{3.157}$$

The following example shows that in the case of ($\varepsilon = +$), a correction term must be added to (3.156): In the definition (3.127) for the commutator spectral density, we replace the operators A and B by

$$\widetilde{A} = A - \langle A\rangle \; ; \quad \widetilde{B} = B - \langle B\rangle . \tag{3.158}$$

Then the spectral density itself does not change at all:

$$S_{AB}^{(+)}\left(t-t'\right) = S_{\widetilde{A}\widetilde{B}}^{(+)}\left(t-t'\right) .$$

Without D, the right-hand side of (3.156) would therefore not change, but the left-hand side would:

$$\left\langle \widetilde{B}\left(t'\right)\widetilde{A}(t)\right\rangle = \left\langle B\left(t'\right)A(t)\right\rangle - \langle B\rangle\langle A\rangle . \tag{3.159}$$

The commutator spectral density thus does not completely determine the correlation function. The reason for this can be read off (3.146), if one decomposes the spectral density into a diagonal and a non-diagonal part:

$$S_{AB}^{(\varepsilon)}(E) = \widehat{S}_{AB}^{(\varepsilon)}(E) + \hbar\,(1-\varepsilon)D\delta(E) . \tag{3.160}$$

Here, the following relations hold:

$$\widehat{S}_{AB}^{(\varepsilon)}(E) = \frac{\hbar}{\Xi}\sum_{n,m}^{E_n \neq E_m} \left\langle E_n \,\middle|\, B \,\middle|\, E_m\right\rangle\left\langle E_m \,\middle|\, A \,\middle|\, E_n\right\rangle \mathrm{e}^{-\beta E_n} \cdot \left(\mathrm{e}^{\beta E}-\varepsilon\right)\delta\left[E-(E_n-E_m)\right] , \tag{3.161}$$

$$D = \frac{1}{\Xi} \sum_{n,m}^{E_n = E_m} \langle E_n \mid B \mid E_m \rangle \langle E_m \mid A \mid E_n \rangle \, \mathrm{e}^{-\beta E_n} \,. \tag{3.162}$$

The diagonal terms contained in D do not appear at all in the commutator spectral density. They are however required for the determination of the correlation functions (3.144) and (3.145). The ($\varepsilon = +$) spectral density alone therefore does not suffice for the determination of the correlations, in the case that the diagonal elements are nonzero. Instead, we then find

$$\langle B(t')\, A(t) \rangle = D + \frac{1}{\hbar} \int_{-\infty}^{+\infty} \mathrm{d}E \, \frac{\widehat{S}_{AB}^{(\varepsilon)}(E)}{\mathrm{e}^{\beta E} - \varepsilon} \exp\left(-\frac{\mathrm{i}}{\hbar} E\, (t - t') \right) , \tag{3.163}$$

$$\langle A(t) B(t') \rangle = D + \frac{1}{\hbar} \int_{-\infty}^{+\infty} \mathrm{d}E \, \frac{\widehat{S}_{AB}^{(\varepsilon)}(E)\mathrm{e}^{\beta E}}{\mathrm{e}^{\beta E} - \varepsilon} \exp\left(-\frac{\mathrm{i}}{\hbar} E\, (t - t') \right) , \tag{3.164}$$

as can be directly read off from the general spectral representations (3.144) and (3.145). Upon insertion into the commutator Green's function $G_{AB}^{(+)}(E)$ or into the commutator spectral density, the constant D is eliminated. The spectral representation of the Green's function defined in (3.151), $G_{AB}(E)$, is then given by:

$$G_{AB}^{(-)}(E) = \\ = \frac{\hbar}{\Xi} \sum_{n,m} \langle E_n \mid B \mid E_m \rangle \langle E_m \mid A \mid E_n \rangle \, \mathrm{e}^{-\beta E_n} \frac{\mathrm{e}^{\beta(E_n - E_m)} + 1}{E - (E_n - E_m)} , \tag{3.165}$$

$$G_{AB}^{(+)}(E) = \\ = \frac{\hbar}{\Xi} \sum_{n,m}^{E_n \neq E_m} \langle E_n \mid B \mid E_m \rangle \langle E_m \mid A \mid E_n \rangle \, \mathrm{e}^{-\beta E_n} \frac{\mathrm{e}^{\beta(E_n - E_m)} - 1}{E - (E_n - E_m)} . \tag{3.166}$$

The following limiting case, which is to be carried out in the complex plane, since $G_{AB}^{(\varepsilon)}(E)$ is defined only there,

$$\lim_{E \to 0} E G_{AB}^{(\varepsilon)}(E) = (1 - \varepsilon) \hbar D \,, \tag{3.167}$$

yields a practical method for the determination of the constant D. The general spectral theorem (3.157) requires the presence of D when using the commutator functions. It can however be determined directly via (3.167) from the associated anticommutator Green's function $G_{AB}^{(-)}(E)$. Two additional important consequences can be seen from (3.167):

1. The commutator Green's function $G_{AB}^{(+)}(E)$ is in every case regular at the origin. This fact can be used as a criterion for approximate solutions.
2. The anticommutator Green's function $G_{AB}^{(-)}(E)$ has a first-order pole at $E = 0$ with the residual $2\hbar D$, in the case that $D \neq 0$. We will discuss simple applications of the fundamental spectral theorem in Sect. 3.3.

3.2.4
Exact Expressions

For realistic problems, Green's functions and spectral densities are, unfortunately, almost never exactly calculable. Approximations thus have to be tolerated. It is then however very useful to have some general, exact expressions at one's disposal (limiting cases, symmetry relations, sum rules etc.), with which one can test the approximations. In this section, we list some of these exact expressions.

One can read directly from the general definitions of the Green's functions the following:

$$G_{AB}^{\text{ret}}\left(t,t'\right)=\varepsilon G_{BA}^{\text{adv}}\left(t',t\right)\ . \tag{3.168}$$

With (3.117) and (3.118), we have namely:

$$\begin{aligned}\langle\langle A(t);B\left(t'\right)\rangle\rangle^{\text{ret}}&=-\mathrm{i}\Theta\left(t-t'\right)\left\langle\left[A(t),B\left(t'\right)\right]_{-\varepsilon}\right\rangle=\\&=+\mathrm{i}\varepsilon\Theta\left(t-t'\right)\left\langle\left[B\left(t'\right),A(t)\right]_{-\varepsilon}\right\rangle=\\&=\varepsilon\langle\langle B\left(t'\right);A(t)\rangle\rangle^{\text{adv}}\ .\end{aligned}$$

After a Fourier transformation, (3.168) becomes:

$$\begin{aligned}&\int\limits_{-\infty}^{+\infty}\mathrm{d}(t-t')\ G_{AB}^{\text{ret}}\left(t-t'\right)\exp\left(\frac{\mathrm{i}}{\hbar}E\left(t-t'\right)\right)=\\&=\varepsilon\int\limits_{-\infty}^{+\infty}\mathrm{d}(t-t')\ G_{BA}^{\text{adv}}\left(t'-t\right)\exp\left(\frac{\mathrm{i}}{\hbar}E\left(t-t'\right)\right)=\\&=\varepsilon\int\limits_{-\infty}^{+\infty}\mathrm{d}(t'-t)\ G_{BA}^{\text{adv}}\left(t'-t\right)\exp\left[\frac{\mathrm{i}}{\hbar}(-E)\left(t'-t\right)\right]\ .\end{aligned}$$

This means that:

$$G_{AB}^{\text{ret}}(E)=\varepsilon G_{BA}^{\text{adv}}(-E)\quad (E\ \text{real})\ . \tag{3.169}$$

If E is complex, then one must take into account the fact that $G_{AB}^{\text{ret}}(E)$ and $G_{AB}^{\text{adv}}(E)$ are analytic only within one half-plane, respectively. When E is complex, the *combined* function (3.151) is to be preferred; it is on the other hand not defined for a real E:

$$G_{AB}(E)=\varepsilon G_{BA}(-E)\qquad (E\ \text{complex})\ . \tag{3.170}$$

For retarded and for advanced Green's functions, the following holds:

$$\left(G_{AB}^{\text{ret, adv}}\left(t,t'\right)\right)^{*}=\varepsilon G_{A^{+}B^{+}}^{\text{ret, adv}}\left(t,t'\right)\ . \tag{3.171}$$

This relation, which follows directly from the definition, has the consequence – in the case that the Green's functions are constructed with Hermitian operators $\left(A=A^{+}\ ,\ B=B^{+}\right)$, as for example in the case of the *response* functions from Sect. 3.1 – that the commutator functions are purely real and the anticommutator functions are purely imaginary.

Using the equation of motion (3.141), we carry out the following rearrangement:

$$\int_{-\infty}^{+\infty} \mathrm{d}E \left\{ E \langle\langle A; B \rangle\rangle_E^{\mathrm{ret}} - \hbar \langle [A, B]_{-\varepsilon} \rangle \right\} =$$

$$= \int_{-\infty}^{+\infty} \mathrm{d}E \, \langle\langle [A, \mathcal{H}]_-; B \rangle\rangle_E^{\mathrm{ret}} =$$

$$= \int_{-\infty}^{+\infty} \mathrm{d}E \, (-\mathrm{i}) \int_0^{\infty} \mathrm{d}t \, \langle [[A, \mathcal{H}]_-(t), B(0)]_{-\varepsilon} \rangle \exp\left(\frac{\mathrm{i}}{\hbar} E t\right) =$$

$$= \hbar \int_0^{\infty} \mathrm{d}t \left\langle \left[\dot{A}(t), B(0)\right]_{-\varepsilon} \right\rangle \int_{-\infty}^{+\infty} \mathrm{d}E \exp\left(\frac{\mathrm{i}}{\hbar} E t\right) =$$

$$= 2\pi \hbar^2 \int_0^{\infty} \mathrm{d}t \left\langle \left[\dot{A}(t), B(0)\right]_{-\varepsilon} \right\rangle \delta(t) \, .$$

Due to

$$\int_0^{\infty} \mathrm{d}x \, \delta(x) f(x) = \frac{1}{2} f(0) \, , \tag{3.172}$$

it finally follows that:

$$\int_{-\infty}^{+\infty} \mathrm{d}E \left\{ E G_{AB}^{\mathrm{ret}}(E) - \hbar \langle [A, B]_{-\varepsilon} \rangle \right\} = \pi \hbar^2 \left\langle \left[\dot{A}(0), B(0)\right]_{-\varepsilon} \right\rangle . \tag{3.173}$$

The analogous relations for the two other Green's functions are given by:

$$\int_{-\infty}^{+\infty} \mathrm{d}E \left\{ E G_{AB}^{\mathrm{adv}}(E) - \hbar \langle [A, B]_{-\varepsilon} \rangle \right\} = -\pi \hbar^2 \left\langle \left[\dot{A}(0), B(0)\right]_{-\varepsilon} \right\rangle , \tag{3.174}$$

$$\int_{-\infty}^{+\infty} \mathrm{d}E \left\{ E G_{AB}^{\mathrm{c}}(E) - \hbar \langle [A, B]_{-\varepsilon} \rangle \right\} = \pi \hbar^2 \left\{ \langle \dot{A}(0) B(0) \rangle + \varepsilon \langle B(0) \dot{A}(0) \rangle \right\} . \tag{3.175}$$

The significance of these relations lies in the following conclusion: The right-hand sides are finite quantities, being expectation values of products of operators (observables). The integrals on the left-hand sides of the equations must therefore converge. The necessary condition for this is:

$$\lim_{E \to \infty} G_{AB}^{\alpha}(E) \approx \frac{\hbar}{E} \langle [A, B]_{-\varepsilon} \rangle \, . \tag{3.176}$$

The expectation value on the right is as a rule directly calculable, so that the high-energy behaviour of the Green's function can be determined in a simple fashion. Consider for example the important response functions $\chi_{ij}^{\alpha\beta}(E)$, $\sigma^{\alpha\beta}(E)$, $\varepsilon(\boldsymbol{q}, E)$ from Sect. 3.1.

For the spectral density $S_{AB}(E)$, there are useful sum rules, which can be obtained independently of the function itself and thus serve as controls for the inevitable approximation procedures. From the definition of the spectral density in (3.127), we have:

$$\left(\mathrm{i}\hbar\frac{\partial}{\partial t}\right)^n \left(2\pi S_{AB}\left(t, t'\right)\right) = \left(\mathrm{i}\hbar\frac{\partial}{\partial t}\right)^n \left\langle\left[A(t), B\left(t'\right)\right]_{-\varepsilon}\right\rangle =$$
$$= \left(\mathrm{i}\hbar\frac{\partial}{\partial t}\right)^n \frac{1}{\hbar}\int\limits_{-\infty}^{+\infty} \mathrm{d}E\, S_{AB}(E) \exp\left(-\frac{\mathrm{i}}{\hbar}E\left(t - t'\right)\right) =$$
$$= \frac{1}{\hbar}\int\limits_{-\infty}^{+\infty} \mathrm{d}E\, S_{AB}(E) E^n \exp\left(-\frac{\mathrm{i}}{\hbar}E\left(t - t'\right)\right) . \tag{3.177}$$

For $t = t'$, we obtain from these expressions the so-called

spectral moments

$$M_{AB}^{(n)} = \frac{1}{\hbar}\int\limits_{-\infty}^{+\infty} \mathrm{d}E\, E^n S_{AB}(E) . \tag{3.178}$$

If we insert the equation of motion (3.27) for Heisenberg operators on the left-hand side of (3.177), and keep in mind that we could have allowed $S_{AB}\left(t, t'\right)$ to be operated upon by $\left(\mathrm{i}\hbar\frac{\partial}{\partial t}\right)^{n-p}\left(-\mathrm{i}\hbar\frac{\partial}{\partial t'}\right)^p$ with $0 \le p \le n$ instead of by $\left(\mathrm{i}\hbar\frac{\partial}{\partial t}\right)^n$, with the same result, then we obtain the following alternative expression for the moments:

$$M_{AB}^{(n)} = \Big\langle\Big[\underbrace{\left[\cdots\left[\left[A, \mathcal{H}\right]_-, \mathcal{H}\right]_- \cdots \mathcal{H}\right]_-}_{(n-p)\text{-fold commutator}},\ \underbrace{\left[\mathcal{H}, \cdots \left[\mathcal{H}, B\right]_- \cdots\right]_-}_{p\text{-fold commutator}}\Big]_{-\varepsilon}\Big\rangle \tag{3.179}$$
$$(0 \le p \le n;\quad n = 0, 1, 2, \ldots)$$

When the Hamiltonian is known, this relation allows us in principle to compute all the moments of the spectral density exactly; this is independent of $S_{AB}(E)$.

With the aid of the spectral moments, an often very useful

high-energy expansion

for the Green's functions can be formulated. We find for the 'combined' Green's function (3.151) the following expression:

$$G_{AB}(E) = \int_{-\infty}^{+\infty} dE' \frac{S_{AB}(E')}{E - E'}$$

$$= \frac{1}{E} \int_{-\infty}^{+\infty} dE' \frac{S_{AB}(E')}{1 - \frac{E'}{E}}$$

$$= \frac{1}{E} \sum_{n=0}^{\infty} \int_{-\infty}^{+\infty} dE' \, S_{AB}(E') \left(\frac{E'}{E}\right)^n .$$

Comparison with (3.178) then yields:

$$G_{AB}(E) = \hbar \sum_{n=0}^{\infty} \frac{M_{AB}^{(n)}}{E^{n+1}} . \tag{3.180}$$

For the extreme high-energy behaviour ($E \to \infty$), this then implies:

$$G_{AB}(E) \approx \frac{\hbar}{E} M_{AB}^{(0)} = \frac{\hbar}{E} \langle [A, \, B]_{-\eta} \rangle . \tag{3.181}$$

The right-hand side can as a rule be readily calculated; therefore, the high-energy behaviour of e.g. the important response functions from Chap. 3.1.2 is already known.

3.2.5 The Kramers-Kronig Relations

According to (3.148) or (3.149), the Green's functions G_{AB}^{ret} and G_{AB}^{adv} are completely determined by the spectral density S_{AB}. On the other hand, the latter can be derived according to (3.154) from the imaginary part of these functions alone. The real and imaginary parts of the Green's functions are therefore not independent of one another.

We consider the integral

$$I_C(E) = \oint_C \mathrm{d}\overline{E} \, \frac{G_{AB}^{\text{ret}}(\overline{E})}{E - \overline{E} - \mathrm{i}0^+} .$$

$G_{AB}^{\text{ret}}(\overline{E})$ is analytic within the entire upper half-plane. This holds, presuming that E is real, for the complete integrand, so that (Fig. 3.3.)

$$I_C(E) = 0$$

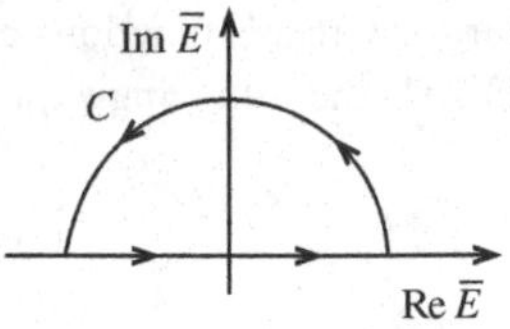

Fig. 3.3 The integration path in the complex E-plane for computing the integral $I_C(E)$

results. If we close the semicircle at infinity, then the integrand along it vanishes due to (3.176). This remains true when we apply the Dirac identity (3.152):

$$0 = \int_{-\infty}^{+\infty} \mathrm{d}\overline{E}\, \frac{G_{AB}^{\mathrm{ret}}(\overline{E})}{E - \overline{E} - \mathrm{i}0^+} = \mathcal{P} \int_{-\infty}^{+\infty} \mathrm{d}\overline{E}\, \frac{G_{AB}^{\mathrm{ret}}(\overline{E})}{E - \overline{E}} + \mathrm{i}\pi G_{AB}^{\mathrm{ret}}(E)\,.$$

From this we find:

$$G_{AB}^{\mathrm{ret}}(E) = \frac{\mathrm{i}}{\pi} \mathcal{P} \int_{-\infty}^{+\infty} \mathrm{d}\overline{E}\, \frac{G_{AB}^{\mathrm{ret}}(\overline{E})}{E - \overline{E}}\,. \tag{3.182}$$

Analogously, one finds on closing the semicircle within the lower half-plane, in which $G_{AB}^{\mathrm{adv}}(\overline{E})$ is analytic, and replacing $-\mathrm{i}0^+$ by $+\mathrm{i}0^+$,

$$G_{AB}^{\mathrm{adv}}(E) = -\frac{\mathrm{i}}{\pi} \mathcal{P} \int_{-\infty}^{+\infty} \mathrm{d}\overline{E}\, \frac{G_{AB}^{\mathrm{adv}}(\overline{E})}{E - \overline{E}}\,. \tag{3.183}$$

Precisely speaking, one requires no knowledge of the complete Green's functions. Determining only the real or only the imaginary part is sufficient. The other part then follows from the relations which we can read off (3.182) and (3.183), i.e. the

Kramers-Kronig relations

$$\mathrm{Re}\, G_{AB}^{\substack{\mathrm{ret}\\\mathrm{adv}}}(E) = \mp \frac{1}{\pi} \mathcal{P} \int_{-\infty}^{+\infty} \mathrm{d}\overline{E}\, \frac{\mathrm{Im}\, G_{AB}^{\substack{\mathrm{ret}\\\mathrm{adv}}}(\overline{E})}{E - \overline{E}}\,, \tag{3.184}$$

$$\mathrm{Im}\, G_{AB}^{\substack{\mathrm{ret}\\\mathrm{adv}}}(E) = \pm \frac{1}{\pi} \mathcal{P} \int_{-\infty}^{+\infty} \mathrm{d}\overline{E}\, \frac{\mathrm{Re}\, G_{AB}^{\substack{\mathrm{ret}\\\mathrm{adv}}}(\overline{E})}{E - \overline{E}}\,. \tag{3.185}$$

If we assume that the spectral density $S_{AB}(E)$ is **real**, then (3.154) holds, and thus:

$$\operatorname{Re} G_{AB}^{\text{ret}}(E) = \operatorname{Re} G_{AB}^{\text{adv}}(E) = \mathcal{P} \int_{-\infty}^{+\infty} \mathrm{d}\overline{E} \, \frac{S_{AB}(\overline{E})}{E - \overline{E}} , \qquad (3.186)$$

$$\operatorname{Im} G_{AB}^{\text{ret}}(E) = - \operatorname{Im} G_{AB}^{\text{adv}}(E) = -\pi S_{AB}(E) . \qquad (3.187)$$

The connection to the causal Green's function is obtained from (3.146) and (3.155):

$$\operatorname{Im} G_{AB}^{\text{c}}(E) = -\pi S_{AB}(E) \frac{\mathrm{e}^{\beta E} + \varepsilon}{\mathrm{e}^{\beta E} - \varepsilon} , \qquad (3.188)$$

$$\operatorname{Re} G_{AB}^{\text{c}}(E) = \operatorname{Re} G_{AB}^{\text{ret, adv}}(E) . \qquad (3.189)$$

Whilst (3.184) and (3.185) remain generally valid, (3.186) to (3.189) require that the spectral density be real. If this the case, as often happens, then these relations can serve as transformation formulas for converting one type of Green's function into another. This is not unimportant, since, as already mentioned, the method of equations of motion uses $G_{AB}^{\text{ret, adv}}$, whilst in contrast, the diagram techniques, which will be treated later, make use of G_{AB}^{c}.

3.2.6 Exercises

Exercise 3.2.1 Prove

$$\frac{\mathrm{d}}{\mathrm{d}t} \Theta \left(t - t'\right) = \delta \left(t - t'\right) = - \frac{\mathrm{d}}{\mathrm{d}t'} \Theta \left(t - t'\right) ,$$

where $\Theta(t - t')$ denotes the step function.

Exercise 3.2.2 Derive the equation of motion of the causal Green's function $G_{AB}^{\text{c}}(t, t')$.

Exercise 3.2.3 Show that

$$\langle B(0) A(t + \mathrm{i}\hbar\beta) \rangle = \langle A(t) B(0) \rangle$$

holds for time-dependent correlation functions when the Hamiltonian is not explicitly time-dependent.

Exercise 3.2.4 Derive the representation (3.147) of the step function:

$$\Theta\left(t-t'\right)=\frac{\mathrm{i}}{2\pi}\int\limits_{-\infty}^{+\infty}\mathrm{d}x\,\frac{\mathrm{e}^{-\mathrm{i}x(t-t')}}{x+\mathrm{i}0^{+}}\,.$$

Exercise 3.2.5 Prove that a complex function $F(E)$ has an analytic continuation in the upper (lower) half-plane, when its Fourier transform $f(t)$ vanishes for $t<0$ $(t>0)$.

Exercise 3.2.6 Calculate the conductivity tensor for the non-interacting electron system, using the result of Ex. 3.1.7.

Exercise 3.2.7 Show that for both retarded and advanced Green's functions, the following relation holds:

$$\left[G_{AB}^{\mathrm{ret\,(adv)}}\left(t,t'\right)\right]^{*}=\varepsilon G_{A^{+}B^{+}}^{\mathrm{ret\,(adv)}}\left(t,t'\right)\,.$$

Exercise 3.2.8 Prove the relation (3.175) for the causal Green's function:

$$\int\limits_{-\infty}^{+\infty}\mathrm{d}E\,\left[EG_{AB}^{c}(E)-\hbar\left\langle[A,B]_{-\varepsilon}\right\rangle\right]=\pi\hbar^{2}\left[\left\langle\dot{A}(0)B(0)\right\rangle+\varepsilon\left\langle B(0)\dot{A}(0)\right\rangle\right]\,.$$

Exercise 3.2.9 Compute all of the spectral moments,

$$M_{k\sigma}^{(n)}=\left\langle\left[\ldots\left[a_{k\sigma},\mathcal{H}\right]_{-},\ldots,\mathcal{H}\right]_{-},\mathcal{H}\right]_{-},a_{k\sigma}^{+}\right]_{+}\right\rangle\,,$$

for a system of non-interacting electrons,

$$H=\sum_{k,\sigma}\varepsilon(k)a_{k\sigma}^{+}a_{k\sigma}\,,$$

and with them the exact spectral density:

$$S_{k\sigma}(E)=-\frac{1}{\pi}\,\mathrm{Im}\left\langle\!\left\langle a_{k\sigma};a_{k\sigma}^{+}\right\rangle\!\right\rangle_{E}^{\mathrm{ret}}\,.$$

Exercise 3.2.10 Consider a free, spinless particle in one dimension:

$$H = \frac{p^2}{2m}, \quad [x, p]_- = \mathrm{i}\,\hbar\,.$$

The (mixed) state of the system is given by the density operator

$$\rho = \mathrm{e}^{-\beta H} \quad \text{(not normalised)}\,.$$

$\beta = \frac{1}{k_B T}$ is a parameter.

1. Calculate the trace of the density operator:

$$\langle \rho \rangle = \int \mathrm{e}^{-\beta H} \mathrm{d}p\,;$$

2. Show that for the expectation value of the energy,

$$\langle H \rangle = \frac{1}{2} k_B T$$

holds.
3. $\langle H \rangle$ is to be calculated from the commutator Green's function $G_p^{(+)}(E) = \langle\!\langle p; p \rangle\!\rangle^{(\varepsilon=+)}$. Solve the equation of motion for $G_p^{(+)}(E)$. (The result is trivial.)
4. Try to determine the expectation value $\langle H \rangle = \frac{1}{2m}\langle p \cdot p \rangle$ from the spectral theorem. Keep in mind the constant D (cf. (3.157))!
5. Compute the constant D using the relation

$$\lim_{E \to 0} E G_p^{(-)}(E) = 2\hbar D$$

from the solution of the equation of motion for the anticommutator Green's function $G_p^{(-)}(E) = \langle\!\langle p; p \rangle\!\rangle^{(\varepsilon=-)}$. Is it possible to determine $\langle H \rangle$?
6. Let an infinitesimal, symmetry-breaking field be defined by

$$H' = \frac{p^2}{2m} + \frac{m}{2}\omega^2 x^2 \quad (\omega \to 0)\,.$$

Set up the equation of motion for the commutator Green's function and solve it for $\omega \neq 0$. (The commutator Green's function $\langle\!\langle x; p \rangle\!\rangle^{(\varepsilon=+)}$ must also be determined.)
7. Find the constant D!
8. Calculate $\langle H \rangle_\omega$ from the spectral theorem for $G_p^{(+)}(E)$ with $\omega \neq 0$.
9. Show that

$$\lim_{\omega \to 0} \langle H' \rangle_\omega = \frac{1}{2} k_B T\,.$$

3.3 First Applications

In this section, we want to demonstrate applications of the abstract Green's function formalism treated in the last section to simple systems. Their properties are naturally known from elementary statistical mechanics. Here, they merely help us to become acquainted with the method.

3.3.1 Non-Interacting Bloch Electrons

As a first example, we discuss a system of electrons in a solid which do not interact with one another, but only with the periodic lattice potential; they are described by the Hamiltonian (2.22):

$$\mathcal{H}_0 = H_0 - \mu\widehat{N} \,, \tag{3.190}$$

$$H_0 = \sum_{k\sigma} \varepsilon(\boldsymbol{k}) a^+_{k\sigma} a_{k\sigma} \,, \tag{3.191}$$

$$\widehat{N} = \sum_{k\sigma} a^+_{k\sigma} a_{k\sigma} \,. \tag{3.192}$$

All of the properties of the electronic system which are of interest can be derived from the so-called

one-electron Green's function

$$G^\alpha_{k\sigma}(E) = \langle\langle a_{k\sigma}; a^+_{k\sigma} \rangle\rangle^\alpha_E \,, \qquad \alpha = \text{ret, adv, c} \,; \quad \varepsilon = - \,. \tag{3.193}$$

Since we are dealing with a purely Fermionic system, the choice of the anticommutator Green's function ($\varepsilon = -$) is preferable, but not necessary.

In solving this simple problem, we proceed exactly as would be required for more complicated cases. The first step is to set up and solve the equation of motion:

$$E G^\alpha_{k\sigma}(E) = \hbar \left\langle \left[a_{k\sigma}, a^+_{k\sigma} \right]_+ \right\rangle + \langle\langle [a_{k\sigma}, \mathcal{H}_0]_- \,; a^+_{k\sigma} \rangle\rangle^\alpha \,. \tag{3.194}$$

Making use of the fundamental commutation relations (2.23) and (2.24) for Fermions, one readily finds:

$$
\begin{aligned}
[a_{k\sigma}, \mathcal{H}_0]_- &= \sum_{k',\sigma'} (\varepsilon(k') - \mu) \left[a_{k\sigma}, a^+_{k'\sigma'} a_{k'\sigma'}\right]_- = \\
&= \sum_{k',\sigma'} (\varepsilon(k') - \mu)\, \delta_{kk'}\, \delta_{\sigma\sigma'}\, a_{k'\sigma'} = \\
&= (\varepsilon(k) - \mu)\, a_{k\sigma} \,.
\end{aligned}
\tag{3.195}
$$

After insertion into (3.194), this leads to the simple equation of motion:

$$
E G^{\alpha}_{k\sigma}(E) = \hbar + (\varepsilon(k) - \mu)\, G^{\alpha}_{k\sigma}(E) \,. \tag{3.196}
$$

Rearrangement and fulfilling the boundary conditions by introducing $+\mathrm{i}0^+$ or $-\mathrm{i}0^+$ yields:

$$
G^{\text{ret, adv}}_{k\sigma}(E) = \frac{\hbar}{E - (\varepsilon(k) - \mu) \pm \mathrm{i}0^+} \,. \tag{3.197}
$$

The singularities of this function clearly correspond to the possible excitation energies of the system. For a complex argument E we employ the *combined* Green's function (3.151):

$$
G_{k\sigma}(E) = \frac{\hbar}{E - (\varepsilon(k) - \mu)} \,. \tag{3.198}
$$

The **one-electron spectral density**

$$
S_{k\sigma}(E) = \hbar\delta\left(E - (\varepsilon(k) - \mu)\right) \tag{3.199}
$$

is important. What is the structure of the associated time-dependent functions? We first consider the retarded Green's function:

$$
G^{\text{ret}}_{k\sigma}\left(t - t'\right) = \frac{1}{2\pi\hbar} \int_{-\infty}^{+\infty} \mathrm{d}E \exp\left(-\frac{\mathrm{i}}{\hbar} E\left(t - t'\right)\right) \frac{\hbar}{E - (\varepsilon(k) - \mu) + \mathrm{i}0^+} \,.
$$

We substitute E by $\overline{E} = E - (\varepsilon(k) - \mu)$:

$$
G^{\text{ret}}_{k\sigma}\left(t - t'\right) = \exp\left[-\frac{\mathrm{i}}{\hbar}(\varepsilon(k) - \mu)\left(t - t'\right)\right] \frac{1}{2\pi} \int_{-\infty}^{+\infty} \mathrm{d}\overline{E}\, \frac{\exp\left(-\frac{\mathrm{i}}{\hbar}\overline{E}\left(t - t'\right)\right)}{\overline{E} + \mathrm{i}0^+} \,.
$$

With (3.147), this gives:

$$
G^{\text{ret}}_{k\sigma}\left(t - t'\right) = -\mathrm{i}\Theta\left(t - t'\right) \exp\left[-\frac{\mathrm{i}}{\hbar}(\varepsilon(k) - \mu)\left(t - t'\right)\right] . \tag{3.200}
$$

The boundary condition (3.133) is thus indeed fulfilled by inserting the infinitesimal $+\mathrm{i}0^+$. In a quite analogous manner, we find the advanced function:

$$G_{k\sigma}^{\mathrm{adv}}\left(t-t'\right)=\mathrm{i}\Theta\left(t'-t\right)\exp\left[-\frac{\mathrm{i}}{\hbar}\left(\varepsilon(\boldsymbol{k})-\mu\right)\left(t-t'\right)\right]. \tag{3.201}$$

In the non-interacting system, the time-dependent Green's functions thus exhibit an oscillatory behaviour with a frequency which corresponds to an exact excitation energy. We shall see later that this remains valid in analogous fashion also for interacting systems. These are then typically characterised by an additional damping factor, which can be interpreted as a finite lifetime of the quasi-particles.

We still want to investigate the causal Green's function, which according to (3.135) must obey the somewhat clumsy boundary conditions

$$G_{k\sigma}^{\mathrm{c}}\left(\left(t-t'\right)=0^+\right)=-\mathrm{i}\left(1-\langle n_{k\sigma}\rangle\right), \tag{3.202}$$

$$G_{k\sigma}^{\mathrm{c}}\left(\left(t-t'\right)=-0^+\right)=+\mathrm{i}\,\langle n_{k\sigma}\rangle\,. \tag{3.203}$$

We therefore write the solution of the equation of motion (3.194) in the following form:

$$G_{k\sigma}^{\mathrm{c}}(E)=\frac{C_1}{E-\left(\varepsilon(\boldsymbol{k})-\mu\right)+\mathrm{i}0^+}+\frac{C_2}{E-\left(\varepsilon(\boldsymbol{k})-\mu\right)-\mathrm{i}0^+}\,.$$

The transformation to the time-dependent function is carried out as for (3.200):

$$\begin{aligned}G_{k\sigma}^{\mathrm{c}}\left(t-t'\right)=&\left(-\mathrm{i}\Theta\left(t-t'\right)\frac{C_1}{\hbar}+\mathrm{i}\Theta\left(t'-t\right)\frac{C_2}{\hbar}\right)\cdot\\&\cdot\exp\left[-\frac{\mathrm{i}}{\hbar}\left(\varepsilon(\boldsymbol{k})-\mu\right)\left(t-t'\right)\right].\end{aligned} \tag{3.204}$$

The boundary conditions (3.202) and (3.203) are thus fulfilled with

$$C_1=\hbar\left(1-\langle n_{k\sigma}\rangle\right);\quad C_2=\hbar\,\langle n_{k\sigma}\rangle\,. \tag{3.205}$$

We can recognise from this simple example that the computational manipulation of the causal Green's function,

$$G_{k\sigma}^{\mathrm{c}}(E)=\frac{\hbar\left(1-\langle n_{k\sigma}\rangle\right)}{E-\left(\varepsilon(\boldsymbol{k})-\mu\right)+\mathrm{i}0^+}+\frac{\hbar\,\langle n_{k\sigma}\rangle}{E-\left(\varepsilon(\boldsymbol{k})-\mu\right)-\mathrm{i}0^+}\,, \tag{3.206}$$

is considerably more complicated than that of the retarded or the advanced functions. In particular, the expectation value $\langle n_{k\sigma}\rangle$ of the number operator must still be determined. The equation of motion method therefore deals almost exclusively with the retarded and the advanced functions.

The time-dependent spectral density can be readily found from (3.199):

$$S_{k\sigma}\left(t-t'\right)=\frac{1}{2\pi}\exp\left[-\frac{\mathrm{i}}{\hbar}(\varepsilon(\boldsymbol{k})-\mu)\left(t-t'\right)\right]. \tag{3.207}$$

The average occupation number $\langle n_{k\sigma}\rangle$ of the $(\boldsymbol{k},\sigma)$ level can be found by inserting (3.199) into the spectral theorem (3.157). We obtain the result which is well known from quantum statistics:

$$\langle n_{k\sigma}\rangle=\frac{1}{\exp\beta\left(\varepsilon(\boldsymbol{k})-\mu\right)+1}. \tag{3.208}$$

This is the **Fermi function**

$$f_{-}(E)=\frac{1}{\mathrm{e}^{\beta(E-\mu)}+1} \tag{3.209}$$

with the value $E=\varepsilon(\boldsymbol{k})$.

Using $\langle n_{k\sigma}\rangle$, and summing over all the wavenumbers $\boldsymbol{k}$ and both spin directions σ, we can fix the total number of electrons N_{e}:

$$\begin{aligned}N_{\mathrm{e}}&=\sum_{k\sigma}\frac{1}{\hbar}\int\limits_{-\infty}^{+\infty}\mathrm{d}E\,S_{k\sigma}(E)\frac{1}{\mathrm{e}^{\beta E}+1}=\\&=\sum_{k\sigma}\frac{1}{\hbar}\int\limits_{-\infty}^{+\infty}\mathrm{d}E\,f_{-}(E)S_{k\sigma}(E-\mu)\,.\end{aligned} \tag{3.210}$$

We denote the density of states per spin by $\rho_\sigma(E)$ for the free Fermion system, for which of course also $\rho_\sigma(E)=\rho_{-\sigma}(E)$ holds; then N_{e} can also be written as follows:

$$N_{\mathrm{e}}=N\sum_{\sigma}\int\limits_{-\infty}^{+\infty}\mathrm{d}E\,f_{-}(E)\rho_\sigma(E)\,. \tag{3.211}$$

N is the number of lattice sites; $\rho_\sigma(E)$ is normalised to 1. Comparison of (3.210) and (3.211) leads to the important definition of the

(quasi-particle) density of states

$$\rho_\sigma(E)=\frac{1}{N\hbar}\sum_{k}S_{k\sigma}(E-\mu)\,. \tag{3.212}$$

The above considerations concerning the electron number N_{e} are of course applicable not only to the non-interacting system, but also hold quite generally. As we will therefore see later, (3.212) already represents the general definition of the quasi-particle density of states for an arbitrary interacting electron system.

For non-interacting electron systems, we can insert (3.199):

$$\rho_\sigma(E) = \frac{1}{N}\sum_{\mathbf{k}} \delta\,(E - \varepsilon(\mathbf{k}))\ . \tag{3.213}$$

If the lattice potential plays no role, i.e.

$$\varepsilon(\mathbf{k}) = \frac{\hbar^2 k^2}{2m}\ ,$$

then $\rho_\sigma(E)$ exhibits the well-known $\sqrt{E}$ dependence:

$$\begin{aligned}
\rho_\sigma(E) &= \frac{1}{N}\sum_{\mathbf{k}} \delta\,(E - \varepsilon(\mathbf{k})) = \frac{V}{N\,(2\pi)^3}\int \mathrm{d}^3k\,\delta\left(E - \frac{\hbar^2k^2}{2m}\right) = \\
&= \frac{V}{2\pi^2 N}\int_0^\infty \mathrm{d}k\,k^2\frac{2m}{\hbar^2}\delta\left(\frac{2mE}{\hbar^2} - k^2\right) = \\
&= \frac{mV}{2\pi^2\hbar^2 N}\int_0^\infty \mathrm{d}k\,k\left[\delta\left(\sqrt{\frac{2mE}{\hbar^2}} - k\right) + \delta\left(\sqrt{\frac{2mE}{\hbar^2}} + k\right)\right] .
\end{aligned}$$

Only the first δ-function contributes:

$$\rho_\sigma(E) = \begin{cases} \dfrac{V}{4\pi^2 N}\left(\dfrac{2m}{\hbar^2}\right)^{3/2}\sqrt{E}\ , & \text{when} \quad E \geq 0\ , \\ 0 & \text{otherwise.} \end{cases} \tag{3.214}$$

The **internal energy** U can be determined in a simple manner as the thermodynamic expectation value of the Hamiltonian using $\langle n_{\mathbf{k}\sigma}\rangle$:

$$\begin{aligned}
U = \langle H_0\rangle &= \sum_{\mathbf{k}\sigma}\varepsilon(\mathbf{k})\,\langle n_{\mathbf{k}\sigma}\rangle = \\
&= \frac{1}{2\hbar}\sum_{\mathbf{k}\sigma}\int_{-\infty}^{+\infty}\mathrm{d}E\ (E + \varepsilon(\mathbf{k}))\, f_-(E) S_{\mathbf{k}\sigma}(E - \mu)\ .
\end{aligned} \tag{3.215}$$

The more complicated expression in the second line will prove to be the generally valid definition of U for interacting electron systems.

From U, we obtain the **free energy** F, and thus finally the entire thermodynamics of the system, with the aid of the following considerations:

Due to

$$F(T,V) = U(T,V) - TS(T,V) = U(T,V) + T\left(\frac{\partial F}{\partial T}\right)_V$$

we have also:

$$U(T,V) = -T^2\left[\frac{\partial}{\partial T}\left(\frac{1}{T}F(T,V)\right)\right]_V . \tag{3.216}$$

Employing the Third Law of Thermodynamics,

$$\lim_{T\to 0}\left[\frac{1}{T}(F(T) - F(0))\right] = \left(\frac{\partial F}{\partial T}\right)_V (T=0) = -S(T=0,V) = 0\,,$$

as well as $F(0,V) = U(0,V)$, we can integrate (3.216):

$$F(T,V) = U(0,V) - T\int_0^T \mathrm{d}T'\,\frac{U(T',V) - U(0,V)}{T'^2} . \tag{3.217}$$

All the other quantities of equilibrium thermodynamics can be derived from $F(T,V)$.

In this section, we have described non-interacting electrons in a solid using the wavenumber-dependent Green's function $G^{\alpha}_{k\sigma}(E)$. We could of course have investigated the single-electron Green's function just as well in the Wannier representation. For H_0, we would then have used (2.33). For

$$G^{\alpha}_{ij\sigma}(E) = \langle\langle a_{i\sigma}; a^{+}_{j\sigma}\rangle\rangle^{\alpha}_{E}\,, \tag{3.218}$$

we find the equation of motion,

$$EG^{\alpha}_{ij\sigma}(E) = \hbar\delta_{ij} + \sum_m (T_{im} - \mu\delta_{im})\,G^{\alpha}_{mj\sigma}(E)\,, \tag{3.219}$$

which is not directly decoupled as is $G^{\alpha}_{k\sigma}(E)$ in (3.196), but which can be readily solved via Fourier transformation:

$$G^{\substack{\text{ret}\\\text{adv}}}_{ij\sigma}(E) = \frac{1}{N}\sum_{k}\frac{\exp\left(\mathrm{i}\boldsymbol{k}\cdot(\boldsymbol{R}_i - \boldsymbol{R}_j)\right)}{E - (\varepsilon(\boldsymbol{k}) - \mu) \pm \mathrm{i}0^+} . \tag{3.220}$$

The physical results which can be derived from this function are of course the same ones which we deduced above from $G^{\alpha}_{k\sigma}$.

3.3.2 Free Spin Waves

As a further, very simple example of an application, we wish to discuss a system of non-interacting Bosons, and we consider in this connection the spin waves of a ferromagnet, which we introduced in Sect. 2.4.4. Our starting point is therefore the Hamiltonian (2.244):

$$H_{\mathrm{SW}} = E_0 + \sum_{\boldsymbol{q}} \hbar\omega(\boldsymbol{q}) a_{\boldsymbol{q}}^{+} a_{\boldsymbol{q}} \,. \tag{3.221}$$

E_0 and $\hbar\omega(q)$ are explained in (2.224) and (2.232), respectively. We define the following

one-magnon Green's function

$$G_{\boldsymbol{q}}^{\alpha}(t, t') = \langle\langle a_{\boldsymbol{q}}(t); a_{\boldsymbol{q}}^{+}(t') \rangle\rangle^{\alpha} \,, \tag{3.222}$$

$\alpha = $ ret, adv, c; $\varepsilon = +$. Since magnons are Bosons, it will prove expedient to use the commutator Green's function.

For magnons, conservation of particle number does not hold. Precisely that number of magnons is excited at a given temperature T, for which the the free energy F is minimised:

$$\left(\frac{\partial F}{\partial N}\right)_{T,V} \overset{!}{=} 0 \,. \tag{3.223}$$

The differential quotient on the left is just the chemical potential μ. It thus follows that:

$$\mu = 0 \,. \tag{3.224}$$

This means that in the equation of motion for the Green's function, we can set $\mathcal{H} = H - \mu N = H$. We require the commutator

$$\begin{aligned} \left[a_{\boldsymbol{q}}, H_{\mathrm{SW}}\right]_{-} &= \sum_{\boldsymbol{q}'} \hbar\omega(\boldsymbol{q}') \left[a_{\boldsymbol{q}}, a_{\boldsymbol{q}'}^{+} a_{\boldsymbol{q}'}\right]_{-} = \\ &= \sum_{\boldsymbol{q}'} \hbar\omega(\boldsymbol{q}') \left[a_{\boldsymbol{q}}, a_{\boldsymbol{q}'}^{+}\right]_{-} a_{\boldsymbol{q}'} = \\ &= \hbar\omega(q) a_{\boldsymbol{q}} \,. \end{aligned} \tag{3.225}$$

The equation of motion then takes on a simple form:

$$E G_{\boldsymbol{q}}^{\alpha}(E) = \hbar + \hbar\omega(\boldsymbol{q}) G_{\boldsymbol{q}}^{\alpha}(E) \,.$$

Rearrangement and taking into account the boundary conditions then leads to:

$$G_q^{\substack{\text{ret}\\\text{adv}}}(E) = \frac{\hbar}{E - \hbar\omega(\boldsymbol{q}) \pm \mathrm{i}0^+} \,. \tag{3.226}$$

The poles again represent the excitation energies, i.e. the energies which must be exchanged on creation or annihilation of a magnon. This is of course just $\hbar\omega(\boldsymbol{q})$.

With (3.154), we find directly from (3.226) the fundamental

one-magnon spectral density

$$S_q(E) = \hbar\delta\left(E - \hbar\omega(\boldsymbol{q})\right) \,. \tag{3.227}$$

The time-dependent Green's function, e.g. the retarded function, represents the undamped harmonic oscillation, as in (3.200) for the free Bloch electrons:

$$G_q^{\text{ret}}\left(t - t'\right) = -\mathrm{i}\Theta\left(t - t'\right) \mathrm{e}^{-\mathrm{i}\omega(\boldsymbol{q})(t - t')} \,. \tag{3.228}$$

The frequency of the oscillations corresponds once again to an exact excitation energy of the system.

Making use of the spectral theorem (3.157) and of the spectral density $S_q(E)$, we obtain the expectation value of the magnon-number operator, the so-called

magnon occupation density

$$m_q = \left\langle a_q^+ a_q \right\rangle = \frac{1}{\exp\left(\beta\hbar\omega(\boldsymbol{q})\right) - 1} + D_q \,. \tag{3.229}$$

Since we started with the commutator Green's function, we still have to determine the constant D_q for the corresponding anticommutator Green's function, as described in detail in Sect. 3.2.3. The fundamental commutation relations for Bosonic systems (1.99) determine the inhomogeneity in the equation of motion:

$$\left\langle \left[a_q, a_q^+\right]_+ \right\rangle = 1 + 2m_q \,. \tag{3.230}$$

Furthermore, the anticommutator Green's function obeys the same equation of motion as the commutator function. We find:

$$G_q^{(-)}(E) = \frac{\hbar\left(1 + 2m_q\right)}{E - \hbar\omega(\boldsymbol{q})} \,. \tag{3.231}$$

For an at least infinitesimal, symmetry-breaking external field ($B_0 \geq 0^+$), the magnon energies are in every case nonzero, and are in fact positive. Then, however, from (3.167) we find:

$$2\hbar D_q = \lim_{E \to 0} E G_q^{(-)}(E) = 0 \,. \tag{3.232}$$

For the occupation density, it thus follows that:

$$m_q = \frac{1}{\mathrm{e}^{\beta\hbar\omega(q)} - 1} \,. \tag{3.233}$$

This is the Bose-Einstein distribution function, and thus the result known from elementary quantum statistics for free Bosonic systems.

The **internal energy** of the spin-wave system corresponds to the expectation value of the Hamiltonian, and is therefore given with (3.221) by

$$U = \langle H \rangle = E_0 + \sum_q \hbar\omega(\boldsymbol{q}) m_q \,. \tag{3.234}$$

The entire equilibrium thermodynamics can finally be derived from the **free energy** F, which is determined from the internal energy as in (3.217). Thus, finally, everything else follows from the magnon occupation density m_q and therefore from the spectral density $S_q(E)$. Even for more complicated interacting systems, as we shall later demonstrate, all the quantities defined in equilibrium thermodynamics become accessible once we have computed the spectral density or, equivalently, one of the Green's functions.

3.3.3 The Two-Spin Problem

As a third example of an application, we wish to treat a model system with interactions, whose partition function can still be calculated exactly, so that all the interesting correlation functions are known in principle. This thus opens up the possibility of comparing the results of the Green's function method with the exact solutions.

The model system in question consists of two spins of magnitudes

$$S_1 = S_2 = \frac{1}{2} \,, \tag{3.235}$$

which are coupled to each other via an exchange interaction J and are presumed to be acted upon by a homogeneous magnetic field. We describe them in terms of the correspondingly simplified Heisenberg model (2.221):

$$H = -J\left(S_1^+ S_2^- + S_1^- S_2^+ + 2S_1^z S_2^z\right) - b\left(S_1^z + S_2^z\right) \,, \tag{3.236}$$

where

$$b = \frac{1}{\hbar} g_J \mu_\mathrm{B} B_0 \,. \tag{3.237}$$

The limitation to $S_1 = S_2 = 1/2$ allows some simplifications:

$$S_i^{\mp} S_i^{\pm} = \frac{\hbar^2}{2} \mp \hbar S_i^z , \tag{3.238}$$

$$S_i^{\pm} S_i^z = -S_i^z S_i^{\pm} = \mp \frac{\hbar}{2} S_i^{\pm} , \tag{3.239}$$

$$\left(S_i^{+}\right)^2 = 0 ; \quad \left(S_i^z\right)^2 = \frac{\hbar^2}{4} . \tag{3.240}$$

For our further discussion, we require several commutators:

$$\begin{aligned} \left[S_1^-, H\right]_- &= -J \left[S_1^-, S_1^+\right]_- S_2^- - 2J \left[S_1^-, S_1^z\right]_- S_2^z - b \left[S_1^-, S_1^z\right]_- = \\ &= 2\hbar J \left(S_1^z S_2^- - S_1^- S_2^z\right) - \hbar b S_1^- . \end{aligned} \tag{3.241}$$

Quite analogously, one finds:

$$\left[S_2^-, H\right]_- = 2\hbar J \left(S_1^- S_2^z - S_1^z S_2^-\right) - \hbar b S_2^- , \tag{3.242}$$

$$\begin{aligned} \left[S_1^z, H\right]_- &= -J \left(\left[S_1^z, S_1^+\right]_- S_2^- + \left[S_1^z, S_1^-\right] S_2^+\right) = \\ &= -\hbar J \left(S_1^+ S_2^- - S_1^- S_2^+\right) = -\left[S_2^z, H\right]_- . \end{aligned} \tag{3.243}$$

Finally, we also require:

$$\begin{aligned} &\left[S_1^z S_2^-, H\right]_- = \\ &= \left[S_1^z, H\right] S_2^- + S_1^z \left[S_2^-, H\right]_- = \\ &= -\hbar J \left(S_1^+ S_2^- - S_1^- S_2^+\right) S_2^- + 2\hbar J \left(S_1^z S_1^- S_2^z - \left(S_1^z\right)^2 S_2^-\right) - \hbar b S_1^z S_2^- = \\ &= \hbar J \left[S_1^- \left(\frac{\hbar^2}{2} + \hbar S_2^z\right)\right] + 2\hbar J \left(-S_1^- S_2^z \frac{\hbar}{2}\right) - 2\hbar J \frac{\hbar^2}{4} S_2^- - \hbar b S_1^z S_2^- = \\ &= \frac{1}{2} \hbar^3 J \left(S_1^- - S_2^-\right) - \hbar b S_1^z S_2^- . \end{aligned} \tag{3.244}$$

If the indices 1 and 2 are exchanged in this expression, which was derived using (3.238) through (3.240), it then follows that:

$$\left[S_2^z S_1^-, H\right]_- = \frac{1}{2} \hbar^3 J \left(S_2^- - S_1^-\right) - \hbar b S_2^z S_1^- . \tag{3.245}$$

Our main goal is the calculation of the magnetisation of the spin system, i.e. the expectation value $\left\langle S_1^z \right\rangle = \left\langle S_2^z \right\rangle \equiv \left\langle S^z \right\rangle$. The corresponding Green's function (retarded or advanced), according to (3.238), is given by:

$$G_{11}^{(+)}\left(t, t'\right) = \langle\!\langle S_1^-(t); S_1^+\left(t'\right) \rangle\!\rangle^{(+)} , \tag{3.246}$$

where the commutator function ($\varepsilon = +$) proves to be expedient. The equation of motion for $G_{11}^{(+)}$ naturally contains new, *higher-order* Green's functions, for which we then can construct further equations of motion. We however obtain a closed system of equations, if we extend $G_{11}^{(+)}$ with the following functions:

$$G_{21}^{(+)}(t,t') = \langle\langle S_2^-(t); S_1^+(t') \rangle\rangle^{(+)} , \tag{3.247}$$

$$\Gamma_{12}^{(+)}(t,t') = \langle\langle (S_1^z S_2^-)(t); S_1^+(t') \rangle\rangle^{(+)} , \tag{3.248}$$

$$\Gamma_{21}^{(+)}(t;t') = \langle\langle (S_2^z S_1^-)(t); S_1^+(t') \rangle\rangle^{(+)} . \tag{3.249}$$

The equations of motion of the energy-dependent Fourier transforms of these functions can be derived using the commutators (3.241) to (3.243), (3.244), and (3.245); we however still require the *inhomogeneities:*

$$\left\langle [S_1^-, S_1^+]_- \right\rangle = -2\hbar \langle S^z \rangle , \tag{3.250}$$

$$\left\langle [S_2^-, S_1^+]_- \right\rangle = 0 , \tag{3.251}$$

$$\left\langle [S_1^z S_2^-, S_1^+]_- \right\rangle = \hbar \langle S_1^+ S_2^- \rangle , \tag{3.252}$$

$$\left\langle [S_2^z S_1^-, S_1^+]_- \right\rangle = -2\hbar \langle S_2^z S_1^z \rangle . \tag{3.253}$$

We define the abbreviations

$$\rho_{12} = \langle S_1^+ S_2^- \rangle + 2 \langle S_1^z S_2^z \rangle , \tag{3.254}$$

$$R^{(+)}(E) = \Gamma_{12}^{(+)}(E) - \Gamma_{21}^{(+)}(E) , \tag{3.255}$$

and thus obtain the following equations of motion:

$$(E + \hbar b) G_{11}^{(+)}(E) = -2\hbar^2 \langle S^z \rangle + 2\hbar J R^{(+)}(E) , \tag{3.256}$$

$$(E + \hbar b) G_{21}^{(+)}(E) = -2\hbar J R^{(+)}(E) , \tag{3.257}$$

$$(E + \hbar b) R^{(+)}(E) = \hbar^2 \rho_{12} + \hbar^3 J \left(G_{11}^{(+)}(E) - G_{21}^{(+)}(E) \right) . \tag{3.258}$$

This system of equations can readily be solved:

$$(E + \hbar b)\left(G_{11}^{(+)}(E) + G_{21}^{(+)}(E) \right) = -2\hbar^2 \langle S^z \rangle , \tag{3.259}$$

$$\left(E + \hbar b - \frac{4\hbar^4 J^2}{E + \hbar b} \right) \left(G_{11}^{(+)}(E) - G_{21}^{(+)}(E) \right) =$$

$$= -2\hbar^2 \langle S^z \rangle + 4\hbar J \frac{\hbar^2 \rho_{12}}{E + \hbar b} . \tag{3.260}$$

However, it is found that the Green's functions have first-order poles at the following energies:

$$E_1 = -\hbar b\,; \quad E_2 = -\hbar b - 2J\hbar^2\,; \quad E_3 = -\hbar b + 2J\hbar^2\,. \tag{3.261}$$

We use this to further rearrange (3.259) and (3.260):

$$G_{11}^{(+)}(E) + G_{21}^{(+)}(E) = \frac{-2\hbar^2 \langle S^z\rangle}{E - E_1}\,,$$

$$G_{11}^{(+)}(E) - G_{21}^{(+)}(E) = -\hbar^2 \langle S^z\rangle \left(\frac{1}{E - E_2} + \frac{1}{E - E_3}\right) - $$

$$- \hbar\rho_{12}\left(\frac{1}{E - E_2} - \frac{1}{E - E_3}\right).$$

Addition or subtraction of these two equations finally leads to:

$$G_{11}^{(+)}(E) = -\frac{\hbar^2 \langle S^z\rangle}{E - E_1} - \frac{\hbar}{2}\frac{\eta_+}{E - E_2} - \frac{\hbar}{2}\frac{\eta_-}{E - E_3}\,, \tag{3.262}$$

$$G_{21}^{(+)}(E) = -\frac{\hbar^2 \langle S^z\rangle}{E - E_1} + \frac{\hbar}{2}\frac{\eta_+}{E - E_2} + \frac{\hbar}{2}\frac{\eta_-}{E - E_3}\,. \tag{3.263}$$

Here, we have also used the abbreviation

$$\eta_\pm = \hbar \langle S^z\rangle \pm \rho_{12}\,. \tag{3.264}$$

The remaining *higher-order* Green's function $R^{(+)}(E)$ can be most simply determined using (3.257):

$$R^{(+)}(E) = -\frac{E - E_1}{2\hbar J} G_{21}^{(+)}(E) =$$

$$= \frac{\hbar}{2J}\langle S^z\rangle - \frac{1}{4J}\left[\eta_+\left(1 - \frac{2J\hbar^2}{E - E_2}\right) + \eta_-\left(1 + \frac{2J\hbar^2}{E - E_3}\right)\right].$$

This gives:

$$R^{(+)}(E) = \frac{\hbar^2}{2}\frac{\eta_+}{E - E_2} - \frac{\hbar^2}{2}\frac{\eta_-}{E - E_3}\,. \tag{3.265}$$

Because of (3.167), these commutator Green's functions must be regular at $E = 0$. This is immediately clear in the presence of a field ($B_0 \neq 0 \Longleftrightarrow b \neq 0 \Longleftrightarrow E_1 \neq 0$). In the absence of a field, it is however assured only by

$$\langle S^z\rangle = 0 \quad \text{for} \quad B_0 = 0 \tag{3.266}$$

From the general analytic properties of the commutator Green's functions, we already obtain the physically important result that in the exchange-coupled two-spin system, there can be no spontaneous magnetisation.

For the complete determination of the Green's functions in (3.262), (3.263) and (3.265), we must determine the expectation values $\langle S^z \rangle$ and ρ_{12} with the aid of the spectral theorem (3.157). The spectral densities associated with the Green's functions can be directly read off (3.262), (3.263) and (3.265) using (3.152) and (3.154):

$$S_{11}^{(+)}(E) = -\hbar^2 \langle S^z \rangle \delta(E - E_1) - \frac{\hbar}{2}\eta_+ \delta(E - E_2) - \frac{\hbar}{2}\eta_- \delta(E - E_3) \,, \qquad (3.267)$$

$$S_{21}^{(+)}(E) = -\hbar^2 \langle S^z \rangle \delta(E - E_1) + \frac{\hbar}{2}\eta_+ \delta(E - E_2) + \frac{\hbar}{2}\eta_- \delta(E - E_3) \,, \qquad (3.268)$$

$$S_r^{(+)}(E) = \frac{\hbar^2}{2}\eta_+ \delta(E - E_2) - \frac{\hbar^2}{2}\eta_- \delta(E - E_3) \,. \qquad (3.269)$$

The spectral theorem (3.157) then gives the following results, where we make use of the abbreviation

$$m_i = \frac{1}{e^{\beta E_i} - 1} \,; \quad i = 1, 2, 3 \,: \qquad (3.270)$$

$$\langle S_1^+ S_1^- \rangle = -\hbar \langle S^z \rangle m_1 - \frac{1}{2}\eta_+ m_2 - \frac{1}{2}\eta_- m_3 + D_{11} \,, \qquad (3.271)$$

$$\langle S_1^+ S_2^- \rangle = -\hbar \langle S^z \rangle m_1 + \frac{1}{2}\eta_+ m_2 + \frac{1}{2}\eta_- m_3 + D_{21} \,, \qquad (3.272)$$

$$\langle S_1^+ S_1^z S_2^- \rangle - \langle S_1^+ S_2^z S_1^- \rangle = \frac{\hbar}{2}\eta_+ m_2 - \frac{\hbar}{2}\eta_- m_3 + D_R \,. \qquad (3.273)$$

With the results

$$\langle S^z \rangle = -\frac{\hbar}{2} + \frac{1}{\hbar} \langle S_1^+ S_1^- \rangle \,, \qquad (3.274)$$

$$\eta_+ = -\frac{2}{\hbar} \left(\langle S_1^+ S_1^z S_2^- \rangle - \langle S_1^+ S_2^z S_1^- \rangle \right) \,, \qquad (3.275)$$

$$\eta_- = -\eta_+ + 2\hbar \langle S^z \rangle \,, \qquad (3.276)$$

$$\langle S_1^z S_2^z \rangle = \frac{1}{4}(\eta_+ - \eta_-) - \frac{1}{2} \langle S_1^+ S_2^- \rangle \qquad (3.277)$$

we have thus determined all the required correlation functions up to the constants D_{11}, D_{21}, and D_R.

According to (3.167), these constants are just the residuals of the $E = 0$ poles of the associated anticommutator Green's functions $G_{11}^{(-)}(E)$, $G_{21}^{(-)}(E)$ and $R^{(-)}(E)$. Their equations of motion differ from those of the commutator functions (3.256) through (3.258) only in terms of the *inhomogeneities* on the right-hand sides of the equations:

$$\left\langle \left[S_1^-, S_1^+ \right]_+ \right\rangle = \hbar^2 ,$$
$$\left\langle \left[S_2^-, S_1^+ \right]_+ \right\rangle = 2 \left\langle S_1^+ S_2^- \right\rangle ,$$
$$\left\langle \left[S_1^z S_2^-, S_1^+ \right]_+ \right\rangle = 0 ,$$
$$\left\langle \left[S_2^z S_1^-, S_1^+ \right]_+ \right\rangle = \hbar^2 \left\langle S^z \right\rangle .$$

We find the following equations of motion:

$$(E + \hbar b) G_{11}^{(-)}(E) = \hbar^3 + 2\hbar J R^{(-)}(E) , \tag{3.278}$$

$$(E + \hbar b) G_{21}^{(-)}(E) = 2\hbar \left\langle S_1^+ S_2^- \right\rangle - 2\hbar J R^{(-)}(E) , \tag{3.279}$$

$$(E + \hbar b) R^{(-)}(E) = -\hbar^3 \left\langle S^z \right\rangle + \hbar^3 J \left(G_{11}^{(-)}(E) - G_{21}^{(-)}(E) \right) . \tag{3.280}$$

They can be further rearranged to:

$$G_{11}^{(-)}(E) + G_{21}^{(-)}(E) = \hbar \frac{\hbar^2 + 2\left\langle S_1^+ S_2^- \right\rangle}{E - E_1} ,$$

$$\left(E + \hbar b - \frac{4\hbar^4 J^2}{E + \hbar b} \right) \left(G_{11}^{(-)}(E) - G_{21}^{(-)}(E) \right) =$$
$$= \hbar^3 - 2\hbar \left\langle S_1^+ S_2^- \right\rangle - 4\hbar^4 J \frac{\left\langle S^z \right\rangle}{E + \hbar b} .$$

This gives:

$$G_{11}^{(-)}(E) - G_{21}^{(-)}(E) =$$
$$= \left(\hbar^3 - 2\hbar \left\langle S_1^+ S_2^- \right\rangle \right) \frac{1}{2} \left(\frac{1}{E - E_2} + \frac{1}{E - E_3} \right) + \hbar^2 \left(\frac{1}{E - E_2} - \frac{1}{E - E_3} \right) \left\langle S^z \right\rangle .$$

Addition or subtraction of these two equations finally leads to:

$$\begin{aligned} G_{11}^{(-)}(E) = &\frac{\hbar}{2} \left(\hbar^2 + 2\left\langle S_1^+ S_2^- \right\rangle \right) \frac{1}{E - E_1} + \\ &+ \frac{\hbar}{2} \left(\frac{\hbar^2}{2} + \hbar \left\langle S^z \right\rangle - \left\langle S_1^+ S_2^- \right\rangle \right) \frac{1}{E - E_2} + \\ &+ \frac{\hbar}{2} \left(\frac{\hbar^2}{2} - \hbar \left\langle S^z \right\rangle - \left\langle S_1^+ S_2^- \right\rangle \right) \frac{1}{E - E_3} , \end{aligned} \tag{3.281}$$

$$G_{21}^{(-)}(E) = \frac{\hbar}{2}\left(\hbar^2 + 2\left\langle S_1^+ S_2^- \right\rangle\right)\frac{1}{E-E_1} - \\ -\frac{\hbar}{2}\left(\frac{\hbar^2}{2} + \hbar\left\langle S^z \right\rangle - \left\langle S_1^+ S_2^- \right\rangle\right)\frac{1}{E-E_2} - \\ -\frac{\hbar}{2}\left(\frac{\hbar^2}{2} - \hbar\left\langle S^z \right\rangle - \left\langle S_1^+ S_2^- \right\rangle\right)\frac{1}{E-E_3} \,. \tag{3.282}$$

For the determination of $R^{(-)}(E)$, we make use of (3.279):

$$R^{(-)}(E) = \frac{1}{2\hbar J}\left[\hbar\left\langle S_1^+ S_2^- \right\rangle - \frac{\hbar^3}{2} + \right. \\ + \frac{\hbar}{2}\left(\frac{\hbar^2}{2} + \hbar\left\langle S^z \right\rangle - \left\langle S_1^+ S_2^- \right\rangle\right)\left(1 - \frac{2J\hbar^2}{E-E_2}\right) + \\ \left. + \frac{\hbar}{2}\left(\frac{\hbar^2}{2} - \hbar\left\langle S^z \right\rangle - \left\langle S_1^+ S_2^- \right\rangle\right)\left(1 + \frac{2J\hbar^2}{E-E_3}\right)\right] .$$

This yields:

$$R^{(-)}(E) = -\frac{\hbar^2}{2}\left(\frac{\hbar^2}{2} + \hbar\left\langle S^z \right\rangle - \left\langle S_1^+ S_2^- \right\rangle\right)\frac{1}{E-E_2} + \\ + \frac{\hbar^2}{2}\left(\frac{\hbar^2}{2} - \hbar\left\langle S^z \right\rangle - \left\langle S_1^+ S_2^- \right\rangle\right)\frac{1}{E-E_3} \,. \tag{3.283}$$

According to (3.167), we can read directly off these expressions:

$$D_{11} = D_{21} = \begin{cases} 0 & \text{for } \; b \neq 0 \,, \\ \dfrac{\hbar^2}{4} + \dfrac{1}{2}\left\langle S_1^+ S_2^- \right\rangle & \text{for } \; b = 0 \,, \end{cases} \tag{3.284}$$

$$D_R \equiv 0 \,. \tag{3.285}$$

Then the equal-time correlation functions (3.271) through (3.275) are completely determined.

The magnetisation $\langle S^z \rangle$ of the two-spin system for $b \neq 0$ is of particular interest; for it, we find the following result, keeping in mind (3.271), (3.274) and (3.284):

$$\left\langle S^z \right\rangle = -\frac{\hbar}{2} - \left\langle S^z \right\rangle m_1 - \frac{\eta_+}{2\hbar} m_2 - \frac{\eta_-}{2\hbar} m_3 \,. \tag{3.286}$$

With (3.276), we obtain from this expression:

$$\left\langle S^z \right\rangle (1 + m_1 + m_3) = -\frac{\hbar}{2} - \frac{\eta_+}{2\hbar}(m_2 - m_3) \,.$$

Equation (3.273) leads to

$$\eta_+ = 2\hbar \langle S^z \rangle \frac{m_3}{1 + m_2 + m_3} \,.$$

Combining these two equations, we obtain the intermediate result:

$$\langle S^z \rangle = -\frac{\hbar}{2} \left(1 + m_1 + m_3 \frac{1 + 2m_2}{1 + m_2 + m_3} \right)^{-1} \,.$$

Inserting the m_i as in (3.270) then gives after some simple rearrangements:

$$\langle S^z \rangle = \frac{\hbar}{2} \frac{\exp(\beta\hbar b) - \exp(-\beta\hbar b)}{1 + \exp\left(-2\beta\hbar^2 J\right) + \exp(\beta\hbar b) + \exp(-\beta\hbar b)} \,. \tag{3.287}$$

For zero applied magnetic field ($B_0 \to 0^+$), the magnetisation of the two-spin system thus vanishes. There is therefore no spontaneous magnetisation, as we had indeed already deduced from the general analytic properties of the commutator Green's function $G_{11}^{(+)}(E)$ in (3.266). When there is no coupling between the two spins, i.e. $J \to 0$, then we obtain the well-known result for the $S = \frac{1}{2}$ paramagnet:

$$\langle S^z \rangle \underset{J \to 0}{\longrightarrow} \frac{\hbar}{2} \tanh\left(\frac{1}{2} g_J \mu_B B_0 \right) \,. \tag{3.288}$$

The exchange coupling of the two spins mediated by J is reflected in particular in the correlations $\langle S_1^+ S_2^- \rangle$ and $\langle S_1^z S_2^z \rangle$, which in the limit $J \to 0$ must be expressed by:

$$\langle S_1^+ S_2^- \rangle \underset{J \to 0}{\longrightarrow} \langle S_1^+ \rangle \langle S_2^- \rangle = 0 \,, \tag{3.289}$$

$$\langle S_1^z S_2^z \rangle \underset{J \to 0}{\longrightarrow} \langle S_1^z \rangle \langle S_2^z \rangle = \langle S^z \rangle^2 \,. \tag{3.290}$$

From (3.271) and (3.272), we first find:

$$\begin{aligned} \langle S_1^+ S_2^- \rangle &= -\langle S_1^+ S_1^- \rangle - 2\hbar \langle S^z \rangle m_1 = \\ &= -\frac{\hbar^2}{2} - \hbar \langle S^z \rangle (1 + 2m_1) \,. \end{aligned} \tag{3.291}$$

In the last step, we made use of (3.238). Inserting (3.287) then leads to

$$\langle S_1^+ S_2^- \rangle = \frac{\hbar^2}{2} \frac{1 - \exp(-2\beta\hbar^2 J)}{1 + (-2\beta\hbar^2 J) + \exp(-\beta\hbar b) + \exp(\beta\hbar b)} \,. \tag{3.292}$$

The limiting case of (3.289) is clearly fulfilled.

The second correlation function, $\langle S_1^z S_2^z \rangle$, can be evaluated as follows: First, from (3.273), (3.275) and (3.276), we obtain:

$$\eta_+ - \eta_- = -2\hbar \langle S^z \rangle \frac{1 + m_2 - m_3}{1 + m_2 + m_3} .$$

This means, according to (3.277):

$$\langle S_1^z S_2^z \rangle = -\frac{\hbar}{2} \langle S^z \rangle \frac{1 + m_2 - m_3}{1 + m_2 + m_3} - \frac{1}{2} \langle S_1^+ S_2^- \rangle .$$

Inserting (3.287) and (3.292) then gives:

$$\langle S_1^z S_2^z \rangle = \frac{\hbar^2}{4} \frac{\exp(\beta\hbar b) + \exp(-\beta\hbar b) - \exp(-2\beta\hbar^2 J) - 1}{1 + \exp(\beta\hbar b) + \exp(-\beta\hbar b) + \exp(-2\beta\hbar^2 J)} . \tag{3.293}$$

Here, again, we can see that the limiting case of $J \to 0$ is reproduced correctly by (3.290).

Thus far, we have assumed that $b \neq 0$, and we must consider separately the special case of $b = 0$. We naturally expect that it can be treated in the limit $b \to 0$ from (3.292) or (3.293). Necessarily, we must have $\langle S^z \rangle = 0$ from (3.266), since the commutator Green's function $G_{11}^{(+)}(E)$ cannot have a pole at $E = 0$. Furthermore, in (3.271) to (3.277), the constants D_{11} and D_{21} are now nonzero (3.284). D_R is however still zero. It follows from (3.271) that:

$$\langle S_1^+ S_1^- \rangle_0 = \frac{\hbar^2}{2} = \frac{1}{2}\rho_{12}(m_3 - m_2) + \frac{\hbar^2}{4} + \frac{1}{2}\langle S_1^+ S_2^- \rangle_0 . \tag{3.294}$$

The same equation results from (3.272), so that for the determination of $\langle S_1^+ S_2^- \rangle$, no other equation is available. The isotropy which is present for $b = 0$ however leads to

$$\langle S_1^+ S_2^- \rangle_0 = 2 \langle S_1^z S_2^z \rangle_0 \tag{3.295}$$

and thus to $\rho_{12} = 2\langle S_1^+ S_2^- \rangle_0$. Equation (3.294) now contains only one unknown:

$$\frac{\hbar^2}{4} = \langle S_1^+ S_2^- \rangle_0 \left(m_3 - m_2 + \frac{1}{2} \right) .$$

In fact, using

$$\langle S_1^+ S_2^- \rangle_{b=0} = \frac{\hbar^2}{2} \frac{1 - \exp\left(-2\beta\hbar^2 J\right)}{3 + \exp\left(-2\beta\hbar J\right)} , \tag{3.296}$$

the limiting case of $b \to 0$ gives the result (3.292), and –due to (3.295)– also that of (3.293). Note that this would **not** have been the case if we had not taken into account the constants D_{11} and D_{12}, which arise from the application of the spectral theorem.

To conclude, we wish to demonstrate the significance of the constants D in the spectral theorem by considering an additional example. The exact result for the correlation $\langle S_1^z S_2^z \rangle$ in (3.293) was obtained finally with the aid of the Green's function $\Gamma_{21}^{(+)}(E)$ defined in (3.249). We could however have obtained the same result with the Green's function

$$P_{21}^{(+)}(E) = \langle\langle S_2^z ; S_1^z \rangle\rangle_E^{(+)} \tag{3.297}$$

and the spectral theorem (3.157). We wish therefore to calculate this Green's function. With the commutator (3.243), its equation of motion is given by:

$$E P_{21}^{(+)}(E) = \hbar J Q^{(+)}(E) \, . \tag{3.298}$$

We denote the Green's function

$$Q^{(+)}(E) = \langle\langle S_1^+ S_2^- - S_2^+ S_1^- ; S_1^z \rangle\rangle_E^{(+)} \tag{3.299}$$

by $Q^{(+)}(E)$. For its equation of motion, we require the following commutator:

$$\begin{aligned} [S_1^+ S_2^-, H]_- &= [S_1^+, H]_- S_2^- + S_1^+ [S_2^-, H]_- = \\ &= \left(-[S_1^-, H]_-\right)^+ S_2^- + S_1^+ [S_2^-, H]_- \, . \end{aligned}$$

We insert (3.241) and (3.242):

$$\begin{aligned} [S_1^+ S_2^-, H]_- = &\left(-2\hbar J \left(S_1^z S_2^+ - S_1^+ S_2^z\right) + \hbar b S_1^+\right) S_2^- + \\ &+ S_1^+ \left(2\hbar J \left(S_1^- S_2^z - S_1^z S_2^-\right) - \hbar b S_2^-\right) , \end{aligned}$$

and use (3.238) and (3.239):

$$\begin{aligned} [S_1^+ S_2^-, H]_- &= -2\hbar J \left[S_1^z \left(\frac{\hbar^2}{2} + \hbar S_2^z \right) - S_1^+ \left(-\frac{\hbar}{2} S_2^- \right) \right] + \\ &\quad + 2\hbar J \left[\left(\frac{\hbar^2}{2} + \hbar S_1^z \right) S_2^z - \left(-\frac{\hbar}{2} S_1^+ \right) S_2^- \right] = \\ &= -J\hbar^3 \left(S_1^z - S_2^z\right) = -[S_2^+ S_1^-, H]_- \, . \end{aligned}$$

With the inhomogeneity

$$\left\langle [S_1^+ S_2^- - S_2^+ S_1^-, S_1^z]_- \right\rangle = \hbar \left\langle -S_1^+ S_2^- - S_2^+ S_1^- \right\rangle = -2\hbar \left\langle S_1^+ S_2^- \right\rangle ,$$

the equation of motion for $Q^{(+)}(E)$ becomes:

$$E Q^{(+)}(E) = -2\hbar^2 \left\langle S_1^+ S_2^- \right\rangle - 2J\hbar^3 \left\{ P_{11}^{(+)}(E) - P_{21}^{(+)}(E) \right\} . \tag{3.300}$$

The corresponding equation of motion for the function

$$P_{11}^{(+)}(E) = \langle\langle S_1^z ; S_1^z \rangle\rangle_E^{(+)} \tag{3.301}$$

is found immediately using (3.243) to be

$$E P_{11}^{(+)}(E) = -\hbar J Q^{(+)}(E) \,. \tag{3.302}$$

From (3.298) and (3.302), we conclude that:

$$P_{11}^{(+)}(E) = -P_{21}^{(+)}(E) \,. \tag{3.303}$$

This leads via (3.300) to

$$E^2 P_{21}^{(+)}(E) = -2\hbar^3 J \langle S_1^+ S_2^- \rangle + 4J^2 \hbar^4 P_{21}^{(+)}(E) \,.$$

We can then readily calculate $P_{21}^{(+)}(E)$:

$$P_{21}^{(+)}(E) = \frac{\hbar}{2} \langle S_1^+ S_2^- \rangle \left(\frac{1}{E + 2\hbar^2 J} - \frac{1}{E - 2\hbar^2 J} \right) \,. \tag{3.304}$$

With the spectral theorem (3.157) and the result (3.292) for $\langle S_1^+ S_2^- \rangle$, we finally obtain:

$$\langle S_1^z S_2^z \rangle = -\frac{\hbar^2}{4} \frac{1 + \exp(-2\beta J \hbar^2)}{1 + \exp(-2\beta \hbar^2 J) + \exp(\beta \hbar b) + \exp(-\beta \hbar b)} + D_{\mathrm{p}} \,. \tag{3.305}$$

Without the constant D_{p}, a contradiction of our earlier result (3.293) would be obtained. D_{p} may therefore under no circumstances be neglected here. To determine D_{p}, we must finally compute the anticommutator Green's function $P_{21}^{(-)}(E)$. With

$$\left\langle [S_2^z, S_1^z]_+ \right\rangle = 2 \langle S_1^z S_2^z \rangle \,,$$

we find its equation of motion:

$$E P_{21}^{(-)}(E) = 2\hbar \langle S_1^z S_2^z \rangle + \hbar J Q^{(-)}(E) \,. \tag{3.306}$$

Because of

$$\left\langle [S_1^+ S_2^-, S_1^z]_+ \right\rangle = \langle S_2^- (S_1^+ S_1^z + S_1^z S_1^+) \rangle = 0 \,,$$

for $Q^{(-)}(E)$, analogously to (3.300), we find:

$$E Q^{(-)}(E) = -2J\hbar^3 \{ P_{11}^{(-)}(E) - P_{21}^{(-)}(E) \} \,. \tag{3.307}$$

Using

$$\left\langle [S_1^z, S_1^z]_+ \right\rangle = 2\left\langle (S_1^z)^2 \right\rangle = \frac{\hbar^2}{2} ,$$

we finally obtain, as in (3.302):

$$E P_{11}^{(-)}(E) = \frac{\hbar^3}{2} - \hbar J Q^{(-)}(E) . \tag{3.308}$$

The Eqs. (3.306), (3.307), and (3.308) form a closed system, which can readily be solved for $P_{21}^{(-)}(E)$:

$$P_{21}^{(-)}(E) = \frac{2\hbar}{E} \left\{ \frac{E^2 - 2\hbar^4 J^2}{E^2 - 4\hbar^4 J^2} \left\langle S_1^z S_2^z \right\rangle - \frac{\hbar^2}{4} \frac{2\hbar^4 J^2}{E^2 - 4\hbar^4 J^2} \right\} . \tag{3.309}$$

In contrast to the commutator Green's function (3.304), the anticommutator-Green's function thus has a first-order pole at $E = 0$. From (3.167), we therefore find:

$$D_{\mathrm{p}} = \frac{1}{2\hbar} \lim_{E \to 0} E P_{21}^{(-)}(E) = \frac{1}{2} \left\langle S_1^z S_2^z \right\rangle + \frac{\hbar^2}{8} . \tag{3.310}$$

Inserting this expression for D_{p} into (3.305), we obtain for $\left\langle S_1^z S_2^z \right\rangle$ the correct result (3.293).

3.3.4 Exercises

Exercise 3.3.1 According to (2.164), the quantised vibrations of the ionic lattice can be described in terms of a non-interacting phonon gas:

$$H = \sum_{q,r} \hbar\omega_r(q) \left(b_{qr}^+ b_{qr} + \frac{1}{2} \right) .$$

One defines the so-called one-phonon Green's function:

$$G_{qr}^{\alpha}(t, t') = \langle\langle b_{qr}(t); b_{qr}^+(t') \rangle\rangle^{\alpha} \qquad (\alpha = \text{ret, adv, c}) .$$

1. Justify the claim that for phonons, the chemical potential μ is equal to zero.
2. Compute $G_{qr}^{\text{ret, adv}}(E)$.
3. Derive the time-dependent Green's function $G_{qr}^{\text{ret, adv}}(t, t')$.
4. Calculate the internal energy U.

Exercise 3.3.2 In Ex. 2.3.5 and Ex. 2.3.6, the BCS theory of superconductivity was treated. The simplified model Hamiltonian,

$$H^* = \sum_{k,\sigma} t(\boldsymbol{k}) a_{k\sigma}^+ a_{k\sigma} - \Delta \sum_{k} \left(b_k + b_k^+ \right) + \frac{1}{V} \Delta^2 ,$$

in which $b_k^+ = a_{k\uparrow}^+ a_{-k\downarrow}^+$ is the Cooper-pair creation operator, and

$$t(\boldsymbol{k}) = \varepsilon(\boldsymbol{k}) - \mu ; \quad t(-\boldsymbol{k}) = t(\boldsymbol{k})$$

was defined, leads to the same expressions for the ground-state energy and for the coefficients u_k and v_k in the BCS *ansatz* $|\mathrm{BCS}\rangle$ (Ex. 2.3.5), if one also chooses

$$\Delta = \Delta^* = V \sum_{k} \langle b_k \rangle = V \sum_{k} \left\langle b_k^+ \right\rangle .$$

1. Calculate the excitation spectrum of the superconductor using the one-electron Green's function

$$G_{k\sigma}^{\mathrm{ret}}(E) = \left\langle\!\left\langle a_{k\sigma} ; a_{k\sigma}^+ \right\rangle\!\right\rangle_E^{\mathrm{ret}} .$$

   Show that it has an energy gap Δ.
2. Derive an equation for determining Δ using the spectral theorem for a suitably defined Green's function. Show that for $T \to 0$, it is equivalent to the gap parameter Δ_k from Ex. 2.3.6, if Δ_k is taken to be independent of $\boldsymbol{k}$.

Exercise 3.3.3

1. Show, using the model Hamiltonian H^* from Ex. 3.3.2, that the following holds for the p-fold commutator of $a_{k\sigma}$ with H^*:

$$\left[\cdots\left[\left[a_{k\sigma}, H^*\right]_-, H^*\right]_- \ldots, H^*\right]_- =$$

$$= \begin{cases} \left(t^2(\boldsymbol{k})+\Delta^2\right)^n a_{k\sigma}\,, & \text{if} \quad p=2n\,, \\ \left(t^2(\boldsymbol{k})+\Delta^2\right)^n \left(t(\boldsymbol{k})a_{k\sigma} - z_\sigma \Delta a^+_{-k-\sigma}\right)\,, & \text{if} \quad p=2n+1\,, \end{cases}$$

$$n = 0, 1, 2, \ldots$$

Compute all of the spectral moments of the one-electron spectral density using this result.

2. Choose a two-pole approach for the one-electron spectral density

$$S_{k\sigma}(E) = \sum_{i=1}^{2} a_{i\sigma}(\boldsymbol{k})\delta\left(E - E_{i\sigma}(\boldsymbol{k})\right)$$

and determine the spectral weights $\alpha_{i\sigma}(\boldsymbol{k})$ as well as the so-called quasi-particle energies $E_{i\sigma}(\boldsymbol{k})$ from the exact results for the first four spectral moments.

Exercise 3.3.4 Investigate the model Hamiltonian H^* for BCS superconductivity,

$$H^* = \sum_k H_k + \frac{\Delta^2}{V},$$

$$H_k = t(\boldsymbol{k})\left(a^+_{k\uparrow}a_{k\uparrow} + a^+_{-k\downarrow}a_{-k\downarrow}\right) - \Delta\left(a^+_{k\uparrow}a^+_{-k\downarrow} + a_{-k\downarrow}a_{k\uparrow}\right).$$

1. Find the energy eigenvalues of H_k.
2. Give the corresponding eigenstates.
3. Give the possible excitation energies of the system.

Exercise 3.3.5

1. Show that the excitations of a BCS superconductor are generated from Ex. 3.3.4 by the operators

$$\rho^+_{k\uparrow} = u_k a^+_{k\uparrow} - v_k a_{-k\downarrow} \; ; \quad \rho^+_{-k\downarrow} = u_k a^+_{-k\downarrow} + v_k a_{k\uparrow} \; .$$

The coefficients u_k and v_k are defined as in Ex. 2.3.6:

$$u_k^2 = \frac{1}{2}\left(1 + \frac{t(\boldsymbol{k})}{\left(t^2(\boldsymbol{k}) + \Delta^2\right)^{1/2}}\right) , \quad v_k^2 = 1 - u_k^2 \; .$$

2. Prove that these operators are purely Fermionic operators.
3. Compute the commutator

$$\left[H^*, \rho^+_{k\uparrow}\right]_- \; .$$

How is this result to be interpreted?

4. Formulate and solve the equation of motion of the retarded Green's function:

$$\widehat{G}^{\text{ret}}_{k\uparrow}(E) = \langle\langle \rho_{k\uparrow}; \rho^+_{k\uparrow} \rangle\rangle^{\text{ret}}_E \; .$$

3.4 The Quasi-Particle Concept

In Sect. 3.3, we discussed relatively simple, exactly solvable model systems, which strictly speaking do not require the Green's function formalism. They were intended merely to introduce the solution **techniques**. The full scope of the method becomes clear only when it is applied to the treatment of interacting systems. In most such cases, to be sure, we will then no longer be able to treat the many-body problem in a mathematically strict manner. Approximations are unavoidable and must be tolerated.

The concept of

quasi-particles

has proven to be extraordinarily useful in this connection, and we shall consider it in detail in the present section. To be more concrete, we will first keep interacting electron systems in mind. The extension to other many-body systems will cause no difficulties.

We wish to investigate which statements about interacting electron systems can be made using Green's functions. To this end, we must first define the operators (or combinations of operators) A and B which are to be used to construct the Green's function being considered. In most practical cases, the type of this function is quite unambiguously fixed by the physical problem and by the representation of the model Hamiltonian used.

3.4.1 One-Electron Green's Functions

As in (2.55), the Hamiltonian of a system of N_e mutually-interacting electrons in the Bloch representation is given by:

$$H = \sum_{k\sigma} \varepsilon(\boldsymbol{k}) a^+_{k\sigma} a_{k\sigma} + \frac{1}{2} \sum_{\substack{kpq \\ \sigma\sigma'}} v_{kp}(\boldsymbol{q}) a^+_{k+q\sigma} a^+_{p-q\sigma'} a_{p\sigma'} a_{k\sigma} \,. \tag{3.311}$$

We limit our considerations to the electrons of a single energy band, so that we can suppress the band indices. For the so-called **Bloch energies** $\varepsilon(\boldsymbol{k})$, from (2.14) and (2.21) we have:

$$\varepsilon(\boldsymbol{k}) = \int \mathrm{d}^3 r \, \psi^*_k(\boldsymbol{r}) \left[-\frac{\hbar^2}{2m} \Delta + V(\boldsymbol{r}) \right] \psi_k(\boldsymbol{r}) \,. \tag{3.312}$$

$\psi_k(\boldsymbol{r})$ is a **Bloch function** and $V(\boldsymbol{r})$ is the periodic lattice potential. We treat $\varepsilon(\boldsymbol{k})$ in the following as a given model parameter. We calculated the Coulomb matrix element in (2.54):

$$v_{kp}(\boldsymbol{q}) = \frac{e^2}{4\pi\varepsilon_0} \iint \mathrm{d}^3 r_1 \, \mathrm{d}^3 r_2 \, \frac{\psi^*_{k+q}(\boldsymbol{r}_1) \, \psi^*_{p-q}(\boldsymbol{r}_2) \, \psi_p(\boldsymbol{r}_2) \, \psi_k(\boldsymbol{r}_1)}{|\boldsymbol{r}_1 - \boldsymbol{r}_2|} \,. \tag{3.313}$$

With a constant lattice potential $V(\boldsymbol{r}) \equiv$ const, this becomes

$$v_{kp}(\boldsymbol{q}) \xrightarrow[V(\boldsymbol{r})=\text{const}]{} v_0(\boldsymbol{q}) = \frac{e^2}{\varepsilon_0 V q^2} \,. \tag{3.314}$$

Frequently, one uses the model Hamiltonian (3.311) also in its **Wannier representation** (cf. e.g. (2.115)):

$$H = \sum_{ij\sigma} T_{ij} a^+_{i\sigma} a_{j\sigma} + \frac{1}{2} \sum_{\substack{ijkl \\ \sigma\sigma'}} v(ij;kl) a^+_{i\sigma} a^+_{j\sigma'} a_{l\sigma'} a_{k\sigma} \,. \tag{3.315}$$

The so-called *hopping integrals*

$$T_{ij} = \int \mathrm{d}^3 r \, \omega^*(\boldsymbol{r} - \boldsymbol{R}_i) \left\{ -\frac{\hbar^2}{2m} \Delta + V(\boldsymbol{r}) \right\} \omega(\boldsymbol{r} - \boldsymbol{R}_j) \tag{3.316}$$

are related to the Bloch energies via a Fourier transform $\varepsilon(\boldsymbol{k})$ (cf. (2.113)). $\omega(\boldsymbol{r} - \boldsymbol{R}_i)$ is the Wannier function centered at $\boldsymbol{R}_i$.

$$\begin{aligned} v(ij;kl) = &\frac{e^2}{4\pi\varepsilon_0} \iint \mathrm{d}^3 r_1 \, \mathrm{d}^3 r_2 \, \omega^*(\boldsymbol{r}_1 - \boldsymbol{R}_i) \, \omega^*(\boldsymbol{r}_2 - \boldsymbol{R}_j) \, \cdot \\ &\cdot \frac{1}{|\boldsymbol{r}_1 - \boldsymbol{r}_2|} \omega(\boldsymbol{r}_2 - \boldsymbol{R}_l) \, \omega(\boldsymbol{r}_1 - \boldsymbol{R}_k) \,. \end{aligned} \tag{3.317}$$

In this section, we show that the **one-electron Green's function**, already introduced in Eq. (3.193),

$$G^{\alpha}_{k\sigma}(E) \equiv \langle\langle a_{k\sigma}; a^{+}_{k\sigma} \rangle\rangle^{\alpha}_{E}\,, \tag{3.318}$$

$$G^{\alpha}_{ij\sigma}(E) \equiv \langle\langle a_{i\sigma}; a^{+}_{j\sigma} \rangle\rangle^{\alpha}_{E}\,, \tag{3.319}$$

$$\alpha = \text{ret, adv, c} \quad (\varepsilon = -)$$

or the equivalent **one-electron spectral density**

$$S_{k\sigma}(E) = \frac{1}{2\pi}\int\limits_{-\infty}^{+\infty} \mathrm{d}(t-t')\,\exp\left(-\frac{\mathrm{i}}{\hbar}E\,(t-t')\right)\left\langle\left[a_{k\sigma}(t), a^{+}_{k\sigma}(t')\right]_{+}\right\rangle, \tag{3.320}$$

$$S_{ij\sigma}(E) = \frac{1}{2\pi}\int\limits_{-\infty}^{+\infty} \mathrm{d}(t-t')\,\exp\left(-\frac{\mathrm{i}}{\hbar}E\,(t-t')\right)\left\langle\left[a_{i\sigma}(t), a^{+}_{j\sigma}(t')\right]_{+}\right\rangle \tag{3.321}$$

determine the entire equilibrium thermodynamics, even for systems of interacting electrons. This of course presumes that one was able in some manner to compute these functions.

To show this, we first write the equation of motion of the $\boldsymbol{k}$-dependent Green's function, for which we require the following commutator:

$$\begin{aligned}[a_{k\sigma}, \mathcal{H}]_{-} &= (\varepsilon(\boldsymbol{k}) - \mu)\, a_{k\sigma} + \\ &\quad + \sum_{\substack{p,q\\ \sigma'}} v_{p,\,k+q}(\boldsymbol{q})\, a^{+}_{p+q\sigma'} a_{p\sigma'} a_{k+q\sigma}\end{aligned} \tag{3.322}$$

(Its derivation can be carried out by the reader as an exercise!). With the *higher-order* Green's function

$$^{\alpha}\Gamma^{\sigma'\sigma}_{pk;\,q}(E) \equiv \langle\langle a^{+}_{p+q\sigma'} a_{p\sigma'} a_{k+q\sigma}; a^{+}_{k\sigma} \rangle\rangle^{\alpha}_{E} \tag{3.323}$$

the equation of motion is then given by:

$$(E - \varepsilon(\boldsymbol{k}) + \mu) G^{\alpha}_{k\sigma}(E) = \hbar + \sum_{p,\,q,\,\sigma'} v_{p,\,k+q}(\boldsymbol{q})\, {}^{\alpha}\Gamma^{\sigma'\sigma}_{pk;\,q}(E)\,. \tag{3.324}$$

The unknown function Γ on the right-hand side prevents a direct solution of this equation. However, we postulate that the following decomposition is permitted:

$$\begin{aligned}\langle\langle [a_{k\sigma}, \mathcal{H} - \mathcal{H}_0]_{-}; a^{+}_{k\sigma} \rangle\rangle^{\alpha}_{E} &= \sum_{p,\,q,\,\sigma'} v_{p,\,k+q}(\boldsymbol{q})\, \Gamma^{\sigma'\sigma}_{pk;\,q}(E) \equiv \\ &\equiv \Sigma^{\alpha}_{\sigma}(\boldsymbol{k}, E) G^{\alpha}_{k\sigma}(E)\,.\end{aligned} \tag{3.325}$$

This equation defines the so-called

self-energy, $\Sigma_\sigma^\alpha(k,E)$,

which we wish to discuss in more detail. With it, the equation of motion (3.324) can be formally solved in a simple fashion:

$$G_{k\sigma}^{\alpha}(E) = \frac{\hbar}{E - \left(\varepsilon(k) - \mu + \Sigma_\sigma^\alpha(k, E)\right)} \,. \tag{3.326}$$

Comparing this expression with that for the non-interacting system (3.197), we see that the entire influence of the particle interactions is contained in the self-energy $\Sigma_\sigma(k, E)$. As a rule, this is a complex-valued function of (k, E), whose real part determines the energy and whose imaginary part the lifetime of the **quasi-particles**, which we have yet to define.

We can reformulate Eq. (3.326) to some extent. We denote the one-electron Green's function of the non-interacting electrons by $G_{k\sigma}^{(0)}(E)$; then from (3.326), it follows that:

$$G_{k\sigma}(E) = \hbar \left\{ \hbar \left[G_{k\sigma}^{(0)}(E) \right]^{-1} - \Sigma_\sigma(k, E) \right\}^{-1}$$

$$\Rightarrow \quad \left\{ \left[G_{k\sigma}^{(0)}(E) \right]^{-1} - \frac{1}{\hbar} \Sigma_\sigma(k, E) \right\} G_{k\sigma}(E) = 1 \,.$$

Here, we have suppressed the index α to make the expressions clearer. Finally, we arrive at the so-called

Dyson equation

$$G_{k\sigma}(E) = G_{k\sigma}^{(0)}(E) + \frac{1}{\hbar} G_{k\sigma}^{(0)}(E) \Sigma_\sigma(k, E) G_{k\sigma}(E) \,. \tag{3.327}$$

Our goal will be to obtain at least an approximate determination of the self-energy. Inserting an approximate expression for $\Sigma_\sigma(k, E)$ into the Dyson equation already implies summation over an *infinite* partial series. We recall however that for the derivation of (3.327), we had to postulate the decomposition (3.325) of the *higher-order* Green's function.

3.4.2 The Electronic Self-Energy

In this section, we wish to obtain an overview of the general structures of the fundamental quantities self-energy, Green's functions, and spectral density. Our starting point is the representation (3.326) of the single-particle Green's function, whereby in the case of the self-energy, we are dealing in general with a complex quantity:

$$\Sigma_\sigma^\alpha(\boldsymbol{k},E) = R_\sigma^\alpha(\boldsymbol{k},E) + \mathrm{i}I_\sigma^\alpha(\boldsymbol{k},E)\,. \tag{3.328}$$

The index α stands for *retarded, advanced*, or *causal*. The corresponding self-energies are quite distinct. Thus, for example, according to (3.186) and (3.187) for a *real* spectral density, we have:

$$\left(G_{\boldsymbol{k}\sigma}^{\mathrm{adv}}(E)\right)^* = G_{\boldsymbol{k}\sigma}^{\mathrm{ret}}(E)\,.$$

This implies

$$\left(\Sigma_\sigma^{\mathrm{adv}}(\boldsymbol{k},E)\right)^* = \Sigma_\sigma^{\mathrm{ret}}(\boldsymbol{k},E)\,. \tag{3.329}$$

The relationship is thus simple. We may limit our considerations without loss of generality to the retarded functions. We then leave off the supplement $+\mathrm{i}0^+$, if $I_\sigma \neq 0$. We omit the index "ret" on the self-energy in the following.

First, we rearrange Eq. (3.326) slightly:

$$G_{\boldsymbol{k}\sigma}^{\mathrm{ret}}(E) = \hbar\frac{\left\{E-(\varepsilon(\boldsymbol{k})-\mu+R_\sigma(\boldsymbol{k},E))\right\}+\mathrm{i}I_\sigma(\boldsymbol{k},E)}{\left\{E-(\varepsilon(\boldsymbol{k})-\mu+R_\sigma(\boldsymbol{k},E))\right\}^2+I_\sigma^2(\boldsymbol{k},E)}\,. \tag{3.330}$$

From (3.154), we then have for the spectral density:

$$S_{\boldsymbol{k}\sigma}(E) = -\frac{\hbar}{\pi}\frac{I_\sigma(\boldsymbol{k},E)}{\left\{E-(\varepsilon(\boldsymbol{k})-\mu+R_\sigma(\boldsymbol{k},E))\right\}^2+I_\sigma^2(\boldsymbol{k},E)}\,. \tag{3.331}$$

With (3.146), we could have given the general spectral representation of the spectral density, which –for the case of the one-electron spectral density which interests us here– becomes:

$$\begin{aligned}S_{\boldsymbol{k}\sigma}(E) = \frac{\hbar}{\Xi}\sum_{n,m}\left|\langle E_n\,|\,a_{\boldsymbol{k}\sigma}^+\,|\,E_m\rangle\right|^2\mathrm{e}^{-\beta E_n}\left(\mathrm{e}^{\beta E}+1\right)\,\cdot\\ \cdot\,\delta\left[E-(E_n-E_m)\right]\,.\end{aligned} \tag{3.332}$$

$S_{\boldsymbol{k}\sigma}(E)$ is thus non-negative for all $(\boldsymbol{k},\sigma,E)$. This, however, according to (3.331), gives

$$I_\sigma(\boldsymbol{k},E) \le 0\,. \tag{3.333}$$

for the imaginary part of the self-energy (retarded!). We now want to investigate the expression (3.331) somewhat more precisely. Without explicit knowledge of $R_\sigma(\boldsymbol{k},E)$ and $I_\sigma(\boldsymbol{k},E)$, we nevertheless expect in the usual case to find more or less prominent maxima in the spectral density at the **resonance energies** $E_{i\sigma}(\boldsymbol{k})$ defined by

$$E_{i\sigma}(\boldsymbol{k}) \stackrel{!}{=} \varepsilon(\boldsymbol{k})-\mu+R_\sigma\left(\boldsymbol{k},E_{i\sigma}(\boldsymbol{k})\right)\;;\quad i=1,2,3,\ldots \tag{3.334}$$

Here, we must distinguish between two cases.

Case A: Let

$$I_\sigma(\boldsymbol{k}, E) \equiv 0 \tag{3.335}$$

hold in a certain energy range, which contains the *resonance* $E_{i\sigma}$. Then, in (3.331), we must go to the limit $I_\sigma \to -0^+$. Using the formulation of the δ-function as a limiting expression,

$$\delta(E - E_0) = \frac{1}{\pi} \lim_{x \to 0} \frac{x}{(E - E_0)^2 + x^2} , \tag{3.336}$$

we then find:

$$S_{\boldsymbol{k}\sigma}(E) = \hbar\delta\left[E - (\varepsilon(\boldsymbol{k}) - \mu + R_\sigma(\boldsymbol{k}, E))\right] . \tag{3.337}$$

We then make use of

$$\delta\left[f(x)\right] = \sum_i \frac{1}{\left|f'(x_i)\right|} \delta(x - x_i) \; ; \quad f(x_i) = 0 , \tag{3.338}$$

which allows us to write the following expression in place of (3.337):

$$S_{\boldsymbol{k}\sigma}(E) = \hbar \sum_{i=1}^{n} \alpha_{i\sigma}(\boldsymbol{k}) \delta(E - E_{i\sigma}(\boldsymbol{k})) , \tag{3.339}$$

$$\alpha_{i\sigma}(\boldsymbol{k}) = \left|1 - \frac{\partial}{\partial E} R_\sigma(\boldsymbol{k}, E)\right|^{-1}_{E = E_{i\sigma}} . \tag{3.340}$$

The sums extend over those resonances $E_{i\sigma}$ which lie within the energy range where (3.335) is valid.

Case B: We assume

$$I_\sigma(\boldsymbol{k}, E) \neq 0 , \tag{3.341}$$

where, however,

$$\left|I_\sigma(\boldsymbol{k}, E)\right| \ll \left|\varepsilon(\boldsymbol{k}) - \mu + R_\sigma(\boldsymbol{k}, E)\right| \tag{3.342}$$

is valid within a certain neighbourhood of the resonance $E_{i\sigma}$. Then we can expect a prominent maximum at the energy $E = E_{i\sigma}$. To see this, we expand the expression

$$F_\sigma(\boldsymbol{k}, E) \equiv \varepsilon(\boldsymbol{k}) - \mu + R_\sigma(\boldsymbol{k}, E)$$

around the resonance position and terminate the series after the linear term:

$$F_\sigma(\boldsymbol{k},E) = F_\sigma(\boldsymbol{k},E_{i\sigma}) + (E-E_{i\sigma})\left.\frac{\partial F_\sigma}{\partial E}\right|_{E=E_{i\sigma}} + \cdots =$$
$$= E_{i\sigma}(\boldsymbol{k}) + (E-E_{i\sigma})\left.\frac{\partial R_\sigma}{\partial E}\right|_{E=E_{i\sigma}} + \cdots$$

This means that:

$$\begin{aligned}(E-\varepsilon(\boldsymbol{k})+\mu-R_\sigma(\boldsymbol{k},E))^2 &\simeq (E-E_{i\sigma})^2\left(1-\left.\frac{\partial R_\sigma}{\partial E}\right|_{E=E_{i\sigma}}\right)^2 = \\ &= \alpha_{i\sigma}^{-2}(\boldsymbol{k})(E-E_{i\sigma}(\boldsymbol{k}))^2 \ .\end{aligned} \tag{3.343}$$

We insert this expression into (3.331). Let us now further assume that $I_\sigma(\boldsymbol{k},E)$ in the neighbourhood of the resonance $E_{i\sigma}$ is a continuous, only weakly varying function of E, so that to a good approximation in the energy region of interest, we can set

$$I_\sigma(\boldsymbol{k},E) \approx I_\sigma(\boldsymbol{k},E_{i\sigma}(\boldsymbol{k})) \equiv I_{i\sigma}(\boldsymbol{k}) \ ; \tag{3.344}$$

then the spectral density can be approximated as follows:

$$S_{\boldsymbol{k}\sigma}^{(i)}(E) \approx -\frac{\hbar}{\pi}\frac{\alpha_{i\sigma}^2(\boldsymbol{k})I_{i\sigma}(\boldsymbol{k})}{(E-E_{i\sigma}(\boldsymbol{k}))^2+(\alpha_{i\sigma}(\boldsymbol{k})I_{i\sigma}(\boldsymbol{k}))^2} \ . \tag{3.345}$$

Under the given assumptions, the spectral density has a Lorentz shape in the neighbourhood of the resonance $E_{i\sigma}$. Note however, that for this to hold, in particular (3.342) must be fulfilled. This condition can however unfortunately be verified only after a complete solution of the problem is at hand. It thus remains speculative for the time being, but is satisfactorily confirmed by a number of examples of concrete applications. However, we shall also encounter systems for which the shape of the spectral density is found to be quite different from the *Lorentz form*, i.e. for which (3.342) is **not** obeyed.

As a rule, however, according to our preliminary considerations, the spectral density will be a linear combination of weighted Lorentz and δ-functions.

This structure of the spectral density will lead us in the next section to the concept of the *quasi-particle*. In this connection, it is interesting to examine the behaviour of the time-dependent spectral density, which one sometimes calls the **propagator**. (Occasionally, this

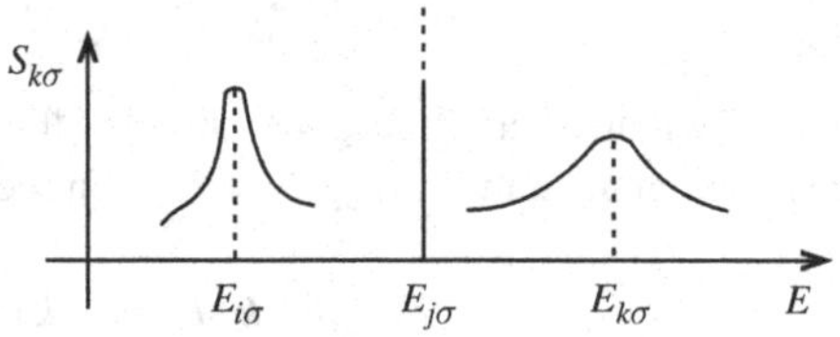

Fig. 3.4 The qualitative shape of the spectral density for a system of interacting Fermions

term is also used for the time-dependent Green's function.) In the case of a non-interacting particle system, it represents a non-damped harmonic oscillation, as one can see from Eq. (3.207). This is true in the system of interacting particles only for case A, when the spectral density can be written as a δ-function (3.337). From (3.339), it then namely follows that:

$$S_{k\sigma}\left(t-t'\right) = \frac{1}{2\pi}\sum_{i=1}^{n}\alpha_{i\sigma}(\boldsymbol{k})\exp\left(-\frac{\mathrm{i}}{\hbar}E_{i\sigma}(\boldsymbol{k})\left(t-t'\right)\right). \tag{3.346}$$

The frequencies of oscillation are then determined by the resonance energies $E_{i\sigma}(\boldsymbol{k})$.

For case B, in contrast, the Lorentz peaks require exponentially damped oscillators. To see this, we assume for the moment that (3.345) holds to a good approximation over the entire energy range. Then we can write:

$$\begin{aligned} S_{k\sigma}^{(i)}\left(t-t'\right) \approx \frac{1}{4\pi^2\mathrm{i}}\int_{-\infty}^{+\infty}\mathrm{d}E\,\exp\left(-\frac{\mathrm{i}}{\hbar}E\left(t-t'\right)\right)\alpha_{i\sigma}(\boldsymbol{k})\,\cdot \\ \cdot\left\{\frac{1}{E-(E_{i\sigma}(\boldsymbol{k})-\mathrm{i}\alpha_{i\sigma}(\boldsymbol{k})I_{i\sigma}(\boldsymbol{k}))} - \right. \\ \left. -\frac{1}{E-(E_{i\sigma}(\boldsymbol{k})+\mathrm{i}\alpha_{i\sigma}(\boldsymbol{k})I_{i\sigma}(\boldsymbol{k}))}\right\}. \end{aligned} \tag{3.347}$$

We solve the integrals using the residual theorem. The spectral weights are positive definite, so that due to (3.333), we must have:

$$\alpha_{i\sigma}(\boldsymbol{k})I_{i\sigma}(\boldsymbol{k}) \le 0\,. \tag{3.348}$$

The first term thus has a pole in the upper half-plane, and the second term has a pole in the lower half-plane. If we choose the following integration paths,

$$t-t'>0:\ \int_{-\infty}^{+\infty}\mathrm{d}E\,\ldots \implies \int_{\smile}\mathrm{d}E\,\ldots\,,$$

$$t-t'<0:\ \int_{-\infty}^{+\infty}\mathrm{d}E\,\ldots \implies \int_{\frown}\mathrm{d}E\,\ldots\,,$$

then the exponential function in (3.347) guarantees that the semicircle which is closed at infinity makes no contribution. According to the residual theorem, for $t-t'>0$ then only the second term makes a contribution to the integral in (3.347), whilst for $t-t'<0$, only the first term contributes. This yields finally:

$$S_{k\sigma}^{(i)}\left(t-t'\right) \approx \frac{1}{2\pi}\alpha_{i\sigma}(\boldsymbol{k})\exp\left(-\frac{\mathrm{i}}{\hbar}E_{i\sigma}(\boldsymbol{k})\left(t-t'\right)\right)\cdot \\ \cdot\exp\left(-\frac{1}{\hbar}\left|\alpha_{i\sigma}(\boldsymbol{k})I_{i\sigma}(\boldsymbol{k})\right|\left|t-t'\right|\right). \tag{3.349}$$

The time-dependent spectral density indeed represents a damped oscillation, whose frequency again corresponds to a resonance $E_{i\sigma}$, whilst the damping is essentially determined by the imaginary part of the self-energy.

Thus, in general for interacting systems, we can expect that $S_{k\sigma}\left(t-t'\right)$ will consist of a superposition of damped and undamped oscillations, whose frequencies correspond to the resonance energies $E_{i\sigma}$. The resulting time dependence can then become quite complicated.

Precisely this qualitative picture of the time dependence of the spectral density will lead us in the next section to the concept of the *quasi-particle*, which is typical of many-body theory.

3.4.3 Quasi-Particles

In this section, we will draw some preliminary conclusions. What is in fact the *new* aspect of the Green's function formalism as compared to conventional methods? What do Green's functions or spectral densities have to do with quasi-particles? What are quasi-particles after all? We presume that they are related to the more or less prominent resonance peaks in the spectral density which we discussed in the last section. We want to clarify this point qualitatively by considering a **special case**

$$T=0\,;\quad |\boldsymbol{k}|>k_{\mathrm{F}}\,;\quad t>t'\,.$$

k_{F} denotes the Fermi wavevector. We presume the system to be in its ground state $\left|E_0\right\rangle$. Upon adding a $(\boldsymbol{k},\sigma)$ electron at the time t, the state

$$\left|\varphi_0(t)\right\rangle = a_{k\sigma}^{+}(t)\left|E_0\right\rangle \tag{3.350}$$

is created, which must not necessarily be an eigenstate of the Hamiltonian. Owing to $|\boldsymbol{k}|>k_{\mathrm{F}}$, only one of the two terms in the definition of the propagator (3.127) $S_{k\sigma}\left(t-t'\right)$ can contribute. We therefore have

$$2\pi\, S_{k\sigma}\left(t-t'\right) = \left\langle\varphi_0(t)\,\middle|\,\varphi_0\left(t'\right)\right\rangle . \tag{3.351}$$

With this, the time-dependent spectral density acquires a clear interpretation.

$2\pi\, S_{k\sigma}\left(t-t'\right)$ is the probability amplitude that the state $\left|\varphi_0\right\rangle$, which was formed from the ground state $\left|E_0\right\rangle$ by adding a $(\boldsymbol{k},\sigma)$ electron at the time t', still exists at a time $t>t'$. $S_{k\sigma}\left(t-t'\right)$ thus characterises the time evolution (**propagation**) of an additional $(\boldsymbol{k},\sigma)$

electron in the N-particle system. If we had presupposed $|\boldsymbol{k}| < k_{\mathrm{F}}$, then $S_{\boldsymbol{k}\sigma}(t-t')$ would describe the propagation of a *hole*.

We must now distinguish between two typical cases:

$$\left|\langle\varphi_0(t)\,|\,\varphi_0(t')\rangle\right|^2 = \text{const} \iff \textbf{stationary}\text{ state,}$$
$$\left|\langle\varphi_0(t)\,|\,\varphi_0(t')\rangle\right|^2 \xrightarrow[t-t'\to\infty]{} 0 \iff \text{state with a }\textbf{finite lifetime.}$$

First, we again consider

1) Non-interacting electrons,
described by

$$\mathcal{H}_0 = \sum_{\boldsymbol{k},\sigma} (\varepsilon(\boldsymbol{k}) - \mu)\, a^+_{\boldsymbol{k}\sigma} a_{\boldsymbol{k}\sigma} .$$

One readily obtains

$$\left[\mathcal{H}_0, a^+_{\boldsymbol{k}\sigma}\right]_- = (\varepsilon(\boldsymbol{k}) - \mu) a^+_{\boldsymbol{k}\sigma} ,$$

with which we find

$$\mathcal{H}_0\left(a^+_{\boldsymbol{k}\sigma}\,|\,E_0\rangle\right) = a^+_{\boldsymbol{k}\sigma}\mathcal{H}_0\,|\,E_0\rangle + \left[\mathcal{H}_0, a^+_{\boldsymbol{k}\sigma}\right]_-\,|\,E_0\rangle =$$
$$= (E_0 + \varepsilon(\boldsymbol{k}) - \mu)\left(a^+_{\boldsymbol{k}\sigma}\,|\,E_0\rangle\right) .$$

In this special case, we thus have $a^+_{\boldsymbol{k}\sigma}\,|\,E_0\rangle$ which is an eigenstate of $\mathcal{H}_0$. Continuing, we obtain:

$$\begin{aligned}|\varphi_0(t)\rangle &= \exp\left(\frac{\mathrm{i}}{\hbar}\mathcal{H}_0 t\right) a^+_{\boldsymbol{k}\sigma} \exp\left(-\frac{\mathrm{i}}{\hbar}\mathcal{H}_0 t\right)|\,E_0\rangle = \\ &= \exp\left(-\frac{\mathrm{i}}{\hbar}E_0 t\right)\exp\left(\frac{\mathrm{i}}{\hbar}\mathcal{H}_0 t\right)\left(a^+_{\boldsymbol{k}\sigma}\,|\,E_0\rangle\right) = \\ &= \exp\left(\frac{\mathrm{i}}{\hbar}(\varepsilon(\boldsymbol{k}) - \mu)\,t\right)\left(a^+_{\boldsymbol{k}\sigma}\,|\,E_0\rangle\right) .\end{aligned} \tag{3.352}$$

Due to $|\boldsymbol{k}| > k_{\mathrm{F}}$ and $\langle E_0\,|\,E_0\rangle = 1$, we also have:

$$\langle E_0\,|\,a_{\boldsymbol{k}\sigma}a^+_{\boldsymbol{k}\sigma}\,|\,E_0\rangle = \langle E_0\,|\,E_0\rangle - \langle E_0\,|\,a^+_{\boldsymbol{k}\sigma}a_{\boldsymbol{k}\sigma}\,|\,E_0\rangle = 1 .$$

We then finally obtain:

$$\langle\varphi_0(t)\,|\,\varphi_0(t')\rangle = \exp\left[-\frac{\mathrm{i}}{\hbar}(\varepsilon(\boldsymbol{k}) - \mu)\,(t-t')\right] . \tag{3.353}$$

The propagator $S_{k\sigma}^{(0)}(t-t')$ thus represents an undamped harmonic oscillation, as we have already seen by another method. Its frequency corresponds to an exact excitation energy of the system, namely $(\varepsilon(\boldsymbol{k})-\mu)$. Because of

$$\left|\left\langle\varphi_0(t)\,\middle|\,\varphi_0\left(t'\right)\right\rangle\right|^2=1\,, \tag{3.354}$$

this is a **stationary** state (see Fig. 3.5).

2) Interacting electron systems
For the propagator $S_{k\sigma}(t-t')$, we obtain from (3.351) by inserting a complete set of eigenstates $|E_n\rangle$ between the two time-dependent creation and annihilation operators:

$$2\pi\,S_{k\sigma}\left(t-t'\right)=\sum_n\left|\left\langle E_n\,\middle|\,a_{k\sigma}^+\,\middle|\,E_0\right\rangle\right|^2\exp\left(-\frac{\mathrm{i}}{\hbar}\left(E_n-E_0\right)\left(t-t'\right)\right)\,. \tag{3.355}$$

In the *free* system, $a_{k\sigma}^+|E_0\rangle$ is an energy eigenstate. Its orthogonality then guarantees that only one term in the sum is nonzero. This no longer holds for an interacting system. In the series expansion

$$|\varphi_0\rangle=a_{k\sigma}^+\,|E_0\rangle=\sum_m c_m\,|E_m\rangle\,, \tag{3.356}$$

several, in general infinitely many expansion coefficients are nonzero. Each summand to be sure still represents a harmonic oscillation; but the superposition of a number of oscillations with different frequencies will guarantee that the sum in (3.355) exhibits a maximum at $t=t'$. For $t-t'>0$, the phase factors $\exp[-(\mathrm{i}/\hbar)(E_n-E_0)(t-t')]$ will gradually distribute themselves over the entire unit circle of the complex plane and thus possibly give rise to

$$\left|\left\langle\varphi_0(t)\,\middle|\,\varphi_0\left(t'\right)\right\rangle\right|^2\xrightarrow[t-t'\to\infty]{}0 \tag{3.357}$$

as a result of destructive interference. The state $|\varphi_0(t')\rangle$ which is formed at the time t' now has a finite lifetime.

Under certain circumstances, it is however possible to represent the erratic time dependence of the propagators as a superposition of damped oscillations with well-defined frequencies:

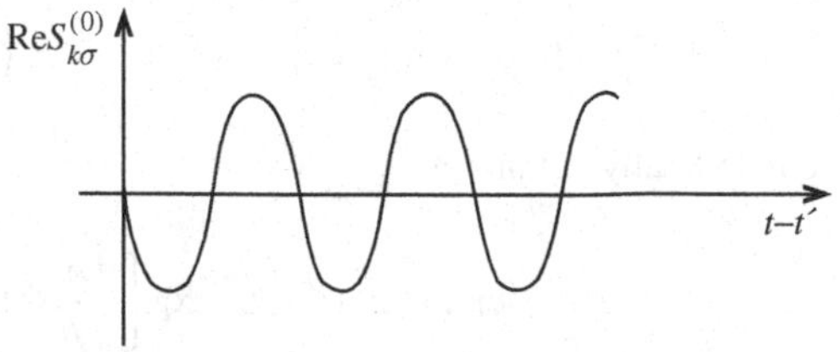

Fig. 3.5 The manifestation of a stationary state in the time-dependent spectral density, in the form of an undamped harmonic oscillation

$$2\pi S_{k\sigma}\left(t-t'\right) = \sum_i \alpha_{i\sigma}(\boldsymbol{k}) \exp\left[-\frac{\mathrm{i}}{\hbar}\left(\eta_{i\sigma}^{\mathrm{QP}}(\boldsymbol{k})\right)\left(t-t'\right)\right]. \tag{3.358}$$

This *ansatz* has formally the same structure as the corresponding expression (3.353) for the *free* system, except that now the *new* single-particle energies are in general complex quantities:

$$\eta_{i\sigma}^{\mathrm{QP}}(\boldsymbol{k}) = \mathrm{Re}\,\eta_{i\sigma}^{\mathrm{QP}}(\boldsymbol{k}) + \mathrm{i}\,\mathrm{Im}\,\eta_{i\sigma}^{\mathrm{QP}}(\boldsymbol{k}). \tag{3.359}$$

The imaginary part $\left(\mathrm{Im}\,\eta_{i\sigma}^{\mathrm{QP}} < 0\right)$ is responsible for the exponential damping of the oscillations. We ascribe these new energies $\eta_{i\sigma}^{\mathrm{QP}}$ to a *fictitious* particle, which we call a

quasi-particle.

It is, roughly speaking, as if the $(\boldsymbol{k},\sigma)$ electron, *implanted* into the N-particle system at time t' as its $(N+1)$-th particle, decays into several quasi-particles, whose energies are determined by the real parts and whose lifetimes by the imaginary parts of $\eta_{i\sigma}^{\mathrm{QP}}$:

$$\begin{aligned} &\text{Quasi-particle energy} \qquad \mathrm{Re}\,\eta_{i\sigma}^{\mathrm{QP}}(\boldsymbol{k}), \\ &\text{Quasi-particle lifetime} \qquad \frac{\hbar}{\left|\mathrm{Im}\,\eta_{i\sigma}^{\mathrm{QP}}(\boldsymbol{k})\right|}. \end{aligned} \tag{3.360}$$

Every quasi-particle is associated with a

spectral weight $\alpha_{i\sigma}(\boldsymbol{k})$,

whereby, due to conservation of the overall number of particles,

$$\sum_i \alpha_{i\sigma}(\boldsymbol{k}) = 1 \tag{3.361}$$

must be fulfilled. Now comparing

$$\begin{aligned} S_{k\sigma}^{(i)}\left(t-t'\right) = \frac{1}{2\pi}\alpha_{i\sigma}(\boldsymbol{k}) \exp\left[-\frac{\mathrm{i}}{\hbar}\left(\mathrm{Re}\,\eta_{i\sigma}^{\mathrm{QP}}(\boldsymbol{k})\right)\left(t-t'\right)\right] \cdot \\ \cdot \exp\left(-\frac{1}{\hbar}\left|\mathrm{Im}\,\eta_{i\sigma}^{\mathrm{QP}}(\boldsymbol{k})\right|\left(t-t'\right)\right) \end{aligned} \tag{3.362}$$

with (3.349), we can see the connection of the quasi-particle properties with the electronic self-energy:

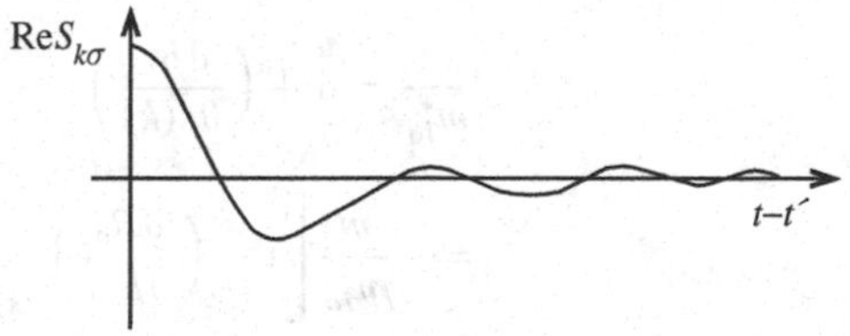

Fig. 3.6 The typical time evolution of the spectral density for the case of a system of interacting particles

Quasi-particle energy: $E_{i\sigma}(\boldsymbol{k})$

$$E_{i\sigma}(\boldsymbol{k}) \overset{!}{=} \varepsilon(\boldsymbol{k}) - \mu + R_\sigma\left(\boldsymbol{k}, E = E_{i\sigma}(\boldsymbol{k})\right) , \tag{3.363}$$

Quasi-particle lifetime: $\tau_{i\sigma}(\boldsymbol{k})$

$$\tau_{i\sigma}(\boldsymbol{k}) = \frac{\hbar}{\left|\alpha_{i\sigma}(\boldsymbol{k}) \cdot I_{i\sigma}(\boldsymbol{k})\right|} . \tag{3.364}$$

The spectral weights $\alpha_{i\sigma}(\boldsymbol{k})$ are, according to (3.340), determined by the real part of the self-energy. Thus the lifetime of the quasi-particles is not influenced solely by the imaginary part, but also by the real part of $\Sigma_\sigma(\boldsymbol{k}, E)$. However, for $I_{i\sigma}(\boldsymbol{k}) = 0$, in every case $\tau_{i\sigma} = \infty$ holds. The Lorentz peaks in the spectral density $S_{\boldsymbol{k}\sigma}(E)$ can also be associated with quasi-particles whose energies are given by the positions and whose lifetimes by the widths of the peaks. δ-functions (3.339) are then special cases, corresponding to quasi-particles with infinitely long lifetimes.

We can finally profit from the analogy with the free system to define an

effective mass $m^*_{i\sigma}(\boldsymbol{k})$

of the quasi-particles. For small wavenumbers, the Bloch energies can always be expanded as follows:

$$\varepsilon(\boldsymbol{k}) = T_0 + \frac{\hbar^2 k^2}{2m} + O\left(k^4\right) . \tag{3.365}$$

T_0 is the lower edge of the respective energy band. Formally, we postulate the same approach as for the quasi-particle energies:

$$E_{i\sigma}(\boldsymbol{k}) = T_{0i\sigma} + \frac{\hbar^2 k^2}{2m^*_{i\sigma}} + O\left(k^4\right) . \tag{3.366}$$

We insert (3.365):

$$E_{i\sigma}(\boldsymbol{k}) = T_{0i\sigma} + \frac{m}{m^*_{i\sigma}}\left(\varepsilon(\boldsymbol{k}) - T_0\right) + \cdots ,$$

and thus obtain:

$$\frac{m}{m^*_{i\sigma}(\boldsymbol{k})} = \frac{\partial E_{i\sigma}(\boldsymbol{k})}{\partial \varepsilon(\boldsymbol{k})} . \tag{3.367}$$

However, from (3.363), this implies that:

$$\frac{m}{m^*_{i\sigma}} = 1 + \left(\frac{\partial R_\sigma}{\partial \varepsilon(\boldsymbol{k})}\right)_{E_{i\sigma}} + \left(\frac{\partial R_\sigma}{\partial E_{i\sigma}}\right)_{\varepsilon(\boldsymbol{k})} \frac{\partial E_{i\sigma}}{\partial \varepsilon(\boldsymbol{k})}$$

$$\Longrightarrow \quad \frac{m}{m^*_{i\sigma}}\left[1 - \left(\frac{\partial R_\sigma}{\partial E_{i\sigma}}\right)_{\varepsilon(\boldsymbol{k})}\right] = 1 + \left(\frac{\partial R_\sigma}{\partial \varepsilon(\boldsymbol{k})}\right)_{E_{i\sigma}} .$$

The real part of the electronic self-energy thus determines the effective mass of the quasi-particles:

$$\frac{m^*_{i\sigma}(\boldsymbol{k})}{m} = \frac{1 - \left(\dfrac{\partial R_\sigma(\boldsymbol{k},\, E_{i\sigma})}{\partial E_{i\sigma}}\right)_{\varepsilon(\boldsymbol{k})}}{1 + \left(\dfrac{\partial R_\sigma(\boldsymbol{k},\, E_{i\sigma})}{\partial \varepsilon(\boldsymbol{k})}\right)_{E_{i\sigma}}} \,. \tag{3.368}$$

We will encounter another important property of the quasi-particles in the following section.

3.4.4 Quasi-Particle Density of States

We will again treat this property in a strict analogy to the free electron gas. For the **average occupation number** $\langle n_{\boldsymbol{k}\sigma}\rangle$ of the $(\boldsymbol{k}, \sigma)$ level, we find using the spectral theorem (3.157) from the one-electron spectral density:

$$\langle n_{\boldsymbol{k}\sigma}\rangle = \left\langle a^+_{\boldsymbol{k}\sigma} a_{\boldsymbol{k}\sigma}\right\rangle = \frac{1}{\hbar}\int\limits_{-\infty}^{+\infty} \mathrm{d}E\, f_-(E) S_{\boldsymbol{k}\sigma}(E-\mu) \,. \tag{3.369}$$

In the non-interacting system, this is the same as (3.208), if we substitute (3.199) for the spectral density. By summing over all wavenumbers $\boldsymbol{k}$ and all spins σ, we can find from $\langle n_{\boldsymbol{k}\sigma}\rangle$ the total number of electrons N_e:

$$N_e = \sum_{\boldsymbol{k}\sigma} \langle n_{\boldsymbol{k}\sigma}\rangle = \frac{1}{\hbar}\sum_{\boldsymbol{k}\sigma}\int\limits_{-\infty}^{+\infty} \mathrm{d}E\, f_-(E) S_{\boldsymbol{k}\sigma}(E-\mu) \,. \tag{3.370}$$

As in the *free* system, N_e must also be accessible through integrating the density of states $\rho_\sigma(E)$ of the interacting system over energy:

$$N_e = N\sum_{\sigma}\int\limits_{-\infty}^{+\infty} \mathrm{d}E\, f_-(E)\rho_\sigma(E) \,. \tag{3.371}$$

N is here the number of lattice sites, and $\rho_\sigma(E)$ is evidently normalised to 1. Comparison with (3.370) then yields the

quasi-particle density of states

$$\rho_\sigma(E) = \frac{1}{N\hbar}\sum_{\boldsymbol{k}} S_{\boldsymbol{k}\sigma}(E-\mu) \,. \tag{3.372}$$

We have already obtained the same expression in Sect. 3.1.1 for non-interacting Bloch electrons. There is thus a close connection between the density of states and the spectral density. All the properties of the spectral density are transferred rather directly to the quasi-particle density of states. If, for example, we insert (3.332) into (3.372), we can see that $\rho_\sigma(E)$, in contrast to the so-called **Bloch density of states** $\rho_0(E)$ ((3.213) or (3.214)) of the non-interacting electron system, will be **temperature dependent**. Furthermore, as we will see from later examples, it is also decisively influenced by the particle number. Since, finally, the spectral density is, according to (3.332), represented by a weighted superposition of δ-functions, whose arguments contain the excitation energies which must be supplied to take on an additional $(\boldsymbol{k}, \sigma)$ electron into the N-particle system or to remove a corresponding electron from it, then $\rho_\sigma(E)$ also has a close connection to experiments (photoemission!).

Because of its fundamental significance, we want to discuss the quasi-particle density of states $\rho_\sigma(E)$ for a relatively simple **special case**. We assume that the electronic self-energy is $\boldsymbol{k}$-independent and real:

$$R_\sigma(\boldsymbol{k}, E) \equiv R_\sigma(E)\,; \quad I_\sigma(\boldsymbol{k}, E) \equiv 0\,. \tag{3.373}$$

Then, from (3.337), we have:

$$S_{\boldsymbol{k}\sigma}(E) = \hbar\delta\,(E - \varepsilon(\boldsymbol{k}) + \mu - R_\sigma(E))\,. \tag{3.374}$$

For the quasi-particle density of states $\rho_\sigma(E)$, this implies that:

$$\rho_\sigma(E) = \frac{1}{N}\sum_{\boldsymbol{k}} \delta\,[E - \varepsilon(\boldsymbol{k}) - R_\sigma(E-\mu)]\,. \tag{3.375}$$

The comparison with the Bloch density of states,

$$\rho_0(E) = \frac{1}{N}\sum_{\boldsymbol{k}} \delta\,(E - \varepsilon(\boldsymbol{k}))\,,$$

finally yields:

$$\rho_\sigma(E) = \rho_0\,[E - R_\sigma(E-\mu)]\,. \tag{3.376}$$

$\rho_\sigma(E)$ is thus nonzero at those energies for which the function $[E - R_\sigma(E-\mu)]$ assumes values lying between the lower and the upper band edge of the *free* Bloch band. If R_σ is simply a weakly varying, smooth function of E, then ρ_σ will not be too different from ρ_0, so that the influence of the particle interactions can be taken into account sufficiently accurately by introducing effective particle masses or similar auxiliary concepts from solid-state physics.

One can, however, readily imagine situations in which $\rho_\sigma(E)$ differs markedly in a qualitative sense from $\rho_0(E)$. This is for example the case when the function $[E - R_\sigma(E-\mu)]$ has a singularity at some point E_0, as indicated in the sketch (see Fig. 3.7). This situation

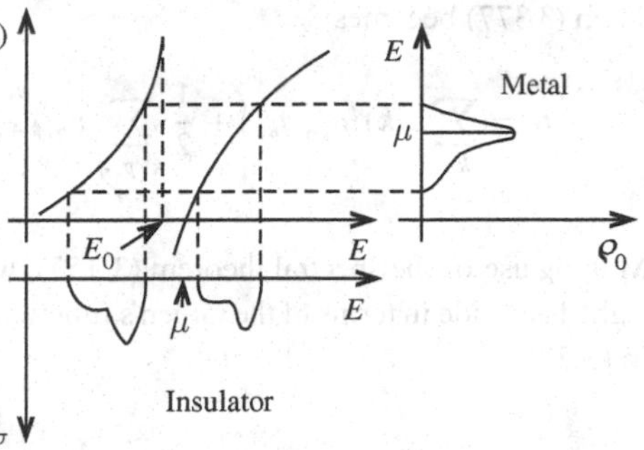

Fig. 3.7 The qualitative energy behaviour of the self-energy for a system of interacting particles, which gives rise to a correlation-related band splitting (Mott insulator)

necessarily leads to band splitting. With a suitable band filling, the chemical potential μ might then fall within the band gap, and as a result, the system would be metallic according to conventional band theory, whilst many-body theory predicts an insulator or a semiconductor. A prominent example of such a situation (*Mott insulator*) is antiferromagnetic NiO.

3.4.5 Internal Energy

In Sects. 3.1.2, 3.1.3, 3.1.4 and 3.1.5, we motivated the use of Green's functions with examples of so-called *response* functions such as

$$\sigma^{\alpha\beta}(E): \quad \text{the electrical conductivity (3.83)},$$

$$\chi^{\alpha\beta}(E): \quad \text{the magnetic susceptibility (3.66)},$$

$$\varepsilon(\boldsymbol{q}, E): \quad \text{the dielectric function (3.104)},$$

which turned out to be retarded commutator Green's functions. These are all *two-or-more-particle functions*, i.e. Green's functions which are constructed from more than one creation and more than one annihilation operator. Naturally, $G_{\boldsymbol{k}\sigma}(E)$ does not belong to this class of functions. However, we want to use the example of the internal energy to show that $G_{\boldsymbol{k}\sigma}(E)$ determines the entire equilibrium thermodynamics of the system of interacting electrons.

For the internal energy, we have initially from (3.311):

$$U = \langle H \rangle = \sum_{\boldsymbol{k}\sigma} \varepsilon(\boldsymbol{k}) \langle a_{\boldsymbol{k}\sigma}^+ a_{\boldsymbol{k}\sigma} \rangle + \frac{1}{2} \sum_{\substack{\boldsymbol{p},\boldsymbol{k},\boldsymbol{q} \\ \sigma\sigma'}} v_{\boldsymbol{k}\boldsymbol{p}}(\boldsymbol{q}) \left\langle a_{\boldsymbol{k}+\boldsymbol{q}\sigma}^+ a_{\boldsymbol{p}-\boldsymbol{q}\sigma'}^+ a_{\boldsymbol{p}\sigma'} a_{\boldsymbol{k}\sigma} \right\rangle . \tag{3.377}$$

We substitute

$$\boldsymbol{q} \rightarrow -\boldsymbol{q}\,, \quad \text{then} \quad \boldsymbol{k} - \boldsymbol{q} \rightarrow \boldsymbol{k}$$

and use from (3.313)

$$v_{\boldsymbol{k}+\boldsymbol{q},\,\boldsymbol{p}}(-\boldsymbol{q}) = v_{\boldsymbol{p},\,\boldsymbol{k}+\boldsymbol{q}}(\boldsymbol{q})\,.$$

Then (3.377) becomes:

$$U = \sum_{k\sigma} \varepsilon(\boldsymbol{k}) \langle a_{k\sigma}^{+} a_{k\sigma} \rangle + \frac{1}{2} \sum_{\substack{k,p,q \\ \sigma,\sigma'}} v_{p,k+q}(\boldsymbol{q}) \left\langle a_{k\sigma}^{+} a_{p+q\sigma'}^{+} a_{p\sigma'} a_{k+q\sigma} \right\rangle . \tag{3.378}$$

Making use of the spectral theorem (3.157), we can express the expectation values on the right-hand side in terms of the Green's functions $G_{k\sigma}(E)$ and $\Gamma_{pk;q}^{\sigma\sigma}(E)$; the latter is defined in (3.323):

$$\begin{aligned} U = \frac{1}{\hbar} \int_{-\infty}^{+\infty} \frac{\mathrm{d}E}{\mathrm{e}^{\beta E}+1} \Bigg\{ & \sum_{k\sigma} \varepsilon(\boldsymbol{k}) \left(-\frac{1}{\pi} \operatorname{Im} G_{k\sigma}^{\mathrm{ret}}(E) \right) + \\ & + \frac{1}{2} \sum_{\substack{kpq \\ \sigma\sigma'}} v_{p,k+q}(\boldsymbol{q}) \left(-\frac{1}{\pi} \operatorname{Im} {}^{\mathrm{ret}}\Gamma_{pk;q}^{\sigma'\sigma}(E) \right) \Bigg\} . \end{aligned} \tag{3.379}$$

Employing the equation of motion (3.324), we can finally replace the *higher-order* Green's function ${}^{\mathrm{ret}}\Gamma_{pk;q}^{\sigma'\sigma}$ by the one-electron Green's function $G_{k\sigma}^{\mathrm{ret}}(E)$:

$$\frac{1}{2} \sum_{\substack{kpq \\ \sigma\sigma'}} v_{p,k+q}(\boldsymbol{q}) {}^{\mathrm{ret}}\Gamma_{pk;q}^{\sigma'\sigma}(E) = \frac{1}{2} \sum_{k\sigma} \left(-\hbar + (E - \varepsilon(\boldsymbol{k}) + \mu) G_{k\sigma}^{\mathrm{ret}}(E) \right) . \tag{3.380}$$

Inserting into (3.379) then leads to:

$$U = \frac{1}{2\hbar} \sum_{k\sigma} \int_{-\infty}^{+\infty} \frac{\mathrm{d}E}{\mathrm{e}^{\beta E}+1} (E + \mu + \varepsilon(\boldsymbol{k})) \left(-\frac{1}{\pi} \operatorname{Im} G_{k\sigma}^{\mathrm{ret}}(E) \right) . \tag{3.381}$$

According to (3.154), on the right-hand side, we have just the single-electron spectral density $S_{k\sigma}(E)$. If we now replace E by $E - \mu$ and insert the Fermi function $f_{-}(E)$ (3.209), then we obtain the notable result,

$$U = \frac{1}{2\hbar} \sum_{k\sigma} \int_{-\infty}^{+\infty} \mathrm{d}E\, f_{-}(E) (E + \varepsilon(\boldsymbol{k}))\, S_{k\sigma}(E - \mu) , \tag{3.382}$$

which we have already encountered in (3.215) in connection with the system of non-interacting electrons. We have, with the aid of (3.382), succeeded in expressing even the contribution of the two-particle Coulomb interaction to the internal energy in terms of the one-electron spectral density. As a result of the generally valid relation (3.217) between the internal energy U and the free energy $F(T, V)$, the latter is likewise completely determined by $S_{k\sigma}(E)$. Thus, our claim that the entire equilibrium thermodynamics of the system of interacting electrons can be derived from the one-electron Green's function or spectral density has been verified.

At the end of this chapter, we want to recall the goals we set out at its beginning. We wished to calculate thermodynamic expectation values and correlation functions. We could do this by solving the Schrödinger equation and then using the eigenstates and eigenvalues of the Hamiltonian $\mathcal{H}$ to construct the partition function. All the desired information can then be derived from it. However, apart from the fact that it would be possible to solve the Schrödinger equation only approximately, this approach would appear to be inefficient in many cases, since possibly a major portion of the tediously calculated terms would cancel out on summing due to the *destructive interference* discussed in Sect. 3.4.3.

The spectral density $S_{k\sigma}(E)$, from which likewise the entire thermodynamical information can be obtained, is to some extent an *all-inclusive* quantity, which already implicitly contains the interference effect mentioned. Only the sufficiently prominent quasi-particle peaks in the spectral density, i.e. those quasi-particles with a sufficiently long lifetime, will make a significant contribution to the various energy integrals. In this sense, the Green's function method, which calculates no states, but instead zooms directly in on the decisive quantities such as the spectral density, appears to be a relatively *efficient* technique. Its **basic idea** consists of replacing an inherently complex interacting many-body system by a *free* gas of quasi-particles. The interaction processes which in fact occur manifest themselves in the renormalised energies and in the possibly finite lifetimes of these quasi-particles.

3.4.6 Exercises

Exercise 3.4.1 For a system of electrons, calculate

$$H = \sum_{k,\sigma} \varepsilon(k) a_{k\sigma}^{+} a_{k\sigma} + \frac{1}{2} \sum_{\substack{k,p,q \\ \sigma,\sigma'}} v_{kp}(q) a_{k+q\sigma}^{+} a_{p-q\sigma'}^{+} a_{p\sigma'} a_{k\sigma} \,,$$

and formulate the equation of motion of the retarded single-particle Green's function. Use it to justify Eq. (3.324) from the text.

Exercise 3.4.2 Let $|E_0\rangle$ be the ground state of the system of non-interacting electrons (Fermi sphere). Calculate the time dependence of the state

$$|\psi_0\rangle = a_{k\sigma}^{+} a_{k'\sigma'} |E_0\rangle \quad (k > k_\mathrm{F},\ k' < k_\mathrm{F}) \ .$$

Is this a stationary state?

Exercise 3.4.3 Let the one-electron Green's function of a system of interacting electrons be given by:

$$G^{\text{ret}}_{k\sigma}(E) = \hbar \left[E - 2\varepsilon(\boldsymbol{k}) + E^2/\varepsilon(\boldsymbol{k}) + \mathrm{i}\gamma\,|E|\right]^{-1}, \quad \gamma > 0\,.$$

1. Find the electronic self-energy $\sum_\sigma(\boldsymbol{k}, E)$.
2. Compute the energies and lifetimes of the quasi-particles.
3. Under which conditions is the quasi-particle concept useful?
4. Calculate the effective masses of the quasi-particles.

Exercise 3.4.4 Assume that the self-energy of a system of interacting electrons has been calculated:

$$\Sigma_\sigma(E) = \frac{a_\sigma\,(E + \mu - b_\sigma)}{(E + \mu - c_\sigma)} \qquad (a_\sigma,\ b_\sigma,\ c_\sigma \quad \text{positive-real}; \quad c_\sigma > b_\sigma)\,.$$

For the density of states of the non-interacting electrons, the following holds:

$$\rho_0(E) = \begin{cases} 1/W & \text{for} \quad 0 \le E \le W\,, \\ 0 & \text{otherwise.} \end{cases}$$

Determine the quasi-particle density of states. Is there a band splitting?

3.5 Self-Examination Questions

For Section 3.1

1. Why is the Schrödinger representation of the time dependence of physical systems also called the *state representation*?
2. Name the most important properties of the time-evolution operator $U_S(t, t_0)$ in the Schrödinger representation.
3. Give the time-evolution operator in the Schrödinger representation for the case that the Hamiltonian is not explicitly time dependent.
4. Give a compact expression for $U_S(t, t_0)$ in the case that $\partial H/\partial t \neq 0$.
5. What is the equation of motion for time-dependent Heisenberg operators?
6. Which relation exists between the operators in the Heisenberg representation and those in the Schrödinger representation?
7. Characterise the Dirac representation.
8. What is the relation between the Dirac and the Schrödinger time-evolution operator?
9. List examples of physically important *response functions*.
10. What is the simplifying assumption of *linear response* theory?
11. What is a double-time, retarded Green's function?
12. Interpret the Kubo formula.
13. Which retarded Green's function determines the tensor of the magnetic susceptibility?
14. How can the susceptibility be used to determine the Curie temperature T_c of the phase transition for para-/ferromagnetism?
15. What is the physical significance of the singularities in the transverse susceptibility $\chi_q^{\pm}(E)$?
16. What relation exists between the dipole-moment operator and the current-density operator?
17. Which Green's function determines the influence of the particle interactions on the electrical conductivity tensor?
18. Sketch the train of argumentation for the derivation of the dielectric function $\varepsilon(\boldsymbol{q}, E)$.
19. What is the physical meaning of the poles of the Green's function $\langle\langle \rho_{\boldsymbol{q}}; \rho_{\boldsymbol{q}}^{+} \rangle\rangle_E^{\text{ret}}$?

For Section 3.2

1. Define retarded, advanced and causal Green's functions.
2. Explain the action of Wick's time-ordering operator.
3. When are the Green's functions homogeneous in time?
4. Give the equation of motion for time-dependent, retarded (advanced, causal) functions. Which boundary conditions do they obey?
5. How does a *chain of equations of motion* arise?
6. Describe the spectral representation of the spectral density $S_{AB}(E)$.
7. What relation exists between the spectral density $S_{AB}(E)$ and the Green's functions $G_{AB}^{\text{ret (adv)}}(E)$?

8. What is the physical significance of the poles of the Green's functions?
9. How do $G^{\text{ret}}_{AB}(E)$ and $G^{\text{adv}}_{AB}(E)$ differ?
10. Formulate the so-called *Dirac identity*.
11. Why is the causal Green's function less suitable for the equation of motion method than the retarded or the advanced function?
12. Formulate and interpret the spectral theorem.
13. Can a commutator Green's function have a pole at $E = 0$?
14. What can you say about the high-energy behaviour ($E \to \infty$) of the Green's function $G^{\alpha}_{AB}(E)$ ($\alpha = $ ret, adv, c)?
15. Explain the relation between the spectral density and the spectral moments.
16. How can one conclude from the spectral representations of the Green's function that the real and imaginary parts of these functions are not independent of each other?
17. What is meant by the Kramers-Kronig relations?
18. What connections exist between the various Green's functions $G^{\text{ret}}_{AB}(E)$, $G^{\text{adv}}_{AB}(E)$ and $G^{\text{c}}_{AB}(E)$ and the spectral density $S_{AB}(E)$?
19. Assume that the causal Green's function $G^{\text{c}}_{AB}(E)$ has been determined in some way. How can you use it to find the retarded Green's function?

For Section 3.3

1. How is the one-electron Green's function defined?
2. What is the one-electron spectral density for non-interacting Bloch electrons?
3. What typical time behaviour is exhibited by the single-particle Green's functions of the *free* electron system?
4. Sketch the derivation of the causal single-particle Green's function for mutually non-interacting Bloch electrons.
5. How does one determine the average occupation number $\langle n_{k\sigma}\rangle$ of the *free* electron system? What is its temperature dependence?
6. What considerations lead to the definition of the quasi-particle density of states $\rho_\sigma(E)$? How is it related to the spectral density?
7. In which manner is the internal energy $U = \langle H_0\rangle$ of the non-interacting electron system determined by the spectral density?
8. How can the free energy $F(T, V)$ be obtained from $U(T, V)$?
9. Give the one-magnon spectral density for a system of non-interacting magnons.
10. Why is the chemical potential μ of magnons equal to zero?
11. Why does it make sense to use the commutator Green's function to describe magnons?
12. Explain $(S_i^+)^2 = 0$ and $(S_i^z)^2 = \hbar^2/4$ for spin-1/2 particles.

For Section 3.4

1. How is the electronic self-energy $\Sigma^{\alpha}_{\sigma}(\boldsymbol{k}, E)$ ($\alpha = $ ret, adv, c) introduced into the equation of motion for the one-electron Green's function?
2. Give the formal solution for $G^{\alpha}_{k\sigma}(E)$ including the self-energy.

3. Formulate and interpret the Dyson equation for the one-electron Green's function.
4. What connection is there between the retarded and the advanced self-energy?
5. Why can the imaginary part of the **retarded** self-energy not be positive?
6. What can you say about the sign of the imaginary part of the **advanced** self-energy?
7. Demonstrate why the real part of the self-energy determines the *resonance values* $E_{i\sigma}(\boldsymbol{k})$ of the one-electron spectral density, at which it exhibits prominent maxima.
8. What is the structure of the spectral density of the interacting electron system in the case that the imaginary part $I_\sigma(\boldsymbol{k}, E)$ of the self-energy is identically zero?
9. What condition on the imaginary part of the self-energy guarantees a prominent maximum in the spectral density?
10. Under which conditions does the spectral density have a Lorentz form in the neighbourhood of a resonance?
11. Which general structure is to be expected in the normal case for the spectral density?
12. What is the structure of the time-dependent spectral density when the imaginary part of the self-energy *vanishes identically*?
13. What is the effect of the Lorentz peaks of $S_{\boldsymbol{k}\sigma}(E)$ on $S_{\boldsymbol{k}\sigma}(t-t')$?
14. Which well-defined physical significance is expressed by the propagator $S_{\boldsymbol{k}\sigma}(t-t')$ in the special case that $T = 0$; $k > k_F$; $t > t'$? How does it change for $k < k_F$?
15. When does one denote $|\varphi_0(t)\rangle$ as a *stationary state*? When does one say that it has a finite lifetime?
16. What is the time dependence of the propagator

$$2\pi S_{\boldsymbol{k}\sigma}(t, t') = \langle\varphi_0(t)\,|\,\varphi_0(t')\,\rangle$$

$\left(|\varphi_0\rangle = a^+_{\boldsymbol{k}\sigma}|E_0\rangle,\ |E_0\rangle\text{: ground state},\ k > k_F\right)$ for the non-interacting electron system?
17. How should this time dependence change for the interacting electron system?
18. Explain how this time dependence leads to the concept of quasi-particles.
19. What is meant by the spectral weight and what is to be understood as the lifetime of a quasi-particle?
20. How do quasi-particle energies and quasi-particle lifetimes relate to the electronic self-energy?
21. How does a quasi-particle manifest itself in the spectral density $S_{\boldsymbol{k}\sigma}(E)$?
22. When does a quasi-particle have an infinitely long lifetime?
23. How is the effective mass of a quasi-particle defined?
24. What close relation is there between the quasi-particle density of states $\rho_\sigma(E)$ and the spectral density of an interacting electron system?
25. Let the self-energy be real and independent of $\boldsymbol{k}$, and assume that it has a singularity at the energy E_0. What does this imply for the quasi-particle density of states?
26. How is the internal energy of an interacting electron system related to the one-electron spectral density?

Systems of Interacting Particles

4

In this section, we want to apply the abstract Green's function formalism introduced in the last chapter to realistic problems in many-body theory, keeping in mind in particular the model systems from Chap. 2. We want to find out on the one hand just which information can be obtained from suitably chosen Green's functions, and on the other, to see how such Green's functions can be calculated in practical cases. We will critically examine the approximations which as a rule must be made in order to make the problems tractable.

4.1 Electrons in Solids

We start by investigating interacting electrons in solids, concentrating here on some typical problems without attempting an exhaustively complete treatment. Two of the models introduced in Sect. 2.1, the jellium model and the Hubbard model, will form the basis for this discussion. We begin with an exactly solvable special case of the Hubbard model.

4.1.1 The Limiting Case of an Infinitely Narrow Band

The Hubbard model describes interacting electrons in relatively narrow energy bands. It is characterised by the Hamiltonian (2.117):

$$\mathcal{H} = \sum_{ij\sigma} \left(T_{ij} - \mu\delta_{ij}\right) a_{i\sigma}^{+} a_{j\sigma} + \frac{1}{2} U \sum_{i,\sigma} n_{i\sigma} n_{i-\sigma} \,. \tag{4.1}$$

We wish to calculate the one-electron Green's function. The form of the Hamiltonian in (4.1) suggests the use of the Wannier formulation (3.319):

$$G_{ij\sigma}^{\alpha}(E) = \langle\!\langle a_{i\sigma}; a_{j\sigma}^{+} \rangle\!\rangle_{E}^{\alpha} \,. \tag{4.2}$$

W. Nolting, *Fundamentals of Many-body Physics*,
DOI 10.1007/978-3-540-71931-1_4, © Springer-Verlag Berlin Heidelberg 2009

For the equation of motion, we require the commutator

$$[a_{i\sigma}, \mathcal{H}]_- = \sum_m (T_{im} - \mu\delta_{im})\, a_{m\sigma} + U n_{i\,-\sigma} a_{i\sigma} \,. \tag{4.3}$$

The second term gives rise to a *higher-order* Green's function:

$$\Gamma^{\alpha}_{ilm;\,j\sigma}(E) = \langle\!\langle a^+_{i\,-\sigma} a_{l\,-\sigma} a_{m\sigma}\,; a^+_{j\sigma} \rangle\!\rangle^{\alpha}_{E} \,. \tag{4.4}$$

Then the equation of motion of the one-electron Green's function is found to be:

$$(E+\mu)G^{\alpha}_{ij\sigma}(E) = \hbar\delta_{ij} + \sum_m T_{im} G^{\alpha}_{mj\sigma}(E) + U\Gamma^{\alpha}_{iii;\,j\sigma}(E) \,, \tag{4.5}$$

which cannot be solved directly on the right-hand side due to the *higher-order* Green's function $\Gamma^{\alpha}_{iii;\,j\sigma}(E)$. Therefore, we write the corresponding equation of motion for this function also:

$$\begin{aligned}
&[n_{i\,-\sigma} a_{i\sigma}, \mathcal{H}_0]_- = \\
&= \sum_{lm\sigma'} (T_{lm} - \mu\delta_{lm}) \left[n_{i\,-\sigma} a_{i\sigma}, a^+_{l\sigma'} a_{m\sigma'}\right]_- = \\
&= \sum_{lm\sigma'} (T_{lm} - \mu\delta_{lm}) \Big\{\delta_{il}\delta_{\sigma\sigma'} n_{i\,-\sigma} a_{m\sigma'} - \\
&\quad - \delta_{il}\delta_{\sigma\,-\sigma'} a^+_{i\,-\sigma} a_{i\sigma} a_{m\sigma'} - \delta_{im}\delta_{\sigma\,-\sigma'} a^+_{l\sigma'} a_{i\,-\sigma} a_{i\sigma}\Big\} = \\
&= \sum_m (T_{im} - \mu\delta_{im}) \left\{n_{i\,-\sigma} a_{m\sigma} + a^+_{i\,-\sigma} a_{m\,-\sigma} a_{i\sigma} - a^+_{m\,-\sigma} a_{i\,-\sigma} a_{i\sigma}\right\} \,,
\end{aligned} \tag{4.6}$$

$$\begin{aligned}
&[n_{i\,-\sigma} a_{i\sigma}, \mathcal{H}_1]_- = \\
&= \frac{1}{2} U \sum_{m,\sigma'} [n_{i\,-\sigma} a_{i\sigma}, n_{m\sigma'} n_{m\,-\sigma'}]_- = \\
&= \frac{1}{2} U \sum_{m,\sigma'} n_{i\,-\sigma} [a_{i\sigma}, n_{m\sigma'} n_{m\,-\sigma'}]_- = \\
&= \frac{1}{2} U \sum_{m,\sigma'} n_{i\,-\sigma} \left\{\delta_{im}\delta_{\sigma\sigma'} a_{m\sigma'} n_{m\,-\sigma'} + \delta_{im}\delta_{\sigma\,-\sigma'} n_{m\sigma'} a_{m\,-\sigma'}\right\} = \\
&= U a_{i\sigma} n_{i\,-\sigma} \,.
\end{aligned} \tag{4.7}$$

In the final step, we made use of the relation (which holds for Fermions):

$$n^2_{i\sigma} = n_{i\sigma} \,.$$

All together, we find for the equation of motion:

$$(E+\mu-U)\Gamma^{\alpha}_{iii;\,j\sigma}(E)=$$
$$=\hbar\delta_{ij}\left\langle n_{i\,-\sigma}\right\rangle+\sum_m T_{im}\left\{\Gamma^{\alpha}_{iim;\,j\sigma}(E)+\Gamma^{\alpha}_{imi;\,j\sigma}(E)-\Gamma^{\alpha}_{mii;\,j\sigma}(E)\right\}. \tag{4.8}$$

In this section, we shall limit ourselves to the relatively simple but quite instructive limiting case of an infinitely narrow band,

$$\varepsilon(\boldsymbol{k})\equiv T_0 \iff T_{ij}=T_0\delta_{ij}\,, \tag{4.9}$$

for which the hierarchy of equations of motion decouples itself. Equation (4.8) is thus simplified to:

$$(E+\mu-U-T_0)\,\Gamma^{\alpha}_{iii;\,j\sigma}(E)=\hbar\delta_{ij}\left\langle n_{-\sigma}\right\rangle . \tag{4.10}$$

Due to translational symmetry, the expectation value of the number operator is independent of the lattice site ($\langle n_{i\sigma}\rangle=\langle n_\sigma\rangle\ \forall i$). We insert the solution of (4.10) into (4.5):

$$(E+\mu-T_0)\,G^{\alpha}_{ii\sigma}=\hbar+\hbar\frac{U\left\langle n_{-\sigma}\right\rangle}{E-(T_0-\mu+U)}\,.$$

Then we finally obtain for the retarded function:

$$G^{\text{ret}}_{ii\sigma}(E)=\frac{\hbar\left(1-\left\langle n_{-\sigma}\right\rangle\right)}{E-(T_0-\mu)+\mathrm{i}0^+}+\frac{\hbar\left\langle n_{-\sigma}\right\rangle}{E-(T_0+U-\mu)+\mathrm{i}0^+}\,. \tag{4.11}$$

$G^{\text{ret}}_{ii\sigma}(E)$ thus has two poles, corresponding to the possible excitation energies:

$$E_{1\sigma}=T_0-\mu=E_{1\,-\sigma}\,, \tag{4.12}$$

$$E_{2\sigma}=T_0+U-\mu=E_{2\,-\sigma}\,. \tag{4.13}$$

The *original* level T_0 splits because of the Coulomb interaction into two spin-independent quasi-particle levels $E_{1\sigma}$ and $E_{2\sigma}$. The spectral density can readily be computed using (3.154) from (4.11):

$$S_{ii\sigma}(E)=\hbar\sum_{j=1}^{2}\alpha_{j\sigma}\delta\left(E-E_{j\sigma}\right). \tag{4.14}$$

The spectral weights

$$\alpha_{1\sigma}=1-\left\langle n_{-\sigma}\right\rangle\,;\quad \alpha_{2\sigma}=\left\langle n_{-\sigma}\right\rangle \tag{4.15}$$

are a measure of the probability that the σ electron encounters a $(-\sigma)$ electron at the same lattice site, $(\alpha_{2\sigma})$, or arrives at an unoccupied site, $(\alpha_{1\sigma})$. In the former case, it must overcome the Coulomb interaction U.

The quasi-particle density of states

$$\begin{aligned} \rho_\sigma(E) &= \frac{1}{N\hbar}\sum_i S_{ii\sigma}(E-\mu) = \frac{1}{\hbar}S_{ii\sigma}(E-\mu) = \\ &= (1-\langle n_{-\sigma}\rangle)\,\delta(E-T_0) + \langle n_{-\sigma}\rangle\,\delta(E-(T_0+U)) \end{aligned} \tag{4.16}$$

consists in this limit of two infinitely narrow *bands* at the energies T_0 and T_0+U. The lower band, which is degenerate to the level, contains $(1-\langle n_{-\sigma}\rangle)$, the upper band $\langle n_{-\sigma}\rangle$ states per atom. The number of states in a quasi-particle subband is thus temperature dependent!

For a complete determination of the quasi-particle density of states, we must find the expectation value $\langle n_{-\sigma}\rangle$ by using the spectral theorem (3.157):

$$\begin{aligned} \langle n_{-\sigma}\rangle &= \frac{1}{\hbar}\int\limits_{-\infty}^{+\infty} \mathrm{d}E\; S_{ii\,-\sigma}(E)\left[\mathrm{e}^{\beta E}+1\right]^{-1} = \\ &= (1-\langle n_\sigma\rangle)\,f_-(T_0) + \langle n_\sigma\rangle\,f_-(T_0+U)\;. \end{aligned}$$

$f_-(E)$ is here again the Fermi function. We insert the corresponding expression for $\langle n_\sigma\rangle$ and thus find:

$$\langle n_{-\sigma}\rangle = \frac{f_-(T_0)}{1+f_-(T_0)-f_-(T_0+U)}\;. \tag{4.17}$$

The complete solution for $\rho_\sigma(E)$ is then given by:

$$\begin{aligned} \rho_\sigma(E) = &\frac{1}{1+f_-(T_0)-f_-(T_0+U)}\Big\{(1-f_-(T_0+U))\,\delta(E-T_0)+ \\ &+ f_-(T_0)\,\delta(E-T_0-U)\Big\} = \\ = &\,\rho_{-\sigma}(E)\;. \end{aligned} \tag{4.18}$$

The quasi-particle density of states is thus spin-independent. A spontaneous magnetisation, i.e. ferromagnetism, is therefore excluded in the limiting case of an infinitely narrow band:

$$\langle n_\sigma\rangle = \langle n_{-\sigma}\rangle = \frac{1}{2}n\;. \tag{4.19}$$

4.1.2 The Hartree-Fock Approximation

In this section, we want to introduce a very simple but also very typical approximation, which is to be sure much too coarse for *exacting* requirements, but frequently can provide a first valuable insight into the physics of the model at hand. The Hartree-Fock approximation of the Hubbard model is known in the literature as the **Stoner model** and as such is referred to in discussing the magnetic behaviour of band electrons.

Our starting point is the following identity for the product of two operators A and B:

$$AB = (A - \langle A \rangle)(B - \langle B \rangle) + A \langle B \rangle + \langle A \rangle B - \langle A \rangle \langle B \rangle \ . \tag{4.20}$$

The simplification consists in a linearisation of this expression. We imagine the product AB as a component of a Green's function, which is indeed defined as a thermodynamic average. The **Hartree-Fock approximation** or also the **molecular-field approximation** neglects the *fluctuations* of the observables around their thermodynamic averages in the Green's functions; it thus makes the replacement

$$AB \underset{\text{HFA}}{\longrightarrow} A \langle B \rangle + \langle A \rangle B - \langle A \rangle \langle B \rangle \ . \tag{4.21}$$

The last term is a pure c-number and does not appear in the equations of motion.

We carry out the approximation (4.21) for the Green's function $\Gamma^{\alpha}_{iii,\, j\sigma}(E)$ in (4.5). As a result of

$$a_{i\sigma} n_{i\,-\sigma} \underset{\text{HFA}}{\longrightarrow} a_{i\sigma} \langle n_{-\sigma} \rangle + \langle a_{i\sigma} \rangle n_{i\,-\sigma} - \langle a_{i\sigma} \rangle \langle n_{-\sigma} \rangle = a_{i\sigma} \langle n_{-\sigma} \rangle \ ,$$

for the *higher-order* Green's function $\Gamma^{\alpha}_{iii;\, j\sigma}(E)$ in the Hartree-Fock approximation, we find:

$$\Gamma^{\alpha}_{iii;\, j\sigma}(E) \underset{\text{HFA}}{\longrightarrow} \langle n_{-\sigma} \rangle \, G^{\alpha}_{ij\sigma}(E) \, . \tag{4.22}$$

$\langle n_{-\sigma} \rangle$, a scalar, can be removed from the Green's function. Then the equation of motion (4.5) can be simplified to

$$(E + \mu - U \langle n_{-\sigma} \rangle) \, G^{\alpha}_{ij\sigma}(E) = \hbar \delta_{ij} + \sum_m T_{im} G^{\alpha}_{mj\sigma}(E) \, ,$$

and by Fourier transformation to wavenumbers, it can be readily solved:

$$\begin{aligned} G_{\boldsymbol{k}\sigma}(E) &= \frac{\hbar}{E - (\varepsilon(\boldsymbol{k}) + U \langle n_{-\sigma} \rangle - \mu)} \equiv \\ &\equiv G^{(0)}_{\boldsymbol{k}\sigma} \left(E - U \langle n_{-\sigma} \rangle\right) \, . \end{aligned} \tag{4.23}$$

In the Hartree-Fock approximation, the one-electron Green's function of the Hubbard model has the same form as that of the *free* system, however with *renormalised*, spin-dependent single-particle energies. Making use of the dimensionless

$$\textbf{magnetisation} \quad m = \frac{1}{2}\left(\langle n_\uparrow\rangle - \langle n_\downarrow\rangle\right) \tag{4.24}$$

and of the

$$\textbf{particle density} \quad n = \langle n_\uparrow\rangle + \langle n_\downarrow\rangle\,, \tag{4.25}$$

the quasi-particle energies are given by:

$$E_\sigma(\boldsymbol{k}) = \left(\varepsilon(\boldsymbol{k}) + \frac{1}{2}Un\right) - z_\sigma m U\,. \tag{4.26}$$

z_σ is a symbol for the sign: $\left(z_\uparrow = +1\,,\ z_\downarrow = -1\right)$. For a complete solution of the problem, we still have to determine the expectation value $\langle n_{-\sigma}\rangle$ in (4.23). This can be done by applying the spectral theorem. From (4.23), we read off directly the following expression

$$S_{\boldsymbol{k}\sigma}(E) = \hbar\delta\left(E - \varepsilon(\boldsymbol{k}) - U\langle n_{-\sigma}\rangle + \mu\right) \tag{4.27}$$

for the spectral density. The spectral theorem (3.157) then yields:

$$\langle n_{-\sigma}\rangle = \frac{1}{N}\sum_i \langle n_{i\,-\sigma}\rangle = \frac{1}{N}\sum_{\boldsymbol{k}}\langle a^+_{\boldsymbol{k}\,-\sigma}a_{\boldsymbol{k}\,-\sigma}\rangle =$$

$$= \frac{1}{N}\sum_{\boldsymbol{k}}\frac{1}{\hbar}\int_{-\infty}^{+\infty} \mathrm{d}E\ f_-(E)S_{\boldsymbol{k}\,-\sigma}(E-\mu)\,.$$

This means that:

$$\langle n_{-\sigma}\rangle = \frac{1}{N}\sum_{\boldsymbol{k}}\left\{1 + \exp\left[\beta\left(\varepsilon(\boldsymbol{k}) + U\langle n_\sigma\rangle - \mu\right)\right]\right\}^{-1}\,. \tag{4.28}$$

This is an implicit functional equation for the average number of particles $\langle n_\sigma\rangle$, $\langle n_{-\sigma}\rangle$, which we can rearrange into a corresponding expression for the magnetisation m:

$$m = \frac{1}{2}\sinh(\beta U m)\frac{1}{N}\sum_{\boldsymbol{k}} g_{\boldsymbol{k}}(\beta, n, m)\,. \tag{4.29}$$

Here, we have used the abbreviation:

$$g_{\boldsymbol{k}}(\beta, n, m) = \left\{\cosh(\beta U m) + \cosh\left[\beta\left(\varepsilon(\boldsymbol{k}) + \frac{1}{2}Un - \mu\right)\right]\right\}^{-1}\,. \tag{4.30}$$

For the particle density, we find:

$$n = \frac{1}{N}\sum_{k}\left\{\exp\left[-\beta\left(\varepsilon(\boldsymbol{k}) + \frac{1}{2}Un - \mu\right)\right] + \cosh(\beta U m)\right\} g_{\boldsymbol{k}}(\beta, n, m) \,. \tag{4.31}$$

We can immediately recognise that the non-magnetic state $m = 0$ always represents a possible solution. In order to see whether there are other solutions, for $m \neq 0$, we rearrange the expression (4.30) slightly using the density of states (3.213) of the non-interacting system,

$$\rho_0(E) = \frac{1}{N}\sum_{k}\delta\left(E - \varepsilon(\boldsymbol{k})\right) ,$$

which we may presume to be known:

$$\begin{aligned}\frac{1}{N}\sum_{k} g_{\boldsymbol{k}}(\beta, n, m) = \int\limits_{-\infty}^{+\infty} \mathrm{d}x\, \rho_0(x)\Bigg\{&\cosh(\beta U m) + \\ &+ \cosh\left[\beta\left(x + \frac{1}{2}Un - \mu\right)\right]\Bigg\}^{-1} .\end{aligned} \tag{4.32}$$

At high temperatures ($T \to \infty \Longleftrightarrow \beta \to 0$), we can expand the hyperbolic functions:

$$\sinh x = x + \frac{1}{3!}x^3 + \cdots ; \quad \cosh x = 1 + \frac{1}{2!}x^2 + \cdots .$$

This means that:

$$\frac{1}{N}\sum_{k} g_{\boldsymbol{k}}(\beta, n, m) \underset{T\to\infty}{\longrightarrow} \frac{1}{2}\int\limits_{-\infty}^{+\infty} \mathrm{d}x\, \rho_0(x) \,, = \frac{1}{2} \tag{4.33}$$

or, with (4.29):

$$m \underset{T\to\infty}{\longrightarrow} \frac{1}{4}\beta U m \,. \tag{4.34}$$

This equation has the single solution $m = 0$. At high temperatures, there is thus no spontaneous magnetisation $m \neq 0$. If a ferromagnetic solution $m \neq 0$ exists at all, then clearly only below a critical temperature T_C (the **Curie temperature**). If we approach T_C *from below*, then m should become very small (second-order phase transition!), so that in (4.32), we can expand the first term. In the neighbourhood of T_C, with (4.25) we should then find:

$$T \overset{<}{\to} T_C : \quad 1 \approx \frac{1}{2}\beta_C U \int\limits_{-\infty}^{+\infty} \mathrm{d}x \frac{\rho_0(x)}{1 + \cosh\left(\beta_C\left(x + \frac{1}{2}Un - \mu\right)\right)} \,. \tag{4.35}$$

For ferromagnetism to be realised in fact, we must require at the least that

$$T_C = 0^+ \quad \left(\beta_C = (k_B T_C)^{-1} \to \infty\right)$$

be fulfilled. With the following representation of the δ-function (proof as exercise (4.1.2)),

$$\delta(x) = \lim_{\beta \to \infty} \frac{1}{2} \frac{\beta}{1 + \cosh(\beta x)}, \tag{4.36}$$

we obtain using (4.35) the criterion

$$1 \approx U\rho_0 \left(\mu - \frac{1}{2} U n\right). \tag{4.37}$$

At $T \approx 0$, the chemical potential μ can be replaced by the Fermi energy E_F. From (4.26), the Fermi energy is related to that of the *free* system (ε_F), due to $m \approx 0$, as $E_F \approx \varepsilon_F + \frac{1}{2} U n$. We then obtain from (4.37) the well-known

Stoner criterion

$$1 \leq U\rho_0(\varepsilon_F) \tag{4.38}$$

for the occurrence of ferromagnetism. If this relation is fulfilled, then the electronic system should exhibit a *spontaneous* magnetisation $m \neq 0$, i.e. one not induced by an external magnetic field. In spite of the various greatly simplifying assumptions which finally led to (4.38), the Stoner criterion has proved to be *correct for predicting trends*.

We finally return to the quasi-particle energies $E_\sigma(\boldsymbol{k})$, (4.26). We have seen that under certain conditions (4.38), the magnetisation can be nonzero, $m \neq 0$. This corresponds to a temperature-dependent **exchange splitting** ΔE_{ex}

$$\Delta E_{ex} = 2Um,$$

which in the present simple example is *rigid*, i.e. $\boldsymbol{k}$-independent. Corresponding to the significance of one-electron Green's functions, the quasi-particle energies $E_\sigma(\boldsymbol{k})$ are just

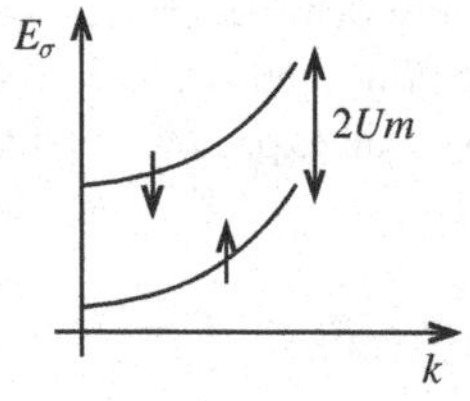

Fig. 4.1 The spin-dependent splitting of the quasi-particle energies in the Stoner model for temperatures below the Curie temperature

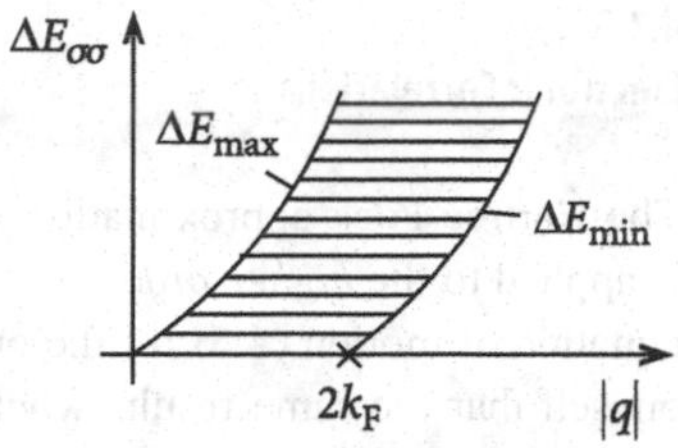

Fig. 4.2 The excitation spectrum in the Stoner model for non-spin-flip transitions

the energies which must be supplied in order to add one $(\boldsymbol{k}, \sigma)$ electron into the N-particle system. For the actual excitation energy within the system, we then find:

$$\Delta E_{\sigma\sigma'}(\boldsymbol{k};\boldsymbol{q}) = E_{\sigma'}(\boldsymbol{k}+\boldsymbol{q}) - E_{\sigma}(\boldsymbol{k}) = \\ = \varepsilon(\boldsymbol{k}+\boldsymbol{q}) - \varepsilon(\boldsymbol{k}) + mU\,(z_\sigma - z_{\sigma'}) \; . \tag{4.39}$$

If an excitation of this type takes place without a spin flip ($\sigma = \sigma'$) within one subband, then because of the rigid band shift, the excitation is identical to a corresponding transition in a non-interacting ($U = 0$) system. In the case of a spherically symmetric Fermi volume, i.e. $\varepsilon(\boldsymbol{k}) = \hbar^2 k^2/2m^*$, the excitation spectrum lies between two curves (see Fig. 4.2):

$$\Delta E_{\max}(\boldsymbol{q}) = \frac{\hbar^2}{2m^*}\left(q^2 + 2k_F\,|\boldsymbol{q}|\right) , \tag{4.40}$$

$$\Delta E_{\min}(\boldsymbol{q}) = \begin{cases} \dfrac{\hbar^2}{2m^*}\left(q^2 - 2k_F\,|\boldsymbol{q}|\right) , & \text{when} \quad |\boldsymbol{q}| > 2k_F \, , \\ 0 & \text{otherwise.} \end{cases} \tag{4.41}$$

Excitations with a spin flip, in contrast, are transitions between the two subbands:

$$\Delta E_{\uparrow\downarrow}(\boldsymbol{k};\boldsymbol{q}) = \varepsilon(\boldsymbol{k}+\boldsymbol{q}) - \varepsilon(\boldsymbol{k}) + 2Um \; . \tag{4.42}$$

One distinguishes between **strong** ($2Um > \varepsilon_F$) and **weak ferromagnetism** ($2Um < \varepsilon_F$) (see Fig. 4.3).

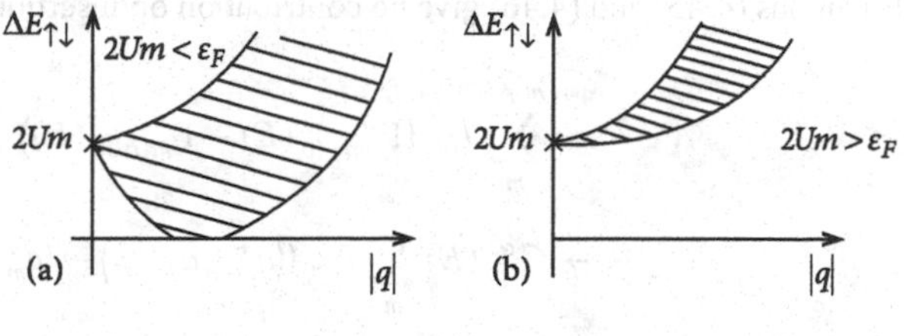

Fig. 4.3 The excitation spectrum in the Stoner model for spin-flip transitions: (**a**) a weak ferromagnet; (**b**) a strong ferromagnet

4.1.3 Electronic Correlations

The Hartree-Fock approximation for the Hubbard model as discussed in the last section is applied to the *higher-order* Green's function $\Gamma^{\alpha}_{iiij\sigma}(E)$ (cf. (4.22)), which appears in the equation of motion (4.5) for the one-electron Green's function. One can readily convince oneself that the same results would have been obtained if the approximation (4.21) had been applied directly to the model Hamiltonian (4.1):

$$\begin{aligned}\mathcal{H} \to \mathcal{H}_S &= \sum_{i,j,\sigma} \left(T_{ij} + (U\langle n_{-\sigma}\rangle - \mu)\,\delta_{ij}\right) a^+_{i\sigma} a_{j\sigma} = \\ &= \sum_{\boldsymbol{k},\sigma} (E_\sigma(\boldsymbol{k}) - \mu)\, a^+_{\boldsymbol{k}\sigma} a_{\boldsymbol{k}\sigma} \,.\end{aligned} \tag{4.43}$$

$\mathcal{H}_S$ defines the actual Stoner model. It is a single-particle operator, for which $G_{\boldsymbol{k}\sigma}(E)$ can readily be calculated exactly, giving agreement with (4.23).

We can obtain additional information by decoupling the chain of equations of motion at a *later* point, e.g. at the new Green's function which appears in the equation of motion (4.8) for $\Gamma^{\alpha}_{iii;\,j\sigma}(E)$. The result can then however no longer be formulated within the framework of a single-particle model. In this connection, one introduces the concept of

particle correlations,

and includes in it all the particle interactions which are **not** describable within a single-particle model and therefore represent genuine

many-body effects.

The *decoupling* mentioned several times above was introduced by Hubbard himself, who thereby suggested an approximate solution to his own model. One applies the Hartree-Fock procedure (4.21) to the Green's functions of Eq. (4.8). Taking into account the conservation of particle number and spin, we thus find:

$$\Gamma^{\alpha}_{iim;j\sigma}(E) \xrightarrow{i\neq m} \langle n_{-\sigma}\rangle\, G^{\alpha}_{mj\sigma}(E), \tag{4.44}$$

$$\Gamma^{\alpha}_{imi;\,j\sigma}(E) \xrightarrow{i\neq m} \left\langle a^+_{i\,-\sigma} a_{m\,-\sigma}\right\rangle G^{\alpha}_{ij\sigma}(E)\,, \tag{4.45}$$

$$\Gamma^{\alpha}_{mii;\,j\sigma}(E) \xrightarrow{i\neq m} \left\langle a^+_{m\,-\sigma} a_{i\,-\sigma}\right\rangle G^{\alpha}_{ij\sigma}(E)\,. \tag{4.46}$$

Equations (4.45) and (4.46) give no contribution on insertion into (4.8),

$$\begin{aligned}&\sum_m^{m\neq i} T_{im}\left(\Gamma^{\alpha}_{imi;\,j\sigma}(E) - \Gamma^{\alpha}_{mii;\,j\sigma}(E)\right) \\ &\longrightarrow G^{\alpha}_{ij\sigma}(E)\sum_m T_{im}\left(\left\langle a^+_{i\,-\sigma} a_{m\,-\sigma}\right\rangle - \left\langle a^+_{m\,-\sigma} a_{i\,-\sigma}\right\rangle\right)\,,\end{aligned} \tag{4.47}$$

if we assume a translationally symmetric lattice as usual:

$$\begin{aligned}
&\sum_m T_{im}\left(\langle a^+_{i-\sigma} a_{m-\sigma}\rangle - \langle a^+_{m-\sigma} a_{i-\sigma}\rangle\right) = \\
&= \frac{1}{N}\sum_i\sum_m T_{im}\left(\langle a^+_{i-\sigma} a_{m-\sigma}\rangle - \langle a^+_{m-\sigma} a_{i-\sigma}\rangle\right) = \\
&= \frac{1}{N}\sum_{i,m}(T_{im} - T_{mi})\langle a^+_{i-\sigma} a_{m-\sigma}\rangle = \\
&= 0\,.
\end{aligned} \tag{4.48}$$

In going from the first to the second line, we made use of translational symmetry; in going from the second line to the third for the second term, we exchanged the summation indices; and in going from the third to the fourth line, we used $T_{im} = T_{mi}$.

With equations (4.47) and (4.48), we have for the equation of motion (4.8):

$$(E + \mu - T_0 - U)\,\Gamma^\alpha_{iii;\,j\sigma}(E) = \hbar\delta_{ij}\,\langle n_{-\sigma}\rangle + \langle n_{-\sigma}\rangle \sum_m^{m\neq i} T_{im} G^\alpha_{mj\sigma}(E)\,.$$

Solving for $\Gamma^\alpha_{iii;\,j\sigma}(E)$ and inserting into (4.5) yields a functional equation for the one-electron Green's function:

$$(E + \mu - T_0)\, G^\alpha_{ij\sigma}(E) = \left(\hbar\delta_{ij} + \sum_m^{m\neq i} T_{im} G^\alpha_{mj\sigma}(E)\right)\cdot \cdot\left(1 + \frac{U\,\langle n_{-\sigma}\rangle}{E + \mu - T_0 - U}\right),$$

which can be solved by Fourier transformation to the wavenumber domain. We define

$$\Sigma_\sigma(E) = U\,\langle n_{-\sigma}\rangle\,\frac{E + \mu - T_0}{E + \mu - U\,(1 - \langle n_{-\sigma}\rangle) - T_0} \tag{4.49}$$

and thus obtain precisely the form expected from the general considerations in Sect. 3.4.1 (cf. (3.326)) for the Green's function of the system of interacting electrons:

$$G_{\boldsymbol{k}\sigma}(E) = \hbar\,[E - (\varepsilon(\boldsymbol{k}) - \mu + \Sigma_\sigma(E))]^{-1}\,. \tag{4.50}$$

One can readily show that for $U \to 0$ (*band limit*) and for $\varepsilon(\boldsymbol{k}) \to T_0$ (*atomic limit*), this solution converges to the exact expressions (3.198) or (4.11). The Hartree-Fock solution (4.23) is correct, in contrast, only in the band limit.

The self-energy is zero when the interaction is *switched off*, ($U = 0$), but also for $\langle n_{-\sigma}\rangle = 0$, because the σ electron then has no interaction partners.

The self-energy is real and independent of $\boldsymbol{k}$ in the *Hubbard solution* (4.49), and thus it fulfils the conditions (3.373) of the special case discussed in Sect. 3.4.4, which implies for the quasi-particle density of states in the representation (3.376) that:

$$\rho_\sigma(E) = \rho_0\left[E - \Sigma_\sigma(E-\mu)\right] . \tag{4.51}$$

The argument $E - \Sigma_\sigma(E-\mu)$ diverges at

$$E_{0\sigma} = U\left(1 - \langle n_{-\sigma}\rangle\right) + T_0 . \tag{4.52}$$

This leads to a band splitting due to electron correlations, which in principle cannot be understood in a single-particle picture.

Finally, we consider the spectral density, for which from (3.374) we have

$$S_{\boldsymbol{k}\sigma}(E) = \hbar\delta\left[E - \varepsilon(\boldsymbol{k}) + \mu - \Sigma_\sigma(E)\right] . \tag{4.53}$$

With the formula (3.338), we can also write

$$S_{\boldsymbol{k}\sigma}(E) = \hbar\sum_{j=1}^{2}\alpha_{j\sigma}(\boldsymbol{k})\delta\left(E + \mu - E_{j\sigma}(\boldsymbol{k})\right) . \tag{4.54}$$

Here, for the quasi-particle energies, we find

$$\begin{aligned} E_{j\sigma}(\boldsymbol{k}) = {} & \frac{1}{2}\left(U + \varepsilon(\boldsymbol{k}) + T_0\right) + \\ & + (-1)^j\sqrt{\frac{1}{4}\left(T_0 + U - \varepsilon(\boldsymbol{k})\right)^2 + U\langle n_{-\sigma}\rangle\left(\varepsilon(\boldsymbol{k}) - T_0\right)} \, ; \end{aligned} \tag{4.55}$$

and for the spectral weights:

$$\alpha_{j\sigma}(\boldsymbol{k}) = (-1)^j\frac{E_{j\sigma}(\boldsymbol{k}) - T_0 - U\left(1 - \langle n_{-\sigma}\rangle\right)}{E_{2\sigma}(\boldsymbol{k}) - E_{1\sigma}(\boldsymbol{k})} . \tag{4.56}$$

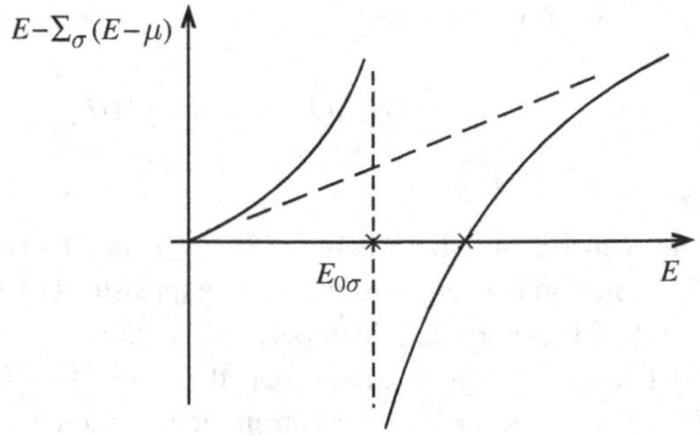

Fig. 4.4 The qualitative energy dependence of the self-energy in the Hubbard solution of the Hubbard model

The band splitting mentioned above manifests itself here in the fact that for each wave-number $\boldsymbol{k}$, two quasi-particle energies exist. These are real, and thus correspond to quasi-particles with infinite lifetimes. If the electron is moving in the upper subband, then it hops mainly onto lattice sites already occupied by another electron from the same energy band with opposite spin. In the lower subband, in contrast, it prefers unoccupied sites. This leads to an energetic spacing of the two subbands of about U, as one can easily verify from (4.55).

The singularity which causes the band splitting, $E_{0\sigma}$, is however also responsible for a serious disadvantage of the Hubbard solution. One should expect that with decreasing U/W (W is the Bloch bandwidth), the initially separated subbands would gradually begin to overlap. We can see however from (4.52) that even for arbitrarily small U/W, a singularity $E_{0\sigma}$ in $(E - \Sigma_\sigma(E-\mu))$ always persists, so that the theory predicts a band gap for all values of the parameters. For small U/W, the Hubbard solution therefore appears questionable.

4.1.4 The Interpolation Method

In this section, we will encounter a very simple approximation method, which can be quite informative for *first estimates*. It is found to be exact in the two extreme limits, the *band limit* (interactions $\to 0$) and the *atomic limit*, ($\varepsilon(\boldsymbol{k}) \to T_0 \ \forall \boldsymbol{k}$), and should therefore also represent a relatively useful approximation in the intermediate range. To explain the method, we begin first with the free system, described by

$$\mathcal{H}_0 = \sum_{\boldsymbol{k},\sigma} (\varepsilon(\boldsymbol{k}) - \mu)\, a^+_{\boldsymbol{k}\sigma}\, a_{\boldsymbol{k}\sigma} \equiv \sum_{ij\sigma} \left(T_{ij} - \mu\delta_{ij}\right) a^+_{i\sigma}\, a_{j\sigma} \,. \tag{4.57}$$

The corresponding *atomic limit* is still simpler:

$$\mathcal{H}_{00} = \sum_{i,\sigma} (T_0 - \mu)\, a^+_{i\sigma}\, a_{i\sigma} \,. \tag{4.58}$$

The single-particle Green's function which applied here will be called the

centroid function $G_{00\sigma}(E)$.

Its equation of motion can be rapidly formulated, giving:

$$G_{00\sigma}(E) = \hbar\,[E - T_0 + \mu]^{-1} \,. \tag{4.59}$$

The single-particle Green's function for $\mathcal{H}_0$ was already derived in (3.198):

$$G^{(0)}_{\boldsymbol{k}\sigma}(E) = \hbar[E - \varepsilon(\boldsymbol{k}) + \mu]^{-1} \,.$$

Clearly, it can be expressed in terms of the centroid function as follows:

$$G^{(0)}_{\boldsymbol{k}\sigma}(E) = \hbar\left[\hbar G^{-1}_{00\sigma}(E) + T_0 - \varepsilon(\boldsymbol{k})\right]^{-1} \,. \tag{4.60}$$

This relation, of course, is exact. We now **postulate** that formally, the same relation between $G_{k\sigma}(E)$ and the centroid function $G_{0\sigma}(E)$ (= solution in the atomic limit!) also holds to a good approximation for **arbitrary model systems**:

Interpolation method

$$G_{k\sigma}(E) = \hbar \left[\hbar G_{0\sigma}^{-1}(E) + T_0 - \varepsilon(\boldsymbol{k})\right]^{-1} . \tag{4.61}$$

This implies for the quasi-particle density of states that:

$$\rho_\sigma(E) = \rho_0 \left[\hbar G_{0\sigma}^{-1}(E-\mu) + T_0\right] . \tag{4.62}$$

Here, $G_{0\sigma}(E)$, as the solution of the atomic limit, can as a rule be relatively simply determined.

We want to evaluate this expression for the Hubbard model. For the *atomic limit* solution, Eq. (4.11) applies:

$$G_{0\sigma}(E) = \hbar \frac{E - T_0 + \mu - U(1 - \langle n_{-\sigma}\rangle)}{(E - T_0 + \mu)(E - T_0 + \mu - U)} .$$

It then follows that:

$$\begin{aligned} \hbar G_{0\sigma}^{-1}(E) &= (E - T_0 + \mu)\left(1 - \frac{U\langle n_{-\sigma}\rangle}{E - T_0 + \mu - U(1 - \langle n_{-\sigma}\rangle)}\right) \\ &= E + \mu - T_0 - \Sigma_\sigma(E) . \end{aligned} \tag{4.63}$$

$\Sigma_\sigma(E)$ is the self-energy (4.49). The interpolation method (4.61) thus gives, with

$$G_{k\sigma}(E) = \hbar\left[E + \mu - \varepsilon(\boldsymbol{k}) - \Sigma_\sigma(E)\right]^{-1} \tag{4.64}$$

for the Hubbard model, exactly the same solution as the *Hubbard decoupling* which we discussed in the last section. By construction, the interpolation method is exact in the *band limit* and in the *atomic limit*.

4.1.5 The Method of Moments

The *Hubbard solution* treated in Sect. 4.1.3 was originally conceived for the description of band magnetism. However, one can readily see that a spontaneous magnetisation is possible in the framework of this theory only under very exceptional, even hardly plausible conditions (e.g. a low particle density n!). We will consider the reasons for this later. Whilst the Stoner model (Sect. 4.1.2) clearly overestimates the occurrence of ferromagnetism – the Stoner criterion (4.30) is too weak –, the *Hubbard solution* gives a criterion which is too restrictive!

We want now to use the example of the Hubbard model to develop a method which is distinctly different from the usual *decoupling procedures* for Green's functions. It has already proven itself as a very effective technique in many-body theory and gives e.g. in the case of the Hubbard model very realistic criteria for the existence of band ferromagnetism.

Our starting point in this case is the one-electron spectral density (3.320) or (3.321):

$$S_{ij(k)\sigma}(E) = \frac{1}{2\pi} \int\limits_{-\infty}^{+\infty} \mathrm{d}(t-t') \exp\left(-\frac{\mathrm{i}}{\hbar} E\,(t-t')\right) \cdot \left\langle \left[a_{i(k)\sigma}(t), a^{+}_{j(k)\sigma}(t') \right]_{+} \right\rangle . \tag{4.65}$$

The procedure consists of two steps. First, one attempts to *guess* the general structure of this fundamental function, guided by exactly solvable limiting cases, spectral representations, approximations which are known to be *reliable*, or general *plausibility considerations*. This leads to a particular *ansatz* for the spectral density, which contains a number of initially unknown parameters. In a second step, these are then adjusted to the exactly calculable spectral moments $M^{(n)}_{k\sigma}$ of the spectral density being sought. The essential point is that for these moments, according to (3.180) and (3.181), there are two equivalent representations. One of them yields the relationship to the spectral density

$$M^{(n)}_{k\sigma} = \frac{1}{\hbar} \int\limits_{-\infty}^{+\infty} \mathrm{d}E\, E^n S_{k\sigma}(E) ; \quad n = 0, 1, 2, \dots , \tag{4.66}$$

whilst via the second relation, all the moments can be calculated exactly, independently of the function being sought, at least in principle:

$$M^{(n)}_{k\sigma} = \frac{1}{N} \sum_{i,j} \mathrm{e}^{-\mathrm{i}k \cdot (R_i - R_j)} \cdot \left\langle \Big[\underbrace{\big[\dots [a_{i\sigma}, \mathcal{H}]_{-} , \dots , \mathcal{H} \big]_{-}}_{(n-p)\text{-fold}} , \underbrace{\big[\mathcal{H}, \dots , \big[\mathcal{H}, a^{+}_{j\sigma} \big]_{-} \dots \big]_{-}}_{p\text{-fold}} \Big]_{+} \right\rangle . \tag{4.67}$$

We are thus seeking an *ansatz* for $S_{k\sigma}(E)$ which contains m free parameters; we will then insert it into (4.66) and finally fix the parameters using the first m moments $M^{(n)}_{k\sigma}$, which can be computed exactly from (4.67). This procedure depends upon two decisive preconditions. For one thing, the *ansatz* must come as close as possible to the correct structure of the spectral density. Secondly, all of the expectation values which occur in the moments must be expressible through $S_{k\sigma}(E)$ in some form by making use of the spectral theorem (3.157), in order to arrive at a closed, self-consistently solvable system of equations. As the order n of the moments increases, the expectation values however become more and more complicated, so that this latter condition sets limits to the number of moments which can be employed.

Now how might a *reasonable ansatz* in the framework of the Hubbard model look? The general considerations in Sect. 3.4.2 have shown that as a rule, the spectral density should have the form of a linear combination of weighted δ- and Lorentz functions. If we are not interested in lifetime effects, we can adopt the version from (3.339):

$$S_{k\sigma}(E) = \hbar \sum_{j=1}^{n_0} \alpha_{j\sigma}(\boldsymbol{k})\delta\left(E + \mu - E_{j\sigma}(\boldsymbol{k})\right) . \tag{4.68}$$

We treat $\alpha_{j\sigma}(\boldsymbol{k})$ and $E_{j\sigma}(\boldsymbol{k})$ as the initially undetermined parameters. Now, the question is: How large is the number n_0 of quasi-particle poles? A hint can be given by the exactly solvable atomic limit, which must naturally also be contained in (4.68) as a limiting case. In this case, however, from (4.14), we have:

$$n_0 = 2 . \tag{4.69}$$

The Hubbard solution (4.54) (or the equivalent result (4.64) of the interpolation method) likewise corresponds to such a *two-pole structure* of the spectral density. It therefore seems attractive to choose as our *ansatz* a sum of two weighted δ-functions. This then contains four undetermined parameters, the two spectral weights $\alpha_{j\sigma}(\boldsymbol{k})$, and the two quasi-particle energies $E_{j\sigma}(\boldsymbol{k})$. These are fixed by the first four exactly calculated spectral moments. Using the model Hamiltonian (4.1) in (4.67), we find after a straightforward but somewhat tedious computation:

$$M_{\boldsymbol{k}\sigma}^{(0)} = 1 , \tag{4.70}$$

$$M_{\boldsymbol{k}\sigma}^{(1)} = (\varepsilon(\boldsymbol{k}) - \mu) + U \langle n_{-\sigma} \rangle , \tag{4.71}$$

$$M_{\boldsymbol{k}\sigma}^{(2)} = (\varepsilon(\boldsymbol{k}) - \mu)^2 + 2U \langle n_{-\sigma} \rangle (\varepsilon(\boldsymbol{k}) - \mu) + U^2 \langle n_{-\sigma} \rangle , \tag{4.72}$$

$$\begin{aligned} M_{\boldsymbol{k}\sigma}^{(3)} &= (\varepsilon(\boldsymbol{k}) - \mu)^3 + 3U \langle n_{-\sigma} \rangle (\varepsilon(\boldsymbol{k}) - \mu)^2 + \\ &+ U^2 \langle n_{-\sigma} \rangle (2 + \langle n_{-\sigma} \rangle)(\varepsilon(\boldsymbol{k}) - \mu) + \\ &+ U^2 \langle n_{-\sigma} \rangle (1 - \langle n_{-\sigma} \rangle)(B_{\boldsymbol{k}-\sigma} - \mu) + U^3 \langle n_{-\sigma} \rangle . \end{aligned} \tag{4.73}$$

Here, as an abbreviation, we have written:

$$\langle n_{-\sigma} \rangle (1 - \langle n_{-\sigma} \rangle) B_{\boldsymbol{k}-\sigma} = B_{S,-\sigma} + B_{W,-\sigma}(\boldsymbol{k}) + T_0 \langle n_{-\sigma} \rangle . \tag{4.74}$$

This term turns out to be decisive for the possibility of a spontaneous spin ordering. It must therefore be carefully considered. Most important is the first term, which gives rise to a **spin-dependent band shift**:

$$B_{S,-\sigma} = \frac{1}{N} \sum_{i,j} T_{ij} \left\langle a_{i-\sigma}^{+} a_{j-\sigma} (2n_{i\sigma} - 1) \right\rangle . \tag{4.75}$$

The second term in (4.74) has an influence in particular on the widths of the quasi-particle bands, due to its $\boldsymbol{k}$ dependence:

$$B_{W,-\sigma}(\boldsymbol{k}) = \frac{1}{N}\sum_{i,j} T_{ij}\,\mathrm{e}^{\mathrm{i}\boldsymbol{k}\cdot(\boldsymbol{R}_i-\boldsymbol{R}_j)} \cdot \Big\{\left\langle n_{i-\sigma}\,n_{j-\sigma}\right\rangle - \langle n_{-\sigma}\rangle^2 - \\ - \left\langle a^+_{j\sigma}\,a^+_{j-\sigma}\,a_{i-\sigma}\,a_{i\sigma}\right\rangle - \left\langle a^+_{j\sigma}\,a^+_{i-\sigma}\,a_{j-\sigma}\,a_{i\sigma}\right\rangle\Big\} . \tag{4.76}$$

The Hubbard model is intended primarily to answer questions concerning magnetism. In this connection, $B_{W,-\sigma}(\boldsymbol{k})$ plays only a minor role. Thus, in the Hartree-Fock approximation, the first two terms on the right-hand side of (4.76) compensate each other. The other two terms are even spin-independent; the following relation holds:

$$\left\langle a^+_{j\sigma}\,a^+_{j-\sigma}\,a_{i-\sigma}\,a_{i\sigma}\right\rangle = \left\langle a^+_{j-\sigma}\,a^+_{j\sigma}\,a_{i\sigma}\,a_{i-\sigma}\right\rangle ,$$

and for real expectation values:

$$\begin{aligned}\left\langle a^+_{j\sigma}\,a^+_{i-\sigma}\,a_{j-\sigma}\,a_{i\sigma}\right\rangle &= \left\langle\left(a^+_{j\sigma}\,a^+_{i-\sigma}\,a_{j-\sigma}\,a_{i\sigma}\right)^+\right\rangle = \\ &= \left\langle a^+_{i\sigma}\,a^+_{j-\sigma}\,a_{i-\sigma}\,a_{j\sigma}\right\rangle = \\ &= \left\langle a^+_{j-\sigma}\,a^+_{i\sigma}\,a_{j\sigma}\,a_{i-\sigma}\right\rangle .\end{aligned}$$

It should therefore be quite sufficient to take $B_{W,-\sigma}(\boldsymbol{k})$ into account merely in the form of an average over all wavenumbers $\boldsymbol{k}$:

$$\begin{aligned}&\frac{1}{N}\sum_{\boldsymbol{k}} B_{W,-\sigma}(\boldsymbol{k}) = \\ &= \frac{1}{N}\sum_{i,j} T_{ij}\left(\frac{1}{N}\sum_{\boldsymbol{k}} \mathrm{e}^{-\mathrm{i}\boldsymbol{k}\cdot(\boldsymbol{R}_i-\boldsymbol{R}_j)}\right)\Big\{\left\langle n_{i-\sigma}\,n_{j-\sigma}\right\rangle - \langle n_{-\sigma}\rangle^2 - \\ &\quad - \left\langle a^+_{j\sigma}\,a^+_{j-\sigma}\,a_{i-\sigma}\,a_{i\sigma}\right\rangle - \left\langle a^+_{j\sigma}\,a^+_{i-\sigma}\,a_{j-\sigma}\,a_{i\sigma}\right\rangle\Big\} = \\ &= \frac{1}{N}\sum_{ij} T_{ij}\delta_{ij}\Big\{\left\langle n_{i-\sigma}\,n_{j-\sigma}\right\rangle - \langle n_{-\sigma}\rangle^2 - \\ &\quad - \left\langle a^+_{j\sigma}\,a^+_{j-\sigma}\,a_{i-\sigma}\,a_{i\sigma}\right\rangle - \left\langle a^+_{j\sigma}\,a^+_{i-\sigma}\,a_{j-\sigma}\,a_{i\sigma}\right\rangle\Big\} = \\ &= T_0\Big\{\langle n_{-\sigma}\rangle\left(1-\langle n_{-\sigma}\rangle\right) - 2\left\langle n_{i-\sigma}\,n_{i\sigma}\right\rangle\Big\} .\end{aligned} \tag{4.77}$$

$B_{k,-\sigma}$ from (4.74) is then absorbed completely into the **band correction** $B_{-\sigma}$:

$$\begin{aligned}\langle n_{-\sigma}\rangle(1-\langle n_{-\sigma}\rangle)B_{-\sigma} &= T_0\,\langle n_{-\sigma}\rangle(1-\langle n_{-\sigma}\rangle)+\\ &+\frac{1}{N}\sum_{i,j}^{i\neq j}T_{ij}\left\langle a^+_{i-\sigma}a_{j-\sigma}(2n_{i\sigma}-1)\right\rangle\,.\end{aligned}\tag{4.78}$$

With the exact spectral moments (4.70), (4.71), (4.72) and (4.73), the free parameters in our *ansatz* (4.68) for the spectral density are fixed via (4.66). For the **quasi-particle energies,** one finds

$$E_{j\sigma}(\boldsymbol{k}) = H_\sigma(\boldsymbol{k})+(-1)^j\sqrt{K_\sigma(\boldsymbol{k})}\,,\tag{4.79}$$

$$H_\sigma(\boldsymbol{k}) = \frac{1}{2}\left(\varepsilon(\boldsymbol{k})+U+B_{-\sigma}\right)\,,\tag{4.80}$$

$$K_\sigma(\boldsymbol{k}) = \frac{1}{4}\left(U+B_{-\sigma}-\varepsilon(\boldsymbol{k})\right)^2+U\,\langle n_{-\sigma}\rangle\left(\varepsilon(\boldsymbol{k})-B_{-\sigma}\right)\,,\tag{4.81}$$

and for the spectral weights:

$$\alpha_{j\sigma}(\boldsymbol{k}) = (-1)^j\,\frac{E_{j\sigma}(\boldsymbol{k})-B_{-\sigma}-U\,(1-\langle n_{-\sigma}\rangle)}{E_{2\sigma}(\boldsymbol{k})-E_{1\sigma}(\boldsymbol{k})}\,.\tag{4.82}$$

These results have the same structure as the *Hubbard solutions* (4.55) and (4.56). New, but very essential, is the band correction $B_{-\sigma}$. If we replace it in the above expressions with its value in the *atomic limit,*

$$B_{-\sigma}\xrightarrow[T_{ij}\to T_0\delta_{ij}]{}T_0\,,\tag{4.83}$$

then we find precisely the same results as the Hubbard solution. In the method of moments, the quasi-particle quantities acquire an additional spin dependence via $B_{-\sigma}$.

To solve the problem completely, we still have to determine the expectation values $\langle n_{-\sigma}\rangle$ and $B_{-\sigma}$, or to express them in terms of $S_{k\sigma}(E)$, in order to arrive at a closed, self-consistently solvable system of equations. For $\langle n_{-\sigma}\rangle$, we can employ the spectral theorem (3.157) directly:

$$\langle n_{-\sigma}\rangle = \frac{1}{N\hbar}\sum_{\boldsymbol{k}}\int\limits_{-\infty}^{+\infty}\mathrm{d}E\;f_-(E)S_{\boldsymbol{k}-\sigma}(E-\mu)\,.\tag{4.84}$$

The band correction $B_{-\sigma}$ is, however, determined essentially by a *higher-order* equal-time correlation function, namely by

$$\left\langle a^+_{i-\sigma}a_{j-\sigma}n_{i\sigma}\right\rangle\,.$$

Fortunately, this term can likewise be expressed in terms of the one-electron spectral density. This however requires some preliminary considerations. First of all, as in (4.3), we have

$$[\mathcal{H}, a^+_{i-\sigma}]_- = \sum_m (T_{mi} - \mu\delta_{mi})\, a^+_{m-\sigma} + U n_{i\sigma} a^+_{i-\sigma}\,, \tag{4.85}$$

and we can then express the desired expectation value as follows:

$$\begin{aligned} \langle a^+_{i-\sigma} a_{j-\sigma} n_{i\sigma}\rangle = &-\frac{1}{U}\sum_m (T_{mi} - \mu\delta_{mi})\langle a^+_{m-\sigma} a_{j-\sigma}\rangle + \\ &+ \frac{1}{U}\Big\langle [\mathcal{H}, a^+_{i-\sigma}]_-\, a_{j-\sigma}\Big\rangle\,. \end{aligned} \tag{4.86}$$

If we now once again use the spectral theorem, as well as the equation of motion (3.27) for time-dependent Heisenberg operators, then for the second term, we can write down the following expressions:

$$\begin{aligned} &\langle [\mathcal{H}, a^+_{i-\sigma}]\, a_{j-\sigma}\rangle = \\ &= \frac{1}{\hbar}\int\limits_{-\infty}^{+\infty} \mathrm{d}E\,\left(\mathrm{e}^{\beta E}+1\right)^{-1} \int\limits_{-\infty}^{+\infty} \mathrm{d}(t-t')\,\cdot \\ &\quad\cdot \exp\left(\frac{\mathrm{i}}{\hbar}E\,(t-t')\right)\left(-\mathrm{i}\hbar\frac{\partial}{\partial t'}\right) S_{ji-\sigma}(t-t') = \\ &= \frac{1}{\hbar}\int\limits_{-\infty}^{+\infty} \mathrm{d}E\,\left(\mathrm{e}^{\beta E}+1\right)^{-1} \int\limits_{-\infty}^{+\infty} \mathrm{d}(t-t')\,\exp\left(\frac{\mathrm{i}}{\hbar}E\,(t-t')\right)\cdot \\ &\quad\cdot \frac{1}{2\pi\hbar}\int\limits_{-\infty}^{+\infty} \mathrm{d}\overline{E}\,\exp\left[-\frac{\mathrm{i}}{\hbar}\overline{E}\,(t-t')\right]\overline{E}\,S_{ji-\sigma}(\overline{E}) = \\ &= \frac{1}{\hbar}\int\limits_{-\infty}^{+\infty} \mathrm{d}E\,\left(\mathrm{e}^{\beta E}+1\right)^{-1} \int\limits_{-\infty}^{+\infty} \mathrm{d}\overline{E}\,\delta(E-\overline{E})\overline{E}\,S_{ji-\sigma}(\overline{E})\,. \end{aligned}$$

This finally leads to:

$$\begin{aligned} \Big\langle [\mathcal{H}, a^+_{i-\sigma}]_-\, a_{j-\sigma}\Big\rangle = &\frac{1}{N\hbar}\sum_{\boldsymbol{k}} \mathrm{e}^{-\mathrm{i}\boldsymbol{k}\cdot(\boldsymbol{R}_i-\boldsymbol{R}_j)}\,\cdot \\ &\cdot \int\limits_{-\infty}^{+\infty} \mathrm{d}E\, f_-(E)(E-\mu) S_{\boldsymbol{k}-\sigma}(E-\mu)\,. \end{aligned} \tag{4.87}$$

For the remaining expectation value, $\langle a^+_{m-\sigma} a_{j-\sigma} \rangle$ in (4.86) , we can make use of the spectral theorem directly:

$$\langle a^+_{m-\sigma} a_{j-\sigma} \rangle = \frac{1}{N\hbar} \sum_{\boldsymbol{k}} \mathrm{e}^{-\mathrm{i}\boldsymbol{k}\cdot(\boldsymbol{R}_m - \boldsymbol{R}_j)} \cdot \int_{-\infty}^{+\infty} \mathrm{d}E\, f_-(E) S_{\boldsymbol{k}-\sigma}(E-\mu)\,. \tag{4.88}$$

We then obtain for the expectation value (4.86):

$$\langle a^+_{i-\sigma} a_{j-\sigma} n_{i\sigma} \rangle = \frac{1}{N\hbar} \sum_{\boldsymbol{k}} \mathrm{e}^{-\mathrm{i}\boldsymbol{k}\cdot(\boldsymbol{R}_i - \boldsymbol{R}_j)} \cdot \int_{-\infty}^{+\infty} \mathrm{d}E\, f_-(E) \frac{1}{U} (E - \varepsilon(\boldsymbol{k}))\, S_{\boldsymbol{k}-\sigma}(E-\mu)\,. \tag{4.89}$$

For the band correction, we require

$$\frac{1}{N} \sum_{i,j} T_{ij} \langle a^+_{i-\sigma} a_{j-\sigma} (2n_{i\sigma} - 1) \rangle =$$

$$= \frac{1}{N\hbar} \sum_{\boldsymbol{k}} \varepsilon(\boldsymbol{k}) \int_{-\infty}^{+\infty} \mathrm{d}E\, f_-(E) \left[\frac{2}{U} (E - \varepsilon(\boldsymbol{k})) - 1 \right] S_{\boldsymbol{k}-\sigma}(E-\mu)\,,$$

from which we must subtract the diagonal term

$$T_0 \langle n_{i-\sigma} (2n_{i\sigma} - 1) \rangle = \frac{T_0}{N\hbar} \sum_{\boldsymbol{k}} \int_{-\infty}^{+\infty} \mathrm{d}E\, f_-(E) \left[\frac{2}{U} (E - \varepsilon(\boldsymbol{k})) - 1 \right] S_{\boldsymbol{k}-\sigma}(E-\mu)\,.$$

If we now make use of our two-pole approach (4.68) for the spectral density, then we find for the band correction which we have been seeking:

$$\begin{aligned} &\langle n_{-\sigma} \rangle (1 - \langle n_{-\sigma} \rangle) B_{-\sigma} = \\ &= \langle n_{-\sigma} \rangle (1 - \langle n_{-\sigma} \rangle) T_0 + \frac{1}{N} \sum_{\boldsymbol{k}} \sum_{j=1}^{2} \alpha_{j-\sigma}(\boldsymbol{k}) (\varepsilon(\boldsymbol{k}) - T_0) f_- \left(E_{j-\sigma}(\boldsymbol{k}) \right) \cdot \\ &\quad \cdot \left[\frac{2}{U} \left(E_{j-\sigma}(\boldsymbol{k}) - \varepsilon(\boldsymbol{k}) \right) - 1 \right]. \end{aligned} \tag{4.90}$$

Clearly, Eqs. (4.79), (4.80), (4.81) and (4.82), (4.84), and (4.90) form a closed system, which can be solved self-consistently. The model parameters are the following:

1. the **temperature**, which enters into the Fermi functions,
2. the **band occupation** $n = \sum_\sigma \langle n_\sigma \rangle$, which determines the chemical potential μ,
3. the **Coulomb interaction** U and
4. the **lattice structure**, which determines the *free* Bloch density of states $\rho_0(E) = \frac{1}{N}\sum_{k} \delta(E - \varepsilon(\boldsymbol{k}))$ or the single-particle energies $\varepsilon(\boldsymbol{k})$ and affects the summation over $\boldsymbol{k}$.

Figure 4.5 illustrates the quasi-particle density of states

$$\rho_\sigma(E) = \frac{1}{N}\sum_{\boldsymbol{k}}\sum_{j=1}^{2} \alpha_{j\sigma}(\boldsymbol{k})\delta\left(E - E_{j\sigma}(\boldsymbol{k})\right) \tag{4.91}$$

for two different band occupations, $n=0.6$ and $n=0.8$, as well as $U=6$eV and $T=0$K. The Bloch density of states used is sketched in Fig. 4.6. We can see that the original band splits into two quasi-particle bands per spin direction.

For the situations indicated, there is an additional shift of the two spin spectra. Since the bands are filled up to the Fermi energies, which are marked by bars, there will be a preferred spin direction and thereby a nonvanishing spontaneous magnetisation m. Finally, the observed band shift is caused by the band correction $B_{-\sigma}$. Once $B_\uparrow \neq B_\downarrow$, it follows

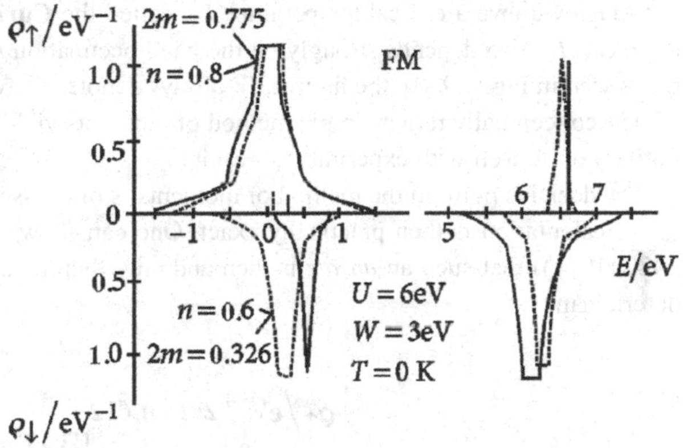

Fig. 4.5 The quasi-particle density of states in the Hubbard model for the ferromagnetic phase as a function of the energy, for two different band occupations, calculated using the method of moments

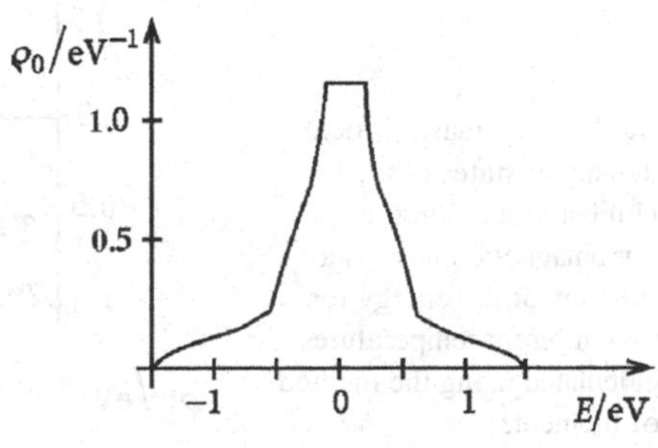

Fig. 4.6 The Bloch density of states of the non-interacting system as a function of the energy

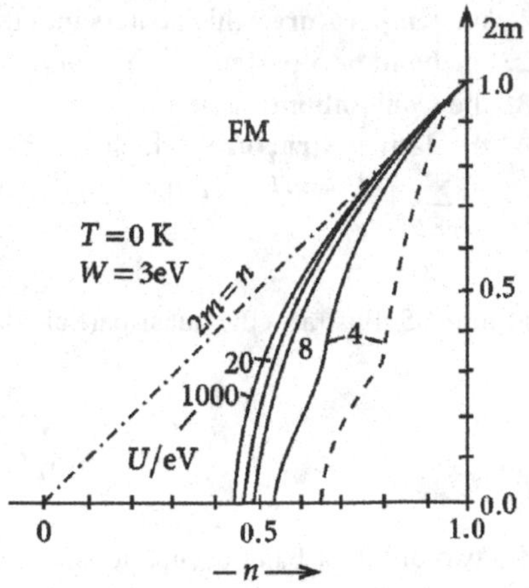

Fig. 4.7 The spontaneous magnetisation m of a system of correlated electrons described by the Hubbard model, as a function of the band occupation n for different values of the Coulomb interaction U, calculated with the method of moments for a Bloch density of states as in Fig. 4.6

that $m \neq 0$. The band correction is lacking in the Hubbard solutions in Sect. 4.1.3, and they therefore do not readily predict the occurrence of ferromagnetism.

Figure 4.7 illustrates the importance of the parameters U and n for the occurrence of ferromagnetism.

The quasi-particle densities of states $\rho_{\uparrow\downarrow}(E)$ are, in contrast to $\rho_0(E)$, noticeably temperature dependent. With increasing temperature, $\rho_\uparrow$ and $\rho_\downarrow$ become increasingly similar, until finally above a critical temperature T_C, called the **Curie temperature**, they become identical. T_C also depends strongly on the band occupation n and the interaction constant U, as seen in Fig. 4.9. (In the figures, W always denotes the width of the *free* Bloch band!)

The conceptually rather simple method of moments yields T_C values which agree qualitatively quite well with experimental results.

The decisive point in the method of moments is of course the *ansatz* in (4.68). The rest of the calculation is then practically exact. One can show (A. Lonke, J. Math. Phys. **12**, 2422 (1971)) that such an *ansatz* is then and only then mathematically precise, when the determinant

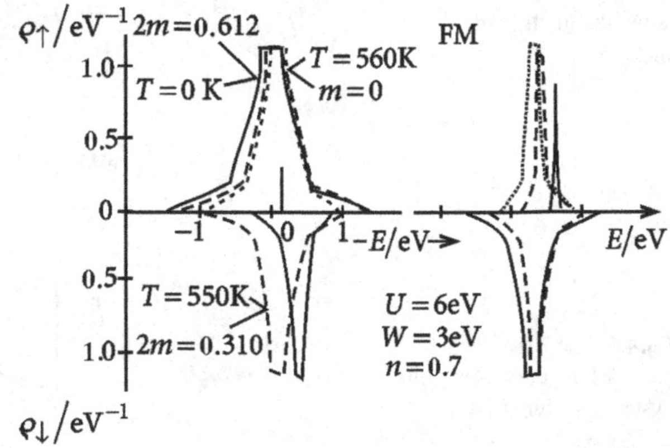

Fig. 4.8 The quasi-particle density of states of the Hubbard model in the ferromagnetic phase as a function of the energy for two different temperatures, calculated using the method of moments

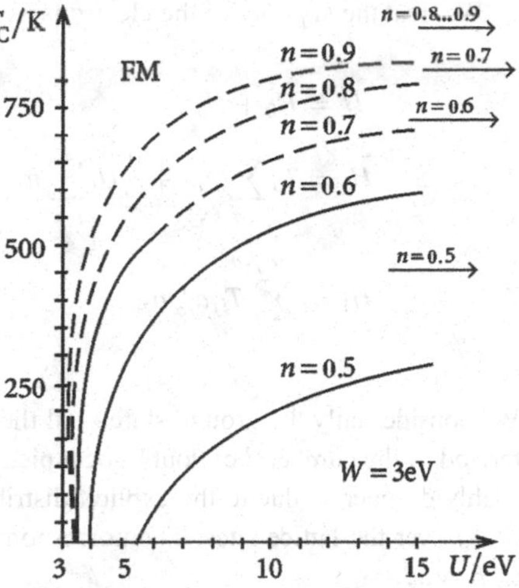

Fig. 4.9 The Curie temperature in the Hubbard model as a function of the Coulomb interaction parameter U for different band occupations n, calculated by the method of moments

$$D_{k\sigma}^{(r)} \equiv \begin{vmatrix} M_{k\sigma}^{(0)} & \cdots & M_{k\sigma}^{(r)} \\ \vdots & & \vdots \\ M_{k\sigma}^{(r)} & \cdots & M_{k\sigma}^{(2r)} \end{vmatrix} \tag{4.92}$$

is zero for $r = n_0$ and nonzero for all lower orders $r = 1, 2, \dots, n_0 - 1$. The elements of the determinant are just the spectral moments (4.67). (As an exercise, one can investigate the atomic limit as solved in Sect. 4.1.1, using (4.92)!)

4.1.6 The Exactly Half-filled Band

Often, valuable physical information can be obtained by transforming the model Hamiltonian for a case of interest to an equivalent effective operator. A rewarding possibility of this type is offered by the Hubbard model for the special case of an exactly half-filled band. In the Hubbard model, the system is described as a lattice of atoms, each of which has a single atomic level which then can be occupied by at most two electrons (of opposite spins). A *half-filled band* here thus means that each atom contributes exactly one electron, i.e. there are just as many electrons as lattice sites ($n = 1$!). In the *atomic limit*, in the ground state, each site is occupied by exactly one electron. The only variable is then the electronic spin. If we now gradually *switch on the hopping*, then the band electrons will still remain strongly localised. *Virtual* site exchanges will however still give rise to an indirect coupling between the electronic spins at the different lattice sites. Such a situation is described as a rule by the Heisenberg model (2.203). We wish to show in this section, using elementary perturbation theory, that in the situation described, i.e. ($n = 1, , U/W \gg 1$), there is an equivalence between the Hubbard and the Heisenberg models.

We treat the *hopping* of the electrons as a perturbation:

$$H = H_0 + H_1 \,, \tag{4.93}$$

$$H_0 = T_0 \sum_{i,\sigma} n_{i\sigma} + \frac{1}{2} U \sum_{i,\sigma} n_{i\sigma} n_{i-\sigma} \,; \quad (n = 1; \quad U/W \gg 1) \,, \tag{4.94}$$

$$H_1 = \sum_{i,j,\sigma}^{i \neq j} T_{ij} a_{i\sigma}^{+} a_{j\sigma} \,. \tag{4.95}$$

We consider only the ground state – all the eigenvalues and eigenstates of H_0 are characterised by the number d of doubly-occupied lattice sites. The states with the same d are still highly degenerate due to the explicit distribution of the N_σ electrons with spin σ ($\sigma = \uparrow$ or $\downarrow$) over the lattice sites. The corresponding enumeration is denoted by Greek letters: $\alpha, \beta, \gamma, \ldots$

$$H_0 \, |d\alpha\rangle^{(0)} = E_d^{(0)} \, |d\alpha\rangle^{(0)} = (N T_0 + d U) \, |d\alpha\rangle^{(0)} \,. \tag{4.96}$$

Since $n = 1$, we have

$$|0\alpha\rangle^{(0)} : \quad 2^N\text{-fold } \textbf{degenerate ground state}.$$

First-order perturbation theory requires the solution of the secular equation,

$$\det \left[{}^{(0)}\langle 0\alpha' \mid H_1 \mid 0\alpha\rangle^{(0)} - E_0^{(1)} \delta_{\alpha\alpha'} \right] \stackrel{!}{=} 0 \,, \tag{4.97}$$

with 2^N solutions $E_{0\alpha}^{(1)}$. Now, one can readily see that

$${}^{(0)}\langle d\alpha' \mid H_1 \mid 0\alpha\rangle^{(0)} \neq 0 \quad \textbf{only} \text{ for } d = 1 \tag{4.98}$$

is allowed, since every term of the operator H_1 produces an empty and a doubly-occupied site. The perturbation matrix in (4.97) thus contains as elements only zeroes. All the energy corrections to first order $E_{0\alpha}^{(1)}$ vanish; the degeneracy remains completely unchanged.

Second-order perturbation theory requires the solution of a system of equations:

$$\sum_{\alpha} C_\alpha \left\{ \sum_{d,\gamma}^{d \neq 0} {}^{(0)}\langle 0\alpha' \mid H_1 \mid d\gamma\rangle^{(0)\,(0)}\langle d\gamma \mid H_1 \mid 0\alpha\rangle^{(0)} \cdot \frac{1}{E_0^{(0)} - E_d^{(0)}} - E_0^{(2)} \delta_{\alpha\alpha'} \right\} \stackrel{!}{=} 0 \,. \tag{4.99}$$

This corresponds to the eigenvalue equation of an **effective Hamiltonian** H_{eff} with the matrix elements:

$$
\begin{aligned}
&{}^{(0)}\langle 0\alpha' | \, H_1 \sum_{d,\gamma}^{d \neq 0} \frac{|d\gamma\rangle^{(0)} \, {}^{(0)}\langle d\gamma|}{E_0^{(0)} - E_d^{(0)}} H_1 \, |0\alpha\rangle^{(0)} = \\
&= -\frac{1}{U} \, {}^{(0)}\langle 0\alpha' | \, H_1 \left(\sum_{d,\gamma} |d\gamma\rangle^{(0)} \, {}^{(0)}\langle d\gamma| \right) H_1 \, |0\alpha\rangle^{(0)} = \\
&= -\frac{1}{U} \, {}^{(0)}\langle 0\alpha' \mid H_1^2 \mid 0\alpha\rangle^{(0)} .
\end{aligned}
\tag{4.100}
$$

In the first step, we made use of (4.98), yielding

$$
\left(E_d^{(0)} - E_0^{(0)} \right) \longrightarrow \left(E_1^{(0)} - E_0^{(0)} \right) = U
$$

and allowing us to leave off the constraint $d \neq 0$. The second step follows from the completeness relation for the *unperturbed* states $|d\gamma\rangle^{(0)}$. Let

P_0: projection operator onto the subspace $d = 0$;

it then follows for our effective Hamiltonian of second order:

$$
H_{\text{eff}} = P_0 \left(-\frac{H_1^2}{U} \right) P_0 . \tag{4.101}
$$

We now rewrite this in terms of spin operators. To do so, we first insert (4.95):

$$
H_{\text{eff}} = -\frac{1}{U} P_0 \left(\sum_{\substack{ij \\ \sigma}}^{i \neq j} \sum_{\substack{mn \\ \sigma'}}^{m \neq n} T_{ij} T_{mn} \, a_{i\sigma}^{+} \, a_{j\sigma} \, a_{m\sigma'}^{+} \, a_{n\sigma'} \right) P_0 . \tag{4.102}
$$

In the multiple sum, only the terms

$$
i = n \quad \text{and} \quad j = m
$$

give nonvanishing contributions. We then have:

$$
\begin{aligned}
H_{\mathrm{eff}} &= -\frac{1}{U} P_0 \left(\sum_{\substack{ij\\ \sigma\sigma'}}^{i\neq j} T_{ij} T_{ji}\, a^+_{i\sigma}\, a_{j\sigma}\, a^+_{j\sigma'}\, a_{i\sigma'} \right) P_0 = \\
&= -\frac{1}{U} P_0 \left(\sum_{\substack{ij\\ \sigma\sigma'}}^{i\neq j} T^2_{ij}\, a^+_{i\sigma}\, a_{i\sigma'} \left(\delta_{\sigma\sigma'} - a^+_{j\sigma'}\, a_{j\sigma} \right) \right) P_0 = \\
&= -\frac{1}{U} P_0 \left(\sum_{ij\sigma}^{i\neq j} T^2_{ij} \left(n_{i\sigma} - n_{i\sigma}\, n_{j\sigma} - a^+_{i\sigma}\, a_{i-\sigma}\, a^+_{j-\sigma}\, a_{j\sigma} \right) \right) P_0 \,.
\end{aligned}
\tag{4.103}
$$

We now introduce the **spin operators**:

$$
S_i^z = \frac{1}{2} \sum_{\sigma} z_\sigma n_{i\sigma} \,, \tag{4.104}
$$

$$
S_i^\sigma = a^+_{i\sigma}\, a_{i-\sigma} \qquad \left(S_i^\uparrow \equiv S_i^+ ,\; S_i^\downarrow \equiv S_i^- \right) . \tag{4.105}
$$

One can readily see that these operators fulfil the elementary commutation relations (2.215) and (2.216) (cf. Ex. 4.1.6). (Remember: $z_\uparrow = +1,\; z_\downarrow = -1$):

$$
\begin{aligned}
P_0 \left(S_i^z S_j^z \right) P_0 &= \frac{1}{4} \sum_{\sigma,\sigma'} z_\sigma\, z_{\sigma'}\, P_0 \left(n_{i\sigma}\, n_{j\sigma'} \right) P_0 = \\
&= \frac{1}{4} \sum_{\sigma} \left\{ P_0 \left(n_{i\sigma}\, n_{j\sigma} \right) P_0 - P_0 \left(n_{i\sigma}\, n_{j-\sigma} \right) P_0 \right\} = \\
&= \frac{1}{4} \sum_{\sigma} \left\{ P_0 \left(n_{i\sigma}\, n_{j\sigma} \right) P_0 - P_0 \left[n_{i\sigma} \left(\mathbf{1} - n_{j\sigma} \right) \right] P_0 \right\} = \\
&= \frac{1}{2} P_0 \left\{ \sum_{\sigma} n_{i\sigma}\, n_{j\sigma} \right\} P_0 - \frac{1}{4} P_0 \left\{ \sum_{\sigma} n_{i\sigma} \right\} P_0 = \\
&= \frac{1}{2} P_0 \left\{ \sum_{\sigma} n_{i\sigma}\, n_{j\sigma} \right\} P_0 - \frac{1}{4} P_0^2 \,.
\end{aligned}
$$

With this we have:

$$
P_0 \left\{ \sum_{\sigma} n_{i\sigma}\, n_{j\sigma} \right\} P_0 = P_0 \left\{ 2 S_i^z S_j^z + \frac{1}{2} \right\} P_0 \,, \tag{4.106}
$$

where in particular we have used

$$P_0 \left\{ \sum_\sigma n_{i\sigma} \right\} P_0 \equiv P_0 \mathbf{1} P_0 , \tag{4.107}$$

a relation which is naturally correct only for our special case $n = 1$. Finally, it follows from (4.105) that:

$$\begin{aligned} P_0 \left\{ \sum_\sigma a^+_{i\sigma} a_{i-\sigma} a^+_{j-\sigma} a_{j\sigma} \right\} P_0 &= P_0 \left\{ \sum_\sigma S_i^\sigma S_j^{-\sigma} \right\} P_0 = \\ &= P_0 \left\{ 2S_i^x S_j^x + 2S_i^y S_j^y \right\} P_0 . \end{aligned} \tag{4.108}$$

Inserting (4.106), (4.107) and (4.108) into (4.103), we obtain an effective operator of the *Heisenberg type*:

$$H_{\text{eff}} = P_0 \left\{ \sum_{i,j}^{i \neq j} \frac{T_{ij}^2}{U} \left(2\boldsymbol{S}_i \cdot \boldsymbol{S}_j - \frac{1}{2} \right) \right\} P_0 . \tag{4.109}$$

The **exchange integrals**

$$J_{ij} = -2 \frac{T_{ij}^2}{U} \tag{4.110}$$

are always negative, which favours an antiferromagnetic ordering of the electronic spins.

We have thus shown that for the half-filled band ($n = 1$), the Hubbard model is equivalent to the Heisenberg model, whereby we are even able to ascribe a microscopic interpretation to the exchange integrals J_{ij}.

The expression (4.100) from second-order perturbation theory describes virtual hopping processes from one site $\boldsymbol{R}_i$ to another, $\boldsymbol{R}_j$, and back again (Fig. 4.10). According to (4.100), these hopping processes lead to a **gain** in energy. The hopping probability is proportional to T_{ij} and is certainly maximal between nearest-neighbour lattice sites. In a ferromagnet, virtual *hopping* is not allowed due to the Pauli principle, since all the spins are parallel. In a paramagnet, the spin directions are statistically distributed over all the possible states. The number of nearest neighbours with antiparallel electronic spins is therefore certainly smaller than in an antiferromagnet. We can therefore indeed expect an antiferromagnetic ground state.

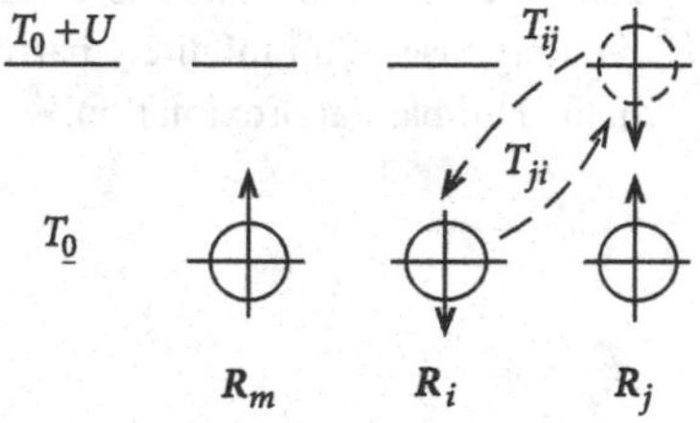

Fig. 4.10 Virtual hopping processes of an electron in the strongly correlated Hubbard model with a half-filled band ($n = 1$)

4.1.7 Exercises

Exercise 4.1.1 What form does the Hubbard Hamiltonian take in the Bloch representation? How is it different from the Hamiltonian of the jellium model?

Exercise 4.1.2 Verify the following formulation of the δ-function:

$$\delta(x) = \frac{1}{2} \lim_{\beta \to \infty} \frac{\beta}{1 + \cosh(\beta x)} \qquad (\beta > 0) .$$

Exercise 4.1.3

1. Carry out the Hartree-Fock approximation for the Hamiltonian of the jellium model. Make use of spin, momentum, and particle-number conservation.
2. Use it to compute the one-electron spectral density.
3. Construct with the aid of the spectral theorem an implicit functional equation for the average occupation number $\langle n_{k\sigma} \rangle$.
4. Calculate the internal energy $U(T)$.
5. Compare $U(T = 0)$ with the perturbation-theoretical result from Sect. 2.1.2.

Exercise 4.1.4 Verify whether

1. the Stoner approximation, and
2. the Hubbard approximation

of the Hubbard model correctly reproduce the exact results for the band limit ($U \to 0$) and for the atomic limit ($\varepsilon(\boldsymbol{k}) \to T_0 \; \forall \boldsymbol{k}$).

Exercise 4.1.5 Calculate the electronic self-energy in the Hubbard model for the limiting case of an infinitely narrow band. Compare the result with the self-energy in the Hubbard approximation.

Exercise 4.1.6

1. Show that the following definition of spin operators makes sense for itinerant band electrons:

$$S_i^z = \frac{\hbar}{2}\left(n_{i\uparrow} - n_{i\downarrow}\right) ; \quad S_i^+ = \hbar a_{i\uparrow}^+ a_{i\downarrow} ; \quad S_i^- = \hbar a_{i\downarrow}^+ a_{i\uparrow} .$$

   Verify the usual commutation relations.

2. Transform the Hubbard Hamiltonian to the spin operators of part 1. Assume the electronic system to be in a static, position-dependent magnetic field:

$$B_0 \exp(-\mathrm{i}\boldsymbol{K} \cdot \boldsymbol{R}_i)\,\boldsymbol{e}_z .$$

3. Compute for the wavenumber-dependent spin operators

$$S^\alpha(\boldsymbol{k}) = \sum_i S_i^\alpha \exp(-\mathrm{i}\boldsymbol{k} \cdot \boldsymbol{R}_i) \qquad (\alpha = x,\, y,\, z,\, +,\, -)$$

   the commutation relations: which are analogous to 1.

Exercise 4.1.7

1. Show, using the result of part 3. in Ex. 4.1.6, that for the Hubbard Hamiltonian in the wavenumber representation, the following holds:

$$H = \sum_{k,\sigma} \varepsilon(k) a^+_{k\sigma} a_{k\sigma} - \frac{2U}{3\hbar^2 N} \sum_{k} \boldsymbol{S}(k) \cdot \boldsymbol{S}(-k) + \frac{1}{2} U \widehat{N} - b S^z(\boldsymbol{K}) ,$$

$$b = \frac{2\mu_B}{\hbar} \mu_0 H , \quad \widehat{N} = \sum_{i\sigma} n_{i\sigma} .$$

2. Prove the following anticommutation relation:

$$\sum_{k} \left[S^-(-k - \boldsymbol{K}), S^+(k + \boldsymbol{K}) \right]_+ = \hbar^2 N \sum_{i} \left(n_{i\uparrow} - n_{i\downarrow} \right)^2$$

$$(\boldsymbol{K} \quad \text{arbitrary!}) .$$

3. Verify the following commutator expressions:

$$\left[S^+(k), \sum_{p} \boldsymbol{S}(p) \boldsymbol{S}(-p) \right]_- = \left[S^+(k), \widehat{N} \right]_- = 0 .$$

4. Calculate the following commutator with the Hubbard Hamiltonian H:

$$\left[S^+(k), H \right]_- = \hbar \sum_{i,j} T_{ij} \left(\mathrm{e}^{-\mathrm{i}k \cdot R_i} - \mathrm{e}^{-\mathrm{i}k \cdot R_j} \right) a^+_{i\uparrow} a_{j\downarrow} + b\hbar S^+(k + \boldsymbol{K}) .$$

5. Confirm the result for the following double commutator:

$$\left[\left[S^+(k), H \right]_- , S^-(-k) \right]_-$$
$$= \hbar^2 \sum_{i,j,\sigma} T_{ij} \left(\mathrm{e}^{-\mathrm{i}k \cdot (R_i - R_j)} - 1 \right) a^+_{i\sigma} a_{j\sigma} + 2b\hbar^2 S^z(\boldsymbol{K}) .$$

Exercise 4.1.8 For a system of interacting electrons in a narrow energy band, one can assume that

$$Q = \frac{1}{N}\sum_{i,j}|T_{ij}|\left(\boldsymbol{R}_i - \boldsymbol{R}_j\right)^2 < \infty ,$$

since the *hopping integrals* T_{ij} decrease as a rule exponentially with increasing distance $|\boldsymbol{R}_i - \boldsymbol{R}_j|$.

1. Set

$$A = S^-(-\boldsymbol{k} - \boldsymbol{K}) ; \quad C = S^+(\boldsymbol{k})$$

and estimate with the help of the partial results from Ex. 4.1.7 the following:

a) $\Sigma_k \left\langle [A, A^+]_+ \right\rangle \le 4\hbar^2 N^2$,

b) $\left\langle [[C, H]_-, C^+]_- \right\rangle \le N\hbar^2 Q k^2 + 2b\hbar^2 \left|\langle S^z(\boldsymbol{K})\rangle\right|$,

c) $\langle [C, A]_- \rangle = 2\hbar \, \langle S^z(-\boldsymbol{K})\rangle$.

Distinguish between commutators $[\ldots,\ldots]_-$ and anticommutators $[\ldots,\ldots]_+$ in 1.a) to 1.c).

2. Define as in the Heisenberg model in Ex. 2.4.7 the magnetisation:

$$M(T, B_0) = \frac{2\mu_B}{\hbar}\frac{1}{N}\sum_i e^{i\boldsymbol{k}\cdot\boldsymbol{R}_i}\left\langle S_i^z\right\rangle$$

Use the results of part 1. to estimate the following using the **Bogoliubov inequaltiy** from Ex. 2.4.5:

$$\beta \ge \frac{M^2}{(2\mu_B)^2}\frac{1}{N}\sum_k \frac{1}{\left|B_0 M\right| + \frac{1}{2}k^2 Q} .$$

3. Show, using the result of part 2., that there can be no **spontaneous** magnetisation in the $d = 1$- and in the $d = 2$-dimensional Hubbard model (**Mermin-Wagner theorem**):

$$M_S(T) = \lim_{B_0 \to 0} M(T, B_0) = 0 \quad \text{for } T \ne 0 \text{ and } d = 1, 2 .$$

Exercise 4.1.9 A system of interacting electrons in a narrow energy band is presumed to be approximately described by the Hubbard model in the *limiting case of an infinitely narrow band*,

$$T_{ij} = T_0 \delta_{ij} \,.$$

1. Verify the following exact representation for the one-electron spectral moments:

$$M_{ii\sigma}^{(n)} = T_0^n + \left[(T_0 + U)^n - T_0^n\right] \langle n_{i-\sigma} \rangle \; ; \quad n = 0, 1, 2, \ldots$$

2. Use *Lonke's theorem* (4.92) to prove that the one-electron spectral density represents a two-pole function, i.e. a linear combination of two δ-functions.
3. Compute the quasi-particle energies and their spectral weights.

Exercise 4.1.10 In Ex. 3.3.2, we have seen that the simplified model Hamiltonian H^*,

$$H^* = \sum_{k,\sigma} t(k) a_{k\sigma}^+ a_{k\sigma} - \Delta \sum_k \left(b_k + b_k^+\right) + \frac{1}{V}\Delta^2 \; ; \quad b_k^+ = a_{k\uparrow}^+ a_{-k\downarrow}^+ \,,$$

describes BCS superconductivity.

1. Give all of the spectral moments of its one-electron spectral density.
2. Show using Lonke's theorem (4.92) that the one-electron spectral density must be a two-pole function.

4.2 Collective Electronic Excitations

All of the results obtained in Sect. 2.1 concerning interacting electrons in solids were found using one-electron Green's functions or one-electron spectral densities. There are, however, also important **collective** electronic excitations such as

charge density waves (plasmons), spin density waves (magnons),

which require other Green's functions for their description. In preparation for their treatment, we first will discuss more or less qualitatively the phenomenon of **screening**, a characteristic consequence of the electron-electron interaction.

4.2.1 Charge Screening (Thomas-Fermi Approximation)

How can collective excitations arise in a system of electrons which are moving in a homogeneously distributed, positively charged *ion sea*?

We begin with the simplest possible assumption, i.e. that the electrons do not mutually interact (Sommerfeld model). One then finds a position-independent particle density n_0 (2.77):

$$n_0 = \frac{k_F^3}{3\pi^2} = \frac{(2m\varepsilon_F)^{3/2}}{3\pi^2\,\hbar^3} = n_0\,(\varepsilon_F)\;. \tag{4.111}$$

We now introduce into the system an additional static electronic charge ($q = -e$), which we may take to be located at the origin of the coordinate system. The electrons interact with this charge. Due to the Coulomb repulsion, they have in the neighbourhood of the *test charge* at $r = 0$ an additional potential energy

$$E_{\text{pot}}(\boldsymbol{r}) = (-e)\varphi(\boldsymbol{r})\,, \tag{4.112}$$

where $\varphi(\boldsymbol{r})$ is the electrostatic potential of the *test charge*. They will thus tend to avoid the neighbourhood of $\boldsymbol{r} = 0$, i.e. the particle density $n(\boldsymbol{r})$ becomes position dependent. In fact, we should solve the Schrödinger equation in order to calculate the particle density,

$$-\frac{\hbar^2}{2m}\Delta\psi_i(\boldsymbol{r}) - e\varphi(\boldsymbol{r})\psi_i(\boldsymbol{r}) = \varepsilon_i\,\psi_i(\boldsymbol{r})\,,$$

and derive the electron density from

$$n(\boldsymbol{r}) = \sum_i \left|\psi_i(\boldsymbol{r})\right|^2 .$$

In the **Thomas-Fermi model**, this procedure is drastically simplified by the assumption that the single-particle energies $\varepsilon(\boldsymbol{k})$ can be written approximately in the presence of the test charge as follows:

$$E(\boldsymbol{k}) \approx \varepsilon(\boldsymbol{k}) - e\varphi(\boldsymbol{r})\,. \tag{4.113}$$

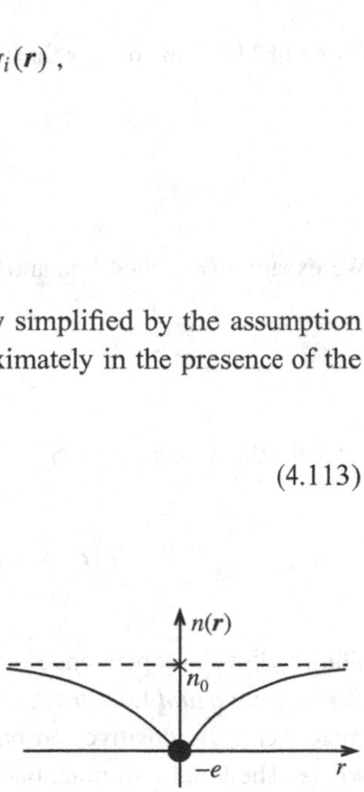

Fig. 4.11 A schematic representation of the position dependence of the particle density in the neighbourhood of a static perturbing charge in the Sommerfeld model

This is naturally not really obvious, since this expression contradicts the uncertainty relation by implying simultaneously a precisely-determined momentum and position for the electron. One must consider the electron to be a wavepacket whose position uncertainty will be of the order of $1/k_F$. In order to accept (4.113), we must then also require that $\varphi(\boldsymbol{r})$ hardly changes over a region of the order of $1/k_F$. If we transform to wavenumber-dependent Fourier components, then the Thomas-Fermi approximation will be realistic only in the region

$$q \ll k_F \ . \tag{4.114}$$

For the unperturbed electron density n_0 (4.111), we have from (3.209):

$$n_0(\varepsilon_F) = \frac{2}{V} \sum_{\boldsymbol{k}} \{\exp[\beta(\varepsilon(\boldsymbol{k}) - \varepsilon_F)] + 1\}^{-1} \ .$$

In order to obtain $n(\boldsymbol{r})$ from n_0, we replace the *unperturbed* single-particle energies $\varepsilon(\boldsymbol{k})$ by the energies $E(\boldsymbol{k})$ from (4.113):

$$\begin{aligned} n(\boldsymbol{r}) &= \frac{2}{V} \sum_{\boldsymbol{k}} \{\exp[\beta(\varepsilon(\boldsymbol{k}) - e\varphi(\boldsymbol{r}) - \varepsilon_F)] + 1\}^{-1} = \\ &= n_0(\varepsilon_F + e\varphi(\boldsymbol{r})) \ . \end{aligned} \tag{4.115}$$

Using (4.111), this means that:

$$n(\boldsymbol{r}) = \frac{[2m(\varepsilon_F + e\varphi(\boldsymbol{r}))]^{3/2}}{3\pi^2 \hbar^3} \ . \tag{4.116}$$

We expand $n(\boldsymbol{r})$ around n_0 and terminate the series under the assumption

$$\varepsilon_F \gg |e\varphi(\boldsymbol{r})|$$

after the linear term:

$$n(\boldsymbol{r}) \approx n_0 + e\varphi(\boldsymbol{r}) \frac{\partial n_0}{\partial \varepsilon_F} = n_0 \left(1 + \frac{3}{2} \frac{e\varphi(\boldsymbol{r})}{\varepsilon_F}\right) \ . \tag{4.117}$$

The resulting r-dependence is shown qualitatively in Fig. 4.11. Around the static charge at $r = 0$, a *virtual* hole forms, which has the same effect as an additional positive charge, since there, the positive ion background charges *show through* more strongly than elsewhere. The Coulomb potential of the test charge is thus shielded, so that the electrons of the system are affected by it only at distances less than a characteristic length, the **screening length**, which we still have to define. We determine this length by using the **Poisson equation**:

$$\Delta\varphi(\boldsymbol{r}) = -\frac{(-e)}{\varepsilon_0}\delta(\boldsymbol{r}) - \frac{(-e)}{\varepsilon_0}\{n(\boldsymbol{r}) - n_0\} \ . \tag{4.118}$$

The first term on the right-hand side represents the charge density of the static point charge. The second term is a result of the now incomplete compensation of the positive ion charges by the electronic charges in the neighbourhood of the perturbing charge. With (4.117), Eq. (4.118) can be simplified to:

$$\left(\Delta - \frac{3}{2}\frac{n_0 e^2}{\varepsilon_0 \varepsilon_\mathrm{F}}\right)\varphi(\boldsymbol{r}) = \frac{e}{\varepsilon_0}\delta(\boldsymbol{r}) \ . \tag{4.119}$$

The solution of this differential equation is most readily obtained by Fourier transformation:

$$\varphi(\boldsymbol{r}) = \frac{V}{(2\pi)^3}\int \mathrm{d}^3q\, \varphi(\boldsymbol{q})\mathrm{e}^{\mathrm{i}\boldsymbol{q}\cdot\boldsymbol{r}} \ ,$$

$$\delta(\boldsymbol{r}) = \frac{1}{(2\pi)^3}\int \mathrm{d}^3q\, \mathrm{e}^{\mathrm{i}\boldsymbol{q}\cdot\boldsymbol{r}} \ .$$

Inserting into (4.119), this yields:

$$\left(-q^2 - \frac{3}{2}\frac{n_0 e^2}{\varepsilon_0 \varepsilon_\mathrm{F}}\right)\varphi(\boldsymbol{q}) = \frac{e}{\varepsilon_0 V} \ .$$

We define

$$q_\mathrm{TF} = \sqrt{\frac{3n_0 e^2}{2\varepsilon_0\varepsilon_\mathrm{F}}} \tag{4.120}$$

and then obtain:

$$\varphi(\boldsymbol{q}) = \frac{-e}{\varepsilon_0 V\left(q^2 + q_\mathrm{TF}^2\right)} \ . \tag{4.121}$$

The reverse transformation makes use of the residual theorem:

$$\begin{aligned}
\varphi(\boldsymbol{r}) &= \frac{-e}{\varepsilon_0(2\pi)^3}\int d^3q\,\frac{e^{i\boldsymbol{q}\cdot\boldsymbol{r}}}{q^2+q_{TF}^2} = \\
&= \frac{-e}{4\pi^2\varepsilon_0}\int_0^{\infty} dq\,\frac{q^2}{q^2+q_{TF}^2}\int_{-1}^{+1} dx\,e^{iqrx} = \\
&= \frac{ie}{4\pi^2\varepsilon_0 r}\int_0^{\infty} dq\,\frac{q}{q^2+q_{TF}^2}\left(e^{iqr}-e^{-iqr}\right) = \\
&= \frac{ie}{4\pi^2\varepsilon_0 r}\int_{-\infty}^{+\infty} dq\,\frac{q\,e^{iqr}}{q^2+q_{TF}^2} = \\
&= \frac{ie}{4\pi^2\varepsilon_0}\frac{1}{r}\int dq\,\frac{q\,e^{iqr}}{(q+iq_{TF})(q-iq_{TF})} = \\
&= \frac{-e}{2\pi\,\varepsilon_0 r}\,\frac{iq_{TF}}{2iq_{TF}}\,e^{-q_{TF}r}\,.
\end{aligned}$$

We find, as expected, a **screened Coulomb potential**

$$\varphi(\boldsymbol{r}) = \frac{-e}{4\pi\varepsilon_0 r}\exp\left(-q_{TF}r\right) \begin{cases} \xrightarrow{\text{small } r} \dfrac{-\varepsilon}{4\pi\varepsilon_0 r} \\ \xrightarrow{r\to\infty} 0 \end{cases} \tag{4.122}$$

(i.e. a *Yukawa potential*). Within the

screening length

$$\lambda_{TF} = q_{TF}^{-1} = \sqrt{\frac{2\varepsilon_0\varepsilon_F}{3n_0 e^2}}\,, \tag{4.123}$$

the potential of the test charge is shielded to $1/e$ of its maximum value. Making use of Eqs. (2.84), (2.85) and (2.86), we can express λ_{TF} in terms of the dimensionless density parameter r_S defined in (2.83):

$$\lambda_{TF} \approx 0.34\,\sqrt{r_S}. \tag{4.124}$$

Typical metallic densities are $2 \le r_S \le 6$. Then λ_{TF} is of the order of the average spacing of the particles. The screening is thus substantial! A characteristic measure of the strength of the screening effect is given by the dielectric function which was introduced in Sect. 3.1.5, $\varepsilon(\boldsymbol{q}, E)$. For the situation discussed here, we have from (3.96):

$$\frac{\rho_{\text{ind}}(\boldsymbol{q}, 0)}{\rho_{\text{ext}}(\boldsymbol{q}, 0)} = \frac{1}{\varepsilon(\boldsymbol{q}, 0)} - 1 \, .$$

Now we find

$$\rho_{\text{ind}}(\boldsymbol{r}) = -e\,(n(\boldsymbol{r}) - n_0)$$

and therefore, from (4.117):

$$\rho^{\text{TF}}_{\text{ind}}(\boldsymbol{q}) = -\frac{3}{2}\frac{e^2}{\varepsilon_{\text{F}}} n_0 \varphi(\boldsymbol{q}) = -\frac{3}{2} e^2 \frac{2\varepsilon_0 q^2_{\text{TF}}}{3e^2} \frac{-e}{\varepsilon_0 V\left(q^2 + q^2_{\text{TF}}\right)} =$$

$$= \frac{e q^2_{\text{TF}}}{V\left(q^2 + q^2_{\text{TF}}\right)} \, .$$

With $\rho_{\text{ext}}(\boldsymbol{q}, 0) = -e/V$, we then obtain for the dielectric function in the Thomas-Fermi approximation the following simple expression:

$$\varepsilon^{\text{TF}}(\boldsymbol{q}) = 1 + \frac{q^2_{\text{TF}}}{q^2} \, . \tag{4.125}$$

The serious disadvantage of the Thomas-Fermi model consists of the assumption that the problem is static. Screening processes should, in contrast, be dynamic processes. If we bring a negative test charge into the electron system, then the negatively-charged electrons will be repelled. They will initially move out past the stationary equilibrium position; this allows the positive background charges to show through more strongly and attracts the electrons again. They flow back, approach the test charge too closely, and are again repelled, etc. The system thus forms a harmonic oscillator and exhibits oscillations in the electron density. This system will then have a **proper frequency**, corresponding to collective excitations referred to as **plasmons**. We will investigate these in the next section. Within the Thomas-Fermi approximation, they are naturally not considered!

4.2.2 Charge Density Waves, Plasmons

In Sect. 3.1.5, we have seen that the dielectric function $\varepsilon(\boldsymbol{q}, E)$ describes the reaction of the electronic system to a time-dependent external perturbation. According to (3.103), we have:

$$\varepsilon^{-1}(\boldsymbol{q}, E) = 1 + \frac{1}{\hbar} v_0(\boldsymbol{q}) \langle\!\langle \hat{\rho}_{\boldsymbol{q}}; \hat{\rho}^+_{\boldsymbol{q}} \rangle\!\rangle^{\text{ret}}_E \, , \tag{4.126}$$

$$v_0(\boldsymbol{q}) = \frac{1}{V} \frac{e^2}{\varepsilon_0 q^2} \, . \tag{4.127}$$

Here, $\widehat{\rho}_q$ is the Fourier component of the density operator:

$$\widehat{\rho}_q = \sum_{k\sigma} a^+_{k\sigma} a_{k+q\sigma} . \tag{4.128}$$

We encountered a first approximation for $\varepsilon(\boldsymbol{q}, E)$ in the preceding section within the framework of the classical Thomas-Fermi model (4.125), which however can be convincing only for static problems ($E = 0$) and $|\boldsymbol{q}| \to 0$.

Via the zeroes of $\varepsilon(\boldsymbol{q}, E)$, we can find the *spontaneous* charge-density fluctuations of the system, which can be excited by an arbitrarily weak perturbative charge. We wish to treat these **proper frequencies** of the system of charged particles in the following. They manifest themselves clearly in the poles of the retarded Green's function,

$$\chi(\boldsymbol{q}, E) = \langle\langle \widehat{\rho}_q ; \widehat{\rho}^+_q \rangle\rangle^{\text{ret}}_E , \tag{4.129}$$

which is also called the **generalised susceptibility** (compare with (3.69), (3.70)). We compute this function initially for the **non-interacting** system. In the process, it is advantageous to begin with the following Green's function,

$$f_{k\sigma}(\boldsymbol{q}, E) = \langle\langle a^+_{k\sigma} a_{k+q\sigma} ; \widehat{\rho}^+_q \rangle\rangle^{\text{ret}}_E , \tag{4.130}$$

which, after summation over $\boldsymbol{k}, \sigma$, yields $\chi(\boldsymbol{q}, E)$. To set up its equation of motion, we require the commutator

$$\begin{aligned}
&\left[a^+_{k\sigma} a_{k+q\sigma}, \mathcal{H}_0\right]_- = \\
&= \sum_{p,\sigma'} (\varepsilon(\boldsymbol{p}) - \mu) \left[a^+_{k\sigma} a_{k+q\sigma}, a^+_{p\sigma'} a_{p\sigma'}\right]_- = \\
&= \sum_{p,\sigma'} (\varepsilon(\boldsymbol{p}) - \mu) \left\{\delta_{\sigma\sigma'} \delta_{p,k+q} a^+_{k\sigma} a_{p\sigma'} - \delta_{\sigma\sigma'} \delta_{p,k} a^+_{p\sigma'} a_{k+q\sigma}\right\} = \\
&= (\varepsilon(\boldsymbol{k}+\boldsymbol{q}) - \varepsilon(\boldsymbol{k})) a^+_{k\sigma} a_{k+q\sigma}
\end{aligned} \tag{4.131}$$

and the inhomogeneity

$$\begin{aligned}
\left[a^+_{k\sigma} a_{k+q\sigma}, \widehat{\rho}^+_q\right]_- &= \sum_{p,\sigma'} \left[a^+_{k\sigma} a_{k+q\sigma}, a^+_{p+q\sigma'} a_{p\sigma'}\right]_- = \\
&= \sum_{p,\sigma'} \left\{\delta_{\sigma\sigma'} \delta_{pk} a^+_{k\sigma} a_{p\sigma'} - \delta_{\sigma\sigma'} \delta_{kp} a^+_{p+q\sigma'} a_{k+q\sigma}\right\} = \\
&= n_{k\sigma} - n_{k+q\sigma} .
\end{aligned} \tag{4.132}$$

We then obtain:

$$\{E - (\varepsilon(\boldsymbol{k}+\boldsymbol{q}) - \varepsilon(\boldsymbol{k}))\} f_{k\sigma}(\boldsymbol{q}, E) = \hbar \left(\langle n_{k\sigma}\rangle^{(0)} - \langle n_{k+q\sigma}\rangle^{(0)}\right) . \tag{4.133}$$

The index "0" refers to averaging within the *free* system. From this we obtain the

susceptibility of the free system

$$\chi_0(\boldsymbol{q}, E) = \hbar \sum_{\boldsymbol{k},\sigma} \frac{\langle n_{\boldsymbol{k}\sigma}\rangle^{(0)} - \langle n_{\boldsymbol{k}+\boldsymbol{q}\sigma}\rangle^{(0)}}{E - (\varepsilon(\boldsymbol{k}+\boldsymbol{q}) - \varepsilon(\boldsymbol{k}))} . \tag{4.134}$$

This is by the way also the generalised susceptibility of the Stoner model, if we substitute $E_\sigma(\boldsymbol{k})$ from (4.26) for $\varepsilon(\boldsymbol{k})$. In the above expression, the summation over σ is purely formal, since the occupation numbers $\langle n_{\boldsymbol{k}\sigma}\rangle^{(0)}$ are of course independent of spins in the free system.

Taking realistic particle interactions into account, we can no longer calculate the susceptibility exactly. We discuss in the following an approximation for the **jellium model**, whose Hamiltonian we will formulate as in (2.72):

$$H = \sum_{\boldsymbol{k}\sigma} \varepsilon(\boldsymbol{k}) a^+_{\boldsymbol{k}\sigma} a_{\boldsymbol{k}\sigma} + \frac{1}{2} \sum_{\boldsymbol{q}}^{\neq 0} v_0(\boldsymbol{q}) \left(\widehat{\rho}_{\boldsymbol{q}} \widehat{\rho}_{-\boldsymbol{q}} - \widehat{N}\right) , \tag{4.135}$$

with $\varepsilon(\boldsymbol{k})$ from (2.64) and $v_0(\boldsymbol{q})$ from (2.127). Our starting point is again the Green's function $f_{\boldsymbol{k}\sigma}(\boldsymbol{q}, E)$, whose equation of motion can be written as follows:

$$\begin{aligned}&[E - (\varepsilon(\boldsymbol{k}+\boldsymbol{q}) - \varepsilon(\boldsymbol{k}))]\, f_{\boldsymbol{k}\sigma}(\boldsymbol{q}, E) = \hbar \left(\langle n_{\boldsymbol{k}\sigma}\rangle - \langle n_{\boldsymbol{k}+\boldsymbol{q}\sigma}\rangle\right) + \\ &+ \frac{1}{2} \sum_{\boldsymbol{q}_1}^{\neq 0} v_0(\boldsymbol{q}_1) \langle\!\langle \left[a^+_{\boldsymbol{k}\sigma} a_{\boldsymbol{k}+\boldsymbol{q}\sigma}, \widehat{\rho}_{\boldsymbol{q}_1} \widehat{\rho}_{-\boldsymbol{q}_1}\right]_- ; \widehat{\rho}^+_{\boldsymbol{q}} \rangle\!\rangle .\end{aligned} \tag{4.136}$$

We now of course average over states of the interacting system. We have already made use of the commutators (4.131) and (4.132) in setting up (4.136). Furthermore, one can readily see that:

$$\left[a^+_{\boldsymbol{k}\sigma} a_{\boldsymbol{k}+\boldsymbol{q}\sigma}, \widehat{N}\right]_- \equiv 0 . \tag{4.137}$$

We further rearrange the equation of motion. We find initially:

$$
\begin{aligned}
&\left[a^+_{k\sigma}\, a_{k+q\sigma}\,,\, \widehat{\rho}_{q_1}\widehat{\rho}_{-q_1}\right]_- = \\
&= \left[a^+_{k\sigma}\, a_{k+q\sigma}\,,\, \widehat{\rho}_{q_1}\right]_- \widehat{\rho}_{-q_1} + \widehat{\rho}_{q_1}\left[a^+_{k\sigma}\, a_{k+q\sigma}\,,\, \widehat{\rho}_{-q_1}\right]_- \,, \\
&\left[a^+_{k\sigma}\, a_{k+q\sigma}\,,\, \widehat{\rho}_{q_1}\right]_- = \\
&= \sum_{p,\sigma'}\left[a^+_{k\sigma}\, a_{k+q\sigma}\,,\, a^+_{p\sigma'} a_{p+q_1\sigma'}\right]_- = \\
&= \sum_{p,\sigma'}\left\{\delta_{\sigma\sigma'}\,\delta_{p,\,k+q}\, a^+_{k\sigma}\, a_{p+q_1\sigma'} - \delta_{\sigma\sigma'}\delta_{k,\,p+q_1}\, a^+_{p\sigma'}\, a_{k+q\sigma}\right\} = \\
&= a^+_{k\sigma}\, a_{k+q+q_1\sigma} - a^+_{k-q_1\sigma}\, a_{k+q\sigma}\,.
\end{aligned}
$$

Analogously, one obtains:

$$
\left[a^+_{k\sigma}\, a_{k+q\sigma}\,,\, \widehat{\rho}_{-q_1}\right]_- = a^+_{k\sigma}\, a_{k+q-q_1\sigma} - a^+_{k+q_1\sigma}\, a_{k+q\sigma}\,.
$$

Using $v_0(\boldsymbol{q}_1) = v_0(-\boldsymbol{q}_1)$, we can then rewrite the equation of motion of the Green's function as follows:

$$
\begin{aligned}
&[E - (\varepsilon(\boldsymbol{k}+\boldsymbol{q}) - \varepsilon(\boldsymbol{k}))]\, f_{k\sigma}(\boldsymbol{q}, E) = \\
&= \hbar\left(\langle n_{k\sigma}\rangle - \langle n_{k+q\sigma}\rangle\right) + \frac{1}{2}\sum_{q_1}^{\neq 0} v_0(\boldsymbol{q}_1)\,\cdot \\
&\cdot\left(\left\langle\!\left\langle \left[\widehat{\rho}_{q_1}, a^+_{k\sigma}\, a_{k+q-q_1\sigma}\right]_+ ; \widehat{\rho}^+_q \right\rangle\!\right\rangle - \left\langle\!\left\langle \left[\widehat{\rho}_{q_1}, a^+_{k+q_1\sigma}\, a_{k+q\sigma}\right]_+ ; \widehat{\rho}^+_q \right\rangle\!\right\rangle\right).
\end{aligned}
\tag{4.138}
$$

These expressions are all still exact. Note that the higher-order Green's functions on the right side now contain only anticommutators! In the next step, we implement the so-called

random phase approximation (RPA):

1. *higher-order* Green's functions are decoupled using the Hartree-Fock method (4.18), whereby momentum conservation must be obeyed. An example:

$$
\begin{aligned}
\widehat{\rho}_{q_1} a^+_{k\sigma}\, a_{k+q-q_1\sigma} \;\xrightarrow{\text{HFA}}\; &\widehat{\rho}_{q_1}\left\langle a^+_{k\sigma}\, a_{k+q-q_1\sigma}\right\rangle + \\
&+ \left\langle\widehat{\rho}_{q_1}\right\rangle a^+_{k\sigma}\, a_{k+q-q_1\sigma} - \\
&- \left\langle\widehat{\rho}_{q_1}\right\rangle\left\langle a^+_{k\sigma}\, a_{k+q-q_1\sigma}\right\rangle = \\
&= \delta_{qq_1}\widehat{\rho}_{q_1}\,\langle n_{k\sigma}\rangle\,.
\end{aligned}
\tag{4.139}
$$

2. Occupation numbers are replaced by those of the *free* system:

$$\langle n_{k\sigma} \rangle \longrightarrow \langle n_{k\sigma} \rangle^{(0)} . \tag{4.140}$$

Thus, the equation of motion (4.138) is now *decoupled*:

$$\begin{aligned}
&[E - (\varepsilon(\boldsymbol{k}+\boldsymbol{q}) - \varepsilon(\boldsymbol{k}))]\, f_{k\sigma}(\boldsymbol{q}, E) = \\
&= \hbar \left(\langle n_{k\sigma} \rangle^{(0)} - \left\langle n_{k+q\sigma} \right\rangle^{(0)} \right) + \\
&+ v_0(\boldsymbol{q}) \left(\langle n_{k\sigma} \rangle^{(0)} - \left\langle n_{k+q\sigma} \right\rangle^{(0)} \right) \langle\langle \widehat{\rho}_q ; \widehat{\rho}_q^+ \rangle\rangle_E .
\end{aligned} \tag{4.141}$$

With $\chi(\boldsymbol{q}, E) \equiv \langle\langle \widehat{\rho}_q ; \widehat{\rho}_q^+ \rangle\rangle_E = \sum_{k\sigma} f_{k\sigma}(\boldsymbol{q}, E)$ and with (4.134), we finally obtain the generalised susceptibility in the RPA:

$$\chi_{\mathrm{RPA}}(\boldsymbol{q}, E) = \frac{\chi_0(\boldsymbol{q}, E)}{1 - \frac{1}{\hbar} v_0(\boldsymbol{q}) \chi_0(\boldsymbol{q}, E)} . \tag{4.142}$$

From (4.126), we find the dielectric function:

$$\begin{aligned}
\varepsilon_{\mathrm{RPA}}(\boldsymbol{q}, E) &= 1 - \frac{1}{\hbar} v_0(\boldsymbol{q}) \chi_0(\boldsymbol{q}, E) = \\
&= 1 - v_0(\boldsymbol{q}) \sum_{k,\sigma} \frac{\langle n_{k\sigma} \rangle^{(0)} - \left\langle n_{k+q\sigma} \right\rangle^{(0)}}{E - (\varepsilon(\boldsymbol{k}+\boldsymbol{q}) - \varepsilon(\boldsymbol{k}))} .
\end{aligned} \tag{4.143}$$

This expression is also called the **Lindhard function**. As shown in Sect. 3.1.5, $\varepsilon(\boldsymbol{q}, E)$ describes the relation between the polarisation $\rho_{\mathrm{ind}}(\boldsymbol{q}, E)$ of the medium, i.e. the fluctuations of the charge density in the electronic system, and an external perturbation $\rho_{\mathrm{ext}}(\boldsymbol{q}, E)$. According to (3.96), we have:

$$\rho_{\mathrm{ind}}(\boldsymbol{q}, E) = \left(\frac{1}{\varepsilon(\boldsymbol{q}, E)} - 1 \right) \rho_{\mathrm{ext}}(\boldsymbol{q}, E) . \tag{4.144}$$

The zeroes of the dielectric function are therefore interesting; they determine the **proper frequencies** of the system. From (4.143), we obtain them by applying the condition

$$f_q(E) \equiv v_0(\boldsymbol{q}) \sum_{k\sigma} \frac{\langle n_{k\sigma} \rangle^{(0)} - \left\langle n_{k+q\sigma} \right\rangle^{(0)}}{E - (\varepsilon(\boldsymbol{k}+\boldsymbol{q}) - \varepsilon(\boldsymbol{k}))} \stackrel{!}{=} 1 . \tag{4.145}$$

The first evaluation of this expression was published by J. Lindhard (1954).

The function $f_q(E)$ exhibits a dense series of poles within the **single-particle continuum**,

$$E_k(\boldsymbol{q}) = \varepsilon(\boldsymbol{k}+\boldsymbol{q}) - \varepsilon(\boldsymbol{k}) . \tag{4.146}$$

Between each pair is an axis crossing $f_q(E) = 1$ (cf. Fig. 4.12). In the thermodynamic limit, these are congruent with the single-particle excitations $E_k(\boldsymbol{q})$ and are thus uninteresting for us here. There is, however, another axis crossing $E_p(\boldsymbol{q})$ outside the continuum, which cannot be a single-particle excitation, but rather represents a **collective mode**:

$$E_p(\boldsymbol{q}) \equiv \hbar\omega_p(\boldsymbol{q}) : \quad \textbf{plasma oscillation, plasmon.}$$

Qualitatively, the excitation spectrum as sketched in Fig. 4.13 is found. Since a long-wave plasma oscillation (q small) represents a correlated motion of a large number of electrons, plasmons have relatively high energies,

$$5\text{eV} \cdots \mathrm{E}_p(\boldsymbol{q}) \cdots 25\text{eV} ,$$

and can therefore not be excited thermally. By injecting high-energy particles into metals, however, it has been possible to excite and observe plasmons.

We now wish to determine the plasmon dispersion relation $\omega_p(\boldsymbol{q})$ approximately for small values of $|\boldsymbol{q}|$. We set

$$\varepsilon(\boldsymbol{k}) = \frac{\hbar^2 k^2}{2m^*} , \tag{4.147}$$

where m^* represents an effective mass of the electrons, which takes into account to first order the otherwise neglected influence of the lattice potential. Because of (4.147), we can then assume that

$$\langle n_{k\sigma} \rangle^{(0)} = \langle n_{-k\sigma} \rangle^{(0)} . \tag{4.148}$$

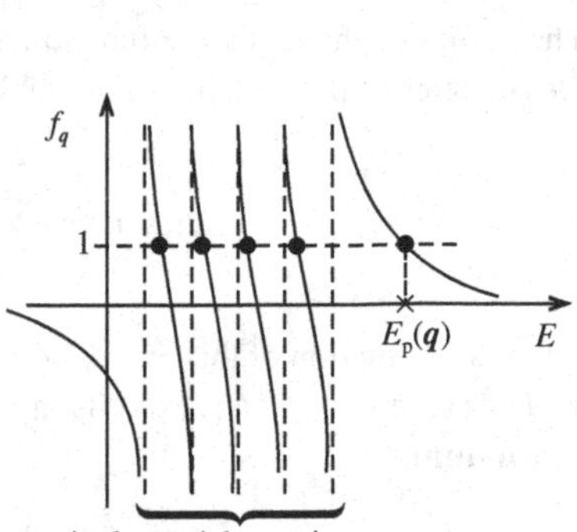

Fig. 4.12 A graphic illustration of the determination of the zeroes of the Lindhard function

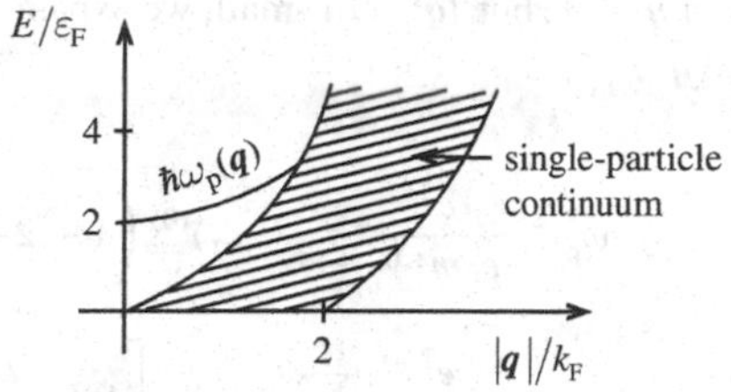

Fig. 4.13 The wavenumber dependence of the zeroes of the Lindhard function (plasmon mode and single-particle continuum)

In the second term of (4.145), we substitute $\boldsymbol{k}$ by $(-\boldsymbol{k}-\boldsymbol{q})$ and then make use of (4.148):

$$\begin{aligned}
1 &= f_{\boldsymbol{q}}(E_p) = \\
&= v_0(\boldsymbol{q}) \sum_{\boldsymbol{k}\sigma} \left\{ \frac{\langle n_{\boldsymbol{k}\sigma}\rangle^{(0)}}{E_p - \varepsilon(\boldsymbol{k}+\boldsymbol{q}) + \varepsilon(\boldsymbol{k})} \frac{\langle n_{-\boldsymbol{k}\sigma}\rangle^{(0)}}{E_p - \varepsilon(-\boldsymbol{k}) + \varepsilon(-\boldsymbol{k}-\boldsymbol{q})} \right\} = \\
&= 2v_0(\boldsymbol{q}) \sum_{\boldsymbol{k}\sigma} \frac{\langle n_{\boldsymbol{k}\sigma}\rangle^{(0)} \left(\varepsilon(\boldsymbol{k}+\boldsymbol{q}) - \varepsilon(\boldsymbol{k})\right)}{E_p^2 - \left(\varepsilon(\boldsymbol{k}+\boldsymbol{q}) - \varepsilon(\boldsymbol{k})\right)^2} .
\end{aligned} \tag{4.149}$$

We next insert (4.147):

$$\omega_p^2 = \frac{e^2}{\varepsilon_0 m^* V q^2} \sum_{\boldsymbol{k}\sigma} \frac{\langle n_{\boldsymbol{k}\sigma}\rangle^{(0)} \left(q^2 + 2\boldsymbol{k}\cdot\boldsymbol{q}\right)}{1 - \frac{\hbar^2}{\omega_p^2}\left(\frac{q^2}{2m^*} + \frac{\boldsymbol{k}\cdot\boldsymbol{q}}{m^*}\right)^2} . \tag{4.150}$$

Let us first investigate the case that $|\boldsymbol{q}| \to 0$. Then we can neglect the expression in parentheses in the denominator, relative to 1. Furthermore, we have:

$$\sum_{\boldsymbol{k},\sigma} \langle n_{\boldsymbol{k}\sigma}\rangle^{(0)} = N_e = n_0 V , \tag{4.151}$$

$$\begin{aligned}
\sum_{\boldsymbol{k},\sigma} \langle n_{\boldsymbol{k}\sigma}\rangle^{(0)} (2\boldsymbol{k}\cdot\boldsymbol{q}) &= \sum_{\boldsymbol{k}',\sigma} \langle n_{-\boldsymbol{k}'\sigma}\rangle^{(0)} \left(-2\boldsymbol{k}'\cdot\boldsymbol{q}\right) = \\
&\overset{(4.148)}{=} -\sum_{\boldsymbol{k}',\sigma} \langle n_{\boldsymbol{k}'\sigma}\rangle^{(0)} \left(2\boldsymbol{k}'\cdot\boldsymbol{q}\right) = \\
&= 0 .
\end{aligned} \tag{4.152}$$

Then, for $|\boldsymbol{q}| = 0$, we obtain the so-called

$$\textbf{plasma frequency:} \quad \omega_p = \omega_p(\boldsymbol{q}=0) = \sqrt{\frac{n_0 e^2}{\varepsilon_0 m^*}} . \tag{4.153}$$

For $\boldsymbol{q} \neq 0$, but $|\boldsymbol{q}|$ still small, we expand the denominator in (4.150) up to quadratic terms in q:

$$\omega_{\rm p}^2 \approx \frac{e^2}{\varepsilon_0 m^* V} \sum_{k\sigma} \langle n_{k\sigma} \rangle^{(0)} \left(1 + 2\frac{\boldsymbol{k}\cdot\boldsymbol{q}}{q^2}\right) \left[1 + \frac{\hbar^2}{\omega_{\rm p}^2} \frac{q^4}{4m^{*2}} \left(1 + 2\frac{\boldsymbol{k}\cdot\boldsymbol{q}}{q^2}\right)\right]^2 =$$

$$= \frac{e^2}{\varepsilon_0 m^* V} \sum_{k\sigma} \langle n_{k\sigma} \rangle^{(0} \left[1 + 2\frac{\boldsymbol{k}\cdot\boldsymbol{q}}{q^2} + \frac{q^4\hbar^2}{4m^{*2}\omega_{\rm p}^2} \cdot \right.$$

$$\left. \cdot \left(1 + 6\frac{\boldsymbol{k}\cdot\boldsymbol{q}}{q^2} + 12\frac{(\boldsymbol{k}\cdot\boldsymbol{q})^2}{q^4} + 8\frac{(\boldsymbol{k}\cdot\boldsymbol{q})^3}{q^6}\right)\right] .$$

The odd powers of $(\boldsymbol{k}\cdot\boldsymbol{q})$ make no contribution, owing to the $\boldsymbol{k}$ summation (cf. (4.152)):

$$\omega_{\rm p}^2(\boldsymbol{q}) \approx \omega_{\rm p}^2(0) + \frac{3e^2\hbar^2}{\varepsilon_0 m^{*3} \omega_{\rm p}^2(\boldsymbol{q})} \frac{1}{V} \sum_{k,\sigma} \langle n_{k\sigma} \rangle^{(0)} (\boldsymbol{k}\cdot\boldsymbol{q})^2 . \tag{4.154}$$

On the right-hand side, we can replace $\omega_{\rm p}^2(\boldsymbol{q})$ by $\omega_{\rm p}^2(0)$, and also, at low temperatures, we can estimate:

$$\frac{1}{V} \sum_{k,\sigma} \langle n_{k\sigma} \rangle^{(0)} k^2 \cos^2\vartheta \approx \frac{2\cdot 2\pi}{(2\pi)^3} \int_{-1}^{+1} {\rm d}\cos\vartheta \, \cos^2\vartheta \int_0^{k_{\rm F}} {\rm d}k \, k^4 =$$

$$\stackrel{(4.111)}{=} \frac{1}{5} n_0 k_{\rm F}^2 .$$

This gives in (4.154) with (4.153):

$$\omega_{\rm p}^2(q) \approx \omega_{\rm p}^2 + \frac{3}{5} \frac{\hbar^2 k_{\rm F}^2}{m^{*2}} q^2 . \tag{4.155}$$

Thus, from the zeroes of the dielectric function $\varepsilon(\boldsymbol{q}, E)$, we have derived the

plasmon dispersion relation:

$$\omega_{\rm p}(\boldsymbol{q}) = \omega_{\rm p} \left(1 + \frac{3}{10} \frac{\hbar^2 k_{\rm F}^2}{m^{*2}\omega_{\rm p}^2} q^2\right) + O\left(q^4\right) . \tag{4.156}$$

In order to compare our general RPA result (4.143) with the semiclassical Thomas-Fermi model from the previous section, we finally evaluate the dielectric function in the **static limit**, $E = 0$. From (4.143) and (4.149), we need to compute:

$$\varepsilon_{\mathrm{RPA}}(\boldsymbol{q},0) = 1 + 2v_0(\boldsymbol{q}) \sum_{\boldsymbol{k},\sigma} \frac{\langle n_{\boldsymbol{k}\sigma} \rangle^{(0)}}{\varepsilon(\boldsymbol{k}+\boldsymbol{q}) - \varepsilon(\boldsymbol{k})} . \tag{4.157}$$

As usual, we replace the $\boldsymbol{k}$ summation by a corresponding integration ($T \approx 0$):

$$\sum_{\boldsymbol{k},\sigma} \langle n_{\boldsymbol{k}\sigma} \rangle^{(0)} \left(\varepsilon(\boldsymbol{k}+\boldsymbol{q}) - \varepsilon(\boldsymbol{k})\right)^{-1} = \frac{2V}{(2\pi)^3} \int\limits_{\mathrm{FS}} \mathrm{d}^3k \left(\varepsilon(\boldsymbol{k}+\boldsymbol{q}) - \varepsilon(\boldsymbol{k})\right)^{-1} \equiv I(\boldsymbol{q}) .$$

FS refers to the *Fermi sphere*. With (4.147), we then find:

$$I(\boldsymbol{q}) = \frac{V}{2\pi^2} \frac{2m^*}{\hbar^2} \int\limits_{-1}^{+1} \mathrm{d}x \int\limits_{0}^{k_{\mathrm{F}}} \mathrm{d}k\, k^2 \frac{1}{2kqx + q^2} =$$

$$= \frac{Vm^*}{\pi^2 \hbar^2} \frac{1}{2q} \int\limits_{0}^{k_{\mathrm{F}}} \mathrm{d}k\, k \ln \left| \frac{q+2k}{q-2k} \right| .$$

The right-hand side contains a standard integral:

$$\int x \ln(a+bx) \mathrm{d}x = \frac{1}{2} \left(x^2 - \frac{a^2}{b^2} \right) \ln(a+bx) - \frac{1}{2} \left(\frac{x^2}{2} - \frac{ax}{b} \right) . \tag{4.158}$$

We then have:

$$I(\boldsymbol{q}) = \frac{Vm^*}{2q\pi^2 \hbar^2} \left[\frac{1}{2} q k_{\mathrm{F}} + \frac{1}{2} \left(k_{\mathrm{F}}^2 - \frac{q^2}{4} \right) \ln \left| \frac{q+2k_{\mathrm{F}}}{q-2k_{\mathrm{F}}} \right| \right] . \tag{4.159}$$

We define the following function:

$$g(u) = \frac{1}{2} \left(1 + \frac{1}{2u} (1-u^2) \ln \left| \frac{1+u}{1-u} \right| \right) . \tag{4.160}$$

Then we can write:

$$I(\boldsymbol{q}) = \frac{1}{2} \frac{Vm^*}{\pi^2 \hbar^2} k_{\mathrm{F}}\, g\left(\frac{q}{2k_{\mathrm{F}}} \right) .$$

With (4.120) and (4.157), the static dielectric function is given by:

$$\varepsilon_{\mathrm{RPA}}(\boldsymbol{q}) = 1 + \frac{q_{\mathrm{TF}}^2}{q^2} g\left(\frac{q}{2k_{\mathrm{F}}} \right) . \tag{4.161}$$

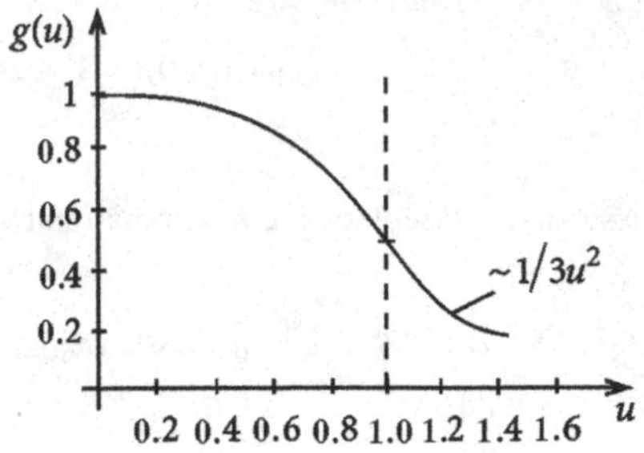

Fig. 4.14 The qualitative behaviour of the Lindhard correction (4.160)

For $g = 1$, this yields the Thomas-Fermi result (4.125). For small q, i.e. long wavelengths, we thus have:

$$\varepsilon_{\mathrm{RPA}}(\boldsymbol{q}) \underset{q \ll k_{\mathrm{F}}}{\approx} \varepsilon^{\mathrm{TF}}(\boldsymbol{q}) \,. \tag{4.162}$$

The so-called **Lindhard correction** $g\,(q/2k_{\mathrm{F}})$ is 1 for $q = 0$ and non-analytic for $q = 2k_{\mathrm{F}}$. There, the first derivative of g exhibits a logarithmic singularity with interesting physical consequences. Using the Poisson equations for the external charge density $\rho_{\mathrm{ext}}(\boldsymbol{r})$, and the overall charge density $\rho(\boldsymbol{r}) = \rho_{\mathrm{ext}}(\boldsymbol{r}) + \rho_{\mathrm{ind}}(\boldsymbol{r})$,

$$q^2\varphi(\boldsymbol{q}) = \frac{1}{\varepsilon_0}\rho(\boldsymbol{q}) \,, \tag{4.163}$$

$$q^2\varphi_{\mathrm{ext}}(\boldsymbol{q}) = \frac{1}{\varepsilon_0}\rho_{\mathrm{ext}}(\boldsymbol{q}) \,, \tag{4.164}$$

we can express the **screened potential** $\varphi(\boldsymbol{q})$ by means of of the static dielectric function $\varepsilon(\boldsymbol{q})$ in terms of the external potential. With (3.96), we find:

$$\varphi(\boldsymbol{q}) = \frac{\varphi_{\mathrm{ext}}(\boldsymbol{q})}{\varepsilon(\boldsymbol{q})} \,. \tag{4.165}$$

If φ_{ext} is the potential of a point charge $(-e)$, i.e.

$$\varphi_{\mathrm{ext}}(\boldsymbol{q}) = \frac{-e}{\varepsilon_0 V q^2} \,,$$

then we obtain –for example with the Thomas-Fermi result (4.125)– just (4.121). However, if we insert the RPA result (4.162) and transform back to real space, then for large distances a term of the form

$$\varphi(\boldsymbol{r}) \sim \frac{1}{r^3}\cos(2k_{\mathrm{F}}r) \tag{4.166}$$

is dominant. The potential thus does not decrease exponentially as in the Thomas-Fermi model, but rather has very long-range oscillations, which are called **Friedel oscillations**.

4.2.3
Spin Density Waves, Magnons

There is another type of collective excitations in a system of interacting band electrons, which arises from the existence of the electronic spin. In Sect. 3.1.3, we introduced the **transverse susceptibility** χ_{ij}^{+-} (3.72), which can be written as follows for band electrons:

$$\chi_{ij}^{+-}(E) = -\gamma \langle\langle a_{i\uparrow}^{+} a_{i\downarrow}; a_{j\downarrow}^{+} a_{j\uparrow} \rangle\rangle_E \,; \quad \left(\gamma = \frac{\mu_0}{V\hbar} g^2 \mu_B^2\right) . \tag{4.167}$$

The poles of the wavelength-dependent Fourier transform,

$$\chi_q^{+-}(E) = \frac{1}{N} \sum_{i,j} \chi_{ij}^{+-}(E) \mathrm{e}^{\mathrm{i}q(R_i - R_j)} = -\gamma \frac{1}{N} \sum_{k,p} \bar{\chi}_{kp}(q) , \tag{4.168}$$

$$\bar{\chi}_{kp}(q) = \langle\langle a_{k\uparrow}^{+} a_{k+q\downarrow}; a_{p\downarrow}^{+} a_{p-q\uparrow} \rangle\rangle_E , \tag{4.169}$$

correspond to spin-wave energies (magnons). The concept of the spin wave was introduced in Sect. 2.4.3 for a system of interacting, localised (!) spins (Heisenberg model). It is a collective excitation which is accompanied by a variation in the z-component of the overall spin by one unit of angular momentum. This spin deviation is not associated with a single electron, but rather is uniformly distributed over the entire spin system. Although it is then not so readily intuitively understandable, the concept of the spin wave can also be applied to itinerant band electrons with their permanent spins. We discuss this point briefly here. We compute $\chi_q^{+-}(E)$ first in the framework of the Stoner model, which is, as in (4.43), described by the Hamiltonian

$$\mathcal{H}_S = \sum_{k,\sigma} (E_\sigma(k) - \mu) a_{k\sigma}^{+} a_{k\sigma} . \tag{4.170}$$

We formulate the equation of motion for the Green's function $\bar{\chi}_{kp}(q)$. To do so, we require the commutator

$$\begin{aligned}
&\left[a_{k\uparrow}^{+} a_{k+q\downarrow}, \mathcal{H}_S\right]_- = \\
&= \sum_{k',\sigma} (E_\sigma(k') - \mu) \left[a_{k\uparrow}^{+} a_{k+q\downarrow}, a_{k'\sigma}^{+} a_{k'\sigma}\right]_- = \\
&= \sum_{k',\sigma} (E_\sigma(k') - \mu) \left(\delta_{\sigma\downarrow} \delta_{k',k+q} a_{k\uparrow}^{+} a_{k'\sigma} - \delta_{\sigma\uparrow} \delta_{k',k} a_{k'\sigma}^{+} a_{k+q\downarrow}\right) = \\
&= (E_\downarrow(k+q) - E_\uparrow(k)) a_{k\uparrow}^{+} a_{k+q\downarrow} = \\
&\overset{(4.39)}{=} \Delta E_{\uparrow\downarrow}(k;q) a_{k\uparrow}^{+} a_{k+q\downarrow} ,
\end{aligned} \tag{4.171}$$

and the *inhomogeneity*:

$$\left\langle \left[a_{k\uparrow}^{+} a_{k+q\downarrow}, a_{p\downarrow}^{+} a_{p-q\uparrow} \right]_{-} \right\rangle = \left(\langle n_{k\uparrow} \rangle^{(S)} - \langle n_{k+q\downarrow} \rangle^{(S)} \right) \delta_{p,\,k+q} \,. \tag{4.172}$$

This yields the simple equation of motion:

$$\left(E - \Delta E_{\uparrow\downarrow}(\boldsymbol{k};\boldsymbol{q}) \right) \overline{\chi}_{kp}(\boldsymbol{q}) = \left(\langle n_{k\uparrow} \rangle^{(S)} - \langle n_{k+q\downarrow} \rangle^{(S)} \right) \delta_{p,\,k+q} \,. \tag{4.173}$$

With (4.168), we find the transverse susceptibility in the Stoner model:

$$\left(\chi_{q}^{+-}(E) \right)^{(S)} = \frac{\gamma}{N} \sum_{k} \frac{\langle n_{k+q\downarrow} \rangle^{(S)} - \langle n_{k\uparrow} \rangle^{(S)}}{E - \Delta E_{\uparrow\downarrow}(\boldsymbol{k};\boldsymbol{q})} \,. \tag{4.174}$$

The poles are identical with the single-particle spin-flip excitation spectrum. In this model, without genuine interactions, there are naturally no collective excitations.

In the next step, we compute the susceptibility within the **Hubbard model**:

$$\mathcal{H} = \sum_{k\sigma} (\varepsilon(\boldsymbol{k}) - \mu)\, a_{k\sigma}^{+} a_{k\sigma} + \frac{U}{N} \sum_{kpq} a_{k\uparrow}^{+} a_{k-q\uparrow} a_{p\downarrow}^{+} a_{p+q\downarrow} \,. \tag{4.175}$$

For the equation of motion of the Green's function $\overline{\chi}_{kp}(\boldsymbol{q})$, we find in comparison to (4.173) an additional term owing to the interactions:

$$\begin{aligned} &\frac{U}{N} \sum_{k'pq'} \left[a_{k\uparrow}^{+} a_{k+q\downarrow}, a_{k'\uparrow}^{+} a_{k'-q'\uparrow} a_{p\downarrow}^{+} a_{p+q'\downarrow} \right]_{-} = \\ &= \frac{U}{N} \sum_{k'pq'} \Big(\delta_{p,\,k+q}\, a_{k\uparrow}^{+} a_{k'\uparrow}^{+} a_{k'-q'\uparrow} a_{p+q'\downarrow} - \\ &\delta_{k,\,k'-q'}\, a_{k'\uparrow}^{+} a_{p\downarrow}^{+} a_{p+q'\downarrow} a_{k+q\downarrow} \Big) = \\ &= \frac{U}{N} \sum_{k'q'} \left(a_{k\uparrow}^{+} a_{k'\uparrow}^{+} a_{k'-q'\uparrow} a_{k+q+q'\downarrow} - a_{k+q'\uparrow}^{+} a_{k'\downarrow}^{+} a_{k'+q'\downarrow} a_{k+q\downarrow} \right) . \end{aligned} \tag{4.176}$$

We thus find in the equation of motion two new *higher-order* Green's functions,

$$H_{kpq}^{k'q'}(E) = \left\langle\!\left\langle a_{k\uparrow}^{+} a_{k'\uparrow}^{+} a_{k'-q'\uparrow} a_{k+q+q'\downarrow} ; a_{p\downarrow}^{+} a_{p-q\uparrow} \right\rangle\!\right\rangle_{E} \,, \tag{4.177}$$

$$K_{kpq}^{k'q'}(E) = \left\langle\!\left\langle a_{k+q'\uparrow}^{+} a_{k'\downarrow}^{+} a_{k'+q'\downarrow} a_{k+q\downarrow} ; a_{p\downarrow}^{+} a_{p-q\uparrow} \right\rangle\!\right\rangle_{E} \,, \tag{4.178}$$

which we simplify by making use of the RPA method, taking care to fulfil momentum and spin conservation:

$$H_{kpq}^{k'q'}(E) \Longrightarrow \langle n_{k'\uparrow}\rangle^{(S)} \delta_{q',0}\overline{\chi}_{kp}(q) - \langle n_{k\uparrow}\rangle^{(S)} \delta_{k,k'-q'}\overline{\chi}_{k+q',p}(q)\,, \tag{4.179}$$

$$K_{kpq}^{k'q'} \Longrightarrow \langle n_{k'\downarrow}\rangle^{(S)} \delta_{q',0}\overline{\chi}_{k,p}(q) - \langle n_{k+q\downarrow}\rangle^{(S)} \overline{\chi}_{k+q',p}(q)\delta_{k',k+q}\,. \tag{4.180}$$

We then find for $\overline{\chi}_{k,p}(q)$ the following simplified equation of motion:

$$\begin{aligned}
&[E-(\varepsilon(k+q)-\varepsilon(k))]\,\overline{\chi}_{kp}(q) = \\
&= \delta_{p,k+q}\left[\langle n_{k\uparrow}\rangle^{(S)} - \langle n_{k+q\downarrow}\rangle^{(S)}\right] + \\
&\quad + \overline{\chi}_{kp}(q)\frac{U}{N}\sum_{k'}\left[\langle n_{k'\uparrow}\rangle^{(S)} - \langle n_{k'\downarrow}\rangle^{(S)}\right] - \\
&\quad - \frac{U}{N}\left[\langle n_{k\uparrow}\rangle^{(S)} - \langle n_{k+q\downarrow}\rangle^{(S)}\right]\sum_{q'}\overline{\chi}_{k+q',p}(E)\,.
\end{aligned} \tag{4.181}$$

With (4.39), it then follows that:

$$\begin{aligned}
\left(E-\Delta E_{\uparrow\downarrow}(k;q)\right)\overline{\chi}_{kp}(q) &= \delta_{p,k+q}\left[\langle n_{k\uparrow}\rangle^{(S)} - \langle n_{k+q\downarrow}\rangle^{(S)}\right] - \\
&\quad - \frac{U}{N}\left[\langle n_{k\uparrow}\rangle^{(S)} - \langle n_{k+q\downarrow}\rangle^{(S)}\right]\sum_{k'}\overline{\chi}_{k'p}(q)\,.
\end{aligned} \tag{4.182}$$

This means, with (4.168) and (4.174):

$$\begin{aligned}
\chi_q^{+-}(E) &= \left(\chi_q^{+-}(E)\right)^{(S)} + \chi_q^{+-}(E)\left[\frac{U}{\gamma}\left(\chi_q^{+-}(E)\right)^{(S)}\right]\,, \\
\chi_q^{+-}(E) &= \frac{\left(\chi_q^{+-}(E)\right)^{(S)}}{1-\gamma^{-1}U\left(\chi_q^{+-}(E)\right)^{(S)}}\,.
\end{aligned} \tag{4.183}$$

This result is very similar to that in the RPA (4.142) for the generalised susceptibility. Its evaluation therefore follows the same scenario as in the preceding section. We shall not repeat the details here.

Qualitatively, we obtain the excitation spectrum shown in Fig. 4.15 for spin-flip processes. $\hbar\omega_m(q)$ is a collective spin-wave mode with

$$\hbar\omega_m(q) \approx Dq^2 \quad (q \to 0)\,. \tag{4.184}$$

The overall spectrum is composed of the single-particle Stoner continuum and the collective mode together. Spin waves in metals were first observed experimentally in iron by inelastic neutron scattering. Their characteristic difference with respect to the spin waves in *localised* spin systems is found from a more detailed analysis to be a T^2 dependence of the magnetisation at low temperatures, instead of the Bloch $T^{3/2}$ law.

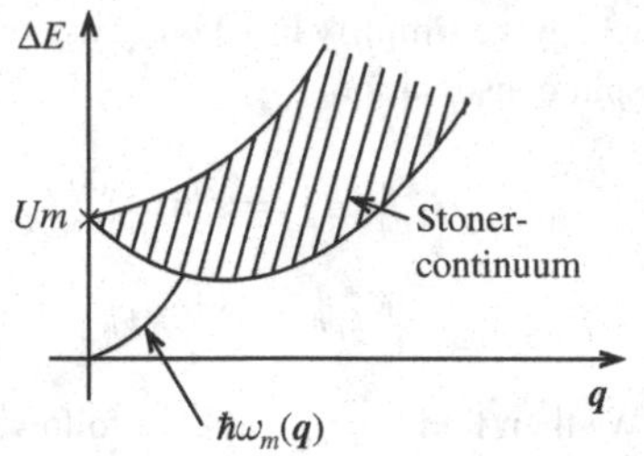

Fig. 4.15 The excitation spectrum of spin-flip processes in a system of band electrons. The *solid line* is the spin-wave dispersion relation

4.2.4 Exercises

Exercise 4.2.1 Prove the following commutator relation:

$$\left[a^+_{k\sigma}a_{k+q\sigma}, \widehat{N}\right] = 0 \quad (\widehat{N}\text{: particle number operator}).$$

Exercise 4.2.2

1. Show that for the Pauli susceptibility of a system of non-interacting electrons, the following holds:

$$\chi_{\text{Pauli}} \simeq 2\mu_B^2\mu_0\rho_0(E_F)\,.$$

   Here, E_F is the Fermi energy, ρ_0 the density of states, μ_B the Bohr magneton and μ_0 the permeability of vacuum. The susceptibility is defined as follows:

$$\chi = \frac{\partial M}{\partial H}\,; \quad M = \mu_B\left(N_\uparrow - N_\downarrow\right) \quad \textit{magnetisation.}$$

   H is a homogeneous magnetic field.
2. Evaluate the *generalised* susceptibility $\chi_0(q)$ of a non-interacting electron system (4.134),

$$\chi_0(q, E=0) = \hbar\sum_{k,\sigma}\frac{\langle n_{k+q\sigma}\rangle^{(0)} - \langle n_{k\sigma}\rangle^{(0)}}{\varepsilon(k+q) - \varepsilon(k)}\,,$$

   at $T = 0$ and compare the result with χ_{Pauli} from Sect. 1.

Exercise 4.2.3

1. Compute the *diagonal* susceptibility of interacting electrons within the Hubbard model:

$$\chi_{\boldsymbol{q}}^{zz}(E) = \frac{1}{N}\sum_{i,j}\chi_{ij}^{zz}(E)\exp\left(\mathrm{i}\boldsymbol{q}\cdot\left(\boldsymbol{R}_i-\boldsymbol{R}_j\right)\right) ,$$

$$\chi_{ij}^{zz}(E) = -\frac{4\mu_{\mathrm{B}}^2\mu_0}{V\hbar^3}\langle\langle\sigma_i^z;\sigma_j^z\rangle\rangle ,$$

$$\sigma_i^z = \frac{\hbar}{2}\left(n_{i\uparrow}-n_{i\downarrow}\right) .$$

   Use an RPA method analogous to that in Sect. 4.2.3.
2. Derive a condition for ferromagnetism with the aid of (3.71) and the result from part 1.

Exercise 4.2.4 *Auger electron spectroscopy (AES)* and *appearance-potential spectroscopy (APS)* have become important experimental methods for the investigation of electronic states in solids. In AES a primary *core hole* is filled by a band electron. The energy released is transferred to another band electron, which is then able to leave the solid. Its kinetic energy is measured. In APS roughly speaking the reverse process takes place. An electron impinges upon a solid and fills an unoccupied band state. The energy released serves to excite a *core electron* into another unoccupied state. The ensuing recombination radiation ($\Longleftrightarrow$ filling of the resulting *core hole*) is used for the detection of the process. Due to the participation of the strictly localised *core state*, the excitation of the two holes or electrons is considered to be intraatomic. We consider a non-degenerate energy band, whose interacting electrons are described within the Stoner model (4.43),

$$\mathcal{H}_S = \sum_{\mathbf{k},\sigma} (E_\sigma(\mathbf{k}) - \mu)\, a^+_{\mathbf{k}\sigma} a_{\mathbf{k}\sigma} \,.$$

An exact description gives the following energy and temperature dependencies for the APS(AES) Intensities:

$$I_{\mathrm{APS}}(E - 2\mu) = \mathrm{e}^{\beta(E - 2\mu)} I_{\mathrm{AES}}(2\mu - E) = \frac{\mathrm{e}^{\beta(E - 2\mu)}}{\mathrm{e}^{\beta(E - 2\mu)} - 1} \left(\frac{1}{\hbar} S^{(2)}_{ii}(E - 2\mu) \right) .$$

Both of these intensities are determined by the same two-particle commutator spectral density:

$$S^{(2)}_{ii}(E) = -\frac{1}{\pi} \operatorname{Im} D_{ii}(E) \,; \quad D_{ij}(E) = \langle\!\langle a_{i-\sigma} a_{i\sigma} ; a^+_{j\sigma} a_{j-\sigma} \rangle\!\rangle^{\mathrm{ret}}_E \,.$$

1. Show that the two-particle spectral density can be expressed as follows by means of the quasi-particle density of states $\rho^{(S)}_\sigma(E)$ in the Stoner model:

$$S^{(2)}_{ii}(E - 2\mu) = \hbar \int \mathrm{d}x\, \rho^{(S)}_\sigma(x) \rho^{(S)}_{-\sigma}(E - x) (1 - f_-(x) - f_-(E - x)) \,,$$

where $f_-(x)$ is the Fermi function.

2. Let W be the width of the energy band of the non-interacting electrons. How wide is the energy range in which $S^{(2)}_{ii}(E - 2\mu)$ is non-vanishing?

Exercise 4.2.5 Some important correlation functions and sum rules can be derived from the intensities I_{APS}, I_{AES} of the AES and APS spectroscopies explained in Ex. 4.2.4. Show using the spectral representation of the two-particle spectral density that the following relations are valid, independently of the single-band model used:

$$\int_{-\infty}^{+\infty} \mathrm{d}E\, I_{\text{APS}}(E - 2\mu) = 1 - n + \langle n_\sigma n_{-\sigma} \rangle \; ; \qquad \int_{-\infty}^{+\infty} \mathrm{d}E\, I_{\text{AES}}(2\mu - E) = \langle n_\sigma n_{-\sigma} \rangle \; .$$

Exercise 4.2.6 Electrons in a non-degenerate energy band (s-band) are to be described by the Hubbard model. Show that for the intensities of the *Auger electron* (AES) and *appearance-potential* (APS) spectroscopies introduced in Ex. 4.2.4, for the case of an **empty** ($n = 0$) energy band, ($\mu \to -\infty$) applies:

$$I_{\text{AES}}(E) = 0 \; ; \quad I_{\text{APS}} = -\frac{1}{\pi} \operatorname{Im} \frac{1}{N} \sum_{\boldsymbol{k}} \frac{\Lambda_{\boldsymbol{k}}^{(0)}(E)}{1 - U \Lambda_{\boldsymbol{k}}^{(0)}(E)} \, ,$$

$$\Lambda_{\boldsymbol{k}}^{(0)}(E) = \frac{1}{N} \sum_{\boldsymbol{p}} \frac{1}{E - \varepsilon(\boldsymbol{k}) - \varepsilon(\boldsymbol{k} - \boldsymbol{p}) + \mathrm{i}0^+} \, .$$

It should be expedient to use a retarded Green's function

$$D_{mn;\,jj}^{\text{ret}}(E) = \langle\!\langle a_{m\sigma} a_{n-\sigma} ; a_{j-\sigma}^{+} a_{j\sigma}^{+} \rangle\!\rangle_{E}^{\text{ret}} \, .$$

Write its equation of motion and show that the *higher-order* Green's functions are greatly simplified, due to $n = 0$! Demonstrate that I_{APS} can be written as a self-folding integral of the Bloch density of states $\rho_0(E)$ for weak electronic correlations, i.e. small U.

Exercise 4.2.7 Calculate as in Ex. 4.2.6 the APS and AES intensities for the case of a **completely occupied** energy band, ($n = 2$).

4.3 Elementary Excitations in Disordered Alloys

4.3.1 Formulation of the Problem

So far, we have investigated the electronic properties of solids with a periodic lattice structure, which are therefore invariant with respect to symmetry operations. They fulfil for example translational symmetry, which we have already used several times, and this guarantees that the single-particle terms of the Hamiltonian are diagonal in $\boldsymbol{k}$ space. The decisive advantage of a periodic solid relative to a disordered system lies in the applicability of Bloch's theorem (2.15), with which one can reduce the entire problem to the solution of the Schrödinger equation for a single microscopic lattice cell. In disordered systems, Bloch's theorem does not apply. In such systems, one must therefore consider a potential of infinite range, which is of course possible with mathematical rigour in only a few, relatively uninteresting limiting cases.

Let us first consider what might be a suitable model Hamiltonian, whose form depends of course essentially on the type of spatial disorder of the system at hand. We want to limit ourselves in the following to the single-particle terms, i.e. we leave the mutual interaction of the elementary excitations out of consideration. Then the model Hamiltonian for all elementary excitations (electrons, phonons, magnons etc.) will have the same formal structure:

$$H = \sum_{i,j}^{i \neq j} \sum_{m,n} T_{ij}^{mn} a_{im}^{+} a_{jn} + \sum_{i,m} \varepsilon_m a_{im}^{+} a_{im} + \sum_{i,j} \sum_{m,n} V_{ij}^{mn} a_{im}^{+} a_{jn} \,. \tag{4.185}$$

The first term describes the *hopping* of the particle from the state $|n\rangle$ at $\boldsymbol{R}_j$ into the state $|m\rangle$ at $\boldsymbol{R}_i$. T_{ij}^{mn} is the corresponding transfer integral. ε_m is the atomic energy of the state $|m\rangle$ in an ideal periodic lattice. The actual problem is to be found in the third term. The **perturbation matrix** V_{ij}^{mn} contains the statistical deviations of the atomic energies and the transfer integrals from the corresponding quantities in the ideal system:

$$V_{ij}^{mn} = (\eta_m - \varepsilon_m)\delta_{ij}\,\delta_{mn} + \left(\widetilde{T}_{ij}^{mn} - T_{ij}^{mn}\right) \,. \tag{4.186}$$

The multiplicity of possible types of disorder can be roughly divided into two classes: **Substitutional disorder** and **structural disorder**. The first category is characterised by a still strictly-periodic arrangement of the lattice building blocks, whereby however for a propagating elementary excitation, the physical conditions vary from one location to another. Examples are alloys and mixed crystals. One refers to structural disorder when an additional deviation of the lattice structure from strict spatial periodicity is present, e.g. in amorphous solids, in glasses, in doped semiconductors, in liquid metals etc.

We shall develop the theoretical concepts on the example of the simplest type of disorder, namely that of

diagonal substitutional disorder.

Here, we consider a periodic lattice in which the atomic building blocks change their character statistically from lattice site to lattice site, whilst the *hopping* integrals are assumed to remain unchanged. From (4.186), we then have:

$$V_{ij}^{mn} = (\eta_m - \varepsilon_m)\,\delta_{ij}\,\delta_{mn}\ . \tag{4.187}$$

This idealised situation can be found to a good approximation in alloys whose pure components have very similar band structures. We could think of Ni-Cu or Ag-Au alloys, for example.

To be more concrete, we consider in the following electrons in a multiple-component alloy with *diagonal* substitutional disorder. It should in the end be clear that the inclusion of band transitions can offer nothing basically new in the context which interests us here. We therefore limit ourselves to a single-band model and suppress the band index:

$$H = \sum_\sigma H_\sigma\ ; \quad H_\sigma = H_{0\sigma} + V_{0\sigma}\ , \tag{4.188}$$

$$H_{0\sigma} = \sum_{i,j} T_{ij}\, a_{i\sigma}^+\, a_{j\sigma}\ , \tag{4.189}$$

$$V_{0\sigma} = \sum_i \eta_{(i)\sigma}\, a_{i\sigma}^+\, a_{i\sigma}\ . \tag{4.190}$$

Let us summarise once more the **assumptions** which we have made:

1. An alloy with α components. Each component is characterised by precisely **one** atomic level ($\eta_{(i)\sigma} = \eta_{m\sigma} - T_{ii} = \widehat{\eta}_{m\sigma}$, in the case that an atom of type m is located at $\boldsymbol{R}_i$),
2. a **diagonal** substitutional disorder (4.187),
3. **no** electron-electron interactions,
4. a statistically independent and homogeneous distribution of the types of atoms.

The constraints 1. and 2. may be reduced. Due to 4., the atomic levels $\eta_{m\sigma}$ become statistically random variables. The concentrations c_m of the alloy components are at the same time the probabilities that an atom of type m is to be found on a particular lattice site.

We are aware of the great information content of the single-particle Green's function,

$$G_{ij\sigma}(E) = \langle\langle a_{i\sigma}; a_{j\sigma}^+ \rangle\rangle_E\ , \tag{4.191}$$

and will therefore try to determine it as precisely as possible. Its equation of motion is readily derived with (4.188):

$$E\,G_{ij\sigma}(E) = \hbar\delta_{ij} + \sum_m \left(T_{im} + \eta_{(i)\sigma}\,\delta_{im}\right) G_{mj\sigma}(E)\ . \tag{4.192}$$

Owing to the lack of translational symmetry, this equation however cannot be solved simply by Fourier transformation.

If we denote the Green's function matrix by $\widehat{G}_\sigma(E)$, whose elements are just the $G_{ij\sigma}(E)$ in the Wannier representation, then we can read (4.192) as a matrix equation, which is simpler to manipulate for many purposes:

$$E\widehat{G}_\sigma(E) = \hbar\mathbf{1} + (H_{0\sigma} + V_{0\sigma})\widehat{G}_\sigma(E) . \tag{4.193}$$

Its formal solution is simple:

$$\widehat{G}_\sigma(E) = \hbar\,[E - H_\sigma]^{-1} . \tag{4.194}$$

The Green's function of the *free* system is given correspondingly by:

$$\widehat{G}_{0\sigma}(E) = \hbar\,[E - H_{0\sigma}]^{-1} . \tag{4.195}$$

Combining the last two equations,

$$\begin{aligned}\widehat{G}_\sigma(E) &= \hbar\left[\hbar\widehat{G}_{0\sigma}^{-1}(E) - V_{0\sigma}\right]^{-1} = \\ &= \left[1 - \frac{1}{\hbar}\widehat{G}_{0\sigma}(E)V_{0\sigma}\right]^{-1}\widehat{G}_{0\sigma}(E) ,\end{aligned}$$

we obtain the **Dyson equation** (3.327):

$$\widehat{G}_\sigma(E) = \widehat{G}_{0\sigma}(E) + \frac{1}{\hbar}\widehat{G}_{0\sigma}(E)V_{0\sigma}\widehat{G}_\sigma(E) . \tag{4.196}$$

If we return for a moment to the Wannier representation, then the Dyson equation reads as follows, using (4.190):

$$G_{ij\sigma}(E) = G_{ij\sigma}^{(0)}(E) + \frac{1}{\hbar}\sum_m G_{im\sigma}^{(0)}(E)\eta_{(m)\sigma}G_{mj\sigma}(E) . \tag{4.197}$$

It will prove to be expedient to introduce at this point the **scattering matrix** (T-matrix) $\widehat{T}_{0\sigma}$. It is defined by the following equation:

$$\widehat{G}_\sigma(E) = \widehat{G}_{0\sigma}(E) + \frac{1}{\hbar}\widehat{G}_{0\sigma}(E)\widehat{T}_{0\sigma}\widehat{G}_{0\sigma}(E) . \tag{4.198}$$

$\widehat{G}_\sigma(E)$ or $G_{ij\sigma}(E)$ depends just like the Hamiltonian H_σ on the actual distribution of the types of atoms $m = 1, 2, \dots, \alpha$ over the lattice. For a given set of concentrations $(c_1, c_2, \dots, c_\alpha)$, there is therefore a whole ensemble of Green's functions. This however means that we encounter a completely unnecessary difficulty, since special configurations are not of interest, as they would hardly be experimentally reproducible. Only the configurationally averaged quantities are important, and we will denote them by angular brackets,

$$\langle \ldots \rangle \quad \Longleftrightarrow \quad \textbf{configurational averaging} \, ,$$

Configurational averaging means taking an ensemble average over all the macroscopically non-distinguishable but microscopically different atomic arrangements which are possible for a given set of concentrations. The practical execution of the averaging process is accomplished as follows: Let F_σ be a functional of the random variables $\eta_{m\sigma}$; then we have:

$$\langle F_\sigma \rangle = \sum_{m=1}^{\alpha} c_m F_\sigma(\eta_{m\sigma}) \, . \tag{4.199}$$

Carrying out this averaging on the Green's function matrix (4.194), we thus define an effective Hamiltonian $H_{\text{eff}}^\sigma(E)$:

$$\left\langle \widehat{G}_\sigma(E) \right\rangle = \left\langle \frac{\hbar}{E - H_\sigma} \right\rangle = \frac{\hbar}{E - H_{\text{eff}}^\sigma(E)} \, . \tag{4.200}$$

H_{eff}^σ now exhibits the full symmetry of the lattice, due to the configurational average that we carried out, but at the price that it is energy dependent and complex under all circumstances. The basic lattice (*free* system) is of course unaffected by the configurational averaging. We thus find:

$$\left\langle \widehat{G}_{0\sigma}(E) \right\rangle \equiv \widehat{G}_{0\sigma}(E) \, , \tag{4.201}$$

and H_{eff}^σ can be written according to (4.188) as:

$$H_{\text{eff}}^\sigma(E) = H_{0\sigma} + \Sigma_{0\sigma}(E) \, . \tag{4.202}$$

The determination of $\Sigma_{0\sigma}(E)$ clearly solves the problem.

Finally, we can write down the Dyson equation (4.196) and the T-matrix equation (4.198) for the configurationally-averaged Green's function:

$$\left\langle \widehat{G}_\sigma(E) \right\rangle = \widehat{G}_{0\sigma}(E) + \frac{1}{\hbar} \widehat{G}_{0\sigma}(E) \Sigma_{0\sigma}(E) \left\langle \widehat{G}_\sigma(E) \right\rangle \, , \tag{4.203}$$

$$\left\langle \widehat{G}_\sigma(E) \right\rangle = \widehat{G}_{0\sigma}(E) + \frac{1}{\hbar} \widehat{G}_{0\sigma}(E) \left\langle \widehat{T}_{0\sigma} \right\rangle \widehat{G}_{0\sigma}(E) \, . \tag{4.204}$$

4.3.2 The Effective-Medium Method

The separation of the model Hamiltonian H_σ as in (4.188) into $H_{0\sigma}$ and $V_{0\sigma}$ is not mandatory in this form. Instead of the *unperturbed* crystal, that is the strictly periodic basic lattice, we could have separated off any suitably chosen **effective medium**, by attributing to each lattice site a fictitious, real or complex potential $v_{K\sigma}$. The effective medium is then defined by the Hamiltonian

$$K_\sigma = \sum_{i,j} T_{ij}\, a_{i\sigma}^+ a_{j\sigma} + \sum_i v_{K\sigma}\, a_{i\sigma}^+ a_{i\sigma} \,. \tag{4.205}$$

It will of course be chosen so that the associated many-body problem can be solved exactly. Since it furthermore exhibits the full symmetry of the basic lattice, the Green's function of the effective medium,

$$\widehat{R}_\sigma(E) = \hbar\,[E - K_\sigma]^{-1} \,, \tag{4.206}$$

is known, and it is diagonal in $\boldsymbol{k}$ space:

$$R_{\boldsymbol{k}\sigma}(E) = \hbar\,(E - \widetilde{\varepsilon}_\sigma(\boldsymbol{k}))^{-1} \,, \tag{4.207}$$

$$\widetilde{\varepsilon}_\sigma(\boldsymbol{k}) = \varepsilon(\boldsymbol{k}) + v_{K\sigma} \,. \tag{4.208}$$

The model Hamiltonian can now be written as:

$$H_\sigma = K_\sigma + V_{K\sigma} \,, \tag{4.209}$$

$$V_{K\sigma} = \sum_i \left(\eta_{(i)\sigma} - v_{K\sigma}\right) a_{i\sigma}^+ a_{i\sigma} \equiv \sum_i \widetilde{\eta}_{(i)\sigma}\, a_{i\sigma}^+ a_{i\sigma} \,. \tag{4.210}$$

$\widetilde{\eta}_{(i)\sigma}$ gives the deviation of the local potential at $\boldsymbol{R}_i$ relative to the effective medium. Equations (4.196), (4.198), (4.203) and (4.204) can be taken over directly. We need only replace $\widehat{G}_{0\sigma}$ by $\widehat{R}_\sigma(E)$. Naturally, the self-energy ($\Sigma_{0\sigma} \Longrightarrow \Sigma_{K\sigma}$) and the T-matrix $\left(\widehat{T}_{0\sigma} \Longrightarrow \widehat{T}_{K\sigma}\right)$ are modified.

$$\widehat{G}_\sigma(E) = \widehat{R}_\sigma(E) + \frac{1}{\hbar}\widehat{R}_\sigma(E) V_{K\sigma} \widehat{G}_\sigma(E) \,, \tag{4.211}$$

$$\widehat{G}_\sigma(E) = \widehat{R}_\sigma(E) + \frac{1}{\hbar}\widehat{R}_\sigma(E) \widehat{T}_{K\sigma} \widehat{R}_\sigma(E) \,, \tag{4.212}$$

$$\left\langle \widehat{G}_\sigma(E) \right\rangle = \widehat{R}_\sigma(E) + \frac{1}{\hbar}\widehat{R}_\sigma(E) \Sigma_{K\sigma}(E) \left\langle \widehat{G}_\sigma(E) \right\rangle \,, \tag{4.213}$$

$$\left\langle \widehat{G}_\sigma(E) \right\rangle = \widehat{R}_\sigma(E) + \frac{1}{\hbar}\widehat{R}_\sigma(E) \left\langle \widehat{T}_{K\sigma} \right\rangle \widehat{R}_\sigma(E) \tag{4.214}$$

Combining the first two equations for the function before averaging, we can express the T-matrix in terms of the statistical potential $V_{K\sigma}$:

$$\widehat{T}_{K\sigma} = V_{K\sigma}\widehat{G}_\sigma \widehat{R}_\sigma^{-1} = V_{K\sigma}\left(1 - \frac{1}{\hbar}\widehat{R}_\sigma V_{K\sigma}\right)^{-1} \widehat{R}_\sigma \widehat{R}_\sigma^{-1} ,$$
$$\widehat{T}_{K\sigma} = V_{K\sigma}\left(1 - \frac{1}{\hbar}\widehat{R}_\sigma V_{K\sigma}\right)^{-1} . \tag{4.215}$$

Quite analogously, by combining (4.213) and (4.214), one obtains:

$$\left\langle \widehat{T}_{K\sigma} \right\rangle = \Sigma_{K\sigma}\left(1 - \frac{1}{\hbar}\widehat{R}_\sigma \Sigma_{K\sigma}\right)^{-1} , \tag{4.216}$$

$$\Sigma_{K\sigma} = \left\langle \widehat{T}_{K\sigma} \right\rangle \left(1 + \frac{1}{\hbar}\left\langle \widehat{T}_{K\sigma} \right\rangle \widehat{R}_\sigma\right)^{-1} . \tag{4.217}$$

We now recall that we have not yet specified the effective medium in any concrete way. We vary it, i.e. we change the type of quasi-particles of the effective medium, until they are no longer scattered on the local potentials. This is the case only when the configuration-averaged T-matrix vanishes:

$$\left\langle \widehat{T}_{K\sigma} \right\rangle \stackrel{!}{=} 0 . \tag{4.218}$$

If we can fulfil this condition, then the entire problem is automatically solved. From (4.217), it then namely follows that:

$$\Sigma_{K\sigma}(E) \equiv 0 \iff \left\langle \widehat{G}_\sigma(E) \right\rangle = \widehat{R}_\sigma(E) . \tag{4.219}$$

The Green's function $\widehat{R}_\sigma$ of the effective medium is however known by construction.

The requirement (4.218) can, however, not as a rule be strictly met, since $\widehat{T}_{K\sigma}$ is not explicitly known. We discuss a suitable approximation method in the next section.

4.3.3 The Coherent Potential Approximation

We first attempt to express the scattering matrix $\widehat{T}_{K\sigma}$ in terms of atomic scattering matrices $\hat{t}^{(i)}_{K\sigma}$. Let $\widehat{G}_{i\sigma}(E)$ be the Green's function for the special case that only the atomic scattering centre at the lattice site $\boldsymbol{R}_i$ is *switched on*, i.e. the sum for $V_{K\sigma}$ in (4.210) contains only one term:

$$V^{(i)}_{K\sigma} = \widetilde{\eta}_{(i)\sigma}\, a^+_{i\sigma} a_{i\sigma} . \tag{4.220}$$

Then it follows from (4.211) that:

$$\begin{aligned}\widehat{G}_{i\sigma}(E) &= \widehat{R}_\sigma(E) + \frac{1}{\hbar}\widehat{R}_\sigma(E)V_{K\sigma}^{(i)}\widehat{R}_\sigma(E) + \\ &+ \frac{1}{\hbar^2}\widehat{R}_\sigma(E)V_{K\sigma}^{(i)}\widehat{R}_\sigma(E)V_{K\sigma}^{(i)}\widehat{R}_\sigma(E) + \cdots = \\ &= \widehat{R}_\sigma(E) + \frac{1}{\hbar}\widehat{R}_\sigma(E)V_{K\sigma}^{(i)}\widehat{G}_{i\sigma}(E)\,.\end{aligned} \tag{4.221}$$

Equation (4.212) then defines the **atomic scattering matrix** $t_{K\sigma}^{(i)}$:

$$\widehat{G}_{i\sigma}(E) = \widehat{R}_\sigma(E) + \frac{1}{\hbar}\widehat{R}_\sigma(E)\hat{t}_{K\sigma}^{(i)}\widehat{R}_\sigma(E)\,. \tag{4.222}$$

Comparison with (4.221) yields:

$$\begin{aligned}\hat{t}_{K\sigma}^{(i)} &= V_{K\sigma}^{(i)} + \frac{1}{\hbar}V_{K\sigma}^{(i)}\widehat{R}_\sigma V_{K\sigma}^{(i)} + \frac{1}{\hbar^2}V_{K\sigma}^{(i)}\widehat{R}_\sigma V_{K\sigma}^{(i)}\widehat{R}_\sigma V_{K\sigma}^{(i)} + \cdots = \\ &= V_{K\sigma}^{(i)}\left(1 - \frac{1}{\hbar}\widehat{R}_\sigma V_{K\sigma}^{(i)}\right)^{-1}\,.\end{aligned} \tag{4.223}$$

This result is naturally consistent with (4.215). We now return to the *full* Dyson equation (4.211) and insert $V_{K\sigma} = \sum_i V_{K\sigma}^{(i)}$ there:

$$\widehat{G}_\sigma = \widehat{R}_\sigma + \frac{1}{\hbar}\widehat{R}_\sigma \sum_i V_{K\sigma}^{(i)}\widehat{R}_\sigma + \frac{1}{\hbar^2}\widehat{R}_\sigma \sum_{i,j} V_{K\sigma}^{(i)}\widehat{R}_\sigma V_{K\sigma}^{(j)}\widehat{R}_\sigma + \cdots$$

With (4.223), we then find:

$$\begin{aligned}\widehat{G}_\sigma(E) &= \widehat{R}_\sigma(E) + \frac{1}{\hbar}\widehat{R}_\sigma(E)\sum_i \hat{t}_{K\sigma}^{(i)}\widehat{R}_\sigma(E) + \\ &+ \frac{1}{\hbar^2}\widehat{R}_\sigma(E)\sum_{i,j}^{i\neq j}\hat{t}_{K\sigma}^{(i)}\widehat{R}_\sigma(E)\hat{t}_{K\sigma}^{(j)}\widehat{R}_\sigma(E) + \\ &+ \frac{1}{\hbar^3}\widehat{R}_\sigma(E)\sum_{i,j,k}^{i\neq j;k\neq j}\hat{t}_{K\sigma}^{(i)}\widehat{R}_\sigma(E)\hat{t}_{K\sigma}^{(j)}\widehat{R}_\sigma(E)\hat{t}_{K\sigma}^{(k)}\widehat{R}_\sigma(E) + \\ &+ \cdots\,.\end{aligned} \tag{4.224}$$

Comparison with (4.212) finally gives the desired result:

$$\widehat{T}_{K\sigma} = \sum_i \hat{t}^{(i)}_{K\sigma} + \frac{1}{\hbar} \sum_{i,j}^{i \neq j} \hat{t}^{(i)}_{K\sigma} \widehat{R}_\sigma(E) \hat{t}^{(j)}_{K\sigma} + \\ + \frac{1}{\hbar^2} \sum_{i,j,k}^{i \neq j;\, j \neq k} \hat{t}^{(i)}_{K\sigma} \widehat{R}_\sigma(E) \hat{t}^{(j)}_{K\sigma} \widehat{R}_\sigma(E) \hat{t}^{(k)}_{K\sigma} + \cdots . \tag{4.225}$$

Up to now, everything is still exact. With a known T-matrix, the entire problem has also been solved, since then via (4.212) the single-particle Green's function is also determined. Every approximate determination of $\widehat{T}_{K\sigma}$ thus leads to an approximate solution of the complete alloy problem.

At this point, it is advisable to formulate the operator relation (4.225) once again for the corresponding matrix elements:

$$\left(\widehat{T}_{K\sigma}\right)_{mn} = t^{(m)}_{K\sigma} \delta_{mn} + \frac{1}{\hbar} t^{(m)}_{K\sigma} \overline{R}_{mn\sigma}(E) t^{(n)}_{K\sigma} + \\ + \frac{1}{\hbar^2} \sum_j t^{(m)}_{K\sigma} \overline{R}_{mj\sigma}(E) t^{(j)}_{K\sigma} \overline{R}_{jn\sigma}(E) t^{(n)}_{K\sigma} + \\ + \cdots . \tag{4.226}$$

Here, we have written as an abbreviation:

$$\overline{R}_{mn\sigma}(E) = R_{mn\sigma}(E)\left(1 - \delta_{mn}\right) . \tag{4.227}$$

We require the configuration-averaged T-matrix. If we carry out the averaging process on the expression (4.225), we obtain terms of the form

$$\left\langle \hat{t}^{(i)}_{K\sigma} \widehat{R}_\sigma \hat{t}^{(j)}_{K\sigma} \widehat{R}_\sigma \cdots \widehat{R}_\sigma \hat{t}^{(n)}_{K\sigma} \right\rangle . \tag{4.228}$$

If the indices $i, j, \ldots, n$ are all pairwise different, then this expression can be factored. The so-called **T-matrix approximation (TMA)** postulates that the propagating electron never returns to a site at which it was already scattered. Owing to the statistical independence of the local scattering potentials, we then find:

$$\left\langle \hat{t}^{(i)}_{K\sigma} \hat{r}_\sigma \hat{t}^{(j)}_{K\sigma} \widehat{R}_\sigma \cdots \widehat{R}_\sigma \hat{t}^{(n)}_{K\sigma} \right\rangle \xrightarrow[\text{TMA}]{} \left\langle \hat{t}^{(i)}_{K\sigma} \right\rangle \widehat{R}_\sigma \left\langle \hat{t}^{(j)}_{K\sigma} \right\rangle \widehat{R}_\sigma \cdots \widehat{R}_\sigma \left\langle \hat{t}^{(n)}_{K\sigma} \right\rangle . \tag{4.229}$$

After the configuration averaging, the atomic scattering matrix is of course independent of the lattice site. Thereby, we can now sum (4.225), if we take into account that

$$\left\langle \hat{t}_{K\sigma}^{(i)} \right\rangle = \langle t_{K\sigma} \rangle \begin{pmatrix} 0 & & 0 \\ & \ddots & \\ & 1 & \\ & & \ddots \\ 0 & & 0 \end{pmatrix} \leftarrow i\text{-th row} .$$

We write

$$\mathbf{1}_i \equiv \begin{pmatrix} 0 & & 0 \\ & \ddots & \\ & 1 & \\ & & \ddots \\ 0 & & 0 \end{pmatrix} \leftarrow i\text{-th row} \qquad (4.230)$$

$$\uparrow$$
$$i\text{-th column}$$

and then obtain:

$$\left\langle \hat{t}_{K\sigma}^{(i)} \right\rangle = \langle t_{K\sigma} \rangle \, \mathbf{1}_i \, . \qquad (4.231)$$

Here, $\langle t_{K\sigma} \rangle$ is now no longer an operator. Instead, with (4.223) and (4.199), we have:

$$\langle t_{K\sigma} \rangle = \sum_{m=1}^{\alpha} c_m \frac{\widehat{\eta}_{m\sigma} - v_{K\sigma}}{1 - \frac{1}{\hbar} R_{ii\sigma}(E) \left(\widehat{\eta}_{m\sigma} - v_{K\sigma}\right)} \, . \qquad (4.232)$$

From (4.225), we then find $\left(\sum_i \left\langle \hat{t}_{K\sigma}^{(i)} \right\rangle = \left\langle \hat{t}_{K\sigma} \right\rangle \mathbf{1}\right)$:

$$\begin{aligned} \left\langle \widehat{T}_{K\sigma} \right\rangle^{\text{TMA}} &= \langle t_{K\sigma} \rangle \, \mathbf{1} \left(1 + \frac{1}{\hbar} \widehat{\widehat{R}}_\sigma(E) \, \langle t_{K\sigma} \rangle + \right. \\ &\quad \left. + \frac{1}{\hbar^2} \widehat{\widehat{R}}_\sigma(E) \, \langle t_{K\sigma} \rangle \, \widehat{\widehat{R}}_\sigma(E) \, \langle t_{K\sigma} \rangle + \cdots \right) , \qquad (4.233) \\ \left\langle \widehat{T}_{K\sigma} \right\rangle^{\text{TMA}} &= \langle t_{K\sigma} \rangle \, \mathbf{1} \left(1 - \frac{1}{\hbar} \widehat{\widehat{R}}_\sigma(E) \, \langle t_{K\sigma} \rangle \right)^{-1} . \end{aligned}$$

Inserting this result into the expression (4.217) for the self-energy, we can then see that it is diagonal within the TMA in the Wannier representation:

$$\left(\Sigma_{K\sigma}^{\text{TMA}}(E) \right)_{ij} = \langle t_{K\sigma} \rangle \, \delta_{ij} \left(1 + \frac{1}{\hbar} R_{ii\sigma}(E) \, \langle t_{K\sigma} \rangle \right)^{-1} . \qquad (4.234)$$

Note that the Green's function $R_{ii\sigma}(E)$ carried the index i only formally, due to the translational symmetry. It is naturally independent of the lattice site:

$$\begin{aligned} R_{ii\sigma}(E) &= \frac{1}{N}\sum_{k} R_{k\sigma}(E) = \\ &= \frac{\hbar}{N}\sum_{k}\left(E - \varepsilon(k) - v_{K\sigma}\right)^{-1} . \end{aligned} \tag{4.235}$$

The self-energy is independent of wavenumber:

$$\Sigma_{K\sigma}^{\mathrm{TMA}}(k;E) \equiv \Sigma_{K\sigma}^{\mathrm{TMA}}(E) . \tag{4.236}$$

The TMA is clearly not a self-consistent procedure, since the solution depends on the arbitrary choice of the effective medium. Intuitively, one would expect that the quality of the TMA is higher, the more similar the effective medium chosen is to the real physical system.

We can now however readily make the procedure self-consistent by making use of the results of the previous section, i.e. by determining the potential of the effective medium through the requirement (4.218). This however means according to (4.233), that through

$$\langle t_{K\sigma}\rangle \overset{!}{=} 0 , \tag{4.237}$$

$v_{K\sigma}$ is fixed. This method is referred to in the literature as the

coherent potential approximation (CPA).

According to (4.232), we solve ($T_{ii} = T_0$, $\widehat{\eta}_{m\sigma} = \eta_{m\sigma} - T_0$):

$$0 \overset{!}{=} \sum_{m=1}^{\alpha} c_m \frac{\eta_{m\sigma} - v_{K\sigma} - T_0}{1 - \frac{1}{\hbar} R_{ii\sigma}(E)\left(\eta_{m\sigma} - v_{K\sigma} - T_0\right)} \tag{4.238}$$

for

$$v_{K\sigma} \to \Sigma_{\sigma}^{\mathrm{CPA}}(E) \tag{4.239}$$

and then, in view of (4.219), we have completely determined the single-particle Green's function

$$\langle G_{k\sigma}(E)\rangle = R_{k\sigma}(E) = \hbar\,\left(E - \varepsilon(k) - \Sigma_{\sigma}^{\mathrm{CPA}}(E)\right)^{-1} . \tag{4.240}$$

Now all the other quantities which are derivable from the single-particle Green's function are known, such as the quasi-particle density of states:

$$\rho_\sigma(E) = -\frac{1}{\pi} \operatorname{Im} \int_{-\infty}^{+\infty} \mathrm{d}x \, \frac{\rho_0(x)}{E - x - \Sigma_\sigma^{\mathrm{CPA}}(E)} \, . \tag{4.241}$$

In this expression, we have replaced the summation over $\boldsymbol{k}$ by an integration over the *free* density of states $\rho_0(x)$.

The equation to be solved, (4.238), is as a rule highly nonlinear and therefore solvable only by numerical methods.

The physical concept of the CPA is clear and relatively simple. Now theories which in some sense follow physical intuition often suffer from the problem that their mathematical structure remains unclear. This is however not the case for the CPA! In spite of its naive concept, it is nevertheless the result of carefully executed mathematics. To demonstrate this, we derive in the next section the CPA formula (4.238) once again in a quite different manner using a diagram technique. In the process, we will encounter a series of other methods which are frequently employed in the theory of disordered systems, and we will experience the power of the CPA.

4.3.4 Diagrammatic Methods

We return once again to the Dyson equation (4.197):

$$\begin{aligned} G_{ij\sigma}(E) &= G_{ij\sigma}^{(0)}(E) + \frac{1}{\hbar} \sum_m G_{im\sigma}^{(0)}(E) \eta_{(m)\sigma} G_{mj\sigma}^{(0)}(E) + \\ &+ \frac{1}{\hbar^2} \sum_{m,n} G_{im\sigma}^{(0)}(E) \eta_{(m)\sigma} G_{mn\sigma}^{(0)}(E) \eta_{(n)\sigma} G_{nj\sigma}^{(0)}(E) + \\ &+ \cdots \end{aligned} \tag{4.242}$$

and carry out the configurational averaging directly on this infinite series:

$$\begin{aligned} \left\langle G_{ij\sigma} \right\rangle &= G_{ij\sigma}^{(0)} + \frac{1}{\hbar} \sum_m G_{im\sigma}^{(0)} G_{mj\sigma}^{(0)} \left\langle \eta_{(m)\sigma} \right\rangle + \\ &+ \frac{1}{\hbar^2} \sum_{m,n} G_{im\sigma}^{(0)} G_{mn\sigma}^{(0)} G_{nj\sigma}^{(0)} \left\langle \eta_{(m)\sigma} \eta_{(n)\sigma} \right\rangle + \\ &+ \cdots \end{aligned} \tag{4.243}$$

Due to the assumed statistical independence of the lattice sites, we naturally have:

$$\left\langle \eta_{(m)\sigma} \eta_{(n)\sigma} \right\rangle = \delta_{mn} \left\langle \eta_{(m)\sigma}^2 \right\rangle + (1 - \delta_{mn}) \left\langle \eta_{(m)\sigma} \right\rangle^2 \, . \tag{4.244}$$

We agree upon the following **diagram rules**

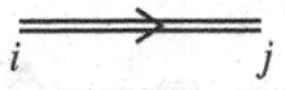

$$\Rightarrow \quad \langle G_{ij\sigma} \rangle$$

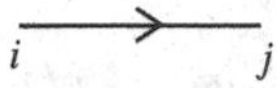

$$\Rightarrow \quad G_{ij\sigma}^{(0)}$$

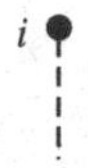

$$\Rightarrow \quad \frac{1}{\hbar}\langle \eta_{(i)\sigma} \rangle = \frac{1}{\hbar}\sum_{m=1}^{\alpha} c_m \widehat{\eta}_{m\sigma}$$

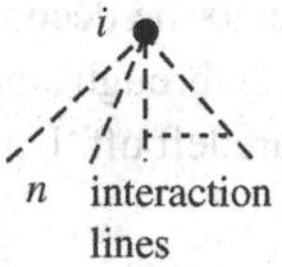

$$\Rightarrow \quad \frac{1}{\hbar}\left\langle \eta_{(i)\sigma}^{n} \right\rangle = \frac{1}{\hbar^n}\sum_{m=1}^{\alpha} c_m \widehat{\eta}_{m\sigma}^{n}$$

With these, we can now represent the above result (4.243) for $\langle G_{ij\sigma} \rangle$ in terms of diagrams:

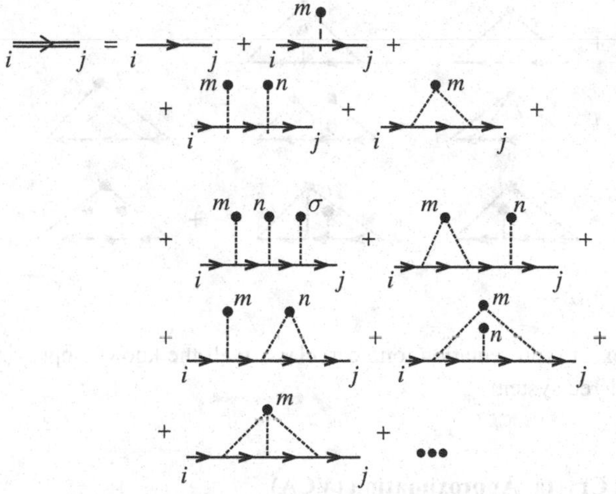

(4.245)

The term **order of a diagram** means the number of **interaction lines** (dashed!). All diagrams up to third order are explicitly plotted in (4.245). In evaluating the diagrams, sums over all the inner indices must be be carried out.

Examples

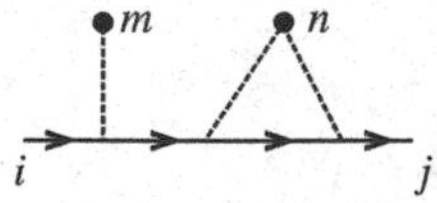

$$\Longleftrightarrow \quad \frac{1}{\hbar^3}\sum_{m,n} G_{im\sigma}^{(0)} G_{mn\sigma}^{(0)} G_{nn\sigma}^{(0)} G_{nj\sigma}^{(0)} \langle \eta_{(m)\sigma} \rangle \left\langle \eta_{(n)\sigma}^{2} \right\rangle .$$

We can now however also represent the **Dyson equation** by diagrams:

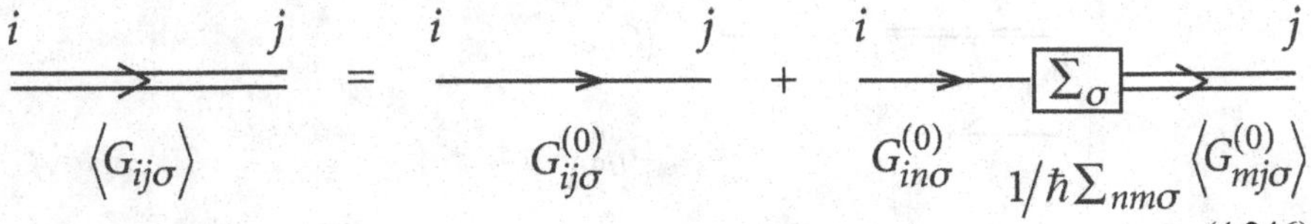

(4.246)

Σ_σ is defined as the sum of all those diagrams which cannot be decomposed into two independent diagrams from the expansion for $\langle G_{ij\sigma}\rangle$ by cutting through a particle propagator (——▸——), whereby the two outer connections at i and j are left off. Up to fourth order, Σ_σ thus consists of the following diagrams:

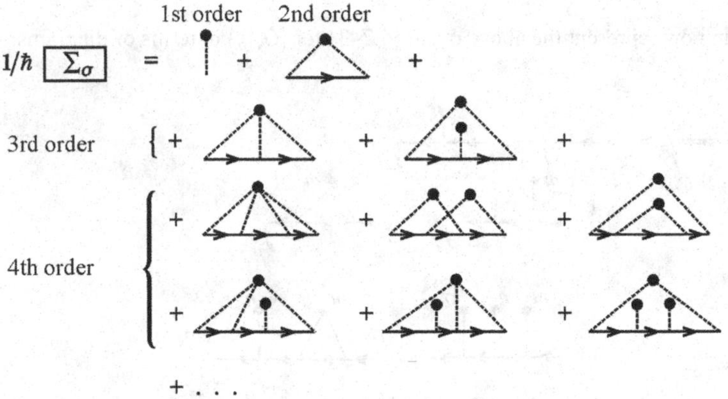

(4.247)

With this still-exact representation, one can classify all the known approximations to the theory of disordered systems.

1) The Virtual Crystal Approximation (VCA)

The simplest approximation is that of the virtual crystal, which includes only the first term from the expansion of $\sum_\sigma$:

$$\frac{1}{\hbar}\Sigma_\sigma^{\mathrm{VCA}} \Longleftrightarrow$$

$$\equiv \delta_{rt}\frac{1}{\hbar}\sum_{m=1}^{\alpha} c_m \widehat{\eta}_{m\sigma} \equiv \delta_{rt}\frac{1}{\hbar}\langle\widehat{\eta}_\sigma\rangle \,. \tag{4.248}$$

For the Green's function, this implies:

$$\langle G_{ij\sigma}(E)\rangle^{\mathrm{VCA}} = G_{ij\sigma}^{(0)}(E) + \frac{1}{\hbar}\langle\widehat{\eta}_\sigma\rangle \sum_m G_{im\sigma}^{(0)} \langle G_{mj\sigma}\rangle^{\mathrm{VCA}} \,.$$

Transformation to wavenumbers then yields:

$$\langle G_{k\sigma}(E)\rangle^{\text{VCA}} = G^{(0)}_{k\sigma}(E - \langle \widehat{\eta}_\sigma \rangle) . \tag{4.249}$$

This corresponds to a molecular-field approximation. The quasi-particle energies $E_\sigma^{\text{VCA}}(\boldsymbol{k})$ are only shifted by a constant amount $\langle \widehat{\eta}_\sigma \rangle$ relative to the free Bloch energies $\varepsilon(\boldsymbol{k})$:

$$E_\sigma^{\text{VCA}}(\boldsymbol{k}) = \varepsilon(\boldsymbol{k}) + \langle \widehat{\eta}_\sigma \rangle - T_0 . \tag{4.250}$$

The VCA is usable, to be sure, only when the atomic levels of the alloying components are not very different from each other.

2) The Single-Site Approximation (SSA)

A somewhat more subtle approximation consists in taking into account all the diagrams from the exact expansion for the self-energy (4.247) which contain just one vertex point, that is the diagrams which are linear in the concentrations c_m. In this case, one neglects correlations between scattering processes at different lattice sites.

$$\frac{1}{\hbar}\Sigma_\sigma^{\text{SSA}} \quad \Longleftrightarrow \tag{4.251}$$

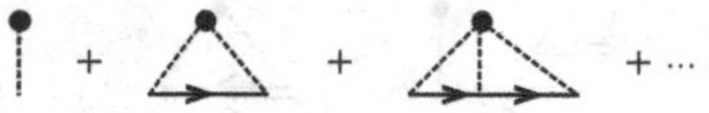

This infinite series can be summed exactly:

$$\frac{1}{\hbar}\left(\Sigma_\sigma^{\text{SSA}}\right)_{rt} =$$

$$= \delta_{rt}\left[\frac{1}{\hbar}\sum_m c_m \widehat{\eta}_{m\sigma} + \frac{1}{\hbar^2}\sum_m c_m \widehat{\eta}^2_{m\sigma} G^{(0)}_{mm\sigma} + \right.$$

$$\left. + \frac{1}{\hbar^3}\sum_m c_m \widehat{\eta}^3_{m\sigma}\left(G^{(0)}_{mm\sigma}\right)^2 + \cdots\right] =$$

$$= \delta_{rt}\frac{1}{\hbar}\sum_m c_m \widehat{\eta}_{m\sigma}\left[1 + \frac{1}{\hbar}\widehat{\eta}_{m\sigma} G^{(0)}_{mm\sigma} + \frac{1}{\hbar^2}\left(\widehat{\eta}_{m\sigma} G^{(0)}_{mm\sigma}\right)^2 + \cdots\right] .$$

The brackets on the right enclose the geometric series:

$$\frac{1}{\hbar}\left(\Sigma_\sigma^{\text{SSA}}\right)_{rt} = \delta_{rt}\frac{1}{\hbar}\sum_m c_m \frac{\widehat{\eta}_{m\sigma}}{1 - \frac{1}{\hbar}\widehat{\eta}_{m\sigma} G^{(0)}_{mm\sigma}} . \tag{4.252}$$

From (4.232), the self-energy in the SSA, which includes multiple scattering only from an isolated local potential, is represented by the atomic T-matrix. The self-energy is independent of wavenumber after Fourier transformation, owing to its *single site* nature:

$$\Sigma_\sigma^{\text{SSA}}(\boldsymbol{k}, E) \equiv \Sigma_\sigma^{\text{SSA}}(E) . \tag{4.253}$$

The full single-particle Green's function is then given by

$$\langle G_{k\sigma}\rangle^{\text{SSA}} = \hbar\left(E - \varepsilon(\boldsymbol{k}) - \Sigma_\sigma^{\text{SSA}}(E)\right)^{-1}, \tag{4.254}$$

and the quasi-particle energies are determined as usual from the poles of this function:

$$\left(E - \varepsilon(\boldsymbol{k}) - \Sigma_\sigma^{\text{SSA}}(E)\right)\Big|_{E=E_\sigma(\boldsymbol{k})} \overset{!}{=} 0\,. \tag{4.255}$$

Because of the $\boldsymbol{k}$-independence, the *full* Green's function can be expressed in terms of that of the *free* system:

$$\langle G_{k\sigma}(E)\rangle^{\text{SSA}} = G_{k\sigma}^{(0)}\left(E - \Sigma_\sigma^{\text{SSA}}(E)\right)\,. \tag{4.256}$$

3) The Modified Propagator Method (MPM)

As a next step, one can make the SSA self-consistent by replacing the free propagator in (4.251) by the *full* Green's function:

$$\frac{1}{\hbar}\Sigma_\sigma^{\text{MPM}} \Longleftrightarrow \tag{4.257}$$

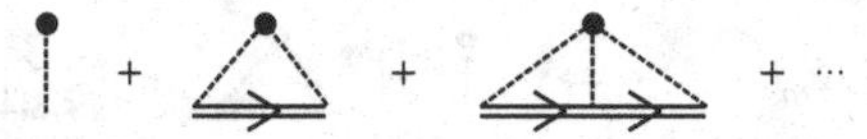

In this way, one reproduces precisely the result of the so-called *modified propagator method*, which however was based originally on a quite different idea:

$$\frac{1}{\hbar}\Sigma_\sigma^{\text{MPM}}(E) = \frac{1}{\hbar}\sum_m c_m \frac{\widehat{\eta}_{m\sigma}}{1 - \frac{1}{\hbar}\widehat{\eta}_{m\sigma} G_{mm\sigma}^{(0)}\left(E - \Sigma_\sigma^{\text{MPM}}(E)\right)}\,. \tag{4.258}$$

One can see immediately that by inserting the full propagator into the SSA diagrams, one brings a large number of diagrams from the exact expansion back into play, which were neglected in the SSA itself. Thus, up to fourth order, from the exact expansion (4.247) for Σ_σ only the *crossed* diagram is still missing; it describes cluster effects.

4) The Average T-Matrix Approximation (ATA)

An obvious defect in the SSA can readily be seen, which in the end is based upon an inaccuracy within our general diagram rules. If, for example, we insert the first term in the expansion of the self-energy (4.247) into the Dyson equation, we obtain among others the diagram sketched here:

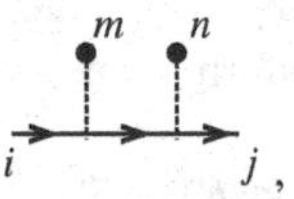

which, due to the two vertex points, yields the contribution

$$\frac{1}{\hbar^2}\langle\widehat{\eta}_\sigma\rangle^2\sum_{m,n}G^{(0)}_{im\sigma}G^{(0)}_{mn\sigma}G^{(0)}_{nj\sigma}\,.$$

According to the rules, we must sum over all m and n. The inaccuracy lies in the diagonal terms $m = n$, since they are included quadratically in the concentrations c, although the correct $m{=}n$ diagram is only linear in c:

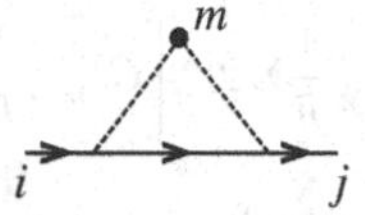

(Note: $\langle\widehat{\eta}^2_\sigma\rangle \neq \langle\widehat{\eta}_\sigma\rangle^2$!)

This obvious error can be eliminated by the following additional prescription: One must subtract from every diagram in the SSA the contributions of all those diagrams which can be constructed from the original diagrams by *breaking off* the interaction lines from vertex points. This is illustrated in the following table:

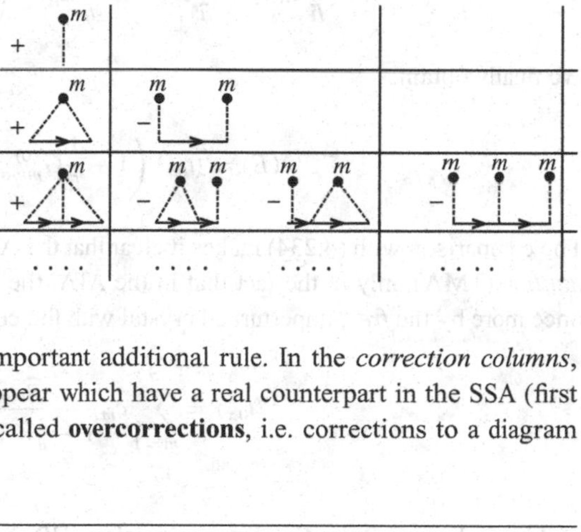

Fig. 4.16 Diagram corrections to the single-site approximation

Finally, we must consider another important additional rule. In the *correction columns*, naturally only those diagrams can appear which have a real counterpart in the SSA (first column). Otherwise we produce so-called **overcorrections**, i.e. corrections to a diagram which does not even exist.

Examples

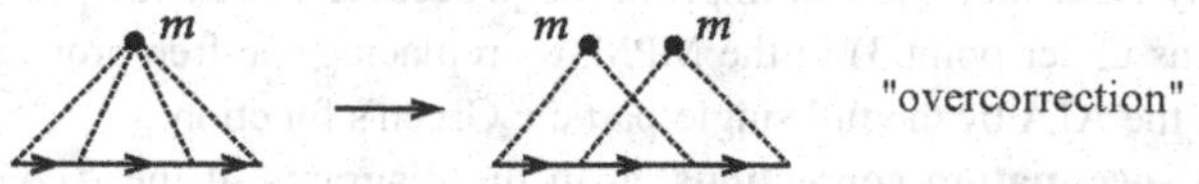

The *crossed* diagram must not be counted. We sum the individual columns separately:

$$\textbf{1st column:}\quad \frac{1}{\hbar}\Sigma_\sigma^{\mathrm{SSA}} = \frac{1}{\hbar}\langle t_{0\sigma}\rangle \qquad \text{(s. (4.252))}\,,$$

$$\textbf{2nd column:}\quad \frac{1}{\hbar}\Sigma_\sigma^{\mathrm{ATA}} G^{(0)}_{mm\sigma}\frac{1}{\hbar}\Sigma_\sigma^{\mathrm{ATA}}\,,$$

$$\vdots$$

$$\textbf{n-th column:}\quad \frac{1}{\hbar}\Sigma_\sigma^{\mathrm{ATA}}\left[G^{(0)}_{mm\sigma}\frac{1}{\hbar}\Sigma_\sigma^{\mathrm{ATA}}\right]^{n-1}.$$

This can be straightforwardly summarised:

$$\begin{aligned}\frac{1}{\hbar}\Sigma_\sigma^{\mathrm{ATA}} &= \frac{1}{\hbar}\langle t_{0\sigma}\rangle - \frac{1}{\hbar}\Sigma_\sigma^{\mathrm{ATA}}\sum_{n=1}^{\infty}\left[G^{(0)}_{mm\sigma}\frac{1}{\hbar}\Sigma_\sigma^{\mathrm{ATA}}\right]^n = \\ &= \frac{1}{\hbar}\langle t_{0\sigma}\rangle + \frac{1}{\hbar}\Sigma_\sigma^{\mathrm{ATA}} - \frac{1}{\hbar}\Sigma_\sigma^{\mathrm{ATA}}\sum_{n=0}^{\infty}\left[G^{(0)}_{mm\sigma}\frac{1}{\hbar}\Sigma_\sigma^{\mathrm{ATA}}\right]^n = \\ &= \frac{1}{\hbar}\langle t_{0\sigma}\rangle + \frac{1}{\hbar}\Sigma_\sigma^{\mathrm{ATA}} - \frac{1}{\hbar}\Sigma_\sigma^{\mathrm{ATA}}\left(1 - G^{(0)}_{mm\sigma}\frac{1}{\hbar}\Sigma_\sigma^{\mathrm{ATA}}\right)^{-1}.\end{aligned}$$

We finally obtain:

$$\Sigma_\sigma^{\mathrm{ATA}}(E) = \langle t_{0\sigma}\rangle\left(1 + \frac{1}{\hbar}G^{(0)}_{mm\sigma}(E)\,\langle t_{0\sigma}\rangle\right)^{-1}. \tag{4.259}$$

The comparison with (4.234) makes it clear that the ATA differs from the *T-matrix approximation* (TMA) only in the fact that in the ATA, the effective medium has been replaced once more by the *free*, unperturbed crystal with the corresponding atomic T-matrix:

$$\langle t_{0\sigma}\rangle = \sum_{m=1}^{\alpha} c_m \frac{\widehat{\eta}_{m\sigma}}{1 - \frac{1}{\hbar}\widehat{\eta}_{m\sigma}G^{(0)}_{ii\sigma}}. \tag{4.260}$$

5) The Coherent Potential Approximation (CPA)

It is immediately clear that one can improve the procedure in one last way by making it self-consistent, as under point 3) in the MPM, by replacing the free propagator $G^{(0)}_{mm\sigma}$ in the diagrams of the ATA by the full single-particle Green's function.

The **multiple-occupation corrections**, as in the diagrams of the ATA, now become somewhat more numerous, although they all have the same origins as explained under point 4). What is new in the CPA is then the inclusion of so-called *nested diagrams*, which allow the propagating particles to be scattered between two scattering events at the same

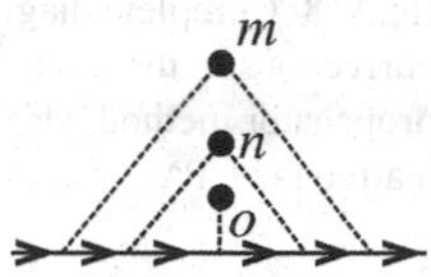

Fig. 4.17 Example of a nested diagram

site m to all the other sites. Difficulties arise again due to the diagonal terms $m = n = o$, which must be subtracted off. This includes for example all of the diagrams in the second column. We again sum the individual columns separately.

1st column:

We find the same result as in the ATA, except that the free propagator must be replaced by the full propagator:

$$\frac{1}{\hbar}\Sigma_\sigma^{(1)} = \frac{1}{\hbar}\sum_{m=1}^{\alpha} c_m \widehat{\eta}_{m\sigma}\left(1 - \frac{1}{\hbar}\widehat{\eta}_{m\sigma}\,\langle G_{ii\sigma}\rangle\right)^{-1} . \tag{4.261}$$

2nd column:

These diagrams are evidently obtained when we interpret Σ_σ as a functional of the propagator γ_σ, which is defined as follows:

$\gamma_\sigma =$ 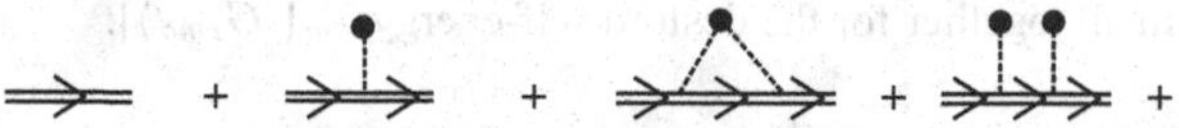

$$= \langle G_{mm\sigma}\rangle + \frac{1}{\hbar}\,\langle G_{mm\sigma}\rangle\,\Sigma_\sigma[\gamma_\sigma]\gamma_\sigma . \tag{4.262}$$

The second column then makes the contribution:

$$-\frac{1}{\hbar}\Big(\Sigma_\sigma[\gamma_\sigma] - \Sigma_\sigma\left[\langle G_{mm\sigma}\rangle\right]\Big) .$$

The first term in (4.262) no longer appears. Corresponding diagrams again have to be extracted.

3rd column:

$$-\frac{1}{\hbar}\Sigma_\sigma[\gamma_\sigma]\,\langle G_{mm\sigma}\rangle\,\frac{1}{\hbar}\Sigma_\sigma[\gamma_\sigma] . \tag{4.263}$$

$$\vdots$$

***n*-th column:**

$$-\frac{1}{\hbar}\Sigma_\sigma[\gamma_\sigma]\left(\langle G_{mm\sigma}\rangle\,\frac{1}{\hbar}\Sigma_\sigma[\gamma_\sigma]\right)^{n-2} . \tag{4.264}$$

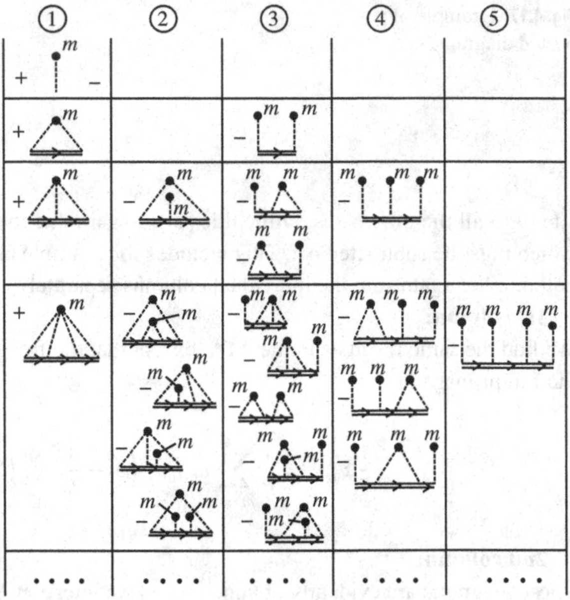

Fig. 4.18 Complete diagram corrections to the modified propagator method which lead to the CPA

We then obtain all together for the desired self-energy $\Sigma_\sigma[\langle G_{mn\sigma}\rangle]$:

$$\Sigma_\sigma\left[\langle G_{mm\sigma}\rangle\right] = \Sigma_\sigma^{(1)} - \Sigma_\sigma[\gamma_\sigma] + \Sigma_\sigma\left[\langle G_{mm\sigma}\rangle\right] + \\ + \Sigma_\sigma[\gamma_\sigma] - \sum_{n=0}^{\infty}\left(\frac{1}{\hbar}\Sigma_\sigma[\gamma_\sigma]\langle G_{mm\sigma}\rangle\right)^n \Sigma_\sigma[\gamma_\sigma]\,.$$

This means in the first instance

$$\Sigma_\sigma^{(1)} = \left(1 - \frac{1}{\hbar}\Sigma_\sigma[\gamma_\sigma]\langle G_{mm\sigma}\rangle\right)^{-1} \Sigma_\sigma[\gamma_\sigma] \tag{4.265}$$

which can be resolved in terms of $\Sigma_\sigma[\gamma_\sigma]$:

$$\Sigma_\sigma[\gamma_\sigma] = \frac{\Sigma_\sigma^{(1)}}{1 + \frac{1}{\hbar}\langle G_{mm\sigma}\rangle \Sigma_\sigma^{(1)}}\,. \tag{4.266}$$

Using (4.262), we rearrange further:

$$\Sigma_\sigma[\gamma_\sigma] = \frac{\Sigma_\sigma^{(1)}\left(1 + \gamma_\sigma \frac{1}{\hbar}\Sigma_\sigma[\gamma_\sigma]\right)}{1 + \gamma_\sigma \frac{1}{\hbar}\Sigma_\sigma[\gamma_\sigma] + \gamma_\sigma \frac{1}{\hbar}\Sigma_\sigma^{(1)}} \\ \Longrightarrow \quad \Sigma_\sigma[\gamma_\sigma]\left(1 + \gamma_\sigma \frac{1}{\hbar}\Sigma_\sigma[\gamma_\sigma]\right) = \Sigma_\sigma^{(1)}\,. \tag{4.267}$$

Now we insert (4.261):

$$\Sigma_\sigma[\gamma_\sigma] = \sum_{m=1}^{\alpha} c_m \widehat{\eta}_{m\sigma} \left[\left(1 - \frac{1}{\hbar}\widehat{\eta}_{m\sigma} \langle G_{ii\sigma}\rangle\right) \left(1 + \gamma_\sigma \frac{1}{\hbar}\Sigma_\sigma[\gamma_\sigma]\right)\right]^{-1} =$$
$$= \sum_{m=1}^{\alpha} c_m \widehat{\eta}_{m\sigma} \left[1 + \gamma_\sigma \frac{1}{\hbar}\Sigma_\sigma[\gamma_\sigma] - \frac{1}{\hbar}\widehat{\eta}_{m\sigma}\gamma_\sigma\right]^{-1} . \tag{4.268}$$

This is a self-consistent functional equation for the self-energy as a functional of γ_σ. The result we actually desired,

$$\Sigma_\sigma^{\mathrm{CPA}}(E) \equiv \Sigma_\sigma\left[\langle\langle G_{mm\sigma}(E)\rangle\rangle\right] , \tag{4.269}$$

is obtained by taking the limit

$$\gamma_\sigma \quad \Longrightarrow \quad \langle G_{mm\sigma}\rangle .$$

This yields the following expression:

$$\Sigma_\sigma^{\mathrm{CPA}}(E) = \sum_m c_m \frac{\widehat{\eta}_{m\sigma}}{1 - \frac{1}{\hbar}\langle G_{ii\sigma}\rangle \left(\widehat{\eta}_{m\sigma} - \Sigma_\sigma^{\mathrm{CPA}}\right)} . \tag{4.270}$$

This result can again be rearranged:

$$\frac{1}{\hbar}\Sigma_\sigma^{\mathrm{CPA}} \langle G_{ii\sigma}\rangle = \sum_m c_m \left(-1 + \frac{1 + \frac{1}{\hbar}\langle G_{ii\sigma}\rangle \Sigma_\sigma^{\mathrm{CPA}}}{1 - \frac{1}{\hbar}\langle G_{ii\sigma}\rangle \left(\widehat{\eta}_{m\sigma} - \Sigma_\sigma^{\mathrm{CPA}}\right)}\right) .$$

With $\sum_m c_m = 1$, it follows that:

$$0 = \left(1 + \frac{1}{\hbar}\Sigma_\sigma^{\mathrm{CPA}} \langle G_{ii\sigma}\rangle\right) \sum_m c_m \left(1 - \frac{1}{1 - \frac{1}{\hbar}\langle G_{ii\sigma}\rangle \left(\widehat{\eta}_{m\sigma} - \Sigma_\sigma^{\mathrm{CPA}}\right)}\right) .$$

This leads finally to

$$0 = \sum_{m=1}^{\alpha} c_m \frac{\eta_{m\sigma} - \Sigma_\sigma^{\mathrm{CPA}}(E) - T_0}{1 - \frac{1}{\hbar}\langle\langle G_{ii\sigma}(E)\rangle\rangle \left(\eta_{m\sigma} - \Sigma_\sigma^{\mathrm{CPA}}(E) - T_0\right)} . \tag{4.271}$$

This self-consistent functional equation for the self-energy $\Sigma_\sigma^{\mathrm{CPA}}(E)$ in the CPA is identical to (4.238), because of

$$\langle\langle G_{ii\sigma}(E)\rangle\rangle = G_{ii\sigma}^{(0)}\left(E - \Sigma_\sigma^{\mathrm{CPA}}(E)\right) . \tag{4.272}$$

Due to the *single-site* nature of the approximation, it is likewise independent of wavenumbers.

We can see that this formally rather simple solution contains three essential parameters:

1. The concentrations of the alloy components,
2. the atomic levels $\eta_{m\sigma}$, and
3. the density of states $\rho_0(E)$ of the unperturbed, pure crystal.

We require $\rho_0(E)$ for the determination of the *free* propagator:

$$G_{ii\sigma}^{(0)}(E) = \int_{-\infty}^{+\infty} \mathrm{d}x \, \frac{\rho_0(x)}{E - x} . \tag{4.273}$$

The considerations in this section have demonstrated that the CPA, among all the methods which make use of the *single-site* aspect, is by far the most successful.

4.3.5 Applications

The most obvious application of the CPA concerns a **binary alloy** containing the atomic species A and B, which are present in the alloy at the concentrations c_A, $c_B = 1 - c_A$. A typical result is sketched in Fig. 4.19. The pure crystals should exhibit the same densities of states, which have the appearance of a *church tower*, whose centres of gravity however lie at different energies. If they are sufficiently far apart, this leads for medium concentrations to a band splitting in the A–B alloy and to clear-cut deformations of the density of states.

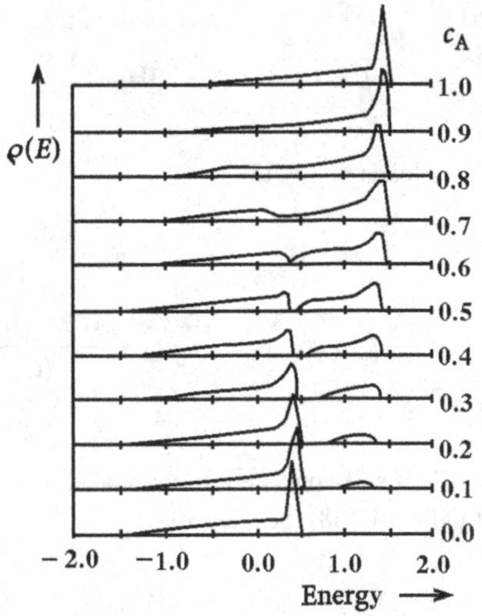

Fig. 4.19 A typical CPA solution for the density of states of a binary A–B alloy where the pure alloying partners A and B have the same densities of states but with shifted centres of gravity. c_A and $c_B = 1 - c_A$ are the concentrations of the alloying partners A and B, resp.

The field of applications of the CPA however extends far beyond just the real alloys. At present, it has matured into a standard procedure in general many-body theory, and especially in connection with the concept of the

alloy analogy.

Think of some electronic many-body model such as for example the Hubbard model (2.117). It is solved in the *atomic limit*, i.e. for the case that all the N energy-band states fall within an N-fold degenerate level T_0. In the Green's function,

$$G_{0\sigma}(E) = \hbar \sum_{m=1}^{p} \frac{\alpha_{m\sigma}}{E - \eta_{m\sigma}} , \tag{4.274}$$

then, the $\eta_{m\sigma}$ are the p quasi-particle levels into which T_0 splits owing to the interactions. The spectral weights α_σ give the degree of degeneracy of the quasi-particle levels:

$N\alpha_{m\sigma}$: **degeneracy of the level** $\eta_{m\sigma}$.

This situation however corresponds to that of a *fictitious* alloy having p components with the

concentrations: $\alpha_{1\sigma}, \alpha_{2\sigma}, \ldots, \alpha_{p\sigma}$

and the

atomic levels: $\eta_{1\sigma}, \eta_{2\sigma}, \ldots, \eta_{p\sigma}$.

These quantities are known. One can insert them into the CPA formula (4.271) and compute the self-energy $\Sigma_\sigma^{\mathrm{CPA}}(E)$ for a crystal lattice defined by $\rho_0(E)$. Thus, for example, the atomic solution (4.11) of the Hubbard model transforms in this way to that of a binary alloy with the atomic levels T_0 and $T_0 + U$ and the *concentrations* $(1 - \langle n_{-\sigma} \rangle)$ and $\langle n_{-\sigma} \rangle$.

This alloy analogy has proven to be an extraordinarily powerful method of many-body theory. In Sect. 4.5.3, we describe a concrete example of an application.

4.4 Spin Systems

In Sect. 2.4.2, we introduced the Heisenberg model as a realistic description of magnetic insulators. In spite of the simple structure of its Hamiltonian (2.203), the associated many-body problem has up to now no exact solution. We must still rely on approximations. We have already encountered one of these in the form of the spin-wave approximation (2.243), which however is limited conceptually to the low-temperature range.

The two-spin system calculated in Sect. 3.3.3 could be solved with mathematical rigour and thus serves as a valuable demonstration example for abstract concepts in many-body theory. For conclusive statements about a macroscopic ferromagnet, it is of course insufficient.

4.4.1 The Tyablikow Approximation

As a model for a ferromagnetic insulator, we consider a system of localised magnetic moments described by spin operators $\boldsymbol{S}_i$, $\boldsymbol{S}_j$ in a homogeneous, constant magnetic field $\boldsymbol{B}_0 = \mu_0 \boldsymbol{H}$:

$$H = -\sum_{i,j} J_{ij} \left(S_i^+ S_j^- + S_i^z S_j^z \right) - \frac{1}{\hbar} g_J \mu_B B_0 \sum_i S_i^z . \tag{4.275}$$

We assume that the spin system orders **ferromagnetically** below a critical temperature T_C, and we are thus primarily interested in its **magnetisation**:

$$M(T, B_0) = \frac{1}{V} g_J \frac{\mu_B}{\hbar} \sum_i \langle S_i^z \rangle_{T, B_0} . \tag{4.276}$$

To calculate it, we need the thermodynamic expectation value of the spin operator S_i^z, which, owing to translational symmetry, will be **in**dependent of the lattice site:

$$\langle S_i^z \rangle_{T, B_0} \equiv \langle S^z \rangle_{T, B_0} . \tag{4.277}$$

Which Green's function can deliver the quantity we are seeking? If we initially for simplicity assume that the localised spins have a magnitude of

$$S = \frac{1}{2} , \tag{4.278}$$

Then the generally-valid relation

$$S_i^{\pm} S_i^{\mp} = \hbar^2 S (S + 1) \pm \hbar S_i^z - \left(S_i^z \right)^2 \tag{4.279}$$

simplifies to

$$S_i^{\pm} S_i^{\mp} = \hbar \left(\hbar S \pm S_i^z \right) \quad \left(\left(S_i^z \right)^2 = \frac{\hbar^2}{4} \mathbf{1} \right) . \tag{4.280}$$

Clearly, then, the following retarded commutator Green's function is a good starting point:

$$G_{ij}^{\text{ret}}(t,t') = \langle\langle S_i^+(t); S_j^-(t') \rangle\rangle^{\text{ret}} . \tag{4.281}$$

We first set up the equation of motion of the energy-dependent Fourier transform and make use of the following commutators:

$$\left[S_i^+, S_j^-\right]_- = 2\hbar\delta_{ij} S_i^z , \tag{4.282}$$

$$\left[S_i^+, H\right]_- = -2\hbar \sum_m J_{mi} \left(S_m^+ S_i^z - S_i^+ S_m^z\right) + g_J \mu_B B_0 S_i^+ . \tag{4.283}$$

For the derivation of the last relation, we were able to make use of $J_{ii} = 0$. The equation of motion for $G_{ij}^{\text{ret}}(E)$ is then given by:

$$\begin{aligned}(E - g_J\mu_B B_0)\, G_{ij}^{\text{ret}}(E) = 2\hbar^2 \delta_{ij} \langle S_i^z \rangle +& \\ + 2\hbar \sum_m J_{im} \Big(\langle\langle S_i^+ S_m^z ; S_j^- \rangle\rangle_E^{\text{ret}} -& \\ - \langle\langle S_m^+ S_i^z ; S_j^- \rangle\rangle_E^{\text{ret}} \Big) .&\end{aligned} \tag{4.284}$$

Due to the *higher-order* Green's functions on the right-hand side, this equation cannot be directly solved. A simple approximation consists of carrying out a Hartree-Fock decoupling (4.21) on the *higher-order* Green's functions. Because of $\langle S_i^+ \rangle \equiv 0$ (conservation of spin!), this means that:

$$\begin{aligned}\langle\langle S_i^+ S_m^z ; S_j^- \rangle\rangle_E^{\text{ret}} &\underset{i\neq m}{\longrightarrow} \langle S^z \rangle\, G_{ij}^{\text{ret}}(E) , \\ \langle\langle S_m^+ S_i^z ; S_j^- \rangle\rangle_E^{\text{ret}} &\underset{i\neq m}{\longrightarrow} \langle S^z \rangle\, G_{mj}^{\text{ret}}(E) .\end{aligned} \tag{4.285}$$

Then the equation of motion (4.284) simplifies to:

$$\begin{aligned}&\left(E - g_J\mu_B B_0 - 2\hbar J_0 \langle S^z \rangle\right) G_{ij}^{\text{ret}}(E) = \\ &= 2\hbar^2 \delta_{ij} \langle S^z \rangle - 2\hbar \langle S^z \rangle \sum_m J_{im} G_{mj}^{\text{ret}}(E) .\end{aligned} \tag{4.286}$$

J_0 is defined in (2.207). This equation can be readily solved by transforming to wavenumbers ((2.212), (2.220)):

$$G_{\boldsymbol{q}}^{\text{ret}}(E) = \frac{2\hbar^2 \langle S^z \rangle}{E - E(\boldsymbol{q}) + \text{i}0^+} . \tag{4.287}$$

The poles of the Green's function correspond to the elementary excitations of the spin system,

$$E(\boldsymbol{q}) = g_J \mu_B B_0 + 2\hbar \left\langle S^z \right\rangle (J_0 - J(\boldsymbol{q})) \ , \tag{4.288}$$

which are found to be temperature dependent, owing to $\langle S^z \rangle$. For $T \to 0$, the approximation (4.285) is exact, and (4.288) then agrees with (2.232) because of $\langle S^z \rangle_{T=0} = \hbar S$ (*ferromagnetic saturation*).

For the complete determination of the quasi-particle energies $E(\boldsymbol{q})$, we still require a functional equation for $\langle S^z \rangle$, which we obtain by making use of the spectral theorem (3.157) and the spectral density

$$S_{\boldsymbol{q}}(E) = -\frac{1}{\pi} \operatorname{Im} G_{\boldsymbol{q}}^{\text{ret}}(E) = 2\hbar^2 \left\langle S^z \right\rangle \delta\left(E - E(\boldsymbol{q})\right) \ . \tag{4.289}$$

It can readily be verified by using the associated anticommutator Green's function that the constant D in the spectral theorem is to be chosen as

$$D \equiv 0 \qquad \text{for } B_0 \geq 0^+ \ . \tag{4.290}$$

We thus find:

$$\begin{aligned} \left\langle S_j^- S_i^+ \right\rangle &= \frac{1}{N} \sum_{\boldsymbol{q}} \mathrm{e}^{-\mathrm{i}\boldsymbol{q}\cdot(\boldsymbol{R}_i - \boldsymbol{R}_j)} \frac{1}{\hbar} \int\limits_{-\infty}^{+\infty} \frac{S_{\boldsymbol{q}}(E)\mathrm{d}E}{\mathrm{e}^{\beta E} - 1} = \\ &= 2\hbar \left\langle S^z \right\rangle \frac{1}{N} \sum_{\boldsymbol{q}} \frac{\mathrm{e}^{-\mathrm{i}\boldsymbol{q}\cdot(\boldsymbol{R}_i - \boldsymbol{R}_j)}}{\mathrm{e}^{\beta E(\boldsymbol{q})} - 1} \ . \end{aligned} \tag{4.291}$$

Due to the stipulated limitation to $S = \frac{1}{2}$, we can relate $\langle S^z \rangle$ directly to $\left\langle S_i^- S_i^+ \right\rangle$ via (4.280),

$$\left\langle S^z \right\rangle = \hbar S - \frac{1}{\hbar} \left\langle S_i^- S_i^+ \right\rangle \ ,$$

from which the following implicit functional equation for $\langle S^z \rangle$ results:

$$\left\langle S^z \right\rangle = \frac{\hbar S}{1 + \frac{2}{N} \sum_{\boldsymbol{q}} \left(\mathrm{e}^{\beta E(\boldsymbol{q})} - 1\right)^{-1}} \ . \tag{4.292}$$

Since according to (4.288), the quasi-particle energies $E(\boldsymbol{q})$ still contain $\langle S^z \rangle$, a general analytic solution is not possible. A numerical solution however poses no problems with a computer.

As the next step, we want to carry out an explicit determination of the Curie temperature T_C from (4.292). To this end, we consider the special case

$$B_0 = 0^+,\ T \overset{<}{\to} T_C \quad \Longleftrightarrow \quad \langle S^z \rangle \overset{>}{\to} 0\,,$$

in which the quasi-particle energies $E(\boldsymbol{q})$ become very small, so that the denominator of (4.292) can be expressed in a series expansion.

$$\langle S^z \rangle \simeq \hbar S \left(\frac{2}{N} \sum_{\boldsymbol{q}} \frac{1}{1 + \beta_C E(\boldsymbol{q}) + \cdots - 1} \right)^{-1} \simeq$$
$$\simeq \hbar S \beta_C \left(\frac{2}{N} \sum_{\boldsymbol{q}} \frac{1}{2\,\langle S^z \rangle\, \hbar\, (J_0 - J(\boldsymbol{q}))} \right)^{-1} .$$

The Curie temperature

$$k_B T_C = \left\{ \frac{1}{NS} \sum_{\boldsymbol{q}} \frac{1}{\hbar^2 (J_0 - J(\boldsymbol{q}))} \right\}^{-1} \tag{4.293}$$

depends of course on the one hand on the exchange integrals, but on the other, it also depends upon the lattice structure, which influences the summation over $\boldsymbol{q}$. The latter can be carried out without difficulties when the exchange integrals are known and the lattice is not too complicated.

One can also show that for low temperatures, the Tyablikow approximation (4.292) correctly reproduces Bloch's $T^{3/2}$ law,

$$1 - \frac{\langle S^z \rangle}{\hbar S} \sim T^{3/2}\,; \tag{4.294}$$

this can be taken as a retroactive validation of the decoupling (4.285), which initially appears somewhat arbitrary. All together, the Tyablikow approximation yields acceptable results over the whole range of temperatures $0 \le T \le T_C$.

Finally, we compute the internal energy U of the spin system as the thermodynamic expectation value of the Hamiltonian:

$$U = \langle H \rangle = -\sum_{i,j} J_{ij} \left(\langle S_i^+ S_j^- \rangle + \langle S_i^z S_j^z \rangle \right) - \frac{1}{\hbar} g_J \mu_B B_0 N \langle S^z \rangle\,. \tag{4.295}$$

The terms $\langle S_i^+ S_j^- \rangle$ and $\langle S^z \rangle$ have already been expressed in terms of the spectral density $S_{\boldsymbol{q}}(E)$. We must still discuss $\langle S_i^z S_j^z \rangle$. We start with the operator identities

$$S_i^- S_i^z = \frac{\hbar}{2} S_i^-\,; \quad S_i^- S_i^+ = \hbar^2 S - \hbar S_i^z\,, \tag{4.296}$$

which hold for $S = \frac{1}{2}$, in order to calculate

$$S_i^- \left[S_i^+, H\right]_- =$$

$$= -2\hbar \sum_m J_{mi} \left\{ \frac{\hbar}{2} S_m^+ S_i^- - \hbar^2 S S_m^z + \hbar S_i^z S_m^z \right\} + g_J \mu_B B_0 \left(\hbar^2 S - \hbar S_i^z\right)$$

with (4.283). This means that:

$$-\sum_{i,j} J_{ij} \left\langle S_i^z S_j^z \right\rangle = \frac{1}{2\hbar^2} \sum_i \left\langle S_i^- [S_i^+, H]_- \right\rangle +$$

$$+ \frac{1}{2} \sum_{i,j} J_{ij} \left\langle S_i^+ S_j^- \right\rangle - \hbar S J_0 N \left\langle S^z \right\rangle -$$

$$- \frac{1}{2\hbar^2} g_J \mu_B B_0 N \left(\hbar^2 S - \hbar \left\langle S^z \right\rangle\right) .$$

We denote the ground-state energy of the ferromagnet (2.224) as E_0,

$$E_0 = -N J_0 \hbar^2 S^2 - N g_J \mu_B B_0 S ,$$

and then find with (4.295) the internal energy U:

$$U = -\frac{1}{2} \sum_{i,j} J_{ij} \left\langle S_i^+ S_j^- \right\rangle + \frac{1}{2\hbar^2} \sum_i \left\langle S_i^- [S_i^+, H]_- \right\rangle +$$

$$+ N S J_0 \left(\left\langle S_i^- S_i^+ \right\rangle - \hbar^2 S\right) - \frac{1}{2} g_J \mu_B B_0 N S +$$

$$+ \frac{1}{2\hbar} g_J \mu_B B_0 N \left\langle S^z \right\rangle - \frac{1}{\hbar} g_J \mu_B B_0 N \left\langle S^z \right\rangle =$$

$$= E_0 + \frac{1}{2\hbar^2} \sum_i \left\langle S_i^- [S_i^+, H]_- \right\rangle +$$

$$+ \frac{1}{2} \sum_{i,j} \left\{ \left(J_0 \delta_{ij} - J_{ij}\right) + \delta_{ij} \frac{1}{\hbar^2} g_J \mu_B B_0 \right\} \left\langle S_j^- S_i^+ \right\rangle .$$

Here, we have used the normalisation $J_{ii} = 0$ several times. After Fourier transformation to wavenumbers, the curly brackets contain just the spin-wave energy $\hbar\omega(\boldsymbol{q})$ of the $S = 1/2$ ferromagnet (2.232):

$$U = E_0 + \frac{1}{2\hbar^3} \sum_{\boldsymbol{q}} \hbar\omega(\boldsymbol{q}) \int_{-\infty}^{+\infty} \mathrm{d}E \, \frac{S_{\boldsymbol{q}}(E)}{\mathrm{e}^{\beta E} - 1} + \tag{4.297}$$

$$+ \frac{1}{2\hbar^2} \sum_i \left\langle S_i^- [S_i^+, H]_- \right\rangle .$$

We can now attempt to express the last term, also, in terms of the spectral density $S_q(E)$:

$$\frac{1}{2\hbar^2}\sum_i \langle S_i^-[S_i^+, H]_-\rangle =$$
$$= \frac{i\hbar}{2\hbar^2}\sum_i \left(\frac{\partial}{\partial t}\langle S_i^-(t')\, S_i^+(t)\rangle\right)_{t=t'} =$$
$$= \frac{i\hbar}{2\hbar^2}\sum_q \left(\frac{1}{\hbar}\frac{\partial}{\partial t}\int_{-\infty}^{+\infty} dE\, \frac{S_q(E)}{e^{\beta E}-1}\exp\left(-\frac{i}{\hbar}E\,(t-t')\right)\right)_{t=t'} = \tag{4.298}$$
$$= \frac{1}{2\hbar^3}\sum_q \int_{-\infty}^{+\infty} dE\, \frac{E\, S_q(E)}{e^{\beta E}-1}\,.$$

Here, we have made use of the equation of motion for time-dependent Heisenberg operators (3.27) and again of the spectral theorem (3.157). Thus, the internal energy U is in the end completely determined by the spectral density $S_q(E)$:

$$U = E_0 + \frac{1}{2\hbar^3}\sum_q \int_{\infty}^{+\infty} dE\, \frac{(E+\hbar\omega(q))}{e^{\beta E}-1} S_q(E)\,. \tag{4.299}$$

If we now insert the expression (4.289) for the spectral density, then the integration over energy can easily be performed. For the $S = (1/2)$ ferromagnet, we then have:

$$U = E_0 + \frac{1}{\hbar}\langle S^z\rangle \sum_q \frac{E(q)+\hbar\omega(q)}{\exp(\beta E(q))-1}\,. \tag{4.300}$$

Using the generally-valid expression (3.217), we can find from $U(T, V)$ the free energy $F(T, V)$, which then determines the complete thermodynamics of the $S = 1/2$ ferromagnet.

Thus far, we have limited our considerations to the special case of $S = 1/2$, since it permits certain simplifications in comparison to the general case of $S \geq 1/2$. Our goal in fact is the determination of $\langle S^z\rangle$ from a suitably chosen Green's function. Due to (4.280), this is immediately possible for $S = 1/2$ with the function (4.281). For $S > 1/2$, however, instead of (4.280) the relation (4.279) must be averaged:

$$\langle S_i^- S_i^+\rangle = \hbar^2 S(S+1) - \hbar\langle S_i^z\rangle - \left\langle (S_i^z)^2\right\rangle\,. \tag{4.301}$$

The term $\langle (S_i^z)^2\rangle$ causes difficulties. It can **not** be expressed by the Green's function (4.281). We therefore choose as our starting point an entire set of Green's functions:

$$G_{ij}^{(n)}(E) = \left\langle\!\left\langle S_i^+; \left(S_j^z\right)^n S_j^- \right\rangle\!\right\rangle_E\,; \quad n = 0, 1, 2, \ldots, 2S-1\,. \tag{4.302}$$

Since the operator to the left of the semicolon is the same as in the case of $S = 1/2$, which we discussed above, the equation of motion is changed only in terms of its inhomogeneity. If we accept the same decouplings as in (4.285), then we can write the solution for the Green's function (4.302) directly:

$$G_{\boldsymbol{q}}^{(n)}(E) = \frac{\hbar \langle [S_i^+, (S_i^z)^n S_i^-]_- \rangle}{E - E(\boldsymbol{q})}. \tag{4.303}$$

The quasi-particle energies $E(\boldsymbol{q})$ are exactly the same as in (4.288). From the spectral theorem, it then follows that:

$$\langle (S_i^z)^n S_i^- S_i^+ \rangle = \langle [S_i^+, (S_i^z)^n S_i^-]_- \rangle \varphi(S) , \tag{4.304}$$

$$\varphi(S) = \frac{1}{N} \sum_{\boldsymbol{q}} \left(e^{\beta E(\boldsymbol{q})} - 1 \right)^{-1} . \tag{4.305}$$

For the expectation value on the left-hand side of (4.304) we can write with (4.301):

$$\begin{aligned} \langle (S_i^z)^n S_i^- S_i^+ \rangle = \hbar^2 S (S+1) \langle (S_i^z)^n \rangle - \\ - \hbar \left\langle (S_i^z)^{n+1} \right\rangle - \left\langle (S_i^z)^{n+2} \right\rangle . \end{aligned} \tag{4.306}$$

We require this equation only for $n = 0, 1, \ldots, 2S - 1$, since the operator identity valid in spin space,

$$\prod_{m_S = -S}^{+S} (S_i^z - \hbar m_S) = 0 , \tag{4.307}$$

guarantees the termination of the series of equations (4.306). For $n = 2S - 1$, the highest power of S_i^z is given by $2S + 1$. This can however be expressed in terms of lower powers of S_i^z by solving the relation (4.307) for $(S_i^z)^{2S+1}$,

$$\left\langle (S_i^z)^{2S+1} \right\rangle = \sum_{n=0}^{2S} \alpha_n(S) \langle (S_i^z)^n \rangle . \tag{4.308}$$

The numbers $\alpha_n(S)$ can for a given spin S be readily derived from (4.307).

As our next step, we prove using complete induction the conjecture

$$S_i^+ (S_i^z)^n = (S_i^z - \hbar)^n S_i^+ . \tag{4.309}$$

For $n = 1$, due to

$$\begin{aligned} S_i^+ S_i^z &= - [S_i^z, S_i^+]_- + S_i^z S_i^+ = \\ &= -\hbar S_i^+ + S_i^z S_i^+ = (S_i^z - \hbar) S_i^+ , \end{aligned}$$

the conjecture is correct. The extrapolation from n to $n+1$ can be made as follows:

$$\begin{aligned}
S_i^+ \left(S_i^z\right)^{n+1} &= S_i^+ \left(S_i^z\right)^n S_i^z = \\
&= \left(S_i^z - \hbar\right)^n S_i^+ S_i^z = \\
&= \left(S_i^z - \hbar\right)^n \left(S_i^z - \hbar\right) S_i^+ = \\
&= \left(S_i^z - \hbar\right)^{n+1} S_i^+ .
\end{aligned}$$

Thus we have proven (4.309). We now use this relation to further evaluate the commutator in (4.304):

$$\begin{aligned}
\left[S_i^+, \left(S_i^z\right)^n S_i^-\right]_- &= \left(S_i^z - \hbar\right)^n S_i^+ S_i^- - \left(S_i^z\right)^n S_i^- S_i^+ = \\
&= \left\{\left(S_i^z - \hbar\right)^n - \left(S_i^z\right)^n\right\} S_i^- S_i^+ + 2\hbar \left(S_i^z - \hbar\right)^n S_i^z .
\end{aligned} \tag{4.310}$$

After averaging and insertion of (4.301), we obtain an expression which –together with (4.306)– converts (4.304) into the following system of equations:

$$\begin{aligned}
&\hbar^2 S(S+1)\left\langle \left(S_i^z\right)^n \right\rangle - \hbar \left\langle \left(S_i^z\right)^{n+1} \right\rangle - \left\langle \left(S_i^z\right)^{n+2} \right\rangle = \\
&= \Big\{ 2\hbar \left\langle \left(S_i^z - \hbar\right)^n S_i^z \right\rangle + \\
&\quad + \left\langle \left(\left(S_i^z - \hbar\right)^n - \left(S_i^z\right)^n\right) \left(\hbar^2 S(S+1) - \hbar S_i^z - \left(S_i^z\right)^2\right) \right\rangle \Big\} \varphi(S) , \\
&\quad n = 0, 1, \dots, 2S - 1 .
\end{aligned} \tag{4.311}$$

These are $2S$ equations, which together with (4.308) allow us to determine the $(2S+1)$ expectation values $\left\langle (S_i^z)^n \right\rangle$, $n = 1, 2, \dots, 2S+1$. This procedure is of course very laborious for large values of S, especially since $\langle S^z \rangle$ reappears in $\varphi(S)$ in a complicated way; but it offers no difficulties in principle.

4.4.2 "Renormalised" Spin Waves

In Sect. 2.4.3, we introduced the concept of the spin wave starting from the fact that the normalised one-magnon state

$$|\boldsymbol{q}\rangle = \frac{1}{\hbar\sqrt{2SN}} S^-(\boldsymbol{q}) \, |S\rangle$$

is an exact eigenstate of the Heisenberg Hamiltonian with the eigenvalue

$$E(\boldsymbol{q}) = E_0 + \hbar\omega(\boldsymbol{q}) .$$

The *linear* spin-wave approximation (Sect. 2.4.4) describes a ferromagnet as a *gas* of non-interacting magnons. It is based on the Holstein-Primakoff transformation ((2.235), (2.236), (2.237) and (2.238)) of the spin operators, which represents an infinite series in the magnon occupation number, that is broken off after the linear term. Such an approximation can, to be sure, be justified only at very low temperatures when the magnon *gas* is still so rarefied that the interactions between the magnons can be neglected. This is no longer allowed at somewhat higher temperatures, and the *ansatz* (2.243) becomes untenable.

In this section, we wish to employ the method of moments, which was demonstrated in Sect. 4.1.5 on the Hubbard model, to *renormalise* the spin-wave energies by including their interactions. We start with the so-called **Dyson-Maleév transformation** of the spin operators:

$$\begin{aligned} S_i^- &= \hbar\sqrt{2S}\alpha_i^+ \,, \\ S_i^+ &= \hbar\sqrt{2S}\left(1-\frac{n_i}{2S}\right)\alpha_i \,, \\ S_i^z &= \hbar\,(S-n_i) \,. \end{aligned} \tag{4.312}$$

α_i^+, α_i are Bosonic operators; they thus fulfil the fundamental commutation relations (1.97), (1.98) and (1.99). $n_i = \alpha_i^+\alpha_i$ can be interpreted as the magnon occupation-number operator. The transformation (4.312) has the advantage relative to the Holstein-Primakoff transformation that no infinite series occur. The Heisenberg Hamiltonian contains a finite number of terms after the transformation:

$$H = E_0 + H_2 + H_4 \,, \tag{4.313}$$

$$H_2 = 2S\hbar^2 \sum_{i,j}\left(J_0\delta_{ij} - J_{ij} + \delta_{ij}\frac{1}{\hbar^2}g_J\mu_B B_0\right)\alpha_i^+\alpha_j \,, \tag{4.314}$$

$$H_4 = -\hbar^2\sum_{i,j} J_{ij}n_i n_j + \hbar^2\sum_{i,j} J_{ij}\alpha_i^+ n_j\alpha_j \,. \tag{4.315}$$

The term H_2 describes *free* spin waves, whilst H_4 contains the interactions between them.

The decisive disadvantage of the transformation (4.312) consists of the fact that S_i^-, S_i^+ are no longer adjoint operators and H is therefore no longer Hermitian. We shall however not treat the resulting complications here (F. J. Dyson, Phys. Rev. **102**, 1217, 1230 (1956)). With

$$\alpha_{\boldsymbol{q}} = \frac{1}{\sqrt{N}}\sum_i e^{-i\boldsymbol{q}\cdot\boldsymbol{R}_i}\alpha_i \,, \tag{4.316}$$

we now rewrite H in terms of wavenumbers:

$$H = E_0 + \sum_{q} \hbar\omega(q)\alpha_q^+\alpha_q + \\ + \frac{\hbar^2}{N} \sum_{q_1 \cdots q_4} \left(J(q_4) - J(q_1 - q_3)\right) \delta_{q_1+q_2,\, q_3+q_4} \alpha_{q_1}^+ \alpha_{q_2}^+ \alpha_{q_3} \alpha_{q_4} \,. \tag{4.317}$$

$\hbar\omega(q)$ again refers to the *bare* spin-wave energies (2.232).

We define the **one-magnon spectral density**:

$$B_q(E) = \frac{1}{2\pi} \int_{-\infty}^{+\infty} \mathrm{d}(t-t') \exp\left(\frac{\mathrm{i}}{\hbar} E\,(t-t')\right) \left\langle \left[\alpha_q(t), \alpha_q^+\,(t')\right]_- \right\rangle . \tag{4.318}$$

The associated spectral moments can be computed from:

$$M_q^{(n)} = \left\langle \left[\underbrace{\left[\ldots \left[[\alpha_q, H]_-, H \right]_-, \ldots, H \right]_-}_{n\text{-fold commutator}}, \alpha_q^+ \right]_- \right\rangle . \tag{4.319}$$

They are related to the spectral density via

$$M_q^{(n)} = \frac{1}{\hbar} \int_{-\infty}^{+\infty} \mathrm{d}E\, E^n B_q(E) \,. \tag{4.320}$$

Which *ansatz* should we choose for $B_q(E)$? Both the spin-wave result

$$B_q^{\mathrm{SW}}(E) = \hbar\delta\left(E - \hbar\omega(q)\right) \tag{4.321}$$

and the Tyablikow approximation (4.289) correspond to one-pole approaches. If we are not particularly interested in lifetime effects, then

$$B_q(E) = b_q \delta\left(E - \hbar\Omega(q)\right) \tag{4.322}$$

represents a physically *reasonable* starting point, where b_q and $\hbar\Omega(q)$ are initially unknown parameters. We now compute using (4.319) the first two spectral moments:

$$M_q^{(0)} = \left\langle \left[\alpha_q, \alpha_q^+\right]_- \right\rangle = 1 \,. \tag{4.323}$$

For the second moment, we require the following commutator:

$$
\begin{aligned}
\left[\alpha_q, H\right]_- &= \hbar\omega(q)\alpha_q + \frac{\hbar^2}{N}\sum_{q_1\cdots q_4}\{J(q_4) - J(q_1 - q_3)\}\,\cdot \\
&\quad\cdot \delta_{q_1+q_2,q_3+q_4}\left(\delta_{q,q_1}\alpha^+_{q_2}\alpha_{q_3}\alpha_{q_4} + \delta_{qq_2}\alpha^+_{q_1}\alpha_{q_3}\alpha_{q_4}\right) = \\
&= \hbar\omega(q)\alpha_q + \frac{\hbar^2}{N}\sum_{q_3,q_4}\{2J(q_4) - J(q - q_3) - J(q_4 - q)\}\,\cdot \\
&\quad\cdot \alpha^+_{q_3+q_4-q}\alpha_{q_3}\alpha_{q_4}\,.
\end{aligned}
\tag{4.324}
$$

With it, we continue the calculation:

$$
\begin{aligned}
&\left[[\alpha_q, H]_-, \alpha_q^+\right]_- = \\
&= \hbar\omega(q) + \frac{\hbar^2}{N}\sum_{q_3,q_4}\Big\{2J(q_4) - J(q - q_3) - J(q_4 - q)\Big\}\alpha^+_{q_3+q_4-q}\,\cdot \\
&\qquad\cdot\left(\delta_{qq_4}\alpha_{q_3} + \delta_{qq_3}\alpha_{q_4}\right) = \\
&= \hbar\omega(q) + 2\frac{\hbar^2}{N}\sum_{\bar{q}}\{J(q) + J(\bar{q}) - J(0) - J(q - \bar{q})\}\,\alpha^+_{\bar{q}}\alpha_{\bar{q}}\,.
\end{aligned}
\tag{4.325}
$$

The second spectral moment is then found as:

$$
M_q^{(1)} = \hbar\omega(q) + 2\frac{\hbar^2}{N}\sum_{\bar{q}}\{J(q) + J(\bar{q}) - J(0) - J(q - \bar{q})\}\left\langle\alpha^+_{\bar{q}}\alpha_{\bar{q}}\right\rangle . \tag{4.326}
$$

The initially unknown parameters in the spectral-density approach (4.322) are uniquely determined by $M_q^{(1)}$ and $M_q^{(2)}$ via (4.320):

$$
b_q \equiv \hbar\,, \tag{4.327}
$$

$$
\hbar\Omega(q) = \hbar\omega(q) + 2\frac{\hbar^2}{N}\sum_{\bar{q}}\frac{J(q) + J(\bar{q}) - J(0) - J(q - \bar{q})}{\exp\left(\beta\hbar\Omega(\bar{q})\right) - 1}\,. \tag{4.328}
$$

For the last equation, we have made use of the spectral theorem:

$$
\left\langle\alpha^+_{\bar{q}}\alpha_{\bar{q}}\right\rangle = \frac{1}{\hbar}\int\limits_{-\infty}^{+\infty}\mathrm{d}E\,\frac{B_{\bar{q}}(E)}{\exp(\beta E) - 1} = \{\exp\left(\beta\hbar\Omega(\bar{q})\right) - 1\}^{-1}\,. \tag{4.329}
$$

The constant D in the general spectral theorem (3.157) is zero here, due to $B_0 \geq 0^+$. It is notable that the *renormalised* spin-wave energies (4.328), calculated using the conceptually simple method of moments, prove to be completely equivalent to those from the well-known Dyson spin-wave theory. The method of moments here again distinguishes itself as a both simple and powerful procedure for finding solutions in many-body theory.

The result in (4.328) can readily be further evaluated for specific systems. The exchange integrals J_{ij} depend only on the lattice spacings $\left|\boldsymbol{R}_i - \boldsymbol{R}_j\right|$. Let z_n be the number of magnetic atoms (spins) in the n-th neighbour shell relative to a chosen atom, and J_n the exchange integral between this atom and its n-th neighbour, and further let

$$\gamma_{\boldsymbol{q}}^{(n)} = \frac{1}{z_n} \sum_{\Delta_n} \mathrm{e}^{\mathrm{i}\boldsymbol{q} \cdot \boldsymbol{R}_{\Delta_n}} , \tag{4.330}$$

where the sum runs over all the *magnetic* lattice sites of the n-th shell; then we can cast the *renormalised* spin-wave energies in the following form:

$$\hbar\Omega(\boldsymbol{q}) = 2S \sum_n \left(1 - \gamma_{\boldsymbol{q}}^{(n)}\right) z_n J_n \left(1 - A_n(T)\right) , \tag{4.331}$$

$$A_n(T) = \frac{1}{NS} \sum_{\boldsymbol{p}} \frac{1 - \gamma_{\boldsymbol{p}}^{(n)}}{\exp\left(\beta\hbar\Omega(\boldsymbol{p})\right) - 1} . \tag{4.332}$$

The prototype of a ferromagnetic Heisenberg spin system is EuO, of which it is known that only exchange with nearest and next-nearest neighbours is significant (J. Als Nielsen et al., Phys. Rev. B **14**, 4908 (1976)):

$$\frac{J_1}{k_{\mathrm{B}}} = 0.625\mathrm{K} ; \quad \frac{J_2}{k_{\mathrm{B}}} = 0.125\mathrm{K} . \tag{4.333}$$

The magnetic $4f$ moments of EuO are strictly localised on the Eu^{2+} lattice sites. They thus occupy a face-centered cubic lattice structure. The summation in (4.332) therefore runs over the first f.c.c. Brillouin zone and can be carried out exactly, so that (4.331) can be self-consistently solved for all temperatures. With the spin-wave energies $\hbar\Omega(\boldsymbol{q})$, we can calculate the magnetisation of the system via

$$\left\langle S^z \right\rangle = \hbar S - \frac{\hbar}{N} \sum_{\boldsymbol{q}} \left\{\exp\left(\beta\hbar\Omega(\boldsymbol{q})\right) - 1\right\}^{-1} \tag{4.334}$$

and compare it with experiment. The result (4.334) should be applicable in the temperature range $0 \leq T \leq 0.8 \cdot T_{\mathrm{C}}$ ($T_{\mathrm{C}}(\mathrm{EuO}) = 69.33\,\mathrm{K}$). From the theory of phase transitions, we know that the magnetisation of a ferromagnet in its critical region $0.9 \cdot T_{\mathrm{C}} \leq T \leq T_{\mathrm{C}}$ can be described by a power law (J. Als Nielsen et al., Phys. Rev. B. **14**, 4908 (1976)):

$$\left\langle S^z \right\rangle = 1.17 \cdot S \cdot \left(1 - \frac{T}{T_{\mathrm{C}}}\right)^{0.36} . \tag{4.335}$$

Combining (4.334) and (4.335) and *"fitting"* the small transition region ($0.8T_{\mathrm{C}} \leq T \leq 0.9T_{\mathrm{C}}$) correctly, we find, as shown in Fig. 4.20, a practically quantitative agreement with the experiments (data points!).

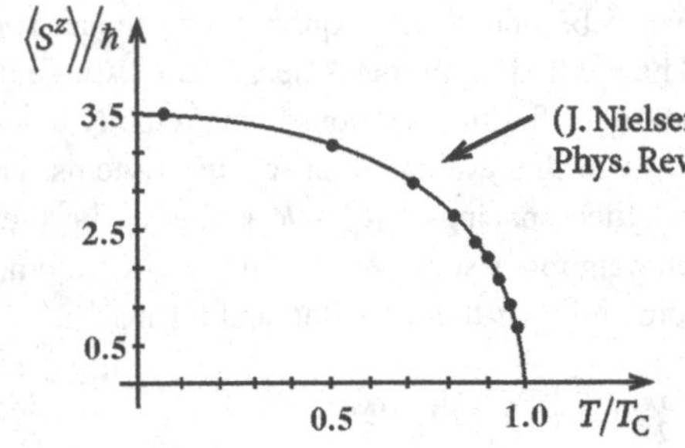

Fig. 4.20 The spontaneous magnetisation of EuO as a function of the temperature, calculated by the method of moments. The points are experimental data

4.4.3 Exercises

Exercise 4.4.1 Show that the Tyablikow approximation for the Heisenberg model obeys Bloch's $T^{3/2}$ law,

$$\frac{1 - \langle S^z \rangle}{\hbar S} \sim T^{3/2} .$$

You can make use of the fact that for small wavenumbers $|\boldsymbol{q}|$, the exchange integrals can be approximated by

$$J_0 - J(\boldsymbol{q}) = \frac{D}{2S\hbar^2} q^2 .$$

Exercise 4.4.2 For a system of localised spins with $S = 1$, derive within the framework of the Heisenberg model the following implicit functional equation:

$$\langle S^z \rangle_{S=1} = \hbar \frac{1 + 2\Phi(1)}{1 + 3\Phi(1) + 3\Phi^2(1)} ,$$

$$\Phi(S) = \frac{1}{N} \sum_{\boldsymbol{q}} [\exp(\beta E(\boldsymbol{q})) - 1]^{-1} ,$$

$$E(\boldsymbol{q}) = 2\hbar \langle S^z \rangle (J_0 - J(\boldsymbol{q})) .$$

Use the Tyablikow approximation for the Green's function defined in Sect. 4.3.2:

$$G_{ij}^{(n)}(E) = \left\langle\!\left\langle S_i^+ ; \left(S_j^z\right)^n S_j^- \right\rangle\!\right\rangle_E^{\text{ret}} ; \quad n = 1, 2 .$$

Compute also $\langle (S^z)^2 \rangle_{S=1}$.

Exercise 4.4.3 Verify the following commutators for spin operators:
1.

$$\left[(S_i^-)^n, S_i^z\right]_- = n\hbar(S_i^-)^n \; ; \quad n = 1, 2, \ldots$$

2.

$$\left[(S_i^-)^n, \left(S_i^z\right)^2\right]_- = n^2\hbar^2(S_i^-)^n + 2n\hbar S_i^z \left(S_i^-\right)^n \; ; \qquad n = 1, 2, \ldots$$

3.

$$\left[S_i^+, (S_i^-)^n\right]_- = \left[2n\hbar S_i^z + \hbar^2 n(n-1)\right]\left(S_i^-\right)^{n-1} \; ; \qquad n = 1, 2, \ldots$$

Exercise 4.4.4 Verify the following operator identity:

$$\left(S_i^-\right)^n \left(S_i^+\right)^n = \prod_{p=1}^{n}\left[\hbar^2 S(S+1) - (n-p)(n-p+1)\hbar^2 - (2n-2p+1)\hbar S_i^z - \left(S_i^z\right)^2\right].$$

Exercise 4.4.5 Find a closed system of equations for the spontaneous magnetisation $\langle S^z \rangle$ of a $S \geq 1/2$ spin ensemble, using the retarded Green's functions

$$G_{ij}^{(n)}(E) = \left\langle\!\left\langle S_i^+ ; (S_j^-)^{n+1}\left(S_j^+\right)^n \right\rangle\!\right\rangle_E^{\mathrm{ret}} ; \quad n = 0, 1, \ldots, 2S-1 \,.$$

Solve the equation of motion by employing the Tyablikow approximation and use some of the results of Exs. 4.4.3 and 4.4.4. Demonstrate the equivalence of the above system of Green's functions to that in (4.302) (see Ex. 4.4.2) explicitly for $S = 1$.

4.5 The Electron-Magnon Interaction

We have already mentioned in a previous section that there are interesting analogies between the lattice vibrations treated in Sect. 2.2 (*phonons*) and the **magnons** introduced in Sect. 2.4.3. Just as the electron-phonon interaction (Sect. 2.3) leads to a series of spectacular phenomena – one only need to think of superconductivity –, so the analogous electron-magnon interaction also has interesting consequences. This is true in particular for those systems in which the magnetic and electronic properties are dominated by electrons from different groups. This, again, is typical of compounds of which rare earths are components. They are therefore the subject of the considerations in the following section.

4.5.1 Magnetic $4f$ Systems (s-f-Model)

The term $4f$ **system** refers to a solid whose electronic properties are due essentially to the existence of partially-filled $4f$ shells. These are thus compounds whose components include the so-called rare earths. The electron configuration of a neutral rare-earth atom corresponds to the stable noble gas configuration of xenon plus additional contributions from the $4f$, $6s$, and often also the $5d$ electrons:

$$[\mathrm{RE}] = [\mathrm{Xe}](4\mathrm{f})^{\mathrm{n}}(5\mathrm{d})^{\mathrm{m}}(6\mathrm{s})^2 \; ; \quad (0 \le \mathrm{n} \le 14\, ;\; \mathrm{m} = 0, 1)\,.$$

In the Periodic Table, the rare earths follow the element lanthanum (La) and are distinguished from it and from each other by the successive filling of the $4f$ shell, i.e. by the number n of their $4f$ electrons. In condensed matter, the $4f$ systems can be insulators, semiconductors, and metals; the rare-earth ions typically exhibit a valence state of 3+,

$$\mathrm{RE} \;\longrightarrow\; (\mathrm{RE})^{3+} + \left\{(6s)^2 + 4f^1\right\} ,$$

whereby the rare-earth atom gives up its two $6s$ electrons and one of the $4f$ electrons. In insulators, e.g. $NdCl_3$, these three electrons participate in the formation of chemical bonds, whilst in metallic $4f$ systems, e.g. Gd, they represent quasi-free conduction electrons. There are a few exceptions to this rule. Ce and Pr can also be tetravalent, Sm, Eu, Tm, and Yb can also be divalent in certain compounds.

An essential property of the $4f$ systems is the strict localisation of the $4f$ electrons. The $4f$ shell is strongly shielded against influences from the environment by the filled electronic shells lying further out in the atom ($5s$, $5p$), so that even in complicated materials, the $4f$ wavefunctions of neighbouring rare-earth ions have practically no overlap. Among other things, this has the result that even in the solid, the $4f$ shell can be described well by Hund's rules from atomic physics. If – according to these rules – the $4f$ electronic angular momenta couple to a total angular momentum $J \neq 0$, then the incompletely filled $4f$ shell produces a permanent magnetic moment, which is likewise strictly localised at the rare-earth site. It is thus not surprising that in certain $4f$ systems, the exchange interaction couples these permanent magnetic moments and produces a collective magnetic ordering, e.g. ferromagnetic order, below a critical temperature which is a characteristic of the material. In this case, we refer to a **magnetic $4f$ system**. Prototypes of this group are the Europium chalcogenides EuX (X = O, S, Se, Te); and the element Gd. The EuX are insulators or semiconductors, whilst Gd is a metal.

The fact that the electronic and magnetic properties of the $4f$ systems are produced by two different types of electrons leads to interesting mutual effects. Thus, one observes for example in ferromagnetic systems a drastic temperature dependence of the structure of the conduction bands which is determined by the state of magnetisation of the $4f$ moments. On the other hand, the system of moments reacts sensitively to changes in the charge-carrier density in the conduction band, which can be produced e.g. by doping with suitable impurities.

The theoretical model with which the magnetic $4f$ systems are to be described is completely uncontroversial. It is the **s-f model**, which we have already introduced in (2.206) and will discuss here in more detail. This model is defined by the following Hamiltonian:

$$H = H_s + H_{ss} + H_f + H_{sf} \; . \tag{4.336}$$

H_s represents the kinetic energy of the conduction electrons and their interactions with the periodic lattice potential:

$$H_s = \sum_{ij\sigma} \left(T_{ij} - \mu\delta_{ij}\right) a_{i\sigma}^+ a_{j\sigma} = \sum_{k\sigma} (\varepsilon(k) - \mu)\, a_{k\sigma}^+ a_{k\sigma} \; . \tag{4.337}$$

This corresponds to the operator $\mathcal{H}_0$ from Eq. (3.190) in Sect. 3.3.1. The symbols in (4.337) have the same meanings as in (3.190).

H_{ss} describes the Coulomb interactions of the conduction electrons, which we for simplicity assume to be strongly shielded, so that they will be of the *Hubbard type*:

$$H_{ss} = \frac{1}{2} U \sum_{i,\sigma} n_{i\sigma} n_{i-\sigma} \; . \tag{4.338}$$

In formulating H_s and H_{ss}, we have presumed that the conduction band is a so-called s band, which can contain at most two electrons per lattice site.

The subsystem of the magnetic $4f$ moments can be very realistically described by the Heisenberg model (2.203), owing to their strict localisation:

$$H_f = -\sum_{i,j} J_{ij} \boldsymbol{S}_i \cdot \boldsymbol{S}_j \; . \tag{4.339}$$

The coupling between the conduction electrons and the $4f$ electrons is now of decisive importance; it is described as an intra-atomic exchange interaction, i.e. a local interction between the spins $\boldsymbol{\sigma}$ of the conduction electrons and the $4f$ spin $\boldsymbol{S}_i$:

$$H_{sf} = -g \sum_i \boldsymbol{\sigma}_i \cdot \boldsymbol{S}_i \; . \tag{4.340}$$

Formally, this operator has the same form as H_f, except that one $4f$ spin is replaced by the conduction electron spin, and the double sum is restricted to its diagonal (intra-atomic) terms. g is here the corresponding s-f exchange constant. For practical purposes, the compact notation in (4.340) is inopportune. One should rather use the formalism of second quantisation for the electronic spins. We therefore transform the spin operators as in (4.104) and (4.105) to creation and annihilation operators:

$$\frac{1}{\hbar}\sigma_i^z = \frac{1}{2}\sum_\sigma z_\sigma n_{i\sigma} \; ; \quad (z_\uparrow = +1; \quad z_\downarrow = -1) \; , \tag{4.341}$$

$$\frac{1}{\hbar}\sigma_i^+ = a_{i\uparrow}^+ a_{i\downarrow} \; , \tag{4.342}$$

$$\frac{1}{\hbar}\sigma_i^- = a_{i\downarrow}^+ a_{i\uparrow} \; . \tag{4.343}$$

The s-f interaction then appears as follows:

$$H_{sf} = -\frac{1}{2} g\hbar \sum_{i,\sigma} \left(z_\sigma S_i^z n_{i\sigma} + S_i^\sigma a_{i-\sigma}^+ a_{i\sigma} \right) \; , \tag{4.344}$$

where we have used the abbreviation

$$S_i^\uparrow \equiv S_i^+ = S_i^x + \mathrm{i}S_i^y \; ; \quad S_i^\downarrow \equiv S_i^- = S_i^x - \mathrm{i}S_i^y \; . \tag{4.345}$$

The interaction is thus composed of two parts, a *diagonal* term between the z components of the spin operators involved, and a *non-diagonal* term, which clearly describes spin exchange processes between the two interaction partners. It is found that it is precisely these spin-flip terms which have a significant influence on the structure of the conduction band.

The s-f model (4.336) has proved to be extraordinarily realistic for describing the magnetic $4f$ systems. It however defines a truly non-trivial many-body problem which cannot be solved in the general case. One can learn much from two limiting cases, which we shall discuss in the next sections.

4.5.2 The Infinitely Narrow Band

We start by disregarding the *dispersion relation* ($\boldsymbol{k}$ dependence) of the energy band states; i.e. as a thought experiment, we allow the lattice constant to become so large that the conduction band becomes degenerate and collapses into a single level T_0. From the model Hamiltonian (4.336), we then obtain:

$$\widehat{H} = (T_0 - \mu) \sum_\sigma n_\sigma + \frac{1}{2} U \sum_\sigma n_\sigma n_{-\sigma} - \frac{1}{2} g\hbar \sum_\sigma \left(z_\sigma S^z n_\sigma + S^\sigma a^+_{-\sigma} a_\sigma \right) . \quad (4.346)$$

Because of $J_{ii} = 0$, H_f is zero in this limit. However, we wish to continue to assume that the localised spin system orders ferromagnetically, i.e. below T_C, it exhibits a finite magnetisation $\langle S^z \rangle$. $\langle S^z \rangle$ cannot of course be derived self-consistently from $\widehat{H}$, and we therefore treat it as a parameter. This will become clear later.

The many-body problem associated with $\widehat{H}$ can be solved exactly with some effort. We define the following operator combinations,

$$d_\sigma = z_\sigma S^z a_\sigma + S^{-\sigma} a_{-\sigma} , \quad (4.347)$$

$$D_\sigma = z_\sigma S^z n_{-\sigma} a_\sigma + S^{-\sigma} n_\sigma a_{-\sigma} , \quad (4.348)$$

$$p_\sigma = n_{-\sigma} a_\sigma , \quad (4.349)$$

and compute the commutators with them:

$$\left[a_\sigma, \widehat{H} \right]_- = T_0 a_\sigma - \frac{1}{2} g\hbar d_\sigma + U p_\sigma , \quad (4.350)$$

$$\left[d_\sigma, \widehat{H} \right]_- = $$

$$= \left(T_0 + \frac{1}{2} g\hbar^2 \right) d_\sigma - \frac{1}{2} g S(S+1) \hbar^3 a_\sigma + \left(U - g\hbar^2 \right) D_\sigma , \quad (4.351)$$

$$\left[D_\sigma, \widehat{H} \right]_- = \left(T_0 + U - \frac{1}{2} g\hbar^2 \right) D_\sigma - \frac{1}{2} g S(S+1) \hbar^3 p_\sigma , \quad (4.352)$$

$$\left[p_\sigma, \widehat{H} \right]_- = (T_0 + U) p_\sigma - \frac{1}{2} g\hbar D_\sigma . \quad (4.353)$$

For the following four Green's functions,

$$G_{a\sigma}(E) = \langle\langle a_\sigma ; a^+_\sigma \rangle\rangle_E , \quad (4.354)$$

$$G_{d\sigma}(E) = \langle\langle d_\sigma ; a^+_\sigma \rangle\rangle_E , \quad (4.355)$$

$$G_{D\sigma}(E) = \langle\langle D_\sigma ; a^+_\sigma \rangle\rangle_E , \quad (4.356)$$

$$G_{p\sigma}(E) = \langle\langle p_\sigma ; a^+_\sigma \rangle\rangle_E , \quad (4.357)$$

we can readily work out the equations of motion by making use of the above commutators:

$$(E - T_0 + \mu)\, G_{a\sigma}(E) = \hbar - \frac{\hbar}{2} g G_{d\sigma}(E) + U G_{p\sigma}(E)\,, \tag{4.358}$$

$$\left(E - T_0 + \mu - \frac{\hbar^2}{2} g\right) G_{d\sigma}(E) =$$

$$= \hbar z_\sigma \left\langle S^z \right\rangle - \frac{\hbar^3}{2} g S(S+1) G_{a\sigma}(E) + \left(U - g\hbar^2\right) G_{D\sigma}(E)\,, \tag{4.359}$$

$$\left(E - T_0 + \mu - U + \frac{1}{2} g\hbar^2\right) G_{D\sigma}(E) =$$

$$= -\hbar^2 \Delta_{-\sigma} - \frac{\hbar^3}{2} g S(S+1) G_{p\sigma}(E)\,, \tag{4.360}$$

$$(E - T_0 + \mu - U)\, G_{p\sigma}(E) = \hbar \left\langle n_{-\sigma} \right\rangle - \frac{\hbar}{2} g G_{D\sigma}(E)\,. \tag{4.361}$$

Here, we have used the abbreviation:

$$\hbar \Delta_\sigma = \left\langle S^\sigma a^+_{-\sigma} a_\sigma \right\rangle + z_\sigma \left\langle S^z n_\sigma \right\rangle\,. \tag{4.362}$$

Equations (4.358), (4.359), (4.360) and (4.361) form a closed system. For their solution, we first insert (4.360) into (4.361):

$$\left(E - T_0 + \mu - U - \frac{\hbar^4/4g^2 S(S+1)}{E - T_0 + \mu - U + \hbar^2/2g}\right) G_{p\sigma}(E) =$$

$$= \hbar \left\langle n_{-\sigma} \right\rangle + \frac{\frac{\hbar^3}{2} g \Delta_{-\sigma}}{E - U + \hbar^2/2g - T_0 + \mu}\,.$$

We abridge:

$$E_3 = T_0 + U - \frac{\hbar^2}{2} g(S+1)\,; \quad E_4 = T_0 + U + \frac{\hbar^2}{2} g S\,. \tag{4.363}$$

$G_{p\sigma}(E)$ is obviously a two-pole function:

$$G_{p\sigma}(E) = \frac{\hbar}{(E - E_3 + \mu)(E - E_4 + \mu)} \cdot$$

$$\cdot \left[\frac{\hbar^2}{2} g \Delta_{-\sigma} + \left\langle n_{-\sigma} \right\rangle \left(E - T_0 + \mu - U + \frac{\hbar^2}{2} g\right)\right].$$

We therefore set

$$G_{p\sigma}(E) = \hbar \left(\frac{\vartheta_{3\sigma}}{E - E_3 + \mu} + \frac{\vartheta_{4\sigma}}{E - E_4 + \mu}\right) \tag{4.364}$$

and determine the spectral weights from:

$$\vartheta_{3\sigma} = \lim_{E \to E_3 - \mu} \frac{1}{\hbar}(E - E_3 + \mu)\, G_{p\sigma}(E) = \\ = \frac{1}{2S+1}\left(S\left\langle n_{-\sigma}\right\rangle - \Delta_{-\sigma}\right) , \tag{4.365}$$

$$\vartheta_{4\sigma} = \lim_{E \to E_4 - \mu} \frac{1}{\hbar}(E - E_4 + \mu)\, G_{p\sigma}(E) = \\ = \frac{1}{2S+1}\left((S+1)\left\langle n_{-\sigma}\right\rangle + \Delta_{-\sigma}\right) . \tag{4.366}$$

Then $G_{p\sigma}(E)$ is completely determined. The sum rule

$$\vartheta_{3\sigma} + \vartheta_{4\sigma} = \left\langle n_{-\sigma}\right\rangle \tag{4.367}$$

can be employed as a consistency check. It becomes clear from (4.361) that $G_{D\sigma}(E)$ must have the same poles as $G_{p\sigma}(E)$:

$$G_{D\sigma}(E) = \hbar^2 \left(\frac{\gamma_{3\sigma}}{E - E_3 + \mu} + \frac{\gamma_{4\sigma}}{E - E_4 + \mu}\right) . \tag{4.368}$$

For the spectral weights, we find as in (4.365) and (4.366):

$$\gamma_{3\sigma} = (S+1)\,\vartheta_{3\sigma} \; ; \quad \gamma_{4\sigma} = -S\vartheta_{4\sigma} . \tag{4.369}$$

Here again, the sum rule (= first spectral moment) is obeyed:

$$\gamma_{3\sigma} + \gamma_{4\sigma} = -\Delta_{-\sigma} . \tag{4.370}$$

We now insert the results for $G_{D\sigma}(E)$ and $G_{p\sigma}(E)$ into (4.358) and (4.359) and solve for $G_{a\sigma}(E)$:

$$\left(E - T_0 + \mu - \frac{\frac{\hbar^4}{4} g^2 S(S+1)}{E - T_0 + \mu - \frac{\hbar^2}{2} g}\right) G_{a\sigma}(E) = \\ = \hbar\left(1 - \frac{\frac{\hbar}{2} g z_\sigma \left\langle S^z\right\rangle}{E - T_0 + \mu - \frac{1}{2} g\hbar^2}\right) + U G_{p\sigma}(E) - \\ - \frac{\frac{1}{2} g\left(U - g\hbar^2\right)\hbar}{E - T_0 + \mu - \frac{1}{2} g} G_{D\sigma}(E) . \tag{4.371}$$

The bracket expression on the left-hand side of the equation can be written as a product $(E - E_1 + \mu)\,(E - E_2 + \mu)$ with

$$E_1 = T_0 - \frac{\hbar^2}{2} g S\,; \quad E_2 = T_0 + \frac{\hbar^2}{2} g(S+1)\,. \tag{4.372}$$

$G_{a\sigma}(E)$ thus evidently represents a four-pole function:

$$G_{a\sigma}(E) = \hbar \sum_{i=1}^{4} \frac{\alpha_{i\sigma}}{E - E_i + \mu}\,. \tag{4.373}$$

We find using (4.371) and

$$\alpha_{i\sigma} = \lim_{E \to E_i - \mu} \frac{1}{\hbar} (E - E_i + \mu)\, G_{a\sigma}(E)$$

the following expressions for the spectral weights:

$$\alpha_{1\sigma} = \frac{1}{2S+1} \left\{ S + 1 + \frac{z_\sigma}{\hbar} \langle S^z \rangle + \Delta_{-\sigma} - (S+1)\, \langle n_{-\sigma} \rangle \right\}\,, \tag{4.374}$$

$$\alpha_{2\sigma} = \frac{1}{2S+1} \left\{ S - \frac{z_\sigma}{\hbar} \langle S^z \rangle - \Delta_{-\sigma} - S\, \langle n_{-\sigma} \rangle \right\}\,, \tag{4.375}$$

$$\alpha_{3\sigma} = \vartheta_{3\sigma}\,; \quad \alpha_{4\sigma} = \vartheta_{4\sigma}\,. \tag{4.376}$$

Now only the Green's function $G_{d\sigma}(E)$ remains; it can be readily determined by employing (4.358):

$$G_{d\sigma}(E) = \hbar^2 \sum_{i=1}^{4} \frac{\beta_{i\sigma}}{E - E_i + \mu}\,. \tag{4.377}$$

For the spectral weights, we now have:

$$\beta_{1\sigma} = S\alpha_{1\sigma}\,; \quad \beta_{2\sigma} = -(S+1)\alpha_{2\sigma}\,; \quad \beta_{3\sigma} = (S+1)\alpha_{3\sigma}\,; \quad \beta_{4\sigma} = -S\alpha_{4\sigma}\,. \tag{4.378}$$

The quantity in which we are principally interested is the one-electron Green's function $G_{a\sigma}(E)$ (4.373), for whose complete determination we still need the expectation values $\Delta_{-\sigma}$ and $\langle n_{-\sigma} \rangle$ as well as the chemical potential μ. We write as an abbreviation

$$f_i(T) = \frac{1}{1 + \exp\left[\beta\,(E_i - \mu)\right]}\,; \quad i = 1, \ldots, 4\,; \tag{4.379}$$

We can then find $\langle n_{-\sigma} \rangle$ from the one-electron spectral density by making use of the spectral theorem:

$$S_{a\sigma}(E) = -\frac{1}{\pi} \operatorname{Im} G_{a\sigma}\left(E + \mathrm{i}0^+\right) = \hbar \sum_{i=1}^{4} \alpha_{i\sigma} \delta\left(E - E_i + \mu\right) . \tag{4.380}$$

We immediately obtain:

$$\langle n_{-\sigma} \rangle = \sum_{i=1}^{4} \alpha_{i-\sigma} f_i(T) . \tag{4.381}$$

The chemical potential μ is determined by the *band* occupation n (= number of electrons per lattice site in the energy *band* considered):

$$n = \sum_{\sigma} \langle n_\sigma \rangle = \sum_{i,\sigma} \alpha_{i\sigma} f_i(T) . \tag{4.382}$$

$\Delta_{-\sigma}$ can likewise be readily derived from the spectral theorem via the *higher-order* spectral density

$$S_{d\sigma}(E) = -\frac{1}{\pi} \operatorname{Im} G_{d\sigma}\left(E + \mathrm{i}0^+\right) = \hbar^2 \sum_{i=1}^{4} \beta_{i\sigma} \delta\left(E - E_i + \mu\right) , \tag{4.383}$$

leading to:

$$\Delta_{-\sigma} = \sum_{i=1}^{4} \beta_{i-\sigma} f_i(T) . \tag{4.384}$$

With (4.381), (4.382) and (4.384), the spectral weights of all four Green's functions are completely determined.

The poles of the one-electron Green's function $G_{a\sigma}(E)$ represent the quasi-particle energies of the interacting system. Due to these interactions, the Bloch band T_0, which had become a degenerate level, splits into four quasi-particle levels E_i, which are listed in (4.363) and (4.372). In contrast to the spectral weights $\alpha_{i\sigma}$, the levels are **in**dependent of the spins, the temperature, and the band occupation.

The $\uparrow$-weights are plotted in Fig. 4.21 as functions of the band occupation n $(0 \le n \le 2)$ as well as the renormalised magnetisation $M = (S - \langle S^z \rangle)/S$ for a realistic set of parameters ($U = 2\text{eV}$; $g = 0.2\text{eV}$; $S = 7/2$). The corresponding $\downarrow$-weights can likewise be read off the figure by using particle-hole symmetry

$$\alpha_{1\sigma}(T, n) = \alpha_{4-\sigma}(T, 2-n) ; \quad \alpha_{2\sigma}(T, n) = \alpha_{3-\sigma}(T, 2-n) . \tag{4.385}$$

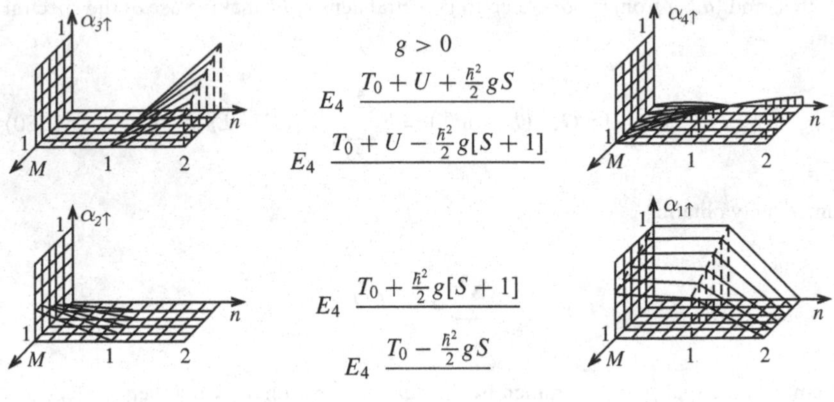

Fig. 4.21 The spectral weights of the exact solution of the s-f model in the limit of an infinitely narrow band are plotted as functions of the particle density n and the renormalised magnetisation $M = (S - \langle S^z \rangle)/S$. E_1 to E_4 are the quasi-particle levels

The temperature dependence of $\alpha_{i\sigma}$ results almost exclusively from the $4f$ magnetisation $\langle S^z \rangle$, which we must regard here as a parameter. This is not completely without problems, since the moment system is naturally also influenced by the conduction electrons via the exchange coupling (4.344). $\langle S^z \rangle$ would in fact have to be determined self-consistently within the framework of the full model. For $H_f = 0$, however, such a self-consistent calculation would yield $\langle S^z \rangle = 0$. We can see from the figure that for each constellation of parameters, at most three of the four levels are in fact found; at least one of them has a vanishing spectral weight. There are in addition a number of special cases in which additional levels are missing, e.g. at $T = 0$ or for $n = 0, 1, 2$.

What is the significance of the spectral weights? When the interactions are *switched off*, the level T_0 is $2N$-fold degenerate (N = number of lattice sites, the factor of 2 enters due to the two spin directions). The $\alpha_{i\sigma}$ now determine how the degeneracy is distributed over the quasi-particle levels when the interactions are again *switched on*. $\alpha_{i\sigma} N$ is the degree of degeneracy of the (i, σ) level. This interpretation now suggests the application to the s-f model of the alloy analogy introduced in Sect. 4.3.5 (CPA), in order to derive from the above *atomic* results statements about the case of finite bandwidths which in fact interests us.

4.5.3 The Alloy Analogy

We imagine a fictitious alloy composed of four components. Each component is characterised by a single level which interests us

$$\eta_{m\sigma} \equiv E_m \quad (m = 1, 2, 3, 4) \tag{4.386}$$

and is statistically distributed throughout the lattice with the

$$\textbf{concentration} \quad c_{m\sigma} \equiv \alpha_{m\sigma}(T, n) \,. \tag{4.387}$$

In the case of very large lattice spacings, the level $\eta_{m\sigma}$ is then all together $(c_{m\sigma} N)$-fold degenerate. This however corresponds precisely to the situation in the *real* system, in which at large spacings each quasi-particle level is $(\alpha_{m\sigma} N)$-fold degenerate.

We choose as the density of states of the energy band under consideration in the undisturbed pure crystal a simple semi-elliptical form,

$$\rho_0(E) = \begin{cases} \dfrac{4}{\pi W}\sqrt{1 - 4\left(\dfrac{E}{W}\right)^2} & \text{when} \quad |E| \leq \dfrac{W}{2}\,, \\ 0 & \text{otherwise}\,, \end{cases} \tag{4.388}$$

and furthermore the precise values of the parameters

$$\hbar^2 g = 0.2\text{eV}\,; \quad U = 2\text{eV}\,; \quad S = \frac{7}{2}\,; \quad W = 1.17\text{eV}\,, \tag{4.389}$$

which can be regarded as realistic for the ferromagnetic $4f$ insulator EuS. We put all this into Eq. (4.271), from which we can then compute the CPA self-energy $\Sigma_\sigma^{\text{CPA}}(E)$ for various temperatures, i.e. for corresponding $4f$ magnetisations, and for various band occupations n self-consistently. With

$$\rho_\sigma(E) = \rho_0\left(E - \Sigma_\sigma^{\text{CPA}}(E)\right) \tag{4.390}$$

we then find the (T, n)-dependent quasi-particle density of states. The actual evaluation must be carried out numerically with a computer. The most noticeable characteristic of the quasi-particle density of states is its multi-subband structure, which in addition exhibits a distinctive (T, n) dependence. Fig. 4.22 shows the dependence on the band occupation n for three different temperatures $T = 0$, $T = 0.8\,T_C$, and $T = T_C = 16.6\,\text{K}$. For $n < 1$, the spectrum in general consists of two low-energy and one higher-energy quasi-particle subbands. These subbands can be roughly and intuitively classified as follows: If the σ electron (for $n < 1$) is moving in the uppermost band, then it is hopping mainly over lattice sites which are already occupied by a $(-\sigma)$ electron. For this to happen, however, the Coulomb interaction energy U must be supplied. This explains why the position of this quasi-particle band is ca. 2eV above the two other bands, in which the electron propagates over empty sites. It is thus clear that the upper subband must vanish for $n = 0$, since then no interaction partners exist, whilst the two lower bands must vanish for $n = 2$, since then there are no empty sites. The two lower-energy bands can be distinguished as follows: In the lowest subband, the electron is moving over lattice sites on which a parallel $4f$ spin is localised; in the second subband, it has an antiparallel orientation relative to the $4f$ spin. With ferromagnetic saturation at $T = 0$, an $\uparrow$-electron can no longer find an antiparallel $4f$ spin, and the second subband therefore does not appear at $T = 0$ in the $\uparrow$ spectrum. In this manner, even details of the notable (T, n) dependence of the quasi-particle density of states $\rho_\sigma(E)$ can be intuitively interpreted. For $n > 1$, i.e. when the Bloch band is more

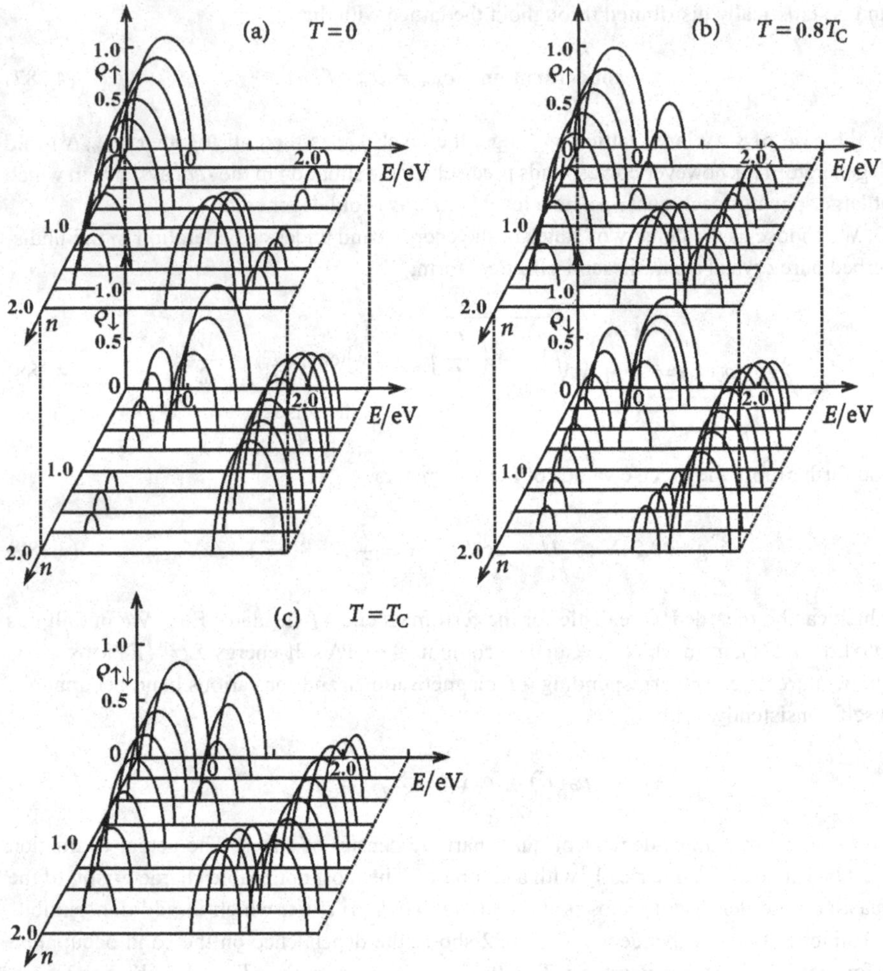

Fig. 4.22 The quasi-particle density of states in the s-f model as a function of the energy for different band occupations n, calculated using the CPA alloy analogy. Figures (**a**), (**b**) and (**c**) represent three different temperatures

than half filled, it is expedient to make use of particle-hole symmetry in the interpretation of the spectra.

4.5.4 The Magnetic Polaron

There is a very informative special case of the s-f model (4.336), that of a single electron (a *test electron*) in an otherwise empty conduction band, which is a thoroughly relevant situation to ferromagnetic insulators such as Eu0 or EuS. This problem is exactly solvable for *ferromagnetic saturation*, i.e. for $T = 0$.

All of the information that interests us can once again be derived from the (retarded or advanced) single-electron Green's function:

$$G_{ij\sigma}(E) \equiv \langle\langle a_{i\sigma}; a^+_{j\sigma} \rangle\rangle_E = \frac{1}{N} \sum_{\boldsymbol{k}} G_{\boldsymbol{k}\sigma}(E) \mathrm{e}^{\mathrm{i}\boldsymbol{k} \cdot (\boldsymbol{R}_i - \boldsymbol{R}_j)} \,. \tag{4.391}$$

We will leave off the index "adv" or "ret" in the following, for simplicity. Three additional, *higher-order* Green's functions will be significant for this problem:

$$D_{ik,j\sigma}(E) = \langle\langle S^z_i a_{k\sigma}; a^+_{j\sigma} \rangle\rangle_E \,, \tag{4.392}$$

$$F_{ik,j\sigma}(E) = \langle\langle S^{-\sigma}_i a_{k-\sigma}; a^+_{j\sigma} \rangle\rangle_E \,, \tag{4.393}$$

$$P_{ik,j\sigma}(E) = \langle\langle n_{i-\sigma} a_{k\sigma}; a^+_{j\sigma} \rangle\rangle_E \,. \tag{4.394}$$

For the equation of motion of the function $G_{ij\sigma}(E)$, we require the commutator:

$$[a_{i\sigma}, H]_- = \sum_m T_{im} a_{m\sigma} + U n_{i-\sigma} a_{i\sigma} - \frac{\hbar}{2} g z_\sigma S^z_i a_{i\sigma} - \frac{\hbar}{2} g S^{-\sigma}_i a_{i-\sigma} \,. \tag{4.395}$$

This yields for the equation of motion:

$$\begin{aligned} &\sum_m (E\delta_{im} - T_{im}) G_{mj\sigma}(E) = \\ &= \hbar\delta_{ij} + U P_{ii,j\sigma}(E) - \frac{\hbar}{2} g \left(z_\sigma D_{ii,j\sigma}(E) + F_{ii,j\sigma}(E)\right) \,. \end{aligned} \tag{4.396}$$

As in (3.325), at this point we introduce the electronic self-energy:

$$\langle\langle [a_{i\sigma}, H - H_s]_-; a^+_{j\sigma} \rangle\rangle_E \equiv \sum_l \Sigma_{il\sigma}(E) G_{lj\sigma}(E) \,. \tag{4.397}$$

The determination of $\Sigma_{il\sigma}(E)$ or of its $\boldsymbol{k}$-dependent Fourier transform $\Sigma_{\boldsymbol{k}\sigma}(E)$ solves this problem. Comparison with (4.396) shows that the self-energy is determined essentially by the *higher-order* Green's functions P, D and F:

$$\sum_l \Sigma_{il\sigma}(E) G_{lj\sigma}(E) = U P_{ii,j\sigma}(E) - \frac{\hbar}{2} g \left(z_\sigma D_{ii,j\sigma}(E) + F_{ii,j\sigma}(E)\right) \,. \tag{4.398}$$

We will now make use of our previous assumption that ($T = 0$, $n = 0$); it implies that we can carry out all the necessary averaging processes of the Green's functions with the ground state $|0\rangle$, which corresponds to an **electron vacuum** and a **magnon vacuum**. For this special case, several obvious simplifications are possible:

$$D_{ik,\,j\sigma}(E) \xrightarrow[T=0,\,n=0]{} \hbar S G_{kj\sigma}(E)\,, \tag{4.399}$$

$$P_{ik,\,j\sigma}(E) \xrightarrow[n=0]{} 0\,. \tag{4.400}$$

The self-energy is thus essentially fixed by the spin-flip function $F_{ik,\,j\sigma}(E)$. The latter becomes particularly simple for $\sigma = \uparrow$. Owing to

$$S_i^+\,|0\rangle = 0 \quad \Longleftrightarrow \quad \langle 0|\,S_i^- = 0\,, \tag{4.401}$$

we have namely:

$$F_{ik,\,j\uparrow}(E) \xrightarrow[T=0,\,n=0]{} 0\,. \tag{4.402}$$

Note that for finite band occupations, $n \neq 0$, due to s-f coupling, the spin system need not necessarily be ferromagnetically saturated. The conclusions in (4.399) and (4.402) are then no longer permitted.

With (4.402), the $\uparrow$ problem can be trivially solved:

$$\begin{aligned} \Sigma_{il\uparrow}^{(0,\,0)}(E) &\equiv -\frac{1}{2} g\hbar^2 S\delta_{il}\,, \\ \Sigma_{k\uparrow}^{(0,\,0)}(E) &\equiv -\frac{1}{2} g\hbar^2 S\,. \end{aligned} \tag{4.403}$$

For the retarded Green's function, we thus have:

$$G_{k\uparrow}^{(0,\,0)}(E) = \hbar \left\{ E - \varepsilon(\boldsymbol{k}) + \frac{1}{2} g\hbar^2 S + \mathrm{i}0^+ \right\}^{-1}\,. \tag{4.404}$$

The $\uparrow$ quasi-particle energies in this special case are merely shifted by a constant amount relative to the free Bloch energies $\varepsilon(\boldsymbol{k})$:

$$E_{\uparrow}^{(0,\,0)}(\boldsymbol{k}) \equiv \varepsilon(\boldsymbol{k}) - \frac{1}{2} g\hbar^2 S\,. \tag{4.405}$$

The $\uparrow$ spectral density is given by a simple δ-function,

$$S_{k\uparrow}^{(0,\,0)}(E) = \hbar\delta\left(E - \varepsilon(\boldsymbol{k}) + \frac{1}{2} g\hbar^2 S \right)\,, \tag{4.406}$$

typical of a quasi-particle with an infinitely long lifetime. The quasi-particle density of states

$$\rho_{\uparrow}^{(0,\,0)}(E) = \frac{1}{N} \sum_{\boldsymbol{k}} S_{k\uparrow}^{(0,\,0)}(E) = \rho_0\left(E + \frac{1}{2} g\hbar^2 S \right) \tag{4.407}$$

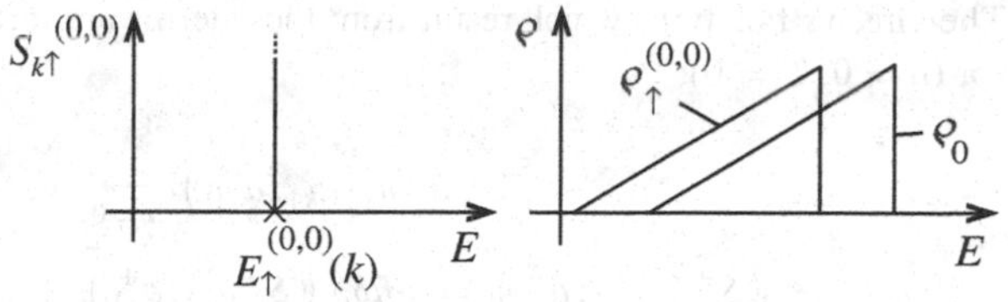

Fig. 4.23 The ↑ spectral density and the ↑ quasi-particle density of states in the exact solution of the ($n = 0, T = 0$) s-f model

remains undistorted relative to the *free* Bloch density of states

$$\rho_0(E) = \frac{1}{N} \sum_{k} \delta\left(E - \varepsilon(\boldsymbol{k})\right) ; \tag{4.408}$$

it is only rigidly shifted by a constant amount.

The ↑ spectrum thus consists of a single quasi-particle band. The CPA results from the last sections therefore prove to be exact for this special case. Physically, these results are simple to understand. At the temperature $T = 0$, the ↑ electron has no possibility to exchange its spin with the completely parallel oriented spin system. The spin-flip terms in the s-f exchange (4.344) are meaningless; only the diagonal part of the s-f interaction is in effect, and it produces a relatively unimportant rigid shift in the quasi-particle spectrum.

The situation becomes more complicated but also more interesting in the case of the ↓ spectrum. A ↓ electron can naturally exchange its spin even at $T = 0$ with one of the antiparallel localised f spins. The spin-flip terms in the s-f exchange in this case will drastically modify the quasi-particle spectrum. We now wish to investigate this point in more detail.

From (4.398), we have:

$$\sum_{l} \Sigma_{il\downarrow}^{(0,0)} G_{lj\downarrow}^{(0,0)} = \frac{1}{2} g\hbar^2 S G_{ij\downarrow}^{(0,0)} - \frac{1}{2} g\hbar F_{ii,j\downarrow}^{(0,0)} . \tag{4.409}$$

We write the equation of motion of the *spin-flip function* $F_{ik,j\downarrow}(E)$. For this, we require the following commutator:

$$\begin{aligned}
\left[S_i^+ a_{k\uparrow}, H\right]_- = & \sum_m T_{km} S_i^+ a_{m\uparrow} + U S_i^+ n_{k\downarrow} a_{k\uparrow} - \\
& - \frac{1}{2} g\hbar \left(S_i^+ S_k^z a_{k\uparrow} + S_i^+ S_k^- a_{k\downarrow}\right) + \\
& + \frac{1}{2} g\hbar^2 \left(n_{i\uparrow} - n_{i\downarrow}\right) S_i^+ a_{k\uparrow} - \\
& - g\hbar^2 S_i^z a_{i\uparrow}^+ a_{i\downarrow} a_{k\uparrow} - \\
& - 2\hbar \sum_m J_{im} \left(S_i^z S_m^+ - S_m^z S_i^+\right) a_{k\uparrow} .
\end{aligned} \tag{4.410}$$

The Green's functions which result from these terms are further simplified to some extent for ($n = 0, T = 0$):

$$\langle\langle S_i^+ n_{k\downarrow} a_{k\uparrow}; a_{j\downarrow}^+ \rangle\rangle \xrightarrow[n=0]{} 0 ,$$

$$\langle\langle S_i^+ S_k^z a_{k\uparrow}; a_{j\downarrow}^+ \rangle\rangle = -\hbar\delta_{ik} \langle\langle S_i^+ a_{k\uparrow}; a_{j\downarrow}^+ \rangle\rangle + \langle\langle S_k^z S_i^+ a_{k\uparrow}; a_{j\downarrow}^+ \rangle\rangle$$

$$\xrightarrow[n=0,\, T=0]{} \hbar(S - \delta_{ik}) F_{ik,j\downarrow}^{(0,0)} ,$$

$$\langle\langle S_i^+ S_k^- a_{k\downarrow}; a_{j\downarrow}^+ \rangle\rangle = 2\hbar\delta_{ik} \langle\langle S_i^z a_{k\downarrow}; a_{j\downarrow}^+ \rangle\rangle + \langle\langle S_k^- S_i^+ a_{k\downarrow}; a_{j\downarrow}^+ \rangle\rangle$$

$$\xrightarrow[n=0,\, T=0]{} 2\hbar^2 S\delta_{ik} G_{ij\downarrow}^{(0,0)} ,$$

$$\langle\langle \left(n_{i\uparrow} - n_{i\downarrow}\right) S_i^+ a_{k\downarrow}; a_{j\downarrow}^+ \rangle\rangle \xrightarrow[n=0]{} 0 ,$$

$$\langle\langle S_i^z a_{i\uparrow}^+ a_{i\downarrow} a_{k\uparrow}; a_{j\downarrow}^+ \rangle\rangle \xrightarrow[n=0]{} 0 ,$$

$$\langle\langle \left(S_i^z S_m^+ - S_m^z S_i^+\right) a_{k\uparrow}; a_{j\downarrow}^+ \rangle\rangle \xrightarrow[n=0,\, T=0]{} \hbar S \left(F_{mk,j\downarrow}^{(0,0)} - F_{ik,j\downarrow}^{(0,0)}\right) .$$

We then find for the equation of motion:

$$\begin{aligned} \left(E + \frac{1}{2} g\hbar^2 (S - \delta_{ik})\right) F_{ik,\,j\downarrow}^{(0,0)}(E) &= \\ = \sum_m T_{km} F_{im,\,j\downarrow}^{(0,0)}(E) &- g\hbar^3 S\delta_{ik} G_{ij\downarrow}^{(0,0)}(E) - \\ - 2\hbar^2 S \sum_m J_{im} &\left(F_{mk,\,j\downarrow}^{(0,0)}(E) - F_{ik,\,j\downarrow}^{(0,0)}(E)\right) . \end{aligned} \tag{4.411}$$

To solve it, we transform the position-dependent functions to $\boldsymbol{k}$-space:

$$G_{ij\sigma}(E) = \frac{1}{N} \sum_{\boldsymbol{k}} \exp\left(\mathrm{i}\boldsymbol{k} \cdot (\boldsymbol{R}_i - \boldsymbol{R}_j)\right) G_{\boldsymbol{k}\sigma}(E) , \tag{4.412}$$

$$F_{ik,\,j\sigma}(E) = \frac{1}{N^{3/2}} \sum_{\boldsymbol{k},\boldsymbol{q}} \exp\left(\mathrm{i}\left(\boldsymbol{q} \cdot \boldsymbol{R}_i + (\boldsymbol{k} - \boldsymbol{q}) \cdot \boldsymbol{R}_k - \boldsymbol{k} \cdot \boldsymbol{R}_j\right)\right) F_{\boldsymbol{k}\boldsymbol{q}\sigma}(E) . \tag{4.413}$$

Then, from (4.411), we obtain after some *simple* rearrangements:

$$\begin{aligned} \left(E + \frac{1}{2} g\hbar^2 S - \varepsilon(\boldsymbol{k} - \boldsymbol{q}) - \hbar\omega(\boldsymbol{q})\right) F_{\boldsymbol{k}\boldsymbol{q}\downarrow}^{(0,0)}(E) &= \\ = \frac{1}{2} g\hbar^2 \frac{1}{N} \sum_{\bar{\boldsymbol{q}}} F_{\boldsymbol{k}\bar{\boldsymbol{q}}\downarrow}^{(0,0)}(E) &- g\hbar^3 S \frac{1}{\sqrt{N}} G_{\boldsymbol{k}\downarrow}^{(0,0)}(E) . \end{aligned} \tag{4.414}$$

The spin-wave energies $\hbar\omega(\boldsymbol{q})$ are as defined in (2.232). We abbreviate by writing:

$$B_{\boldsymbol{k}}(E) = \frac{1}{N}\sum_{\boldsymbol{q}}\left\{E + \frac{1}{2}g\hbar^2 S - \varepsilon(\boldsymbol{k}-\boldsymbol{q}) - \hbar\omega(\boldsymbol{q})\right\}^{-1} . \tag{4.415}$$

Then we have from (4.414):

$$\frac{1}{\sqrt{N}}\sum_{\boldsymbol{q}} F^{(0,0)}_{\boldsymbol{kq}\downarrow}(E) = -\frac{g\hbar^3 S B_{\boldsymbol{k}}(E)}{1-\frac{1}{2}g\hbar^2 B_{\boldsymbol{k}}(E)} G^{(0,0)}_{\boldsymbol{k}\downarrow}(E) . \tag{4.416}$$

The equation of motion of the single-particle Green's function is given, from (4.396), (4.399), (4.400), (4.412) and (4.413), by:

$$\left(E - \frac{1}{2}g\hbar^2 S - \varepsilon(\boldsymbol{k})\right) G^{(0,0)}_{\boldsymbol{k}\downarrow}(E) = \hbar - \frac{1}{2}g\hbar\frac{1}{\sqrt{N}}\sum_{\boldsymbol{q}} F^{(0,0)}_{\boldsymbol{kq}\downarrow}(E) . \tag{4.417}$$

Into this equation, we insert (4.416):

$$\left(E - \frac{1}{2}g\hbar^2 S - \varepsilon(\boldsymbol{k})\right) G^{(0,0)}_{\boldsymbol{k}\downarrow}(E) = \hbar + \frac{\frac{1}{2}g^2\hbar^4 S}{1-\frac{1}{2}g\hbar^2 B_{\boldsymbol{k}}(E)} B_{\boldsymbol{k}}(E) G^{(0,0)}_{\boldsymbol{k}\downarrow}(E) .$$

Comparison with

$$G^{(0,0)}_{\boldsymbol{k}\downarrow}(E) = \hbar\left\{E - \varepsilon(\boldsymbol{k}) - \Sigma^{(0,0)}_{\boldsymbol{k}\downarrow}(E)\right\}^{-1} \tag{4.418}$$

finally yields the $\downarrow$ self-energy:

$$\Sigma^{(0,0)}_{\boldsymbol{k}\downarrow}(E) = \frac{1}{2}g\hbar^2 S\left(1 + \frac{g\hbar^2 B_{\boldsymbol{k}}(E)}{1-\frac{1}{2}g\hbar^2 B_{\boldsymbol{k}}(E)}\right) . \tag{4.419}$$

The problem is thereby completely and exactly solved.

We want to try to interpret this result. First of all, we can achieve a significant simplification in its evaluation if we suppress the magnon energies $\hbar\omega(\boldsymbol{q})$ in (4.415). This is certainly permitted, since they are always several orders of magnitude smaller than other typical energies such as the Bloch band width W or the s-f coupling constant g. With this simplification, the in general complex *propagator* $B_{\boldsymbol{k}}(E)$ becomes **in**dependent of wavenumbers:

$$B_{\boldsymbol{k}}(E) \equiv B(E) = R_{\mathrm{B}}(E) + \mathrm{i}I_{\mathrm{B}}(E) . \tag{4.420}$$

Here, the imaginary part $I_B(E)$ is practically identical to the ↑ density of states (4.407):

$$\begin{aligned} I_B(E) &= -\frac{\pi}{N}\sum_{q}\delta\left(E+\frac{1}{2}g\hbar^2 S-\varepsilon(k-q)\right)= \\ &= -\frac{\pi}{N}\sum_{\hat{q}}\delta\left(E+\frac{1}{2}g\hbar^2 S-\varepsilon(\hat{q})\right)= \\ &= -\pi\rho_0\left(E+\frac{1}{2}g\hbar^2 S\right)=-\pi\rho_{\uparrow}^{(0,0)}(E)\,. \end{aligned} \tag{4.421}$$

The real part is a principal-value integral:

$$R_B(E)=\mathcal{P}\int \mathrm{d}x\,\frac{\rho_0(x)}{E+\frac{1}{2}g\hbar^2 S-x}\,. \tag{4.422}$$

The electronic self-energy (4.419) will in general be a complex quantity, which, owing to the above stipulation that we will neglect $\hbar\omega(q)$, likewise becomes independent of wavenumbers:

$$\Sigma_{k\downarrow}^{(0,0)}(E)\equiv\Sigma_{\downarrow}^{(0,0)}(E)=R_{\downarrow}(E)+\mathrm{i}I_{\downarrow}(E)\,. \tag{4.423}$$

If we insert (4.420) into (4.419), we obtain as our concrete result:

$$R_{\downarrow}(E)=\frac{1}{2}g\hbar^2 S\left(1+g\hbar^2\frac{R_B(E)\left(1-\frac{1}{2}g\hbar^2 R_B(E)\right)-\frac{1}{2}g\hbar^2 I_B^2(E)}{\left(1-\frac{1}{2}g\hbar^2 R_B(E)\right)^2+\frac{1}{4}g^2\hbar^4 I_B^2(E)}\right), \tag{4.424}$$

$$I_{\downarrow}(E)=\frac{1}{2}g^2\hbar^4 S\frac{I_B(E)}{\left(1-\frac{1}{2}g\hbar^2 R_B(E)\right)^2+\frac{1}{4}g^2\hbar^4 I_B^2(E)}\,. \tag{4.425}$$

Comparison with (4.421) shows that the imaginary part of the electronic ↓ self-energy is then and only then nonzero, when the ↑ density of states $\rho_{\uparrow}^{(0,0)}(E)$ takes on finite values. $I_{\downarrow}\neq 0$ means that the lifetime of the corresponding quasi-particle is finite. It is clearly limited by spin-flip processes. If we recall that we strictly speaking should have included also the magnon energies in the above expressions, it becomes clear that the original ↓ electron reverses its spin on emission of a magnon and thereby becomes a ↑ electron. This is of course only then possible, if suitable ↑ states are available which can accept the originally ↓ electron.

If the Green's function already has a pole outside the range $\rho_{\uparrow}^{(0,0)}(E)\neq 0$, i.e. if

$$E=\varepsilon(k)+R_{\downarrow}(E)$$

can be fulfilled there, then an additional quasi-particle appears, now however with an infinite lifetime. The spectral density

$$S_{k\downarrow}^{(0,0)}(E) = -\frac{1}{\pi}\,\mathrm{Im}\,G_{k\downarrow}^{(0,0)}(E+\mathrm{i}0^+)$$

will thus as a rule be composed of two terms which correspond to two different elementary processes (see Fig. 4.24):

$$S_{k\downarrow}^{(0,0)}(E) = \\ = \begin{cases} -\dfrac{\hbar}{\pi}\dfrac{I_\downarrow(E)}{\left(E-\varepsilon(\boldsymbol{k})-R_\downarrow(E)\right)^2+I_\downarrow^2(E)}\,, & \text{for}\quad \varepsilon_0 \le E+\frac{1}{2}g\hbar^2S \le \varepsilon_0+W\,,\\ \hbar\delta\left(E-\varepsilon(\boldsymbol{k})-R_\downarrow(E)\right) & \text{otherwise}\,. \end{cases} \tag{4.426}$$

(ε_0 is the lower band edge and W the width of the Bloch band.) The original $\downarrow$ electron can exchange its spin with the localised spin system through magnon emission and thereby become a $\uparrow$ electron. This leads to the first term in (4.426), yielding a **scattering spectrum** which is always a few eV in width and occupies the same energy range as the $\uparrow$ density of states. The $\downarrow$ electron can however also form a bound state with an antiparallel $4f$ spin. As long as its energy lies outside that of the scattering spectrum, it will give rise to a quasi-particle with an infinitely long lifetime, which is referred to as a **magnetic polaron**.

At the conclusion of this chapter, we want to discuss the exact $T = 0$ results by referring to the specific Bloch density of states for a simple cubic lattice, which can be calculated in the "tight-binding approximation" with the energies from (2.110). The details of such a calculation are not important here.

Figure 4.25 shows the spectral density $S_{k\downarrow}^{(0,0)}(E)$ for several values of the $\boldsymbol{k}$ vector within the first Brillouin zone and for three different coupling strengths $g\hbar^2$. With weak coupling ($g\hbar^2 = 0.05$eV), the spectral density consists of a narrow peak whose position is $\boldsymbol{k}$-dependent and lies near the energy $\varepsilon(\boldsymbol{k}) + \frac{1}{2}g\hbar^2S$, which corresponds to a molecular-

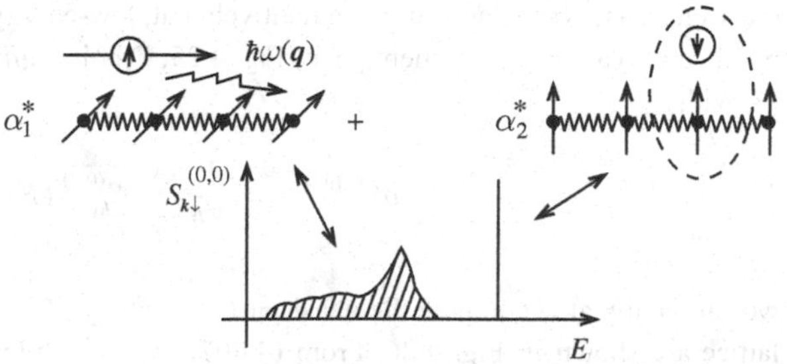

Fig. 4.24 A schematic illustration of the elementary processes which contribute to the exact $\downarrow$ spectral density of the (n=0, T=0) s-f model; on the *left*: magnon emission, on the *right*: formation of a stable magnetic polaron

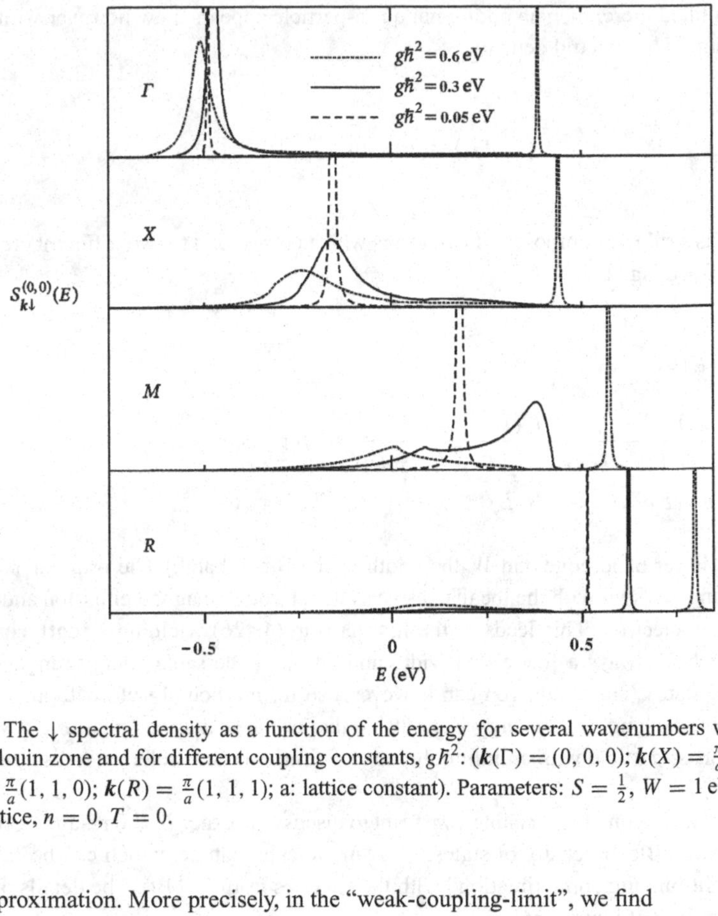

Fig. 4.25 The ↓ spectral density as a function of the energy for several wavenumbers within the first Brillouin zone and for different coupling constants, $g\hbar^2$: ($\boldsymbol{k}(\Gamma) = (0, 0, 0)$; $\boldsymbol{k}(X) = \frac{\pi}{a}(1, 0, 0)$; $\boldsymbol{k}(M) = \frac{\pi}{a}(1, 1, 0)$; $\boldsymbol{k}(R) = \frac{\pi}{a}(1, 1, 1)$; a: lattice constant). Parameters: $S = \frac{1}{2}$, $W = 1$ eV, simple cubic lattice, $n = 0$, $T = 0$.

field approximation. More precisely, in the "weak-coupling-limit", we find

$$E_{\downarrow}(\boldsymbol{k}) \approx \varepsilon(\boldsymbol{k}) + \frac{1}{2} g\hbar^2 S + \frac{g^2\hbar^4 S}{2N} \sum_{\boldsymbol{q}} \frac{1}{\varepsilon(\boldsymbol{k}) + g\hbar^2 S - \varepsilon(\boldsymbol{q})} . \tag{4.427}$$

With stronger coupling, the picture changes completely. As already indicated schematically in Fig. 4.24, a sharp, high-energy peak splits off, which corresponds to the stable magnetic polaron. The scattering spectrum, which arises from magnon emission through the ↓ electron, is seen as a rule as a relatively flat, low-energy structure, but it is sometimes bundled into a fairly prominent peak (Fig. 4.25; Γ point; $g\hbar^2 = 0.6$eV).

With

$$\rho_{\sigma}^{(0,0)}(E) = \frac{1}{N\hbar} \sum_{\boldsymbol{k}} S_{\boldsymbol{k}\sigma}^{(0,0)}(E) , \tag{4.428}$$

we can finally also compute the quasi-particle density of states. Results for a simple cubic lattice are shown in Fig. 4.26. From (4.407), $\rho_{\uparrow}(E)$ is identical to the Bloch density of

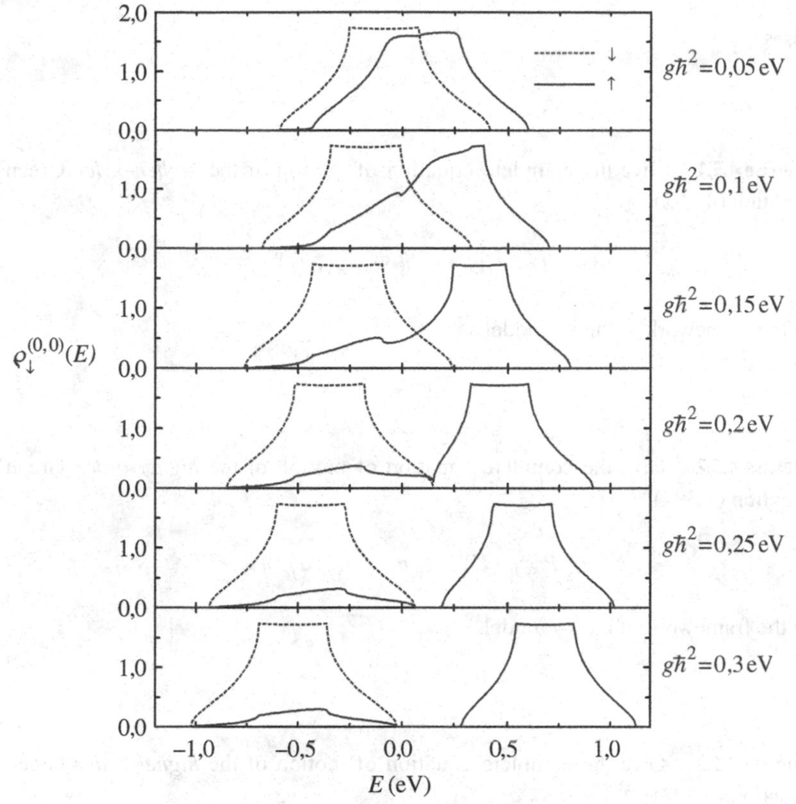

Fig. 4.26 The quasi-particle density of states $\rho_\sigma(E)$ as a function of the energy E for different coupling strengths $g\hbar^2$. The *solid lines* are for $\sigma = \downarrow$, the *dashed lines* for $\sigma = \uparrow$. Parameters: $S = \frac{7}{2}$, $W = 1$ eV, simple cubic lattice, $n = 0$, $T = 0$

states, and is merely rigidly shifted by a constant energy of $-\frac{1}{2}g\hbar^2 S$. Considerably more structure is shown by $\rho_\downarrow(E)$. The two elementary processes shown lead already at moderate coupling strengths to a splitting of the original Bloch band into two quasi-particle bands. The lower band is formed as a result of magnon emission. Since the $\downarrow$ electron reverses its spin in this process, there must be unoccupied $\uparrow$ states on which the $\downarrow$ electron can then "land". This explains why the "scattering band" occupies the same energy region as $\rho_\uparrow(E)$. The upper quasi-particle band consists of polaron states.

The many-body correlations thus give rise here to a phenomenon which could not be explained by conventional single-particle theory.

4.5.5 Exercises

Exercise 4.5.1 Give the complete equation of motion of the *higher-order* Green's function (4.392):

$$D_{ik,\,j\sigma}(E) = \langle\langle S_i^z a_{k\sigma}; a_{j\sigma}^+ \rangle\rangle_E$$

in the framework of the s-f model.

Exercise 4.5.2 Give the complete equation of motion of the *higher-order* Green's function (4.394):

$$P_{ik,\,j\sigma}(E) = \langle\langle n_{i\,-\sigma} a_{k\sigma}; a_{j\sigma}^+ \rangle\rangle_E$$

in the framework of the s-f model.

Exercise 4.5.3 Give the complete equation of motion of the *higher-order* Green's function (4.393):

$$F_{ik,\,j\sigma}(E) = \langle\langle S_i^{-\sigma} a_{k\,-\sigma}; a_{j\sigma}^+ \rangle\rangle_E$$

in the framework of the s-f model.

Exercise 4.5.4 Discuss the special case within the s-f model of a single hole in an otherwise fully occupied conduction band. For a ferromagnetically saturated f spin system, this situation can be treated with mathematical rigour.

1. Show that the one-electron Green's function for $\sigma = \downarrow$ electrons takes on the following simple form:

$$G_{k\downarrow}^{(n=2,\,T=0)}(E) = \hbar \left(E - \varepsilon(\boldsymbol{k}) - U - \frac{1}{2} g\hbar^2 S + \mathrm{i}0^+ \right)^{-1} .$$

2. Compute the electronic $\sigma = \uparrow$ self-energy. Compare the result with the *magnetic polaron* discussed in Sect. 4.5.4.

Exercise 4.5.5 Apply the Hartree-Fock approximation to the equation of motion of the one-electron Green's function in the s-f model. Test the result by comparing with the exact cases of the atomic limit and the empty or the completely filled conduction band at $T = 0$. What would you see as the principal disadvantage of this approximation?

4.6 Self-Examination Questions

For Section 4.1

1. How is the Hubbard Hamiltonian formulated in the limiting case of an infinitely narrow band?
2. What structures do the one-electron Green's function and spectral density have in this limiting case?
3. Can ferromagnetism occur in the case of an infinitely narrow band?
4. What is referred to as the Hartree-Fock or molecular-field approximation of a Green's function?
5. Which form does the one-electron Green's function of the Hubbard model take on in the Hartree-Fock approximation?
6. What is the relation between the Stoner and the Hubbard models?
7. What are the quasi-particle energies of the Stoner model?
8. Explain the Stoner criterion for the occurrence of ferromagnetism.
9. When does one speak of strong, and when of weak ferromagnetism?
10. What is meant by *particle correlations*?
11. To what extent can the so-called *Hubbard decouplings* also be interpreted as a molecular field approximation?
12. How can one readily see from the self-energy that the *Hubbard approximation* for the Hubbard model leads to a splitting into two quasi-particle bands?
13. What is the lifetime of the quasi-particles in the Hubbard approximation?
14. Name a significant disadvantage of the *Hubbard solution*.
15. What is the relationship within the interpolation method between the Green's function of a model system and the associated solution in the atomic limit?
16. Compare within the Hubbard model the solutions of the one-electron Green's function by the interpolation method with those from Hubbard's decoupling method.
17. Sketch the method of moments.
18. Justify the two-pole *ansatz* for the spectral density in the Hubbard model.
19. How do the quasi-particle energies in the Hubbard approximation differ from those in the method of moments?
20. Why are the solutions in the method of moments more realistic for the description of magnetic electron systems than those resulting from the Hubbard decouplings?
21. Which physical quantities determine the actual form of the quasi-particle density of states in the Hubbard model?
22. What are the preconditions for an equivalence between the Hubbard model and the Heisenberg model?
23. Can you explain why the Hubbard model for a half-filled energy band ($n = 1$) favours antiferromagnetism over ferromagnetism?

For Section 4.2

1. What is the simplifying assumption of the Thomas-Fermi approximation?
2. What is meant by the *screening length*?
3. Which simple structure is assumed by the dielectric function $\varepsilon(\boldsymbol{q})$ in the Thomas-Fermi approximation?
4. What are *plasmons*? Which Green's function determines them through its poles?
5. Can one describe charge-density waves (plasmons) by a one-electron Green's function?
6. Which form does the susceptibility $\chi_0(\boldsymbol{q}, E)$ of the non-interacting electron system take?
7. How is the susceptibility in the *random phase approximation* (RPA) related to $\chi_0(\boldsymbol{q}, E)$?
8. Sketch the determination of the plasmon dispersion relation $\hbar\omega_{\mathrm{p}}(\boldsymbol{q})$ via the *Lindhard function* graphically.
9. What is the order of magnitude of plasmon energies?
10. How is the *plasma frequency* defined?
11. Give the wave-number dependence of the plasmon dispersion relation $\omega_{\mathrm{p}}(\boldsymbol{q})$ for small $|\boldsymbol{q}|$.
12. What is the *Lindhard correction*? What is its relation to the *Friedel oscillations* of the shielded Coulomb potential of a *perturbation charge density* $\rho_{\mathrm{ext}}(\boldsymbol{r})$?
13. Which Green's function is suitable for the determination and discussion of spin-density waves and magnons in the Hubbard model?

For Section 4.3

1. Define the concepts of structural disorder, substitutional disorder, and diagonal substitutional disorder.
2. What is the decisive advantage of a periodic solid as compared to a disordered system for the theoretical description?
3. How do the T-matrix equation and the Dyson equation differ?
4. What is meant by configurational averaging in a disordered system? How is it carried out in practice?
5. Explain the effective-medium method.
6. Which equation defines the **atomic** scattering matrix?
7. What simplification makes use of the so-called *T-matrix approximation* (TMA)?
8. One describes the TMA as *non-self-consistent*. What does this mean?
9. How does one go from the TMA to the *coherent potential approximation* (CPA)?
10. The CPA is considered – in contrast to the TMA – to be *self-consistent*. Why?
11. Formulate the diagram rules for the single-particle Green's function of disordered systems.
12. What is meant by the order of a diagram?
13. Describe the diagram representation of the Dyson equation.
14. Characterise the *virtual crystal approximation* (VCA). When can it be applied?

15. What does the *single site approximation* (SSA) neglect?
16. Which form does the self-energy have in the SSA?
17. How is the *modified propagator method* (MPM) derived from the SSA?
18. Which diagram corrections were applied in order to go from the SSA to the *average T-matrix approximation* (ATA)? What is meant in this connection by *overcorrections*?
19. How do the self-energies in the TMA differ from those in the ATA?
20. How does one obtain the CPA from the ATA? Which *multiple-occupation corrections* have to be considered?
21. Why is the CPA self-energy independent of the wavenumber?
22. Which parameters determine the CPA self-energy?
23. What is meant by the concept of "alloy analogy" in connection with the CPA?
24. Formulate the CPA alloy analogy in the Hubbard model.

For Section 4.4

1. Which Green's function is expedient for the calculation of the magnetisation of a *spin-(1/2) system* within the Heisenberg model?
2. What is meant by the *Tyablikow approximation*? How well does it work for low temperatures ($T \to 0$)?
3. Does the *Tyablikow approximation* obey Bloch's $T^{3/2}$ law?
4. What difficulties occur in calculating the magnetisation of $S > 1/2$ *systems*?
5. Formulate the Dyson-Maleév transformation of the spin operators.
6. Which advantage and which disadvantage does the Dyson-Maleév transformation have in comparison to the Holstein-Primakoff transformation?
7. Which simple approach for the magnon spectral density yields Dyson's full spin-wave result via the method of moments?

For Section 4.5

1. What is meant by a $4f$ system?
2. Give the Hamiltonian of the s-f model. Which solids are typically described by this model?
3. How many poles does the one-electron Green's function of the s-f model have in the limiting case of an infinitely narrow band? Characterise them.
4. Formulate the CPA alloy analogy in the s-f model.
5. Try to give a physical interpretation of the different quasi-particle bands in the CPA solution of the s-f model.
6. Describe the $\uparrow$ quasi-particle energies for an electron in an otherwise empty conduction band at $T = 0$. Why do they have such a simple form in this limit?
7. Is the CPA solution for the special case described in 6. correct?
8. Why is the imaginary part of the $\downarrow$ self-energy in this special case nonzero just when the $\uparrow$ density of states assumes finite values?

9. Which physical processes determine the lifetimes of the $\downarrow$ quasi-particles?
10. What does a δ-function imply for the lifetime of the corresponding quasi-particle?
11. Which elementary processes give rise in the above exactly solvable special case to a splitting of the $\downarrow$ quasi-particle density of states into two subbands?

Perturbation Theory ($T = 0$) 5

The general considerations in Chap. 3 have shown that we can express everything that we need for the description of physical systems with suitably defined Green's functions. With just this statement, however, we have not yet solved any many-body problem. We need to find procedures for determining such Green's functions. Several of these we encountered in Chap. 4 in connection with specific problems in solid-state physics. The goal of the present chapter is to develop a

diagrammatic perturbation theory,

whereby we first want to presuppose generally

$$T = 0: \quad \langle \ldots \rangle \quad \Longrightarrow \quad \langle E_0 \,|\, \ldots \,|\, E_0 \rangle \,.$$

All average values are to be carried out over the ground state $|E_0\rangle$ of the interacting system.

5.1 Causal Green's Functions

5.1.1 "Conventional" Time-dependent Perturbation Theory

We decompose the Hamiltonian $\mathcal{H}$,

$$\mathcal{H} = \mathcal{H}_0 + V \,, \tag{5.1}$$

as usual into an *unperturbed* part $\mathcal{H}_0$ and a *perturbation* V. We presume that this decomposition is carried out in such a way that the eigenvalue problem for $\mathcal{H}_0$ can be regarded as solved:

W. Nolting, *Fundamentals of Many-body Physics*,
DOI 10.1007/978-3-540-71931-1_5, © Springer-Verlag Berlin Heidelberg 2009

$$\mathcal{H}_0 \left| \eta_n \right\rangle = \eta_n \left| \eta_n \right\rangle . \tag{5.2}$$

We seek the ground state of the complete problem:

$$\mathcal{H} \left| E_0 \right\rangle = E_0 \left| E_0 \right\rangle . \tag{5.3}$$

Often, one splits off a coupling constant λ from the perturbation V, which as a rule is a particle interaction,

$$V = \lambda v , \tag{5.4}$$

and then attempts to expand the quantities sought, i.e. E_0, $\left| E_0 \right\rangle$, in powers of λ. If λ is sufficiently small, one will then be able to terminate the series after a finite number of terms. If this precondition is not fulfilled, one will instead try to sum infinite series containing the dominant terms.

With (5.2) and (5.3), we initially have:

$$\left\langle \eta_0 \right| V \left| E_0 \right\rangle = \left\langle \eta_0 \right| (\mathcal{H} - \mathcal{H}_0) \left| E_0 \right\rangle = (E_0 - \eta_0) \left\langle \eta_0 \middle| E_0 \right\rangle .$$

This yields the still-exact

level shift

$$\Delta E_0 \equiv E_0 - \eta_0 = \frac{\left\langle \eta_0 \right| V \left| E_0 \right\rangle}{\left\langle \eta_0 \middle| E_0 \right\rangle} . \tag{5.5}$$

We of course cannot make much use of this shift, since $\left| E_0 \right\rangle$ is still unknown. We define the *projection operator*

$$P_0 \equiv \left| \eta_0 \right\rangle \left\langle \eta_0 \right| . \tag{5.6}$$

For the *orthogonal projector* Q, we find:

$$\begin{aligned} Q_0 \equiv \mathbf{1} - P_0 &= \sum_{n=0}^{\infty} \left| \eta_n \right\rangle \left\langle \eta_n \right| - \left| \eta_0 \right\rangle \left\langle \eta_0 \right| = \\ &= \sum_{n=1}^{\infty} \left| \eta_n \right\rangle \left\langle \eta_n \right| . \end{aligned} \tag{5.7}$$

We now return to the exact eigenvalue equation (5.3), for which we assume the ground state $\left| E_0 \right\rangle$ to be non-degenerate. With an arbitrary real constant D, we can write:

$$(D - \mathcal{H}_0) \left| E_0 \right\rangle = (D - \mathcal{H} + V) \left| E_0 \right\rangle = (D - E_0 + V) \left| E_0 \right\rangle .$$

The operator $(D - \mathcal{H}_0)$ has a unique inversion, as long as $\mathcal{H}_0$ does not have just the constant D itself as an eigenvalue:

$$\left| E_0 \right\rangle = \frac{1}{D - \mathcal{H}_0} (D - E_0 + V) \left| E_0 \right\rangle .$$

We now make use of the projectors introduced above:

$$\left| E_0 \right\rangle = P_0 \left| E_0 \right\rangle + Q_0 \left| E_0 \right\rangle = \left| \eta_0 \right\rangle \left\langle \eta_0 \middle| E_0 \right\rangle + Q_0 \left| E_0 \right\rangle .$$

With the definition

$$\left| \widetilde{E}_0 \right\rangle = \frac{\left| E_0 \right\rangle}{\left\langle \eta_0 \middle| E_0 \right\rangle} , \tag{5.8}$$

we then obtain an equation for $| \widetilde{E}_0 \rangle$,

$$\left| \widetilde{E}_0 \right\rangle = \left| \eta_0 \right\rangle + \frac{1}{D - \mathcal{H}_0} Q_0 (D - E_0 + V) \left| \widetilde{E}_0 \right\rangle , \tag{5.9}$$

which clearly can be iterated. In (5.9), we have already made use of the fact that Q_0 commutes with $\mathcal{H}_0$. From the definition (5.6), it namely follows immediately that:

$$[P_0, \mathcal{H}_0]_- = 0 \tag{5.10}$$

and thus also:

$$[Q_0, \mathcal{H}_0]_- = 0 . \tag{5.11}$$

By iteration of (5.9), we obtain the

Fundamental formula of perturbation theory

$$| \widetilde{E}_0 \rangle = \sum_{m=0}^{\infty} \left\{ \frac{1}{D - \mathcal{H}_0} Q_0 (D - E_0 + V) \right\}^m \left| \eta_0 \right\rangle . \tag{5.12}$$

On the right-hand side, only the *unperturbed* ground state occurs, to be sure accompanied by the eigenvalue E_0, which still has to be determined. We still have the constant D as a free parameter. For the level shift ΔE_0, we find with (5.12) in (5.5):

$$\begin{aligned} \Delta E_0 &= \left\langle \eta_0 \middle| V \middle| \widetilde{E}_0 \right\rangle = \\ &= \sum_{m=0}^{\infty} \left\langle \eta_0 \right| V \left\{ \frac{1}{D - \mathcal{H}_0} Q_0 (D - E_0 + V) \right\}^m \left| \eta_0 \right\rangle . \end{aligned} \tag{5.13}$$

Specific choices of D yield different versions of time-independent perturbation theory.

1. **The Schrödinger perturbation theory**
   If we choose

$$D = \eta_0 \,, \tag{5.14}$$

then we obtain:

$$\left|\widetilde{E}_0\right\rangle = \sum_{m=0}^{\infty} \left\{ \frac{1}{\eta_0 - \mathcal{H}_0} Q_0 (V - \Delta E_0) \right\}^m \left|\eta_0\right\rangle \,, \tag{5.15}$$

$$\Delta E_0 = \sum_{m=0}^{\infty} \left\langle \eta_0 \right| V \left\{ \frac{1}{\eta_0 - \mathcal{H}_0} Q_0 (V - \Delta E_0) \right\}^m \left|\eta_0\right\rangle . \tag{5.16}$$

For a practical evaluation, these general results must now be ordered according to powers of the coupling constant λ. To do this, we evaluate the leading terms of the level shift explicitly:

$$\Delta E_0(m = 0) = \left\langle \eta_0 \middle| \lambda v \middle| \eta_0 \right\rangle \sim \lambda \,, \tag{5.17}$$

$$\begin{aligned} \Delta E_0(m = 1) &= \left\langle \eta_0 \right| V \frac{1}{\eta_0 - \mathcal{H}_0} Q_0 \left(V - \Delta E_0\right) \left|\eta_0\right\rangle = \\ &= \left\langle \eta_0 \right| V \frac{1}{\eta_0 - \mathcal{H}_0} \sum_{n=1}^{\infty} \left|\eta_n\right\rangle \left\langle \eta_n \middle| V \middle| \eta_0 \right\rangle = \\ &= \sum_{n=1}^{\infty} \frac{\left| \left\langle \eta_0 \middle| \lambda v \middle| \eta_n \right\rangle \right|^2}{\eta_0 - \eta_n} \sim \lambda^2 . \end{aligned} \tag{5.18}$$

These are the well-known results of the Schrödinger perturbation theory. Up to m=1, the perturbation expansion runs parallel to the powers of λ, that is,

$$\Delta E_0(m) \sim \lambda^{m+1} \quad (m = 0, 1) . \tag{5.19}$$

This however already no longer holds for the $m = 2$ term.

$$\begin{aligned} &\Delta E_0(m = 2) = \\ &= \left\langle \eta_0 \right| V \frac{1}{\eta_0 - \mathcal{H}_0} Q_0 \left(V - \Delta E_0\right) \frac{1}{\eta_0 - \mathcal{H}_0} Q_0 \left(V - \Delta E_0\right) \left|\eta_0\right\rangle = \\ &= \sum_{n=1}^{\infty} \left\langle \eta_0 \right| V \frac{1}{\eta_0 - \mathcal{H}_0} Q_0 \left(V - \Delta E_0\right) \left|\eta_n\right\rangle \left\langle \eta_n \middle| V \middle| \eta_0 \right\rangle \frac{1}{\eta_0 - \eta_n} = \\ &= \sum_{n=1}^{\infty} \sum_{m=1}^{\infty} \frac{\left\langle \eta_0 \middle| V \middle| \eta_m \right\rangle \left\langle \eta_m \middle| V \middle| \eta_n \right\rangle \left\langle \eta_n \middle| V \middle| \eta_0 \right\rangle}{(\eta_0 - \eta_m)(\eta_0 - \eta_n)} - \\ &- \Delta E_0 \sum_{n=1}^{\infty} \frac{\left| \left\langle \eta_0 \middle| V \middle| \eta_n \right\rangle \right|^2}{(\eta_0 - \eta_n)^2} . \end{aligned} \tag{5.20}$$

The first term is proportional to λ^3, the second contains all powers of $\lambda \geq 3$ due to ΔE_0. The ordering process becomes more and more tedious with increasing m. It is for example not possible to formulate the general energy correction proportional to λ^n in a concrete and clearly-arranged form. This proves to be a considerable disadvantage when the physical problem requires the summation of an infinite partial series. In that case, one needs perturbation expansions which directly yield the corrections proportional to λ^n. We will encounter such expansions in the next section. First, however, we ask whether the

2. **Brillouin-Wigner perturbation theory**
   is more suitable in the above sense than Schrödinger's perturbation theory. Here, we set

$$D = E_0 \tag{5.21}$$

and then obtain:

$$\left|\widetilde{E}_0\right\rangle = \sum_{m=0}^{\infty} \left\{ \frac{1}{E_0 - \mathcal{H}_0} Q_0 V \right\}^m \left|\eta_0\right\rangle , \tag{5.22}$$

$$\Delta E_0 = \sum_{m=0}^{\infty} \left\langle \eta_0 \right| V \left\{ \frac{1}{E_0 - \mathcal{H}_0} Q_0 V \right\}^m \left|\eta_0\right\rangle . \tag{5.23}$$

One can readily see that the desired ordering in terms of powers of λ will give rise to the same difficulties here as in 1.

A trick can help at this point: The in reality time-independent problem is converted artificially into a time-dependent one. This makes it possible to use the time-evolution operator, which from (3.18) or (3.40) consists of terms which are ordered according to powers of λ, to construct the ground state of the interacting system from that of the non-interacting system.

5.1.2
"Switching on" the Interaction Adiabatically

We make the Hamiltonian (5.1) *artificially* time dependent by replacing it with

$$\mathcal{H}_\alpha = \mathcal{H}_0 + V e^{-\alpha |t|} ; \quad \alpha > 0 . \tag{5.24}$$

Beginning with the unperturbed system ($\mathcal{H}_0$) at $t = -\infty$, we switch on the interaction slowly, so that it has reached full strength at $t = 0$, and then is switched off in the same manner for $t \to \infty$:

$$\lim_{t \to \pm\infty} \mathcal{H}_\alpha = \mathcal{H}_0 ; \quad \lim_{t \to 0} \mathcal{H}_\alpha = \mathcal{H} . \tag{5.25}$$

At the end of the calculation, the limit $\alpha \to 0$ will be carried out, i.e. the interaction is switched on and off with infinite slowness (i.e. *adiabatically*). If now the ground state $\left|\eta_0\right\rangle$

of the free system is not degenerate, and furthermore the overlap $\langle \eta_0 \mid E_0 \rangle$ is finite, then it appears at least plausible that the ground state $|E_0\rangle$ of the interacting system evolves during this *adiabatic* switching process continuously out of $|\eta_0\rangle$. We want to investigate this question more quantitatively in the following.

It proves expedient to formulate the operators of interest in the Dirac representation. From (3.34), we have for the interaction operator,

$$V_D(t)\exp(-\alpha\,|t|) = \exp\left(\frac{i}{\hbar}\mathcal{H}_0 t\right) V \exp\left(-\frac{i}{\hbar}\mathcal{H}_0 t\right) \exp\left(-\alpha\,|t|\right) \tag{5.26}$$

which determines the time-evolution operator in (3.40) and (3.18):

$$\begin{aligned} U_\alpha^D(t, t_0) = \sum_{n=0}^{\infty} \frac{1}{n!}\left(-\frac{i}{\hbar}\right)^n \int_{t_0}^{t} \cdots \int dt_1 \cdots dt_n\, e^{-\alpha(|t_1|+\cdots+|t_n|)}\, \cdot \\ \cdot\, T_D\left\{ V_D(t_1)\cdots V_D(t_n)\right\} . \end{aligned} \tag{5.27}$$

Each term belongs to a particular power of the coupling constant λ. The expansion is thus favourably ordered in the sense of the considerations in the previous section.

The action of the time-evolution operator is clear from (3.30):

$$\left|\psi_\alpha^D(t)\right\rangle = U_\alpha^D(t, t_0)\left|\psi_\alpha^D(t_0)\right\rangle . \tag{5.28}$$

The equation of motion (3.37),

$$i\hbar\left|\dot{\psi}_\alpha^D(t)\right\rangle = e^{-\alpha|t|}\, V_D(t)\left|\psi_\alpha^D(t)\right\rangle ,$$

implies for $\alpha > 0$ that:

$$i\hbar\left|\dot{\psi}_\alpha^D(t \to \pm\infty)\right\rangle = 0 .$$

In the interaction representation in this limit, the state thus becomes time independent. We take

$$\left|\psi_\alpha^D(t \to -\infty)\right\rangle = \left|\eta_0\right\rangle , \tag{5.29}$$

since at $T = 0$, $\left|\psi_\alpha^D(t \to -\infty)\right\rangle$ differs from the ground state of the free system only in terms of a phase factor. The latter can be set to 1 without loss of generality. Then, however, the phase for the corresponding limiting state at $t \to +\infty$ is no longer free:

$$\left|\psi_\alpha^D(t \to +\infty)\right\rangle = e^{i\varphi}\left|\eta_0\right\rangle . \tag{5.30}$$

With (5.28), we then obtain for the time evolution of the Dirac state:

$$\left|\psi_\alpha^{\mathrm{D}}(t)\right\rangle = U_\alpha^{\mathrm{D}}(t, -\infty)\left|\eta_0\right\rangle . \tag{5.31}$$

At $t = 0$, the interaction is fully switched on. It can of course not be excluded that the state $\left|\psi_\alpha^{\mathrm{D}}(0)\right\rangle$ still depends on α. α indeed determines the speed of the switching-on process. If the latter is however carried out *adiabatically*, ($\alpha \to 0$), then it seems clear that at every given time t, the ground state which corresponds to the current value of the interaction strength will be found. Thus the desired exact ground state should be calculable from (5.31) and

$$\left|E_0^{\mathrm{D}}\right\rangle \stackrel{?}{=} \lim_{\alpha \to 0}\left|\psi_\alpha^{\mathrm{D}}(0)\right\rangle . \tag{5.32}$$

Since, however, we must assume explicitly that $\alpha > 0$ for (5.29), it is by no means certain that the limit $\alpha \to 0$ exists in fact.

Gell-Mann–Low theorem
If the state

$$\lim_{\alpha \to 0} \frac{U_\alpha^{\mathrm{D}}(0, -\infty)\left|\eta_0\right\rangle}{\left\langle \eta_0\right| U_\alpha^{\mathrm{D}}(0, -\infty)\left|\eta_0\right\rangle} = \lim_{\alpha \to 0} \frac{\left|\psi_\alpha^{\mathrm{D}}(0)\right\rangle}{\left\langle \eta_0 \mid \psi_\alpha^{\mathrm{D}}(0)\right\rangle} \tag{5.33}$$

exists for every order of perturbation theory, then it is an exact eigenstate of $\mathcal{H}$. The limiting value (5.32), on the other hand, *does not exist!*

This theorem fixes the eigenstate which evolves from the *free* ground state during the adiabatic switching-on of the perturbation. This need not necessarily be the ground state of the interacting system. Therefore, we will later have to *postulate* the additional assumption that no *crossings* of the states occur during their evolution from the free states. This will as a rule be correct; however this additional assumption naturally excludes phenomena such as superconductivity. In that case, the interactions lead to a new type of ground state with a different symmetry and a lower energy than the *adiabatic* ground state.

We will briefly sketch the proof of the Gell-Mann–Low theorem: The starting point is the relation

$$\begin{aligned}(\mathcal{H}_0 - \eta_0)\left|\psi_\alpha^{\mathrm{D}}(0)\right\rangle &= (\mathcal{H}_0 - \eta_0)\, U_\alpha^{\mathrm{D}}(0, -\infty)\left|\eta_0\right\rangle = \\ &= \left[\mathcal{H}_0, U_\alpha^{\mathrm{D}}(0, -\infty)\right]_- \left|\eta_0\right\rangle .\end{aligned} \tag{5.34}$$

Inserting (5.27), we see that the following commutators must be evaluated:

$$\begin{aligned}
&[\mathcal{H}_0, V_D(t_1)\cdots V_D(t_n)]_- = \\
&= [\mathcal{H}_0, V_D(t_1)]_-\ V_D(t_2)\cdots V_D(t_n) + \\
&+ V_D(t_1)\,[\mathcal{H}_0, V_D(t_2)]_-\ V_D(t_3)\cdots V_D(t_n) + \\
&+\cdots+ \\
&+ V_D(t_1)\,V_D(t_2)\cdots[\mathcal{H}_0, V_D(t_n)]_- = \\
&= -\mathrm{i}\hbar\left\{\frac{\partial}{\partial t_1}+\frac{\partial}{\partial t_2}+\cdots+\frac{\partial}{\partial t_n}\right\}V_D(t_1)\cdots V_D(t_n)\,.
\end{aligned} \tag{5.35}$$

In the last step, we made use of the equation of motion (3.35). It is immediately clear that from (5.35), it follows that:

$$[\mathcal{H}_0, T_D(V_D(t_1)\cdots V_D(t_n))]_- = -\mathrm{i}\hbar\left(\sum_{j=1}^{n}\frac{\partial}{\partial t_j}\right)T_D\big(V_D(t_1)\cdots V_D(t_n)\big)\,. \tag{5.36}$$

We insert this along with (5.27) into (5.34):

$$\begin{aligned}
&(\mathcal{H}_0-\eta_0)\left|\psi_\alpha^D(0)\right\rangle = \\
&= -\sum_{n=1}^{\infty}\frac{1}{n!}\left(-\frac{\mathrm{i}}{\hbar}\right)^{n-1}\int_{-\infty}^{0}\cdots\int \mathrm{d}t_1\cdots\mathrm{d}t_n\,\cdot \\
&\quad\cdot\ \mathrm{e}^{-\alpha(|t_1|+\cdots+|t_n|)}\left(\sum_{j=1}^{n}\frac{\partial}{\partial t_j}\right)T_D\big(V_D(t_1)\cdots V_D(t_n)\big)\left|\eta_0\right\rangle\,.
\end{aligned} \tag{5.37}$$

Because of the subsequent integrations, the n time derivatives naturally make the same contribution; one need only suitably reorder the time indices. We can then agree upon the replacement

$$\left(\sum_{j=1}^{n}\frac{\partial}{\partial t_j}\right) \longrightarrow n\frac{\partial}{\partial t_n}$$

in (5.37). Now, however,

$$\int_{-\infty}^{0} \mathrm{d}t_n\, \mathrm{e}^{+\alpha t_n} \frac{\partial}{\partial t_n} T_\mathrm{D}\big(V_\mathrm{D}(t_1)\cdots V_\mathrm{D}(t_n)\big) =$$

$$= \Big[\mathrm{e}^{\alpha t_n} T_\mathrm{D}\big(V_\mathrm{D}(t_1)\cdots V_\mathrm{D}(t_n)\big)\Big]_{-\infty}^{0} - \int_{-\infty}^{0} \mathrm{d}t_n\, \alpha\, \mathrm{e}^{\alpha t_n} T_\mathrm{D}\big(V_\mathrm{D}(t_1)\cdots V_\mathrm{D}(t_n)\big) =$$

$$= V_\mathrm{D}(0) T_\mathrm{D}\big(V_\mathrm{D}(t_1)\cdots V_\mathrm{D}(t_{n-1})\big) - \alpha \int_{-\infty}^{0} \mathrm{d}t_n\, \mathrm{e}^{-\alpha |t_n|} T_\mathrm{D}\big(V_\mathrm{D}(t_1)\cdots V_\mathrm{D}(t_n)\big)\,.$$

For (5.37), this means that:

$$\begin{aligned}
&(\mathcal{H}_0 - \eta_0)\,\big|\psi_\alpha^\mathrm{D}(0)\big\rangle \\
&= -V_\mathrm{D}(0) \sum_{n=1}^{\infty} \frac{1}{(n-1)!}\left(-\frac{\mathrm{i}}{\hbar}\right)^{n-1} \cdot \\
&\quad \cdot \int_{-\infty}^{0}\cdots\int \mathrm{d}t_1\cdots \mathrm{d}t_{n-1}\, \mathrm{e}^{-\alpha(|t_1|+\cdots+|t_{n-1}|)} T_\mathrm{D}\big(V_\mathrm{D}(t_1)\cdots V_\mathrm{D}(t_{n-1})\big)\,\big|\eta_0\big\rangle + \\
&\quad + \alpha \sum_{n=1}^{\infty} \frac{1}{(n-1)!}\left(-\frac{\mathrm{i}}{\hbar}\right)^{n-1} \int_{-\infty}^{0}\cdots\int \mathrm{d}t_1\cdots \mathrm{d}t_n\, \cdot \\
&\quad \cdot\, \mathrm{e}^{-\alpha(|t_1|+\cdots+|t_n|)} T_\mathrm{D}\big(V_\mathrm{D}(t_1)\cdots V_\mathrm{D}(t_n)\big)\,\big|\eta_0\big\rangle\,.
\end{aligned} \tag{5.38}$$

Due to (5.4), we have:

$$T_\mathrm{D}\big(V_\mathrm{D}(t_1)\cdots V_\mathrm{D}(t_n)\big) \sim \lambda^n\,.$$

In the second term in (5.38), we then have an expression of the form

$$\alpha \left(-\frac{\mathrm{i}}{\hbar}\right)^{n-1} \frac{1}{(n-1)!}\lambda^n = \alpha \mathrm{i}\hbar\lambda \frac{\partial}{\partial\lambda}\left[\left(-\frac{\mathrm{i}}{\hbar}\right)^{n} \frac{1}{n!}\lambda^n\right]. \tag{5.39}$$

We can thus combine the two terms in (5.38):

$$\begin{aligned}
&(\mathcal{H}_0 - \eta_0)\,\big|\psi_\alpha^\mathrm{D}(0)\big\rangle \\
&= \left(-V_\mathrm{D}(0) + \mathrm{i}\hbar\alpha\lambda\frac{\partial}{\partial\lambda}\right) \sum_{n=0}^{\infty} \frac{1}{n!}\left(-\frac{\mathrm{i}}{\hbar}\right)^{n} \int_{-\infty}^{0}\cdots\int \mathrm{d}t_1\cdots \mathrm{d}t_n\, \mathrm{e}^{-\alpha(|t_1|+\cdots+|t_n|)}\, \cdot \\
&\quad \cdot\, T_\mathrm{D}\big(V_\mathrm{D}(t_1)\cdots V_\mathrm{D}(t_n)\big)\,\big|\eta_0\big\rangle = \\
&= \left(-V_\mathrm{D}(0) + \mathrm{i}\hbar\alpha\lambda\frac{\partial}{\partial\lambda}\right)\,\big|\psi_\alpha^\mathrm{D}(0)\big\rangle\,.
\end{aligned} \tag{5.40}$$

At $t = 0$, the interaction in the Dirac representation is identical with that of the Schrödinger representation. We can thus now combine $\mathcal{H}_0$ with $V_{\mathrm{D}}(0)$:

$$(\mathcal{H} - \eta_0)\left|\psi_\alpha^{\mathrm{D}}(0)\right\rangle = \mathrm{i}\hbar\alpha\lambda\frac{\partial}{\partial\lambda}\left|\psi_\alpha^{\mathrm{D}}(0)\right\rangle . \tag{5.41}$$

We rearrange:

$$\begin{aligned}
&\left(\mathcal{H} - \eta_0 - \mathrm{i}\hbar\alpha\lambda\frac{\partial}{\partial\lambda}\right)\frac{\left|\psi_\alpha^{\mathrm{D}}(0)\right\rangle}{\left\langle\eta_0 \mid \psi_\alpha^{\mathrm{D}}(0)\right\rangle} = \\
&= (\mathcal{H} - \eta_0)\frac{\left|\psi_\alpha^{\mathrm{D}}(0)\right\rangle}{\left\langle\eta_0 \mid \psi_\alpha^{\mathrm{D}}(0)\right\rangle} - \frac{\mathrm{i}\hbar\alpha\lambda}{\left\langle\eta_0 \mid \psi_\alpha^{\mathrm{D}}(0)\right\rangle}\frac{\partial}{\partial\lambda}\left|\psi_\alpha^{\mathrm{D}}(0)\right\rangle + \\
&+ \frac{\mathrm{i}\hbar\alpha\left|\psi_\alpha^{\mathrm{D}}(0)\right\rangle}{\left(\left\langle\eta_0 \mid \psi_\alpha^{\mathrm{D}}(0)\right\rangle\right)^2}\left\langle\eta_0\right|\lambda\frac{\partial}{\partial\lambda}\left|\psi_\alpha^{\mathrm{D}}(0)\right\rangle = \\
&\overset{(5.41)}{=} (\mathcal{H} - \eta_0)\frac{\left|\psi_\alpha^{\mathrm{D}}(0)\right\rangle}{\left\langle\eta_0 \mid \psi_\alpha^{\mathrm{D}}(0)\right\rangle} - (\mathcal{H} - \eta_0)\frac{\left|\psi_\alpha^{\mathrm{D}}(0)\right\rangle}{\left\langle\eta_0 \mid \psi_\alpha^{\mathrm{D}}(0)\right\rangle} + \\
&+ \frac{\left|\psi_\alpha^{\mathrm{D}}(0)\right\rangle}{\left\langle\eta_0 \mid \psi_\alpha^{\mathrm{D}}(0)\right\rangle}\frac{\left\langle\eta_0\right|(\mathcal{H} - \eta_0)\left\langle\psi_\alpha^{\mathrm{D}}(0)\right|}{\left\langle\eta_0 \mid \psi_\alpha^{\mathrm{D}}(0)\right\rangle} = \\
&= \frac{\left|\psi_\alpha^{\mathrm{D}}(0)\right\rangle}{\left\langle\eta_0 \mid \psi_\alpha^{\mathrm{D}}(0)\right\rangle}\left\{\frac{\left\langle\eta_0\right|\mathcal{H}\left|\psi_\alpha^{\mathrm{D}}(0)\right\rangle}{\left\langle\eta_0 \mid \psi_\alpha^{\mathrm{D}}(0)\right\rangle} - \eta_0\right\} .
\end{aligned}$$

All together, we have now found the result:

$$\left\{\mathcal{H} - \frac{\left\langle\eta_0 \mid \mathcal{H} \mid \psi_\alpha^{\mathrm{D}}(0)\right\rangle}{\left\langle\eta_0 \mid \psi_\alpha^{\mathrm{D}}(0)\right\rangle} - \mathrm{i}\hbar\alpha\lambda\frac{\partial}{\partial\lambda}\right\}\frac{\left|\psi_\alpha^{\mathrm{D}}(0)\right\rangle}{\left\langle\eta_0 \mid \psi_\alpha^{\mathrm{D}}(0)\right\rangle} = 0 . \tag{5.42}$$

Now, by construction, the state on the right next to the brackets must also exist for $\alpha \to 0$ in every order of perturbation theory, i.e. in every order in the coupling constant λ. This remains true when we differentiate this expression with respect to λ. If we now go to the limit $\alpha \to 0$ in (5.42), then the third term within the brackets vanishes:

$$\left\{\mathcal{H} - \frac{\left\langle\eta_0 \mid \mathcal{H} \mid \psi_0^{\mathrm{D}}(0)\right\rangle}{\left\langle\eta_0 \mid \psi_0^{\mathrm{D}}(0)\right\rangle}\right\}\frac{\left|\psi_0^{\mathrm{D}}(0)\right\rangle}{\left\langle\eta_0 \mid \psi_0^{\mathrm{D}}(0)\right\rangle} = 0 . \tag{5.43}$$

This proves the assertion of the Gell-Mann–Low theorem. We have shown that the state (5.33) is an exact eigenstate under the conditions assumed. In agreement with previous considerations, we make the additional assumption that it is also the ground state:

$$\frac{\left|\psi_0^{\mathrm{D}}(0)\right\rangle}{\left\langle\eta_0 \mid \psi_0^{\mathrm{D}}(0)\right\rangle} \overset{!}{=} \frac{\left|E_0^{\mathrm{D}}(0)\right\rangle}{\left\langle\eta_0 \mid E_0^{\mathrm{D}}(0)\right\rangle} = \left|\widetilde{E}_0\right\rangle . \tag{5.44}$$

Finally, we show to conclude this section that the numerator and the denominator of (5.33), each by itself, do *not* exist in the limit $\alpha \to 0$. To do so, we consider the following expression:

$$\begin{aligned}
&\mathrm{i}\hbar\alpha\lambda\frac{\partial}{\partial\lambda}\ln\left\langle\eta_0 \mid \psi_\alpha^{\mathrm{D}}(0)\right\rangle = \\
&= \frac{1}{\left\langle\eta_0 \mid \psi_\alpha^{\mathrm{D}}(0)\right\rangle}\mathrm{i}\hbar\alpha\lambda\frac{\partial}{\partial\lambda}\left\langle\eta_0 \mid \psi_\alpha^{\mathrm{D}}(0)\right\rangle = \\
&\overset{(5.41)}{=} \frac{1}{\left\langle\eta_0 \mid \psi_\alpha^{\mathrm{D}}(0)\right\rangle}\left\langle\eta_0\right|(\mathcal{H}-\eta_0)\left|\psi_\alpha^{\mathrm{D}}(0)\right\rangle = \\
&= \frac{\left\langle\eta_0\right|V_{\mathrm{D}}(0)\left|\psi_\alpha^{\mathrm{D}}(0)\right\rangle}{\left\langle\eta_0 \mid \psi_\alpha^{\mathrm{D}}(0)\right\rangle} \underset{\alpha\to 0}{\overset{(5.5)}{\longrightarrow}} \Delta E_0(\lambda)\,.
\end{aligned}$$

It then follows that:

$$\frac{\partial}{\partial\lambda}\ln\left\langle\eta_0 \mid \psi_\alpha^{\mathrm{D}}(0)\right\rangle \underset{\alpha\to 0}{\longrightarrow} \frac{1}{\mathrm{i}\hbar}\frac{\Delta E_0(\lambda)}{\lambda}\frac{1}{\alpha}\,.$$

The integration over λ leads to an expression of the form

$$\ln\left\langle\eta_0 \mid \psi_\alpha^{\mathrm{D}}(0)\right\rangle \underset{\alpha\to 0}{\longrightarrow} \frac{-\mathrm{i}f(\lambda)}{\alpha}$$

and thus

$$\left\langle\eta_0 \mid \psi_\alpha^{\mathrm{D}}(0)\right\rangle \underset{\alpha\to 0}{\longrightarrow} \exp\left(-\mathrm{i}\frac{f(\lambda)}{\alpha}\right)\,. \tag{5.45}$$

The state $\left|\psi_\alpha^{\mathrm{D}}(0)\right\rangle$ thus has a phase which diverges as $1/\alpha$ for $\alpha \to 0$. The limit in (5.32) therefore does not exist. This divergent phase apparently cancels out in the state (5.33).

5.1.3 Causal Green's Functions

Green's functions are, according to the definition in Sect. 3.2.1, expectation values of time-dependent Heisenberg operators. Since in this section, we are generally considering the limiting case $T = 0$, these expectation values are to be taken over the ground state. The Heisenberg representation is not suitable for a perturbation-theory calculation; the Dirac representation is more convenient. We therefore first investigate the corresponding transformations.

In (5.44), we found the ground state of the interacting system:

$$\left|\widetilde{E}_0\right\rangle = \lim_{\alpha\to 0}\frac{U_\alpha^{\mathrm{D}}(0,-\infty)\left|\eta_0\right\rangle}{\left\langle\eta_0\right|U_\alpha^{\mathrm{D}}(0,-\infty)\left|\eta_0\right\rangle}\,. \tag{5.46}$$

Since the interaction is again switched off for times greater than zero in the same manner as it was switched on for negative times starting from $t = -\infty$, we could have proved the Gell-Mann–Low theorem just as well for the state

$$|\widetilde{E}_0'\rangle = \lim_{\alpha \to 0} \frac{U_\alpha^{\mathrm{D}}(0, +\infty)\,|\,\eta_0\rangle}{\langle \eta_0\,|\,U_\alpha^{\mathrm{D}}(0, +\infty)\,|\,\eta_0\rangle}\,. \tag{5.47}$$

Due to the fact that $|\eta_0\rangle$ is by construction non-degenerate, $|\widetilde{E}_0\rangle$ and $|\widetilde{E}_0'\rangle$ can differ at most by a phase. Owing to

$$\langle \eta_0\,|\,\widetilde{E}_0\rangle = \langle \eta_0\,|\,\widetilde{E}_0'\rangle = 1\,, \tag{5.48}$$

we can even write:

$$|\widetilde{E}_0\rangle \equiv |\widetilde{E}_0'\rangle\,. \tag{5.49}$$

The *normalised* ground state which is the same in all representations at the time $t = 0$ is then given by:

$$|E_0\rangle = \frac{|\widetilde{E}_0\rangle}{\left(\langle \widetilde{E}_0\,|\,\widetilde{E}_0\rangle\right)^{1/2}} = \frac{|\widetilde{E}_0'\rangle}{\left(\langle \widetilde{E}_0'\,|\,\widetilde{E}_0'\rangle\right)^{1/2}}\,. \tag{5.50}$$

For the time-evolution operator in the Dirac representation, from (3.33) we then obtain

$$U_\alpha^{\mathrm{D}}(t, t') = \exp\left(\frac{\mathrm{i}}{\hbar}\mathcal{H}_0 t\right) U_\alpha^{\mathrm{S}}(t, t') \exp\left(-\frac{\mathrm{i}}{\hbar}\mathcal{H}_0 t'\right), \tag{5.51}$$

and thus for an arbitrary operator A in the Heisenberg representation:

$$\begin{aligned} A_\alpha^{\mathrm{H}}(t) &= U_\alpha^{\mathrm{S}}(0, t) A_\alpha^{\mathrm{S}} U_\alpha^{\mathrm{S}}(t, 0) = \\ &= U_\alpha^{\mathrm{S}}(0, t) \exp\left(-\frac{\mathrm{i}}{\hbar}\mathcal{H}_0 t\right) A^{\mathrm{D}}(t) \exp\left(\frac{\mathrm{i}}{\hbar}\mathcal{H}_0 t\right) U_\alpha^{\mathrm{S}}(t, 0) = \\ &= U_\alpha^{\mathrm{D}}(0, t) A^{\mathrm{D}}(t) U_\alpha^{\mathrm{D}}(t, 0)\,. \end{aligned} \tag{5.52}$$

We can now find the expectation value of a Heisenberg observable in the ground state:

$$\begin{aligned} \langle E_0\,|\,A^{\mathrm{H}}(t)\,|\,E_0\rangle &= \frac{\langle \widetilde{E}_0\,|\,A^{\mathrm{H}}_{\alpha\to 0}(t)\,|\,\widetilde{E}_0\rangle}{\langle \widetilde{E}_0\,|\,\widetilde{E}_0\rangle} = \\ &\overset{(5.49)}{=} \frac{\langle \widetilde{E}_0'\,|\,A^{\mathrm{H}}_{\alpha\to 0}(t)\,|\,\widetilde{E}_0\rangle}{\langle \widetilde{E}_0'\,|\,\widetilde{E}_0\rangle} = \\ &= \lim_{\alpha\to 0} \frac{\langle \eta_0\,|\,U_\alpha^{\mathrm{D}}(+\infty, 0) A_\alpha^{\mathrm{H}}(t) U_\alpha^{\mathrm{D}}(0, -\infty)\,|\,\eta_0\rangle}{\langle \eta_0\,|\,U_\alpha^{\mathrm{D}}(\infty, 0) U_\alpha^{\mathrm{D}}(0, -\infty)\,|\,\eta_0\rangle\rangle}\,. \end{aligned}$$

We define the

$$\textbf{scattering matrix:}\quad S_\alpha = U_\alpha^{\mathrm{D}}(+\infty, -\infty)\,, \tag{5.53}$$

and can then rewrite the expectation value in terms of the *interacting* ground state, with the aid of (5.52), as an expression which refers to the ground state $|\eta_0\rangle$ of the free system:

$$\langle E_0 \mid A^{\mathrm{H}}(t) \mid E_0 \rangle = \lim_{\alpha \to 0} \frac{\langle \eta_0 \mid U_\alpha^{\mathrm{D}}(\infty, t) A^{\mathrm{D}}(t) U_\alpha(t, -\infty) \mid \eta_0 \rangle}{\langle \eta_0 \mid S_\alpha \mid \eta_0 \rangle} . \tag{5.54}$$

Together with (5.51), this relation can be generalised immediately to several operators:

$$\langle E_0 \mid A^{\mathrm{H}}(t) B^{\mathrm{H}}(t') \mid E_0 \rangle = \\ = \lim_{\alpha \to 0} \frac{\langle \eta_0 \mid U_\alpha^{\mathrm{D}}(\infty, t) A^{\mathrm{D}}(t) U_\alpha^{\mathrm{D}}(t, t') B^{\mathrm{D}}(t') U_\alpha^{\mathrm{D}}(t', -\infty) \mid \eta_0 \rangle}{\langle \eta_0 \mid S_\alpha \mid \eta_0 \rangle} . \tag{5.55}$$

Using this expression, we now wish to cast the causal Green's function defined in (3.119) in a form which is tractable for perturbation theory. To do so, we insert U_α^{D} from (5.27), henceforth leaving off the index "D", since we will work only with the Dirac representation in the following.

In the definition of the causal Green's function (3.119), Wick's time-ordering operator T_ε appears. It orders operators at later times to the left, whereby each permutation introduces a factor of $\varepsilon = +1$ for Bosonic operators and $\varepsilon = -1$ for Fermionic operators. In U_α^{D}, in contrast, Dyson's time-ordering operator T_{D} (3.15) occurs. It orders similarly to T_ε, but without the factor ε. T_{D} acts on the interaction $V(t)$. It always consists however for Fermions of an even number of creation or annihilation operators, so that the replacement

$$T_{\mathrm{D}} \Longrightarrow T_\varepsilon$$

in the time-evolution operator is always permitted. We now assert that the expectation value (5.54) can be written as follows:

$$\langle E_0 \mid A^{\mathrm{H}}(t) \mid E_0 \rangle = \\ = \lim_{\alpha \to 0} \frac{1}{\langle \eta_0 \mid S_\alpha \mid \eta_0 \rangle} \sum_{\nu=0}^{\infty} \frac{1}{\nu!} \left(-\frac{\mathrm{i}}{\hbar} \right)^\nu \int_{-\infty}^{+\infty} \cdots \int \mathrm{d}t_1 \cdots \mathrm{d}t_\nu \cdot \\ \cdot \mathrm{e}^{-\alpha(|t_1| + \cdots + |t_\nu|)} \langle \eta_0 | T_\varepsilon \{ V(t_1) \cdots V(t_\nu) A(t) \} | \eta_0 \rangle . \tag{5.56}$$

For the **proof**, we consider a *snapshot* of the ν-th term:

$$n \text{ times} \quad t_1, t_2, \ldots, t_n > t , \\ m \text{ times} \quad \bar{t}_1, \bar{t}_2, \ldots, \bar{t}_m < t ,$$

with $m + n = \nu$. In this case, we have:

$$T_\varepsilon\{\ldots\} = T_\varepsilon\{ V(t_1) \cdots V(t_n) \} A(t) T_\varepsilon \{ V(\bar{t}_1) \cdots V(\bar{t}_m) \} .$$

This situation can, for ν independent times, give rise to

$$\frac{\nu!}{n!\, m!}$$

possibilities which all make the same contribution. We then take **all** of the possibilities into account by summation over all the *conceivable* values of n and m, with $\nu = n + m$ as a boundary condition:

$$\sum_{\nu=0}^{\infty} \frac{1}{\nu!}\left(-\frac{\mathrm{i}}{\hbar}\right)^{\nu} \int_{-\infty}^{+\infty} \cdots \int \mathrm{d}t_1 \cdots \mathrm{d}t_\nu \, \mathrm{e}^{-\alpha(|t_1|+\cdots+|t_\nu|)} \cdot$$
$$\cdot \, T_\varepsilon\left\{V(t_1)\cdots V(t_\nu)\, A(t)\right\} =$$
$$= \sum_{\nu=0}^{\infty} \frac{1}{\nu!} \sum_{n,m}^{0\ldots\infty} \left(-\frac{\mathrm{i}}{\hbar}\right)^{\nu} \frac{\nu!}{n!\,m!} \delta_{\nu,n+m} \int_{t}^{\infty} \cdots \int \mathrm{d}t_1 \cdots \mathrm{d}t_n \cdot$$
$$\cdot \, \mathrm{e}^{-\alpha(|t_1|+\cdots+|t_n|)} T_\varepsilon\left\{V(t_1)\cdots V(t_n)\right\} A(t) \cdot$$
$$\cdot \int_{-\infty}^{t} \cdots \int \mathrm{d}\bar{t}_1 \cdots \mathrm{d}\bar{t}_m \, \mathrm{e}^{-\alpha(|\bar{t}_1|+\cdots+|\bar{t}_m|)} T_\varepsilon\left\{V(\bar{t}_1)\cdots V(\bar{t}_m)\right\} =$$
$$= \left[\sum_{n=0}^{\infty} \frac{1}{n!}\left(-\frac{\mathrm{i}}{\hbar}\right)^{n} \int_{t}^{\infty} \cdots \int \mathrm{d}t_1 \cdots \mathrm{d}t_n \, \mathrm{e}^{-\alpha(|t_1|+\cdots+|t_n|)} \cdot \right.$$
$$\left. \cdot \, T_\varepsilon\left\{V(t_1)\cdots V(t_n)\right\}\right] A(t) \left[\sum_{m=0}^{\infty} \frac{1}{m!}\left(-\frac{\mathrm{i}}{\hbar}\right)^{m} \cdot \right.$$
$$\left. \cdot \int_{-\infty}^{t} \cdots \int \mathrm{d}\bar{t}_1 \cdots \mathrm{d}\bar{t}_m \, \mathrm{e}^{-\alpha(|\bar{t}_1|+\cdots+|\bar{t}_m|)} T_\varepsilon\left\{V(\bar{t}_1)\cdots V(\bar{t}_m)\right\}\right] .$$

The comparison with (5.27) then yields:

$$\sum_{\nu=0}^{\infty} \frac{1}{\nu!}\left(-\frac{\mathrm{i}}{\hbar}\right)^{\nu} \int_{-\infty}^{+\infty} \cdots \int \mathrm{d}t_1 \cdots \mathrm{d}t_\nu \, \mathrm{e}^{-\alpha(|t_1|+\cdots+|t_\nu|)} \cdot$$
$$\cdot \, T_\varepsilon\left\{V(t_1)\cdots V(t_\nu)\, A(t)\right\} =$$
$$= U_\alpha(\infty, t) A(t) U_\alpha(t, -\infty) . \tag{5.57}$$

Together with (5.54), this relation proves the assertion of (5.56).

The same train of thought allows us to rearrange (5.55), also. We merely have to divide the integration variables into three groups. This leads to:

$$\begin{aligned}&\left\langle E_0 \middle| T_\varepsilon\left\{A^{\mathrm{H}}(t)B^{\mathrm{H}}\left(t'\right)\right\} \middle| E_0\right\rangle = \\ &= \lim_{\alpha\to 0} \frac{1}{\left\langle \eta_0 \mid S_\alpha \mid \eta_0\right\rangle} \sum_{\nu=0}^{\infty} \frac{1}{\nu!}\left(-\frac{\mathrm{i}}{\hbar}\right)^\nu \int_{-\infty}^{+\infty}\cdots\int \mathrm{d}t_1\cdots\mathrm{d}t_\nu \,\cdot \\ &\cdot\, \mathrm{e}^{-\alpha(|t_1|+\cdots+|t_\nu|)}\left\langle \eta_0 \middle| T_\varepsilon\left\{V\left(t_1\right)\cdots V\left(t_\nu\right)A(t)B\left(t'\right)\right\} \middle| \eta_0\right\rangle .\end{aligned} \tag{5.58}$$

With this, we can now specifically cast the causal $T = 0$ Green's function in a form which will prove to be convenient for a diagrammatic perturbation theory:

Causal one-electron Green's function
($T = 0$):

$$\begin{aligned}&\mathrm{i}G^{\mathrm{c}}_{k\sigma}\left(t, t'\right) = \\ &= \lim_{\alpha\to 0} \frac{1}{\left\langle \eta_0 \mid S_\alpha \mid \eta_0\right\rangle} \sum_{\nu=0}^{\infty} \frac{1}{\nu!}\left(-\frac{\mathrm{i}}{\hbar}\right)^\nu \int_{-\infty}^{+\infty}\cdots\int \mathrm{d}t_1\cdots\mathrm{d}t_\nu \,\cdot \\ &\cdot\, \mathrm{e}^{-\alpha(|t_1|+\cdots+|t_\nu|)}\left\langle \eta_0 \middle| T_\varepsilon\left\{V\left(t_1\right)\cdots V\left(t_\nu\right)a_{k\sigma}(t)a^+_{k\sigma}\left(t'\right)\right\} \middle| \eta_0\right\rangle .\end{aligned} \tag{5.59}$$

The denominator $\langle \eta_0 \mid S_\alpha \mid \eta_0\rangle$ has here a form analogous to that of the numerator, except that the operators $a_{k\sigma}$, $a^+_{k\sigma}$ are lacking in the argument of the T_ε operator.

5.1.4 Exercises

Exercise 5.1.1 Let $\mathcal{P} = |\eta\rangle\langle\eta|$ be the projection operator onto the eigenstate $|\eta\rangle$ of the Hamiltonian H. Show that $\mathcal{P}$ and the orthogonal projector $\mathcal{Q} = \mathbf{1} - \mathcal{P}$ commute with H.

Exercise 5.1.2 If

$$H = H_0 + \lambda v = H(\lambda)$$

is the Hamiltonian of a system of particles with interactions, then the (normalised) ground state $|E_0\rangle$ and the ground-state energy E_0 will be functions of the coupling constant λ. Show that for the level shift of the *unperturbed* ground state $|\eta_0\rangle$, the following relation holds due to the interaction λv:

$$\Delta E_0(\lambda) = E_0(\lambda) - \eta_0 = \int_0^\lambda \frac{\mathrm{d}\lambda'}{\lambda}\left\langle E_0\left(\lambda'\right) \middle| \lambda' v \middle| E_0\left(\lambda'\right)\right\rangle .$$

Exercise 5.1.3 Electrons in a valence band which are interacting with an antiferromagnetically ordered, localised spin system can be described by the following simplified s-f model:

$$H = H_0 + H_1\,; \quad H_0 = \sum_{\substack{k,\sigma \\ \alpha,\beta}} \varepsilon_{\alpha\beta}(\boldsymbol{k}) a^+_{\boldsymbol{k}\sigma\alpha} a_{\boldsymbol{k}\sigma\beta}\,;$$

$$H_1 = -\frac{1}{2} g \sum_{\boldsymbol{k},\sigma,\alpha} z_\sigma \left\langle S^z_\alpha \right\rangle a^+_{\boldsymbol{k}\sigma\alpha} a_{\boldsymbol{k}\sigma\alpha}\,.$$

$\alpha = \mathrm{A}, \mathrm{B}$ denotes the two chemically equivalent ferromagnetic sublattices A and B:

$$\left\langle S^z_{\mathrm{A}} \right\rangle = -\left\langle S^z_{\mathrm{B}} \right\rangle = \left\langle S^z \right\rangle.$$

The Bloch energies

$$\varepsilon_{\mathrm{AA}}(\boldsymbol{k}) = \varepsilon_{\mathrm{BB}}(\boldsymbol{k}) = \varepsilon(\boldsymbol{k}); \qquad \varepsilon_{\mathrm{AB}}(\boldsymbol{k}) = \varepsilon^*_{\mathrm{BA}}(\boldsymbol{k}) = t(\boldsymbol{k})$$

are assumed to be known, where $\boldsymbol{k}$ is a wavenumber within the first Brillouin zone of one of the two equivalent sublattices.

1. Compute the eigenvalues and eigenstates of the *unperturbed* operator H_0.
2. Calculate the energy corrections to first and second order in the Schrödinger perturbation theory.
3. Calculate the energy corrections to first and second order in the Brillouin-Wigner perturbation theory.
4. Compare the results of part 5.2. and 5.3. with the exact energy eigenvalues.

5.2 Wick's Theorem

5.2.1 The Normal Product

In order to be specific, we shall concentrate in this section exclusively on

Fermi systems,

which are subject to a **pair interaction** of the form

$$V(t) = \frac{1}{2} \sum_{\substack{kl \\ mn}} v(kl; nm) a^+_k(t) a^+_l(t) a_m(t) a_n(t)\,. \tag{5.60}$$

The most obvious realisation would be the electron-electron Coulomb interaction with $k \equiv (\boldsymbol{k}, \sigma)$,... The operators are given in their Dirac representation, whereby the time dependence is in fact trivial. According to the **Baker-Hausdorff theorem** (see Ex. 3.1.2), we have:

$$a_k(t) = \exp\left(\frac{\mathrm{i}}{\hbar}\mathcal{H}_0 t\right) a_k \exp\left(-\frac{\mathrm{i}}{\hbar}\mathcal{H}_0 t\right) = \sum_{n=0}^{\infty} \frac{1}{n!}\left(-\frac{\mathrm{i}}{\hbar}t\right)^n L^n(\mathcal{H}_0)\, a_k\,, \tag{5.61}$$

$$L(\mathcal{H}_0)\, a_k = [a_k, \mathcal{H}_0]_- = \left(\varepsilon(\boldsymbol{k}) - \mu\right) a_k\,, \tag{5.62}$$

$$\begin{aligned} L^n(\mathcal{H}_0)\, a_k &= \Big[\underbrace{\ldots[[a_k, \mathcal{H}_0]_-, \mathcal{H}_0]_- \ \ldots, \mathcal{H}_0}_{n\text{-fold}}\Big]_- = \\ &= \left(\varepsilon(\boldsymbol{k}) - \mu\right)^n a_k\,. \end{aligned} \tag{5.63}$$

This implies that:

$$a_k(t) = \exp\left(-\frac{\mathrm{i}}{\hbar}\left(\varepsilon(\boldsymbol{k}) - \mu\right)t\right) a_k\,, \tag{5.64}$$

$$a_k^+(t) = \exp\left(\frac{\mathrm{i}}{\hbar}\left(\varepsilon(\boldsymbol{k}) - \mu\right)t\right) a_k^+\,. \tag{5.65}$$

According to (5.59), our task consists of finding expectation values of the following form:

$$\begin{aligned} &\langle \eta_0 |\, T_\varepsilon \big\{ a_{k_1}^+(t_1)\, a_{l_1}^+(t_1)\, a_{m_1}(t_1)\, a_{n_1}(t_1) \cdots \\ &\cdots a_{k_n}^+(t_n)\, a_{l_n}^+(t_n)\, a_{m_n}(t_n)\, a_{n_n}(t_n)\, a_{k\sigma}(t) a_{k\sigma}^+(t') \big\} \,|\, \eta_0 \rangle\,. \end{aligned} \tag{5.66}$$

We shall try to rewrite these products in such a way that the application of the operators to the ground state $|\eta_0\rangle$ of the non-interacting system becomes generally feasible. To this end, we introduce new operators. In the ground state $|\eta_0\rangle$, all the levels within the *Fermi sphere* (of radius k_F in $\boldsymbol{k}$-space) are occupied. The operator

$$\gamma_{k\sigma}^+ = \begin{cases} a_{k\sigma}^+ & \text{for} \quad |\boldsymbol{k}| > k_F\,, \\ a_{k\sigma} & \text{for} \quad |\boldsymbol{k}| \le k_F \end{cases} \tag{5.67}$$

thus creates a particle outside the Fermi sphere, and a hole within it. The corresponding annihilation is effected by the operator $\gamma_{k\sigma}$:

$$\gamma_{k\sigma} = \begin{cases} a_{k\sigma} & \text{for} \quad |\boldsymbol{k}| > k_F\,, \\ a_{k\sigma}^+ & \text{for} \quad |\boldsymbol{k}| \le k_F\,, \end{cases} \tag{5.68}$$

For the γ's, the same fundamental commutation relations naturally hold as for the a's. When $|\boldsymbol{k}| > k_F$, $\gamma_{k\sigma}$ and $\gamma_{k\sigma}^+$ are creation and annihilation operators for particles; when $|\boldsymbol{k}| \le k_F$, they apply to holes. Due to

$$\gamma_{k\sigma}\, |\eta_0\rangle = 0\,, \tag{5.69}$$

$|\eta_0\rangle$ is also referred to as the **Fermi vacuum** or the *vacuum state*. We now introduce the so-called

normal product $N\left(\cdots \gamma_k^+ \cdots \gamma_l\right)$

of a series of such creation and annihilation operators γ^+, γ relative to the Fermi vacuum by means of the prescription that all the creation operators γ^+ must stand to the left of all the annihilation operators γ. Each permutation of two operators required for this prescription introduces a factor of (-1). The ordering of the γ's among themselves and of the γ^+'s among themselves is irrelevant.

Examples

$$
\begin{aligned}
N\left(\gamma_1\gamma_2^+\gamma_3\right) &= (-1)\gamma_2^+\gamma_1\gamma_3 = \\
&= (-1)^2\gamma_2^+\gamma_3\gamma_1 = \\
&= N\left(\gamma_2^+\gamma_3\gamma_1\right) = \\
&= (-1)^3 N\left(\gamma_3\gamma_2^+\gamma_1\right).
\end{aligned}
$$

If the "original" a and a^+ are found in the argument of N, they are to be interpreted according to (5.67) and (5.68) as γ or γ^+, respectively.

It is important for our purposes that:

$$\langle\eta_0|\, N\left(\cdots\gamma_k^+\cdots\gamma_l\cdots\right)|\eta_0\rangle = 0\,. \tag{5.70}$$

A decomposition of T products into N products in (5.66) would thus be desirable, and we therefore will try to achieve it in the following. We define the

(contraction)

$$\underbrace{A(t)B}\left(t'\right) \equiv T_\varepsilon\left\{A(t)B\left(t'\right)\right\} - N\left\{A(t)B\left(t'\right)\right\}. \tag{5.71}$$

If the operators A and B are both creation operators or both annihilation operators, then the contraction is clearly zero. Thus, only the following two cases are interesting:

$$\underbrace{\gamma_k(t)\gamma_{k'}^+}\left(t'\right) = \begin{cases} \gamma_k(t)\gamma_{k'}^+\left(t'\right) + \gamma_{k'}^+\left(t'\right)\gamma_k(t) & \text{for} \quad t > t'\,, \\ 0 & \text{for} \quad t < t'\,, \end{cases} \tag{5.72}$$

$$\underbrace{\gamma_{k'}^+\left(t'\right)\gamma_k}(t) = \begin{cases} -\gamma_k(t)\gamma_{k'}^+\left(t'\right) - \gamma_{k'}^+\left(t'\right)\gamma_k(t) & \text{for} \quad t > t'\,, \\ 0 & \text{for} \quad t < t'\,. \end{cases} \tag{5.73}$$

Since the time dependence of the γ's is trivial due to (5.64) and (5.65), and furthermore for $k \neq k'$ all creation and annihilation operators anticommute, the contractions given above are all zero for $k \neq k'$.

We formulate (5.72) and (5.73) once more explicitly in terms of the *original* operators a_k, a_k^+:

$|\boldsymbol{k}| > k_\mathrm{F}$:

$$\underbrace{a_k(t)a_{k'}^+}\left(t'\right) = \delta_{kk'} \begin{cases} \exp\left[-\frac{\mathrm{i}}{\hbar}\left(\varepsilon(\boldsymbol{k}) - \mu\right)\left(t - t'\right)\right] & \text{for} \quad t > t'\,, \\ 0 & \text{for} \quad t < t'\,, \end{cases} \tag{5.74}$$

$$\underbrace{a_{k'}^+\left(t'\right)a_k}(t) = \delta_{kk'} \begin{cases} -\exp\left[-\frac{\mathrm{i}}{\hbar}\left(\varepsilon(\boldsymbol{k}) - \mu\right)\left(t - t'\right)\right] & \text{for} \quad t > t'\,, \\ 0 & \text{for} \quad t < t'\,, \end{cases} \tag{5.75}$$

$|\boldsymbol{k}| \leq k_{\mathrm{F}}$:

$$\underbracket{a_k(t)a^+_{k'}}\left(t'\right) = \delta_{kk'} \begin{cases} 0 & \text{for} \quad t > t'\,, \\ -\exp\left[-\frac{\mathrm{i}}{\hbar}\left(\varepsilon(\boldsymbol{k}) - \mu\right)\left(t - t'\right)\right] & \text{for} \quad t < t'\,, \end{cases} \tag{5.76}$$

$$\underbracket{a^+_{k'}\left(t'\right) a_k}(t) = \delta_{kk'} \begin{cases} 0 & \text{for} \quad t > t'\,, \\ \exp\left[-\frac{\mathrm{i}}{\hbar}\left(\varepsilon(\boldsymbol{k}) - \mu\right)\left(t - t'\right)\right] & \text{for} \quad t < t'\,, \end{cases} \tag{5.77}$$

Now, from (3.204), for the *free*, causal Green's function ($\varepsilon = -1, T = 0$), we find:

$$\begin{aligned} \mathrm{i}G^{0,\,\mathrm{c}}_{k\sigma}\left(t - t'\right) = &\left\{\Theta\left(t - t'\right)\left(1 - \langle n_{k\sigma}\rangle^{(0)}\right) - \Theta\left(t' - t\right)\langle n_{k\sigma}\rangle^{(0)}\right\} \cdot \\ &\cdot \exp\left[-\frac{\mathrm{i}}{\hbar}\left(\varepsilon(\boldsymbol{k}) - \mu\right)\left(t - t'\right)\right]. \end{aligned} \tag{5.78}$$

Comparison yields:

$$\underbracket{a_k(t)a^+_k}\left(t'\right) = \mathrm{i}G^{0,\,\mathrm{c}}_k\left(t - t'\right), \tag{5.79}$$

$$\underbracket{a^+_k\left(t'\right) a_k}(t) = -\mathrm{i}G^{0,\,\mathrm{c}}_k\left(t - t'\right). \tag{5.80}$$

The T_ε products of two time dependent creation and annihilation operators thus decompose into normal products which vanish on taking averages over the free ground state $|\eta_0\rangle$, and contractions which are simply free, causal Green's functions. (The argument of the Green's function always contains the *annihilation time* minus the *creation time*). Contractions are here c numbers, that is they are not operators:

$$\left\langle \eta_0 \middle| \underbracket{a_k(t)a^+_k}\left(t'\right) \middle| \eta_0 \right\rangle = \underbracket{a_k(t)a^+_k}\left(t'\right). \tag{5.81}$$

These important relations hold for $t \neq t'$. For $t = t'$, the causal functions are **not** defined.

The problem of *simultaneity* always arises when in a typical term such as (5.66), a creation operator and an annihilation operator which stem from the same interaction potential $V(t)$ are contracted with each other. By convention, the T_ε products leave the operators in their *natural* order, i.e. the creation operator is to the left of the annihilation operator. This means that

$$t_{\text{annihilation operator}} - t_{\text{creation operator}} = 0^-$$

and thus:

$$\underbracket{a_k(t)a^+_k}(t) = \mathrm{i}G^{0,\,\mathrm{c}}_k(0^-); \quad \underbracket{a^+_k(t)a_k}(t) = -\mathrm{i}G^{0,\,\mathrm{c}}_k(0^-), \tag{5.82}$$

where, from (5.78), we have:

$$\mathrm{i}G^{0,\,\mathrm{c}}_k(0^-) = -\langle n_k\rangle^{(0)}. \tag{5.83}$$

5.2.2 Wick's Theorem

Equation (5.71) yields the desired decomposition of a T_ε product of two factors into normal products and contractions. In general, this follows from Wick's theorem for a T_ε product of n factors.

Theorem 5.2.1

$$U, V, W, \ldots, X, Y, Z: \quad \text{Fermionic operators}\,.$$

We then have:

$$
\begin{aligned}
T_\varepsilon(UVW\ldots XYZ) = N(UVW\ldots XYZ)+ & \\
\begin{pmatrix}\text{N-products}\\ \text{with one}\\ \text{contraction}\end{pmatrix} \quad & +N(\underbracket{UV}W\ldots XYZ)+ \\
& +N(\underbracket{UVW}\ldots XYZ)+ \\
& +\cdots+ \\
\begin{pmatrix}\text{N-products}\\ \text{with two}\\ \text{contractions}\end{pmatrix} \quad & +N(\underbracket{UV}W\ldots X\underbracket{YZ})+ \\
& +N(\underbracket{UVW}\ldots X\underbracket{YZ})+ \\
& +\cdots+ \\
& +\{\textit{total pairing}\}
\end{aligned}
\tag{5.84}
$$

The term **total pairing** refers to the the complete decomposition of the operator product $UVW\ldots XYZ$ in all possible ways into contractions, which naturally presupposes an even number of operators.

Wick's theorem, as quoted above, is an operator identity. However, its true usefulness only becomes clear on carrying out an averaging process over the ground state $|\eta_0\rangle$, when all the normal products vanish:

$$\langle\eta_0|\,T_\varepsilon(UVW\ldots XYZ)\,|\eta_0\rangle = \{\textit{total pairing}\}\,. \tag{5.85}$$

Here, according to (5.79) through (5.82), a *total pairing* represents a sum of products of causal Green's functions of the free systems.

Before we prove the theorem, let us consider as an *example* the T_ε product of four operators:

$$
\begin{aligned}
T_\varepsilon(UVWX) = N(UVWX)+ & \\
&+\Big(\underbracket{UV}\,N(WX)+\underbracket{VW}\,N(UX)+\\
&+\underbracket{WX}\,N(UV)+\underbracket{UX}\,N(VW)-\\
&-\underbracket{UW}\,N(VX)-\underbracket{VX}\,N(UW)\Big)+\\
&+\Big\{\underbracket{UV}\,\underbracket{WX}-\underbracket{UW}\,\underbracket{VX}+\underbracket{UX}\,\underbracket{VW}\Big\}\,.
\end{aligned}
\tag{5.86}
$$

Contractions, as c numbers, can of course be extracted from the N product as prefactors. To do this, we agree upon a **sign convention**, that the operators to be contracted must be adjacent in the N product. The required permutations of pairs of Fermion operators each yield a factor of (-1).

For the **proof** of the theorem, we can permit ourselves the assumption that the operators $U, V, W, \dots, X, Y, Z$ are already time-ordered within the argument of the T_ε operator. If this is not yet the case, then we reorder the argument of the T_ε operator correspondingly. This contributes for p permutations a factor $(-1)^p$ to the left-hand side of (5.84). The same factor occurs also in each summand on the right-hand side when we reorder the arguments of the N products correspondingly.

The decisive point for the proof of Wick's theorem is the following:

Lemma 5.2.1 Let Z be an operator at an **earlier** time than $U, V, W, \dots, X, Y$. Then we have:

$$
\begin{aligned}
N(UV\dots XY)Z = N(UV\dots X&\underbracket{YZ})+\\
&+N(UV\dots \underbracket{XYZ})+\\
&+\dots+\\
&+N(\underbracket{UV\dots XYZ})+\\
&+N(UV\dots XYZ)\,.
\end{aligned}
\tag{5.87}
$$

Proof

1. Z: **annihilation operator** on the Fermi vacuum. Then:

$$
\underbracket{UZ} = T_\varepsilon(UZ) - N(UZ) = UZ - UZ = 0\,.
$$

(5.87) is thus clearly proven, provided that

$$
N(UV\dots XY)Z = N(UV\dots XYZ)
$$

can be assumed. This is however certainly the case, since Z is presumed to be an annihilation operator.

2. Z: **Creation operator** on the Fermi vacuum.

We can assume that the operators $U, V, \ldots, X, Y$ in (5.87) are already normally ordered. If this is not the case, we reorder them suitably in the arguments of the N products, which introduces into *each* summand a factor of $(-1)^m$. If, however, the normal order is already present, then we can take the operators $U, V, \ldots\, X, Y$ all together to be annihilation operators on the Fermi vacuum without loss of generality. We can namely later *fill in creation operators* from the left, without changing the normal ordering. The additional terms which then appear on the right contain contractions of the creation operator Z with one of the *filled-in* creation operators, and are thus all equal to zero.

We therefore must prove the lemma only for the case that Z is a creation operator and $U, V, \ldots, X, Y$ are annihilation operators. We carry out the proof by **complete induction:**

$n = 2$:

We wish to show that:

$$N(Y)Z = N(\underbracket{YZ}) + N(YZ).$$

The contraction is a c number. Therefore, the relation is proved if and only if

$$YZ = \underbracket{YZ} + N(YZ)$$

holds. With the precondition $t_y > t_z$, this is however just the definition (5.71) of the contraction:

$$T_\varepsilon(YZ) = \underbracket{YZ} + N(YZ).$$

$n \Longrightarrow n + 1$:

We multiply (5.87) from the left by an additional annihilation operator D: since $U, V, \ldots, X, Y$ are also annihilation operators, it follows initially that:

$$DN(UV \ldots XY)Z = N(DUV \ldots XY)Z.$$

From the condition of the induction, (5.87) is valid for n operators:

$$\begin{aligned}
DN(UV \ldots XY)Z &= DN(UV \ldots X\underbracket{YZ}) + DN(UV \ldots \underbracket{XYZ}) + \\
&\quad + \cdots + DN(\underbracket{UV \ldots XYZ}) + DN(U \ldots Z) = \\
&= N(DUV \ldots X\underbracket{YZ}) + N(DUV \ldots \underbracket{XYZ}) + \\
&\quad + \cdots + N(D\underbracket{UV \ldots XYZ}) + DN(UV \ldots XYZ).
\end{aligned}$$

In the last step, we made use of the fact that the only creation operator in the arguments of the N products, the operator Z, occurs only within contractions, i.e. it no longer acts as an operator. We can thus include the annihilation operator D, as indicated, within the N products. Eq. (5.87) is then proved, if

$$DN(UV\dots XYZ) = N(\underbrace{DU\dots YZ}) + N(DU\dots YZ)$$

can be demonstrated to hold. Now, however, we have:

$$\begin{aligned}
DN(UV\dots YZ) &= (-1)^n DZUV\dots Y = \\
&\overset{(t_D > t_Z)}{=} (-1)^n \underbrace{DZ}UV\dots Y + (-1)^n N(DZ)UV\dots Y = \\
&= (-1)^{2n} N(\underbrace{DUV\dots YZ}) + (-1)^{n+1} N(ZD)UV\dots Y = \\
&= N(\underbrace{DUV\dots YZ}) + (-1)^{n+1} N(ZDUV\dots Y) = \\
&= N(\underbrace{DUV\dots YZ}) + (-1)^{n+1}(-1)^{n+1} N(DUV\dots YZ) = \\
&= N(\underbrace{DUV\dots YZ}) + N(DUV\dots YZ) .
\end{aligned}$$

This proves the lemma (5.87).

One can immediately see that (5.87) remains valid even if the normal product on the left-hand side already contains one or more contractions:

$$\begin{aligned}
N(U\underbrace{V\dots X}Y)Z &= N(U\underbrace{V\dots X}\underbrace{YZ}) + \dots + \\
&+ \dots + N(\underbrace{U\underbrace{V\dots X}YZ}) + N(U\underbrace{V\dots X}YZ) .
\end{aligned} \tag{5.88}$$

We can now proceed to prove Wick's theorem (5.84) once again, using complete induction. We employ the fact explained above that the operators can be assumed to be already time-ordered.

$\boxed{n = 2 :}$

$$T_\varepsilon(UV) = UV = N(UV) + \underbrace{UV} .$$

This is nothing other than the definition of a contraction.

$\boxed{n \Longrightarrow n+2:}$

We multiply (5.84) from the right with AB, whereby

$$t_U > t_V > t_W > \cdots > t_X > t_Y > t_Z > t_A > t_B$$

can be assumed, and then we apply the lemma (5.87) twice.

$$\begin{aligned}
T_\varepsilon(UV\ldots YZ)AB = T_\varepsilon(UV\ldots YZAB) =& \\
&= N(UV\ldots YZ)AB + N(\underbracket{UV}\ldots YZ)AB + \\
&\quad + \cdots = \\
&= N(UV\ldots YZA)B + N(UV\ldots Y\underbracket{ZA})B + \\
&\quad + \cdots + \\
&\quad + N(\underbracket{UV}\ldots YZA)B + N(\underbracket{UV}\ldots Y\underbracket{ZA})B + \\
&\quad + \cdots + \\
&\quad + \{\mathit{total\ pairing}\}_n \big(\underbracket{AB} + N(AB)\big) = \\
&= N(UV\ldots YZAB) + N(UV\ldots YZ\underbracket{AB}) + \\
&\quad + \ldots + \\
&\quad + N(UV\ldots Y\underbracket{ZA}B) + N(UV\ldots \underbracket{Y\underbracket{ZA}B}) + \\
&\quad + \ldots + \\
&\quad + \{\mathit{total\ pairing}\}_n N(AB) + \\
&\quad + \{\mathit{total\ pairing}\}_n \underbracket{AB} = \\
&= N(UV\ldots YZAB) + N(UV\ldots YZ\underbracket{AB}) + \\
&\quad + N(UV\ldots Y\underbracket{ZAB}) + \cdots + \\
&\quad + N(\underbracket{UV}\ldots YZ\underbracket{AB}) + N(\underbracket{UV}\ldots Y\underbracket{ZAB}) + \cdots + \\
&\quad + \cdots + \\
&\quad + \{\mathit{total\ pairing}\}_{n+2}\,.
\end{aligned}$$

The $\{total\ pairing\}_{n+2}$ results from $\{total\ pairing\}_n \underset{\sqcup}{AB}$ and from those terms for which in the n-th step all the operators are paired except for two in the argument of the N products.

Thus we have proved the fundamental Wick theorem!

5.2.3 Exercises

Exercise 5.2.1

1. Write the time-ordered product

$$T_\varepsilon\left\{a_{k\sigma}(t_1)\, a^+_{l\sigma'}(t_2)\, a_{m\sigma}(t_3)\, a^+_{n\sigma'}(t_3)\right\}$$

   in terms of normal products and suitable contractions.
2. Express the expectation value of the time-ordered product in part 1. in the ground state $|\eta_0\rangle$ of the unperturbed system in terms of products of the free causal Green's functions.

Exercise 5.2.2 Evaluate explicitly the expectation value of the time-ordered product in part 1. of Ex. 5.2.1 in the ground state $|\eta_0\rangle$ of the unperturbed system for the special case $\boldsymbol{k} = \boldsymbol{l} = \boldsymbol{m} = \boldsymbol{n}$, $\sigma = \sigma'$ for

1. $t_1 > t_2 > t_3$,
2. $t_1 > t_3 > t_2$.

Check the results by direct calculation of the expectation values, i.e. without using Wick's theorem.

5.3 Feynman Diagrams

Wick's theorem shows the way to construct perturbation expansions for the various expectation values. The main task consists of forming all imaginable contractions from the given products of creation and annihilation operators, whereby these are related according to (5.79) and (5.82) directly to the unperturbed causal Green's functions. This task requires as a rule considerable effort, which however can be effectively reduced by introducing Feynman graphs.

We start with the expectation value of the time-evolution operator,

$$\langle \eta_0 | U_\alpha(t, t') | \eta_0 \rangle ,$$

which is also called the **vacuum amplitude**. Other examples will then follow quite naturally.

5.3.1 Perturbation Expansion for the Vacuum Amplitude

According to (5.27), we must calculate the following:

$$\langle \eta_0 | U_\alpha(t, t') | \eta_0 \rangle = 1 + \sum_{n=1}^{\infty} \langle \eta_0 | U_\alpha^{(n)}(t, t') | \eta_0 \rangle , \tag{5.89}$$

$$\langle \eta_0 | U_\alpha^{(n)}(t, t') | \eta_0 \rangle = \frac{1}{n!}\left(-\frac{\mathrm{i}}{\hbar}\right)^n \int_{t'}^{t} \cdots \int \mathrm{d}t_1 \cdots \mathrm{d}t_n \cdot \\ \cdot \mathrm{e}^{-\alpha(|t_1| + \cdots + |t_n|)} \langle \eta_0 | T_\varepsilon \{ V(t_1) \cdots V(t_n) \} | \eta_0 \rangle . \tag{5.90}$$

$V(t)$ is taken to be a pair interaction of the type (5.60). It will later prove expedient to insert a trivial integration into it:

$$V(t_1) = \frac{1}{2} \sum_{\substack{kl \\ mn}} v(kl; nm) \int_{-\infty}^{+\infty} \mathrm{d}t_1' \, \delta(t_1 - t_1') \, a_k^+(t_1) \, a_l^+(t_1') \, a_m(t_1') \, a_n(t_1) \tag{5.91}$$

As an example, we consider the first term of the perturbation expansion (5.89):

$$\langle \eta_0 | U_\alpha^{(1)}(t, t') | \eta_0 \rangle = -\frac{\mathrm{i}}{2\hbar} \int_{t'}^{t} \mathrm{d}t_1 \, \mathrm{e}^{-\alpha|t_1|} \sum_{klmn} \int_{-\infty}^{+\infty} \mathrm{d}t_1' \, \delta(t_1 - t_1') \cdot \\ \cdot v(kl; nm) \langle \eta_0 | T_\varepsilon \{ a_k^+(t_1) \, a_l^+(t_1') \, a_m(t_1') \, a_n(t_1) \} | \eta_0 \rangle .$$

To evaluate the matrix element, we make use of Wick's theorem:

$$\langle \eta_0 | T_\varepsilon \{ \ldots \} | \eta_0 \rangle = \underbrace{a_k^+(t_1) \, a_n(t_1)}\underbrace{a_l^+(t_1') \, a_m(t_1')} - \\ - \underbrace{a_k^+(t_1) \, a_m(t_1')}\underbrace{a_l^+(t_1') \, a_n(t_1)} = \\ = \left[-\mathrm{i} G_k^{0,\mathrm{c}}(0^-) \delta_{kn} \right] \left[-\mathrm{i} G_l^{0,\mathrm{c}}(0^-) \delta_{lm} \right] - \\ - \left[-\mathrm{i} G_k^{0,\mathrm{c}}(t_1' - t_1) \, \delta_{km} \right] \left[-\mathrm{i} G_l^{0,\mathrm{c}}(t_1 - t_1') \, \delta_{ln} \right] .$$

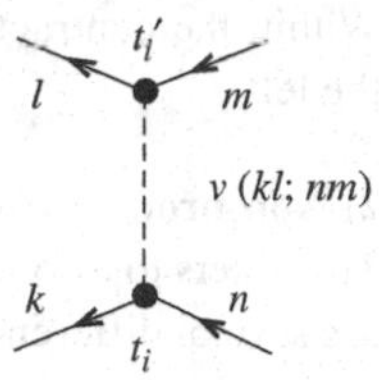

Fig. 5.1 The annotation of a vertex as a basic element of a Feynman diagram

With (5.83), this yields after insertion:

$$\left\langle \eta_0 \right| U_\alpha^{(1)}\left(t, t'\right) \left| \eta_0 \right\rangle =$$
$$= \frac{\mathrm{i}}{2\hbar} \int\limits_{t'}^{t} \mathrm{d}t_1\, \mathrm{e}^{-\alpha|t_1|} \sum_{k,l} \langle n_k \rangle^{(0)} \langle n_l \rangle^{(0)} \left(v(kl; lk) - v(kl; kl) \right). \tag{5.92}$$

We want to visualise this result using diagrams. In the following, we wish to work out step by step a unique set of translation rules for the complicated terms in the perturbation expansion into so-called

Feynman graphs.

Vertex. The interaction is indicated by a dashed line (Fig. 5.1). The time indices t_i, t_i' serve only to distinguish the ends of the interaction line. Due to $\delta(t_i - t_i')$ in the integrands of (5.91), both points of course finally denote the same time. A line which enters a vertex point symbolises an annihilation operator, and a line which emerges from a vertex symbolises a creation operator.

A **contraction** is represented by a solid line with an arrow which connects two vertex points. We imagine a time axis with a time index which increases from left to right. We distinguish between:

1) Propagating Lines

The time argument of the Green's function, as per our previous convention, always contains *(annihilation time − creation time)*.

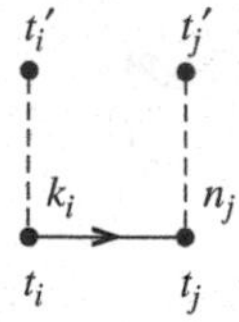

$$\Longleftrightarrow \quad \underbrace{a_{k_i}^+ (t_i)\, a_{n_j} (t_j)} =$$
$$= -\mathrm{i}G_{k_i}^{0,\,\mathrm{c}} \left(t_j - t_i\right) \delta_{k_i n_j}, \tag{5.93}$$

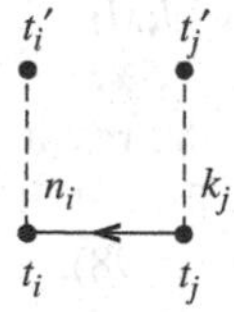

$$\Longleftrightarrow \quad \underbrace{a_{n_i} (t_i)\, a_{k_j}^+ (t_j)} =$$
$$= \mathrm{i}G_{k_j}^{0,\,\mathrm{c}} \left(t_i - t_j\right) \delta_{n_i k_j}. \tag{5.94}$$

Within the contraction, the operator appears before the time, which is placed further to the left.

2) Non-propagating Lines
This refers to a solid line which emerges from and reenters one and the same vertex. There are several different possibilities for this:

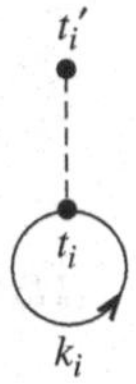

$$\Longleftrightarrow \quad \underbrace{a^+_{k_i}(t_i) a_{k_i}(t_i)} = \\ = -\mathrm{i}G^{0,\,c}_{k_i}(0^-)\delta_{k_i n_i} = \langle n_{k_i} \rangle^{(0)} \delta_{k_i n_i} \,. \tag{5.95}$$

This assignment is per convention; the arrow on the **bubble** is therefore in fact superfluous.

$$\Longleftrightarrow \quad \underbrace{a^+_{k_i}(t_i) a_{m_i}(t_i')} = \\ = -\mathrm{i}G^{0,\,c}_{k_i}(0^-)\delta_{k_i m_i} = \langle n_{k_i} \rangle^{(0)} \delta_{k_i m_i} \,. \tag{5.96}$$

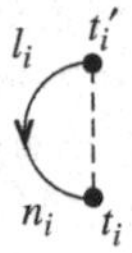

$$\Longleftrightarrow \quad \underbrace{a_{n_i}(t_i) a^+_{l_i}(t_i')} = \\ = \mathrm{i}G^{0,\,c}_{n_i}(0^-)\delta_{n_i l_i} = -\langle n_{l_i} \rangle^{(0)} \delta_{l_i n_i} \,. \tag{5.97}$$

We agree upon the convention that in the T_ε product, the contractions are always to be sorted in such a way that for same times, the operators with the "primed" times will be placed to the right of those with the "unprimed" times. Combining (5.96) with (5.97), we can see that within a contraction, the "primed" times can be permuted with the "unprimed" times. We will make use of this later.

The first term in the perturbation expansion for $U_\alpha(t, t')$ has only a single vertex. The solid lines can therefore be only non-propagating lines. The fourfold sum thus becomes a double sum:

$$\langle \eta_0 | U^{(1)}_\alpha(t, t') | \eta_0 \rangle = -\frac{\mathrm{i}}{2\hbar} \int\limits_{t'}^{t} \mathrm{d}t_1\, \mathrm{e}^{-\alpha|t_1|} \int\limits_{-\infty}^{+\infty} \mathrm{d}t_1' \delta(t - t_1) \,\cdot \\ \cdot \sum_{k,l} \left\{ \quad v(kl;kl) \quad + \quad v(kl;kl) \right\} . \tag{5.98}$$

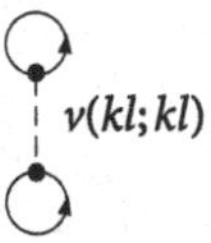

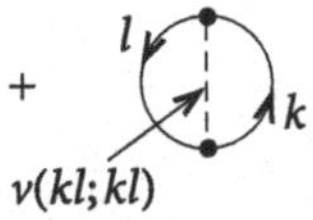

Applying the diagram rules listed above, we find directly the result (5.92). Since every vertex must be entered by two lines and two must emerge from it, it is clear that to first order, no additional diagrams are possible besides the two in (5.98).

For the first term in the perturbation expansion, the diagram representation is child's play; it becomes useful only for higher-order terms and for partial summations.

How many different graphs are possible with n vertices? We can see the answer as follows: for n vertices, there are $2n$ outgoing arrows. The first outgoing arrow then has $2n$ possibilities to end on a vertex as an ingoing arrow, whilst the second arrow then has only $(2n - 1)$ possibilities, the third has $(2n - 2)$ etc.:

n vertices $\iff$ $(2n)!$ different graphs for the vacuum amplitude.

However, not all of them must always be explicitly counted. Graphs which are related to each other simply by a permutation of the indices on an interaction line are naturally identical, since later, a summation will be carried out over **all** wavenumbers. Furthermore, those diagrams which differ only in the arrangement of the time indices are the same, since the integration is performed independently over all times. We shall later need to systematise this description to some extent.

Before we formulate the general diagram rules, we wish as an exercise to investigate the second term of the perturbation expansion in somewhat more detail:

$$\begin{aligned}\langle\eta_0|\,U_\alpha^{(2)}(t,t')\,|\eta_0\rangle &= \\ &= \frac{1}{2^2 2!}\left(-\frac{\mathrm{i}}{\hbar}\right)^2 \int_{t'}^{t}\!\cdots\!\int \mathrm{d}t_1\,\mathrm{d}t_1'\,\mathrm{d}t_2\,\mathrm{d}t_2'\; \mathrm{e}^{-\alpha(|t_1|+|t_2|)}\,\delta(t_1-t_1')\,\cdot \\ &\quad \cdot\,\delta(t_2-t_2')\sum_{k_1l_1m_1n_1}\;\sum_{k_2l_2m_2n_2} v(k_1l_1;n_1m_1)\,v(k_2l_2;n_2m_2)\,\cdot \\ &\quad \cdot\,\langle\eta_0|\,T_\varepsilon\{(2)\}\,|\eta_0\rangle\,.\end{aligned} \tag{5.99}$$

The *total pairing* of the time-ordered product in $\langle\eta_0 \mid T_\varepsilon\{(2)\} \mid \eta_0\rangle$ contains 24 terms:

$$a^+_{k_1}(t_1)a^+_{l_1}(t_1')a_{m_1}(t')a_{n_1}(t_1)a^+_{k_2}(t_2)a^+_{2l}(t_2')a_{m_2}(t_2')a_{n_2}(t_2)$$

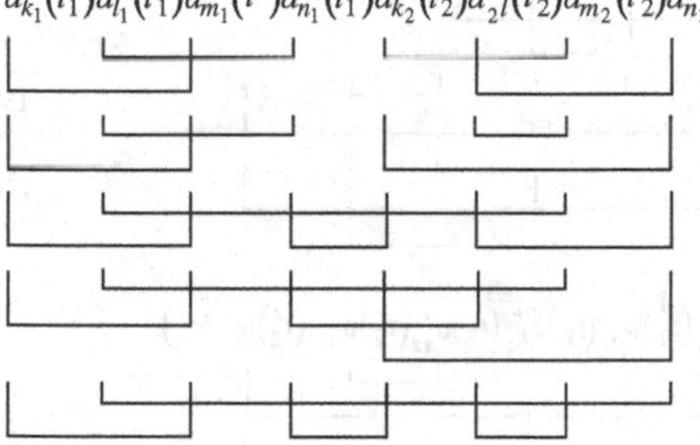

1)
2)
3)
4)
5)

$$a^+_{k_1}(t_1)a^+_{l_1}(t'_1)a_{m_1}(t')a_{n_1}(t_1)a^+_{k_2}(t_2)a^+_{2l}(t'_2)a_{m_2}(t'_2)a_{n_2}(t_2)$$

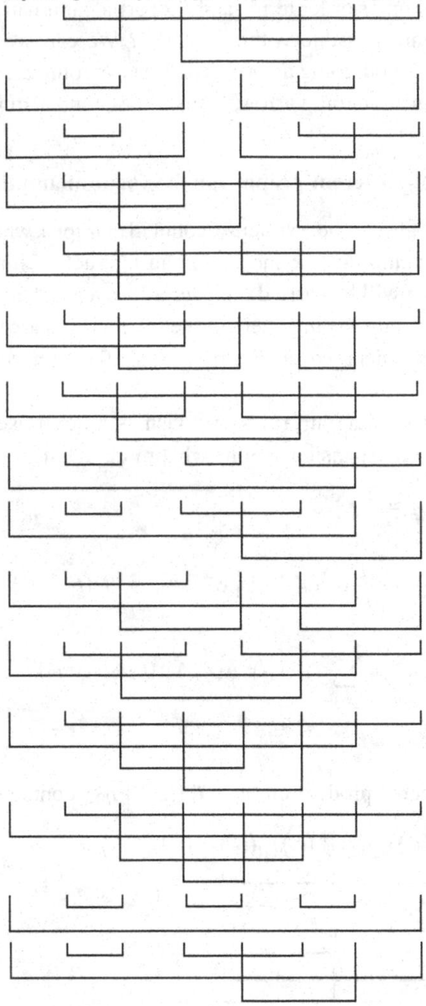

6)
7)
8)
9)
10)
11)
12)
13)
14)
15)
16)
17)
18)
19)
20)

$$a^+_{k_1}(t_1)a^+_{l_1}(t'_1)a_{m_1}(t'_1)a_{n_1}(t_1)a^+_{k_2}(t_2)a^+_{l_2}(t'_2)a_{m_2}(t'_2)a_{n_2}(t_2)$$

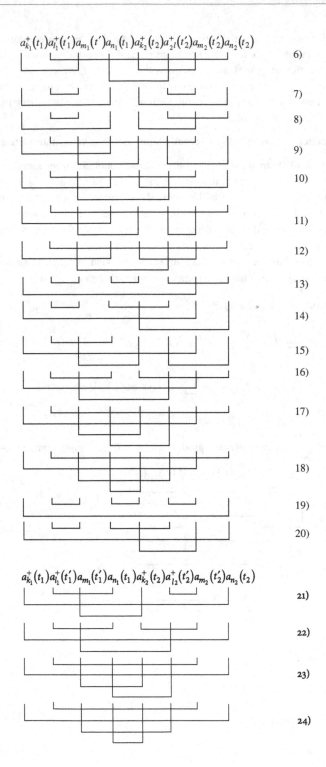

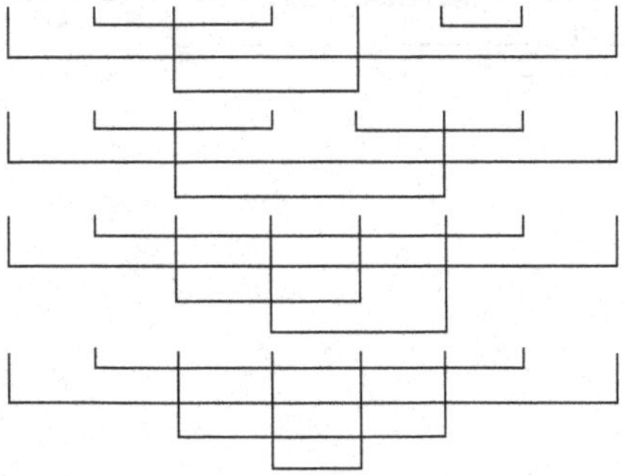

21)
22)
23)
24)

In evaluating the contractions indicated here, one must keep in mind that the operators to be contracted must be adjacent to one another. The pairwise permutations required to achieve this each introduce a factor of (-1). Furthermore, we agreed upon the convention that within a contraction, the operators must be arranged in such a way that the operator with the smaller time index stands to the left, and when the times are the same, the operator with the *unprimed* time is on the left. This sounds rather complicated, but it can be greatly simplified by using the **loop rule**, which we shall prove later.

We translate the above contributions to the *total pairing* into the *diagram language*:

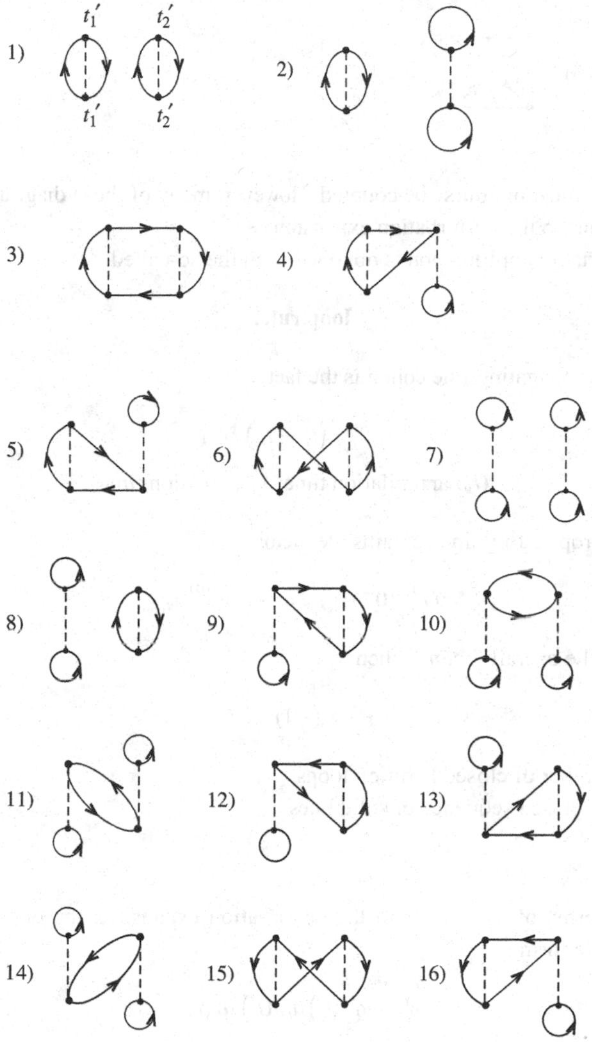

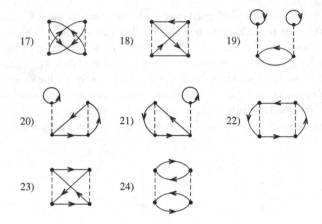

17) 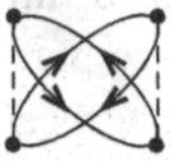 18)  19) 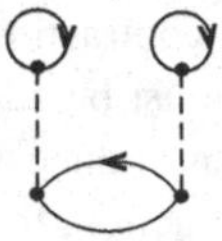

20) 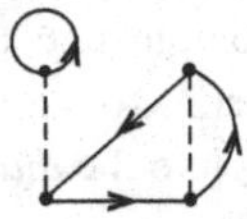 21) 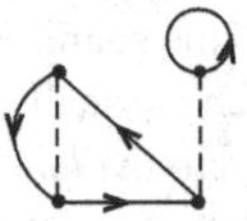 22) 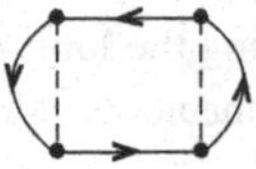

23) 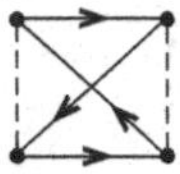 24) 

All 24 diagrams must of course be counted. However, many of these diagrams make identical contributions to the perturbation expansion.

A first important simplification is obtained from the so-called

loop rule.

1. Every solid propagating line contains the factor

$$\mathrm{i}G^{0,\,\mathrm{c}}_{k_\nu}\left(t_\nu - t_\mu\right)\delta_{k_\nu,k_\mu}$$

(t_ν: annihilation time; t_μ : creation time).

2. Every non-propagating line contains the factor

$$\mathrm{i}G^{0,\,\mathrm{c}}_{k_\nu}(0^-)\delta_{k_\nu k_\mu} = -\left\langle \eta_{k_\nu}\right\rangle^{(0)}\delta_{k_\nu k_\mu}\ . \tag{5.100}$$

3. The sign of the overall factor is then

$$(-1)^S\,,$$

with S = number of closed Fermion loops,
and a *loop* is closed sequence of solid lines.

Proof In the terms of n-th order in the perturbation expansion, the operators occur as four-tuples of the form

$$a_k^+(t)a_l^+\left(t'\right)a_m\left(t'\right)a_n(t)\,.$$

We can rewrite these without change of sign to

$$a_k^+(t)a_n(t)a_l^+\left(t'\right)a_m\left(t'\right)\,,$$

since within the T_ε product, this requires two permutations in each case. *Same-time* operators always enter a loop in the form

$$a_k^+(t)\,a_n(t)\;,$$

as long as it is not a bubble, which we will treat separately. Such operator products can be moved arbitrarily through the T_ε product without changes of sign. Then a loop can always be arranged as follows:

$$a^+(t_1)\,a\,(t_1)\,a^+\,(t_2)a\,(t_2)\cdots\,a^+(t_{n-1})a(t_{n-1})a^+\,(t_n)a\,(t_n)\,.$$

In this process, $a^+\,(t_1)\,a\,(t_1)$ is held fixed, whilst $a^+\,(t_n)\,a\,(t_n)$ is moved to the extreme right, $a^+(t_{n-1})a(t_{n-1})$ follows, etc. If now all the time indices in a contraction are different, then it corresponds to a propagating line. The inner contractions in the expression above then have an operator ordering which leads according to (5.94) to a contribution of the form 1. If *same-time* operators with *primed* and *unprimed* times are contracted, this corresponds to a non-propagating line of the form (5.97), which makes a contribution as in 2. This holds again for the inner contractions. The only exception is an outer contraction in which the operators to be contracted are arranged in the *wrong* order. Evaluating the whole loop using the prescriptions 1 and 2 will then lead to an additional factor of (-1).

If the diagram term consists of several loops, then one can reorder the operators in the T_ε product from the outset in such a way that in the *total pairing*, the loops are directly factored. This can be accomplished without sign changes, since each loop is of course constructed from an even number of operators.

A bubble $a_k^+(t)a_k(t)$ represents a special case of a loop.

From 2, it makes the contribution $-\,\langle n_k\rangle^{(0)}$, and from 3 an additional factor of (-1) occurs; thus all together $+\,\langle n_k\rangle^{(0)}$. This indeed agrees with (5.95). We have thus proved the loop rule.

"Preliminary" diagram rules. Terms of n-th order in the perturbation expansion for $\langle \eta_0 \mid U_\alpha(t,t') \mid \eta_0\rangle$:

All (!) of the diagrams with n vertices are to be drawn, whose end points are joined pairwise by solid, directed lines. The contribution of a diagram can then be computed as follows:

1. Vertex $i \Longleftrightarrow v(k_i l_i;\, n_i m_i)$.
2. Propagating line $\Longleftrightarrow \mathrm{i}G_{k_i}^{0,\,c}(t_i - t_j)\delta_{k_i,k_j}$.
3. Non-propagating line $\Longleftrightarrow -\left\langle n_{k_i}\right\rangle^{(0)}\,\delta_{k_i,k_j}$.
4. Factor $(-1)^S$; S = number of Fermion loops.
5. Summation over all wavenumbers and possibly spins $\ldots,k_i,l_i,m_i,n_i,\ldots$.
6. Insert δ-functions $\delta(t_i - t_i')$; and the switching-on factor $\exp\left[-\alpha(|t_1| + \cdots + |t_n|)\right]$.
7. Integrate over all t_i, t_i' from t' to t.
8. Include a factor $\frac{1}{n!}\left(-\frac{\mathrm{i}}{2\hbar}\right)^n$.

Examples Diagram 3)

$$
\begin{aligned}
3) &= \frac{1}{2!}\left(-\frac{i}{2\hbar}\right)^2 \int_{t'}^{t}\cdots\int dt_1\, dt_1'\, dt_2\, dt_2'\, \delta\left(t_1 - t_1'\right)\delta\left(t_2 - t_2'\right) \cdot \\
&\quad \cdot e^{-\alpha(|t_1|+|t_2|)} \sum_{\substack{k_1,\dots,n_1\\ k_2,\dots,n_2}} v\left(k_1\dots\right) v\left(k_2\dots\right)(-1) \cdot \\
&\quad \cdot \left(iG_{l_1}^{0,c}\left(t_2' - t_1'\right)\right)\left(iG_{n_1}^{0,c}\left(t_1 - t_2\right)\right)\left(-\left\langle n_{k_1}\right\rangle^{(0)}\right)\left(-\left\langle n_{n_2}\right\rangle^{(0)}\right) \cdot \\
&\quad \cdot \delta_{l_1, m_2}\delta_{n_1, k_2}\delta_{k_1, m_1}\delta_{n_2, l_2} = \\
&= \frac{+1}{2!}\left(-\frac{i}{2\hbar}\right)^2 \iint_{t'}^{t} dt_1\, dt_2\, e^{-\alpha(|t_1|+|t_2|)} \sum_{k_1, l_1, n_1, n_2} v\left(k_1 l_1; n_1 k_1\right) \cdot \\
&\quad \cdot v\left(n_1 n_2; n_2 l_1\right) G_{l_1}^{0,c}\left(t_2 - t_1\right) G_{n_1}^{0,c}\left(t_1 - t_2\right)\left\langle n_{k_1}\right\rangle^{(0)}\left\langle n_{n_2}\right\rangle^{(0)} .
\end{aligned}
$$

5.3.2 The Linked-Cluster Theorem

The procedure which we have thus far developed still seems to be too complicated. We want to simplify it further by making use of topology. What is in fact the meaning of

"all" diagrams with n vertices

in the rules given above? Among these, there are a number of diagrams which each make the same contribution to the perturbation expansion:

Diagrams **with the same structure** are those which can be converted into one another by exchanging their vertices and permuting the times at their vertices. With n vertices, there are $n!$ possible permutations of the vertices among themselves and 2^n permutations of *above* and *below* on the individual vertices. For a given diagram type with n vertices, there are thus

$2^n n!$ diagrams of the same structure,

which each make the same contribution to the perturbation expansion, since an independent summation over all wavenumbers and integration over all times will later be performed. The indices on the wavenumbers and on the times are only an aid to characterising the variables.

Examples

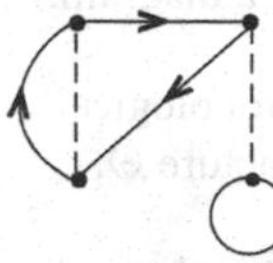

We can find all the diagrams with the *same structure* as the one sketched for example by carrying out the following prescription: Leave off the arrows and construct all diagrams by permutation of the *right* and *left* as well as of *above* and *below*:

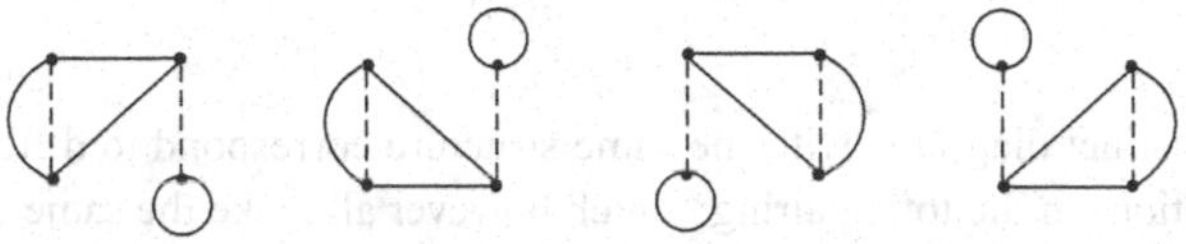

For each of these diagrams, there are now still two possibilities for the sense of rotation. All together, we thus have eight diagrams ($2^2 2!$) of the same structure.

Among the $2^n n!$ diagrams of the same structure, however, there are some which are already

topologically equivalent.

These are diagrams with certain symmetries, which mean that a permutation of certain vertices or permutation of the vertex points yields identical diagrams. Thus, the diagram

1.

is invariant with respect to a permutation of above and below.

The diagram

2.

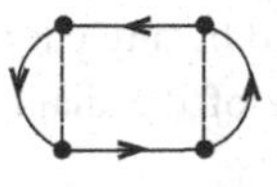

remains invariant when the two vertices are exchanged and simultaneously *above* and *below* are permuted on both vertices.

We introduce the following notation:

Θ : Structure of a diagram,

$h(\Theta)$: Number of topologically **equivalent** diagrams within a structure Θ,

$A_n(\Theta)$: Number of topologically **distinct** diagrams within a structure

$$A_n(\Theta) = \frac{2^n n!}{h(\Theta)} . \tag{5.101}$$

Topologically distinct diagrams with the same structure correspond to different combinations of contractions in the total pairing, which however all make the same contribution to the perturbation term.

One thus chooses from each of the pairwise different structures

$$\Theta_1, \Theta_2, \ldots, \Theta_\nu, \ldots$$

one representative $D_\nu^{(n)}$ and computes its contribution $U(D_\nu^{(n)})$ according to the diagram rules from the preceding section. Then the overall contribution from the structure Θ_ν is:

$$U(\Theta_\nu) = A_n(\Theta_\nu) U\left(D_\nu^{(n)}\right) = a_n(\Theta_\nu) U^*\left(D_\nu^{(n)}\right) . \tag{5.102}$$

Here, $U^*(D_\nu^{(n)})$ is the contribution of the diagram $D_\nu^{(n)}$ **without** the factor required by rule 8., i.e.

$$a_n(\Theta_\nu) = \frac{1}{h(\Theta_\nu)} \left(-\frac{\mathrm{i}}{\hbar}\right)^n . \tag{5.103}$$

Finally, one sums over all the structures:

$$\left\langle \eta_0 \middle| U_\alpha^{(n)}(t, t') \middle| \eta_0 \right\rangle = \sum_\nu a_n(\Theta_\nu) U^*\left(D_\nu^{(n)}\right) . \tag{5.104}$$

We now define

connected diagrams

as those which cannot be decomposed by any cut into two independent diagrams of lower order without cutting through a line of the diagram. The diagrams 1), 2), 7) and 8) in Sect. 5.3.1 are clearly not connected.

Now let $D^{(n)}$ be a diagram with the structure Θ, which can be decomposed into the two connected diagrams $D_1^{(n_1)}$ and $D_2^{(n_2)}$ with the structures Θ_1 and Θ_2, and is thus not itself connected. Then for $\Theta_1 \neq \Theta_2$, we have:

$$h(\Theta) = h(\Theta_1)\, h(\Theta_2)\,, \tag{5.105}$$

since for each diagram from Θ_1, there are $h(\Theta_2)$ topologically equivalent diagrams with the structure Θ_2. The overall contribution of the structure Θ is then given by:

$$U(\Theta) = \frac{\left(-\frac{\mathrm{i}}{\hbar}\right)^{n_1+n_2}}{h(\Theta_1)\, h(\Theta_2)} U^*\left(D^{(n)}\right) .$$

Non-connected diagrams have no common integration or summation variables in their substructures. Therefore, the overall contribution $U^*(D^{(n)})$ can be factored:

$$U^*\left(D^{(n)}\right) = U^*\left(D_1^{(n_1)}\right) U^*\left(D_2^{(n_2)}\right) . \tag{5.106}$$

However, this also signifies that:

$$U(\Theta) = U(\Theta_1)\, U(\Theta_2) \quad (\Theta_1 \neq \Theta_2)\,. \tag{5.107}$$

With the same structures ($\Theta_1 = \Theta_2$), we have instead of (5.105):

$$h(\Theta) = h(\Theta_1)\, h(\Theta_2)\, 2! = 2!\, h^2(\Theta_1)\,, \tag{5.108}$$

since a permutation of the same structures yields further topologically equivalent diagrams:

$$U(\Theta) = \frac{1}{2!} U^2(\Theta_1) \quad (\Theta_1 = \Theta_2)\,. \tag{5.109}$$

These considerations can readily be generalised to arbitrary structures Θ. Assume that

$$\Theta = p_1\Theta_1 + \cdots + p_n\Theta_n\,; \quad p_\nu \in \mathbf{N}\,, \tag{5.110}$$

where Θ_ν are connected structures. Then for the overall contribution of this structure, we have:

$$U(\Theta) = \frac{1}{p_1!} U^{p_1}(\Theta_1) \frac{1}{p_2!} U^{p_2}(\Theta_2) \cdots \frac{1}{p_n!} U^{p_n}(\Theta_n)\,. \tag{5.111}$$

Let us now consider the full perturbation expansion of the time-evolution operator $U_\alpha(t, t')$:

$$\begin{aligned}
\langle \eta_0 | U_\alpha(t, t') | \eta_0 \rangle &= 1 + \sum_\Theta U(\Theta) = \\
&= 1 + \underbrace{U(\Theta_1) + U(\Theta_2) + \cdots +}_{\substack{\text{Contributions of all the} \\ \text{connected diagrams}}} \\
&\quad + \frac{1}{2!} U^2(\Theta_1) + U(\Theta_1)\, U(\Theta_2) + U(\Theta_1)\, U(\Theta_3) + \cdots + \\
&\quad + \frac{1}{2!} U^2(\Theta_2) + U(\Theta_2)\, U(\Theta_3) + \cdots + \\
&\quad \vdots
\end{aligned}$$

$$+\underbrace{\frac{1}{2!}U^2(\Theta_n) + U(\Theta_n)\,U(\Theta_{n+1}) + \cdots +}_{\substack{\text{Contributions of all the non-connected diagrams}\\ \text{which are decomposable into \textbf{two} connected diagrams}}}$$

$$+\frac{1}{3!}U^3(\Theta_1) + \frac{1}{2!}U^2(\Theta_1)\,U(\Theta_2) + \cdots +$$

$$+\,U(\Theta_1)\,U(\Theta_2)\,U(\Theta_3) + \cdots +$$

$$+\underbrace{\frac{1}{3!}U^3(\Theta_2) + \frac{1}{2!}U^2(\Theta_2)\,U(\Theta_1) + \cdots +}_{\substack{\text{Contributions of all the non-connected diagrams}\\ \text{which are decomposable into \textbf{three} connected diagrams}}}$$

$$+\cdots =$$

$$= 1 + \left(\sum_\nu^{\text{conn}} U(\Theta_\nu)\right) +$$

$$+\frac{1}{2!}\Big(U^2(\Theta_1) + 2U(\Theta_1)\,U(\Theta_2) + \cdots +$$

$$+\,U^2(\Theta_2) + 2U(\Theta_2)\,U(\Theta_3) + \cdots\Big) +$$

$$+\frac{1}{3!}\Big(U^3(\Theta_1) + 3U^2(\Theta_1)\,U(\Theta_2) + 6U(\Theta_1)\,U(\Theta_2)\,U(\Theta_3) + \cdots\Big) +$$

$$+\cdots =$$

$$= 1 + \left\{\sum_\nu^{\text{conn}} U(\Theta_\nu)\right\} + \frac{1}{2!}\left\{\sum_\nu^{\text{conn}} U(\Theta_\nu)\right\}^2 + \cdots .$$

We have thus derived the important

linked-cluster theorem

$$\langle \eta_0 |\, U_\alpha(t, t')\, | \eta_0 \rangle = \exp\left\{\sum_\nu^{\text{conn}} U(\Theta_\nu)\right\} \tag{5.112}$$

with the notable consequence that we now have to sum only the connected diagrams which exhibit pairwise different structures.

We can now update the diagram rules of the preceding section:

Perturbation-theoretical calculation of the vacuum amplitude

$$\langle \eta_0 |\, U_\alpha(t, t')\, | \eta_0 \rangle .$$

One finds all the **connected** diagrams with pairwise different structures and computes the contribution of a diagram of n-th order as follows:

1. Vertex $\Longleftrightarrow$ $v(kl;nm)$.
2. Propagating line $\Longleftrightarrow$ $\mathrm{i}G^{0,\mathrm{c}}_{k_\nu}(t_\nu - t_\mu)\delta_{k_\nu,k_\mu}$.
3. Non-propagating line $\Longleftrightarrow$ $-\left\langle n_{k_\nu}\right\rangle^{(0)} \delta_{k_\nu,k_\mu}$.
4. Summation over all $\ldots, k_i, l_i, m_i, n_i, \ldots$
5. Multiplication by $\exp\big(-\alpha(|t_1| + \cdots + |t_n|)\big)\delta(t_1 - t_1')\cdots\delta(t_n - t_n')$, then integration over all t_i', t_i from t' to t.
6. Factor $\left(-\frac{\mathrm{i}}{\hbar}\right)^n \frac{(-1)^S}{h(\Theta)}$.

Finally, one inserts the resulting contribution $U(\Theta)$ into (5.112).

5.3.3 The Principal Theorem of Connected Diagrams

Up to now, we have been considering the development of diagrams for the vacuum amplitude

$$\left\langle \eta_0 \middle| U_\alpha\left(t, t'\right) \middle| \eta_0 \right\rangle ,$$

which becomes the scattering matrix S_α (5.53) for $t = +\infty$ and $t' = -\infty$. In fact, however, we are interested in expressions of the form (5.58):

$$\left\langle E_0 \middle| T_\varepsilon\left\{A^{\mathrm{H}}(t)B^{\mathrm{H}}\left(t'\right)\right\} \middle| E_0 \right\rangle =$$

$$= \lim_{\alpha \to 0} \frac{1}{\left\langle \eta_0 \middle| S_\alpha \middle| \eta_0 \right\rangle} \sum_{\nu=0}^{\infty} \frac{1}{\nu!}\left(-\frac{\mathrm{i}}{\hbar}\right)^\nu \int_{-\infty}^{+\infty} \cdots \int \mathrm{d}t_1 \cdots \mathrm{d}t_\nu \cdot \tag{5.113}$$

$$\cdot\ \mathrm{e}^{-\alpha(|t_1| + \cdots + |t_\nu|)} \left\langle \eta_0 \middle| T_\varepsilon\left\{V(t_1) \cdots V(t_\nu)\, A(t) B\left(t'\right)\right\} \middle| \eta_0 \right\rangle ,$$

in which the operators are on the right in the Dirac representation and $A(t)$, $B(t')$ are supposed to be products of Fermionic creation and annihilation operators. The perturbation expansion of the numerator on the right-hand side is carried out quite analogously to that of the vacuum amplitude which we have been discussing:

1 Wick's theorem: total pairing of the creation and annihilation operators which occur.
2 Summations over all the *inner* k_i, l_i,... **No** summation is carried out over the *outer* indices of the operators occurring in A and B.
3 Integrations over all the *inner* time variables from $-\infty$ to $+\infty$, but **not** over t and t'.

$A(t)$ is supposed to contain $\bar{n}$ creation and annihilation operators, $B(t')$ $\bar{m}$, where $\bar{m} + \bar{n}$ is an even number. A diagram of n-th order can then be represented symbolically as in Fig. 5.2:

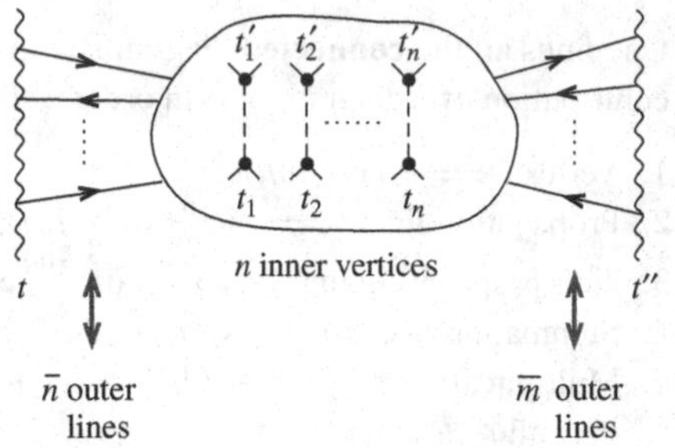

Fig. 5.2 The general structure of an open diagram of n-th order

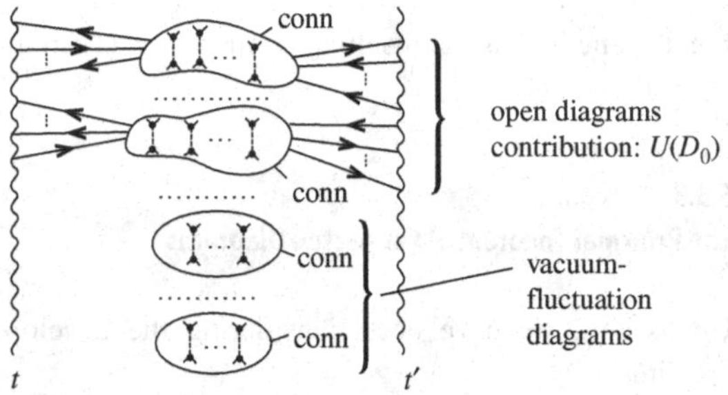

Fig. 5.3 The general structure of the contribution to the vacuum amplitude of an arbitrary diagram

We distinguish among:

"open" diagrams	= diagrams **with** outer lines,
"closed" diagrams; *and* **vacuum fluctuation diagrams**	= diagrams **without** outer lines.

We then clearly expect that:

Every open diagram consists of open, connected diagrams plus connected vacuum fluctuation diagrams.

One obtains **all** such diagrams by adding **all** of the possible vacuum fluctuation diagrams to **every** combination D_0 of open, connected diagrams. The former contribute, as in (5.112), a factor of

$$\exp\left\{\sum_{\nu}^{\text{conn}} U\left(\Theta_{\nu}\right)\right\} .$$

All diagrams with the same combination D_0 of open, connected diagrams thus contribute to the denominator in (5.113) with

$$U\left(D_0\right) \exp\left\{\sum_{\nu}^{\text{conn}} U\left(\Theta_{\nu}\right)\right\} .$$

It then follows that:

The overall contribution of all diagrams to the perturbation expansion is:

$$\left(\sum_{D_0} U(D_0)\right) \exp\left\{\sum_{\nu}^{\text{conn}} U(\Theta_\nu)\right\}. \tag{5.114}$$

The summation runs over all combinations of open, connected diagrams. This is the

principal theorem of connected diagrams,

without which every diagram expansion would be illusory. If we insert this theorem into (5.113), then the contribution from the vacuum fluctuation diagrams just cancels out with $\langle \eta_0 \mid S_\alpha \mid \eta_0 \rangle$:

$$\left\langle E_0 \middle| T_\varepsilon\left\{A^{\mathrm{H}}(t) B^{\mathrm{H}}(t')\right\} \middle| E_0 \right\rangle = \lim_{\alpha \to 0} \sum_{D_0} U(D_0). \tag{5.115}$$

Here, D_0 is thus a combination of open, connected diagrams with all together $\overline{n}$ attached outer lines at t and $\overline{m}$ at t', where $\overline{n}$ and $\overline{m}$ are the numbers of Fermionic operators in $A(t)$ and $B(t')$.

Quite analogously, we find for the simpler expression (5.56):

$$\left\langle E_0 \middle| A^{\mathrm{H}}(t) \middle| E_0 \right\rangle = \lim_{\alpha \to 0} \sum_{\overline{D}_0} U(\overline{D}_0). \tag{5.116}$$

In this expression, $\overline{D}_0$ is now a combination of open, connected diagrams with as many solid lines attached at t as there are Fermionic operators contained in $A(t)$.

For both cases, (5.115) and (5.116), we discuss examples of applications in the following sections.

5.3.4 Exercises

Exercise 5.3.1 Evaluate the vacuum amplitude $\langle \eta_0 \mid U_\alpha(t, t') \mid \eta_0 \rangle$ in first order perturbation theory for the

1. Hubbard model and the
2. jellium model.

Exercise 5.3.2 In second-order perturbation theory for the vacuum amplitude, we find the diagram:

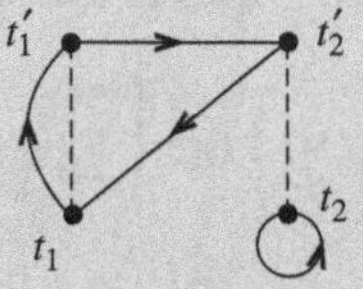

1. Calculate the contribution of this diagram.
2. What does it contribute in the Hubbard model?
3. What contribution does it make in the jellium model?

Exercise 5.3.3 Find out which of the diagrams of second order for the vacuum amplitude as listed in Sect. 5.3.1 are topologically distinct, but have the same structure; i.e. they correspond to different terms of the *total pairing* which make the same contributions to the perturbation expansion. How many of the 24 diagrams accordingly must be evaluated explicitly?

5.4 Single-Particle Green's Functions

5.4.1 Diagrammatic Perturbation Expansions

An important application of diagrammatic perturbation theory concerns the causal single-particle Green's function:

$$\mathrm{i}G^{\mathrm{c}}_{k\sigma}\left(t-t'\right) = \left\langle E_0 \middle| T_\varepsilon\left\{a_{k\sigma}(t)a^+_{k\sigma}\left(t'\right)\right\}\middle| E_0\right\rangle . \tag{5.117}$$

This corresponds to the case of (5.115), i.e. it is to be summed over all the pairwise distinct structures of connected diagrams with two solid outer lines, corresponding to the operators $a_{k\sigma}(t)$ and $a^+_{k\sigma}(t')$.

We insert here a remark on interactions. We consider pairwise interactions, $((v(|\boldsymbol{r}_1 - \boldsymbol{r}_2|))$. The entire system is assumed to have translational symmetry. Then the momenta at a vertex are not arbitrary, but instead

momentum conservation at a vertex

must be required:

$$\boldsymbol{k} - \boldsymbol{n} = \boldsymbol{m} - \boldsymbol{l} = \boldsymbol{q} . \tag{5.118}$$

The sum of the *inward-pointing* momenta is equal to the sum of the *outward-pointing* momenta. At each vertex **point**, we furthermore require conservation of spin:

$$\sigma_k = \sigma_n\,; \quad \sigma_l = \sigma_m\,. \tag{5.119}$$

Due to (5.118) and (5.119), the number of summations is again greatly reduced. We shall make use of this at a suitable juncture.

We now come to the diagram expansion for the Green's function. In

zeroth order,

we find merely a line propagating from t' to t:

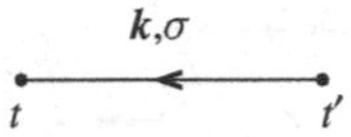

this corresponds to the contribution:

$$\mathrm{i}G^{0,\,\mathrm{c}}_{k\sigma}\left(t - t'\right)\,.$$

In

first order

we must evaluate:

$$\frac{1}{1!}\frac{1}{2}\left(-\frac{\mathrm{i}}{\hbar}\right)\int\limits_{-\infty}^{+\infty}\mathrm{d}t_1\int\limits_{-\infty}^{+\infty}\mathrm{d}t_1'\,\delta\left(t_1 - t_1'\right)\mathrm{e}^{-\alpha|t_1|}\sum_{\substack{\boldsymbol{k}_1\boldsymbol{l}_1\boldsymbol{m}_1\boldsymbol{n}_1\\ \sigma_1\sigma_1'}} v\left(\boldsymbol{k}_1\boldsymbol{l}_1;\boldsymbol{n}_1\boldsymbol{m}_1\right)\,\cdot$$

$$\cdot\left\langle\eta_0\middle|T_\varepsilon\{a_{\boldsymbol{k}\sigma}(t)a^+_{\boldsymbol{k}\sigma}(t')a^+_{\boldsymbol{k}_1\sigma_1}(t_1)a_{\boldsymbol{n}_1\sigma_1}(t_1)a^+_{\boldsymbol{l}_1\sigma_1'}(t_1')a_{\boldsymbol{m}_1\sigma_1'}(t_1')\}\middle|\eta_0\right\rangle$$

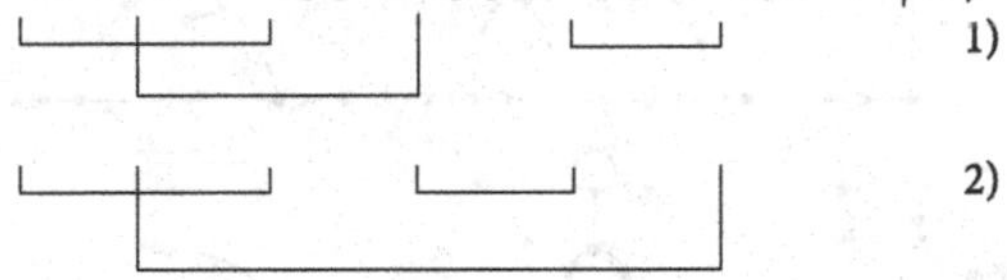

Only open, connected diagrams need be considered.

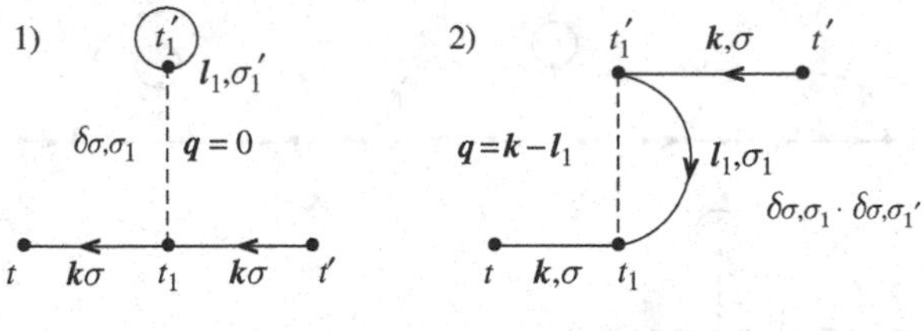

At the vertex, one can naturally also exchange *upper* and *lower*. This yields topologically distinct diagrams of the same structure, which are taken into account by inserting the factor $2^1 \cdot 1!$.

Second order

We have to count the following open, connected diagrams:

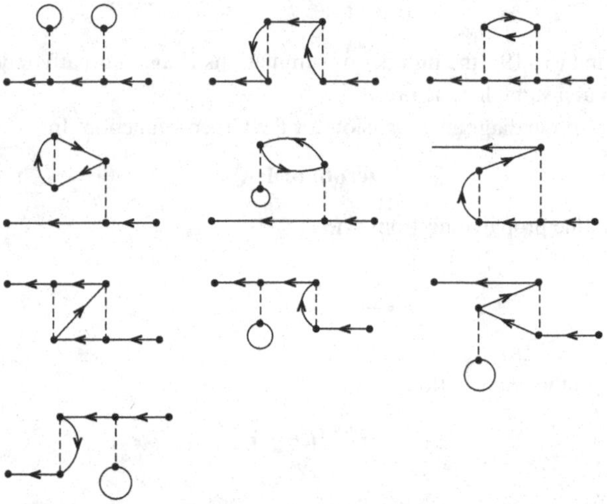

For each of these diagrams there are again $2^2 \cdot 2! = 8$ topologically distinct diagrams of the same structure which make the same contributions. Topologically equivalent diagrams do **not** occur owing to the *outer* attachments. For the Green's function diagrams, one sometimes chooses a somewhat modified representation by *stretching* the propagating lines, but not necessarily drawing in the vertices as perpendicular lines. The above diagrams are then drawn as follows:

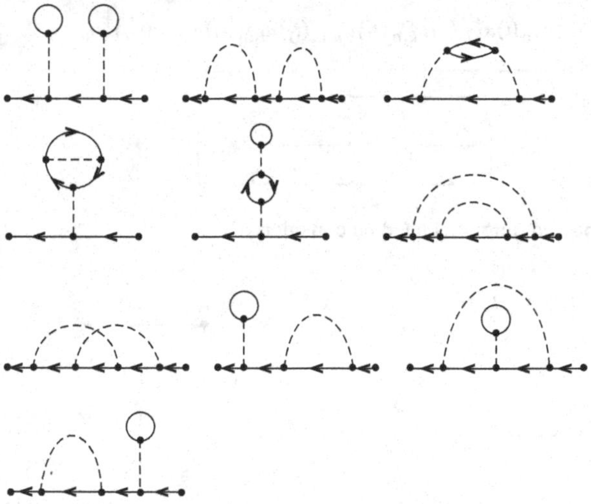

"stretched" diagrams

The rules for the evaluation of these not-numbered diagrams are obtained immediately from those in Sect. 5.3.2 for the vacuum amplitude:

We draw a member from each structure Θ of connected diagrams with two outer attachments. Each diagram of n-th order contains n vertices and $(2n+1)$ solid lines, among them two outer ones. The contribution of such a diagram is then computed as follows:

1. Vertex $\Longleftrightarrow$ $v(kl;nm)$.
2. Propagating line $\Longleftrightarrow$ $\mathrm{i}G^{0,\mathrm{c}}_{k_\nu}(t_\nu - t_\mu)\delta_{k_\nu,k_\mu}$.
3. Non-propagating line $\Longleftrightarrow$ $\mathrm{i}G^{0,\mathrm{c}}_{k_\nu}(0^-)\delta_{k_\nu,k_\mu}$.
4. Momentum conservation at the vertex; spin conservation at the vertex point.
5. Multiplication by $\mathrm{e}^{-\alpha(|t_1|+\ldots+|t_n|)}\delta(t_1 - t_1')\cdots\delta(t_n - t_n')$.
6. Summation over all the *inner* wavenumbers and spins $\ldots, k_i, l_i, m_i, n_i, \ldots$ as well as integration over all the *inner* times t_i, t_i' from $-\infty$ to $+\infty$.
7. The factor $\left(-\frac{\mathrm{i}}{\hbar}\right)^n(-1)^S$; $S =$ number of loops $(\hat{n}(\Theta) \equiv 1)$.

In 2 and 3, k_ν and k_μ refer to the indices (wavenumber, spin) which connect the propagators $\mathrm{i}G^{0,\mathrm{c}}$ to each other.

The evaluation of the diagrams using these rules can be somewhat tedious, since as seen in (5.78), the causal one-electron Green's function exhibits an unfavourable time dependence. One is thus well advised to use the Fourier transform:

$$G^{0,\mathrm{c}}_k\left(t-t'\right) = \frac{1}{2\pi\hbar}\int\limits_{-\infty}^{+\infty} \mathrm{d}E\, G^{0,\mathrm{c}}_k(E)\exp\left(-\frac{\mathrm{i}}{\hbar}E\left(t-t'\right)\right). \tag{5.120}$$

In the diagrams, the transformation is carried out as follows:

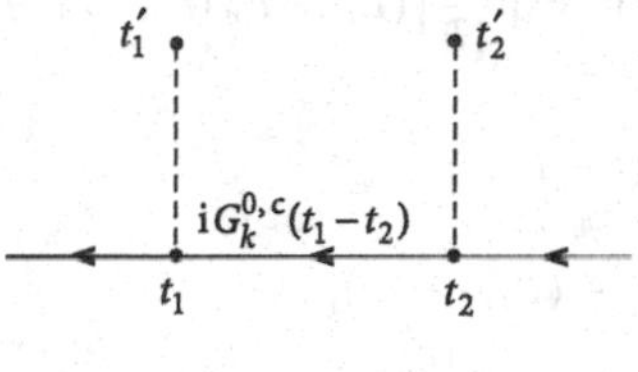

$$\frac{\exp\left(-\frac{\mathrm{i}}{\hbar}Et_1\right)}{\sqrt{2\pi\hbar}}\left(\mathrm{i}G^{0,\mathrm{c}}_k(E)\right)\frac{\exp\left(\frac{\mathrm{i}}{\hbar}Et_2\right)}{\sqrt{2\pi\hbar}}$$

The outgoing line at t_2 is associated with the additional factor

$$\frac{\exp\left(\frac{\mathrm{i}}{\hbar}Et_2\right)}{\sqrt{2\pi\hbar}}.$$

The ingoing line at t_1, in contrast, produces a term

$$\frac{\exp\left(-\frac{\mathrm{i}}{\hbar}Et_1\right)}{\sqrt{2\pi\hbar}}.$$

It is thus advisable to index the ingoing and outgoing lines at a vertex additionally with the energies. The whole vertex is then associated, apart from the matrix element $v(kl;nm)$, with a factor (Fig. 5.4):

$$\frac{1}{(2\pi\hbar)^2}\exp\left\{\frac{\mathrm{i}}{\hbar}(E_k - E_n)t + \frac{\mathrm{i}}{\hbar}(E_l - E_m)t' - \alpha\,|t|\right\}\delta\left(t-t'\right).$$

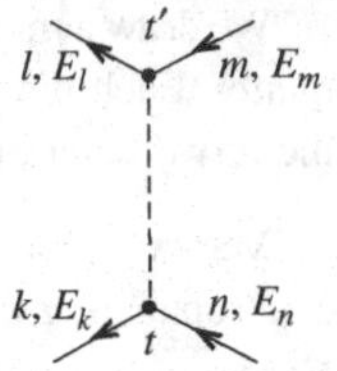

Fig. 5.4 The annotation of a vertex in a diagram for an energy-dependent one-electron Green's function

The subsequent integration over time is readily carried out:

$$\int_{-\infty}^{+\infty} \mathrm{d}t \int_{-\infty}^{+\infty} \mathrm{d}t' \; \exp\left\{\frac{\mathrm{i}}{\hbar}\left[(E_k - E_n)\,t + (E_l - E_m)\,t'\right]\right\} \exp(-\alpha\,|t|\,)\,\delta\left(t - t'\right) =$$

$$= \int_0^{+\infty} \mathrm{d}t \; \exp\left(\frac{\mathrm{i}}{\hbar}\overline{E}t - \alpha t\right) + \int_{-\infty}^{0} \mathrm{d}t \; \exp\left(\frac{\mathrm{i}}{\hbar}\overline{E}t + \alpha t\right) =$$

$$= \frac{-1}{\frac{\mathrm{i}}{\hbar}\overline{E} - \alpha} + \frac{1}{\frac{\mathrm{i}}{\hbar}\overline{E} + \alpha} = \frac{2\alpha}{\left(\frac{1}{\hbar}\overline{E}\right)^2 + \alpha^2}\,,$$

$$\overline{E} = (E_k - E_n) + (E_l - E_m)\,.$$

Taking the limit $\alpha \to 0$ (*adiabatic switching on*) then makes this expression into a δ-function:

$$\lim_{\alpha \to 0} \frac{1}{(2\pi\hbar)^2} \iint_{-\infty}^{+\infty} \mathrm{d}t\,\mathrm{d}t' \; \exp\left\{\frac{\mathrm{i}}{\hbar}\left[(E_k - E_n)\,t + (E_l - E_m)\,t'\right]\right\} \cdot$$

$$\cdot \; \exp(-\alpha\,|t|\,)\,\delta\left(t - t'\right) = \tag{5.121}$$

$$= \frac{1}{2\pi\hbar}\,\delta\left[(E_k + E_l) - (E_m + E_n)\right]\,.$$

This however simply guarantees

conservation of energy at the vertex.

The outer lines take on a certain special role:

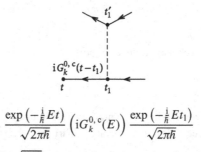

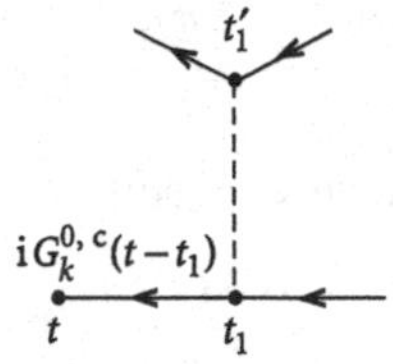

$$\frac{\exp\left(-\frac{\mathrm{i}}{\hbar}Et\right)}{\sqrt{2\pi\hbar}} \left(\mathrm{i}G_k^{0,\,\mathrm{c}}(E)\right) \frac{\exp\left(-\frac{\mathrm{i}}{\hbar}Et_1\right)}{\sqrt{2\pi\hbar}}$$

The factor $\exp\left[(\mathrm{i}/\hbar)Et_1\right]/\sqrt{2\pi\hbar}$, as described above, is taken into the vertex at t_1 and contributes after the integration over t_1 to the corresponding δ-function (5.121). Then the term

$$\frac{\exp\left(-\frac{\mathrm{i}}{\hbar}Et\right)}{\sqrt{2\pi\hbar}}\left(\mathrm{i}G_k^{0,\mathrm{c}}(E)\right)$$

still remains; it is finally integrated over all E in order to obtain $G_k^{\mathrm{c}}(t-t')$.

For the line which enters the diagram from the right of t', an analogous factor applies:

$$\mathrm{i}G_k^{0,\mathrm{c}}(E')\,\frac{\exp\left(\frac{\mathrm{i}}{\hbar}E't'\right)}{\sqrt{2\pi\hbar}} .$$

If the *inner* summations and integrations all together yield the numerical value I, then we have for the overall diagram:

$$\begin{aligned}
\mathrm{i}\widetilde{G}_k\left(t-t'\right) &= \frac{\mathrm{i}}{2\pi\hbar}\iint \mathrm{d}E\,\mathrm{d}E'\left(\mathrm{i}G_k^{0,\mathrm{c}}(E)\right)\left(\mathrm{i}G_k^{0,\mathrm{c}}(E')\right)\cdot\\
&\quad\cdot\exp\left[\frac{\mathrm{i}}{\hbar}\left(E't'-Et\right)\right]\overset{!}{=}\\
&\overset{!}{=}\mathrm{i}\widetilde{G}_k\left((t+t_0)-(t'+t_0)\right)=\\
&=\frac{\mathrm{i}}{2\pi\hbar}\iint \mathrm{d}E\mathrm{d}E'\left(\mathrm{i}G_k^{0,\mathrm{c}}(E)\right)\left(\mathrm{i}G_k^{0,\mathrm{c}}(E')\right)\cdot\\
&\quad\cdot\exp\left[\frac{\mathrm{i}}{\hbar}\left(E't'-Et\right)\right]\exp\left[\frac{\mathrm{i}}{\hbar}\left(E'-E\right)t_0\right].
\end{aligned}$$

Since from (3.129), the Green's function depends only upon the time difference, it follows that:

$$\mathrm{i}\widetilde{G}_k\left(t-t'\right)=\frac{\mathrm{i}}{2\pi\hbar}\int\mathrm{d}E\left(\mathrm{i}G_k^{0,\mathrm{c}}(E)\right)^2\exp\left(\frac{\mathrm{i}}{\hbar}E\left(t'-t\right)\right).$$

For the Fourier transform, this implies that:

$$\mathrm{i}\widetilde{G}_k(E)=I\left(\mathrm{i}G_k^{0,\mathrm{c}}(E)\right)^2 . \tag{5.122}$$

The two outer attachments of a diagram for the single-particle Green's function $G_k^{\mathrm{c}}(E)$ thus have not only the same wavenumber and spin $k=(\boldsymbol{k},\sigma)$, but also the same energy E, in the last analysis a consequence of energy conservation at each vertex.

If we now recall the structure of the free, energy-dependent causal $T=0$ Green's function according to (3.206),

$$\begin{aligned}
&G_{\boldsymbol{k}\sigma}^{0,\mathrm{c}}(E)=\frac{\hbar}{E-(\varepsilon(\boldsymbol{k})-\varepsilon_F)\pm\mathrm{i}0^+}\\
&(+\quad\text{for}\quad|\boldsymbol{k}|>k_\mathrm{F}\,,\quad-\quad\text{for}\quad|\boldsymbol{k}|<k_\mathrm{F})\,,
\end{aligned} \tag{5.123}$$

then we have everything we need in order to formulate the diagram rules for $\mathrm{i}G_{\boldsymbol{k}\sigma}^{\mathrm{c}}(E)$,

diagram rules for $\mathrm{i}G^{\mathrm{c}}_{k\sigma}(E)$:

One member from each structure Θ of connected diagrams with two outer attachments must be found. A diagram of n-th order (with n vertices, $(2n + 1)$ solid lines) then yields the following contribution:

1. Vertex $\Longleftrightarrow$ $\frac{1}{2\pi\hbar} v(kl, nm)\delta\left[(E_k + E_l) - (E_m + E_n)\right]$.
2. Propagating **and** non-propagating line $\Longleftrightarrow$ $\mathrm{i}G^{0,\mathrm{c}}_{k_\nu}(E_{k_\nu})\delta_{k_\nu k_\mu}$.
3. Factor: $(-1)^{\mathrm{S}}\left(-\frac{\mathrm{i}}{\hbar}\right)^n$.
4. Summation over all the *inner* indices $k_i, l_i, \ldots$; integration over all the *inner* energies $E_{k_i}, E_{l_i}, \ldots$
5. *Outer* attachments: $\mathrm{i}G^{0,\mathrm{c}}_k(E)$.

5.4.2 The Dyson Equation

As we already mentioned earlier, summing of a finite number of terms of a perturbation expansion is not always expedient, e.g. when the perturbation is not really small or when divergences occur in the individual perturbation terms. It is then often preferable to formulate an approximation by summing an infinite partial series. Such a possibility is opened up by the **Dyson equation**, which we have already encountered in (3.327). We now wish to reconstruct it with the aid of our diagram techniques.

Definition 5.4.1: Self-energy contribution = that portion of a diagram which is connected with the remainder of the diagram by two propagating lines.

Examples

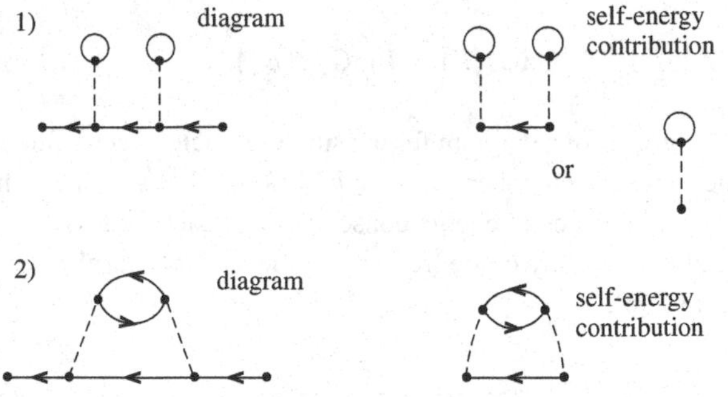

A self-energy contribution is thus a diagram part with two outer *attachments*, for one outgoing and one incoming propagating line.

Definition 5.4.2: **The proper (irreducible) self-energy contribution** = that portion of the self-energy part which **cannot** be decomposed into two independent self-energy contributions by removal of a propagating line.

Examples

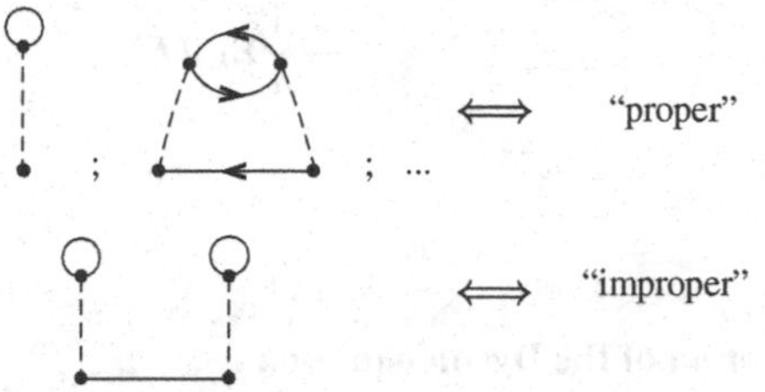

Up to second order, the causal Green's function has the following irreducible self-energy contributions:

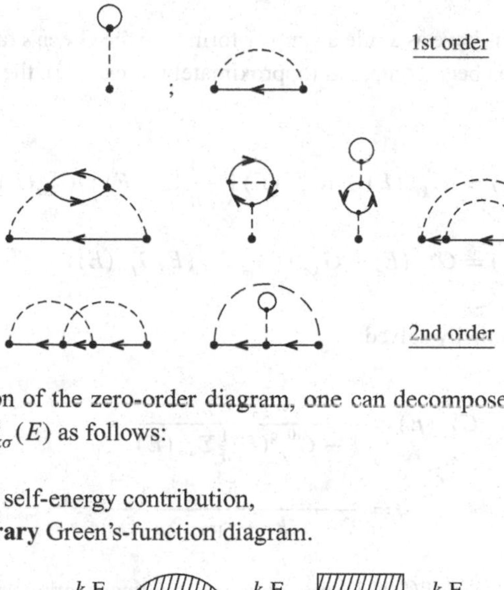

With the exception of the zero-order diagram, one can decompose **every** diagram which contributes to $iG^{c}_{k\sigma}(E)$ as follows:

(I): $iG^{0,c}_{k}(E)$,

(II): the proper self-energy contribution,

(III): **any arbitrary** Green's-function diagram.

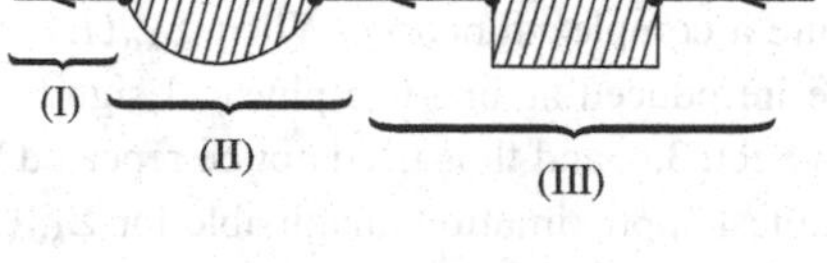

Due to conservation of energy and momentum at each vertex, the middle propagator has the same fixed indices k, E as the incoming and outgoing propagating lines.

Evidently, one obtains **all** the diagrams by summing in (II) over **all** the proper self-energy contributions and in (III) over **all** the Green's function diagrams.

Definition 5.4.3: **Self-energy** $(\Sigma_{k\sigma}(E)) = \mathrm{i}\hbar \cdot$ Sum over all the proper self-energy contributions. Using the notation

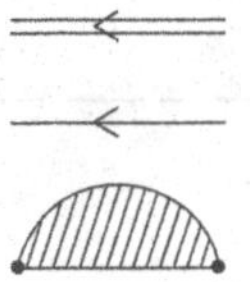

$$: \mathrm{i}G^{\mathrm{c}}_{k\sigma}(E)$$

$$: \mathrm{i}G^{0,\mathrm{c}}_{k\sigma}(E)$$

$$: -\frac{\mathrm{i}}{\hbar}\Sigma_{k\sigma}(E)\,,$$

we find the

Diagram representation of the Dyson equation

The self-energy diagrams have as a rule a simpler form than the Green's function diagrams. Once the self-energy has been computed (approximately or exactly), the Green's function is also determined:

$$\begin{aligned}
\mathrm{i}G^{\mathrm{c}}_{k\sigma}(E) &= \mathrm{i}G^{0,\mathrm{c}}_{k\sigma}(E) + \mathrm{i}G^{0,\mathrm{c}}_{k\sigma}(E)\left(-\frac{\mathrm{i}}{\hbar}\Sigma_{k\sigma}(E)\right)\mathrm{i}G^{\mathrm{c}}_{k\sigma}(E)\,,\\
G^{\mathrm{c}}_{k\sigma}(E) &= G^{0,\mathrm{c}}_{k\sigma}(E) + G^{0,\mathrm{c}}_{k\sigma}(E)\frac{1}{\hbar}\Sigma_{k\sigma}(E)G^{\mathrm{c}}_{k\sigma}(E)\,.
\end{aligned} \tag{5.124}$$

This equation can be formally solved:

$$G^{\mathrm{c}}_{k\sigma}(E) = \frac{G^{0,\mathrm{c}}_{k\sigma}(E)}{1 - G^{0,\mathrm{c}}_{k\sigma}(E)\frac{1}{\hbar}\Sigma_{k\sigma}(E)} = \tag{5.125}$$

$$= \frac{\hbar}{E - \varepsilon(\boldsymbol{k}) + \varepsilon_{\mathrm{F}} - \Sigma_{k\sigma}(E)}\,. \tag{5.126}$$

In the last step, we made use of (5.123), including the imaginary infinitesimal $\pm\mathrm{i}0^+$ within $\Sigma_{k\sigma}(E)$, which is as a rule a complex function. When $\Sigma_{k\sigma}(E)$ is real, then $\pm\mathrm{i}0^+$ in the sense of (5.123) must be introduced again. The physical significance of the self-energy was discussed in detail in Sect. 3.4, and thus need not be repeated here.

Note that even the simplest approximation imaginable for $\Sigma_{k\sigma}(E)$ according to (5.124) requires summing of an **infinite** partial series:

If within the jellium model (Sect. 2.1.2), we assume

$$v(\boldsymbol{q} = \boldsymbol{0}) = 0\,, \tag{5.127}$$

then all the diagrams with *bubbles* make no contribution, since they correspond to a momentum transfer of $q = 0$. Then, as the simplest approximation, we have:

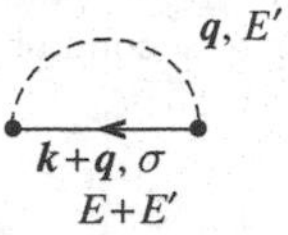

$$\hat{=} -\frac{\mathrm{i}}{\hbar}\Sigma^{(1)}_{k\sigma}(E)\,.$$

The evaluation of this expression follows immediately from the diagram rules in Sect. 5.4.1:

$$-\frac{\mathrm{i}}{\hbar}\Sigma^{(1)}_{k\sigma}(E) = -\frac{\mathrm{i}}{\hbar}\frac{1}{2\pi\hbar}\sum_{q}^{q\neq 0} v(\boldsymbol{q})\int \mathrm{d}E'\,\mathrm{i}G^{0,\,\mathrm{c}}_{k+q\sigma}\left(E+E'\right) =$$

$$= \frac{-\mathrm{i}}{2\pi\hbar^2}\sum_{q}^{q\neq 0} v(\boldsymbol{q})\mathrm{i}2\pi\hbar G^{0,\,\mathrm{c}}_{k+q\sigma}(0^-)\,.$$

Here, we assumed specifically $v(\boldsymbol{kl};\boldsymbol{nm}) = v(\boldsymbol{k}-\boldsymbol{n}) = v(\boldsymbol{m}-\boldsymbol{l}) = v(\boldsymbol{q})$ as in the jellium model. We then have

$$\Sigma^{(1)}_{k\sigma}(E) = -\sum_{q} v(\boldsymbol{q})\left\langle n_{k+q,\sigma}\right\rangle^{(0)}\,. \tag{5.128}$$

5.4.3 Exercises

Exercise 5.4.1 Within the Hubbard model, calculate the first-order contribution to the self-energy of the one-electron Green's function. Which approximation of the equation of motion method (Chap. 4) corresponds to this self-energy?

Exercise 5.4.2 Discuss the self-energy diagrams of second order for the Hubbard model which can yield a nonzero contribution to the one-electron Green's function.

Exercise 5.4.3 Within the Hubbard model, calculate the one-electron Green's function in first-order perturbation theory and compare the result with the Green's function which is found by computing the self-energy to first order. (Ex. 5.4.1).

Exercise 5.4.4 Even the lowest-order approximation for the self-energy requires the summation of an infinite partial series of diagrams for the one-electron Green's function. Give all of the diagrams which occur up to second order.

Exercise 5.4.5 Which approximation is obtained if, in the self-energy to first order, the *free* Green's function propagators are replaced by the **full** propagators:

$$\widehat{\Sigma}^{(1)}_{k\sigma}(E) =$$

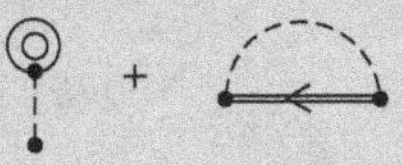

Give examples of new diagrams which occur as a result of this so-called **renormalisation** of the particle propagators as compared to the approximation in Ex. 5.4.4.

5.5 The Ground-State Energy of the Electron Gas (Jellium Model)

5.5.1 First-Order Perturbation Theory

Following the one-electron Green's functions, we now discuss another application of the diagram techniques developed in Sect. 5.3. We consider the ground-state energy of the interacting electron gas, which we describe within the jellium model (Sect. 2.1.2). It is characterised by the interaction operator

$$V(t) = \frac{1}{2} \sum_{klmn} v(kl;nm) a_k^+(t) a_l^+(t) a_m(t) a_n(t) \,, \qquad (5.129)$$

$$k \equiv (\boldsymbol{k}, \sigma_k) \,,$$

where for the matrix element

$$v(kl;nm) = v(\boldsymbol{k} - \boldsymbol{n}) \delta_{\boldsymbol{k}+\boldsymbol{l},\, \boldsymbol{m}+\boldsymbol{n}} \delta_{\sigma_k \sigma_n} \delta_{\sigma_m \sigma_l} \qquad (5.130)$$

we find:

$$v(\boldsymbol{q}) = \frac{e^2}{\varepsilon_0 V q^2} \,; \quad v(\boldsymbol{0}) = 0 \,. \qquad (5.131)$$

The $\boldsymbol{q} = \boldsymbol{0}$-matrix element is exactly compensated by the homogeneously distributed positively-charged ionic background.

For the ground-state energy and the level shift, we have from (5.5) and (5.43):

$$E_0 = \lim_{\alpha \to 0} \frac{\langle \eta_0 \mid \mathcal{H}_\alpha \mid \psi_\alpha^{\mathrm{D}}(0) \rangle}{\langle \eta_0 \mid \psi_\alpha^{\mathrm{D}}(0) \rangle}, \tag{5.132}$$

$$\Delta E_0 = E_0 - \eta_0 = \lim_{\alpha \to 0} \frac{\langle \eta_0 \mid V(t=0) U_\alpha(0,-\infty) \mid \eta_0 \rangle}{\langle \eta_0 \mid U_\alpha(0,-\infty) \mid \eta_0 \rangle}. \tag{5.133}$$

All the operators are of course here again taken to be in their Dirac representations. The denominator is already familiar from Sect. 5.3. It is the vacuum amplitude for $t' = -\infty$ and $t = 0$. We still have to evaluate the following expression:

$$\begin{aligned} \Delta E_0 = \lim_{\alpha \to 0} & \frac{1}{\langle \eta_0 \mid U_\alpha(0,-\infty) \mid \eta_0 \rangle} \sum_{n=0}^{\infty} \frac{1}{n!} \left(-\frac{\mathrm{i}}{\hbar} \right)^n \cdot \\ & \cdot \int_{-\infty}^{0} \cdots \int \mathrm{d}t_1 \cdots \mathrm{d}t_n \, \mathrm{e}^{-\alpha(|t_1| + \cdots + |t_n|)} \cdot \\ & \cdot \langle \eta_0 | V(t=0) T_\varepsilon \{ V(t_1) \cdots V(t_n) \} | \eta_0 \rangle . \end{aligned} \tag{5.134}$$

We can include the operator $V(t=0)$ as indicated within the T_ε product, since the times $t_1, \ldots, t_n$ are all ≤ 0. The diagram expansion of this expression corresponds to the situation in (5.116):

$$\Delta E_0 = \lim_{\alpha \to 0} \sum_{\widehat{D}_0} U\left(\widehat{D}_0\right). \tag{5.135}$$

The sum runs over all the combinations $\widehat{D}_0$ of open, connected diagrams with four solid lines attached at $t = t' = 0$. According to Wick's theorem, we have to construct the *total pairing* from typical terms of the perturbation expansion such as

$$\begin{aligned} T_\varepsilon \Big\{ & a_k^+(t=0) a_l^+(t'=0)\, a_m(t'=0)\, a_n(t=0) a_{k_1}^+(t_1)\, a_{l_1}^+(t_1') \cdot \\ & \cdot a_{m_1}(t')\, a_{n_1}(t_1) \cdots a_{k_n}^+(t_n)\, a_{l_n}^+(t_n')\, a_{m_n}(t_n')\, a_{n_n}(t_n) \Big\} . \end{aligned}$$

The Feynman diagrams have formally the same structures as those for the vacuum amplitude in Sect. 5.3.1, with the exception that the left vertex is fixed at $t = t' = 0$. We integrate or sum over the times, momenta, and spins of the *inner* vertices. At this point, we can already contemplate how many topologically distinct diagrams of the same structure

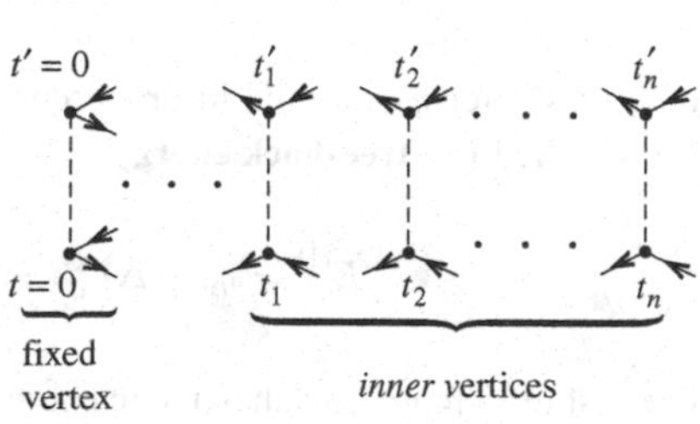

Fig. 5.5 Outer and inner vertices for the computation of the ground-state energy within the jellium model

there can be for a given order. Due to the integrations and summations, we can permute the *inner* vertices among themselves and exchange *above* and *below* on them. The left vertex is fixed. However, here, too, *above* and *below* can be exchanged:

$$A(\Theta_n) = \frac{2^{n+1} n!}{h(\Theta_n)} . \tag{5.136}$$

$h(\Theta_n)$ is the number of topologically equivalent diagrams.

The diagrams in first-order perturbation theory ($n = 0$) contain no *inner* vertices:

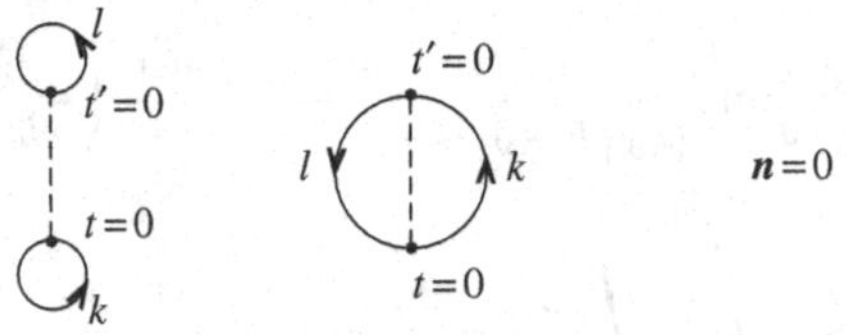

A *bubble* as in the left diagram makes no contribution within the jellium model due to

$$v(\boldsymbol{k} - \boldsymbol{n}) = v(\boldsymbol{0}) = 0 .$$

Thus only the second diagram remains. Exchanging *above* and *below* yields a topologically equivalent diagram. We thus find $h(\Theta_0) = 2;\ A(\Theta_0) = 1$. The rules of Sect. 5.3 then yield:

$$\begin{aligned} \Delta E_0^{(1)} &= -\frac{1}{2} \sum_{klmn} v(kl; nm) \langle n_k \rangle^{(0)} \langle n_l \rangle^{(0)} \delta_{km} \delta_{ln} = \\ &= -\frac{1}{2} \sum_{\substack{\boldsymbol{k}\boldsymbol{l} \\ \sigma_k, \sigma_l}} v(\boldsymbol{k} - \boldsymbol{l}) \delta^2_{\sigma_k \sigma_l} \left\langle n_{\boldsymbol{k}\sigma_k} \right\rangle^{(0)} \left\langle n_{\boldsymbol{l}\sigma_l} \right\rangle^{(0)} . \end{aligned}$$

With $\boldsymbol{l} = \boldsymbol{k} + \boldsymbol{q}$ and $\sigma_k = \sigma_l = \sigma$, we then must still calculate:

$$\Delta E_0^{(1)} = -\sum_{\boldsymbol{k}, \boldsymbol{q}}^{\boldsymbol{q} \neq \boldsymbol{0}} v(\boldsymbol{q}) \Theta(k_{\mathrm{F}} - k) \Theta(k_{\mathrm{F}} - |\boldsymbol{k} + \boldsymbol{q}|) . \tag{5.137}$$

A similar expression was already evaluated in Sect. 2.1.2. From (2.96), we have:

$$\Delta E_0^{(1)} = -\frac{0.916}{r_s} N [\mathrm{ryd}] . \tag{5.138}$$

With (2.87) for η_0, we find in first-order perturbation theory for the ground-state energy the so-called **Hartree-Fock energy**:

$$E_0^{(1)} = \eta_0 + \Delta E_0^{(1)} = N \left(\frac{2.21}{r_s^2} - \frac{0.916}{r_s} \right) [\mathrm{ryd}] . \tag{5.139}$$

The first term represents the kinetic energy and the second the **exchange energy**.

5.5.2
Second-Order Perturbation Theory

Now, how do the diagrams for second-order perturbation theory look? According to (5.134), the following expression must be evaluated:

$$\Delta E_0^{(2)} = \lim_{\alpha \to 0} \frac{1}{\langle \eta_0 \,|\, U_\alpha(0, -\infty) \,|\, \eta_0 \rangle} \left(-\frac{\mathrm{i}}{\hbar}\right) \int\limits_{-\infty}^{+\infty} \mathrm{d}t' \, \delta\left(t'\right) \int\limits_{-\infty}^{0} \mathrm{d}t_1 \cdot$$

$$\cdot \int\limits_{-\infty}^{+\infty} \mathrm{d}t_1' \, \delta\left(t_1 - t_1'\right) \mathrm{e}^{-\alpha |t_1|} \frac{1}{4} \sum_{klmn} \sum_{k_1 l_1 m_1 n_1} v(kl;nm) v\left(k_1 l_1; n_1 m_1\right) \cdot \tag{5.140}$$

$$\cdot \left\langle \eta_0 \middle| T_\varepsilon \left\{ a_k^+(0) a_l^+\left(t'\right) a_m\left(t'\right) a_n(0) a_{k_1}^+\left(t_1\right) a_{l_1}^+\left(t_1'\right) a_{m_1}\left(t_1'\right) a_{n_1}\left(t_1\right) \right\} \middle| \eta_0 \right\rangle .$$

Only connected, open diagrams need be considered. For each diagram structure, there are from (5.136)

$$A\left(\Theta_1\right) = \frac{4}{h\left(\Theta_1\right)} \tag{5.141}$$

topologically distinct diagrams of the same structure. The following structures occur:

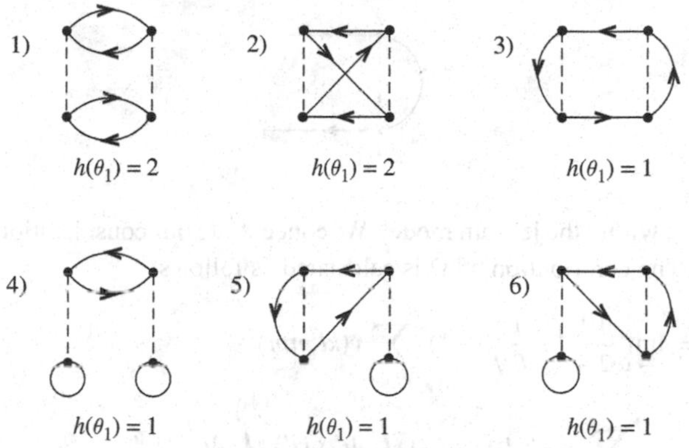

Due to the *fixed* vertex on the left, diagrams 5) and 6) – in contrast to the case of the vacuum amplitude – do *not* have the same structure.

Owing to (5.131), all the diagrams with *bubbles* make no contribution. The structures 4), 5) and 6) therefore need not be evaluated. It can be readily seen that this also holds for diagrams of type 3):

To see this, we examine the exact annotation of the diagram:

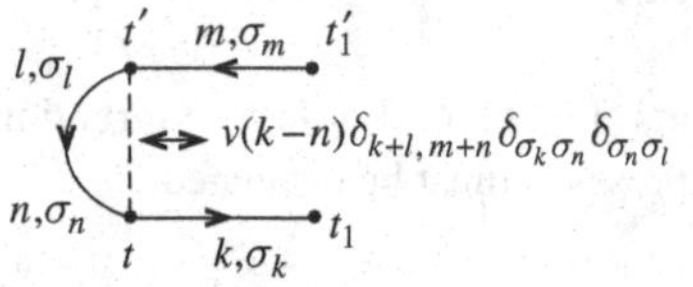

$$\Downarrow$$

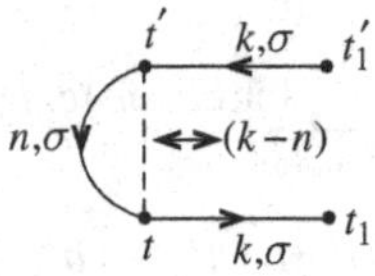

Because of $t' > t_1'$, the propagator *above*, $\mathrm{i}G^{0,\mathrm{c}}_{k\sigma}(t' - t_1')$, is only nonzero when $k > k_\mathrm{F}$, according to (5.78); the propagator *below*, $\mathrm{i}G^{0,\mathrm{c}}_{k\sigma}(t_1 - t)$, is however nonzero only when $k < k_\mathrm{F}$, due to $t_1 < t$. The two cases cannot occur simultaneously. Thus the net contribution of this diagram is zero. Since the t_i, t_i' are always less than the *fixed* times t, t' (after carrying out the trivial integrations), this also holds for all the higher orders. Diagrams of the type

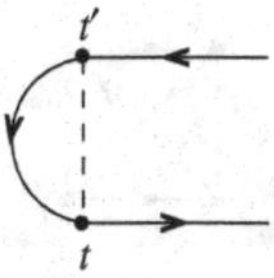

cannot contribute within the jellium model. We concentrate our considerations on the structures 1) and 2). The contribution of 1) is calculated as follows:

$$U_{(1)}(\Theta_1) = \lim_{\alpha \to 0} \frac{4}{2}\frac{1}{4}\left(-\frac{\mathrm{i}}{\hbar}\right)(-1)^2 \sum_{klmn} v(kl;nm) \cdot$$

$$\cdot \sum_{k_1 l_1 m_1 n_1} v\left(k_1 l_1; n_1 m_1\right) \int\limits_{-\infty}^{+\infty} \mathrm{d}t'\, \delta\left(t'\right) \int\limits_{-\infty}^{0} \mathrm{d}t_1\, \mathrm{e}^{-\alpha |t_1|} \cdot$$

$$\cdot \int\limits_{-\infty}^{+\infty} \mathrm{d}t_1'\, \delta\left(t_1 - t_1'\right) \left(\mathrm{i}G^{0,\mathrm{c}}_{k}\left(t_1 - 0\right) \delta_{kn_1}\right) \left(\mathrm{i}G^{0,\mathrm{c}}_{n}\left(0 - t_1\right) \delta_{nk_1}\right) \cdot$$

$$\cdot \left(\mathrm{i}G^{0,\mathrm{c}}_{l}\left(t_1' - t'\right) \delta_{lm_1}\right) \left(\mathrm{i}G^{0,\mathrm{c}}_{m}\left(t' - t_1'\right) \delta_{ml_1}\right) =$$

$$= \lim_{\alpha \to 0} \left(-\frac{\mathrm{i}}{2\hbar}\right) \sum_{klmn} v(kl;nm) v(nm;kl) \cdot$$

$$\cdot \int\limits_{-\infty}^{0} \mathrm{d}t_1\, \mathrm{e}^{-\alpha |t_1|} \left(\mathrm{i}G^{0,\mathrm{c}}_{k}(t_1)\right) \left(\mathrm{i}G^{0,\mathrm{c}}_{n}(-t_1)\right) \left(\mathrm{i}G^{0,\mathrm{c}}_{l}(t_1)\right) \left(\mathrm{i}G^{0,\mathrm{c}}_{m}(-t_1)\right) .$$

In this expression, we now insert the free, causal Green's functions from (5.78):

$$U_{(1)}(\Theta_1) = \lim_{\alpha \to 0} \left(-\frac{\mathrm{i}}{2\hbar}\right) \sum_{\substack{klmn \\ \sigma_k\sigma_l\sigma_m\sigma_n}}^{|l|,\,|k|<k_F<|n|,\,|m|} v(\boldsymbol{k}-\boldsymbol{n})\delta_{\boldsymbol{m}+\boldsymbol{n},\,\boldsymbol{k}+\boldsymbol{l}} \cdot$$

$$\cdot\ \delta_{\sigma_k\sigma_n}\delta_{\sigma_m\sigma_l} v(\boldsymbol{n}-\boldsymbol{k})\delta_{\boldsymbol{k}+\boldsymbol{l},\,\boldsymbol{m}+\boldsymbol{n}}\delta_{\sigma_n\sigma_k}\delta_{\sigma_m\sigma_l} \cdot$$

$$\cdot \int_{-\infty}^{0} \mathrm{d}t_1 \exp\left[\alpha t_1 - \frac{\mathrm{i}}{\hbar}\Big(\varepsilon(\boldsymbol{k})-\mu+\varepsilon(\boldsymbol{l})-\mu-\varepsilon(\boldsymbol{n})+\mu-\varepsilon(\boldsymbol{m})+\mu\Big)t_1\right] =$$

$$= \frac{1}{2}4 \sum_{klmn}^{|l|,\,|k|<k_F<|n|,\,|m|} v^2(\boldsymbol{n}-\boldsymbol{k}) \frac{\delta_{\boldsymbol{k}+\boldsymbol{l},\,\boldsymbol{m}+\boldsymbol{n}}}{\varepsilon(\boldsymbol{k})+\varepsilon(\boldsymbol{l})-\varepsilon(\boldsymbol{n})-\varepsilon(\boldsymbol{m})}\,.$$

With $\boldsymbol{n} = \boldsymbol{k}+\boldsymbol{q}$, $\boldsymbol{l} = \boldsymbol{p}$, $\boldsymbol{m} = \boldsymbol{p}-\boldsymbol{q}$, it finally follows that:

$$U_{(1)}(\Theta_1) = 2 \sum_{\substack{\boldsymbol{k},\,\boldsymbol{p},\,\boldsymbol{q} \\ \left(\substack{p,\,k<k_F \\ |\boldsymbol{k}+\boldsymbol{q}|,\,|\boldsymbol{p}-\boldsymbol{q}|>k_F}\right)}}^{\boldsymbol{q}\neq 0} \frac{v^2(\boldsymbol{q})}{\varepsilon(\boldsymbol{k})+\varepsilon(\boldsymbol{p})-\varepsilon(\boldsymbol{k}+\boldsymbol{q})-\varepsilon(\boldsymbol{p}-\boldsymbol{q})}\,. \tag{5.142}$$

As we shall later show, this contribution diverges owing to the Coulomb interaction $v^2(\boldsymbol{q})$. This is not the case for the structure 2):

$$U_{(2)}(\Theta_1) = \lim_{\alpha \to 0} \frac{4}{2}\frac{1}{4}\left(-\frac{\mathrm{i}}{\hbar}\right)(-1)\sum_{klmn} v(kl;nm)\ \cdot$$

$$\cdot \sum_{k_1l_1m_1n_1} v(k_1l_1;n_1m_1) \int_{-\infty}^{0} \mathrm{d}t_1\, \mathrm{e}^{-\alpha|t_1|}\left(\mathrm{i}G_k^{0,\mathrm{c}}(t_1)\,\delta_{km_1}\right)\cdot$$

$$\cdot \left(\mathrm{i}G_n^{0,\mathrm{c}}(-t_1)\,\delta_{nk_1}\right)\left(\mathrm{i}G_l^{0,\mathrm{c}}(t_1)\,\delta_{ln_1}\right)\left(\mathrm{i}G_m^{0,\mathrm{c}}(-t_1)\,\delta_{ml_1}\right) =$$

$$= \frac{\mathrm{i}}{2\hbar}\sum_{\substack{klmn \\ \sigma_k\sigma_l\sigma_m\sigma_n}} v(\boldsymbol{k}-\boldsymbol{n})\delta_{\boldsymbol{k}+\boldsymbol{l},\,\boldsymbol{m}+\boldsymbol{n}}\delta_{\sigma_k\sigma_n}\delta_{\sigma_m\sigma_l}\ \cdot$$

$$\cdot\ v(\boldsymbol{n}-\boldsymbol{l})\delta_{\boldsymbol{n}+\boldsymbol{m},\,\boldsymbol{l}+\boldsymbol{k}}\delta_{\sigma_n\sigma_l}\delta_{\sigma_m\sigma_k}\ \cdot$$

$$\cdot \left[-\frac{\mathrm{i}}{\hbar}\left(\varepsilon(\boldsymbol{k})+\varepsilon(\boldsymbol{l})-\varepsilon(\boldsymbol{n})-\varepsilon(\boldsymbol{m})\right)\right]^{-1}\Bigg|_{\substack{k,\,l<k_F \\ n,\,m>k_F}} =$$

$$= -\sum_{klmn}^{k,\,l<k_F<n,\,m} \frac{\delta_{\boldsymbol{k}+\boldsymbol{l},\,\boldsymbol{n}+\boldsymbol{m}}\,v(\boldsymbol{k}-\boldsymbol{n})v(\boldsymbol{n}-\boldsymbol{l})}{\varepsilon(\boldsymbol{k})+\varepsilon(\boldsymbol{l})-\varepsilon(\boldsymbol{n})-\varepsilon(\boldsymbol{m})}\,.$$

The structure 2) thus makes the contribution:

$$U_{(2)}(\Theta_1) = -\sum_{\substack{k,\,p,\,q \\ \left(\substack{p,\,k < k_F \\ |k+q|,\,|p-q| > k_F}\right)}}^{q \neq 0} \frac{v(\boldsymbol{q})v(\boldsymbol{k}+\boldsymbol{q}-\boldsymbol{p})}{\varepsilon(\boldsymbol{k})+\varepsilon(\boldsymbol{p})-\varepsilon(\boldsymbol{k}+\boldsymbol{q})-\varepsilon(\boldsymbol{p}-\boldsymbol{q})} . \tag{5.143}$$

We now want to prove by explicit evaluation the assertion made above that the contribution $U_{(1)}(\Theta_1)$ diverges:

$$\varepsilon(\boldsymbol{k})+\varepsilon(\boldsymbol{p})-\varepsilon(\boldsymbol{k}+\boldsymbol{q})-\varepsilon(\boldsymbol{p}-\boldsymbol{q}) = \frac{\hbar^2}{m}\boldsymbol{q}\cdot(\boldsymbol{p}-\boldsymbol{k}-\boldsymbol{q}) .$$

We normalise the wavenumbers

$$\bar{q} = -\frac{\boldsymbol{q}}{k_F}; \quad \bar{k} = -\frac{\boldsymbol{k}}{k_F}; \quad \bar{p} = \frac{\boldsymbol{p}}{k_F}$$

and as usual replace the sums by integrals:

$$\sum_{k} \Longrightarrow \frac{V}{(2\pi)^3}\int d^3k .$$

We then still have to evaluate the expression:

$$U_{(1)}(\Theta_1) = \frac{-2V^3}{(2\pi)^9} k_F^3 \frac{e^4}{\varepsilon_0^2 V^2}\int \frac{d^3\bar{q}}{\bar{q}^4}\iint d^3\bar{k}\, d^3\bar{p}\, \frac{m/\hbar^2}{\bar{q}\cdot(\bar{p}+\bar{k}+\bar{q})} .$$

Here, we are still using the energy unit "ryd" (2.35):

$$1\text{ryd} = \frac{me^4}{2\hbar^2(4\pi\varepsilon_0)^2} , \tag{5.144}$$

$$U_{(1)}(\Theta_1) = -\frac{3N}{8\pi^5}\iiint_{\left(\substack{\bar{p},\,\bar{k} < 1 \\ |\bar{k}+\bar{q}|,\,|\bar{p}+\bar{q}| > 1}\right)} d^3\bar{q}\, d^3\bar{k}\, d^3\bar{p}\, \frac{1}{\bar{q}^4}\frac{1}{\bar{q}\cdot(\bar{p}+\bar{k}+\bar{q})} . \tag{5.145}$$

We have made use of $k_F^3 = 3\pi^2 N/V$ in this expression. We abbreviate

$$x_p = \frac{\bar{p}\cdot\bar{q}}{\bar{p}\bar{q}}; \quad x_k = \frac{\bar{k}\cdot\bar{q}}{\bar{k}\bar{q}}$$

and consider the integral

$$I(\bar{q}) = \iint d^3\bar{p}\, d^3\bar{k}\, \frac{1}{\bar{q}\bar{p}x_p + \bar{q}\bar{k}x_k + \bar{q}^2} . \tag{5.146}$$

The range of integration is determined by

$$\bar{k} < 1 < |\bar{k}+\bar{q}| ; \quad \bar{p} < 1 < |\bar{p}+\bar{q}| .$$

We estimate these expressions for small values of $\bar{q}$.

$$
\begin{aligned}
|\bar{\boldsymbol{k}}+\bar{\boldsymbol{q}}| &= \sqrt{\bar{k}^2+\bar{q}^2+2\bar{k}\bar{q}x_k} = \bar{k}\left(1+2x_k\frac{\bar{q}}{\bar{k}}+\frac{\bar{q}^2}{\bar{k}^2}\right)^{1/2} = \\
&= \bar{k}+\bar{q}x_k+O(\bar{q}^2)\,, \\
|\bar{\boldsymbol{p}}+\bar{\boldsymbol{q}}| &= \bar{p}+\bar{q}x_p+O(\bar{q}^2)\,.
\end{aligned}
$$

For the range of integration, this implies:

$$1-\bar{q}x_k<\bar{k}<1\,;\quad 1-\bar{q}x_p<\bar{p}<1\,.$$

We define the polar axis to be parallel to $\bar{\boldsymbol{q}}$ and then have to evaluate:

$$I(\bar{q}) \approx 4\pi^2 \int\limits_{-1}^{+1} \mathrm{d}x_k \int\limits_{-1}^{+1} \mathrm{d}x_p \int\limits_{1-\bar{q}x_k}^{1} \mathrm{d}\bar{k} \int\limits_{1-\bar{q}x_p}^{1} \mathrm{d}\bar{p}\, \frac{\bar{k}^2\bar{p}^2}{\bar{q}\bar{p}x_p+\bar{q}\bar{k}x_k+\bar{q}^2}\,.$$

For $\bar{q}\to 0$, we can assume in the denominator of the integrand that $\bar{k}, \bar{p} = 1+O(\bar{q})$:

$$
\begin{aligned}
I(\bar{q}) &\approx 4\pi^2 \int\limits_{-1}^{+1} \mathrm{d}x_k \int\limits_{-1}^{+1} \mathrm{d}x_p \int\limits_{1-\bar{q}x_k}^{1} \mathrm{d}\bar{k} \int\limits_{1-\bar{q}x_p}^{1} \mathrm{d}\bar{p}\, \frac{\bar{k}^2\bar{p}^2}{\bar{q}\,(x_p+x_k)} = \\
&= \frac{4\pi^2}{9} \iint\limits_{-1}^{+1} \mathrm{d}x_k\,\mathrm{d}x_p\, \frac{\left\{1-(1-\bar{q}x_k)^3\right\}\left\{1-(1-\bar{q}x_p)^3\right\}}{\bar{q}\,(x_p+x_k)} \approx \\
&\approx \alpha\bar{q}+O\left(\bar{q}^2\right)\,.
\end{aligned}
\tag{5.147}
$$

Here,

$$\alpha = 4\pi^2 \iint\limits_{-1}^{+1} \mathrm{d}x_k\,\mathrm{d}x_p\, \frac{x_k x_p}{x_k+x_p}$$

is a simple numerical value. Inserting this result into (5.145),

$$U_{(1)}(\Theta_1) \approx -\frac{3N}{2\pi^4}\alpha \int\limits_0^{?} \frac{\mathrm{d}\bar{q}}{\bar{q}}\,, \tag{5.148}$$

We can see that the integral diverges at its lower bound. At the upper bound, our estimate is not valid; however, due to the $1/\bar{q}^4$ term, no irregularities occur there.

For the structure 2), from (5.143), we find:

$$U_{(2)}(\Theta_1) =$$
$$= -\frac{V^3}{(2\pi)^9}\frac{e^4}{\varepsilon_0^2 V^2}\frac{m}{\hbar^2} \iiint\limits_{\left(\substack{k,\,p<k_F \\ |k+q|,\,|p-q|>k_F}\right)} d^3q\, d^3k\, d^3p \frac{1}{q^2\,|\boldsymbol{k}+\boldsymbol{q}-\boldsymbol{p}|^2\boldsymbol{q}\cdot(\boldsymbol{p}-\boldsymbol{k}-\boldsymbol{q})} =$$
$$= -\frac{V k_F^3}{16\pi^7} \iiint\limits_{\left(\substack{\bar{k},\,\bar{p}<1 \\ |\bar{k}+\bar{q}|,\,|\bar{p}-\bar{q}|>1}\right)} d^3\bar{q}\, d^3\bar{k}\, d^3\bar{p} \frac{1}{\bar{q}^2\,|\bar{p}-\bar{k}-\bar{q}|^2\bar{q}\cdot(\bar{p}-\bar{k}-\bar{q})}[\text{ryd}]\,.$$

We now replace $\bar{q}$ by $-\bar{q}$ and $\bar{k}$ by $-\bar{k}$ and then obtain an expression which can be integrated analytically (L. Onsager *et al.*, Annalen der Physik **18**, 71 (1966)):

$$U_{(2)}(\Theta_1) = \frac{3N}{16\pi^5}\int \frac{d^3\bar{q}}{\bar{q}^2}\int\limits_{\left(\substack{\bar{k},\,\bar{p}<1 \\ |\bar{k}+\bar{q}|,\,|\bar{p}+\bar{q}|>1}\right)} d^3\bar{k}\int d^3\bar{p}\, \frac{1}{|\bar{p}+\bar{k}+\bar{q}|^2\bar{q}\cdot(\bar{p}+\bar{k}+\bar{q})}[\text{ryd}] = \tag{5.149}$$
$$= 0.0484\cdot N[\text{ryd}]\,.$$

The origin of the divergence of the structure 1) lies in the factor $v^2(\boldsymbol{q})$. This also holds for all the higher orders, each of which contains a diagram of the type 1), which contributes a factor $v^{n+1}(\boldsymbol{q})$ that produces the divergence. Such diagrams are called *ring diagrams*; they represent continuing sequences of *structural elements*. These contribute the same momentum transfer $\boldsymbol{q}$ at *every* interaction line.

Within the jellium model, we make the strange observation that first-order perturbation theory gives good results (5.138), but every additional term of the perturbation expansion diverges. If one however sums the infinite series, then the contributions of the ring diagrams cancel, so that a finite value is obtained.

Fig. 5.6 The definition of a ring diagram

5.5.3 The Correlation Energy

The so-called *Hartree-Fock solution* (5.139) for the ground-state energy of the interacting electron gas, which we derived here using first-order perturbation theory, yielded finally the expectation value of the Coulomb interaction in the **unperturbed** ground state $|\eta_0\rangle$. It takes the Pauli principle into account, which guarantees that electrons with parallel spins cannot approach each other too closely. This leads to a reduction in the ground-state energy, since it keeps like-charged particles at a certain distance. Due to the repulsive electron-electron interactions, it should however be improbable that electrons with antiparallel spins approach each other closely. This fact, that even particles with opposite spins are **correlated** with each other, is **not** taken into account in the Hartree-Fock approximation. One therefore refers to the deviation of the exact ground-state energy from its Hartree-Fock value as the

correlation energy,

which we want to estimate more precisely in this section following a procedure of M. Gell-Mann and K. A. Brueckner (Phys. Rev. **106**, 364 (1957)) in the limiting case of high electron densities. According to the Rayleigh-Ritz variational principle, the perturbation-theoretical result (5.139) already represents an upper limit for the true ground-state energy. Taking the correlations into account should therefore lead to a further reduction of the energy.

As a measure of the electron density, we make use of the dimensionless density parameter r_s, which was defined by (2.83):

$$\frac{V}{N} = \frac{4\pi}{3}(a_B r_s)^3 \; ; \quad a_B = \frac{4\pi\,\varepsilon_0\,\hbar^2}{me^2} \, .$$

a_B is the first Bohr radius. High electron densities correspond to low values of r_s.

In estimating the higher-order perturbation corrections, it can prove to be reasonable to reverse the transition from (3.14) to (3.18) for the time-evolution operator. We can then, instead of (5.134), also use the following formula for the ground-state energy:

$$\Delta E_0 = \lim_{\alpha \to 0} \frac{1}{\langle \eta_0 \,|\, U_\alpha(0, -\infty) \,|\, \eta_0 \rangle} \sum_{n=0}^{\infty} \left(-\frac{\mathrm{i}}{\hbar}\right)^n \cdot$$
$$\cdot \int\limits_{-\infty}^{0} \mathrm{d}t_1 \int\limits_{-\infty}^{t_1} \mathrm{d}t_2 \cdots \int\limits_{-\infty}^{t_{n-1}} \mathrm{d}t_n \, \mathrm{e}^{-\alpha(|t_1| + \cdots + |t_n|)} \cdot \tag{5.150}$$
$$\cdot \langle \eta_0 | \, V(t=0) V(t_1) \, V(t_2) \cdots V(t_n) \, | \eta_0 \rangle \, .$$

The operators are in this case already time-ordered. T_ε thus acts as an identity and can be left off. Note that in (5.150), as compared to (5.134), the factor $1/n!$ is missing. In counting the topologically distinct diagrams of the same structure, we must be careful, since the vertices cannot be permuted arbitrarily any longer owing to the fixed time ordering. This just explains the factor $1/n!$. Instead of (5.136), we now have:

$$A^*(\Theta_n) = \frac{2^{n+1}}{h(\Theta_n)} = 2^{n+1} \, . \tag{5.151}$$

Because of the fixed ordering of the vertices, there are now no longer any connected topologically equivalent diagrams.

We first consider those corresponding ring diagrams from third-order perturbation theory ($n = 2$) which we want to evaluate with (5.150) (Fig. 5.7).

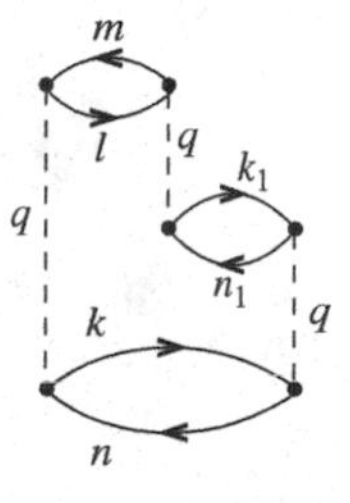

Fig. 5.7 A third-order ring diagram

$$U_{\mathrm{Ring}}(\Theta_2) =$$
$$= \lim_{\alpha\to 0} 8\frac{1}{8}\left(-\frac{\mathrm{i}}{\hbar}\right)^2 (-1)^3 \sum_{klmn}\sum_{k_1\ldots}\sum_{k_2\ldots} v(kl,nm)v(k_1l_1,n_1m_1)\,v(k_2l_2,n_2m_2)\,\cdot$$
$$\cdot \int\limits_{-\infty}^{0} \mathrm{d}t_1 \int\limits_{-\infty}^{t_1} \mathrm{d}t_2\, \mathrm{e}^{-\alpha(|t_1|+|t_2|)} \cdot \left(\mathrm{i}G_k^{0,\mathrm{c}}(t_2)\,\delta_{kn_2}\right)\left(\mathrm{i}G_n^{0,\mathrm{c}}(-t_2)\,\delta_{nk_2}\right)\,\cdot$$
$$\cdot\left(\mathrm{i}G_l^{0,\mathrm{c}}(t_1)\,\delta_{l,m_1}\right)\left(\mathrm{i}G_m^{0,\mathrm{c}}(-t_1)\,\delta_{m,l_1}\right)\left(\mathrm{i}G_{k_1}^{0,\mathrm{c}}(t_2-t_1)\,\delta_{k_1m_2}\right)\left(\mathrm{i}G_{n_1}^{0,\mathrm{c}}(t_1-t_2)\,\delta_{n_1l_2}\right).$$

Following (5.130), we insert the explicit Coulomb matrix elements and carry out the summation over spins, which yields trivial numerical values, since the free Green's functions are independent of spin:

$$= -8\left(-\frac{\mathrm{i}}{\hbar}\right)^2 \lim_{\alpha\to 0}\sum_{\substack{klmn\\k_1}} v(\boldsymbol{k}-\boldsymbol{n})v(\boldsymbol{n}-\boldsymbol{k})\,\cdot$$
$$\cdot\, v(\boldsymbol{n}-\boldsymbol{k})\int\limits_{-\infty}^{0}\mathrm{d}t_1\int\limits_{-\infty}^{t_1}\mathrm{d}t_2\,\mathrm{e}^{-\alpha(|t_1|+|t_2|)}\delta_{\boldsymbol{k}+\boldsymbol{l},\,\boldsymbol{m}+\boldsymbol{n}}\left(\mathrm{i}G_{\boldsymbol{k}}^{0,\mathrm{c}}(t_2)\right)\left(\mathrm{i}G_{\boldsymbol{n}}^{0,\mathrm{c}}(-t_2)\right)\,\cdot$$
$$\cdot\left(\mathrm{i}G_{\boldsymbol{l}}^{0,\mathrm{c}}(t_1)\right)\left(\mathrm{i}G_{\boldsymbol{m}}^{0,\mathrm{c}}(-t_1)\right)\left(\mathrm{i}G_{\boldsymbol{k}_1}^{0,\mathrm{c}}(t_2-t_1)\right)\left(\mathrm{i}G_{\boldsymbol{k}_1+\boldsymbol{k}-\boldsymbol{n}}^{0,\mathrm{c}}(t_1-t_2)\right).$$

We write

$$\boldsymbol{k}\to\boldsymbol{k}_1\,;\quad \boldsymbol{l}\to\boldsymbol{k}_2\,;\quad \boldsymbol{m}\to\boldsymbol{k}_2+\boldsymbol{q}\,;\quad \boldsymbol{n}\to\boldsymbol{k}_1-\boldsymbol{q}\,;\quad \boldsymbol{k}_1\to\boldsymbol{k}_3$$

and then obtain after insertion of (5.78):

$$U_{\mathrm{Ring}}(\Theta_2) = 8\left(-\frac{\mathrm{i}}{\hbar}\right)^2 \lim_{\alpha\to 0}\sum_{\substack{\boldsymbol{k}_1,\boldsymbol{k}_2,\boldsymbol{k}_3\\ \boldsymbol{q}}}^{\boldsymbol{q}\neq 0} v^3(\boldsymbol{q})\int\limits_{-\infty}^{0}\mathrm{d}t_1\int\limits_{-\infty}^{t_1}\mathrm{d}t_2\,\mathrm{e}^{\alpha(t_1+t_2)}\,\cdot$$
$$\cdot\,\Theta(k_{\mathrm{F}}-k_1)\,\Theta(|\boldsymbol{k}_1-\boldsymbol{q}|-k_{\mathrm{F}})\,\cdot$$
$$\cdot\,\Theta(k_{\mathrm{F}}-k_2)\,\Theta(|\boldsymbol{k}_2+\boldsymbol{q}|-k_{\mathrm{F}})\,\cdot$$
$$\cdot\,\Theta(k_{\mathrm{F}}-k_3)\,\Theta(|\boldsymbol{k}_3+\boldsymbol{q}|-k_{\mathrm{F}})\,\cdot$$
$$\cdot\,\exp\!\left[-\frac{\mathrm{i}}{\hbar}\Big(\varepsilon(\boldsymbol{k}_2)-\varepsilon(\boldsymbol{k}_2+\boldsymbol{q})-\varepsilon(\boldsymbol{k}_3)+\varepsilon(\boldsymbol{k}_3+\boldsymbol{q})\Big)t_1\right]\cdot$$
$$\cdot\,\exp\!\left[-\frac{\mathrm{i}}{\hbar}\Big(\varepsilon(\boldsymbol{k}_1)-\varepsilon(\boldsymbol{k}_1-\boldsymbol{q})+\varepsilon(\boldsymbol{k}_3)-\varepsilon(\boldsymbol{k}_3+\boldsymbol{q})\Big)t_2\right]=$$

$$= 8 \sum_{\substack{q \neq 0 \\ k_1, k_2, k_3, q \\ \begin{pmatrix} k_1 < k_F < |k_1 - q| \\ k_2 < k_F < |k_2 + q| \\ k_3 < k_F < |k_3 + q| \end{pmatrix}}} \frac{v^3(q)}{(\varepsilon(k_1) - \varepsilon(k_1 - q) + \varepsilon(k_3) - \varepsilon(k_3 + q))} \cdot$$

$$\cdot \frac{1}{(\varepsilon(k_2) - \varepsilon(k_2 + q) + \varepsilon(k_1) - \varepsilon(k_1 - q))} .$$

We substitute

$$q \to -\frac{q}{k_F}; \quad k_2 \to -\frac{k_2}{k_F}; \quad k_3 \to -\frac{k_3}{k_F}$$

and again use the energy unit ryd (5.144). With

$$k_F a_B = \frac{\alpha}{r_s}; \quad \alpha = \left(\frac{9\pi}{4}\right)^{1/3} \quad \text{(cf. (2.86))},$$

we then find the intermediate result:

$$U_{\text{Ring}}(\Theta_2) = \frac{3N}{4\pi^7 \alpha} r_s \int \frac{d^3\bar{q}}{\bar{q}^6} \iiint\limits_{\substack{\bar{k}_i < 1 < |\bar{k}_i + \bar{q}| \\ i=1,2,3}} d^3\bar{k}_1 d^3\bar{k}_2 d^3\bar{k}_3 \cdot \tag{5.152}$$

$$\cdot \frac{1}{[\bar{q} \cdot (\bar{k}_1 + \bar{k}_3 + \bar{q})][\bar{q} \cdot (\bar{k}_1 + \bar{k}_2 + \bar{q})]} [\text{ryd}]$$

We want to demonstrate, as in (5.79), that this contribution also diverges. To this end, we first investigate the triple integral over the $\bar{k}_i$

$$I_{(2)}(\bar{q}) \equiv \iiint\limits_{\substack{\bar{k}_i < 1 < |\bar{k}_i + \bar{q}| \\ i=1,2,3}} d^3\bar{k}_1 \, d^3\bar{k}_2 \, d^3\bar{k}_3 \left\{ \left(\bar{q}\bar{k}_1 x_1 + \bar{q}\bar{k}_3 x_3 + \bar{q}^2\right) \cdot \right. \tag{5.153}$$

$$\left. \cdot \left(\bar{q}\bar{k}_1 x_1 + \bar{q}\bar{k}_2 x_2 + \bar{q}^2\right) \right\}^{-1}$$

for small values of $\bar{q}$. We again have abbreviated

$$x_i = \frac{\bar{k}_i \cdot \bar{q}}{\bar{k}_i \bar{q}}; \quad i = 1, 2, 3 .$$

The range of integration can be estimated as from (5.146) to be

$$1 - \bar{q} x_i < \bar{k}_i < 1; \quad i = 1, 2, 3 .$$

Within these ranges, however, $\bar{k}_i = 1 + O(\bar{q})$. The polar axis is chosen to be parallel to $\bar{\boldsymbol{q}}$:

$$I_{(2)}(\bar{\boldsymbol{q}}) \approx 8\pi^3 \iiint\limits_{-1}^{+1} \mathrm{d}x_1\, \mathrm{d}x_2\, \mathrm{d}x_3 \iiint\limits_{1-\bar{q}x_i}^{+1} \mathrm{d}\bar{k}_1\, \mathrm{d}\bar{k}_2\, \mathrm{d}\bar{k}_3\, \bar{k}_1^2 \bar{k}_2^2 \bar{k}_3^2 \cdot$$

$$\cdot \left\{\bar{q}^2 \left[(x_1 + x_3)(x_1 + x_2) + O(\bar{q})\right]\right\}^{-1} =$$

$$= \frac{8\pi^3}{9} \iiint\limits_{-1}^{+1} \mathrm{d}x_1\, \mathrm{d}x_2\, \mathrm{d}x_3 \left[1 - (1 - \bar{q}x_1)^3\right]\left[1 - (1 - \bar{q}x_2)^3\right] \cdot$$

$$\cdot \left[1 - (1 - \bar{q}x_3)^3\right]\left\{\bar{q}^2 (x_1 + x_3)(x_1 + x_2) + O\left(\bar{q}^3\right)\right\}^{-1} .$$

This yields

$$I_{(2)}(\bar{\boldsymbol{q}}) = \alpha_{(2)}\bar{q} + O\left(\bar{q}^2\right) \tag{5.154}$$

with a simple numerical factor $\alpha_{(2)}$. The $\bar{q}$-dependence is thus the same as in (5.147). For the remaining contribution (5.152) of the ring diagram, we then have

$$U_{\mathrm{Ring}}\left(\Theta_2\right) \approx \gamma_{(2)} r_s \int\limits_0^{?} \frac{\mathrm{d}\bar{q}}{\bar{q}^3} , \tag{5.155}$$

which obviously diverges at the lower limit of the integration.

In $(n + 1)$-th order perturbation theory, the integral which is analogous to (5.153), $I_{(n)}(\bar{q})$, can be estimated for small values of $\bar{q}$. The integrations over the *inner* times $t_1, t_2, \ldots, t_n$ each contribute a factor of $\bar{q}^{-1}$ with the above estimate. At each vertex, we have in fact three independent summations over $\boldsymbol{k}$. In the case of ring diagrams, each *inner* vertex, except for the last, yields only *one* additional, independent summation over wavenumbers. With n *inner* vertices, this gives $(n - 1)$ summations. The *fixed* vertex at the left contributes three summations, one of them over $\bar{q}$. All together, after taking the thermodynamic limit in $I_{(n)}(\bar{q})$, we have $(n + 1)\,\bar{k}_i$ integrations, each from $1 - \bar{q}x_i$ to 1. Each one contributes a factor of $\bar{q}$ (after expansion for small $\bar{q}$, as shown just before (5.154)). This yields all together

$$I_{(n)}(\bar{q}) \approx \alpha_{(n)}\bar{q} + O\left(\bar{q}^2\right) . \tag{5.156}$$

The contribution U of the ring diagram in $(n + 1)$-th order perturbation theory therefore diverges due to the factor $v^{n+1}(\bar{\boldsymbol{q}}) \sim \bar{q}^{-(2n+2)}$ as

$$U_{\mathrm{Ring}}\left(\Theta_n\right) \sim \int\limits_0^{?} \frac{\mathrm{d}\bar{q}}{\bar{q}^{2n-1}} . \tag{5.157}$$

Compare the particular results (5.148) for $n = 1$ and (5.155) for $n = 2$.

Now, the dependencies of the contributions of the individual diagrams on the density parameter r_s are important; they can be readily estimated for arbitrary diagrams, i.e. not only for ring diagrams. In second-order perturbation theory ($n = 1$), we computed all of the possible diagrams for the jellium model in Sect. 5.5.2 exactly. They proved to be independent of r_s. In every order which increases by $\Delta n = 1$, an additional factor of

$$v\left(\boldsymbol{k}_i - \boldsymbol{k}_j\right) \sim \frac{1}{|\boldsymbol{k}_i - \boldsymbol{k}_j|^2}$$

occurs, which after scaling contributes a factor k_{F}^{-2}. Each new *inner* vertex furthermore yields an additional time variable over which we must integrate. This adds an additional energy denominator $\left\{\varepsilon(k_i) + \cdots\right\}^{-1}$, which after scaling, owing to $\varepsilon(k_i) \sim k_i^2$, likewise contributes a factor k_{F}^{-2}. Every additional *inner* vertex requires *one* further $\boldsymbol{k}$-summation, which, after taking the thermodynamic limit,

$$\sum_{\boldsymbol{k}} \longrightarrow \frac{V}{(2\pi)^3} \int \mathrm{d}^3k\,,$$

leads to a scaling factor k_{F}^3. All together, each increase in the order of perturbation theory by Δn=1 gives a factor of k_{F}^{-1} and thus, due to

$$k_{\mathrm{F}} a_{\mathrm{B}} = \frac{\alpha}{r_s}\,,$$

a factor r_s. For the contribution of $U(\Theta_n)$ to the $(n+1)$-th-order perturbation theory, we thus have:

$$U\left(\Theta_n\right) \sim r_s^{n-1}\,; \quad n = 0, 1, 2, \dots \tag{5.158}$$

This is naturally extremely favourable for a perturbation theory in the region of high electron densities ($r_s \to 0$). The perturbation expansion could be terminated after a finite number of terms, if the divergent $q \to 0$ behaviour were not present in certain terms; this in turn results from the long range of the Coulomb interaction.

The correlation energy must naturally be finite in the end, i.e. the Coulomb interaction is finally screened by the electron gas itself. Thus, the divergent terms in the perturbation expansion must compensate each other in the summation to give a finite value. We shall therefore attempt to carry out infinite partial sums over the *critical* diagrams, whilst the non-divergent terms, due to (5.158), need to be taken into account only up to a finite order. We must however be aware that not only the ring diagrams exhibit singularities. We therefore first justify why we may nevertheless limit ourselves essentially to the ring diagrams in carrying out the evaluation. The actual ring diagrams look like those in Fig. 5.8.

In Fig. 5.9, a divergent non-ring diagram is shown as an example. It belongs to $U\left(\Theta_2\right)$ and yields a term $v^2(q)$; it is therefore *less divergent* than the corresponding ring diagram, but *just as divergent* as $U_{\mathrm{Ring}}\left(\Theta_1\right)$. It is thus not immediately apparent just why it may be neglected. It diverges as

$$\int_0^{?} \frac{\mathrm{d}\bar{q}}{\bar{q}^{4-3}}\, g_2^{(3)}(\bar{q})\,,$$

where $g_2^{(3)}(\bar{q})$ is a *harmless* factor which remains finite for $\bar{q} \to 0$. The upper index indicates the order $(n+1)$ of perturbation theory, and the lower index ($m = 2$) gives the number of equal momentum transfers of the structure being considered, whereby a factor of

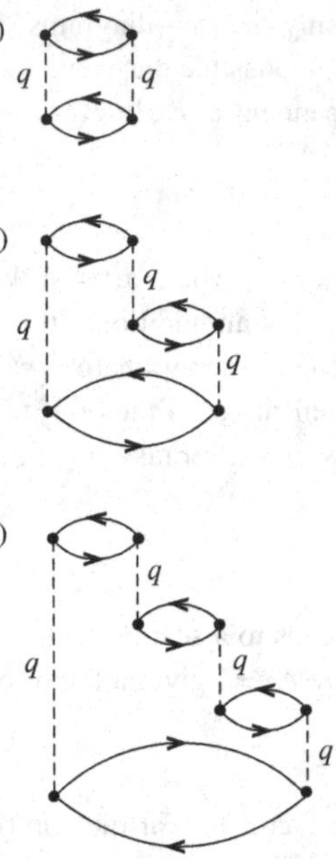

Fig. 5.8 (a–c) Low-order ring diagrams

$\bar{q}^{-2m}$ is brought into play. In general, one can write for the contribution of the n-th order diagram:

$$U(\Theta_n) = r_s^{n-1} \sum_{m=1}^{n+1} \int_0^{?} \mathrm{d}\bar{q}\, \bar{q}^{3-2m} g_m^{(n+1)}(\bar{q}), \tag{5.159}$$

$$g_m^{(n+1)}(\bar{q}) \underset{\bar{q} \to 0}{\longrightarrow} \text{const}.$$

All the contributions except for $m = 1$ are divergent. The actual ring diagrams correspond to $m = n + 1$ (see (5.157)). Why need one include only the ring diagrams in the evaluation?

As already mentioned, the physical origin of the divergences is the long range of the Coulomb potential. The reaction of the electron gas to the potential leads to a screening effect, so that only those wavenumbers q are relevant whose contribution exceeds a minimum value k_m. If we take as a measure of k_m the reciprocal of the **Thomas-Fermi screening length** (4.123),

$$k_m \sim r_s^{-1/2} \quad \Longrightarrow \quad \bar{k}_m = \frac{k_m}{k_\mathrm{F}} \sim \frac{r_s^{-1/2}}{k_\mathrm{F}} \sim r_s^{+1/2},$$

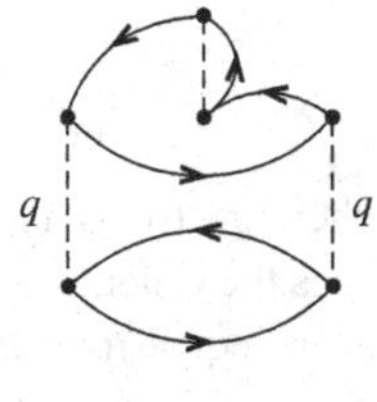

Fig. 5.9 An example of a divergent non-ring diagram

then we obtain the estimate:

$$m = 2: \quad \int\limits_{\bar{k}_m}^{?} \frac{\mathrm{d}\bar{q}}{\bar{q}} g_2^{(n+1)}(\bar{q}) \sim \ln \bar{k}_m \sim \ln r_s\,,$$

$$m > 2: \quad \int\limits_{\bar{k}_m}^{?} \mathrm{d}\bar{q} \cdot \bar{q}^{3-2m} g_m^{(n+1)}(\bar{q}) \sim \bar{k}_m^{4-2m} \sim r_s^{2-m}\,.$$

The contribution of a diagram with n *inner* vertices then, due to (5.159), scales as

$$U(\Theta_n) \sim r_s^{n+1-m} \quad (m > 2)\,. \tag{5.160}$$

For the ring diagrams ($m = n + 1$), this means that

$$U_{\text{Ring}}(\Theta_n) \sim r_s^0 \quad (m > 2) \tag{5.161}$$

and for the contributions of all the other diagrams:

$$U(\Theta_n) \sim r_s^t \underset{r_s \to 0}{\longrightarrow} 0 \quad (t > 0)\,.$$

A special case is $n = 1$, which we calculated explicitly in the last section:

$$\begin{aligned} U_{\text{Ring}}(\Theta_1) &\sim \ln r_s\,, \\ U_{(2)}(\Theta_1) &\sim r_s^0 \quad (\text{s. } (5.149))\,. \end{aligned} \tag{5.162}$$

For high electron densities (r_s small!), we then find the following expression to be a reasonable approximation for the correlation energy:

$$E_{\text{corr}} \approx \sum_{n=1}^{\infty} U_{\text{Ring}}(\Theta_n) + U_{(2)}(\Theta_1)\,. \tag{5.163}$$

As our next step, we therefore try to sum the ring diagrams.
Our starting point is the following representation for $U_{\text{Ring}}(\Theta)$:

$$U_{\text{Ring}}(\Theta_{n-1}) = (-1)^{n-1} \frac{3N}{8\pi^5} \left(\frac{r_s}{\alpha\pi^2}\right)^{n-2} \int \frac{I_{(n-1)}(\bar{\boldsymbol{q}})}{\bar{q}^{2n}} \mathrm{d}^3\bar{q}\,[\text{ryd}]\,, \tag{5.164}$$

$$I_{(n-1)}(\bar{q}) = \frac{1}{n} \int_{-\infty}^{+\infty} \cdots \int \mathrm{d}t_1 \cdots \mathrm{d}t_n\, F_{\bar{q}}(t_1) \cdots F_{\bar{q}}(t_n)\, \delta(t_1 + \cdots + t_n)\,, \tag{5.165}$$

$$F_{\bar{q}}(t) = \int\limits_{\bar{p} < 1 < |\bar{p} + \bar{q}|} \mathrm{d}^3\bar{p}\, \exp\left(-\left(\frac{1}{2}\bar{q}^2 + \bar{\boldsymbol{q}} \cdot \bar{\boldsymbol{p}}\right)|t|\right)\,. \tag{5.166}$$

For a general proof of this assertion, we must refer to the original literature (M. Gell-Mann, K. A. Brueckner, Phys. Rev. **106**, 364 (1957)). However, we shall examine the case of $n = 2$ explicitly:

$$I_{(1)}(\bar{\boldsymbol{q}}) =$$

$$= \frac{1}{2} \iint\limits_{-\infty}^{+\infty} \mathrm{d}t_1\, \mathrm{d}t_2\, F_{\bar{q}}(t_1)\, F_{\bar{q}}(t_2)\, \delta(t_1 + t_2) =$$

$$= \frac{1}{2} \int\limits_{-\infty}^{+\infty} \mathrm{d}t_1\, F_{\bar{q}}(t_1)\, F_{\bar{q}}(-t_1) =$$

$$= \iint\limits_{\substack{\bar{p}_i < 1 < |\bar{\boldsymbol{p}}_i + \bar{\boldsymbol{q}}| \\ i=1,2}} \mathrm{d}^3\bar{p}_1\, \mathrm{d}^3\bar{p}_2 \int\limits_0^{\infty} \mathrm{d}t\, \exp - \left(\left(\frac{1}{2}\bar{q}^2 + \bar{\boldsymbol{q}} \cdot \bar{\boldsymbol{p}}_1 \right) t \right) \exp - \left(\left(\frac{1}{2}\bar{q}^2 + \bar{\boldsymbol{q}} \cdot \bar{\boldsymbol{p}}_2 \right) t \right).$$

The exponents

$$\frac{1}{2}\bar{q}^2 + \bar{\boldsymbol{q}} \cdot \bar{\boldsymbol{p}}_i = \frac{1}{2}\left[(\bar{\boldsymbol{q}} + \bar{\boldsymbol{p}}_i)^2 - \bar{\boldsymbol{p}}_i^2 \right] \tag{5.167}$$

are positive within the range of integration, so that the above integrals converge in any case:

$$I_{(1)}(\bar{\boldsymbol{q}}) = \iint\limits_{\bar{p}_i < 1 < |\bar{\boldsymbol{p}}_i + \bar{\boldsymbol{q}}|\, i=1,2} \mathrm{d}^3\bar{p}_1 \mathrm{d}^3\bar{p}_2 \frac{1}{\bar{\boldsymbol{q}} \cdot (\bar{\boldsymbol{q}} + \bar{\boldsymbol{p}}_1 + \bar{\boldsymbol{p}}_2)} \quad \text{(s. (5.146))}.$$

If we insert this result into (5.164), we indeed find precisely (5.145). We recommend the verification of the case $n = 3$ as an exercise for the reader.

We now insert the following expression for the δ-function into (5.164):

$$\delta(t) = \frac{1}{2\pi} \int\limits_{-\infty}^{+\infty} \mathrm{d}\omega\, \mathrm{e}^{\mathrm{i}\omega t} = \frac{\bar{q}}{2\pi} \int\limits_{-\infty}^{+\infty} \mathrm{d}\omega\, \mathrm{e}^{\mathrm{i}\bar{q}\omega t},$$

and thus obtain:

$$\begin{aligned} I_{(n-1)}(\bar{q}) &= \frac{\bar{q}}{2\pi n} \int\limits_{-\infty}^{+\infty} \mathrm{d}\omega \int \cdots \int\limits_{-\infty}^{+\infty} \mathrm{d}t_1 \cdots \mathrm{d}t_n\, F_{\bar{q}}(t_1) \cdots F_{\bar{q}}(t_n) \cdot \\ &\quad \cdot \exp\left[i\bar{q}\omega(t_1 + \cdots + t_n) \right] = \\ &= \frac{\bar{q}}{2\pi n} \int\limits_{-\infty}^{+\infty} \mathrm{d}\omega \left[\int\limits_{-\infty}^{+\infty} \mathrm{d}t\, F_{\bar{q}}(t) \mathrm{e}^{\mathrm{i}\bar{q}\omega t} \right]^n. \end{aligned} \tag{5.168}$$

We further evaluate the expression in square brackets:

$$R_{\bar{q}}(\omega) \equiv \int_{-\infty}^{+\infty} \mathrm{d}t\, F_{\bar{q}}(t)\mathrm{e}^{\mathrm{i}\bar{q}\omega t} =$$

$$= \int\limits_{\bar{p}<1<|\bar{p}+\bar{q}|} \mathrm{d}^3\bar{p} \int_{-\infty}^{+\infty} \mathrm{d}t\, \exp(\mathrm{i}\bar{q}\omega t)\exp\left(-\left(\frac{1}{2}\bar{q}^2 + \bar{\boldsymbol{q}}\cdot\bar{\boldsymbol{p}}\right)|t|\right) = \tag{5.169}$$

$$= \int \mathrm{d}^3\bar{p}\,\Theta(1-\bar{p})\,\Theta(|\bar{\boldsymbol{p}}+\bar{\boldsymbol{q}}|-1)\,\frac{2\left(\frac{1}{2}\bar{q}^2 + \bar{\boldsymbol{q}}\cdot\bar{\boldsymbol{p}}\right)}{\bar{q}^2\omega^2 + \left(\frac{1}{2}\bar{q}^2 + \bar{\boldsymbol{q}}\cdot\bar{\boldsymbol{p}}\right)^2}\,.$$

Due to (5.167), the quotient in the integrand is antisymmetrical with respect to the permutation $\bar{\boldsymbol{p}} \rightleftharpoons \bar{\boldsymbol{p}} + \bar{\boldsymbol{q}}$. In the product of the step functions

$$\Theta(1-\bar{p})\,\Theta(|\bar{\boldsymbol{p}}+\bar{\boldsymbol{q}}|-1) = \Theta(1-\bar{p})\left\{1-\Theta(1-|\bar{\boldsymbol{p}}+\bar{\boldsymbol{q}}|)\right\},$$

in contrast, the second term is symmetrical with respect to this permutation, so that all together, we have:

$$R_{\bar{q}}(\omega) = 2\int \mathrm{d}^3\bar{p}\,\Theta(1-\bar{p})\,\frac{\frac{1}{2}\bar{q}^2 + \bar{\boldsymbol{q}}\cdot\bar{\boldsymbol{p}}}{\bar{q}^2\omega^2 + \left(\frac{1}{2}\bar{q}^2 + \bar{\boldsymbol{q}}\cdot\bar{\boldsymbol{p}}\right)^2}\,.$$

We first carry out the integrations over angles. Using

$$\int_{-1}^{+1} \frac{\mathrm{d}x}{\pm\mathrm{i}\bar{q}\omega + \frac{1}{2}\bar{q}^2 + \bar{q}\bar{p}x} = \frac{1}{\bar{q}\bar{p}}\ln\frac{\pm\mathrm{i}\bar{q}\omega + \frac{1}{2}\bar{q}^2 + \bar{q}\bar{p}}{\pm\mathrm{i}\bar{q}\omega + \frac{1}{2}\bar{q}^2 - \bar{q}\bar{p}}\,,$$

we find the intermediate result:

$$R_{\bar{q}}(\omega) = \frac{2\pi}{\bar{q}}\int_0^1 \mathrm{d}\bar{p}\,\bar{p}\ln\frac{\left(\frac{1}{2}\bar{q}+\bar{p}\right)^2 + \omega^2}{\left(\frac{1}{2}\bar{q}-\bar{p}\right)^2 + \omega^2}\,.$$

Defining

$$x = \bar{p} \pm \frac{1}{2}\bar{q}\,,$$

we must still evaluate:

$$R_{\bar{q}}(\omega) = \frac{2\pi}{\bar{q}}\left[\int_{\frac{1}{2}\bar{q}}^{1+\frac{1}{2}\bar{q}} \mathrm{d}x\left(x-\frac{1}{2}\bar{q}\right)\ln(x^2+\omega^2) - \int_{-\frac{1}{2}\bar{q}}^{1-\frac{1}{2}\bar{q}} \mathrm{d}x\left(x+\frac{1}{2}\bar{q}\right)\ln\left(x^2+\omega^2\right)\right].$$

The integrals are elementary:

$$\int \mathrm{d}x\,\ln\left(x^2+\omega^2\right) = x\ln\left(x^2+\omega^2\right) + 2\omega\arctan\frac{x}{\omega} - 2x + C_1\,, \tag{5.170}$$

$$\int \mathrm{d}x\,x\ln\left(x^2+\omega^2\right) = \frac{1}{2}\left(x^2+\omega^2\right)\ln\left(x^2+\omega^2\right) - \frac{x^2}{2} + C_2\,. \tag{5.171}$$

We thus obtain:

$$R_{\bar{q}}(\omega) = 2\pi \left\{ 1 - \omega \left(\arctan \frac{1 + \frac{1}{2}\bar{q}}{\omega} + \arctan \frac{1 - \frac{1}{2}\bar{q}}{\omega} \right) + \right.$$
$$\left. + \frac{1 - \frac{1}{4}\bar{q}^2 + \omega^2}{2\bar{q}} \ln \frac{\left(1 + \frac{1}{2}\bar{q}\right)^2 + \omega^2}{\left(1 - \frac{1}{2}\bar{q}\right)^2 + \omega^2} \right\} . \tag{5.172}$$

We first insert this into (5.168), in order to evaluate (5.164):

$$\Delta E_{\text{Ring}} = \sum_{n=2}^{\infty} \Delta E_{\text{Ring}}^{(n)} = \sum_{n=2}^{\infty} U_{\text{Ring}}(\Theta_{n-1}) =$$
$$= -\frac{3N}{8\pi^5} \left(\frac{\alpha\pi^2}{r_s} \right)^2 \int d^3\bar{q} \, \frac{\bar{q}}{2\pi} \int_{-\infty}^{+\infty} d\omega \sum_{n=2}^{\infty} \frac{(-1)^n}{n} \left(r_s \frac{R_{\bar{q}}(\omega)}{\alpha\pi^2\bar{q}^2} \right)^n . \tag{5.173}$$

The expansion converges in the case that

$$-1 < r_s \frac{R_{\bar{q}}(\omega)}{\alpha\pi^2\bar{q}^2} < +1 \tag{5.174}$$

can be assumed, which however certainly becomes questionable for small values of $\bar{q}$:

$$\Delta E_{\text{Ring}} = \frac{3N}{16\pi^6} \left(\frac{\alpha\pi^2}{r_s} \right)^2 \int d^3\bar{q}\,\bar{q} \int_{-\infty}^{+\infty} d\omega \left[\ln \left(1 + r_s \frac{R_{\bar{q}}(\omega)}{\alpha\pi^2\bar{q}^2} \right) - r_s \frac{R_{\bar{q}}(\omega)}{\alpha\pi^2\bar{q}^2} \right] . \tag{5.175}$$

We are now in principle finished. The remaining multiple integrals must be evaluated numerically. The use of the logarithm proves always to be correct, in spite of (5.174) (K. Sawada: Phys. Rev. **106**, 372 (1957); K. Sawada, K. Brueckner, N. Fukuda, R. Brout: Phys. Rev. **108**, 507 (1957)):

$$\Delta E_{\text{Ring}} = N \left[\frac{2}{\pi^2}(1 - \ln 2) \ln r_s - 0.142 + O\left(r_s \ln r_s\right) \right] [\text{ryd}] . \tag{5.176}$$

With (5.149) and (5.176) inserted into (5.163), we finally arrive at the correlation energy:

$$\frac{1}{N} E_{\text{corr}} = \left[0.0622 \ln r_s - 0.094 + O\left(r_s \ln r_s\right) \right] [\text{ryd}] . \tag{5.177}$$

Higher-order corrections correspond to higher powers of r_s, which we can neglect for the case of high electron densities. Only for high densities is (5.177) also acceptable. Note, however, that for typical metallic electron densities, r_s lies in the range $1 < r_s < 6$.

5.6 Diagrammatic Partial Sums

In Sect. 5.4.2, we derived the Dyson equation for the one-electron Green's function. A central point in the derivation was the introduction of the concept of self energy (5.124).

Every approximation to the self energy $\Sigma_{k\sigma}(E)$, no matter how simple, already corresponds to an infinite partial sum. The self-energy concept is however not the only possibility to form partial sums. In this section, we wish to introduce additional variants. The summing of such infinite partial series is often very important, sometimes even unavoidable. Quasi-particle lifetime effects can, for example, be calculated only in this way. In the last section, we saw that divergences in the individual terms of the perturbation expansion for the ground-state energy will compensate each other through the use of suitable partial sums to give finite values.

For many diagram expansions, a considerable reduction in the number of diagrams considered can be achieved if one includes only those diagrams which contain a self-energy part in **none** of their particle lines ("skeleton diagrams"), and instead in the remaining diagrams, one replaces every free propagator by the full propagator. In a similar manner, interaction lines also may be renormalised (*dressing*). Some of the most important methods of this type will be discussed in outline form in the following sections, whereby we will limit our considerations to the concrete example of the jellium model.

5.6.1 The Polarisation Propagator

In connection with the dielectric function $\varepsilon(\boldsymbol{q}, E)$, in Sect. 3.1.5 we introduced the so-called **density correlation**. This is a two-particle Green's function:

$$D_{\boldsymbol{q}}\left(t,t'\right) = \langle\langle \rho_{\boldsymbol{q}}(t);\rho_{\boldsymbol{q}}^{+}\left(t'\right)\rangle\rangle = -\mathrm{i}\left\langle T_{\varepsilon}\left(\rho_{\boldsymbol{q}}(t)\rho_{\boldsymbol{q}}^{+}\left(t'\right)\right)\right\rangle = \\ = \frac{1}{2\pi\hbar}\int\limits_{-\infty}^{+\infty} \mathrm{d}E\, D_{\boldsymbol{q}}(E)\exp\left(-\frac{\mathrm{i}}{\hbar}E\left(t-t'\right)\right). \tag{5.178}$$

We have already encountered the density operators $\rho_{\boldsymbol{q}}(t)$ in (3.97):

$$\rho_{\boldsymbol{q}}(t) = \sum_{\boldsymbol{k},\sigma} a_{\boldsymbol{k}\sigma}^{+}(t)a_{\boldsymbol{k}+\boldsymbol{q}\sigma}(t); \qquad \rho_{\boldsymbol{q}}^{+}(t) = \rho_{-\boldsymbol{q}}(t). \tag{5.179}$$

The computation of the expression (5.178),

$$\mathrm{i}D_{\boldsymbol{q}}(E) = \int\limits_{-\infty}^{+\infty} \mathrm{d}\left(t-t'\right)\exp\left(\frac{\mathrm{i}}{\hbar}E\left(t-t'\right)\right)\cdot \\ \cdot\sum_{\substack{\boldsymbol{k},\boldsymbol{p}\\ \sigma,\sigma'}}\left\langle E_0\left|T_{\varepsilon}\left\{a_{\boldsymbol{k}\sigma}^{+}(t)a_{\boldsymbol{k}+\boldsymbol{q}\sigma}(t)a_{\boldsymbol{p}\sigma'}^{+}\left(t'\right)a_{\boldsymbol{p}-\boldsymbol{q}\sigma'}\left(t'\right)\right\}\right|E_0\right\rangle, \tag{5.180}$$

corresponds to the task formulated in (5.115). The sums run over all combinations of open, connected diagrams, which all together exhibit four outer lines, two attached at t and two at t':

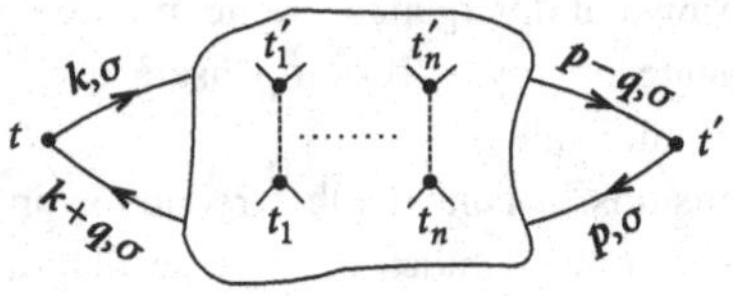

If we once again assume conservation of spin at each vertex point (no spin exchange between the interacting particles!), then it becomes immediately clear that (5.180) is nonzero only for $\sigma = \sigma'$. This is already taken into account in the figure. Let us think in particular of the jellium model; all open diagrams of the above type are, due to $v(\mathbf{0}) = 0$, themselves already connected. Then for example a diagram contribution of the form

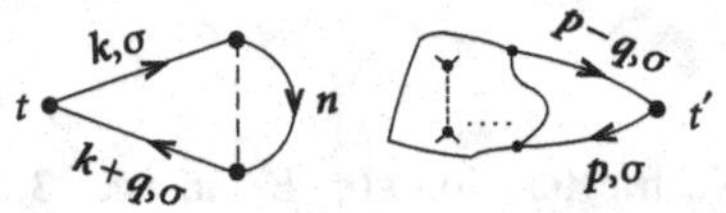

requires from conservation of momentum at the vertex that $\boldsymbol{k} + \boldsymbol{q} + \boldsymbol{n} = \boldsymbol{k} + \boldsymbol{n} \iff \boldsymbol{q} = 0$, and thus makes no contribution. Thus, in (5.180), we need not sum over *combinations* of open, connected diagrams with *all together* four outer lines, but rather only over the connected diagrams themselves.

The zeroth order contains no vertex. Therefore, for $\boldsymbol{q} \neq 0$, only one diagram is possible.

$\boxed{n = 0:}$

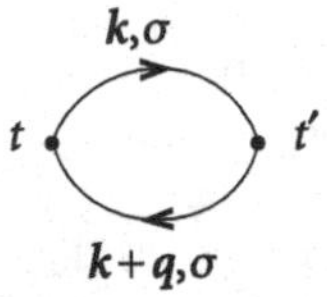

$$\iff (-1)\left(\mathrm{i}G^{0,\,c}_{k\sigma}\left(t'-t\right)\right)\delta_{\boldsymbol{k},\,\boldsymbol{p}-\boldsymbol{q}}\left(\mathrm{i}G^{0,\,c}_{\boldsymbol{k}+\boldsymbol{q},\sigma}\left(t'-t\right)\right)\delta_{\boldsymbol{p},\,\boldsymbol{k}+\boldsymbol{q}}\,.$$

The factor (-1) results from the loop rule.
To first order, we find the following diagram structures:

$\boxed{n = 1 :}$

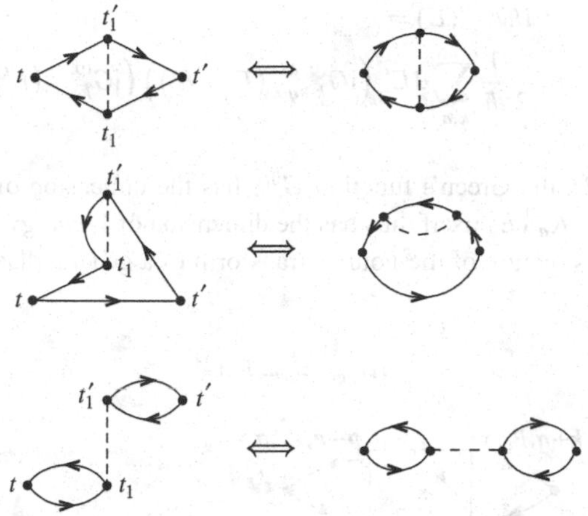

The two representations are of course completely equivalent; the right-hand one is the more usual.

As in the case of the single-particle Green's functions in Sect. 5.4.1, we can again employ the energy-dependent Fourier transforms. The most important consequence of this is then energy conservation at each vertex. Aside from this, the diagram rules formulated following (5.123) can be adopted practically without change. For a quantitative analysis, however, a careful evaluation of the trivial factors is indispensable. The latter represent a genuinely serious source of errors. We shall therefore formulate the diagram rules for the two-particle Green's function $D_{\boldsymbol{q}}(t, t')$ or $D_{\boldsymbol{q}}(E)$ explicitly once again.

Let us first consider the diagrams of order $n = 0$:

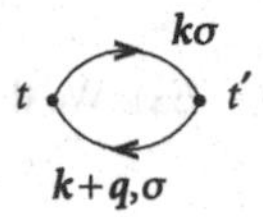

$$\equiv \mathrm{i}\hbar\Lambda_{\boldsymbol{q}}^{(0)}(t, t') . \qquad (5.181)$$

Corresponding to (5.180), we find:

$$\mathrm{i}\hbar\Lambda_{\boldsymbol{q}}^{(0)}(t, t') = -\sum_{\substack{\boldsymbol{k},\sigma\\ \boldsymbol{p}}} \delta_{\boldsymbol{k}+\boldsymbol{q},\boldsymbol{p}} \left(\mathrm{i}G_{\boldsymbol{k}+\boldsymbol{q}\sigma}^{0,\mathrm{c}}(t - t')\right)\left(\mathrm{i}G_{\boldsymbol{k}\sigma}^{0,\mathrm{c}}(t' - t)\right) =$$

$$= -\frac{1}{(2\pi\hbar)^2}\sum_{\boldsymbol{k},\sigma}\iint \mathrm{d}E\,\mathrm{d}E' \exp\left(-\frac{\mathrm{i}}{\hbar}E(t - t')\right) \cdot$$

$$\cdot \left(\mathrm{i}G_{\boldsymbol{k}+\boldsymbol{q}\sigma}^{0,\mathrm{c}}(E + E')\right)\left(\mathrm{i}G_{\boldsymbol{k}\sigma}^{0,\mathrm{c}}(E')\right) .$$

This implies for the energy-dependent Fourier transform:

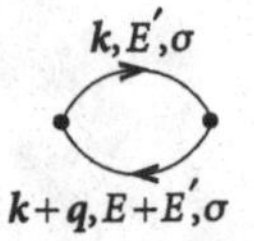

$$\equiv \mathrm{i}\hbar\Lambda_q^{(0)}(E) =$$

$$= \frac{-1}{2\pi\hbar}\sum_{k,\sigma}\mathrm{d}E'\left(\mathrm{i}G_{k+q\sigma}^{(0,\,c)}\left(E+E'\right)\right)\left(\mathrm{i}G_{k\sigma}^{(0,\,c)}\left(E'\right)\right). \tag{5.182}$$

According to (5.123), the Green's function $G^{0,\,\mathrm{c}}$ has the dimension of a *time*. This then holds also for $\hbar\Lambda_q^{(0)}$. $\Lambda_q^{(0)}(E)$ itself thus has the dimension of $1/$*energy*.

What is then the structure of the Fourier transform of a general diagram of the perturbation expansion?

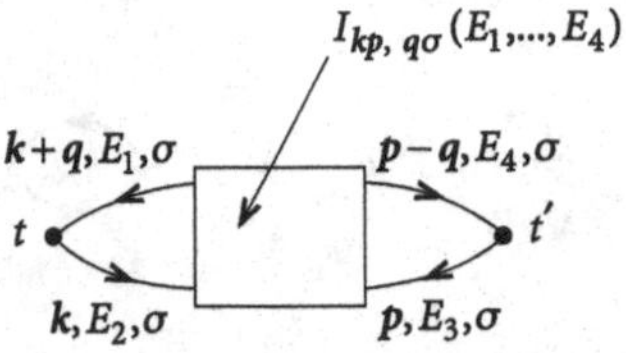

$$\equiv \mathrm{i}\widetilde{D}_q\left(t,t'\right).$$

Following the considerations in Sect. 5.4.1, the four outer lines make the following contributions:

$$(k+q,\,E_1):\quad \frac{1}{\sqrt{2\pi\hbar}}\exp\left(-\frac{\mathrm{i}}{\hbar}E_1t\right)\left(\mathrm{i}G_{k+q\sigma}^{0,\,\mathrm{c}}\left(E_1\right)\right),$$

$$(k,\,E_2):\quad \frac{1}{\sqrt{2\pi\hbar}}\exp\left(\frac{\mathrm{i}}{\hbar}E_2t\right)\left(\mathrm{i}G_{k\sigma}^{0,\,\mathrm{c}}\left(E_2\right)\right),$$

$$(p,\,E_3):\quad \frac{1}{\sqrt{2\pi\hbar}}\exp\left(\frac{\mathrm{i}}{\hbar}E_3t'\right)\left(\mathrm{i}G_{p\sigma}^{0,\,\mathrm{c}}\left(E_3\right)\right),$$

$$(p-q,\,E_4):\quad \frac{1}{\sqrt{2\pi\hbar}}\exp\left(-\frac{\mathrm{i}}{\hbar}E_4t'\right)\left(\mathrm{i}G_{p-q\sigma}^{0,\,\mathrm{c}}\left(E_4\right)\right).$$

The *kernel* of the diagram contributes $I_{kp,q\sigma}(E_1\cdots E_4)$. We then have all together:

$$\mathrm{i}\widetilde{D}_q\left(t,t'\right) = \frac{-1}{(2\pi\hbar)^2}\sum_{\substack{k,\,p\\ \sigma}}\int\cdots\int\mathrm{d}E_1\cdots\mathrm{d}E_4\,I_{kp,\,q\sigma}\left(E_1\cdots E_4\right)\cdot$$

$$\cdot\exp\left\{-\frac{\mathrm{i}}{\hbar}\left[\left(E_1-E_2\right)t-\left(E_3-E_4\right)t'\right]\right\}\left(\mathrm{i}G_{k+q,\,\sigma}^{0,\,\mathrm{c}}\left(E_1\right)\right)\cdot$$

$$\cdot\left(\mathrm{i}G_{k\sigma}^{0,\,\mathrm{c}}\left(E_2\right)\right)\left(\mathrm{i}G_{p\sigma}^{0,\,\mathrm{c}}\left(E_3\right)\right)\left(\mathrm{i}G_{p-q\sigma}^{0,\,\mathrm{c}}\left(E_4\right)\right).$$

Green's functions depend only on time differences (3.128). Therefore, we can assume that

$$E_1 - E_2 = E_3 - E_4 \equiv E.$$

With $E_1 = E' + E$, $E_2 = E'$, $E_3 = E''$, $E_4 = E'' - E$, we then obtain after Fourier transformation:

$$\begin{aligned} \mathrm{i}\widetilde{D}_{\boldsymbol{q}}(E) = \frac{-1}{2\pi\hbar} \sum_{\boldsymbol{k},\,\boldsymbol{p},\,\sigma} \iint \mathrm{d}E'\,\mathrm{d}E''\, I_{\boldsymbol{kp},\,\boldsymbol{q}\sigma}\left(E', E''; E\right) \cdot \\ \cdot \left(\mathrm{i}G^{0,\,\mathrm{c}}_{\boldsymbol{k}+\boldsymbol{q}\sigma}\left(E'+E\right)\right)\left(\mathrm{i}G^{0,\,\mathrm{c}}_{\boldsymbol{k}\sigma}\left(E'\right)\right) \cdot \\ \cdot \left(\mathrm{i}G^{0,\,\mathrm{c}}_{\boldsymbol{p}\sigma}\left(E''\right)\right)\left(\mathrm{i}G^{0,\,\mathrm{c}}_{\boldsymbol{p}-\boldsymbol{q}\sigma}\left(E''-E\right)\right). \end{aligned} \tag{5.183}$$

The factor (-1) results from the loop rule. We can now formulate the diagram rules for the two-particle Green's function,

diagram rules for $\mathrm{i}D_{\boldsymbol{q}}(E)$:

Consider all the diagrams with four outer attachments as in Fig. 5.10. A diagram of n-th order (n vertices!) is then to be evaluated as follows (cf. Sect. 5.4.1):

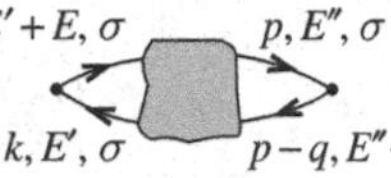

Fig. 5.10 The general diagram structure of the density correlation

1 Vertex $\Longleftrightarrow \frac{1}{2\pi\hbar}\left(-\frac{\mathrm{i}}{\hbar}\right) v(kl, nm)\delta\left[(E_k + E_l) - (E_m + E_n)\right]$.
2 Propagator (= *solid* line): $\mathrm{i}G^{0,\,\mathrm{c}}_{\boldsymbol{k}_\nu}\left(E_{k_\nu}\right)\delta_{\boldsymbol{k}_\nu \boldsymbol{k}_\mu}$.
3 Factor: $(-1)^S$; S = number of Fermion loops.
4 *Outer* attachments:

$$\text{left:} \quad \frac{1}{\sqrt{2\pi\hbar}}\left(\mathrm{i}G^{0,\,\mathrm{c}}_{\boldsymbol{k}+\boldsymbol{q}}\left(E'+E\right)\right)\left(\mathrm{i}G^{0,\,\mathrm{c}}_{\boldsymbol{k}}\left(E'\right)\right),$$

$$\text{right:} \quad \frac{1}{\sqrt{2\pi\hbar}}\left(\mathrm{i}G^{0,\,\mathrm{c}}_{\boldsymbol{p}-\boldsymbol{q}}\left(E''-E\right)\right)\left(\mathrm{i}G^{0,\,\mathrm{c}}_{\boldsymbol{p}}\left(E''\right)\right)$$

5 Summation or integration over all the *inner* wavenumbers, spins and energies. Among these are also $\boldsymbol{k}$, $\boldsymbol{p}$, σ, E', E'', *not* however E and $\boldsymbol{q}$.

With this, we can systematically develop the perturbation expansion for the two-particle Green's function $\mathrm{i}D_{\boldsymbol{q}}(E)$.

Just as with the Dyson equation for the single-electron Green's function in Sect. 5.4.2, we can see that the diagram expansion for $\mathrm{i}D_{\boldsymbol{q}}(E)$, according to the above rules, also contains an infinite partial series which can be separated off.

Definition 5.6.1:

The polarisation part = diagram contributing to $\mathrm{i}D_{\boldsymbol{q}}(E)$ with two fixed outer attachments, into each of which an external line enters and from each of which an external line emerges.

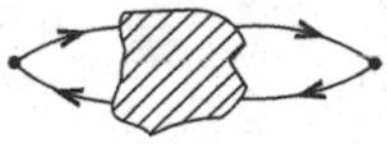

We are thus dealing here with a diagram from the expansion of

$$\Longleftrightarrow \quad \mathrm{i}D_q(E)\,.$$

Examples

1)  2) 

Definition 5.6.2: **The proper, irreducible polarisation part** = the polarisation part which **cannot** be decomposed into two independent diagrams by separating off an interaction line.

Examples

1)  ,  , ... irreducible

2) 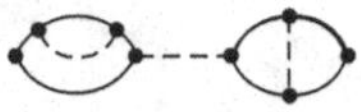 , ... reducible

---

Every diagram which is not itself already an irreducible polarisation part can then evidently be decomposed as follows:

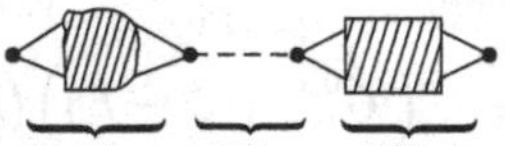

(I): Irreducible polarisation part,
(II): interaction line,
(III): Diagram from the expansion of $\mathrm{i}D_q(E)$.

We clearly obtain **all** the diagrams if in (I) we sum over **all** the irreducible polarisation parts and in (III) over **all** possible diagrams and furthermore add on the sum of all irreducible polarisation parts themselves.

Definition 5.6.3:

$$\mathrm{i}\hbar\Lambda_q(E) \Longleftrightarrow$$ 

= sum of all irreducible polarisation parts.

with this, we can formulate the following diagrammatic **Dyson equation**:

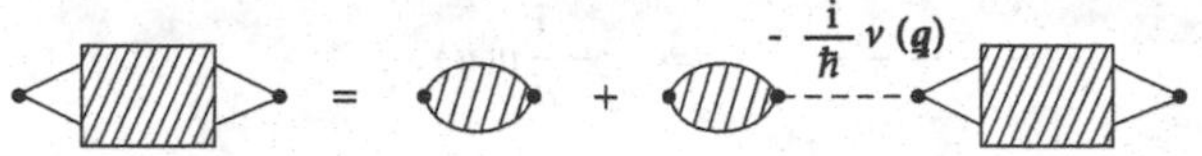

(5.184)

$iD_q(E)$ $\quad$ $i\hbar\Lambda_q(E)$ $\quad$ $i\hbar\Lambda_q(E)$ $\quad$ $iD_q(E)$

We still have to consider the contribution of the interaction line, which deviates somewhat from point 1 of the above diagram rules. Conservation of energy, which is guaranteed by the δ-function in 1, is already provided for in the case of **this** interaction line through the attachments at the left and the right. If in the evaluation of the second summand in (5.184) we take the *right-hand* attachments of $i\hbar\Lambda_q(E)$ and the *left-hand* attachments of $iD_q(E)$ to be *outer* attachments as in rule 5.6.1., then the integration over time (5.121) at the vertex can naturally be dispensed with and thus the factor $(1/2\pi\hbar)\delta\left[(E_k + E_l) - (E_m + E_n)\right]$ drops out. This becomes clear in the derivation of (5.183).

The formal solution of (5.184) is given by

$$D_q(E) = \frac{\hbar\Lambda_q(E)}{1 - \Lambda_q(E)v(q)} . \tag{5.185}$$

$D_q(E)$ is thus completely determined by the polarisation propagator $\Lambda_q(E)$, which is to be computed according to the above diagram rules and has a considerably simpler structure than the *original* two-particle Green's function $D_q(E)$.

5.6.2 Effective Interactions

Using the polarisation propagator $\Lambda_q(E)$ which was introduced in the previous section, we can develop another very useful concept, that of the **effective (dressed) interaction**. In the diagram expansion for the single-particle Green's function $iG^c_{k\sigma}(E)$, which we carried out in Sect. 5.4.1, a series of graphs occurs which contain an irreducible polarisation part in an interaction line.

Example

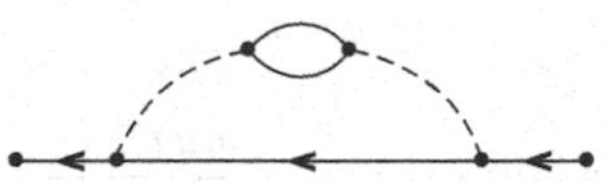

On summing over all possible diagrams, we will encounter a number of such graphs which differ from the one in the sketch only in that the loops in the interaction line are replaced by something more complex. We want to combine the ensemble of all such diagrams by introducing the effective interaction $v_{\text{eff}}(q, E)$. We adopt by convention the following symbols:

$$= = = = = = \quad \Longleftrightarrow \quad -\frac{\mathrm{i}}{\hbar} v_{\mathrm{eff}}(\boldsymbol{q}, E)$$

$$- - - - - - \quad \Longleftrightarrow \quad -\frac{\mathrm{i}}{\hbar} v(\boldsymbol{q}) \, .$$

The prefactor $(-i/\hbar)$ corresponds to the diagram rule 5.6.1. for $\mathrm{i}D_{\boldsymbol{q}}(E)$. Energy conservation, which leads to the term

$$\frac{1}{2\pi\hbar} \delta \left[(E_k + E_l) - (E_m + E_n) \right] \, ,$$

must be taken into account both for $v_{\mathrm{eff}}(\boldsymbol{q}, E)$ as well as for the *undressed* interaction $v(\boldsymbol{q})$, and will therefore not be explicitly considered in the following.

Now, in the diagram expansion for $\mathrm{i}G^{\mathrm{c}}_{k\sigma}(E)$, one suppresses all the diagrams which contain a polarisation part in at least one interaction line. In the remaining diagrams, one then replaces the simple (*undressed*) interaction lines $v(\boldsymbol{q})$ by the effective ones, whereby $v_{\mathrm{eff}}(\boldsymbol{q}, E)$ results from the following expansion:

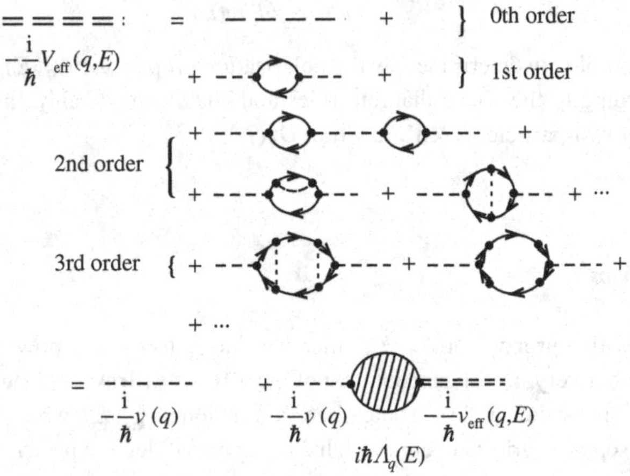

The n-th order in the expansion for $v_{\mathrm{eff}}(\boldsymbol{q}, E)$ is characterised by $(n + 1)$ interaction lines $v(\boldsymbol{q})$. Obviously, in the form given, the diagrams can be combined into a Dyson equation with the formal solution:

$$v_{\mathrm{eff}}(\boldsymbol{q}, E) = \frac{v(\boldsymbol{q})}{1 - v(\boldsymbol{q})\Lambda_{\boldsymbol{q}}(E)} \, . \tag{5.186}$$

Due to the polarisation propagator $\Lambda_{\boldsymbol{q}}(E)$, the effective interaction is also uniquely determined.

Those diagrams which remain in the expansion of $\mathrm{i}G^{\mathrm{c}}_{k\sigma}(E)$ after the introduction of the effective interaction as described above are referred to as *skeleton diagrams*.

In Sect. 3.1.5, we discussed the dielectric function $\varepsilon(\boldsymbol{q}, E)$. Via the relation (3.104), we can now relate it to the effective interaction or the polarisation propagator:

$$\frac{1}{\varepsilon^{c}(\boldsymbol{q}, E)} = 1 + \frac{1}{\hbar} v(\boldsymbol{q}) \langle\langle \rho_{\boldsymbol{q}} ; \rho_{\boldsymbol{q}}^{+} \rangle\rangle_{E}^{c} = = 1 + \frac{1}{\hbar} v(\boldsymbol{q}) D_{\boldsymbol{q}}(E) .$$

The superscript c is meant to indicate that the diagram techniques used here refer to the causal Green's function. The considerations in Sect. 3.1.5, in contrast, referred to the corresponding retarded function. Owing to (3.188) and (3.189) for $T \to 0$,

$$\begin{aligned} \operatorname{Re} \varepsilon^{\mathrm{ret}}(\boldsymbol{q}, E) &= \operatorname{Re} \varepsilon^{c}(\boldsymbol{q}, E) , \\ \operatorname{Im} \varepsilon^{\mathrm{ret}}(\boldsymbol{q}, E) &= \operatorname{sign}(E) \operatorname{Im} \varepsilon^{c}(\boldsymbol{q}, E) , \end{aligned} \tag{5.187}$$

the transformation is however simple. With (5.185), it follows that:

$$\varepsilon^{c}(\boldsymbol{q}, E) = 1 - v(\boldsymbol{q}) \Lambda_{\boldsymbol{q}}(E) . \tag{5.188}$$

Inserting this into (5.186),

$$v_{\mathrm{eff}}(\boldsymbol{q}, E) = \frac{v(\boldsymbol{q})}{\varepsilon^{c}(\boldsymbol{q}, E)} , \tag{5.189}$$

we can see that the dielectric function expresses the screening of the *undressed* interaction due to the polarisation of the Fermi sea by the particle interactions.

We want to consider at this point the discussion of an important special case which will allow us to establish a direct relationship between the self-energy of the Dyson equation and the polarisation propagator. Our starting point is the simplest approximation $\Lambda_{\boldsymbol{q}}^{(0)}(E)$ for the polarisation propagator, which we formulated in (5.182). For the effective interaction, it requires the summation of ring diagrams.

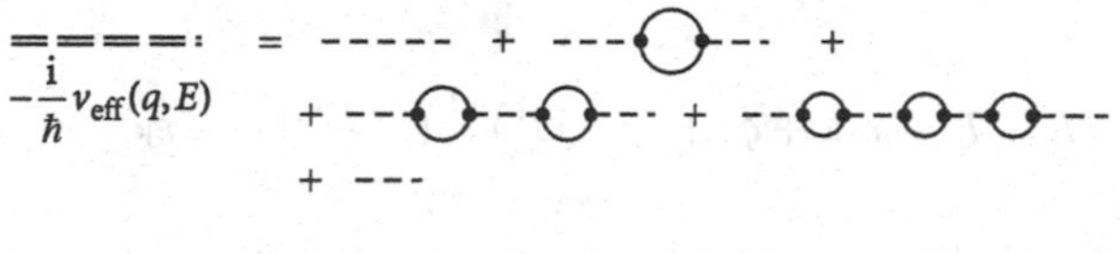

In this way, we now construct on the other hand an infinite partial series for the self-energy $\Sigma_{\boldsymbol{k}\sigma}(E)$. In the example of an application at the end of Sect. 5.4.2, we replace the undressed interaction in $\left(\Sigma_{\boldsymbol{k}\sigma}^{(1)}(E)\right)$ by the effective interaction:

$$-\frac{i}{\hbar}\widetilde{\Sigma}^{(1)}_{k\sigma}(E) \equiv$$

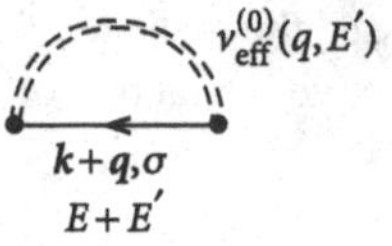

$$-\frac{\mathrm{i}}{\hbar}\widetilde{\Sigma}^{(1)}_{k\sigma}(E) = -\frac{\mathrm{i}}{\hbar}\frac{1}{2\pi\hbar}\sum_{q}\int \mathrm{d}E'\, v^{(0)}_{\text{eff}}(\boldsymbol{q}, E')\,\mathrm{i}G^{0,\,\mathrm{c}}_{k+q,\sigma}(E+E')\,. \tag{5.190}$$

This leads to the so-called **RPA (Random Phase Approximation)**, which is characterised by the lowest-order diagram for $\Lambda_{q}(E)$ and already by **all** the ring diagrams for $\Sigma_{k\sigma}(E)$:

$$\widetilde{\Sigma}^{(1)}_{k\sigma}(E) = \frac{1}{2\pi\hbar}\sum_{q}\int_{-\infty}^{+\infty}\mathrm{d}E'\left(\mathrm{i}G^{0,\,\mathrm{c}}_{k+q\sigma}(E+E')\right)\frac{v(\boldsymbol{q})}{1 - v(\boldsymbol{q})\Lambda^{(0)}_{q}(E')}\,. \tag{5.191}$$

We can see that the lowest order of perturbation theory for a *higher-order* Green's function already corresponds to an infinite partial sum for a Green's function of lower order. This is, by the way, a typical property of the diagram techniques.

We want to evaluate the RPA result for the polarisation propagator explicitly in one case. Our starting point is the intermediate result (5.182), into which we insert (5.123):

$$\Lambda^{(0)}_{q}(E) = \frac{-\mathrm{i}}{2\pi}2\sum_{k}\int_{-\infty}^{+\infty}\mathrm{d}E'\frac{1}{E + E' - (\varepsilon(\boldsymbol{k}+\boldsymbol{q}) - \varepsilon_{\mathrm{F}}) + \mathrm{i}0_{k+q}}\,\cdot$$
$$\cdot\frac{1}{E' - (\varepsilon(\boldsymbol{k}) - \varepsilon_{\mathrm{F}}) + \mathrm{i}0_{k}}\,.$$

0_k is positive, as long as $|\boldsymbol{k}| > k_{\mathrm{F}}$, and negative when $|\boldsymbol{k}| < k_{\mathrm{F}}$. We solve the integral by integration within the complex plane. If both poles lie in the same semi-plane, we can close the integration path in the other semi-plane. Within the area enclosed by the integration path, there are no poles; thus, the integral vanishes. We still need to discuss two cases:

1. $|\boldsymbol{k}+\boldsymbol{q}| > k_{\mathrm{F}};\ |\boldsymbol{k}| < k_{\mathrm{F}}$

   In this case, the following integral must be solved:

$$\int \mathrm{d}E'\,\left(E + E' - \varepsilon(\boldsymbol{k}+\boldsymbol{q}) + \varepsilon_{\mathrm{F}} + \mathrm{i}0^{+}\right)^{-1}\left(E' - \varepsilon(\boldsymbol{k}) + \varepsilon_{\mathrm{F}} - \mathrm{i}0^{+}\right)^{-1} =$$

$$= 2\pi\mathrm{i}\left(E + \varepsilon(\boldsymbol{k}) - \varepsilon_{\mathrm{F}} - \varepsilon(\boldsymbol{k}+\boldsymbol{q}) + \varepsilon_{\mathrm{F}} + \mathrm{i}0^{+}\right)^{-1} =$$
$$= 2\pi\mathrm{i}\left(E + \varepsilon(\boldsymbol{k}) - \varepsilon(\boldsymbol{k}+\boldsymbol{q}) + \mathrm{i}0^{+}\right)^{-1}\,.$$

2. $|\boldsymbol{k}+\boldsymbol{q}| < k_{\mathrm{F}}$; $|\boldsymbol{k}| > k_{\mathrm{F}}$

$$\int \mathrm{d}E' \left(E + E' - \varepsilon(\boldsymbol{k}+\boldsymbol{q}) + \varepsilon_{\mathrm{F}} - \mathrm{i}0^+\right)^{-1} \left(E' - \varepsilon(\boldsymbol{k}) + \varepsilon_{\mathrm{F}} + \mathrm{i}0^+\right)^{-1} =$$

$$= 2\pi\mathrm{i}\left(-E + \varepsilon(\boldsymbol{k}+\boldsymbol{q}) - \varepsilon_{\mathrm{F}} - \varepsilon(\boldsymbol{k}) + \varepsilon_{\mathrm{F}} + \mathrm{i}0^+\right)^{-1} =$$
$$= -2\pi\mathrm{i}\left(E + \varepsilon_{\boldsymbol{k}} - \varepsilon(\boldsymbol{k}+\boldsymbol{q}) - \mathrm{i}0^+\right)^{-1}.$$

The final result for $\Lambda_{\boldsymbol{q}}^{(0)}(E)$ is then:

$$\Lambda_{\boldsymbol{q}}^{(0)}(E) = 2\sum_{\boldsymbol{k}} \left\{ \frac{\left(1 - \langle n_{\boldsymbol{k}+\boldsymbol{q}}\rangle^{(0)}\right)\langle n_{\boldsymbol{k}}\rangle^{(0)}}{E + \varepsilon(\boldsymbol{k}) - \varepsilon(\boldsymbol{k}+\boldsymbol{q}) + \mathrm{i}0^+} - \frac{\left(1 - \langle n_{\boldsymbol{k}}\rangle^{(0)}\right)\langle n_{\boldsymbol{k}+\boldsymbol{q}}\rangle^{(0)}}{E + \varepsilon(\boldsymbol{k}) - \varepsilon(\boldsymbol{k}+\boldsymbol{q}) - \mathrm{i}0^+} \right\}. \tag{5.192}$$

Finally, we can express the ground-state energy of the interacting electron gas in terms of the polarisation propagator. Because of $v(\boldsymbol{0}) = 0$ (jellium model), all the ground-state diagrams are connected and open. With the fixed vertex at $t = t' = 0$, four outer lines (two creation operators and two annihilation operators) are joined:

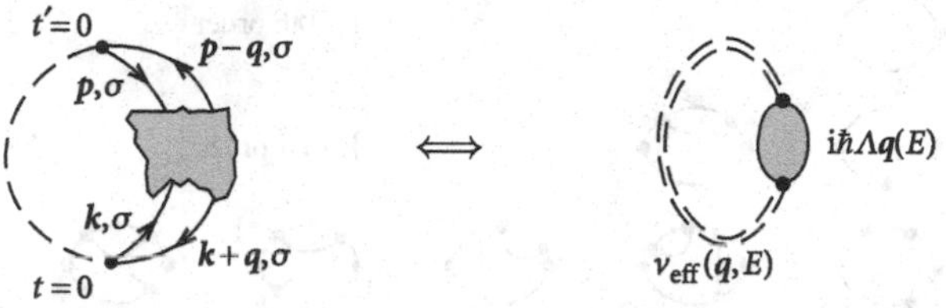

The remaining diagram attached to the vertex corresponds to the general diagram of the density correlation. Therefore, an irreducible polarisation part can again be separated off. What remains is then a diagram of the effective interaction

$$\Delta E_0 = 2\sum_{\boldsymbol{q}} \int_{-\infty}^{+\infty} \frac{\mathrm{d}E}{2\pi\hbar} v_{\mathrm{eff}}(\boldsymbol{q}, E)\mathrm{i}\hbar\Lambda_{\boldsymbol{q}}(E). \tag{5.193}$$

The factors 2 and $1/2\pi\hbar$ result from the summation over spins and from energy conservation at the vertex. The factor $-\mathrm{i}/\hbar$, which is related according to the diagram rule 1. to $\mathrm{i}D_{\boldsymbol{q}}(E)$ with one vertex, in the end results from the perturbation expansion of the time-evolution operator, and thus appears only for the *inner* vertices. The effective interaction thus enters (5.193) without this factor. With (5.188) and (5.189), we can finally express the level shifts in terms of the dielectric function:

$$\Delta E_0 = \frac{\mathrm{i}}{\pi}\sum_{\boldsymbol{q}} \int_{-\infty}^{+\infty} \mathrm{d}E \left\{\frac{1}{\varepsilon^{\mathrm{c}}(\boldsymbol{q}, E)} - 1\right\}. \tag{5.194}$$

ΔE_0 is of course real; therefore, we must have:

$$\Delta E_0 = -\frac{1}{\pi} \sum_q \int_{-\infty}^{+\infty} \mathrm{d}E \; \mathrm{Im} \frac{1}{\varepsilon^{\mathrm{c}}(\boldsymbol{q}, E)} \,. \tag{5.195}$$

As we discussed in detail in the last section, in particular the ring diagrams are important for the ground-state energy. We find these precisely when we insert the lowest-order approximation $\Lambda_{\boldsymbol{q}}^{(0)}(E)$ for the polarisation propagator in (5.193):

$$\Delta E_0^{\mathrm{RPA}} = \frac{\mathrm{i}}{\pi} \sum_q \int_{-\infty}^{+\infty} \mathrm{d}E \, \frac{v(\boldsymbol{q})\Lambda_{\boldsymbol{q}}^{(0)}(E)}{1 - v(\boldsymbol{q})\Lambda_{\boldsymbol{q}}^{(0)}(E)} \,.$$

5.6.3 Vertex Function

The method of simplifying diagram expansions by introducing certain *diagram blocks* can be extended in a variety of ways. Thus far, we have introduced the self-energy of the single-particle Green's function, the polarisation propagator of the density correlation, and the effective interaction. Let us look once again at the polarisation propagator $\Lambda_{\boldsymbol{q}}(E)$ somewhat more closely. In the lowest orders, we find the following diagram contributions:

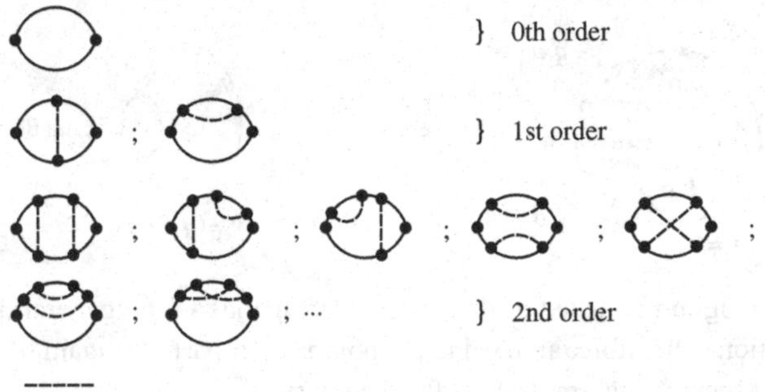

We define:

Definition 5.6.4: The **vertex part**
= diagram contribution with **two** attachments for particle lines and **one** attachment for an interaction line.

Examples

Definition 5.6.5: Irreducible vertex part
= a vertex part from which **no** independent self-energy diagram can be split off by separation of a propagating line, nor any independent polarisation diagram by separation of an interaction line.

Examples

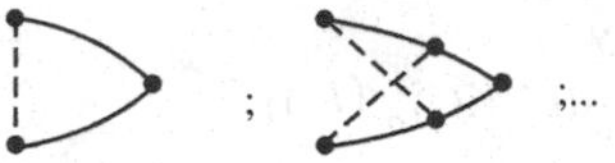

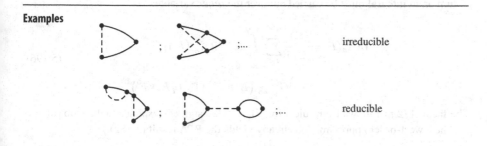

irreducible

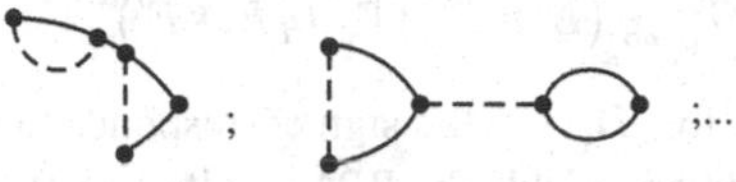

reducible

Finally, we also define:

Definition 5.6.6: The **vertex function**

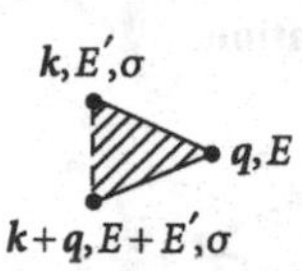

$\Longleftrightarrow \quad \Gamma_\sigma(\boldsymbol{q}E;\boldsymbol{k}E') =$ sum of all irreducible vertex parts.

We list the lowest orders:

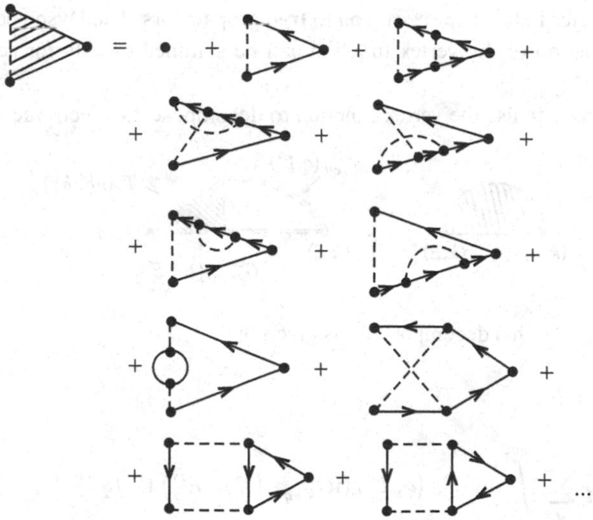

We can use the vertex function for the representation of the polarisation propagator:

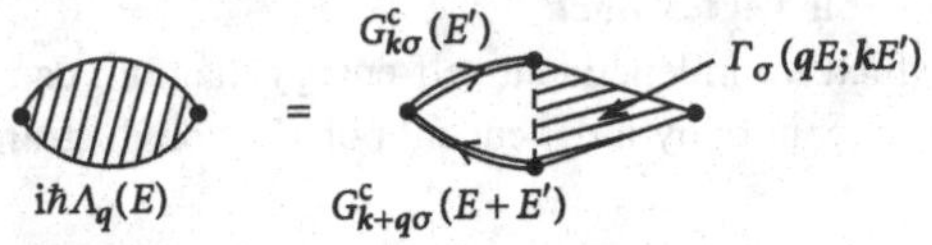

A sum or an integral must be carried out over the *inner* variable:

$$i\hbar\Lambda_q(E) = \frac{-1}{2\pi\hbar}\sum_{k\sigma}\int\limits_{-\infty}^{+\infty} dE' \left(iG^{c}_{k\sigma}(E')\right) \cdot \cdot \left(iG^{c}_{k+q\sigma}(E+E')\right)\Gamma_\sigma(qE;kE') \,. \tag{5.196}$$

The factor $1/2\pi\hbar$ follows from rule 4 for $iD_q(E)$; the sign corresponds to the loop rule.

The lowest-order approximation already yields the RPA result (5.182):

$$\textbf{RPA:}\quad G^{c}_{k\sigma} \longrightarrow G^{0,c}_{k\sigma}\,, \qquad \Gamma_\sigma \longrightarrow 1\,. \tag{5.197}$$

Physically, this means neglecting all the scattering processes of the particle-hole pair. These are better taken into account in the so-called **ladder approximation**:

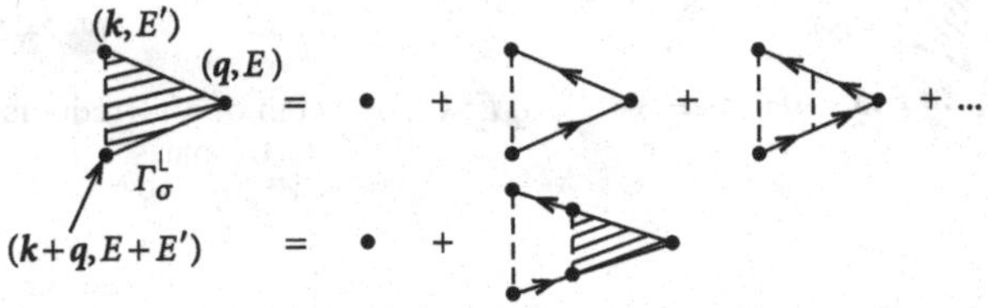

(5.198)

The solid lines refer in this approximation to **free** propagators. The Dyson equation for the ladder approximation of the vertex function can be summed exactly for certain types of interactions.

Finally, we can also use the vertex function to decompose the electronic self-energy:

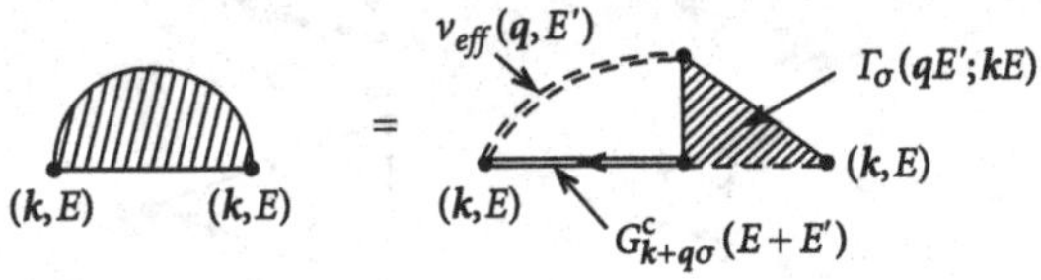

Written as a formula, this decomposition is given by:

$$-\frac{i}{\hbar}\Sigma_{k\sigma}(E) = = \frac{-i}{2\pi\hbar^2}\sum_{q}\int\limits_{-\infty}^{+\infty} dE'\, v_{\text{eff}}(q,E')\left(iG^{c}_{k+q\sigma}(E+E')\right)\Gamma_\sigma(qE';kE)\,. \tag{5.199}$$

If we use the simplest approximation (5.197) together with $v_{\text{eff}} \to v$, we obtain the *Hartree-Fock approximation* (5.128).

5.6.4 Exercises

Exercise 5.6.1 Calculate the transverse susceptibility approximately, using suitable partial sums within the Hubbard model:

$$\chi_q^{\pm}(E) = -\gamma \int\limits_{-\infty}^{+\infty} \mathrm{d}\left(t-t'\right) \exp\left[\frac{\mathrm{i}}{\hbar} E\left(t-t'\right)\right] \frac{1}{N} \sum_{k,p} \Big\{ -\mathrm{i} \langle E_0 | T_\varepsilon \left(a_{k\uparrow}^{+}(t) a_{k+q\downarrow}(t) \cdot \right.$$

$$\left. \cdot\, a_{p\downarrow}^{+}\left(t'\right) a_{p-q\uparrow}\left(t'\right)\right) | E_0 \rangle \Big\} = -\frac{\gamma}{N} \widehat{\chi}_q^{\pm}(E).$$

It can be treated diagrammatically in complete analogy to the density-density Green's function $D_q(E)$ which was discussed in Sect. 5.6.

1. Show, using the Dyson equation, that $\widehat{\chi}_q^{\pm}(E)$ is completely determined by the suitably defined polarisation propagator.
2. Calculate the vertex function in the ladder approximation.
3. Give a representation of the transverse susceptibility in terms of the *full* one-electron Green's function and the vertex function.
4. In the exact expression for the transverse susceptibility from 3, replace the *full* propagators by the *free* propagators and use the ladder approximation as in 2 for the vertex function. Compare the result for the transverse susceptibility with that from Sect. 4.2.3.

Exercise 5.6.2 The T-matrix 5.6.4. introduced in Sect. 4.3.1 can be defined as follows:

$$\left(-\frac{\mathrm{i}}{\hbar} T_{k\sigma}(E)\right) = \text{sum over all the proper and improper self-energy parts.}$$

Find, making use of the T-matrix, an exact diagrammatic representation of the one-electron Green's function. Derive the connection between the T-matrix and the self energy.

Exercise 5.6.3 In Ex. 4.2.4, it was shown that the so-called *appearance-potential spectroscopy (APS)* and *Auger electron spectroscopy* (AES) are completely determined by the two-particle spectral density

$$S^{(2)}_{ii\sigma}(E) = -\frac{1}{\pi} \operatorname{Im} \langle\langle a_{i-\sigma} a_{i\sigma} ; a^{+}_{i\sigma} a^{+}_{i-\sigma} \rangle\rangle_E \, .$$

1. Develop an appropriate diagrammatic representation.
2. Describe as in 4.2.6 the interacting electron system within the Hubbard model and develop an approximation which describes exactly the direct interactions of the two excited particles (*direct correlations*), but neglects the interactions with the rest of the system (*indirect correlations*).
3. How could one extend the approximation in 2 with a previously-determined *full* one-electron Green's function in order to take into account the *indirect* correlations, at least approximately?

5.7 Self-Examination Questions

For Section 5.1

1. What is the fundamental formula of perturbation theory?
2. How are the Schrödinger perturbation theory and the Brillouin-Wigner perturbation theory obtained from the fundamental formula?
3. What is the disadvantage of *conventional*, time-independent perturbation theory?
4. What is meant by *adiabatic switching on* of an interaction?
5. Formulate and interpret the Gell-Mann–Low theorem.
6. How does the normalised ground state of the interacting system develop according to the Gell-Mann–Low theorem from that of the free system?
7. Describe the expectation value of an arbitrary, time-dependent Heisenberg observable in the ground state using the *trick* of adiabatic switching-on in terms of an expression which refers to the ground state $|\eta_0\rangle$ of the free system.
8. Discuss the structure of the causal one-electron Green's function, as appropriate for diagrammatic perturbation theory.

For Section 5.2

1. What is meant by the *Fermi vacuum*?
2. How is the *normal product* defined? For what purpose was it introduced?
3. What is meant by a *contraction*?
4. To what does the contraction of two annihilation operators give rise?
5. Why is the contraction $\underbrace{a_k(t)a_{k'}^{+}(t')}$ not an operator?

   How is it related to the causal $T = 0$ Green's function? What is the convention for equal times $t = t'$?
6. Formulate Wick's theorem.
7. What does *total pairing* mean?

For Section 5.3

1. Which expectation value is referred to as a *vacuum amplitude*?
2. Which diagrams contain the first term of the perturbation expansion for the vacuum amplitude?
3. What is meant by a *vertex*?
4. How many different graphs with **four** vertices contribute to the vacuum amplitude?
5. Formulate the *loop rule*.
6. Which diagrams are referred to as having *the same structure*?
7. How many diagrams of the same structure with n vertices are there for the vacuum amplitude?

8. What are *topologically equivalent* diagrams?
9. What is a *connected* diagram?
10. Formulate and interpret the *linked cluster theorem*.
11. What is an *open* diagram? What is a *vacuum-fluctuation diagram*?
12. What does the *principal theorem of connected diagrams* state?

For Section 5.4

1. What does conservation of momentum at a vertex mean?
2. What is meant by *stretched* diagrams of the single-particle Green's function?
3. Which considerations lead to conservation of energy at a vertex?
4. What are the diagram rules for the causal one-electron Green's function $i G^{c}_{k\sigma}(E)$?
5. What is a self-energy part? When is it called *proper* or irreducible?
6. How is the self energy defined?
7. Describe the diagram representation of the Dyson equation.

For Section 5.5

1. Which expression must be evaluated for the calculation of the level shifts $\Delta E_0 = E_0 - \eta_0$?
2. How do the Feynman diagrams for ΔE_0 differ from those for the vacuum amplitude?
3. How many topologically distinct diagrams of the same structure Θ_n are there to order n in the expansion for ΔE_0?
4. What is found in first-order perturbation theory for the ground-state energy in the jellium model?
5. Why do diagrams with *bubbles* make no contribution in the jellium model?
6. Which diagram structures contribute to second-order perturbation theory for the ground-state energy?
7. Which type of diagrams causes divergences even in second-order perturbation theory for the ground-state energy in the jellium model?
8. What is meant by *ring diagrams*?
9. Interpret the concept of *correlation energy*.
10. How does a ground-state diagram of n-th order depend on the density parameter r_s?
11. Why can one limit oneself in the approximate determination of the ground-state energy in the jellium model for high electron densities to non-divergent diagrams of low order, whilst in contrast the ring diagrams must be included as infinite partial sums?
12. What is the physical cause of the divergence of a ring diagram?

For Section 5.6

1. Formulate the diagram rules for the two-particle Green's function $\mathrm{i}D_q(E)$ (*density-density correlation*).
2. What is a polarisation part? When is it referred to as *proper* or irreducible?

3. What is meant by a polarisation propagator? Which form does it assume in the so-called RPA?
4. Formulate the Dyson equation for $\mathrm{i}D_{\boldsymbol{q}}(E)$ using the polarisation propagator.
5. How can one define an *effective interaction* $v_{\text{eff}}(\boldsymbol{q}, E)$ using the polarisation propagator?
6. What are *skeleton diagrams*?
7. What is the relation between the effective interaction $v_{\text{eff}}(\boldsymbol{q}, E)$ and the dielectric function?
8. How can the so-called RPA of the self-energy $\Sigma_{\boldsymbol{k}\sigma}(E)$ be represented diagrammatically using the *dressed* interaction $v_{\text{eff}}(\boldsymbol{q}, E)$?
9. Give the ground-state energy in the jellium model in terms of the polarisation propagator and the effective interaction.
10. Express the level shift ΔE_0 in terms of the dielectric function. Which result is found in the RPA?
11. What is a vertex part? When is it irreducible?
12. How is the vertex function defined?
13. Represent the polarisation propagator using the vertex function.
14. What is meant by the *ladder approximation*?
15. How can one determine the self-energy $\Sigma_{\boldsymbol{k}\sigma}(E)$ using the vertex function?

6 Perturbation Theory at Finite Temperatures

6.1 The Matsubara Method

Up to now, we have concerned ourselves exclusively with perturbation theory methods which apply at $T = 0$. Experiments, however, are carried out at finite temperatures. Since every theory in the end has the goal of explaining experimental results or of predicting them, the extension to $T > 0$ is unavoidable. We should at least investigate whether the $T = 0$ methods of the preceding chapter can be carried over to the case that $T \neq 0$ in some form. Considerations of this type are closely connected with the name Matsubara (T. Matsubara, Progr. Theoret. Phys. **14**, 351 (1955)). We therefore refer to the procedures discussed in the present section as **Matsubara methods**.

As we showed in Chap. 3, the retarded Green's function has a direct relation to experiments (e.g. response functions, quasi-particle densities of states, correlation functions, excitation energies). There are a series of methods for its approximate determination (equation of motion method, method of moments, CPA,...); a perturbation-theoretical diagram technique in the sense of Chap. 5, however, cannot be formulated. For the retarded function, there is no Wick's theorem, although a Dyson equation as in (3.313) can be constructed. The retarded Green's function is thus not so readily accessible via perturbation theory. For perturbation theory, at least at $T = 0$, the causal function is eminently suitable. The special form (5.85) of Wick's theorem makes it possible to apply quite effective diagram techniques. Since, on the other hand, at $T = 0$ the transformation from the causal to the retarded function is very simple ((3.174), (3.175)), it is worthwhile to develop the perturbation theory for the causal Green's function as in Chap. 5.

At finite temperatures, however, the conditions for the application of Wick's theorem in the special form of (5.85) are no longer fulfilled. We can no longer take averages using only the ground state $|\eta_0\rangle$ of the non-interacting system, and the vanishing of the normal products as in (5.70) can no longer be exploited. The **Matsubara functions** which we now introduce are actually none other than suitably generalised causal Green's functions which permit the application of a modified Wick's theorem and thus become accessible to perturbation theory. Furthermore, we will be able to show that the transition from the causal functions to the presently more interesting retarded Green's functions is quite simple.

W. Nolting, *Fundamentals of Many-body Physics*,
DOI 10.1007/978-3-540-71931-1_6, © Springer-Verlag Berlin Heidelberg 2009

6.1.1 Matsubara Functions

Making use of Eqs. (3.118) and (3.119), we have derived the fact that when the Hamiltonian is not explicitly time dependent, the correlation functions $\langle A(t)B(t')\rangle$, $\langle B(t')A(t)\rangle$ and thereby all three Green's functions depend only on the time difference $t - t'$:

$$\langle A(t)B(t')\rangle = \langle A(t-t')B(0)\rangle = \langle A(0)B(t'-t)\rangle \, ,$$
$$\langle B(t')A(t)\rangle = \langle B(t'-t)A(0)\rangle = \langle B(0)A(t-t')\rangle \, .$$

The two correlation functions which construct the Green's functions $G^{\alpha}_{AB}(t, t')$ (α = ret, adv, c) are not mutually independent if one allows the time variable formally to assume **complex** values:

$$\begin{aligned}
&\Xi\langle A(t-\mathrm{i}\hbar\beta)B(t')\rangle = \\
&= \mathrm{Tr}\left\{\exp(-\beta\mathcal{H})\exp\left[\frac{\mathrm{i}}{\hbar}\mathcal{H}(t-\mathrm{i}\hbar\beta)\right]A(0)\exp\left[-\frac{\mathrm{i}}{\hbar}\mathcal{H}(t-\mathrm{i}\hbar\beta)\right]B(t')\right\} = \\
&= \mathrm{Tr}\left\{B(t')\exp(-\beta\mathcal{H})\exp(+\beta\mathcal{H})\exp\left(\frac{\mathrm{i}}{\hbar}\mathcal{H}t\right)A(0)\exp\left(-\frac{\mathrm{i}}{\hbar}\mathcal{H}t\right)\exp(-\beta\mathcal{H})\right\} = \\
&= \mathrm{Tr}\left\{\exp(-\beta\mathcal{H})B(t')A(t)\right\} .
\end{aligned}$$

Here, we have made use of the cyclic invariance of the trace several times. The resulting relation

$$\langle A(t-\mathrm{i}\hbar\beta)B(t')\rangle = \langle B(t')A(t)\rangle \tag{6.1}$$

makes it reasonable to extend the definitions of the Green's functions also to complex times. This would have an additional advantage: Every *normal* perturbation theory is based on the assumption that the Hamiltonian $\mathcal{H}$ of the system can be decomposed according to $\mathcal{H} = \mathcal{H}_0 + V$, where it is assumed that the eigenvalue problem for $\mathcal{H}_0$ is exactly solvable. For averaged quantities at $T \neq 0$ (3.120), the perturbation V then appears in two distinct places, first in the Heisenberg representation of the time-dependent operators via $\exp(\pm\frac{\mathrm{i}}{\hbar}\mathcal{H}t)$, and secondly in the density operator $\exp(-\beta\mathcal{H})$ of the grand canonical averaging procedure. For both, we would then in fact have to carry out a perturbation expansion. The effort required can be reduced if one takes $\hbar\beta$ to be the real or the imaginary part of a complex time. The two exponential functions can then be combined.

The Matsubara method presumes purely imaginary times and introduces the real quantity

$$\tau = -\mathrm{i}t \, . \tag{6.2}$$

This leads to a modified Heisenberg representation for the operators:

$$A(\tau) = \exp\left(\frac{1}{\hbar}\mathcal{H}\tau\right) A(0) \exp\left(-\frac{1}{\hbar}\mathcal{H}\tau\right). \tag{6.3}$$

In making use of this representation, we must take some care, since the operator $\exp\left(\frac{1}{\hbar}\mathcal{H}\tau\right)$ that produces the imaginary time shifts is **not** unitary. The equation of motion is then given by:

$$-\hbar\frac{\partial}{\partial\tau}A(\tau) = \left[A(\tau), \mathcal{H}\right]_- . \tag{6.4}$$

With the step function

$$\Theta(\tau) = \begin{cases} 1\,, & \text{if} \quad \tau > 0 \iff t \quad \text{negative imaginary,} \\ 0\,, & \text{if} \quad \tau < 0 \iff t \quad \text{positive imaginary}\,, \end{cases} \tag{6.5}$$

we can define a **time-ordering operator**:

$$T_\tau\left\{A(\tau)B\left(\tau'\right)\right\} = \Theta\left(\tau - \tau'\right) A(\tau)B\left(\tau'\right) + \varepsilon^p\Theta\left(\tau' - \tau\right) B\left(\tau'\right) A(\tau)\,. \tag{6.6}$$

p is the number of permutations of creation and annihilation operators which are necessary in order to bring the second term again into the same operator ordering as the first term. ε is as usual defined by ($\varepsilon = +1$: Bosons; and $\varepsilon = -1$: Fermions). However, for simplicity in the following we assume that A and B are purely Fermionic or purely Bosonic operators, so that we can set $p = 1$.

After these preliminary considerations, we now define the

Matsubara function

$$G^M_{AB}\left(\tau, \tau'\right) \equiv \langle\langle A(\tau); B\left(\tau'\right)\rangle\rangle^M = -\left\langle T_\tau\left(A(\tau)B\left(\tau'\right)\right)\right\rangle . \tag{6.7}$$

It follows immediately from the definition that the equation of motion of this function, inserting (6.4) and (6.6), is given by:

$$-\hbar\frac{\partial}{\partial\tau}G^{\mathrm{M}}_{AB}\left(\tau, \tau'\right) = \hbar\delta\left(\tau - \tau'\right)\langle[A, B]_{-\varepsilon}\rangle + \left\langle\left\langle\left[A(\tau), \mathcal{H}\right]_- ; B\left(\tau'\right)\right\rangle\right\rangle^{\mathrm{M}} . \tag{6.8}$$

We wish to list some of its important properties. Due to

$$
\begin{aligned}
&\Xi\left\langle A(\tau)B\left(\tau'\right)\right\rangle = \\
&= \mathrm{Tr}\left\{\exp(-\beta\mathcal{H})\exp\left(\frac{1}{\hbar}\mathcal{H}\tau\right)A\exp\left(-\frac{1}{\hbar}\mathcal{H}\left(\tau-\tau'\right)\right)B\exp\left(-\frac{1}{\hbar}\mathcal{H}\tau'\right)\right\} = \\
&= \mathrm{Tr}\left\{\exp(-\beta\mathcal{H})\exp\left(\frac{1}{\hbar}\mathcal{H}\left(\tau-\tau'\right)\right)A\exp\left(-\frac{1}{\hbar}\mathcal{H}\left(\tau-\tau'\right)\right)B\right\} = \\
&= \Xi\left\langle A\left(\tau-\tau'\right)B\right\rangle ,
\end{aligned}
$$

the Matsubara function also depends only on time differences:

$$
G^{\mathrm{M}}_{AB}\left(\tau,\tau'\right) = G^{\mathrm{M}}_{AB}\left(\tau-\tau',0\right) = G^{\mathrm{M}}_{AB}\left(0,\tau'-\tau\right) . \tag{6.9}
$$

Another quite important property concerns its periodicity. Let

$$
\hbar\beta > \tau-\tau'+n\hbar\beta > 0 \quad n\in\mathbf{Z} ;
$$

then we have:

$$
\begin{aligned}
&\Xi G^{\mathrm{M}}_{AB}\underbrace{\left(\tau-\tau'+n\hbar\beta\right)}_{>0} = \\
&= -\mathrm{Tr}\left\{\exp(-\beta\mathcal{H})T_\tau\left(A\left(\tau-\tau'+n\hbar\beta\right)B(0)\right)\right\} = \\
&= -\mathrm{Tr}\left\{\exp(-\beta\mathcal{H})A\left(\tau-\tau'+n\hbar\beta\right)B(0)\right\} = \\
&= -\mathrm{Tr}\left\{\exp(-\beta\mathcal{H})\exp\left(\frac{1}{\hbar}\mathcal{H}\left(\tau-\tau'+n\hbar\beta\right)\right)A(0)\cdot\right. \\
&\qquad\left.\cdot\exp\left(-\frac{1}{\hbar}\mathcal{H}\left(\tau-\tau'+n\hbar\beta\right)\right)B(0)\right\} = \\
&= -\mathrm{Tr}\left\{\exp\left(\frac{1}{\hbar}\mathcal{H}\left(\tau-\tau'+(n-1)\hbar\beta\right)\right)A(0)\cdot\right. \\
&\qquad\left.\cdot\exp\left(-\frac{1}{\hbar}\mathcal{H}\left(\tau-\tau'+(n-1)\hbar\beta\right)\right)\exp(-\beta\mathcal{H})B(0)\right\} = \\
&= -\mathrm{Tr}\left\{\exp(-\beta\mathcal{H})B(0)A\underbrace{\left(\tau-\tau'+(n-1)\hbar\beta\right)}_{<0}\right\} = \\
&= -\mathrm{Tr}\left\{\exp(-\beta\mathcal{H})T_\tau\left(B(0)A\left(\tau-\tau'+(n-1)\hbar\beta\right)\right)\right\} = \\
&= -\varepsilon\mathrm{Tr}\left\{\exp(-\beta\mathcal{H})T_\tau\left(A\left(\tau-\tau'+(n-1)\hbar\beta\right)B(0)\right)\right\} .
\end{aligned}
$$

This yields the important relation:

$$\begin{aligned}&\hbar\beta > \tau - \tau' + n\hbar\beta > 0 :\\ &G^{\mathrm{M}}_{AB}\left(\tau - \tau' + n\hbar\beta\right) = \varepsilon G^{\mathrm{M}}_{AB}\left(\tau - \tau' + (n-1)\hbar\beta\right) .\end{aligned} \tag{6.10}$$

In particular, for $n = 1$ we find:

$$\begin{aligned}&G^{\mathrm{M}}_{AB}\left(\tau - \tau' + \hbar\beta\right) = \varepsilon G^{\mathrm{M}}_{AB}\left(\tau - \tau'\right) ,\\ &\text{when} \quad -\hbar\beta < \tau - \tau' < 0 .\end{aligned} \tag{6.11}$$

The Matsubara function is thus periodic with a periodicity interval of $2\hbar\beta$. We can therefore limit our considerations to the time interval $-\hbar\beta < \tau - \tau' < 0$.

Owing to this periodicity, we can make use of a **Fourier expansion** for the Matsubara function:

$$G^{\mathrm{M}}(\tau) = \frac{1}{2}a_0 + \sum_{n=1}^{\infty}\left[a_n \cos\frac{n\pi}{\hbar\beta}\tau + b_n \sin\frac{n\pi}{\hbar\beta}\tau\right] ,$$

$$a_n = \frac{1}{\hbar\beta}\int\limits_{-\hbar\beta}^{+\hbar\beta} \mathrm{d}\tau\, G^{\mathrm{M}}(\tau)\cos\left(\frac{n\pi}{\hbar\beta}\tau\right) ,$$

$$b_n = \frac{1}{\hbar\beta}\int\limits_{-\hbar\beta}^{+\hbar\beta} \mathrm{d}\tau\, G^{\mathrm{M}}(\tau)\sin\left(\frac{n\pi}{\hbar\beta}\tau\right) .$$

We define

$$E_n = \frac{n\pi}{\beta} ; \quad G^{\mathrm{M}}\left(E_n\right) = \frac{1}{2}\hbar\beta\left(a_n + \mathrm{i}b_n\right) \tag{6.12}$$

and can then write:

$$G^{\mathrm{M}}(\tau) = \frac{1}{\hbar\beta}\sum_{n=-\infty}^{+\infty} \exp\left(-\frac{\mathrm{i}}{\hbar}E_n\tau\right)G^{\mathrm{M}}\left(E_n\right) , \tag{6.13}$$

$$G^{\mathrm{M}}\left(E_n\right) = \frac{1}{2}\int\limits_{-\hbar\beta}^{+\hbar\beta} \mathrm{d}\tau\, G^{\mathrm{M}}(\tau)\exp\left(\frac{\mathrm{i}}{\hbar}E_n\tau\right) . \tag{6.14}$$

This can be somewhat further simplified:

$$\begin{aligned}G^{\mathrm{M}}\left(E_n\right) &= \frac{1}{2}\int\limits_0^{\hbar\beta} \ldots + \frac{1}{2}\int\limits_{-\hbar\beta}^{0} \ldots =\\ &= \frac{1}{2}\int\limits_0^{\hbar\beta} \mathrm{d}\tau\, G^{\mathrm{M}}(\tau)\exp\left(\frac{\mathrm{i}}{\hbar}E_n\tau\right) +\end{aligned}$$

$$+\frac{1}{2}\int\limits_0^{\hbar\beta} \mathrm{d}\tau' \, G^{\mathrm{M}}\left(\tau' - \hbar\beta\right) \exp\left(\frac{\mathrm{i}}{\hbar} E_n \tau'\right) \exp(-\mathrm{i} E_n \beta) =$$

$$\left(\tau' = \tau + \hbar\beta\right)$$

$$= \left[1 + \varepsilon \exp(-\mathrm{i}\beta E_n)\right] \frac{1}{2} \int\limits_0^{\hbar\beta} \mathrm{d}\tau \, G^{\mathrm{M}}(\tau) \exp\left(\frac{\mathrm{i}}{\hbar} E_n \tau\right) .$$

The expression in brackets vanishes for Fermions ($\varepsilon = -1$), in the case that n is even, and for Bosons ($\varepsilon = +1$) when n is odd. We then have:

$$G^{\mathrm{M}}(\tau) = \frac{1}{\hbar\beta} \sum_{n=-\infty}^{+\infty} \exp\left(-\frac{\mathrm{i}}{\hbar} E_n \tau\right) G^{\mathrm{M}}\left(E_n\right) , \tag{6.15}$$

$$G^{\mathrm{M}}\left(E_n\right) = \int\limits_0^{\hbar\beta} \mathrm{d}\tau \, G^{\mathrm{M}}(\tau) \exp\left(\frac{\mathrm{i}}{\hbar} E_n \tau\right) , \tag{6.16}$$

$$E_n = \begin{cases} 2n\pi/\beta : & \text{Bosons,} \\ (2n+1)\pi/\beta : & \text{Fermions.} \end{cases} \tag{6.17}$$

For these Matsubara functions, we shall be able to formulate Wick's theorem later on. In order to show that they also have a direct connection to experiments, we demonstrate their relation to the retarded Green's function. This is possible using the **spectral representation** (notation as in Sect. 3.2.2):

$$\begin{aligned} &\langle A(\tau) B(0) \rangle \\ &= \frac{1}{\Xi} \sum_n \langle E_n | \, A(\tau) B(0) \, | E_n \rangle \exp(-\beta E_n) = \\ &= \frac{1}{\Xi} \sum_{n,m} \langle E_n \, | A | \, E_m \rangle \langle E_m \, | B | \, E_n \rangle \exp(-\beta E_n) \exp\left[\frac{1}{\hbar}\left(E_n - E_m\right) \tau\right] . \end{aligned}$$

We derived the following expression for the spectral density $S_{AB}(E)$ using (3.134):

$$\begin{aligned} S_{AB}(E) = \frac{\hbar}{\Xi} \sum_{n,m} \langle E_n \, | A | \, E_m \rangle \langle E_m \, | B | \, E_n \rangle \, \mathrm{e}^{-\beta E_n} \cdot \\ \cdot \left(1 - \varepsilon \mathrm{e}^{-\beta E}\right) \delta\left[E - (E_m - E_n)\right] . \end{aligned}$$

We thus find:

$$\langle A(\tau)B(0)\rangle = \frac{1}{\hbar}\int\limits_{-\infty}^{+\infty} \mathrm{d}E \, \frac{S_{AB}(E)}{1-\varepsilon \exp(-\beta E)} \exp\left(-\frac{1}{\hbar}E\tau\right) . \tag{6.18}$$

Within the integration range in (6.16), τ always remains positive, so that for the Matsubara function, we need evaluate only:

$$G^{\mathrm{M}}_{AB}(E_n) = -\int\limits_{0}^{\hbar\beta} \mathrm{d}\tau \, \exp\left(\frac{\mathrm{i}}{\hbar}E_n\tau\right) \langle A(\tau)B(0)\rangle . \tag{6.19}$$

We insert

$$\int\limits_{0}^{\hbar\beta} \mathrm{d}\tau \, \exp\left(\frac{1}{\hbar}(\mathrm{i}E_n - E)\,\tau\right) = \frac{\hbar}{\mathrm{i}E_n - E}\Big[\exp(\mathrm{i}\beta E_n)\exp(-\beta E) - 1\Big] =$$

$$= \frac{\hbar}{\mathrm{i}E_n - E}\Big[\varepsilon \exp(-\beta E) - 1\Big]$$

together with (6.18) into (6.19):

$$G^{\mathrm{M}}_{AB}(E_n) = \int\limits_{-\infty}^{+\infty} \mathrm{d}E' \, \frac{S_{AB}(E')}{\mathrm{i}E_n - E'} . \tag{6.20}$$

Comparison with (3.136) verifies the formal agreement with the spectral representation of the retarded Green's functions after making the replacement

$$\mathrm{i}E_n \longrightarrow E + \mathrm{i}0^+ . \tag{6.21}$$

We thus obtain the retarded Green's function from the Matsubara function quite simply by an analytic continuation of the imaginary axis to the real E axis. For completeness, we mention that the advanced Green's function can be obtained from the Matsubara function (6.20) via the transition $\mathrm{i}E_n \to E - \mathrm{i}0^+$.

6.1.2 The Grand Canonical Partition Function

The following considerations concern systems containing Fermions or Bosons which may be subject to a pairwise interaction, as usual:

$$\mathcal{H} = \mathcal{H}_0 + V , \tag{6.22}$$

$$\mathcal{H}_0 = \sum_k (\varepsilon(\boldsymbol{k}) - \mu) a_k^+ a_k \,, \tag{6.23}$$

$$V = \frac{1}{2} \sum_{klmn} v(kl;nm) a_k^+ a_l^+ a_m a_n \,. \tag{6.24}$$

In the case of $S = 1/2$-Fermions, $k \equiv (\boldsymbol{k}, \sigma)$; for $S = 0$-Bosons $k = \boldsymbol{k}$ is to be read. In the end, the goal will be to compute expectation values of time-ordered operator products, whereby the averaging process is to be carried out over the grand canonical ensemble:

$$\left\langle T_\tau \left(\cdots I\,(\tau_i) \cdots J\,(\tau_j) \cdots \right) \right\rangle = \frac{1}{\Xi} \mathrm{Tr} \left\{ \mathrm{e}^{-\beta \mathcal{H}} T_\tau \left(\cdots I\,(\tau_i) \cdots J\,(\tau_j) \cdots \right) \right\} . \tag{6.25}$$

Ξ is the **grand canonical partition function**, which we have often used already.

$$\Xi = \mathrm{Tr} \left\{ \mathrm{e}^{-\beta \mathcal{H}} \right\} . \tag{6.26}$$

As we shall see, this important function will play a similar role to that of the vacuum amplitude in the $T = 0$ formalism.

For the construction of a $T = 0$ perturbation theory, we found the Dirac or interaction representation to be particularly favourable. This is true in modified form also of the Matsubara formalism. The following considerations are thus developed for the most part parallel to those in Sect. 3.1.1. We first define the transition to the Dirac representation, in analogy to (3.34), for an arbitrary operator A_S from the Schrödinger representation as follows:

$$A_D(\tau) = \exp\left(\frac{1}{\hbar} \mathcal{H}_0 \tau \right) A_S \exp\left(-\frac{1}{\hbar} \mathcal{H}_0 \tau \right) . \tag{6.27}$$

For the transformation into the Heisenberg representation, we have from (6.3):

$$A_H(\tau) = \exp\left(\frac{1}{\hbar} \mathcal{H} \tau \right) A_S \exp\left(-\frac{1}{\hbar} \mathcal{H} \tau \right) . \tag{6.28}$$

A_S is at most explicitly time dependent. We define in analogy to Dirac's time-evolution operator (3.33):

$$U_D\left(\tau, \tau'\right) = \exp\left(\frac{1}{\hbar} \mathcal{H}_0 \tau \right) \exp\left(-\frac{1}{\hbar} \mathcal{H}\left(\tau - \tau'\right) \right) \exp\left(-\frac{1}{\hbar} \mathcal{H}_0 \tau' \right) . \tag{6.29}$$

This operator is, to be sure, **not unitary**, but like its analog in (3.33) for real times, it has the following properties:

$$U_D\left(\tau_1, \tau_2\right) U_D\left(\tau_2, \tau_3\right) = U_D\left(\tau_1, \tau_3\right) \,, \tag{6.30}$$

$$U_D(\tau, \tau) = \mathbf{1} \,. \tag{6.31}$$

Via U_D, the Dirac and the Heisenberg representations can be related to each other:

$$A_\mathrm{H}(\tau) = \exp\left(\frac{1}{\hbar}\mathcal{H}\tau\right)\exp\left(-\frac{1}{\hbar}\mathcal{H}_0\tau\right)A_\mathrm{D}(\tau)\exp\left(\frac{1}{\hbar}\mathcal{H}_0\tau\right)\exp\left(-\frac{1}{\hbar}\mathcal{H}\tau\right) =$$
$$= U_\mathrm{D}(0,\tau)A_\mathrm{D}(\tau)U_\mathrm{D}(\tau,0)\,. \tag{6.32}$$

Using (6.29), we can readily derive the equation of motion of the time-evolution operator:

$$-\hbar\frac{\partial}{\partial\tau}U_\mathrm{D}\left(\tau,\tau'\right) =$$
$$= -\exp\left(\frac{1}{\hbar}\mathcal{H}_0\tau\right)(\mathcal{H}_0 - \mathcal{H})\exp\left(-\frac{1}{\hbar}\mathcal{H}\left(\tau-\tau'\right)\right)\exp\left(-\frac{1}{\hbar}\mathcal{H}_0\tau'\right) =$$
$$= \exp\left(\frac{1}{\hbar}\mathcal{H}_0\tau\right)V\exp\left(-\frac{1}{\hbar}\mathcal{H}_0\tau\right)\exp\left(\frac{1}{\hbar}\mathcal{H}_0\tau\right)\exp\left(-\frac{1}{\hbar}\mathcal{H}\left(\tau-\tau'\right)\right)$$
$$\exp\left(-\frac{1}{\hbar}\mathcal{H}_0\tau'\right),$$
$$-\hbar\frac{\partial}{\partial\tau}U_\mathrm{D}\left(\tau,\tau'\right) = V_\mathrm{D}(\tau)U_\mathrm{D}\left(\tau,\tau'\right)\,. \tag{6.33}$$

$V_\mathrm{D}(\tau)$ is the interaction in the Dirac representation. With (6.31) as boundary condition, the formal solution of the equation of motion is given by:

$$U_\mathrm{D}\left(\tau,\tau'\right) = \mathbf{1} - \frac{1}{\hbar}\int_{\tau'}^{\tau}\mathrm{d}\tau''\,V_\mathrm{D}\left(\tau''\right)U_\mathrm{D}\left(\tau'',\tau'\right)\,. \tag{6.34}$$

This agrees, apart from unimportant factors, with (3.12). We thus find as a result of the same considerations as those that led to (3.13) and (3.17):

$$U_\mathrm{D}\left(\tau,\tau'\right) = \sum_{n=0}^{\infty}\frac{1}{n!}\left(-\frac{1}{\hbar}\right)^n\int_{\tau'}^{\tau}\mathrm{d}\tau_1\cdots\int_{\tau'}^{\tau}\mathrm{d}\tau_n T_\tau\left(V_\mathrm{D}\left(\tau_1\right)\cdots V_\mathrm{D}\left(\tau_n\right)\right)\,. \tag{6.35}$$

With the same justification as for (5.56), we were able to replace Dyson's time-ordering operator T_D (3.15), which in fact appears in the expansion (6.35) and *sorts* without the factor ε, by the operator T_τ from (6.6). This is permissible, since from (6.24), the interaction V is constructed with an even number of creation and annihilation operators.

Equation (6.35) is the starting point for a $T > 0$ perturbation theory. We can draw a first important conclusion for the grand canonical partition function. It follows from (6.29) that:

$$\exp\left(-\frac{1}{\hbar}\mathcal{H}\tau\right) = \exp\left(-\frac{1}{\hbar}\mathcal{H}_0\tau\right)U_\mathrm{D}(\tau,0)\,.$$

If we choose in particular $\tau = \hbar\beta$,

$$e^{-\beta\mathcal{H}} = e^{-\beta\mathcal{H}_0} U_D(\hbar\beta, 0) , \tag{6.36}$$

then we can relate the partition function to U_D:

$$\begin{aligned} \Xi &= \mathrm{Tr}\{e^{-\beta\mathcal{H}_0} U_D(\hbar\beta, 0)\} = \\ &= \sum_{n=0}^{\infty} \frac{1}{n!}\left(-\frac{1}{\hbar}\right)^n \int_0^{\hbar\beta} \cdots \int d\tau_1 \cdots d\tau_n \,\mathrm{Tr}\Big\{e^{-\beta\mathcal{H}_0} T_\tau\big(V_D(\tau_1)\cdots V_D(\tau_n)\big)\Big\} . \end{aligned} \tag{6.37}$$

6.1.3 The Single-Particle Matsubara Function

The **single-particle Matsubara function** will be of particular interest:

$$G_k^M(\tau) = -\big\langle T_\tau\big(a_k(\tau)a_k^+(0)\big)\big\rangle . \tag{6.38}$$

We will show later that it obeys a Dyson equation:

$$\begin{aligned} G_k^M(\tau) &= \frac{1}{\hbar\beta}\sum_{n=-\infty}^{+\infty} \exp\left(-\frac{i}{\hbar}E_n\tau\right) G_k^M(E_n) , \\ G_k^M(E_n) &= \frac{\hbar}{iE_n - \big(\varepsilon(\boldsymbol{k}) - \mu\big) - \Sigma^M(k, E_n)} . \end{aligned} \tag{6.39}$$

Here, the self-energy $\Sigma^M(k, E_n)$ depends upon the retarded self-energy, which we have already encountered and which takes the influence of the particle interactions into account, via the following transition:

$$\Sigma^M(k, E_n) \xrightarrow[iE_n \to E + i0^+]{} \Sigma^{ret}(k, E) = R^{ret}(k, E) + iI^{ret}(k, E) . \tag{6.40}$$

R^{ret} and I^{ret}, according to (3.317), directly determine the single-particle spectral density $S_k(E)$, whose significance and direct relation to experiments were emphasized in Chap. 3.

For the perturbation theory which we wish to describe in the following, we require the Matsubara function $G_k^{0,M}(\tau)$ for the system of **non**-interacting particles defined by $\mathcal{H}_0$, which of course can be calculated exactly. We first derive explicitly the time evolution of the *Heisenberg operator* $a_k(\tau)$. The relation

$$a_k\mathcal{H}_0^n = \big(\varepsilon(\boldsymbol{k}) - \mu + \mathcal{H}_0\big)^n a_k \tag{6.41}$$

is proved by complete induction. Due to

$$[a_k, \mathcal{H}_0]_- = (\varepsilon(\boldsymbol{k}) - \mu)a_k\,,$$

the proposition is clearly correct for $n = 1$:

$$a_k\mathcal{H}_0 = [a_k, \mathcal{H}_0]_- + \mathcal{H}_0 a_k = \big(\varepsilon(\boldsymbol{k}) - \mu + \mathcal{H}_0\big)a_k\,.$$

The extension from n to $n+1$ is accomplished as follows:

$$\begin{aligned}
a_k\mathcal{H}_0^{n+1} &= (a_k\mathcal{H}_0^n)\mathcal{H}_0 = \\
&= \big(\varepsilon(\boldsymbol{k}) - \mu + \mathcal{H}_0\big)^n a_k\mathcal{H}_0 = \\
&= \big(\varepsilon(\boldsymbol{k}) - \mu + \mathcal{H}_0\big)^n \big(\varepsilon(\boldsymbol{k}) - \mu + \mathcal{H}_0\big)a_k = \\
&= \big(\varepsilon(\boldsymbol{k}) - \mu + \mathcal{H}_0\big)^{n+1} a_k \quad \text{q. e. d.}
\end{aligned}$$

With (6.27), we furthermore have:

$$\begin{aligned}
&\exp\left(\frac{1}{\hbar}\mathcal{H}_0\tau\right)a_k\exp\left(-\frac{1}{\hbar}\mathcal{H}_0\tau\right) = \\
&= \exp\left(\frac{1}{\hbar}\mathcal{H}_0\tau\right)\sum_{n=0}^{\infty}\frac{1}{n!}\left(-\frac{\tau}{\hbar}\right)^n a_k\mathcal{H}_0^n = \\
&= \exp\left(\frac{1}{\hbar}\mathcal{H}_0\tau\right)\sum_{n=0}^{\infty}\frac{1}{n!}\left(-\frac{1}{\hbar}\big(\varepsilon(\boldsymbol{k}) - \mu + \mathcal{H}_0\big)\tau\right)^n a_k = \\
&= \exp\left(\frac{1}{\hbar}\mathcal{H}_0\tau\right)\exp\left(-\frac{1}{\hbar}\big(\varepsilon(\boldsymbol{k}) - \mu + \mathcal{H}_0\big)\tau\right)a_k\,.
\end{aligned}$$

This means that:

$$a_k(\tau) = a_k\exp\left(-\frac{1}{\hbar}\big(\varepsilon(\boldsymbol{k}) - \mu\big)\tau\right). \tag{6.42}$$

Quite analogously, one proves that:

$$a_k^+(\tau) = a_k^+\exp\left(\frac{1}{\hbar}\big(\varepsilon(\boldsymbol{k}) - \mu\big)\tau\right). \tag{6.43}$$

We can see that in the *modified* Heisenberg representation $a_k(\tau)$ and $a_k^+(\tau)$ are no longer mutually adjoint for $\tau \neq 0$.

Using (6.42) and (6.43), the *free* single-particle Matsubara function can be readily computed:

$$\begin{aligned}
G_k^{0,\mathrm{M}} &= -\left\langle T_\tau\left(a_k(\tau)a_k^+(0)\right)\right\rangle^{(0)} = \\
&= -\Theta(\tau)\left\langle a_k(\tau)a_k^+(0)\right\rangle^{(0)} - \varepsilon\Theta(-\tau)\left\langle a_k^+(0)a_k(\tau)\right\rangle^{(0)} = \\
&= -\exp\left(-\frac{1}{\hbar}(\varepsilon(\boldsymbol{k})-\mu)\tau\right)\left\{\Theta(\tau)\left\langle a_k a_k^+\right\rangle^{(0)} + \varepsilon\Theta(-\tau)\left\langle a_k^+ a_k\right\rangle^{(0)}\right\} , \qquad (6.44)\\
G_k^{0,\mathrm{M}}(\tau) &= -\exp\left(-\frac{1}{\hbar}(\varepsilon(\boldsymbol{k})-\mu)\tau\right)\left\{\Theta(\tau)\left(1+\varepsilon\,\langle n_k\rangle^{(0)}\right) + \Theta(-\tau)\varepsilon\,\langle n_k\rangle^{(0)}\right\} .
\end{aligned}$$

This result strongly reminds us of the representation (3.190) for the causal function. The expectation value of the number operator $\langle n_k\rangle^{(0)}$ is determined with the aid of (6.18):

$$\begin{aligned}
\left\langle a_k a_k^+\right\rangle^{(0)} &= \frac{1}{\hbar}\int\limits_{-\infty}^{+\infty} \mathrm{d}E\,\frac{S_k^{(0)}(E)}{1-\varepsilon \mathrm{e}^{-\beta E}} \overset{(3.3.185)}{=} \frac{1}{1-\varepsilon \mathrm{e}^{-\beta(\varepsilon(\boldsymbol{k})-\mu)}} = \\
&= \frac{\mathrm{e}^{\beta(\varepsilon(\boldsymbol{k})-\mu)}}{\mathrm{e}^{\beta(\varepsilon(\boldsymbol{k})-\mu)}-\varepsilon} = 1 + \frac{\varepsilon}{\mathrm{e}^{\beta(\varepsilon(\boldsymbol{k})-\mu)}-\varepsilon} = 1+\varepsilon\,\langle n_k\rangle^{(0)} .
\end{aligned}$$

This yields the result which is well known from quantum statistics (the Fermi-Dirac or the Bose-Einstein function):

$$\langle n_k\rangle^{(0)} = \left\{\mathrm{e}^{\beta(\varepsilon(\boldsymbol{k})-\mu)}-\varepsilon\right\}^{-1} . \qquad (6.45)$$

The energy-dependent Matsubara function can be quickly computed by insertion of (3.185) into (6.20):

$$G_k^{0,\mathrm{M}}(E_n) = \frac{\hbar}{\mathrm{i}E_n - \left(\varepsilon(\boldsymbol{k})-\mu\right)} . \qquad (6.46)$$

Of course, we could also have inserted (6.44) into (6.16) and transformed directly. – The temperature dependence is here contained only in the energies $E_n \sim \beta^{-1}$. We shall see later how the mean occupation numbers enter back into the equations when diagrams and correlation functions are explicitly evaluated.

We now want to bring the single-particle function of the interacting system (6.38) into a suitable form for perturbation theory:

$$G_k^{\mathrm{M}}(\tau_1,\tau_2) = -\left\langle T_\tau\left(a_k(\tau_1)\,a_k^+(\tau_2)\right)\right\rangle . \qquad (6.47)$$

The operators are given here still in their modified Heisenberg representation. The time differences $\tau_1-\tau_2$ are limited to the range

$$-\hbar\beta < \tau_1-\tau_2 < +\hbar\beta .$$

We can therefore assume for τ_1 and τ_2 that

$$0 < \tau_1, \tau_2 < \hbar\beta \,. \tag{6.48}$$

Equation (6.47) can be further rearranged using (6.36) and (6.32), whereby we initially assume that $\tau_1 > \tau_2$:

$$\begin{aligned}
G_k^{\mathrm{M}}(\tau_1, \tau_2) &= -\frac{1}{\Xi}\mathrm{Tr}\left\{\mathrm{e}^{-\beta\mathcal{H}} T_\tau\left(a_k(\tau_1)\, a_k^+(\tau_2)\right)\right\} = \\
&= -\frac{1}{\Xi}\mathrm{Tr}\left\{\mathrm{e}^{-\beta\mathcal{H}} a_k(\tau_1)\, a_k^+(\tau_2)\right\} = \\
&= -\frac{1}{\Xi}\mathrm{Tr}\left\{\mathrm{e}^{-\beta\mathcal{H}_0} U_{\mathrm{D}}(\hbar\beta, 0) U_{\mathrm{D}}(0, \tau_1) \cdot \right. \\
&\qquad \left. \cdot\, a_k^{\mathrm{D}}(\tau_1)\, U_{\mathrm{D}}(\tau_1, 0)\, U_{\mathrm{D}}(0, \tau_2)\, a_k^{+\mathrm{D}}(\tau_2)\, U_{\mathrm{D}}(\tau_2, 0)\right\} = \\
&= -\frac{1}{\Xi}\mathrm{Tr}\left\{\mathrm{e}^{-\beta\mathcal{H}_0} U_{\mathrm{D}}(\hbar\beta, \tau_1) a_k^{\mathrm{D}}(\tau_1)\, U_{\mathrm{D}}(\tau_1, \tau_2)\, a_k^{+\mathrm{D}}(\tau_2)\, U_{\mathrm{D}}(\tau_2, 0)\right\}.
\end{aligned}$$

Since, from (6.48), $\hbar\beta$ is the *latest time*, the operators in the trace are already time-ordered. We can therefore once again introduce the time-ordering operator T_τ, and in the argument of T_τ, we can factor the operators U_{D} without a sign change past a_k^{D} or $a_k^{+\mathrm{D}}$, since according to (6.35) and (6.24), they are composed of an even number of creation and annihilation operators:

$$\begin{aligned}
&G_k^{\mathrm{M}}(\tau_1, \tau_2) = \\
&= -\frac{1}{\Xi}\mathrm{Tr}\left\{\mathrm{e}^{-\beta\mathcal{H}_0} T_\tau\left(U_{\mathrm{D}}(\hbar\beta, \tau_1)\, a_k^{\mathrm{D}}(\tau_1)\, U_{\mathrm{D}}(\tau_1, \tau_2)\, a_k^{+\mathrm{D}}(\tau_2)\, U_{\mathrm{D}}(\tau_2, 0)\right)\right\} = \\
&= -\frac{1}{\Xi}\mathrm{Tr}\left\{\mathrm{e}^{-\beta\mathcal{H}_0} T_\tau\left(U_{\mathrm{D}}(\hbar\beta, \tau_1)\, U_{\mathrm{D}}(\tau_1, \tau_2)\, U_{\mathrm{D}}(\tau_2, 0)\, a_k^{\mathrm{D}}(\tau_1)\, a_k^{+\mathrm{D}}(\tau_2)\right)\right\} = \\
&= -\frac{1}{\Xi}\mathrm{Tr}\left\{\mathrm{e}^{-\beta\mathcal{H}_0} T_\tau\left(U_{\mathrm{D}}(\hbar\beta, 0) a_k^{\mathrm{D}}(\tau_1)\, a_k^{+\mathrm{D}}(\tau_2)\right)\right\}.
\end{aligned}$$

In the final step, we once again made use of (6.30). We now have to investigate the other case, that $\tau_1 < \tau_2$:

$$\begin{aligned}
&G_k^{\mathrm{M}}(\tau_1, \tau_2) \\
&= -\frac{\varepsilon}{\Xi}\mathrm{Tr}\left\{\mathrm{e}^{-\beta\mathcal{H}} a_k^+(\tau_2)\, a_k(\tau_1)\right\} = \\
&= -\frac{\varepsilon}{\Xi}\mathrm{Tr}\left\{\mathrm{e}^{-\beta\mathcal{H}_0} U_{\mathrm{D}}(\hbar\beta, 0) U_{\mathrm{D}}(0, \tau_2)\, a_k^{+\mathrm{D}}(\tau_2) \cdot \right. \\
&\qquad \left. \cdot\, U_{\mathrm{D}}(\tau_2, 0)\, U_{\mathrm{D}}(0, \tau_1)\, a_k^{\mathrm{D}}(\tau_1)\, U_{\mathrm{D}}(\tau_1, 0)\right\} =
\end{aligned}$$

$$= -\frac{\varepsilon}{\Xi}\mathrm{Tr}\Big\{\mathrm{e}^{-\beta\mathcal{H}_0}U_\mathrm{D}(\hbar\beta,\tau_2)\,a_k^{+\mathrm{D}}(\tau_2)\,U_\mathrm{D}(\tau_2,\tau_1)\,a_k^\mathrm{D}(\tau_1)\,U_\mathrm{D}(\tau_1,0)\Big\} =$$

$$= -\frac{\varepsilon}{\Xi}\mathrm{Tr}\Big\{\mathrm{e}^{-\beta\mathcal{H}_0}T_\tau\Big(U_\mathrm{D}(\hbar\beta,\tau_2)\,a_k^{+\mathrm{D}}(\tau_2)\,U_\mathrm{D}(\tau_2,\tau_1)\,a_k^\mathrm{D}(\tau_1)\,U_\mathrm{D}(\tau_1,0)\Big)\Big\} =$$

$$= -\frac{\varepsilon}{\Xi}\mathrm{Tr}\Big\{\mathrm{e}^{-\beta\mathcal{H}_0}T_\tau\big(U_\mathrm{D}(\hbar\beta,0)a_k^{+\mathrm{D}}(\tau_2)\,a_k^\mathrm{D}(\tau_1)\big)\Big\} =$$

$$= -\frac{1}{\Xi}\mathrm{Tr}\Big\{\mathrm{e}^{-\beta\mathcal{H}_0}T_\tau\big(U_\mathrm{D}(\hbar\beta,0)a_k^\mathrm{D}(\tau_1)\,a_k^{+\mathrm{D}}(\tau_2)\big)\Big\}\,.$$

Both cases, $\tau_1 > \tau_2$ and $\tau_1 < \tau_2$, thus lead to the same result. If we suppress the index D on the operators, since now **all** the operators are given in their Dirac representation, we can express this result as:

$$G_k^\mathrm{M}(\tau_1,\tau_2) = -\frac{\mathrm{Tr}\big\{\mathrm{e}^{-\beta\mathcal{H}_0}T_\tau\big(U(\hbar\beta,0)a_k(\tau_1)\,a_k^+(\tau_2)\big)\big\}}{\mathrm{Tr}\big\{\mathrm{e}^{-\beta\mathcal{H}_0}U(\hbar\beta,0)\big\}}\,. \tag{6.49}$$

If we now insert the time-evolution operator U as in (6.35), so we can recognise a clear analogy with the causal $T = 0$ Green's function (5-59). It is therefore not surprising that we will be able to use practically the same procedure for the evaluation of (6.49) as in Chap. 5. Important differences are that the time integrations are carried out over **finite** ranges, and that **no** *switching-on factors* occur. We have at no point had to employ the hypothesis of *adiabatic switching-on* (cf. Sect. 5.1.2). The partition function Ξ takes on roughly the same role in the Matsubara formalism which the vacuum amplitude (5.89) played in the $T = 0$ formalism. This will become more clear in the following section.

6.2 Diagrammatic Perturbation Theory

6.2.1 Wick's Theorem

For a diagrammatic analysis of the time-ordered products in (6.49), we need a tool which can assume the function of Wick's theorem (5.85) in the $T = 0$ formalism for the causal function. We shall call this tool, which we will now develop, the **generalised Wick theorem**. We have to evaluate expressions of the following form:

$$\mathrm{Tr}\big\{\mathrm{e}^{-\beta\mathcal{H}_0}T_\tau(UVW\cdots XYZ)\big\} = \Xi_0\,\langle T_\tau(UVW\cdots XYZ)\rangle^{(0)}\,.$$

Ξ_0 is the grand canonical partition function of the non-interacting system. $U, V, W\ldots$ are creation and annihilation operators in the Dirac representation, each of which acts at some particular time τ. We define:

a contraction

$$\underset{\sqcup}{UV} = \langle T_\tau(UV)\rangle^{(0)} = \varepsilon \underset{\sqcup}{VU} \; . \tag{6.50}$$

Since U and V are presumed to be creation and annihilation operators, the contraction, in analogy to the case of $T = 0$, will be essentially the single-particle Matsubara function. We now prove the

generalised Wick theorem

$$\begin{aligned}\langle T_\tau(UVW\cdots XYZ)\rangle^{(0)} &= (\underbrace{UV}\underbrace{W\cdots X}\underbrace{YZ}) + (UVW\cdots XYZ) + \cdots = \\ &= \{\textit{total pairing}\} \; .\end{aligned} \tag{6.51}$$

Note that this theorem does **not** imply an operator identity. Under the term **total pairing**, we mean (as in Sect. 5.2.2) the complete partitioning of the operator products $UVW\cdots XYZ$ into products of contractions in all possible ways, which of course presumes an even number of operators. The latter will however always be the case. $\mathcal{H}_0$ namely commutes with the particle number operator $\widehat{N}$; the number of particles is therefore a conserved quantity. An expectation value of the form $\langle UV \cdots YZ\rangle^{(0)}$ is thus only then nonzero when the product contains the same number of creation and annihilation operators. All together, we thus always have an even number of operators. We now introduce for (6.51) the **sign convention**: the operators to be contracted are first to be brought into neighbouring positions. Each transposition which is required to achieve this contributes a factor of ε.

We can initially assume, as in the proof of Wick's theorem in Sect. 5.2.2, that the operators are already time-ordered on the left side of (6.51). If this were not the case, the corresponding permutations would imply for **each** term in (6.51) the same factor ε^m. We can thus assume without loss of generality for the proof that

$$\tau_U > \tau_V > \tau_W > \cdots > \tau_X > \tau_Y > \tau_Z \; . \tag{6.52}$$

Due to (6.42) and (6.43), the time dependence of the creation and annihilation operators is very simple. We write:

$$U = \gamma_U(\tau_U)\,\alpha_U \; ; \quad \alpha_U = a_U^+ \text{ or } a_U \; , \tag{6.53}$$

$$\gamma_U(\tau_U) = \exp\!\left(\sigma_U \frac{1}{\hbar}(\varepsilon(\boldsymbol{U}) - \mu)\tau_U\right) ; \quad \sigma_U = \begin{cases} - , & \text{when} \quad \alpha_U = a_U \; , \\ + , & \text{when} \quad \alpha_U = a_U^+ \; . \end{cases} \tag{6.54}$$

Let us first consider the contraction

$$\underset{\llcorner\!\lrcorner}{UV} = \langle T_\tau(UV)\rangle^{(0)} = \langle UV\rangle^{(0)} = \gamma_U(\tau_U)\gamma_V(\tau_V)\,\langle\alpha_U\alpha_V\rangle^{(0)}\,. \tag{6.55}$$

Since the averaging is carried out over the *free* system, we can further conclude that:

$$\begin{aligned}&\langle\alpha_U\alpha_V\rangle^{(0)} \neq 0 \quad \text{only in the case that}\\ &1)\ \ \alpha_U = a_U\,, \quad \alpha_V = a_U^+\,,\\ &2)\ \ \alpha_U = a_U^+\,, \quad \alpha_V = a_U\,.\end{aligned}$$

From this, it follows with (6.45) that:

1.

$$\begin{aligned}\left\langle a_U a_U^+\right\rangle^{(0)} &= 1 + \varepsilon\,\langle n_U\rangle^{(0)} =\\ &= 1 + \frac{\varepsilon}{\mathrm{e}^{\beta(\varepsilon(U)-\mu)} - \varepsilon} = \frac{1}{1 - \varepsilon\,\mathrm{e}^{-\beta(\varepsilon(U)-\mu)}} =\\ &= \frac{\left[a_U, a_U^+\right]_{-\varepsilon}}{1 - \varepsilon\gamma_U(\hbar\beta)}\,.\end{aligned}$$

2.

$$\begin{aligned}\left\langle a_U^+ a_U\right\rangle^{(0)} &= \langle n_U\rangle^{(0)} = \frac{1}{\gamma_U(\hbar\beta) - \varepsilon} =\\ &= \frac{-\varepsilon}{1 - \varepsilon\gamma_U(\hbar\beta)} = \frac{\left[a_U^+, a_U\right]_{-\varepsilon}}{1 - \varepsilon\gamma_U(\hbar\beta)}\,.\end{aligned}$$

We can obviously combine the two cases:

$$\underset{\llcorner\!\lrcorner}{UV} = \gamma_U(\tau_U)\gamma_V(\tau_V)\,\frac{[\alpha_U, \alpha_V]_{-\varepsilon}}{1 - \varepsilon\gamma_U(\hbar\beta)}\,. \tag{6.56}$$

We now come to the actual proof of (6.51). We first have:

$$\langle UV\cdots YZ\rangle^{(0)} = \gamma_U\gamma_V\cdots\gamma_Y\gamma_Z\,\langle\alpha_U\alpha_V\cdots\alpha_Y\alpha_Z\rangle^{(0)}\,.$$

We now attempt to *pull* the operator α_U all the way to the right:

$$\begin{aligned}\frac{\langle UV\cdots YZ\rangle^{(0)}}{\gamma_U\gamma_V\cdots\gamma_Y\gamma_Z} &= \left\langle[\alpha_U, \alpha_V]_{-\varepsilon}\,\alpha_W\cdots\alpha_Z\right\rangle^{(0)} +\\ &+ \varepsilon\left\langle\alpha_V\,[\alpha_U, \alpha_W]_{-\varepsilon}\cdots\alpha_Z\right\rangle^{(0)} +\\ &+ \cdots +\\ &+ \varepsilon^{p-2}\left\langle\alpha_V\alpha_W\cdots[\alpha_U, \alpha_Z]_{-\varepsilon}\right\rangle^{(0)} +\\ &+ \varepsilon^{p-1}\,\langle\alpha_V\alpha_W\cdots\alpha_Y\alpha_Z\alpha_U\rangle^{(0)}\,.\end{aligned} \tag{6.57}$$

p is the number of operators in the expectation value. Since p must be an even number, we have $\varepsilon^{p-1} = \varepsilon$. We rearrange the last term in (6.57) once more. For this, (6.41) is helpful:

$$
\begin{aligned}
a_U \, \mathrm{e}^{-\beta \mathcal{H}_0} &= \sum_{n=0}^{\infty} \frac{1}{n!} (-\beta)^n a_U \mathcal{H}_0^n = \\
&= \sum_{n=0}^{\infty} \frac{1}{n!} \big(-\beta(\varepsilon(\boldsymbol{U}) - \mu + \mathcal{H}_0)\big)^n a_U = \\
&= \mathrm{e}^{-\beta(\varepsilon(\boldsymbol{U}) - \mu + \mathcal{H}_0)} a_U = \\
&= \gamma_U(\hbar\beta) \mathrm{e}^{-\beta \mathcal{H}_0} a_U \, .
\end{aligned}
$$

Analogously, one finds

$$
a_U^+ \mathrm{e}^{-\beta \mathcal{H}_0} = \mathrm{e}^{+\beta(\varepsilon(\boldsymbol{U}) - \mu) - \beta \mathcal{H}_0} a_U^+ = \gamma_U(\hbar\beta) \mathrm{e}^{-\beta \mathcal{H}_0} a_U^+ \, ,
$$

so that we can summarise as:

$$
\alpha_U \, \mathrm{e}^{-\beta \mathcal{H}_0} = \gamma_U(\hbar\beta) \mathrm{e}^{-\beta \mathcal{H}_0} \alpha_U \, . \tag{6.58}
$$

Making use of the cyclic invariance of the trace and Eq. (6.58), we find for the last term in (6.57):

$$
\begin{aligned}
\langle \alpha_V \alpha_W \cdots \alpha_Z \alpha_U \rangle^{(0)} &= \frac{1}{\Xi_0} \mathrm{Tr} \left\{ \mathrm{e}^{-\beta \mathcal{H}_0} \alpha_V \alpha_W \cdots \alpha_Z \alpha_U \right\} = \\
&= \frac{1}{\Xi_0} \mathrm{Tr} \left\{ \alpha_U \, \mathrm{e}^{-\beta \mathcal{H}_0} \alpha_V \alpha_W \cdots \alpha_Z \right\} = \\
&= \frac{\gamma_U(\hbar\beta)}{\Xi_0} \mathrm{Tr} \left\{ \mathrm{e}^{-\beta \mathcal{H}_0} \alpha_U \alpha_V \cdots \alpha_Z \right\} = \\
&= \gamma_U(\hbar\beta) \, \langle \alpha_U \alpha_V \cdots \alpha_Z \rangle^{(0)} \, .
\end{aligned}
$$

Inserting into (6.57), this yields:

$$
\begin{aligned}
&\frac{\langle U V \cdots Y Z \rangle^{(0)}}{\gamma_U \gamma_V \cdots \gamma_Z} \big(1 - \varepsilon \gamma_U(\hbar\beta)\big) = \\
&= \left\langle [\alpha_U, \alpha_V]_{-\varepsilon} \, \alpha_W \cdots \alpha_Z \right\rangle^{(0)} + \\
&+ \varepsilon \left\langle \alpha_V \, [\alpha_U, \alpha_W]_{-\varepsilon} \cdots \alpha_Z \right\rangle^{(0)} + \\
&+ \cdots + \\
&+ \varepsilon^{p-2} \left\langle \alpha_V \alpha_W \cdots [\alpha_U, \alpha_Z]_{-\varepsilon} \right\rangle^{(0)} \, .
\end{aligned}
$$

Finally, with (6.56), it follows that:

$$\begin{aligned}
\langle UVW\cdots XYZ\rangle^{(0)} &= \big\langle \underbrace{UV}W\cdots XYZ\big\rangle^{(0)}+\\
&+\varepsilon\big\langle V\underbrace{UW}\cdots XYZ\big\rangle^{(0)}+\\
&+\cdots+\\
&+\varepsilon^{p-2}\big\langle VW\cdots \underbrace{UZ}\big\rangle^{(0)}=\\
&=\big\langle \underbrace{UV}W\cdots XYZ\big\rangle^{(0)}+\\
&+\big\langle \underbrace{UVW}\cdots XYZ\big\rangle^{(0)}+\\
&+\cdots+\\
&+\big\langle \underbrace{UVW\cdots XYZ}\big\rangle^{(0)}.
\end{aligned} \tag{6.59}$$

The contraction itself is a c-number, and can therefore be factored out of the expectation value. We can again apply (6.59) to the remaining mean value. Finally, we obtain the *total pairing*. With (6.52), we have then proven the *generalised* Wick theorem (6.51).

6.2.2 Diagram Analysis of the Grand-Canonical Partition Function

We start with the analysis of the grand canonical partition function Ξ, from which all the macroscopic thermodynamics of the system of interacting particles can be derived. We presume the grand canonical partition function Ξ_0 of the non-interacting system to be known, and then, according to (6.37), we must calculate the following:

$$\frac{\Xi}{\Xi_0}=\frac{1}{\Xi_0}\mathrm{Tr}\left\{\mathrm{e}^{-\beta\mathcal{H}_0}U(\hbar\beta,0)\right\}=\langle U(\hbar\beta,0)\rangle^{(0)}\ . \tag{6.60}$$

The associated perturbation expansion is likewise given in (6.37). Its n-th term is given by:

$$\begin{aligned}
&\frac{1}{n!}\left(-\frac{1}{\hbar}\right)^n\int_0^{\hbar\beta}\cdots\int \mathrm{d}\tau_1\cdots\mathrm{d}\tau_n\,\big\langle T_\tau\big(V(\tau_1)\cdots V(\tau_n)\big)\big\rangle^{(0)}=\\
&=\frac{1}{n!}\left(-\frac{1}{\hbar}\right)^n\frac{1}{2^n}\sum_{\substack{k_1l_1\\ m_1n_1}}\cdots\sum_{\substack{k_nl_n\\ m_nn_n}} v\left(k_1l_1;n_1m_1\right)\cdots v(k_n\cdots)\cdot
\end{aligned} \tag{6.61}$$

$$\cdot \int_0^{\hbar\beta} \cdots \int d\tau_1 \cdots d\tau_n \left\langle T_\tau \left\{ a^+_{k_1}(\tau_1)\, a^+_{l_1}(\tau_1)\, a_{m_1}(\tau_1)\, a_{n_1}(\tau_1) \cdot \right.\right.$$

$$\left.\left. \cdots \cdot a^+_{k_n}(\tau_n)\, a^+_{l_n}(\tau_n)\, a_{m_n}(\tau_n)\, a_{n_n}(\tau_n) \right\} \right\rangle^{(0)} .$$

Apart from the *switching-on factors* and $(-1/\hbar)^n$ instead of $(-\mathrm{i}/\hbar)^n$, this expression is identical to (5.5.90), the expansion for the vacuum amplitude. Since the algebraic structure of the generalised Wick theorems (6.51) is the same as in the $T = 0$ case, when we average (5.84) over the ground state $|\eta_0\rangle$ of the free system, we can directly adopt practically all of the rules and laws derived in Sect. 5.3. The Feynman diagrams will have the same structures as in the case of $T = 0$. We could therefore repeat the treatment of the vacuum amplitude in Sect. 5.3 nearly intact; we shall do this however only in outline form.

Previously, we separated the time arguments from a four-tuple:

$$a^+_k(\tau)a^+_l(\tau)a_m(\tau)a_n(\tau)$$

into τ and τ', in order to be able to formally distinguish between *below* and *above* at a vertex. We will dispense with this distinction here, but adopt the convention that a^+_k and a_n are attached to the same vertex point and a^+_l and a_m to the other one. Every combination of contractions from the *total pairing* will be represented by a Feynman diagram. The number of vertices corresponds to the *order* of the diagram.

We adopt the same notation as in Sect. 5.3.2. Thus, Θ denotes the structure of a diagram. All the *topologically distinct* diagrams of the same structure make the same contribution to the perturbation expansion, but they belong to different combinations of contractions and must therefore be counted separately. We can give their number. As in (5.101), we have:

$$A_n(\Theta) = \frac{2^n n!}{h(\Theta)} . \tag{6.62}$$

The factor 2^n results from permutations of *below* and *above* at a vertex, and $n!$ from the permutation of the vertices. $h(\Theta)$ is the number of topologically **equivalent** diagrams within the structure Θ. These correspond to identical combinations of contractions, and thus are to be counted only once.

We again denote by **connected diagrams** those which cannot be decomposed into two independent diagrams of lower order my means of a cut. For the contribution of all connected diagrams with the structure Θ of order n, we write:

$$U_n(\Theta) = \frac{2^n}{h(\Theta)} \left(-\frac{1}{\hbar}\right)^n U^*_n\left(D^{(n)}\right) , \tag{6.63}$$

$$U^*_n\left(D^{(n)}\right) = \int_0^{\hbar\beta} \cdots \int d\tau_1 \cdots d\tau_n \left\langle T_\tau \left(V(\tau_1) \cdots V(\tau_n)\right)\right\rangle^{(0)}_{\mathrm{cont}} . \tag{6.64}$$

We now consider a non-connected diagram which consists of p connected diagram components with $n_1, n_2, \ldots, n_p$ vertices ($n_1 + n_2 + \cdots + n_p = n$). Non-connected diagrams have no common integration or summation variables in their substructures. Therefore, the overall contribution can be factored as in (6.64):

$$U_n^*\left(D^{(n)}\right) = U_{n_1}^*\left(D^{(n_1)}\right) \cdots U_{n_p}^*\left(D^{(n_p)}\right) .$$

If, among the p connected diagram components, there are $p_1, \ldots, p_\nu$ which are the same, with the structures $\Theta_1, \ldots, \Theta_\nu$,

$$\Theta = p_1\Theta_1 + p_2\Theta_2 + \cdots + p_\nu\Theta_\nu ; \quad p_1 + p_2 + \cdots + p_\nu = p ,$$

then in (6.63) we must set

$$h(\Theta) = p_1!\, h^{p_1}(\Theta_1)\, p_2!\, h^{p_2}(\Theta_2) \cdots p_\nu!\, h^{p_\nu}(\Theta_\nu) . \tag{6.65}$$

The factorials $p_1!, p_2!, \ldots, p_\nu!$ result from the fact that a permutation of the p_μ diagram components among themselves leads to topologically equivalent diagrams. We then obtain for the overall contribution of the structure:

$$U_n(\Theta) = \frac{1}{p_1!} U^{p_1}(\Theta_1) \frac{1}{p_2!} U^{p_2}(\Theta_2) \cdots \frac{1}{p_\nu!} U^{p_\nu}(\Theta_\nu) . \tag{6.66}$$

We can now easily formulate the overall perturbation expansion for Ξ/Ξ_0 in compact form:

$$\frac{\Xi}{\Xi_0} = \sum_{p_1, p_2, \ldots}^{0 \ldots \infty} \frac{1}{p_1!} U^{p_1}(\Theta_1) \frac{1}{p_2!} U^{p_2}(\Theta_2) \cdots \tag{6.67}$$

All the pairwise distinct connected diagram structures occur in the product on the right. Every p_ν runs from 0 to ∞. This means that

$$\frac{\Xi}{\Xi_0} = \exp\big(U(\Theta_1)\big) \exp\big(U(\Theta_2)\big) \cdots$$

or

$$\frac{\Xi}{\Xi_0} = \exp\left\{ \sum_\nu^{\text{conn}} U(\Theta_\nu) \right\} . \tag{6.68}$$

This corresponds to the **linked-cluster theorem** (5.112). We can thus immediately limit ourselves to the connected diagrams for the evaluation of the grand canonical partition function.

The analysis of a diagram component is carried out in complete analogy to the special case of $T = 0$.

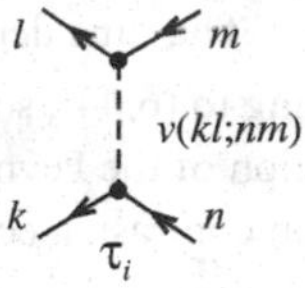

Fig. 6.1 The annotation of a vertex in the diagram analysis of Matsubara functions

The vertices are denoted by times τ_i, whose indices increase in going from left to right. Every vertex is associated with a factor $v(kl; nm)$. The factors $1/2$ cancel out in the overall expression with the term 2^n in (6.63).

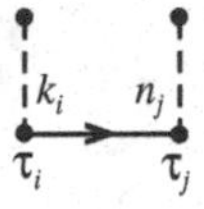

$$= \underbrace{a^+_{k_i}(\tau_i)\, a_{n_j}(\tau_j)} = \varepsilon G^{0,\mathrm{M}}_{k_i}(\tau_j - \tau_i)\, \delta_{k_i n_j} \,, \tag{6.69}$$

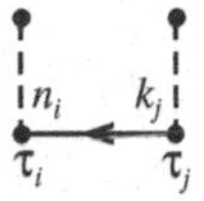

$$= \underbrace{a_{n_i}(\tau_i)\, a^+_{k_j}(\tau_j)} = -G^{0,\mathrm{M}}_{n_i}(\tau_i - \tau_j)\, \delta_{n_i k_j} \,. \tag{6.70}$$

When the times are equal, ($\tau_i = \tau_j = \tau$), we assume the convention as in the case of $T = 0$ that the time-ordering operator should move the creation operator to the left:

$$\begin{aligned} \underbrace{a^+_k(\tau) a_k(\tau)} &\overset{!}{=} \left\langle T_\tau\left(a^+_k(\tau) a_k\left(\tau - 0^+\right)\right)\right\rangle^{(0)} = \\ &= -\varepsilon G^{0,\mathrm{M}}_k\left(-0^+\right) = \\ &= \varepsilon \underbrace{a_k(\tau) a^+_k(\tau)} \,. \end{aligned} \tag{6.71}$$

With (6.44), this means that:

$$\underbrace{a^+_k(\tau) a_k(\tau)} = \langle n_k \rangle^{(0)} \,. \tag{6.72}$$

We are thereby in a position to formulate the

diagram rules for the grand canonical partition function Ξ/Ξ_0:

All the connected diagrams with pairwise differing structures are sought and the contribution of the diagrams of n-th order (n vertices, $2n$ propagators) is computed as follows:

1. Vertex $\Leftrightarrow v(kl; nm)$.
2. Propagating line $\Leftrightarrow -G^{0,\mathrm{M}}_{k_\nu}(\tau_\nu - \tau_\mu)\delta_{k_\nu, k_\mu}$ (from τ_μ to τ_ν).
3. Non-propagating line (equal times) $\Leftrightarrow -G^{0,\mathrm{M}}_{k_\nu}(-0^+)\delta_{k_\nu k_\mu}$.
4. Summation over all the $\ldots, k_i, l_i, m_i, n_i, \ldots$
5. Integration over all the $\tau_1, \ldots, \tau_n$ from 0 to $\hbar\beta$.
6. Factor: $\left(-\frac{1}{\hbar}\right)^n \frac{\varepsilon^s}{h(\Theta)}$; S = number of loops.

The loop rule, which in rule 6.2.2. leads to the factor ε^s, is proved as in the $T = 0$ case; cf. (5.100).

The time dependence of the *free* single-particle Matsubara function $G_k^{0,\mathrm{M}}(\tau)$ is, according to (6.44), somewhat clumsy. The required distinction between $\tau \gtrless 0$ makes the evaluation of the Feynman diagrams relatively complicated. The energy-dependent function has, in contrast, a considerably simpler structure (6.46). From (6.15) we see that:

$$G_k^{0,\mathrm{M}}(\tau_2 - \tau_1) = \frac{1}{\hbar\beta} \sum_n \exp\left(-\frac{\mathrm{i}}{\hbar} E_n (\tau_2 - \tau_1)\right) G_k^{0,\mathrm{M}}(E_n) \,. \tag{6.73}$$

In the diagrams, we adopt the following assignment:

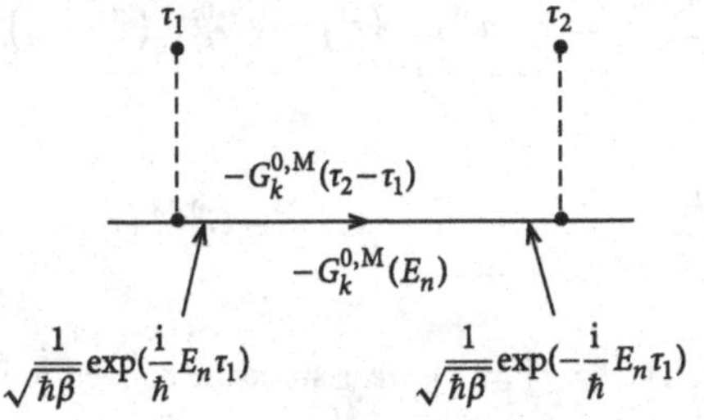

Every line which emerges from a vertex point yields an additional factor

$$\frac{\exp\left(\frac{\mathrm{i}}{\hbar} E_n \tau_1\right)}{\sqrt{\hbar\beta}} \,.$$

The corresponding line which enters at τ_2 contributes a factor

$$\exp\left(-\frac{\mathrm{i}}{\hbar} E_n \tau_2\right) / \sqrt{\hbar\beta} \,.$$

The time dependencies are concentrated exclusively in these exponential functions.

The vertex at τ then contains the factors sketched in Fig. 6.2, with which one can readily carry out the integrations over time:

$$\frac{1}{(\hbar\beta)^2} \int_0^{\hbar\beta} \mathrm{d}\tau \, \exp\left(\frac{\mathrm{i}}{\hbar}(E_k + E_l - E_m - E_n)\tau\right) =$$

$$= \frac{1}{(\hbar\beta)^2} \frac{\exp\left[\frac{\mathrm{i}}{\hbar}(E_k + E_l - E_m - E_n)\tau\right]}{\frac{\mathrm{i}}{\hbar}(E_k + E_l - E_m - E_n)} \Bigg|_0^{\hbar\beta} =$$

$$= \frac{1}{\hbar\beta} \begin{cases} 0\,, & \text{when} \quad (E_k + E_l) \neq (E_m + E_n)\,, \\ 1\,, & \text{when} \quad (E_k + E_l) = (E_m + E_n)\,. \end{cases}$$

The combination $(E_k + E_l - E_m - E_n)$ is, from (6.17), an even-integral multiple of π/β for both Fermions and Bosons. The integrations over time thus lead to energy conservation at the vertex. We can now reformulate the

diagram rules for the grand canonical partition function Ξ/Ξ_0:

All the connected diagrams with pairwise differing structures are sought and the contribution of the diagrams of n-th order (n vertices, $2n$ propagators) is computed according to the following prescriptions:

Fig. 6.2 Vertex annotation for the transition from the time-dependent to the energy-dependent Matsubara functions

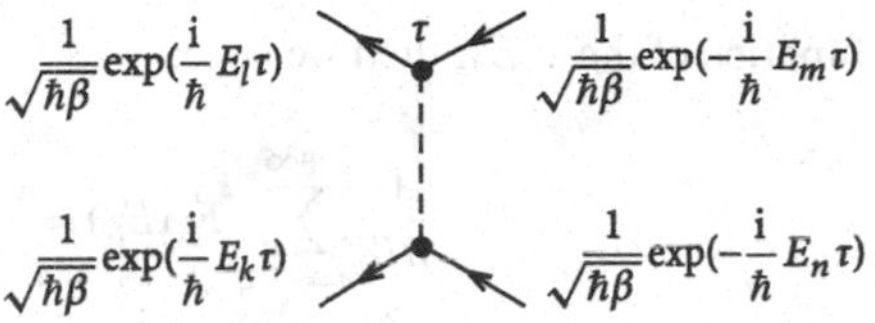

1. Vertex $\Leftrightarrow v(kl;nm)\frac{1}{\hbar\beta}\delta_{E_k+E_l,E_m+E_n}$.
2. Solid line (propagating or non-propagating):

$$-G_k^{0,\mathrm{M}}(E_{n_k}) = \frac{-\hbar}{\mathrm{i}E_{n_k} - (\varepsilon(\boldsymbol{k}) - \mu)} \,.$$

3. In addition, for non-propagating lines:

$$\exp\left(\frac{\mathrm{i}}{\hbar}E_{n_k}0^+\right) .$$

4. Summations over all $\ldots, k_i, l_i, m_i, n_i, \ldots$ and over all E_{n_i}.
5. Factor: $\left(-\frac{1}{\hbar}\right)^n \frac{\varepsilon^s}{h(\Theta)}$; S = number of loops.

The convergence-inducing factor for non-propagating lines in rule 3. can be read directly off (6.73). It follows from our convention of taking the limit

$$\tau_2 \longrightarrow \tau_1 - 0^+$$

for equal times. This, however, means that we must also associate a factor from rule 3. with a solid line which begins and ends at the same vertex, in addition to the part 2.

The summations over wavenumbers which are required by rule 4. will practically always be replaced by the corresponding integrations on going to the thermodynamic limit:

$$\sum_k \Rightarrow \frac{V}{(2\pi)^3}\int \mathrm{d}^3k \,. \tag{6.74}$$

The summations over the Matsubara energies E_n, for which (6.17) holds, are a new feature. These summations can also be converted to integrations. Let $F = F(\mathrm{i}E_n)$ be some

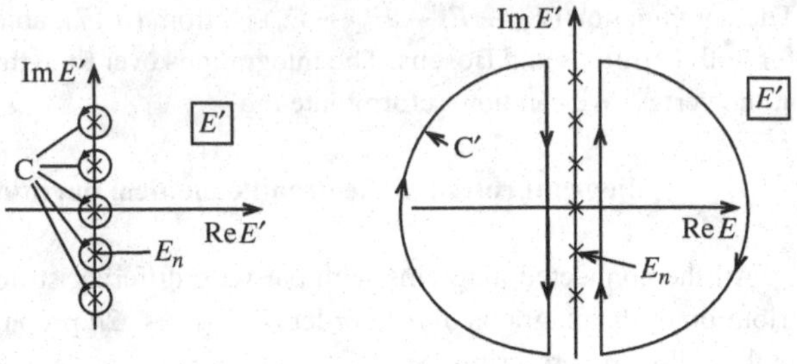

Fig. 6.3 Integration paths in the complex energy plane for carrying out the summations over the Matsubara energies

function of these E_n; then we have:

$$\frac{1}{\hbar\beta}\sum_{n=-\infty}^{+\infty} F(\mathrm{i}E_n) = \frac{-1}{2\pi\hbar\mathrm{i}}\int_C \mathrm{d}E'\,\frac{F(E')}{1-\varepsilon\mathrm{e}^{\beta E'}} \,. \tag{6.75}$$

C refers to the path in the complex E'-plane which is sketched in Fig. 6.3. If the function $F(E')$ vanishes at infinity more rapidly than $1/E'$, then we may later replace C by the contour C'.

For the proof of (6.75), we rearrange the right-hand side as follows:

$$I = \frac{-1}{2\pi\hbar\mathrm{i}}\sum_n \int_{C_n} \mathrm{d}E'\,\frac{F(E')}{E'-\mathrm{i}E_n} f\left(E'\right) \,.$$

Here, we expect that

$$f\left(E'\right) = \frac{E'-\mathrm{i}E_n}{1-\varepsilon\mathrm{e}^{\beta E'}}$$

holds. $f(E')$ remains finite for $E' = iE_n$:

$$\begin{aligned} f\left(E'=\mathrm{i}E_n\right) &= \lim_{E'\to\mathrm{i}E_n}\frac{\frac{\mathrm{d}}{\mathrm{d}E'}\left(E'-\mathrm{i}E_n\right)}{\frac{\mathrm{d}}{\mathrm{d}E'}\left(1-\varepsilon\mathrm{e}^{\beta E'}\right)} = \\ &= \lim_{E'\to\mathrm{i}E_n}\frac{1}{-\varepsilon\beta\mathrm{e}^{\beta E'}} = -\frac{1}{\beta} \,. \end{aligned}$$

The integrand of I thus has a first-order pole at $E' = \mathrm{i}E_n$ with the residual $-\frac{1}{\beta}F(\mathrm{i}E_n)$. According to the residual theorem, it then follows for I that:

$$I = \frac{1}{\hbar\beta}\sum_n F\left(\mathrm{i}E_n\right) \,.$$

This proves the contention (6.75).

In the next section, we will practice the application of the diagram rules on some specific examples.

6.2.3 Ring Diagrams

As demonstrated in Sect. 5.5 for the ground-state energy of the jellium model, likewise in the case of the grand canonical partition function Ξ, the ring diagrams play a decisive role. They can be summed exactly. We shall demonstrate this again for the jellium model, i.e. for a system of Fermions. As an example, let us consider a third-order ring diagram:

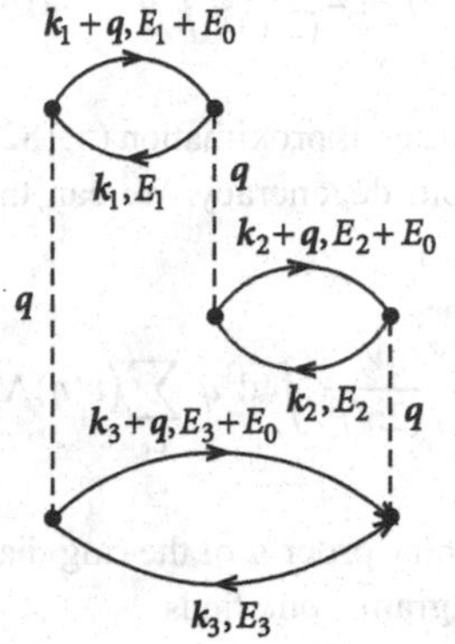

The conservation of energy and momentum have already been taken into account in the annotation of the diagram. The E_ν's are, by construction, Fermi quanta,

$$E_\nu = (2n_\nu + 1)\,\frac{\pi}{\beta}\,;$$

as a result, the energy transfer E_0 must be a Bose quantum:

$$E_0 = 2n_0\frac{\pi}{\beta}.$$

Due to conservation of spin at the vertex, we must still take three independent summations over spins into account. The term I_3 is then found to contribute:

$$I_3 = \underbrace{2^3\left(-\frac{1}{\hbar}\right)^3 \frac{(-1)^3}{h(\Theta_3)}}_{(5)}\ \underbrace{\frac{V^4}{(2\pi)^{12}}\int\cdots\int d^3q\, d^3k_1\, d^3k_2\, d^3k_3 \sum_{E_0E_1E_2E_3}}_{(4)} \cdot$$

$$\cdot \underbrace{\frac{1}{(\hbar\beta)^3}v^3(q)}_{(1)}\ \underbrace{\prod_{\nu=1}^{3}\left(\frac{-\hbar}{\mathrm{i}E_\nu - \varepsilon(\boldsymbol{k}_\nu) + \mu}\ \frac{-\hbar}{\mathrm{i}\,(E_\nu + E_0) - \varepsilon(\boldsymbol{k}_\nu + \boldsymbol{q}) + \mu}\right)}_{(2)}\,.$$

We factor out the term

$$\widetilde{\Lambda}_{\boldsymbol{q}}(\boldsymbol{k};E_0)=\frac{1}{\hbar\beta}\sum_{E_n}\frac{-\hbar}{\mathrm{i}E_n-\varepsilon(\boldsymbol{k})+\mu}\,\frac{-\hbar}{\mathrm{i}\,(E_n+E_0)-\varepsilon(\boldsymbol{k}+\boldsymbol{q})+\mu} \tag{6.76}$$

and use the following definition:

$$\hbar\Lambda_{\boldsymbol{q}}^{(0)}(E_0)=2\frac{V}{(2\pi)^3}\int \mathrm{d}^3k\,\widetilde{\Lambda}_{\boldsymbol{q}}(\boldsymbol{k};E_0)\;. \tag{6.77}$$

This corresponds to the lowest-order approximation (5.182) of the polarisation propagator. The factor 2 results from the spin degeneracy. We can then write the diagram contribution I_3 as follows:

$$I_3=\frac{1}{h\,(\Theta_3)}\frac{V}{(2\pi)^3}\int \mathrm{d}^3q\,\sum_{E_0}\left(v(\boldsymbol{q})\Lambda_{\boldsymbol{q}}^{(0)}(E_0)\right)^3\;. \tag{6.78}$$

This can be extended to an arbitrary order n of the ring diagram. For the number $h(\Theta_n)$ of topologically equivalent ring diagrams, one finds

$$h\,(\Theta_n)=2n\;. \tag{6.79}$$

Topologically equivalent diagrams are obtained e.g. by **cyclic** permutations of the vertices. This gives n different possibilities. Permutation of *above* and *below* simultaneously at the vertices yields once more a factor of 2. For the contribution of n-th order, we thus have:

$$I_n=\frac{1}{2}\frac{V}{(2\pi)^3}\int \mathrm{d}^3q\,\sum_{E_0}\frac{1}{n}\left(v(\boldsymbol{q})\Lambda_{\boldsymbol{q}}^{(0)}(E_0)\right)^n,\quad n\geq 2\;. \tag{6.80}$$

The contribution from $n=1$ has been left out. This is an *equal-time* diagram, which according to rule 3. receives an additional factor $\exp\left(\frac{\mathrm{i}}{\hbar}E_0 0^+\right)$. We shall see that without this factor, the $n=1$ contribution to the grand-canonical partition function would be divergent. Since, on the other hand, it causes no *disturbance* in the $n\geq 2$ terms, we include it in (6.80) generally for all n:

$$\sum_{n=1}^{\infty}I_n=\frac{1}{2}\frac{V}{(2\pi)^3}\int \mathrm{d}^3q\,\sum_{E_0}\left[\sum_{n=1}^{\infty}\frac{1}{n}\left(v(\boldsymbol{q})\Lambda_{\boldsymbol{q}}^{(0)}(E_0)\exp\left(\frac{\mathrm{i}}{\hbar}E_0 0^+\right)\right)^n\right].$$

With

$$\ln(1-x)=-\sum_{n=1}^{\infty}\frac{x^n}{n}\,,$$

the *ring-diagram approximation* for the grand canonical partition function can then be written as follows, also taking into account the fact that the $n = 0$ term contributes just 1:

$$\left(\frac{\Xi}{\Xi_0}\right)_{\text{Ring}} = 1 - \frac{1}{2}\frac{V}{(2\pi)^3}\int d^3q \sum_{E_0} \ln\left(1 - v(\boldsymbol{q})\Lambda_{\boldsymbol{q}}^{(0)}(E_0)\right) . \tag{6.81}$$

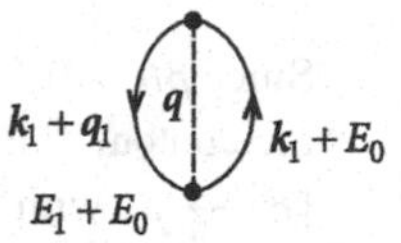

Fig. 6.4 A first-order diagram in the ring-diagram approximation for the grand canonical partition function

We have not written the convergence-inducing factor explicitly; we merely add it in *as needed*.

Finally, $\Lambda_{\boldsymbol{q}}^{(0)}(E_0)$ still remains to be evaluated. To this end, we first compute $\widetilde{\Lambda}_{\boldsymbol{q}}(\boldsymbol{k}; E_0)$ using (6.76). With (6.75), we find:

$$\widetilde{\Lambda}_{\boldsymbol{q}}(\boldsymbol{k}; E_0) = \frac{-1}{2\pi\hbar i}\int_C \frac{dE'}{1 + e^{\beta E'}} \frac{\hbar^2}{\left(E' - \varepsilon(\boldsymbol{k}) + \mu\right)\left(E' + iE_0 - \varepsilon(\boldsymbol{k} + \boldsymbol{q}) + \mu\right)} .$$

The fraction on the right vanishes as $1/E'^2$ at infinity. We can thus replace the path C by C' (cf. Fig. 6.3). Two first-order poles lie within the region bounded by C'; they are each circumscribed mathematically in a negative direction. With the residual theorem, it then follows that:

$$\widetilde{\Lambda}_{\boldsymbol{q}}(\boldsymbol{k}; E_0) = \frac{\hbar}{\left(1 + e^{\beta(\varepsilon(\boldsymbol{k}) - \mu)}\right)\left(\varepsilon(\boldsymbol{k}) - \varepsilon(\boldsymbol{k} + \boldsymbol{q}) + iE_0\right)} + \\ + \frac{\hbar}{\left(1 + e^{\beta(\varepsilon(\boldsymbol{k}+\boldsymbol{q}) - \mu)} e^{-i\beta E_0}\right)} .$$

E_0 is a Bosonic quantum, and therefore $e^{i\beta E_0} = +1$. We still have:

$$\widetilde{\Lambda}_{\boldsymbol{q}}(\boldsymbol{k}; E_0) = \hbar \frac{\langle n_{\boldsymbol{k}}\rangle^{(0)} - \langle n_{\boldsymbol{k}+\boldsymbol{q}}\rangle^{(0)}}{\varepsilon(\boldsymbol{k}) - \varepsilon(\boldsymbol{k} + \boldsymbol{q}) + iE_0} . \tag{6.82}$$

The result

$$\Lambda_{\boldsymbol{q}}^{(0)}(E_0) = \frac{2V}{(2\pi)^3}\int d^3k \frac{\langle n_{\boldsymbol{k}}\rangle^{(0)} - \langle n_{\boldsymbol{k}+\boldsymbol{q}}\rangle^{(0)}}{\varepsilon(\boldsymbol{k}) - \varepsilon(\boldsymbol{k} + \boldsymbol{q}) + iE_0} \tag{6.83}$$

can be compared to (5.192).

If we insert this result into (6.81), then we can again convert the summation over the Matsubara frequencies E_0 into an integration. In the series expansion for the logarithm, the $n = 1$ term would then give rise to difficulties without the convergence-inducing factor

$\exp\left(\frac{i}{\hbar}E_0 0^+\right)$, since $\Lambda_q^{(0)}$ behaves *only* as $1/E_0$ at infinity. With this factor, two cases can be distinguished:

1. $|E| \to \infty$ with $\mathrm{Re}\, E > 0$:
   The overall integrand behaves asymptotically as
   $$\exp\frac{\left(\frac{1}{\hbar}\mathrm{Re}\; E0^+\right)}{\exp(\beta E)|E|} \sim \frac{1}{|E|}\exp\left(-(\beta\hbar - 0^+)\frac{\mathrm{Re}\; E}{\hbar}\right).$$
   Since $\beta\hbar > 0^+ > 0$, the preconditions for Jordan's lemma are met. We can thus replace the contour C by C'.
2. $|E| \to \infty$ with $\mathrm{Re}\; E < 0$:
   The integrand now behaves asymptotically as
   $$\frac{\exp\left(\frac{1}{\hbar}\mathrm{Re}\; E0^+\right)}{1}\frac{1}{|E|}$$
   and thus likewise meets the preconditions.

6.2.4 The Single-Particle Matsubara Function

We have already carried out the the most important preparations for the diagrammatic analysisof the single-particle Matsubara function in Sect. 6.1.3. The following considerations are based on (6.49) and run generally parallel to those in Sect. 5.3.3. The diagrams which contribute to the numerator of the perturbation expansion multiplied by $1/\Xi_0$ in (6.49) are all open. They contain two extended *outer* lines, of which one begins at τ_2 and the other enters at τ_1. If a diagram component is attached to one of these two *outer* connections, then necessarily it is also attached to the other. This is required by particle number conservation. An equal number of creators and of annihilators contributes to each combination of contractions.

Every open diagram of this form consists of an open, connected diagram plus combinations of closed, connected diagrams from the expansion for Ξ. One therefore obtains **all** the diagrams if one adds to **each** open, connected diagram D_0 with two *outer* attachments all the possible Ξ/Ξ_0 diagrams. The latter contribute, as seen from (6.68), the factor

$$\exp\left\{\sum_{\nu}^{\mathrm{conn}} U\left(\Theta_\nu\right)\right\}.$$

The overall contribution of all diagrams in the perturbation expansion to the numerator multiplied by $1/\Xi_0$ in (6.49) is then:

$$\left\{\sum_{D_0} U(D_0)\right\}\exp\left\{\sum_{\nu}^{\mathrm{conn}} U\left(\Theta_\nu\right)\right\}.$$

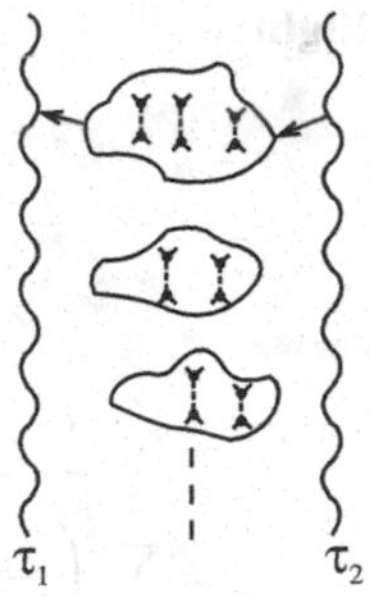

Fig. 6.5 The general structure of an open diagram for the single-particle Matsubara function

The last factor just cancels out with the denominator multiplied by $1/\Xi_0$ in (6.49), so that for the Matsubara function, we have:

$$G_k^{\mathrm{M}}(\tau_1, \tau_2) = -\left\langle T_\tau\left(U(\hbar\beta, 0)a_k(\tau_1)\, a_k^+(\tau_2)\right)\right\rangle_{\substack{\text{conn}\\ \text{open}}}^{(0)} . \tag{6.84}$$

The diagram rules for the energy-dependent function can be derived directly from those for the grand canonical partition function which we formulated following (6.72), whereby the sums and integrations are to be limited merely to *inner* variables.

We therefore go immediately to the energy-dependent function. Initially, one readily sees that every connected diagram with two outer lines has no topologically equivalent diagram of the same structure:

$$h\left(\Theta_n\right) = 1 \quad \forall n . \tag{6.85}$$

We can adopt practically all the rules from Sect. 6.2.2; only the outer lines require a certain special treatment.

Owing to energy conservation at each vertex, the two outer lines carry the same energy E_n:

Left:

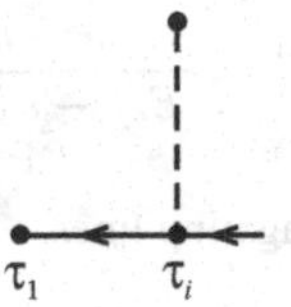

$$-\,G_k^{0,\mathrm{M}}(\tau_1 - \tau_i) =$$

$$= \sum_n \underbrace{\left(\frac{1}{\sqrt{\hbar\beta}}\exp\left(-\frac{\mathrm{i}}{\hbar}E_n\tau_1\right)\right)}_{(1)} \underbrace{\left(-G_k^{0,\mathrm{M}}(E_n)\right)}_{(2)} \underbrace{\left(\frac{1}{\sqrt{\hbar\beta}}\exp\left(\frac{\mathrm{i}}{\hbar}E_n\tau_i\right)\right)}_{(3)} .$$

The contribution (3) is associated to the vertex and guarantees that energy is conserved. (2) is taken over by the solid outer line. (1) is required for the entire Fourier decomposition.

Right:

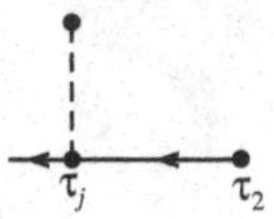

$$-G_k^{0,\mathrm{M}}(\tau_j-\tau_2)=$$
$$=\sum_n \underbrace{\left(\frac{1}{\sqrt{\hbar\beta}}\exp\left(-\frac{\mathrm{i}}{\hbar}E_n\tau_j\right)\right)}_{(1)}\underbrace{\left(-G_k^{0,\mathrm{M}}(E_n)\right)}_{(2)}\underbrace{\left(\frac{1}{\sqrt{\hbar\beta}}\exp\left(\frac{1}{\hbar}E_n\tau_2\right)\right)}_{(3)}.$$

(1) goes into the vertex, (2) is associated with the outer line, and (3) appears in the Fourier decomposition.

If the inner lines all together make the contribution I, then it follows that:

$$-G_k^{\mathrm{M}}(\tau_1-\tau_2)=\frac{1}{\hbar\beta}\sum_n I\left(-G_k^{0,\mathrm{M}}(E_n)\right)^2\exp\left(-\frac{\mathrm{i}}{\hbar}E_n(\tau_1-\tau_2)\right), \tag{6.86}$$

$$-G_k^{\mathrm{M}}(E_n)=I\left(G_k^{0,\mathrm{M}}(E_n)\right)^2. \tag{6.87}$$

with this, we now have the

diagram rules for $-G_k^{\mathrm{M}}(E_n)$.

All the connected diagrams with pairwise differing structures and two outer lines are sought. A diagram of n-th order (n vertices, $2n+1$ extended lines, of which 2 are outer lines) is evaluated according to the following prescription:

1. Vertex: $v(kl;nm)\frac{1}{\hbar\beta}\delta_{E_k+E_l,E_m+E_n}$.
2. Extended (propagating and non-propagating) lines:

$$-G_{k_i}^{0,\mathrm{M}}=\frac{-\hbar}{\mathrm{i}E_{n_{k_i}}-\varepsilon(\boldsymbol{k}_i)+\mu}.$$

3. An additional factor for non-propagating lines

$$\exp\left(\frac{\mathrm{i}}{\hbar}E_{n_k}0^+\right).$$

4. Summation over all the *inner* k_i, l_i and all the *inner* Matsubara energies.
5. Extended *outer* lines: $G_k^{0,\mathrm{M}}(E_n)$.
6. Factor: $\left(-\frac{1}{\hbar}\right)^n\varepsilon^s$; S = number of loops.

To conclude, we wish to calculate explicitly the diagrams of zeroth and first order for the system of interacting electrons ($\varepsilon=-1$) as an example of an application:

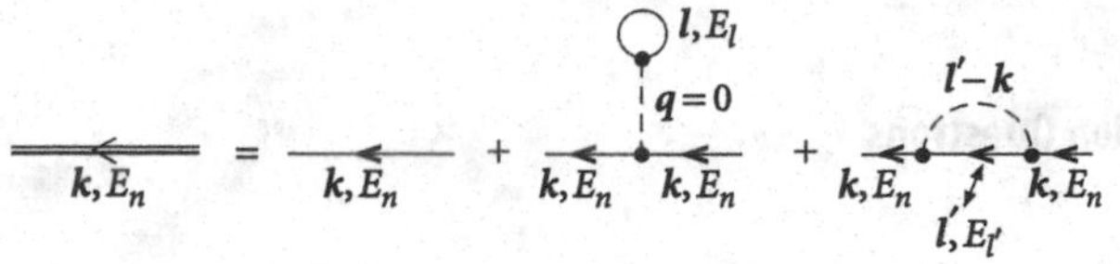

The above rules prescribe the following contribution:

$$\left(G^{\mathrm{M}}_{k\sigma}(E_n)\right)^{(1)} = G^{0,\mathrm{M}}_{k\sigma}(E_n) + \frac{2}{\hbar^2\beta}\frac{V}{(2\pi)^3}\int d^3l \sum_{n_l} G^{0,\mathrm{M}}_{l\sigma}(E_l)\cdot$$
$$\cdot\, v(\mathbf{0})\exp\left(\frac{\mathrm{i}}{\hbar}E_l 0^+\right)\left(G^{0,\mathrm{M}}_{k\sigma}(E_n)\right)^2 +$$
$$+\left(-\frac{1}{\hbar^2\beta}\right)\frac{V}{(2\pi)^3}\int d^3l' \sum_{n_{l'}} G^{0,\mathrm{M}}_{l'\sigma}(E_{l'})\, v(\mathbf{l}'-\mathbf{k})\cdot$$
$$\cdot\exp\left(\frac{\mathrm{i}}{\hbar}E_{l'}0^+\right)\left(G^{0,\mathrm{M}}_{k\sigma}(E_n)\right)^2 .$$

This can be summarised as follows:

$$\left(G^{\mathrm{M}}_{k\sigma}(E_n)\right)^{(1)} = G^{0,\mathrm{M}}_{k\sigma}(E_n) + G^{0,\mathrm{M}}_{k\sigma}(E_n)\frac{1}{\hbar}\Sigma^{(1)}_{k\sigma}(E_n)\, G^{0,\mathrm{M}}_{k\sigma}(E_n)\,, \tag{6.88}$$

$$\frac{1}{\hbar}\Sigma^{(1)}_{k\sigma}(E_n) = \frac{1}{\hbar^2\beta}\frac{V}{(2\pi)^3}\int d^3l \sum_{n_l}(2v(\mathbf{0}) - v(\mathbf{l}-\mathbf{k}))\cdot$$
$$\cdot\exp\left(\frac{\mathrm{i}}{\hbar}E_l 0^+\right)G^{0,\mathrm{M}}_{l\sigma}(E_l)\,. \tag{6.89}$$

In this approximation, the self-energy is energy-**in**dependent. However, we can see from this example that the $T \neq 0$ Matsubara formalism also allows the definition of a self-energy. This is of course not surprising, since the $T = 0$ and $T \neq 0$ diagrams are structurally identical. All the Dyson equations in the fifth chapter can be taken over directly.

The $T \neq 0$ Matsubara formalism does not at any point require the hypothesis of an adiabatic switching-on, which is afflicted with a degree of uncertainty. The Gell-Mann–Low theorem guarantees only that the adiabatically switched-on state be an eigenstate of the full Hamiltonian. It need not, starting from the ground state of the *free* system, necessarily yield the ground state of the interacting system after switching on the interactions. This will, to be sure, as a rule be the case, but one can also *land* in an excited state. The limit of $T \to 0$ in the Matsubara formalism, in contrast, yields the ground state in all cases (W. Kohn, J. M. Luttinger: Phys. Rev. **118**, 41 (1960)).

6.3 Self-Examination Questions

For Section 6.1

1. Why is the retarded Green's function not suitable for perturbation theory?
2. Why can Wick's theorem as in Sect. 5.2.2 not be used also for $T \neq 0$ problems?
3. What close connection exists between the correlation functions $\langle A(t)B(t')\rangle$ and $\langle B(t')A(t)\rangle$, if one formally allows complex values for the time variables?
4. The Matsubara method presumes purely imaginary times ($\tau = -it$ is real!). What form does the modified Heisenberg representation for time-dependent operators take in this case?
5. How is the Matsubara function defined? Give its equation of motion.
6. What periodicity is exhibited by the Matsubara function?
7. How does one obtain the retarded Green's function, which is actually the function of interest, from the Matsubara function?
8. Write the equation of motion and further properties of the time-evolution operator in the Dirac representation for purely imaginary times.
9. What is the formal solution of the time-evolution operator $U_D(\tau, \tau'')$?
10. Express $e^{-\beta\mathcal{H}}$ in terms of U_D.
11. How is the single-particle Matsubara function $G_k^M(\tau)$ defined?
12. What form is taken by the single-particle Matsubara function for the non-interacting system?
13. Which form of the single-particle Matsubara function of an interacting system is suitable for diagrammatic perturbation theory?
14. Is the hypothesis of *adiabatic switching-on* of the perturbation also required in the Matsubara formalism?
15. Which quantity within the Matsubara formalism takes on the role of the vacuum amplitude from the $T = 0$ theory?

For Section 6.2

1. In the Matsubara formalism, how is a *contraction* defined?
2. Formulate the so-called *generalised Wick theorem*! Is it an operator identity?
3. How do the diagram rules for the grand canonical partition function Ξ differ from those for the $T = 0$ vacuum amplitude?
4. Is there a *linked-cluster theorem* for Ξ?
5. How does one carry out summations over Matsubara energies?
6. Describe the *ring-diagram approximation* for the grand-canonical partition function.
7. Explain why that for the single-particle Matsubara function, only open, connected diagrams with two outer lines need be summed over.
8. Formulate the diagram rules for the energy-dependent single-particle Matsubara function.
9. Can Dyson equations also be formulated in the Matsubara formalism?

Solutions of the Exercises

Section 1.4

Solution 1.4.1

1. Hamiltonian of the two-particle system:

$$H = H_1 + H_2 = -\frac{\hbar^2}{2m}\left(\Delta_1 + \Delta_2\right) + V\left(x_1\right) + V\left(x_2\right) .$$

Non-symmetrised eigenstate:

$$|\varphi_{\alpha_1}\varphi_{\alpha_2}\rangle = |\varphi_{\alpha_1}^{(1)}\rangle \, |\varphi_{\alpha_2}^{(2)}\rangle .$$

Real-space representation:

$$\langle x_1 x_2 \mid \varphi_{\alpha_1}\varphi_{\alpha_2}\rangle = \varphi_n\left(x_1\right)\varphi_m\left(x_2\right)\chi_S\left(m_S^{(1)}\right)\chi_S\left(m_{S'}^{(2)}\right) ,$$

χ_S : Spin function (identical particles have the same spins S)

$\alpha_1 = (n, m_S)\ ; \quad \alpha_2 = (m, m_{S'})\ .$

2. Solution of the single-particle problem:

$$\left(-\frac{\hbar^2}{2m}\Delta + V(x)\right)\varphi(x) = E\varphi(x) .$$

Initially, we have:

$$\varphi(x) \equiv 0 \quad \text{for} \quad x < 0 \quad \text{and} \quad x > a .$$

For $0 \le x \le a$, we need to solve:

$$-\frac{\hbar^2}{2m}\Delta\varphi(x) = E\varphi(x)\,.$$

Trial solution:

$$\varphi(x) = c\sin(\gamma_1 x + \gamma_2)\,.$$

Boundary conditions:

$$\begin{aligned} \varphi(0) = 0 \quad &\Longrightarrow \quad \gamma_2 = 0\,, \\ \varphi(a) = 0 \quad &\Longrightarrow \quad \gamma_1 = n\frac{\pi}{a}\,; \quad n = 1, 2, 3, \dots\,. \end{aligned}$$

Energy eigenvalues:

$$E = \frac{\hbar^2}{2m}\gamma_1^2 \quad \Longrightarrow \quad E_n = \frac{\hbar^2\pi^2}{2ma^2}n^2\,; \quad n = 1, 2, \dots\,.$$

Eigenfunctions:

$$\varphi_n(x) = c\sin\left(n\frac{\pi}{a}x\right),$$

$$1 \overset{!}{=} c^2 \int_0^a \sin^2\left(n\frac{\pi}{a}x\right)\mathrm{d}x \quad \Longrightarrow \quad c = \sqrt{\frac{2}{a}}\,,$$

$$\varphi_n(x) = \begin{cases} \sqrt{\frac{2}{a}}\sin\left(n\frac{\pi}{a}x\right) & \text{for} \quad 0 \le x \le a\,, \\ 0 & \text{otherwise.} \end{cases}$$

3. Two-particle problem:

$$\begin{aligned} |\varphi_{\alpha_1}\varphi_{\alpha_2}\rangle^{(\pm)} \quad \longrightarrow \quad \frac{1}{\sqrt{2}}\Big\{&\varphi_n(x_1)\,\varphi_m(x_2)\,\chi_S\left(m_S^{(1)}\right)\chi_S\left(m_{S'}^{(2)}\right) \pm \\ &\pm\varphi_n(x_2)\,\varphi_m(x_1)\,\chi_S\left(m_S^{(2)}\right)\chi_S\left(m_{S'}^{(1)}\right)\Big\}\,, \end{aligned}$$

(+): Bosons,
(−): Fermions: $(n, m_S) \neq (m, m_{S'})$ due to the Pauli principle.

4. Ground-state energy of the N-particle system:

Bosons:
All particles in the $n = 1$ state:

$$E_0 = N\frac{\hbar^2\pi^2}{2ma^2}\,.$$

Fermions:

$$E_0 = 2\sum_{n=1}^{N/2} \frac{\hbar^2\pi^2}{2ma^2} n^2 \approx \frac{\hbar^2\pi^2}{2ma^2}\frac{N^3}{24}$$

with

$$\sum_{n=1}^{N/2} n^2 \underset{N\gg 1}{\approx} \int_1^{N/2} n^2 \mathrm{d}n = \frac{1}{3}\left(\frac{N^3}{8} - 1\right) \approx \frac{N^3}{24}.$$

Solution 1.4.2

1.

$$P_{12}\,|0,0\rangle_t = -\,|0,0\rangle_t \qquad \text{antisymmetric,}$$

$$P_{12}\,\big|1,M_S\big\rangle_t = \big|1,M_S\big\rangle_t \qquad \text{symmetric.}$$

$$(M_S = 0, \pm 1)$$

2. We carry out the proof in terms of components:

$$P_{12}S_1^z P_{12}\left|m_{S_1}^{(1)}, m_{S_2}^{(2)}\right\rangle =$$

$$= P_{12}S_1^z\left|m_{S_1}^{(2)}, m_{S_2}^{(1)}\right\rangle = \hbar m_{S_2} P_{12}\left|m_{S_1}^{(2)}, m_{S_2}^{(1)}\right\rangle =$$

$$= \hbar m_{S_2}\left|m_{S_1}^{(1)}, m_{S_2}^{(2)}\right\rangle = S_2^z\left|m_{S_1}^{(1)}, m_{S_2}^{(2)}\right\rangle.$$

This expression is valid for arbitrary two-particle states, and thus also for the symmetrised basis states of $\mathcal{H}_2^{(\pm)}$. In $\mathcal{H}_2^{(\pm)}$, the operator identity:

$$P_{12}S_1^z P_{12} = S_2^z$$

then holds. Analogously, one can show that:

$$P_{12}S_2^z P_{12} = S_1^z.$$

Now the x- and y-components remain to be dealt with:

$$S_j^x = \frac{1}{2}\left(S_j^+ + S_j^-\right); \quad S_j^y = \frac{1}{2\mathrm{i}}\left(S_j^+ - S_j^-\right); \quad j = 1, 2.$$

We obtain:

$$\begin{aligned}
P_{12}S_1^{\pm}P_{12}\left|m_{S_1}^{(1)}m_{S_2}^{(2)}\right\rangle &= P_{12}S_1^{\pm}\left|m_{S_1}^{(2)}m_{S_2}^{(1)}\right\rangle = \\
&= \hbar\sqrt{\left(\frac{1}{2}\mp m_{S_2}\right)\left(\frac{1}{2}\pm m_{S_2}+1\right)}P_{12}\left|m_{S_1}^{(2)},(m_{S_2}\pm 1)^{(1)}\right\rangle = \\
&= \hbar\sqrt{\left(\frac{1}{2}\mp m_{S_2}\right)\left(\frac{1}{2}\pm m_{S_2}+1\right)}\left|m_{S_1}^{(1)},(m_{S_2}\pm 1)^{(2)}\right\rangle = \\
&= S_2^{\pm}\left|m_{S_1}^{(1)},m_{S_2}^{(2)}\right\rangle .
\end{aligned}$$

Conclusions as above:

$$P_{12}S_{1,2}^{\pm}P_{12} = S_{2,1}^{\pm} .$$

With this, it also follows that:

$$P_{12}S_{1,2}^{x,y}P_{12} = S_{2,1}^{x,y} .$$

This proves the proposition.

3. $\boldsymbol{S}_1\cdot\boldsymbol{S}_2 = S_1^zS_2^z + \frac{1}{2}\left(S_1^+S_2^- + S_1^-S_2^+\right)$.

$\boxed{m_{S_1}=m_{S_2}=m_S}$

$$\begin{aligned}
S_1^{\pm}S_2^{\mp}\left|m_S^{(1)},m_S^{(2)}\right\rangle &= 0 , \\
S_1^zS_2^z\left|m_S^{(1)},m_S^{(2)}\right\rangle &= \frac{\hbar^2}{4}\left|m_S^{(1)},m_S^{(2)}\right\rangle , \\
\frac{1}{2}\left(\mathbf{1}+\frac{4}{\hbar^2}\boldsymbol{S}_1\cdot\boldsymbol{S}_2\right)\left|m_S^{(1)},m_S^{(2)}\right\rangle &= \left|m_S^{(1)},m_S^{(2)}\right\rangle = \left|m_S^{(2)},m_S^{(1)}\right\rangle .
\end{aligned}$$

$\boxed{m_{S_1}\neq m_{S_2}}$

$$\begin{aligned}
S_1^+S_2^-\left|m_{S_1}^{(1)},m_{S_2}^{(2)}\right\rangle &= \hbar^2\delta_{m_{S_1},-(1/2)}\delta_{m_{S_2},(1/2)}\left|(m_{S_1}+1)^{(1)},(m_{S_2}-1)^{(2)}\right\rangle = \\
&= \hbar^2\delta_{m_{S_1},-(1/2)}\delta_{m_{S_2},(1/2)}\left|m_{S_2}^{(1)},m_{S_1}^{(2)}\right\rangle = \\
&= \hbar^2\delta_{m_{S_1},-(1/2)}\delta_{m_{S_2},(1/2)}\left|m_{S_1}^{(2)},m_{S_2}^{(1)}\right\rangle .
\end{aligned}$$

Analogously:

$$\begin{aligned}
S_1^-S_2^+\left|m_{S_1}^{(1)},\,m_{S_2}^{(2)}\right\rangle &= \hbar^2\delta_{m_{S_2},-(1/2)}\delta_{m_{S_1},(1/2)}\left|m_{S_1}^{(2)},m_{S_2}^{(1)}\right\rangle , \\
\frac{1}{2}\left(S_1^+S_2^- + S_1^-S_2^+\right)\left|m_{S_1}^{(1)},m_{S_2}^{(2)}\right\rangle &= \frac{\hbar^2}{2}\left|m_{S_1}^{(2)},m_{S_2}^{(1)}\right\rangle .
\end{aligned}$$

Furthermore, we have:

$$S_1^z S_2^z \left| m_{S_1}^{(1)}, m_{S_2}^{(2)} \right\rangle = -\frac{\hbar^2}{4} \left| m_{S_1}^{(1)}, m_{S_2}^{(2)} \right\rangle .$$

All together, the result is thus:

$$\begin{aligned}
&\frac{1}{2}\left(\mathbf{1} + \frac{4}{\hbar^2}\mathbf{S}_1 \cdot \mathbf{S}_2\right) \left| m_{S_1}^{(1)}, m_{S_2}^{(2)} \right\rangle = \\
&= \frac{1}{2}\left(1 - \frac{4}{\hbar^2}\frac{\hbar^2}{4}\right) \left| m_{S_1}^{(1)}, m_{S_2}^{(2)} \right\rangle + \frac{1}{2}\left(\frac{4}{\hbar^2}\frac{\hbar^2}{2}\right) \left| m_{S_1}^{(2)}, m_{S_2}^{(1)} \right\rangle = \\
&= \left| m_{S_1}^{(2)}, m_{S_2}^{(1)} \right\rangle .
\end{aligned}$$

With this, it is clear that quite generally:

$$P_{12} \left| m_{S_1}^{(1)}, m_{S_2}^{(2)} \right\rangle = \left| m_{S_1}^{(2)}, m_{S_2}^{(1)} \right\rangle .$$

Solution 1.4.3

Proof through complete induction:

Initiation of induction:

$N = 1:$

$$\begin{aligned}
\langle 0 \mid a_{\beta_1} a_{\alpha_1}^+ \mid 0 \rangle &= \langle 0 | \left(\delta(\beta_1 - \alpha_1) + \varepsilon a_{\alpha_1}^+ a_{\beta_1}\right) | 0 \rangle = \\
&= \delta(\beta_1 - \alpha_1) \langle 0 \mid 0 \rangle + \varepsilon \langle 0 \mid a_{\alpha_1}^+ a_{\beta_1} \mid 0 \rangle = \\
&= \delta(\beta_1 - \alpha_1) .
\end{aligned}$$

$N = 2:$

$$\begin{aligned}
&\langle 0 \mid a_{\beta_2} a_{\beta_1} a_{\alpha_1}^+ a_{\alpha_2}^+ \mid 0 \rangle = \\
&= \langle 0 | a_{\beta_2} \left(\delta(\beta_1 - \alpha_1) + \varepsilon a_{\alpha_1}^+ a_{\beta_1}\right) a_{\alpha_2}^+ | 0 \rangle = \\
&= \delta(\beta_1 - \alpha_1) \langle 0 | \left(\delta(\beta_2 - \alpha_2) + \varepsilon a_{\alpha_2}^+ a_{\beta_2}\right) | 0 \rangle + \\
&\quad + \varepsilon \langle 0 | a_{\beta_2} a_{\alpha_1}^+ \left(\delta(\beta_1 - \alpha_2) + \varepsilon a_{\alpha_2}^+ a_{\beta_1}\right) | 0 \rangle = \\
&= \delta(\beta_1 - \alpha_1)(\beta_2 - \alpha_2) + \varepsilon (\beta_1 - \alpha_2) \langle 0 | \left((\beta_2 - \alpha_1) + \varepsilon a_{\alpha_1}^+ a_{\beta_2}\right) | 0 \rangle = \\
&= (\beta_1 - \alpha_1)(\beta_2 - \alpha_2) + \varepsilon (\beta_1 - \alpha_2)(\beta_2 - \alpha_1) .
\end{aligned}$$

Conclusion from induction for $N-1 \to N$:

$$\langle 0|\, a_{\beta_N} \cdots a_{\beta_1} a^+_{\alpha_1} \cdots a^+_{\alpha_N}\, |0\rangle \overset{\substack{\text{"pull through"! } \alpha_{\beta_1} \\ \text{to the right}}}{\downarrow=}$$

$$= (\beta_1 - \alpha_1)\, \langle 0|\, a_{\beta_N} \cdots a_{\beta_2} a^+_{\alpha_2} \cdots a^+_{\alpha_N}\, |0\rangle +$$

$$+ \varepsilon\, (\beta_1 - \alpha_2)\, \langle 0|\, a_{\beta_N} \cdots a_{\beta_2} a^+_{\alpha_1} a^+_{\alpha_3} \cdots a^+_{\alpha_N}\, |0\rangle +$$

$$+ \cdots +$$

$$+ \varepsilon^{N-1} \delta\, (\beta_1 - \alpha_N)\, \langle 0|\, a_{\beta_N} \cdots a_{\beta_2} a^+_{\alpha_1} a^+_{\alpha_2} \cdots a^+_{\alpha_{N-1}}\, |0\rangle =$$

Precondition for induction

$$\overset{\downarrow}{=} (\beta_1 - \alpha_1) \sum_{\mathcal{P}_\alpha} \varepsilon^{\mathcal{P}_\alpha} \mathcal{P}_\alpha \left[\delta\,(\beta_2 - \alpha_2) \cdots \delta\,(\beta_N - \alpha_N)\right] +$$

$$+ \varepsilon\delta(\beta_1 - \alpha_2) \sum_{\mathcal{P}_\alpha} \varepsilon^{\mathcal{P}_\alpha} \mathcal{P}_\alpha \left[\delta(\beta_2 - \alpha_1)\delta\,(\beta_3 - \alpha_3) \cdots \delta\,(\beta_N - \alpha_N)\right] +$$

$$+ \cdots +$$

$$+ \varepsilon^{N-1} \delta\,(\beta_1 - \alpha_N) \sum_{\mathcal{P}_\alpha} \varepsilon^{\mathcal{P}_\alpha} \mathcal{P}_\alpha \left[\delta\,(\beta_2 - \alpha_1)\,\delta\,(\beta_3 - \alpha_2) \cdots \delta\,(\beta_N - \alpha_{N-1})\right] =$$

$$= \sum_{\mathcal{P}_\alpha} \varepsilon^{\mathcal{P}_\alpha} \mathcal{P}_\alpha \left[\delta\,(\beta_1 - \alpha_1)\,\delta\,(\beta_2 - \alpha_2) \cdots \delta\,(\beta_N - \alpha_N)\right] \quad \text{q. e. d.}$$

Solution 1.4.4

One-particle basis:

$$|\boldsymbol{k}\rangle \quad \Longleftrightarrow \quad \langle \boldsymbol{r} \mid \boldsymbol{k}\rangle = \varphi_{\boldsymbol{k}}(\boldsymbol{r}) = (2\pi)^{-3/2} \mathrm{e}^{\mathrm{i}\boldsymbol{k}\cdot\boldsymbol{r}}$$

plane wave.

Operator for the kinetic energy:

$$\sum_{i=1}^{N} \frac{p_i^2}{2m} \quad \Longrightarrow \quad \iint \mathrm{d}^3k\mathrm{d}^3k'\, \langle \boldsymbol{k}|\, \frac{p^2}{2m}\, |\boldsymbol{k}'\rangle\, a^+_{\boldsymbol{k}} a_{\boldsymbol{k}'}\,.$$

Matrix element:

$$\langle \boldsymbol{k}|\, \frac{p^2}{2m}\, |\boldsymbol{k}'\rangle = \int \mathrm{d}^3r\, \langle \boldsymbol{k} \mid \boldsymbol{r}\rangle\, \langle \boldsymbol{r}|\, \frac{p^2}{2m}\, |\boldsymbol{k}'\rangle = \int \mathrm{d}^3r\, \varphi^*_{\boldsymbol{k}}(\boldsymbol{r}) \left(-\frac{\hbar^2}{2m}\Delta\right) \varphi_{\boldsymbol{k}'}(\boldsymbol{r}) =$$

$$= \frac{\hbar^2 k'^2}{2m} (2\pi)^{-3} \int \mathrm{d}^3r\, \mathrm{e}^{-\mathrm{i}(\boldsymbol{k}-\boldsymbol{k}')\cdot\boldsymbol{r}} = \frac{\hbar^2 k'^2}{2m} \delta\left(\boldsymbol{k} - \boldsymbol{k}'\right).$$

Single-particle operator:

$$\sum_{i=1}^{N} \frac{p_i^2}{2m} \quad \Longrightarrow \quad \int \mathrm{d}k \, \frac{\hbar^2 k^2}{2m} a_k^+ a_k \, .$$

Operator for the Coulomb interaction:

$$\frac{1}{2} \sum_{i,j}^{i \neq j} V_{ij} \quad \Longrightarrow \quad \frac{1}{2} \int \cdots \int \mathrm{d}^3k_1 \, \mathrm{d}^3k_2 \, \mathrm{d}^3k_3 \, \mathrm{d}^3k_4 \, \langle \boldsymbol{k}_1 \boldsymbol{k}_2 | \, V_{12} \, | \boldsymbol{k}_3 \boldsymbol{k}_4 \rangle a_{\boldsymbol{k}_1}^+ a_{\boldsymbol{k}_2}^+ a_{\boldsymbol{k}_4} a_{\boldsymbol{k}_3} \, .$$

The matrix element may be symmetrised, but it may also be non-symmetrised:

$$M \equiv \left\langle \boldsymbol{k}_1^{(1)} \right| \left\langle \boldsymbol{k}_2^{(2)} \right| V_{12} \left| \boldsymbol{k}_3^{(1)} \right\rangle \left| \boldsymbol{k}_4^{(2)} \right\rangle .$$

The real-space representation is expedient, since then V_{12} is diagonal:

$$\begin{aligned}
M = & \int \cdots \int \mathrm{d}^3r_1 \cdots \mathrm{d}^3r_4 \left(\left\langle \boldsymbol{k}_1^{(1)} \middle| \boldsymbol{r}_1^{(1)} \right\rangle \left\langle \boldsymbol{r}_1^{(1)} \right| \right) \left(\left\langle \boldsymbol{k}_2^{(2)} \middle| \boldsymbol{r}_2^{(2)} \right\rangle \left\langle \boldsymbol{r}_2^{(2)} \right| \right) \cdot \\
& \cdot V \left(\left| \hat{\boldsymbol{r}}^{(1)} - \hat{\boldsymbol{r}}^{(2)} \right| \right) \left(\left| \boldsymbol{r}_3^{(1)} \right\rangle \left\langle \boldsymbol{r}_3^{(1)} \middle| \boldsymbol{k}_3^{(1)} \right\rangle \right) \left(\left| \boldsymbol{r}_4^{(2)} \right\rangle \left\langle \boldsymbol{r}_4^{(2)} \middle| \boldsymbol{k}_4^{(2)} \right\rangle \right) = \\
= & \int \cdots \int \mathrm{d}^3r_1 \cdots \mathrm{d}^3r_4 \, V \left(\left| \boldsymbol{r}_3 - \boldsymbol{r}_4 \right| \right) \left\langle \boldsymbol{k}_1^{(1)} \middle| \boldsymbol{r}_1^{(1)} \right\rangle \left\langle \boldsymbol{r}_1^{(1)} \middle| \boldsymbol{r}_3^{(1)} \right\rangle \cdot \\
& \cdot \left\langle \boldsymbol{k}_2^{(2)} \middle| \boldsymbol{r}_2^{(2)} \right\rangle \left\langle \boldsymbol{r}_2^{(2)} \middle| \boldsymbol{r}_4^{(2)} \right\rangle \left\langle \boldsymbol{r}_3^{(1)} \middle| \boldsymbol{k}_3^{(1)} \right\rangle \left\langle \boldsymbol{r}_4^{(2)} \middle| \boldsymbol{k}_4^{(2)} \right\rangle = \\
= & (2\pi)^{-6} \iint \mathrm{d}^3r_1 \, \mathrm{d}^3r_2 \, V \left(\left| \boldsymbol{r}_1 - \boldsymbol{r}_2 \right| \right) \mathrm{e}^{-\mathrm{i}(\boldsymbol{k}_1 - \boldsymbol{k}_3) \cdot \boldsymbol{r}_1} \mathrm{e}^{-\mathrm{i}(\boldsymbol{k}_2 - \boldsymbol{k}_4) \cdot \boldsymbol{r}_2} \, .
\end{aligned}$$

Making the coordinate transformation

$$\begin{aligned}
\boldsymbol{r} = \boldsymbol{r}_1 - \boldsymbol{r}_2 \, ; \quad \boldsymbol{R} = \frac{1}{2} (\boldsymbol{r}_1 + \boldsymbol{r}_2) \quad \Longrightarrow \quad & \boldsymbol{r}_1 = \frac{1}{2} \boldsymbol{r} + \boldsymbol{R} \, , \\
& \boldsymbol{r}_2 = \boldsymbol{R} - \frac{1}{2} \boldsymbol{r}
\end{aligned}$$

it then follows that:

$$\begin{aligned}
M = & (2\pi)^{-6} \int \mathrm{d}^3R \, \mathrm{e}^{-\mathrm{i}(\boldsymbol{k}_1 - \boldsymbol{k}_3 + \boldsymbol{k}_2 - \boldsymbol{k}_4) \cdot \boldsymbol{R}} \, \cdot \\
& \cdot \int \mathrm{d}^3r \, V(r) \mathrm{e}^{-\frac{1}{2}(\boldsymbol{k}_1 - \boldsymbol{k}_3 - \boldsymbol{k}_2 + \boldsymbol{k}_4) \cdot \boldsymbol{r}} = \\
= & (2\pi)^{-3} \delta \left(\boldsymbol{k}_1 - \boldsymbol{k}_3 + \boldsymbol{k}_2 - \boldsymbol{k}_4 \right) \int \mathrm{d}^3r \, V(r) \mathrm{e}^{-(\boldsymbol{k}_1 - \boldsymbol{k}_3) \cdot \boldsymbol{r}} = \\
= & V(\boldsymbol{k}_1 - \boldsymbol{k}_3) \delta \left(\boldsymbol{k}_1 - \boldsymbol{k}_3 + \boldsymbol{k}_2 - \boldsymbol{k}_4 \right) .
\end{aligned}$$

Substitution:

$$\boldsymbol{k}_1 \to \boldsymbol{k}+\boldsymbol{q}\,, \quad \boldsymbol{k}_2 \to \boldsymbol{p}-\boldsymbol{q}\,, \quad \boldsymbol{k}_3 \to \boldsymbol{k}\,.$$

Result:

$$\frac{1}{2}\sum_{i,j}^{i\neq j} V_{ij} \to \frac{1}{2}\iiint \mathrm{d}^3k\,\mathrm{d}^3p\,\mathrm{d}^3q\; V(\boldsymbol{q}) a^+_{\boldsymbol{k}+\boldsymbol{q}} a^+_{\boldsymbol{p}-\boldsymbol{q}} a_{\boldsymbol{p}} a_{\boldsymbol{k}} \quad \text{q. e. d.}$$

Solution 1.4.5

$$\begin{aligned}
H &= \widehat{T} + \widehat{V}\,,\\
\widehat{T} &= \int \mathrm{d}^3k\, \frac{\hbar^2k^2}{2m} a^+_{\boldsymbol{k}} a_{\boldsymbol{k}}\,,\\
\widehat{V} &= \frac{1}{2}\iiint \mathrm{d}^3k\,\mathrm{d}^3p\,\mathrm{d}^3q\; V(\boldsymbol{q}) a^+_{\boldsymbol{k}+\boldsymbol{q}} a^+_{\boldsymbol{p}-\boldsymbol{q}} a_{\boldsymbol{p}} a_{\boldsymbol{k}}\,.
\end{aligned}$$

1.

$$\begin{aligned}
\left[\widehat{N}, \widehat{T}\right]_- &= \int \mathrm{d}^3p \int \mathrm{d}^3k\, \frac{\hbar^2k^2}{2m} \left[\hat{n}_{\boldsymbol{p}}, \hat{n}_{\boldsymbol{k}}\right]_-\,,\\
\left[\hat{n}_{\boldsymbol{p}}, \hat{n}_{\boldsymbol{k}}\right]_- &= a^+_{\boldsymbol{p}} a_{\boldsymbol{p}} a^+_{\boldsymbol{k}} a_{\boldsymbol{k}} - \hat{n}_{\boldsymbol{k}}\hat{n}_{\boldsymbol{p}} =\\
&= a^+_{\boldsymbol{p}} \left(\delta(\boldsymbol{p}-\boldsymbol{k}) + \varepsilon a^+_{\boldsymbol{k}} a_{\boldsymbol{p}}\right) a_{\boldsymbol{k}} - \hat{n}_{\boldsymbol{k}}\hat{n}_{\boldsymbol{p}} =\\
&= \delta(\boldsymbol{p}-\boldsymbol{k}) a^+_{\boldsymbol{p}} a_{\boldsymbol{k}} + \varepsilon^2 a^+_{\boldsymbol{k}} a^+_{\boldsymbol{p}} a_{\boldsymbol{p}} a_{\boldsymbol{k}} - \hat{n}_{\boldsymbol{k}}\hat{n}_{\boldsymbol{p}} =\\
&= \delta(\boldsymbol{p}-\boldsymbol{k}) \hat{n}_{\boldsymbol{k}} + \varepsilon a^+_{\boldsymbol{k}} a^+_{\boldsymbol{p}} a_{\boldsymbol{k}} a_{\boldsymbol{p}} - \hat{n}_{\boldsymbol{k}}\hat{n}_{\boldsymbol{p}} =\\
&= \delta(\boldsymbol{p}-\boldsymbol{k}) \hat{n}_{\boldsymbol{k}} + \varepsilon a^+_{\boldsymbol{k}} \left(\varepsilon a_{\boldsymbol{k}} a^+_{\boldsymbol{p}} - \varepsilon\delta(\boldsymbol{p}-\boldsymbol{k})\right) a_{\boldsymbol{p}} - \hat{n}_{\boldsymbol{k}}\hat{n}_{\boldsymbol{p}} =\\
&= \delta(\boldsymbol{p}-\boldsymbol{k}) \hat{n}_{\boldsymbol{k}} + \hat{n}_{\boldsymbol{k}}\hat{n}_{\boldsymbol{p}} - \delta(\boldsymbol{p}-\boldsymbol{k}) a^+_{\boldsymbol{k}} a_{\boldsymbol{p}} - \hat{n}_{\boldsymbol{k}}\hat{n}_{\boldsymbol{p}} =\\
&= 0\\
\Longrightarrow \quad \left[\widehat{N}, \widehat{T}\right]_- &= 0\,.
\end{aligned}$$

2.

$$\begin{aligned}
\left[\widehat{N}, \widehat{V}\right] &= \frac{1}{2}\iiiint \mathrm{d}^3\bar{q}\,\mathrm{d}^3k\,\mathrm{d}^3p\,\mathrm{d}^3q\; V(\boldsymbol{q}) \left[a^+_{\bar{\boldsymbol{q}}} a_{\bar{\boldsymbol{q}}}, a^+_{\boldsymbol{k}+\boldsymbol{q}} a^+_{\boldsymbol{p}-\boldsymbol{q}} a_{\boldsymbol{p}} a_{\boldsymbol{k}}\right]_- =\\
&= \frac{1}{2}\iiiint \mathrm{d}^3\bar{q}\,\mathrm{d}^3k\,\mathrm{d}^3p\,\mathrm{d}^3q\; V(\boldsymbol{q}) \Big\{\delta(\bar{\boldsymbol{q}}-\boldsymbol{k}-\boldsymbol{q}) a^+_{\bar{\boldsymbol{q}}} a^+_{\boldsymbol{p}-\boldsymbol{q}} a_{\boldsymbol{p}} a_{\boldsymbol{k}} +
\end{aligned}$$

$$+\varepsilon\delta(\bar{q}-p+q)a_{\bar{q}}^{+}a_{k+q}^{+}a_p a_k - \varepsilon^5\delta(\bar{q}-p)a_{k+q}^{+}a_{p-q}^{+}a_k a_{\bar{q}} -$$

$$-\varepsilon^6\delta(\bar{q}-k)a_{k+q}^{+}a_{p-q}^{+}a_p a_{\bar{q}}\Big\} =$$

$$= \frac{1}{2}\iiint d^3k\, d^3p\, d^3q\; V(q)\Big\{a_{k+q}^{+}a_{p-q}^{+}a_p a_k + \varepsilon a_{p-q}^{+}a_{k+q}^{+}a_p a_k -$$

$$-\varepsilon a_{k+q}^{+}a_{p-q}^{+}a_k a_p - a_{k+q}^{+}a_{p-q}^{+}a_p a_k\Big\} =$$

$$= \frac{1}{2}\iiint d^3k\, d^3p\, d^3q\; V(q)\Big\{2a_{k+q}^{+}a_{p-q}^{+}a_p a_k - 2a_{k+q}^{+}a_{p-q}^{+}a_p a_k\Big\} =$$

$$= 0$$

$$\Longrightarrow \quad \big[\widehat{N}, \widehat{V}\big]_- = 0\,.$$

Solution 1.4.6

According to Sect. 1.2, we have for the relation between field operators and general creation and annihilation operators the following:

$$\widehat{\psi}^{+}(r) = \int d\alpha\; \varphi_\alpha^*(r) a_\alpha^{+}\,,$$

$$\widehat{\psi}(r) = \int d\alpha\; \varphi_\alpha(r) a_\alpha\,.$$

In the k representation with plane waves, this means that:

$$\widehat{\psi}^{+}(r) = (2\pi)^{-3/2}\int d^3k\; e^{-ik\cdot r} a_k^{+}\,,$$

$$\widehat{\psi}(r) = (2\pi)^{-3/2}\int d^3k\; e^{ik\cdot r} a_k\,.$$

We first discuss the *kinetic energy:*

$$\widehat{T} = \int d^3r\, \widehat{\psi}^{+}(r)\left\{-\frac{\hbar^2}{2m}\Delta_r\right\}\widehat{\psi}(r) =$$

$$= (2\pi)^{-3}\iiint d^3r\, d^3k\, d^3k'\; e^{-ik\cdot r}\left\{-\frac{\hbar^2}{2m}\Delta_r\right\} e^{ik'\cdot r} a_k^{+} a_{k'} =$$

$$= \iint d^3k\, d^3k' \left(\frac{\hbar^2 k'^2}{2m}\right) a_k^{+} a_{k'} \underbrace{(2\pi)^{-3}\int d^3r\; e^{-i(k-k')\cdot r}}_{\delta(k-k')} =$$

$$= \int d^3k \left(\frac{\hbar^2 k^2}{2m}\right) a_k^{+} a_k\,.$$

The *potential energy* requires somewhat more effort:

$$\widehat{V} = \frac{1}{2}\iint d^3r\, d^3r'\, \widehat{\psi}^+(\boldsymbol{r})\widehat{\psi}^+(\boldsymbol{r}')\, V\left(|\boldsymbol{r}-\boldsymbol{r}'|\right)\widehat{\psi}(\boldsymbol{r}')\,\widehat{\psi}(\boldsymbol{r}) =$$

$$= \frac{1}{2}(2\pi)^{-6}\iint d^3r\, d^3r'\, V\left(|\boldsymbol{r}-\boldsymbol{r}'|\right)\int\cdots\int d^3k_1\cdots d^3k_4\;\cdot$$

$$\cdot\; a^+_{\boldsymbol{k}_1}a^+_{\boldsymbol{k}_2}a_{\boldsymbol{k}_3}a_{\boldsymbol{k}_4}\,e^{-i(\boldsymbol{k}_1\boldsymbol{r}+\boldsymbol{k}_2\boldsymbol{r}')}e^{i(\boldsymbol{k}_3\cdot\boldsymbol{r}'+\boldsymbol{k}_4\cdot\boldsymbol{r})}\,.$$

Centre-of-mass and relative coordinates:

$$\bar{\boldsymbol{r}} = \boldsymbol{r}-\boldsymbol{r}'\,;\qquad \boldsymbol{R} = \frac{1}{2}\left(\boldsymbol{r}+\boldsymbol{r}'\right)$$

$$\Longrightarrow\quad \boldsymbol{r} = \frac{1}{2}\bar{\boldsymbol{r}}+\boldsymbol{R}\,;\quad \boldsymbol{r}' = \boldsymbol{R}-\frac{1}{2}\bar{\boldsymbol{r}}\,.$$

This means that:

$$\widehat{V} = \frac{1}{2}\int\cdots\int d^3k_1\cdots d^3k_4\, a^+_{\boldsymbol{k}_1}a^+_{\boldsymbol{k}_2}a_{\boldsymbol{k}_3}a_{\boldsymbol{k}_4}\;\cdot$$

$$\cdot\;(2\pi)^{-3}\int d^3R\, e^{-i(\boldsymbol{k}_1+\boldsymbol{k}_2-\boldsymbol{k}_3-\boldsymbol{k}_4)\boldsymbol{R}}\;\cdot$$

$$\cdot\;(2\pi)^{-3}\int d^3\bar{r}\, V(\bar{r})e^{i\frac{1}{2}(-\boldsymbol{k}_1+\boldsymbol{k}_2-\boldsymbol{k}_3+\boldsymbol{k}_4)\bar{\boldsymbol{r}}} =$$

$$= \frac{1}{2}\int\cdots\int d^3k_1\cdots d^3k_4\, a^+_{\boldsymbol{k}_1}a^+_{\boldsymbol{k}_2}a_{\boldsymbol{k}_3}a_{\boldsymbol{k}_4}\,\delta\left(\boldsymbol{k}_1+\boldsymbol{k}_2-\boldsymbol{k}_3-\boldsymbol{k}_4\right)\;\cdot$$

$$\cdot\;(2\pi)^{-3}\int d^3\bar{r}\, V(\bar{r})e^{i\frac{1}{2}(-\boldsymbol{k}_1+\boldsymbol{k}_2-\boldsymbol{k}_3+\boldsymbol{k}_4)\cdot\bar{\boldsymbol{r}}} =$$

$$= \frac{1}{2}\iiint d^3k_1\, d^3k_2\, d^3k_3\, a^+_{\boldsymbol{k}_1}a^+_{\boldsymbol{k}_2}a_{\boldsymbol{k}_3}a_{\boldsymbol{k}_1+\boldsymbol{k}_2-\boldsymbol{k}_3}\;\cdot$$

$$\cdot\;(2\pi)^{-3}\int d^3\bar{r}\, V(\bar{r})e^{i(\boldsymbol{k}_2-\boldsymbol{k}_3)\cdot\bar{\boldsymbol{r}}}\,.$$

Setting

$$\boldsymbol{k}_1\to\boldsymbol{k}+\boldsymbol{q}\,;\quad \boldsymbol{k}_2\to\boldsymbol{p}-\boldsymbol{q}\,;\quad \boldsymbol{k}_3\to\boldsymbol{p}\,,$$

then, with $V(\boldsymbol{q}) = V(-\boldsymbol{q})$, we obtain:

$$\widehat{V} = \frac{1}{2}\iiint d^3k\, d^3p\, d^3q\, V(\boldsymbol{q})a^+_{\boldsymbol{k}+\boldsymbol{q}}a^+_{\boldsymbol{p}-\boldsymbol{q}}a_{\boldsymbol{p}}a_{\boldsymbol{k}}\quad \text{q. e. d.}$$

Solution 1.4.7

1.

$$\left[\hat{n}_\alpha, a_\beta^+\right]_- = \hat{n}_\alpha a_\beta^+ - a_\beta^+ \hat{n}_\alpha = a_\alpha^+ a_\alpha a_\beta^+ - a_\beta^+ \hat{n}_\alpha =$$
$$= a_\alpha^+ \left(\delta_{\alpha\beta} + \varepsilon a_\beta^+ a_\alpha\right) - a_\beta^+ \hat{n}_\alpha = \delta_{\alpha\beta} a_\alpha^+ + \varepsilon^2 a_\beta^+ a_\alpha^+ a_\alpha - a_\beta^+ \hat{n}_\alpha =$$
$$= \delta_{\alpha\beta} a_\alpha^+ .$$

2.

$$\left[\hat{n}_\alpha, a_\beta\right]_- = \hat{n}_\alpha a_\beta - a_\beta \hat{n}_\alpha = \varepsilon a_\alpha^+ a_\beta a_\alpha - a_\beta \hat{n}_\alpha =$$
$$= \left(a_\beta a_\alpha^+ - \delta_{\alpha\beta}\right) a_\alpha - a_\beta \hat{n}_\alpha = -\delta_{\alpha\beta} a_\alpha .$$

3.

$$\left[\hat{N}, a_\alpha^+\right]_- = \sum_\gamma \left[\hat{n}_\gamma, a_\alpha^+\right]_- \stackrel{1.}{=} \sum_\gamma \delta_{\gamma\alpha} a_\alpha^+ = a_\alpha^+ .$$

4.

$$\left[\widehat{N}, a_\alpha\right]_- = \sum_\gamma \left[\hat{n}_\gamma, a_\alpha\right]_- = \sum_\gamma \delta_{\gamma\alpha}(-a_\alpha) = -a_\alpha .$$

Solution 1.4.8

1. $[a_\alpha, a_\beta]_+ = 0$.
   From this, it follows in particular for $\alpha = \beta$:

$$0 = [a_\alpha, a_\alpha]_+ = (a_\alpha)^2 + (a_\alpha)^2 = 2\,(a_\alpha)^2 \quad \Longrightarrow \quad (a_\alpha)^2 = 0 .$$

   Owing to the Pauli principle, two Fermions can not have exactly the same set of quantum numbers. Therefore, two *equivalent Fermions* cannot be annihilated. Analogously, it follows that:

$$0 = \left[a_\alpha^+, a_\alpha^+\right]_+ \quad \Longleftrightarrow \quad (a_\alpha^+)^2 = 0 .$$

2.

$$(\hat{n}_\alpha)^2 = a_\alpha^+ a_\alpha a_\alpha^+ a_\alpha = a_\alpha^+ \left(1 + a_\alpha^+ a_\alpha\right) a_\alpha =$$
$$= \hat{n}_\alpha + \left(a_\alpha^+\right)^2 (a_\alpha)^2 = \hat{n}_\alpha \quad \textit{(Interpretation?)} .$$

3.

$$a_\alpha \hat{n}_\alpha = a_\alpha a_\alpha^+ a_\alpha = \left(1 + a_\alpha^+ a_\alpha\right) a_\alpha = a_\alpha + a_\alpha^+ (a_\alpha)^2 = a_\alpha \,,$$
$$a_\alpha^+ \hat{n}_\alpha = \left(a_\alpha^+\right)^2 a_\alpha = 0 \,.$$

4.

$$\hat{n}_\alpha a_\alpha = a_\alpha^+ (a_\alpha)^2 = 0 \,,$$
$$a_\alpha \hat{n}_\alpha = \left(1 + a_\alpha^+ a_\alpha\right) a_\alpha = a_\alpha + a_\alpha^+ (a_\alpha)^2 = a_\alpha \,.$$

Solution 1.4.9

1. Non-interacting, identical Bosons or Fermions:

$$H = \sum_{i=1}^{N} H_1^{(i)} \,.$$

Eigenvalue equation:

$$H_1^{(i)} \left| \varphi_r^{(i)} \right\rangle = \varepsilon_r \left| \varphi_r^{(i)} \right\rangle , \quad \left\langle \varphi_r^{(i)} \,\middle|\, \varphi_s^{(i)} \right\rangle = \delta_{rs} \,.$$

The single-particle operator in the second quantisation:

$$H = \sum_{r,\,s} \langle \varphi_r \mid H_1 \mid \varphi_s \rangle a_r^+ a_s = \sum_{r,\,s} \varepsilon_s \delta_{rs} a_r^+ a_s$$
$$\Longrightarrow \quad H = \sum_r \varepsilon_r a_r^+ a_r = \sum_r \varepsilon_r \hat{n}_r \,.$$

2. Non-normalised density matrix for the grand canonical ensemble:

$$\rho = \exp\left[-\beta \left(H - \mu \widehat{N}\right)\right],$$
$$\widehat{N} = \sum_r \hat{n}_r \,.$$

The normalised Fock states

$$\left| N; n_1 n_2 \cdots n_i \cdots \right\rangle^{(\varepsilon)}$$

are eigenstates of $\hat{n}_r$ and thus also of $\widehat{N}$ and H:

$$H\,|N; n_1 \cdots\rangle^{(\varepsilon)} = \left(\sum_r \varepsilon_r n_r\right) |N; n_1 \cdots\rangle^{(\varepsilon)},$$

$$\widehat{N}\,|N; n_1 \cdots\rangle^{(\varepsilon)} = N\,|N; n_1 \cdots\rangle^{(\varepsilon)}.$$

Taking the trace is therefore expedient with these Fock states:

$$^{(\varepsilon)}\langle N; n_1 n_2 \cdots|\exp\left[-\beta\left(H - \mu\widehat{N}\right)\right]|N; n_1 n_2 \cdots\rangle^{(\varepsilon)} =$$

$$= \exp\left[-\beta \sum_r (\varepsilon_r - \mu)\, n_r\right] \qquad \text{with} \quad \sum_r n_r = N\,.$$

From this, it follows that:

$$\mathrm{Tr}\,\rho = \sum_{N=0}^{\infty} \sum_{\substack{\{n_r\} \\ (\sum n_r = N)}} \exp\left[-\beta \sum_r (\varepsilon_r - \mu)\, n_r\right] =$$

$$= \sum_{N=0}^{\infty} \sum_{\substack{\{n_r\} \\ (\sum n_r = N)}} \prod_r \mathrm{e}^{-\beta(\varepsilon_r - \mu) n_r} =$$

$$= \sum_{n_1}\sum_{n_2} \cdots \sum_{n_r} \cdots \prod_r \mathrm{e}^{-\beta(\varepsilon_r - \mu) n_r} =$$

$$= \left(\sum_{n_1} \mathrm{e}^{-\beta n_1(\varepsilon_1 - \mu)}\right)\left(\sum_{n_2} \mathrm{e}^{-\beta n_2(\varepsilon_2 - \mu)}\right) \cdots .$$

Grand canonical partition function:

$$\Xi(T, V, \mu) = \mathrm{Tr}\,\rho = \prod_r \left(\sum_{n_r} \mathrm{e}^{-\beta n_r(\varepsilon_r - \mu)}\right).$$

Bosons ($n_r = 0, 1, 2, \ldots$):

$$\Xi_{\mathrm{B}}(T, V, \mu) = \prod_r \frac{1}{1 - \varepsilon^{-\beta(\varepsilon_r - \mu)}}\,.$$

Fermions ($n_r = 0, 1$):

$$\Xi_{\mathrm{F}}(T, V, \mu) = \prod_r \left(1 + \mathrm{e}^{-\beta(\varepsilon_r - \mu)}\right).$$

3. Expectation value of the particle number:

$$\langle\widehat{N}\rangle = \frac{1}{\Xi}\,\mathrm{Tr}(\rho\widehat{N})\,.$$

Taking the trace is to be recommended for the Fock states, since they are eigenstates of $\widehat{N}$:

$$\langle\widehat{N}\rangle = \frac{1}{\Xi}\sum_{N=0}^{\infty}\sum_{\substack{\{n_r\}\\(\sum n_r = N)}}\left\{N\exp\left[-\beta\sum_r(\varepsilon_r-\mu)\,n_r\right]\right\} =$$

$$= \frac{1}{\beta}\frac{\partial}{\partial\mu}\ln\Xi\,.$$

With part 2.:

$$\frac{\partial}{\partial\mu}\ln\Xi_{\mathrm{B}} = \frac{\partial}{\partial\mu}\left\{-\sum_r\ln\left[1-\mathrm{e}^{-\beta(\varepsilon_r-\mu)}\right]\right\} =$$

$$= -\sum_r\frac{-\beta\mathrm{e}^{-\beta(\varepsilon_r-\mu)}}{1-\mathrm{e}^{-\beta(\varepsilon_r-\mu)}} = \beta\sum_r\frac{1}{\mathrm{e}^{\beta(\varepsilon_r-\mu)}-1}\,,$$

$$\frac{\partial}{\partial\mu}\ln\Xi_{\mathrm{F}} = \frac{\partial}{\partial\mu}\left\{\sum_r\ln\left[1+\mathrm{e}^{-\beta(\varepsilon_r-\mu)}\right]\right\} =$$

$$= \beta\sum_r\frac{\mathrm{e}^{-\beta(\varepsilon_r-\mu)}}{1+\mathrm{e}^{-\beta(\varepsilon_r-\mu)}} = \beta\sum_r\frac{1}{\mathrm{e}^{\beta(\varepsilon_r-\mu)}+1}\,.$$

This means that:

$$\langle\widehat{N}\rangle = \begin{cases}\displaystyle\sum_r\frac{1}{\mathrm{e}^{\beta(\varepsilon_r-\mu)}-1}: & \textit{Bosons},\\[2ex] \displaystyle\sum_r\frac{1}{\mathrm{e}^{\beta(\varepsilon_r-\mu)}+1}: & \textit{Fermions}.\end{cases}$$

4. Internal energy:

$$U = \langle H\rangle = \frac{1}{\Xi}\,\mathrm{Tr}(\rho H)\,.$$

Fock states are also eigenstates of H, and thus are again suitable for taking the trace, as is required here!

$$U = \frac{1}{\Xi} \sum_{N=0}^{\infty} \sum_{\substack{\{n_r\} \\ (\sum n_r = N)}} \left[\left(\sum_i \varepsilon_i n_i \right) \mathrm{e}^{-\beta \sum_r (\varepsilon_r - \mu) n_r} \right] =$$

$$= -\frac{\partial}{\partial \beta} \ln \Xi + \mu \left\langle \widehat{N} \right\rangle ,$$

$$-\frac{\partial}{\partial \beta} \ln \Xi_{\mathrm{B}} = \sum_r \frac{(\varepsilon_r - \mu)\, \mathrm{e}^{-\beta(\varepsilon_r - \mu)}}{1 - \mathrm{e}^{-\beta(\varepsilon_r - \mu)}} = -\mu \left\langle \widehat{N} \right\rangle + \sum_r \frac{\varepsilon_r}{\mathrm{e}^{\beta(\varepsilon_r - \mu)} - 1},$$

$$-\frac{\partial}{\partial \beta} \ln \Xi_{\mathrm{F}} = -\sum \frac{-(\varepsilon_r - \mu)\, \mathrm{e}^{-\beta(\varepsilon_r - \mu)}}{1 + \mathrm{e}^{-\beta(\varepsilon_r - \mu)}} = -\mu \left\langle \widehat{N} \right\rangle + \sum_r \frac{\varepsilon_r}{\mathrm{e}^{\beta(\varepsilon_r - \mu)} + 1} .$$

Finally, this yields:

$$U = \begin{cases} \sum_r \dfrac{\varepsilon_r}{\mathrm{e}^{\beta(\varepsilon_r - \mu)} - 1} : & \textit{Bosons,} \\ \sum_r \dfrac{\varepsilon_r}{\mathrm{e}^{\beta(\varepsilon_r - \mu)} + 1} : & \textit{Fermions.} \end{cases}$$

5. Fock states are also eigenstates of the occupation-number operator:

$$\langle \hat{n}_i \rangle = \frac{1}{\Xi} \mathrm{Tr}(\rho \hat{n}_i) = \frac{1}{\Xi} \sum_{N=0}^{\infty} \sum_{\substack{\{n_r\} \\ (\sum n_r = N)}} \left[n_i \, \mathrm{e}^{-\beta \sum_r (\varepsilon_r - \mu) n_r} \right] =$$

$$= -\frac{1}{\beta} \frac{\partial}{\partial \varepsilon_i} \ln \Xi ,$$

$$-\frac{1}{\beta} \frac{\partial}{\partial \varepsilon_i} \ln \Xi_{\mathrm{B}} = +\frac{1}{\beta} \sum_r \frac{+\beta \mathrm{e}^{-\beta(\varepsilon_r - \mu)}}{1 - \mathrm{e}^{-\beta(\varepsilon_r - \mu)}} \frac{\partial \varepsilon_r}{\partial \varepsilon_i} =$$

$$= \frac{1}{\mathrm{e}^{\beta(\varepsilon_i - \mu)} - 1} \quad \textit{(Bose distribution function)},$$

$$-\frac{1}{\beta} \frac{\partial}{\partial \varepsilon_i} \ln \Xi_{\mathrm{F}} = -\frac{1}{\beta} \sum_r \frac{-\beta \mathrm{e}^{-\beta(\varepsilon_r - \mu)}}{1 + \mathrm{e}^{-\beta(\varepsilon_r - \mu)}} \frac{\partial \varepsilon_r}{\partial \varepsilon_i} =$$

$$= \frac{1}{\mathrm{e}^{\beta(\varepsilon_i - \mu)} + 1} \quad \textit{(Fermi distribution function)}.$$

It follows that:

$$\langle \hat{n}_i \rangle = \begin{cases} \left\{ \exp\left[\beta\left(\varepsilon_i - \mu\right)\right] - 1 \right\}^{-1} & \textit{Bosons,} \\ \left\{ \exp\left[\beta\left(\varepsilon_i - \mu\right)\right] + 1 \right\}^{-1} & \textit{Fermions.} \end{cases}$$

By comparison with the preceding exercises, one can immediately see that:

$$\langle \widehat{N} \rangle = \sum_r \langle \hat{n}_r \rangle \; ; \quad U = \sum_r \varepsilon_r \langle \hat{n}_r \rangle \; .$$

Solution 1.4.10

1.

$$\widehat{N} = \sum_\sigma (\hat{n}_{1\sigma} + \hat{n}_{2\sigma}) \; ,$$

$$H = \sum_\sigma \left(\varepsilon_{1\sigma} \hat{n}_{1\sigma} + \varepsilon_{2\sigma} \hat{n}_{2\sigma} + V \left(c^+_{1\sigma} c_{2\sigma} + c^+_{2\sigma} c_{1\sigma} \right) \right) .$$

As usual, we have:

$$\left[\hat{n}_{i\sigma}, \hat{n}_{j\sigma'} \right]_- = 0 \, , \quad i, j \in \{1, 2\} \, .$$

This implies:

$$[\widehat{N}, H]_- = \left[\sum_\sigma (\hat{n}_{1\sigma} + \hat{n}_{2\sigma}) , \; V \sum_{\sigma'} \left(c^+_{1\sigma'} c_{2\sigma'} + c^+_{2\sigma'} c_{1\sigma'} \right) \right]_- =$$

$$= V \sum_\sigma \sum_{\sigma'} \Big\{ \left[\hat{n}_{1\sigma}, c^+_{1\sigma'} c_{2\sigma'} \right]_- + \left[\hat{n}_{1\sigma}, c^+_{2\sigma'} c_{1\sigma'} \right]_- +$$

$$+ \left[\hat{n}_{2\sigma}, c^+_{1\sigma'} c_{2\sigma'} \right]_- + \left[\hat{n}_{2\sigma}, c^+_{2\sigma'} c_{1\sigma'} \right]_- \Big\} \, .$$

We make use of the generally-valid relation:

$$\left[\widehat{A}, \widehat{B}\widehat{C} \right]_- = \left[\widehat{A}, \widehat{B} \right]_- \widehat{C} + \widehat{B} \left[\widehat{A}, \widehat{C} \right]_- \, .$$

Moreover, one can readily see that (cf. Ex. 1.4.7.):

$$\left[\hat{n}_{i\sigma}, c_{j\sigma'} \right]_- = -\delta_{ij} \delta_{\sigma\sigma'} c_{i\sigma} \, ,$$

$$\left[\hat{n}_{i\sigma}, c^+_{j\sigma'} \right]_- = \delta_{ij} \delta_{\sigma\sigma'} c^+_{i\sigma} \, .$$

We then finally obtain:

$$[\widehat{N}, H]_- = V \sum_\sigma \left\{ c^+_{1\sigma} c_{2\sigma} - c^+_{2\sigma} c_{1\sigma} - c^+_{1\sigma} c_{2\sigma} + c^+_{2\sigma} c_{1\sigma} \right\} = 0 \, .$$

2. Fock states:

$$|N; F\rangle = \left| N; n_{1\uparrow} n_{1\downarrow}; n_{2\uparrow} n_{2\downarrow} \right\rangle^{(-)} .$$

The eigenvalue equation:

$$
\begin{aligned}
& H\,|E\rangle = |E\rangle \\
\Longrightarrow\quad & \langle N;F \mid H \mid E\rangle = E\,\langle N;F \mid E\rangle \\
\Longrightarrow\quad & \sum_{N'}\sum_{F'} \langle N;F \mid H \mid N';F'\rangle\,\langle N';F' \mid E\rangle = E\,\langle N;F \mid E\rangle\,.
\end{aligned}
$$

According to 1., H conserves the particle number. This means that:

$$\langle N;F \mid H \mid N';F'\rangle \sim \delta_{NN'}\,.$$

We thus have to solve the following homogeneous system of equations for $N = 0, 1, 2, 3, 4$:

$$\sum_{F'}\Big(\langle N;F \mid H \mid N;F'\rangle - E\delta_{FF'}\Big)\,\langle N;F' \mid E\rangle = 0\,.$$

The eigenvalues are determined from the condition for a nontrivial solution:

$$\det\left(H^{(N)}_{FF'} - E\delta_{FF'}\right) \stackrel{!}{=} 0\,,\quad H^{(N)}_{FF'} = \langle N;F \mid H \mid N;F'\rangle\,.$$

3. $\boxed{N=0}$

$$|0;F\rangle = |0;00;00\rangle^{(-)}$$

is clearly an eigenstate with $E^{(0)} = 0$.

$\boxed{n=1}$

Four possible Fock states:

$$|1;F\rangle = |1;10;00\rangle^{(-)};\ |1;01;00\rangle^{(-)};\ |1;00;10\rangle^{(-)};\ |1;00;01\rangle^{(-)}\,.$$

$$
\begin{aligned}
H\,|1;10;00\rangle^{(-)} &= \varepsilon_1\,|1;10;00\rangle^{(-)} + V\,|1;00;10\rangle^{(-)}\,,\\
H\,|1;01;00\rangle^{(-)} &= \varepsilon_1\,|1;01;00\rangle^{(-)} + V\,|1;00;01\rangle^{(-)}\,,\\
H\,|1;00;10\rangle^{(-)} &= \varepsilon_2\,|1;00;10\rangle^{(-)} + V\,|1;10;00\rangle^{(0)}\,,\\
H\,|1;00;01\rangle^{(-)} &= \varepsilon_2\,|1;00;01\rangle^{(-)} + V\,|1;01;00\rangle^{(-)}\,,
\end{aligned}
$$

$$\left(H^{(1)}_{FF'}\right) \equiv \begin{pmatrix} \varepsilon_1 & 0 & V & 0 \\ 0 & \varepsilon_1 & 0 & V \\ V & 0 & \varepsilon_2 & 0 \\ 0 & V & 0 & \varepsilon_2 \end{pmatrix}$$

Condition for a solution:

$$0 \overset{!}{=} \begin{vmatrix} \varepsilon_1 - E & 0 & V & 0 \\ 0 & \varepsilon_1 - E & 0 & V \\ V & 0 & \varepsilon_2 - E & 0 \\ 0 & V & 0 & \varepsilon_2 - E \end{vmatrix} =$$

$$= - \begin{vmatrix} \varepsilon_1 - E & 0 & V & 0 \\ V & 0 & \varepsilon_2 - E & 0 \\ 0 & \varepsilon_1 - E & 0 & V \\ 0 & V & 0 & \varepsilon_2 - E \end{vmatrix} =$$

$$= \begin{vmatrix} \varepsilon_1 - E & V & 0 & 0 \\ V & \varepsilon_2 - E & 0 & 0 \\ 0 & 0 & \varepsilon_1 - E & V \\ 0 & 0 & V & \varepsilon_2 - E \end{vmatrix} =$$

$$= \left\{(\varepsilon_1 - E)(\varepsilon_2 - E) - V^2\right\}^2$$

$$\Longrightarrow \quad E_1^{(1)} = E_2^{(1)} = E_+ ; \quad E_3^{(1)} = E_4^{(1)} = E_- ,$$

$$E_\pm = \frac{1}{2}\left(\varepsilon_1 + \varepsilon_2 \pm \sqrt{(\varepsilon_1 - \varepsilon_2)^2 + 4V^2}\right) .$$

4. $\boxed{N = 2}$

Six possible Fock states:

$$|2; F\rangle = |2; 11; 00\rangle^{(-)}, \ |2; 00; 11\rangle^{(-)} ;$$
$$|2; 10; 10\rangle^{(-)}; \ |2; 10; 01\rangle^{(-)} ;$$
$$|2; 01; 10\rangle^{(-)}; \ |2; 01; 01\rangle^{(-)} .$$

Two of the Fock states are already eigenstates of H:

$$H\,|2; 10; 10\rangle^{(-)} = (\varepsilon_1 + \varepsilon_2)\,|2; 10; 10\rangle^{(-)} ,$$
$$H\,|2; 01; 01\rangle^{(-)} = (\varepsilon_1 + \varepsilon_2)\,|2; 01; 01\rangle^{(-)}$$

$$\Longrightarrow \quad E_{1,2}^{(2)} = \varepsilon_1 + \varepsilon_2 ,$$

$$H\,|2;11;00\rangle^{(-)} = 2\varepsilon_1\,|2;11;00\rangle^{(-)} - V\,|2;01;10\rangle^{(-)} + V\,|2;10;01\rangle^{(-)}\,,$$
$$H\,|2;00;11\rangle^{(-)} = 2\varepsilon_2\,|2;00;11\rangle^{(-)} + V\,|2;10;01\rangle^{(-)} - V\,|2;01;10\rangle^{(-)}\,,$$
$$H\,|2;10;01\rangle^{(-)} = (\varepsilon_1+\varepsilon_2)\,|2;10;01\rangle^{(-)} + V\,|2;00;11\rangle^{(-)} + V\,|2;11;00\rangle^{(-)}\,,$$
$$H\,|2;01;10\rangle^{(-)} = (\varepsilon_1+\varepsilon_2)\,|2;01;10\rangle^{(-)} - V\,|2;11;00\rangle^{(-)} - V\,|2;00;11\rangle^{(-)}\,.$$

We therefore still have to solve the following 4×4 secular determinant:

$$0 \stackrel{!}{=} \begin{vmatrix} 2\varepsilon_1 - E & 0 & V & -V \\ 0 & 2\varepsilon_2 - E & V & -V \\ V & V & \varepsilon_1+\varepsilon_2-E & 0 \\ -V & -V & 0 & \varepsilon_1+\varepsilon_2-E \end{vmatrix} =$$

$$= \begin{vmatrix} 2\varepsilon_1 - E & -2\varepsilon_2 + E & 0 & 0 \\ 0 & 2\varepsilon_2 - E & V & -V \\ V & V & \varepsilon_1+\varepsilon_2-E & 0 \\ 0 & 0 & \varepsilon_1+\varepsilon_2-E & \varepsilon_1+\varepsilon_2-E \end{vmatrix} =$$

$$= (\varepsilon_1+\varepsilon_2-E)\begin{vmatrix} 2\varepsilon_1 - E & -2\varepsilon_2 + E & 0 \\ 0 & 2\varepsilon_2 - E & V \\ V & V & \varepsilon_1+\varepsilon_2-E \end{vmatrix} -$$

$$-(\varepsilon_1+\varepsilon_2-E)\begin{vmatrix} 2\varepsilon_1 - E & -2\varepsilon_2 + E & 0 \\ 0 & 2\varepsilon_2 - E & -V \\ V & V & 0 \end{vmatrix}\,.$$

From this, we can read off another possible solution:

$$E_3^{(2)} = \varepsilon_1 + \varepsilon_2\,.$$

It remains to calculate:

$$\begin{aligned} 0 &= (2\varepsilon_1 - E)(2\varepsilon_2 - E)(\varepsilon_1+\varepsilon_2-E) - V^2(2\varepsilon_2 - E) - \\ &\quad - V^2(2\varepsilon_1 - E) - V^2(2\varepsilon_2 - E) - V^2(2\varepsilon_1 - E) = \\ &= (2\varepsilon_1 - E)(2\varepsilon_2 - E)(\varepsilon_1+\varepsilon_2-E) - \\ &\quad - 2V^2(2\varepsilon_2 - E + 2\varepsilon_1 - E)\,. \end{aligned}$$

This immediately yields the next solution:

$$E_4^{(2)} = \varepsilon_1 + \varepsilon_2 .$$

Finally, we have only a quadratic equation:

$$0 = (2\varepsilon_1 - E)(2\varepsilon_2 - E) - 4V^2 \quad \Longrightarrow \quad E_{5,6}^{(2)} = 2E_\pm .$$

5. $\boxed{N = 3}$

$$|3; F\rangle = |3; 01; 11\rangle^{(-)}, \ |3; 10, 11\rangle^{(-)}, \ |3; 11, 01\rangle^{(-)}, \ |3; 11, 10\rangle^{(-)},$$

$$H\,|3; 01, 11\rangle^{(-)} = (\varepsilon_1 + 2\varepsilon_2)\,|3; 01, 11\rangle^{(-)} - V\,|3; 11, 01\rangle^{(-)},$$

$$H\,|3; 10, 11\rangle^{(-)} = (\varepsilon_1 + 2\varepsilon_2)\,|3; 10, 11\rangle^{(-)} - V\,|3; 11, 10\rangle^{(-)},$$

$$H\,|3; 11, 01\rangle^{(-)} = (2\varepsilon_1 + \varepsilon_2)\,|3; 11, 01\rangle^{(-)} - V\,|3; 01, 11\rangle^{(-)},$$

$$H\,|3; 11, 10\rangle^{(-)} = (2\varepsilon_1 + \varepsilon_2)\,|3; 11, 10\rangle^{(-)} - V\,|3; 10, 11\rangle^{(-)}.$$

The secular determinant:

$$0 \stackrel{!}{=} \begin{vmatrix} (\varepsilon_1 + 2\varepsilon_2) - E & 0 & -V & 0 \\ 0 & (\varepsilon_1 + 2\varepsilon_2) - E & 0 & -V \\ -V & 0 & (2\varepsilon_1 + \varepsilon_2) - E & 0 \\ 0 & -V & 0 & (2\varepsilon_1 + \varepsilon_2) - E \end{vmatrix} =$$

$$= - \begin{vmatrix} (\varepsilon_1 + 2\varepsilon_2) - E & 0 & -V & 0 \\ -V & 0 & (2\varepsilon_1 + \varepsilon_2) - E & 0 \\ 0 & (\varepsilon_1 + 2\varepsilon_2) - E & 0 & -V \\ 0 & -V & 0 & (2\varepsilon_1 + \varepsilon_2) - E \end{vmatrix} =$$

$$= \begin{vmatrix} (\varepsilon_1 + 2\varepsilon_2) - E & -V & 0 & 0 \\ -V & (2\varepsilon_1 + \varepsilon_2) - E & 0 & 0 \\ 0 & 0 & (\varepsilon_1 + 2\varepsilon_2) - E & -V \\ 0 & 0 & -V & (2\varepsilon_1 + \varepsilon_2) - E \end{vmatrix} =$$

$$= \left\{ [(\varepsilon_1 + 2\varepsilon_2) - E]\,[(2\varepsilon_1 + \varepsilon_2) - E] - V^2 \right\}^2$$

$$\Longrightarrow \quad E_{1,2}^{(3)} = \widetilde{E}_+ ; \quad E_{3,4}^{(3)} = \widetilde{E}_- ,$$

$$\widetilde{E}_\pm = E_\pm + (\varepsilon_1 + \varepsilon_2) .$$

$\boxed{N = 4}$

$$
\begin{aligned}
|4; F\rangle &= |4; 11, 11\rangle^{(-)} , \\
H\,|4; F\rangle &= 2\,(\varepsilon_1 + \varepsilon_2)\,|4; 11, 11\rangle^{(-)} \\
\Longrightarrow \quad E^{(4)} &= 2\,(\varepsilon_1 + \varepsilon_2) \,.
\end{aligned}
$$

Section 2.1.4

Solution 2.1.1

$$
\boldsymbol{k} = \frac{2\pi}{L}\left(n_x, n_y, n_z\right) , \quad n_{x,\,y,\,z} = 0, \pm 1, \pm 2, \ldots, \pm\left(\frac{N'}{2} - 1\right), \frac{N'}{2} ,
$$

$$
L = N' a_x = N' a_y = N' a_z \,, \quad a_{x,\,y,\,z} = a \,.
$$

We can see immediately that:

$$
\sum_{\boldsymbol{k}} \mathrm{e}^{\mathrm{i}\boldsymbol{k}\cdot(\boldsymbol{R}_i - \boldsymbol{R}_j)} = N'^3 = N \,, \quad \text{when} \quad i = j \,.
$$

We thus need to discuss only the case $i \neq j$:

$$
\begin{aligned}
&\sum_{\boldsymbol{k}} \mathrm{e}^{\mathrm{i}\boldsymbol{k}\cdot(\boldsymbol{R}_i - \boldsymbol{R}_j)} = \\
&= \sum_{k_x} \mathrm{e}^{\mathrm{i}k_x(R_{ix} - R_{jx})} \sum_{k_y} \mathrm{e}^{\mathrm{i}k_y(R_{iy} - R_{jy})} \sum_{k_z} \mathrm{e}^{\mathrm{i}k_z(R_{iz} - R_{jz})} = \\
&= \sum_{n_x} \mathrm{e}^{\mathrm{i}\frac{2\pi}{N'}n_x(i_x - j_x)} \sum_{n_y} \mathrm{e}^{\mathrm{i}\frac{2\pi}{N'}n_y(i_y - j_y)} \sum_{n_z} \mathrm{e}^{\mathrm{i}\frac{2\pi}{N'}n_z(i_z - j_z)} \,.
\end{aligned}
$$

We compute the first factor as an example:

$$
i_x, j_x \in \mathbf{Z} \quad \text{with} \quad -\frac{N'}{2} < i_x, j_x \leq +\frac{N'}{2} \,.
$$

$$
\begin{aligned}
&\sum_{n_x} \mathrm{e}^{\mathrm{i}\frac{2\pi}{N'}n_x(i_x - j_x)} = \\
&= \sum_{n_x=0}^{N'/2} \mathrm{e}^{\mathrm{i}\frac{2\pi}{N'}n_x(i_x - j_x)} + \sum_{n_x=-\frac{N'}{2}+1}^{-1} \mathrm{e}^{\mathrm{i}\frac{2\pi}{N'}n_x(i_x - j_x)} =
\end{aligned}
$$

$$= \sum_{n_x=0}^{N'/2} e^{i\frac{2\pi}{N'}n_x(i_x - j_x)} + \sum_{n_x=\frac{N'}{2}-1}^{N'-1} e^{i\frac{2\pi}{N'}n_x(i_x - j_x)} \underbrace{e^{-i\frac{2\pi}{N'}N'(i_x - j_x)}}_{=+1} =$$

$$= \sum_{n_x=0}^{N'-1} e^{i\frac{2\pi}{N'}n_x(i_x - j_x)} = \frac{1 - e^{i2\pi\,(i_x - j_x)}}{1 - e^{i\frac{2\pi}{N'}(i_x - j_x)}} .$$

For $i_x \neq j_x$, the numerator is zero and the denominator is finite. Completely analogous expressions are found for the y- and z-components. We have thereby proved the contention:

$$\frac{1}{N}\sum_{k} e^{ik\cdot(R_i - R_j)} = \delta_{ij} \; !$$

Solution 2.1.2

$$a_{i\sigma} = \frac{1}{\sqrt{N}} \sum_{k}^{1.\mathrm{BZ}} e^{ik\cdot R_i} a_{k\sigma}$$

$$\Longrightarrow \quad \left[a_{i\sigma}, a_{j\sigma'}\right]_+ = \frac{1}{N}\sum_{k}\sum_{k'} e^{i(k\cdot R_i + k'\cdot R_j)} \underbrace{[a_{k\sigma}, a_{k'\sigma'}]_+}_{=0} = 0.$$

In complete analogy, one finds:

$$\left[a^+_{i\sigma}, a^+_{j\sigma'}\right]_+ = 0 .$$

It remains to determine:

$$\left[a_{i\sigma}, a^+_{j\sigma'}\right]_+ = \frac{1}{N}\sum_{k}\sum_{k'} e^{i(k\cdot R_i - k'\cdot R_j)} \left[a_{k\sigma}, a^+_{k'\sigma'}\right]_+ =$$

$$= \frac{1}{N}\sum_{k}\sum_{k'} e^{i(k\cdot R_i - k'\cdot R_j)} \delta_{\sigma\sigma'}\delta_{kk'} =$$

$$= \delta_{\sigma\sigma'}\frac{1}{N}\sum_{k} e^{ik\cdot(R_i - R_j)} =$$

$$= \delta_{\sigma\sigma'}\delta_{ij} \quad \text{(cf. Ex. 2.1.1)} .$$

Solution 2.1.3

1.

$$p(y) = \int_{-\infty}^{y} dx\, g(x) \quad \Longleftrightarrow \quad g(y) = \frac{dp(y)}{dy} .$$

Through integration by parts, we find that:

$$\int_{-\infty}^{+\infty} \mathrm{d}x\, g(x) f_-(x) = p(x) f_-(x)\Big|_{-\infty}^{+\infty} - \int_{-\infty}^{+\infty} \mathrm{d}x\, p(x) \frac{\partial f_-(x)}{\partial x}\,.$$

The first term vanishes at the upper bound of the integration, since $f_-(x)$ goes to zero more quickly than p diverges. At the lower bound, $f_-(x) = 1$ and $p(x) = 0$. We have therefore found that:

$$I(T) = -\int_{-\infty}^{+\infty} \mathrm{d}x\, p(x) \frac{\partial f_-(x)}{\partial x}\,.$$

The second factor on the right-hand side differs notably from zero only within the thin Fermi layer ($\mu \pm 4k_{\mathrm{B}}T$)!

2. Taylor series expansion:

$$p(x) = p(\mu) + \sum_{n=1}^{\infty} \frac{(x-\mu)^n}{n!} \left(\frac{\mathrm{d}^n p(x)}{\mathrm{d}x^n} \right)_{x=\mu}\,.$$

The first term makes the following contribution:

$$J_0(\mu) = -p(\mu) \int_{-\infty}^{+\infty} \mathrm{d}x\, \frac{\partial f_-(x)}{\partial x} = p(\mu)\,.$$

In the sum, only even powers of $(x - \mu)$ contribute, since

$$\frac{\partial f_-(x)}{\partial x} = -\beta \frac{\mathrm{e}^{\beta(x-\mu)}}{\left[\mathrm{e}^{\beta(x-\mu)} + 1\right]^2} = \frac{-\beta}{\left[\mathrm{e}^{(1/2)\beta(x-\mu)} + \mathrm{e}^{-(1/2)\beta(x-\mu)}\right]^2}$$

is an *even* function of $(x - \mu)$.

We insert this expansion for $p(x)$ into the integral $I(T)$:

$$I(T,\mu) = J_0(\mu) + \beta \sum_{n=1}^{\infty} \frac{1}{(2n)!} \left(\frac{\mathrm{d}^{2n-1}}{\mathrm{d}x^{2n-1}} g(x) \right)_{x=\mu} J_{2n}(T)\,.$$

Here, we have made use of the following abbreviation:

$$J_{2n}(T) = \int_{-\infty}^{+\infty} \mathrm{d}x\, (x-\mu)^{2n} \frac{\mathrm{e}^{\beta(x-\mu)}}{\left[\mathrm{e}^{\beta(x-\mu)} + 1\right]^2}\,.$$

We can evaluate this expression further:

$$J_{2n}(T) = \beta^{-(2n+1)} \int_{-\infty}^{+\infty} \mathrm{d}y\, y^{2n} \frac{\mathrm{e}^y}{(\mathrm{e}^y + 1)^2} =$$

$$= -2\beta^{-(2n+1)} \left(\frac{\mathrm{d}}{\mathrm{d}\lambda} \int_0^\infty \mathrm{d}y \frac{y^{2n-1}}{\mathrm{e}^{\lambda y} + 1} \right)_{\lambda = 1} =$$

$$= -2\beta^{-(2n+1)} \left(\frac{\mathrm{d}}{\mathrm{d}\lambda} \lambda^{-2n} \int_0^\infty \mathrm{d}u \, \frac{u^{2n-1}}{\mathrm{e}^u + 1} \right)_{\lambda = 1} =$$

$$= 4n\beta^{-(2n+1)} \left(1 - 2^{1-2n}\right) \Gamma(2n) \zeta(2n) .$$

Riemann's ζ function:

$$\zeta(n) = \sum_{p=1}^{\infty} \frac{1}{p^n} = \frac{1}{\left(1 - 2^{1-n}\right) \Gamma(n)} \int_0^\infty \mathrm{d}u \, \frac{u^{n-1}}{\mathrm{e}^u + 1} ,$$

$$n \in \mathbf{N} ; \quad \text{then:} \quad \Gamma(n) = (n-1)!$$

In particular:

$$\zeta(2) = \frac{\pi^2}{6} ; \quad \zeta(4) = \frac{\pi^4}{90} ; \quad \zeta(6) = \frac{\pi^6}{945} ; \quad \ldots$$

This implies that:

$$I(T, \mu) = p(\mu) + 2 \sum_{n=1}^{\infty} \left(1 - 2^{1-2n}\right) (k_{\mathrm{B}} T)^{2n} \zeta(2n) g^{(2n-1)}(\mu) .$$

3. The value of this so-called

Sommerfeld expansion

becomes particularly clear for functions $g(x)$ for which

$$g^{(n)}(x)\big|_{x=\mu} \approx \frac{g(\mu)}{\mu^n}$$

is valid, such as e.g. the density of states $\rho_0(x) \sim \sqrt{x}$ in the *Sommerfeld model* (see Ex. 2.1.4). Then the series converges in fact very quickly, since the ratios of subsequent terms are of the order of

$$\left(\frac{k_{\mathrm{B}} T}{\mu}\right)^2 \quad \left(\approx 10^{-4} \quad \text{for metals at room temperature}\right) .$$

As a rule, one needs to include only the first few terms in the sum:

$$\int_{-\infty}^{+\infty} dx\, g(x) f_-(x) = \int_{-\infty}^{\mu} dx\, g(x) + \frac{\pi^2}{6}(k_B T)^2 g'(\mu) + \frac{7\pi^4}{360}(k_B T)^4 g'''(\mu) + O\left[(k_B T/\mu)^6\right] .$$

Solution 2.1.4

1. The Schrödinger equation:

$$-\frac{\hbar^2}{2m}\Delta\psi_k(r) = \varepsilon(k)\psi_k(r) ,$$

$$\Delta = \frac{d^2}{dx^2} + \frac{d^2}{dy^2} + \frac{d^2}{dz^2} .$$

Trial solution:

$\psi_k(r) = \alpha e^{ik \cdot r}$, $|\psi_k(r)|^2 d^3r =$ Probability that the electron is to be found within the volume element d^3r at the position r.

Normalisation:

$$1 \stackrel{!}{=} \int_V d^3r\, |\psi_k(r)|^2 \implies \alpha = \frac{1}{\sqrt{V}} .$$

Eigenfunctions:

$$\psi_k(r) = \frac{1}{\sqrt{V}} e^{ik \cdot r} .$$

This solution takes *no* boundary conditions into account. If the electrons remain within the crystal, we would in fact have to require that $\psi \equiv 0$ at the crystal boundaries. This, however, proves not to be expedient. Periodic boundary conditions are easier to work with and can be justified in the thermodynamic limit ($N \to \infty$, $V \to \infty$, $N/V \to$ const).

$$\psi_k(x+L, y, z) \stackrel{!}{=} \psi_k(x, y+L, z) \stackrel{!}{=} \psi_k(x, y, z+L) \stackrel{!}{=} \psi_k(x, y, z)$$

$$\implies k_{x,y,z} = \frac{2\pi}{L} n_{x,y,z} , \quad n_{x,y,z} \in \mathbf{Z} .$$

"Grid volume" = volume per state in $\boldsymbol{k}$-space:

$$\Delta k = \frac{(2\pi)^3}{L^3} = \frac{(2\pi)^3}{V} \,.$$

Energy eigenvalues:

$$\varepsilon(\boldsymbol{k}) = \frac{\hbar^2 k^2}{2m} = \frac{\hbar^2}{2m}\left(k_x^2 + k_y^2 + k_z^2\right) = \frac{2\hbar^2\pi^2}{mL^2}\left(n_x^2 + n_y^2 + n_z^2\right) .$$

From this, discrete energy levels result from the boundary conditions!

2. In the ground state, the electrons occupy all the states with

$$\varepsilon(\boldsymbol{k}) \le \varepsilon_F = \frac{\hbar^2 k_F^2}{2m} \,,$$

ε_F: Fermi energy, k_F: Fermi wavevector.

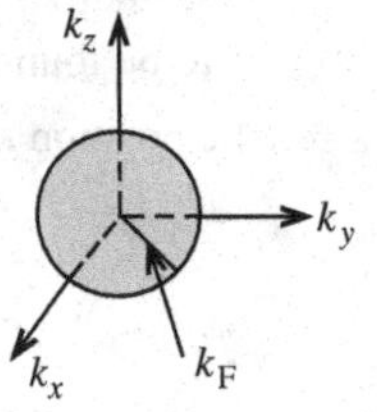

Fig. A.1

Overall number of electrons:

$$N = 2 \sum_{\boldsymbol{k}}^{k \le k_F} 1 \,.$$

The factor 2 appears because of spin degeneracy:

$$N = \frac{2}{\Delta k} \int\limits_{k \le k_F} d^3k = \frac{2V}{(2\pi)^3} \frac{4\pi}{3} k_F^3$$

$$\Longrightarrow \quad k_F = \left(3\pi^2 n\right)^{1/3} ; \quad \varepsilon_F = \frac{\hbar^2}{2m}\left(3\pi^2 n\right)^{2/3} .$$

3.

$$\bar{\varepsilon} = \frac{2}{N}\sum_{k}^{k \le k_F} \frac{\hbar^2 k^2}{2m} = \frac{2}{N}\frac{\hbar^2}{2m} 4\pi \int_0^{k_F} dk\, k^4 \frac{1}{\Delta k} =$$

$$= \frac{2}{N}\frac{4\pi\hbar^2}{2m}\frac{k_F^5}{5}\frac{V}{(2\pi)^3} = \frac{2}{N}\frac{4\pi\hbar^2}{2m}\frac{k_F^2}{5} 3\pi^2 n \frac{V}{8\pi^3} =$$

$$= \frac{3}{5}\frac{\hbar^2 k_F^2}{2m} = \frac{3}{5}\varepsilon_F .$$

4.

$$\rho_0(E)\mathrm{d}E = \text{the number of states within the energy interval} \quad [E; E + \mathrm{d}E],$$

$$\rho_0(E)\mathrm{d}E = \frac{2}{\Delta k} \int_{\substack{\text{shell} \\ [E;E+\mathrm{d}E]}} \mathrm{d}^3 k .$$

The integration is carried out over a shell in $\boldsymbol{k}$-space, which contains the $\boldsymbol{k}$-vectors belonging to energies between E and $E + \mathrm{d}E$. With the phase-space volume

$$\varphi(E) = \int_{\varepsilon(\boldsymbol{k}) \le E} \mathrm{d}^3 k = \frac{4\pi}{3} k^3 \bigg|_{\varepsilon(\boldsymbol{k}) = E} = \frac{4\pi}{3}\left(\frac{2m}{\hbar^2}E\right)^{3/2},$$

it follows that

$$\frac{\mathrm{d}\varphi(E)}{\mathrm{d}E} = 2\pi \left(\frac{2m}{\hbar^2}\right)^{3/2} \sqrt{E} .$$

Furthermore, we have:

$$\rho_0(E)\mathrm{d}E = \frac{2V}{(2\pi)^3}\left(\frac{\mathrm{d}\varphi(E)}{\mathrm{d}E}\right)\mathrm{d}E$$

$$\Longrightarrow \quad \rho_0(E) = \begin{cases} d\sqrt{E}, & \text{when} \quad E \ge 0, \\ 0 & \text{otherwise,} \end{cases}$$

$$d = \frac{V}{2\pi^2}\left(\frac{2m}{\hbar^2}\right)^{3/2} = \frac{3N}{2\varepsilon_F^{3/2}} .$$

5.

$n = N/V$: average electron density,

$v = 1/n$: mean volume per electron,

$$v = \frac{4\pi}{3}(a_{\mathrm{B}}r_s)^3\,,$$

$$a_{\mathrm{B}} = \frac{4\pi\,\varepsilon_0\hbar^2}{me^2}: \quad \textit{first Bohr radius,}$$

$$1\,\mathrm{ryd} \equiv \frac{1}{4\pi\,\varepsilon_0}\frac{e^2}{2a_{\mathrm{B}}} = \frac{me^4}{2\hbar^2\,(4\pi\,\varepsilon_0)^2}\,,$$

$$a_{\mathrm{B}}r_s = \left(\frac{3v}{4\pi}\right)^{1/3} = \left(\frac{3}{4\pi n}\right)^{1/3} \quad \Longrightarrow \quad k_{\mathrm{F}}a_{\mathrm{B}}r_s = \left(\frac{9\pi}{4}\right)^{1/3} = \alpha\,.$$

With this, we find:

$$\varepsilon_{\mathrm{F}} = \frac{\hbar^2 k_{\mathrm{F}}^2}{2m} = \frac{\hbar^2}{2m}(k_{\mathrm{F}}a_{\mathrm{B}}r_s)^2\,\frac{1}{a_{\mathrm{B}}^2 r_s^2} =$$

$$= \frac{\alpha^2}{r_s^2}\left(\frac{\hbar^2}{2m}\frac{m^2e^4}{(4\pi\,\varepsilon_0)^2\,\hbar^4}\right) = \frac{\alpha^2}{r_s^2}\left(\frac{me^4}{2\hbar^2\,(4\pi\,\varepsilon_0)^2}\right).$$

And thus:

$$\varepsilon_{\mathrm{F}} = \frac{\alpha^2}{r_s^2}[\mathrm{ryd}] \quad (\alpha = 1.92)$$

$$\Longrightarrow \quad E_0 = N\bar{\varepsilon} = N\frac{3}{5}\varepsilon_{\mathrm{F}} = N\frac{2.21}{r_s^2}[\mathrm{ryd}]\,.$$

Solution 2.1.5

1. From Ex. 1.4.9, we have:

$$\langle\hat{n}_i\rangle = \Big\{\exp\left[\beta\,(\varepsilon_i - \mu)\right] + 1\Big\}^{-1}\,.$$

This is the probability that the state of energy ε_i is occupied at a temperature T! In the Sommerfeld model for the electrons in a metal, this implies that:

$$\langle\hat{n}_{k\sigma}\rangle = f_-[E = \varepsilon(\boldsymbol{k})]\,,$$

$$f_-(E) = \frac{1}{\mathrm{e}^{\beta(E-\mu)} + 1}\,.$$

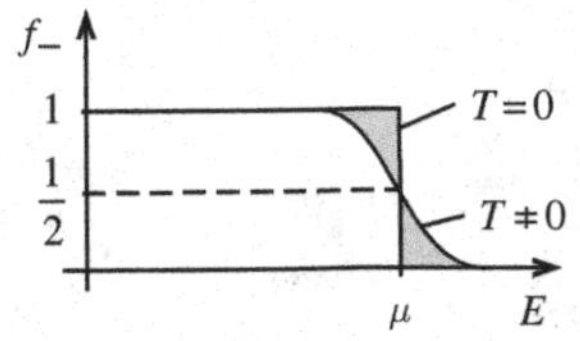

Fig. A.2

At $T \neq 0$, the kinetic energy of the electrons increases. Some of them leave the levels at $\varepsilon < \varepsilon_F$ for higher-lying levels which are not occupied at $T = 0$. However, for **all** temperatures, we have:

$$f_-(E = \mu) = \frac{1}{2} .$$

The *broadening* of the Fermi edge with increasing temperature occurs symmetrically:

$$f_-(\mu + \Delta E) = 1 - f_-(\mu - \Delta E) ,$$

$$\frac{\mathrm{d}f_-(E)}{\mathrm{d}E} = -\beta \frac{\mathrm{e}^{\beta(E-\mu)}}{\left(\mathrm{e}^{\beta(E-\mu)} + 1\right)^2} \underset{E \to \mu}{\longrightarrow} -\frac{1}{4k_\mathrm{B}T} .$$

The width of the broadened *Fermi layer* can thus be estimated to be approximately $4k_\mathrm{B}T$!

$$\mu = \mu(T) \quad \text{(cf. part 3.)}; \quad \mu(T = 0) = \varepsilon_F .$$

Numerical values:

$$k_\mathrm{B}T[\mathrm{eV}] = \frac{T[\mathrm{K}]}{11605} ,$$

$$\varepsilon_F = 1 \cdots 10\,\mathrm{eV} \quad \text{(typical of metals)}$$

$$\Longrightarrow \quad \frac{k_\mathrm{B}T}{\varepsilon_F} \geq \frac{1}{40} \quad (\text{at} \quad T = 290\mathrm{K}) .$$

Thus, at usual temperatures, only a narrow region around the Fermi edge is *broadened*. The high-energy tail of the distribution

$$E - \mu \gg k_\mathrm{B}T ; \quad f_-(E) \approx \exp\left[-\beta(E - \mu)\right] ,$$

corresponds to the classical Boltzmann distribution.

2. $f_-(E)\rho_0(E)$ = density of the *occupied* states:

$$N = 2 \int_{-\infty}^{+\infty} \mathrm{d}E\, f_-(E)\rho_0(E) ,$$

$$U(T) = 2 \int_{-\infty}^{+\infty} \mathrm{d}E\, E f_-(E)\rho_0(E) .$$

Fig. A.3

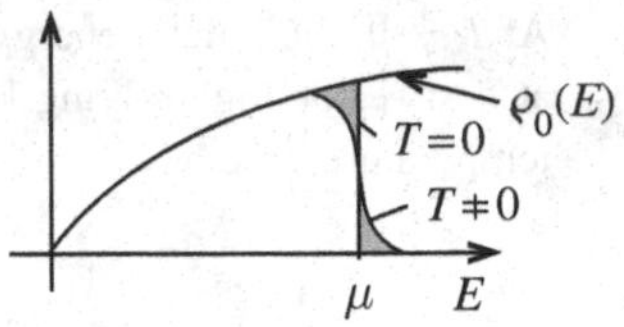

More formally:

$$N = \sum_{k\sigma} \langle \hat{n}_{k\sigma} \rangle = 2 \sum_{k} f_-(\varepsilon(\mathbf{k}))\,,$$

$$U(T) = \langle H \rangle = \sum_{k\sigma} \varepsilon(\mathbf{k}) \langle \hat{n}_{k\sigma} \rangle = 2 \sum_{k} \varepsilon(\mathbf{k}) f_-(\varepsilon(\mathbf{k}))\,,$$

$$\rho_0(E) = \sum_{k} \delta(E - \varepsilon(\mathbf{k}))$$

$$\Longrightarrow \quad N = \int_{-\infty}^{+\infty} \mathrm{d}E\, f_-(E)\rho_0(E)\,,$$

$$U(T) = \int_{-\infty}^{+\infty} \mathrm{d}E\, E f_-(E)\rho_0(E)\,.$$

The particle number N is in fact naturally not dependent on the temperature!

3.

$$\rho_0(E) = \begin{cases} \frac{3N}{2\varepsilon_F^{3/2}} \sqrt{E} & \text{for} \quad E \geq 0\,, \\ 0 & \text{otherwise.} \end{cases}$$

fulfils the conditions for the Sommerfeld expansion!

$$N \approx \int_{-\infty}^{\mu} \mathrm{d}E\, \rho_0(E) + \frac{\pi^2}{6} (k_B T)^2 \rho_0'(\mu) + \cdots =$$

$$= \frac{3N}{2\varepsilon_F^{3/2}} \left(\frac{2}{3} \mu^{3/2} + \frac{\pi^2}{6} (k_B T)^2 \frac{1}{2} \mu^{-1/2} + \cdots \right)$$

$$\Longrightarrow \quad 1 \approx \left(\frac{\mu}{\varepsilon_F} \right)^{3/2} \Bigg[1 + \frac{\pi^2}{8} \underbrace{\left(\frac{k_B T}{\mu} \right)^2}_{\text{typically} \approx 10^{-4}!} + \cdots \Bigg]$$

$$\Longrightarrow \quad \frac{\mu}{\varepsilon_F} \approx 1 - \frac{2}{3} \frac{\pi^2}{8} \left(\frac{k_B T}{\mu} \right)^2$$

$$\Longrightarrow \quad \mu \approx \varepsilon_F \left[1 - \frac{\pi^2}{12} \left(\frac{k_B T}{\varepsilon_F} \right)^2 \right].$$

The temperature dependence is thus as a rule very weak!

4.

$$U(T) \approx \int_0^{\mu} \mathrm{d}E\, E\rho_0(E) + \frac{\pi^2}{6}(k_\mathrm{B}T)^2\left[\mu\rho_0'(u) + \rho_0(\mu)\right] =$$

$$= \frac{2}{5}\mu^2\rho_0(\mu) + \frac{\pi^2}{4}(k_\mathrm{B}T)^2\rho_0(\mu) =$$

$$= \frac{3N}{2\varepsilon_\mathrm{F}^{3/2}}\left[\frac{2}{5}\mu^{5/2} + \frac{\pi^2}{4}(k_\mathrm{B}T)^2\mu^{1/2}\right] =$$

$$= \frac{3}{5}N\varepsilon_\mathrm{F}\left[\left(\frac{\mu}{\varepsilon_\mathrm{F}}\right)^{5/2} + \frac{5\pi^2}{8}\left(\frac{k_\mathrm{B}T}{\varepsilon_\mathrm{F}}\right)^2\left(\frac{\mu}{\varepsilon_\mathrm{F}}\right)^{1/2}\right],$$

$$\left(\frac{\mu}{\varepsilon_\mathrm{F}}\right)^n \approx 1 - n\frac{\pi^2}{12}\left(\frac{k_\mathrm{B}T}{\varepsilon_\mathrm{F}}\right)^2,$$

$$U(T) = U(0)\left[1 + \frac{5\pi^2}{8}\left(\frac{k_\mathrm{B}T}{\varepsilon_\mathrm{F}}\right)^2 - \frac{5\pi^2}{8}\left(\frac{k_\mathrm{B}T}{\varepsilon_\mathrm{F}}\right)^2 + \cdots\right]$$

$$\Longrightarrow \quad U(T) - U(0) = U(0)\frac{5\pi^2}{12}\left(\frac{k_\mathrm{B}T}{\varepsilon_\mathrm{F}}\right)^2 + O\left[\left(\frac{k_\mathrm{B}T}{\varepsilon_\mathrm{F}}\right)^4\right].$$

Specific heat:

$$c_V = \left(\frac{\partial U}{\partial T}\right)_V = \gamma T$$

$$\gamma = U(0)\frac{5\pi^2}{6}\frac{k_\mathrm{B}^2}{\varepsilon_\mathrm{F}^2} = \frac{1}{2}N\pi^2\frac{k_\mathrm{B}^2}{\varepsilon_\mathrm{F}} = \frac{1}{3}\pi^2 k_\mathrm{B}^2\rho_0(\varepsilon_\mathrm{F}).$$

5. The grand canonical ensemble:
Entropy:

$$S = k_\mathrm{B}\frac{\partial}{\partial T}(T\ln\Xi),$$

$$S = \frac{\partial}{\partial T}\left\{k_\mathrm{B}T\sum_{\boldsymbol{k}\sigma}\ln\left(1 + \mathrm{e}^{-\beta(\varepsilon(\boldsymbol{k})-\mu)}\right)\right\} =$$

$$= k_\mathrm{B}\sum_{\boldsymbol{k}\sigma}\ln\left(1 + \mathrm{e}^{-\beta(\varepsilon(\boldsymbol{k})-\mu)}\right) +$$

$$+ k_\mathrm{B}T\frac{1}{k_\mathrm{B}T^2}\sum_{\boldsymbol{k}\sigma}\frac{\mathrm{e}^{-\beta(\varepsilon(\boldsymbol{k})-\mu)}}{1 + \mathrm{e}^{-\beta(\varepsilon(\boldsymbol{k})-\mu)}}(\varepsilon(\boldsymbol{k}) - \mu)$$

$$\left(\frac{\partial\mu}{\partial T} \approx 0\right),$$

$$\Longrightarrow \quad S = \sum_{k\sigma} \left\{ k_B \ln \left(\frac{1}{1 - \langle \hat{n}_{k\sigma} \rangle} \right) + k_B \beta (\varepsilon(\boldsymbol{k}) - \mu) \langle \hat{n}_{k\sigma} \rangle \right\} ,$$

$$-\beta\,(\varepsilon(\boldsymbol{k}) - \mu) = \ln \langle \hat{n}_{k\sigma} \rangle + \ln\left(1 + \mathrm{e}^{-\beta(\varepsilon(\boldsymbol{k}) - \mu)}\right) = \\ = \ln \langle \hat{n}_{k\sigma} \rangle - \ln\left(1 - \langle \hat{n}_{k\sigma} \rangle\right) .$$

With this, we then have:

$$S = -k_B \sum_{k\sigma} \Big[\ln\left(1 - \langle \hat{n}_{k\sigma} \rangle\right) + \langle \hat{n}_{k\sigma} \rangle \ln \langle n_{k\sigma} \rangle - \langle \hat{n}_{k\sigma} \rangle \ln\left(1 - \langle \hat{n}_{k\sigma} \rangle\right) \Big] ,$$

$$S = -k_B \sum_{k\sigma} \Big[\underbrace{\langle \hat{n}_{k\sigma} \rangle \ln \langle \hat{n}_{k\sigma} \rangle}_{\text{Contribution of the electrons}} + \underbrace{\left(1 - \langle \hat{n}_{k\sigma} \rangle\right) \ln\left(1 - \langle \hat{n}_{k\sigma} \rangle\right)}_{\text{Contribution of the holes}} \Big] .$$

Behaviour for $T \to 0$:

$$\varepsilon(\boldsymbol{k}) > \mu \quad \Longrightarrow \quad \langle \hat{n}_{k\sigma} \rangle \underset{T \to 0}{\longrightarrow} 0 ; \quad \ln\left(1 - \langle \hat{n}_{k\sigma} \rangle\right) \underset{T \to 0}{\longrightarrow} 0 ,$$

$$\varepsilon(\boldsymbol{k}) < \mu \quad \Longrightarrow \quad \langle \hat{n}_{k\sigma} \rangle \underset{T \to 0}{\longrightarrow} 1 ; \quad \ln \langle \hat{n}_{k\sigma} \rangle \underset{T \to 0}{\longrightarrow} 0 .$$

From all this, the validity of the Third Law follows:

$$S \underset{T \to 0}{\longrightarrow} 0 .$$

Solution 2.1.6

1. Operator for the electron density:

$$\widehat{\rho}(\boldsymbol{r}) = \sum_{i=1}^{N} \delta\left(\boldsymbol{r} - \hat{\boldsymbol{r}}_i\right) .$$

Second quantisation with Wannier states $|i\sigma\rangle$:

$$\widehat{\rho}(\boldsymbol{r}) = \sum_{\substack{ij \\ \sigma\sigma'}} \langle i\sigma \mid \delta\left(\boldsymbol{r} - \hat{\boldsymbol{r}}'\right) \mid j\sigma' \rangle\, a_{i\sigma}^{+} a_{j\sigma} .$$

Matrix element:

$$\langle i\sigma \mid \delta\left(\boldsymbol{r}-\hat{\boldsymbol{r}}'\right) \mid j\sigma'\rangle = \sum_{\sigma''} \int \mathrm{d}^3r'' \, \langle i\sigma \mid \delta\left(\boldsymbol{r}-\hat{\boldsymbol{r}}'\right) \mid \boldsymbol{r}''\sigma''\rangle \langle \boldsymbol{r}''\sigma'' \mid j\sigma'\rangle =$$

$$= \sum_{\sigma''} \int \mathrm{d}^3r'' \, \delta\left(\boldsymbol{r}-\boldsymbol{r}''\right) \langle i\sigma \mid \boldsymbol{r}''\sigma''\rangle \langle \boldsymbol{r}''\sigma'' \mid j\sigma'\rangle =$$

$$= \sum_{\sigma''} \delta_{\sigma\sigma''}\delta_{\sigma''\sigma'} \langle i \mid \boldsymbol{r}\rangle \langle \boldsymbol{r} \mid j\rangle =$$

$$= \delta_{\sigma\sigma'}\omega^*\left(\boldsymbol{r}-\boldsymbol{R}_i\right)\omega\left(\boldsymbol{r}-\boldsymbol{R}_j\right)$$

$$\Longrightarrow \quad \widehat{\rho}(\boldsymbol{r}) = \sum_{ij\sigma} \left(\omega^*\left(\boldsymbol{r}-\boldsymbol{R}_i\right)\omega\left(\boldsymbol{r}-\boldsymbol{R}_j\right)\right) a_{i\sigma}^+ a_{j\sigma} \, .$$

2.

$$\int \mathrm{d}^3r \, \omega^*\left(\boldsymbol{r}-\boldsymbol{R}_i\right)\omega\left(\boldsymbol{r}-\boldsymbol{R}_j\right) = \delta_{ij}$$

$$\Longrightarrow \quad \int \mathrm{d}^3r \, \widehat{\rho}(\boldsymbol{r}) = \sum_{ij\sigma} \delta_{ij} a_{i\sigma}^+ a_{j\sigma} = \sum_{i\sigma} \hat{n}_{i\sigma}$$

$$\Longrightarrow \quad \widehat{N} = \int \mathrm{d}^3r \, \widehat{\rho}(\boldsymbol{r}) \, .$$

3. In the jellium model: Bloch functions $\Longrightarrow$ plane waves.

   Wannier functions:

$$\omega\left(\boldsymbol{r}-\boldsymbol{R}_i\right) = \frac{1}{\sqrt{N}} \sum_{\boldsymbol{k}} \mathrm{e}^{-\mathrm{i}\boldsymbol{k}\cdot\boldsymbol{R}_i} \frac{1}{\sqrt{V}} \mathrm{e}^{\mathrm{i}\boldsymbol{k}\cdot\boldsymbol{r}} \, .$$

   This implies that:

$$\langle i\sigma \mid \delta\left(\boldsymbol{r}-\hat{\boldsymbol{r}}'\right) \mid j\sigma'\rangle = \delta_{\sigma\sigma'} \frac{1}{VN} \sum_{\boldsymbol{k}\boldsymbol{k}'} \mathrm{e}^{-\mathrm{i}\boldsymbol{k}(\boldsymbol{r}-\boldsymbol{R}_i)} \mathrm{e}^{\mathrm{i}\boldsymbol{k}'(\boldsymbol{r}-\boldsymbol{R}_j)} =$$

$$= \delta_{\sigma\sigma'} \frac{1}{V} \sum_{\boldsymbol{q}} \mathrm{e}^{\mathrm{i}\boldsymbol{q}\cdot\boldsymbol{r}} \frac{1}{N} \sum_{\boldsymbol{k}} \mathrm{e}^{\mathrm{i}\boldsymbol{k}\cdot\boldsymbol{R}_i} \mathrm{e}^{-\mathrm{i}(\boldsymbol{k}+\boldsymbol{q})\cdot\boldsymbol{R}_j} \, .$$

   It then follows that:

$$\widehat{\rho}(\boldsymbol{r}) = \frac{1}{V} \sum_{\boldsymbol{q}} \widehat{\rho}_{\boldsymbol{q}} \mathrm{e}^{\mathrm{i}\boldsymbol{q}\cdot\boldsymbol{r}} \, ,$$

$$\widehat{\rho}_{\boldsymbol{q}} = \frac{1}{N} \sum_{ij\sigma} \sum_{\boldsymbol{k}} \mathrm{e}^{\mathrm{i}\boldsymbol{k}\cdot\boldsymbol{R}_i} \mathrm{e}^{-\mathrm{i}(\boldsymbol{k}+\boldsymbol{q})\cdot\boldsymbol{R}_j} a_{i\sigma}^+ a_{j\sigma} = \sum_{\boldsymbol{k},\sigma} a_{\boldsymbol{k}\sigma}^+ a_{\boldsymbol{k}+\boldsymbol{q}\sigma} \, .$$

Solution 2.1.7

$$\begin{aligned}\widehat{\rho}(\boldsymbol{r}) &= \sum_{\sigma',\sigma''}\iint \mathrm{d}^3r'\mathrm{d}^3r''\, \langle \boldsymbol{r}'\sigma' | \delta(\boldsymbol{r}-\hat{\boldsymbol{r}}) | \boldsymbol{r}''\sigma''\rangle \widehat{\psi}^+_{\sigma'}(\boldsymbol{r}')\,\widehat{\psi}_{\sigma''}(\boldsymbol{r}'') = \\ &= \sum_{\sigma',\sigma''}\iint \mathrm{d}^3r'\mathrm{d}^3r''\,\delta(\boldsymbol{r}-\boldsymbol{r}'')\,\langle \boldsymbol{r}'\sigma' \mid \boldsymbol{r}''\sigma''\rangle \widehat{\psi}^+_{\sigma'}(\boldsymbol{r}')\,\widehat{\psi}_{\sigma''}(\boldsymbol{r}'') = \\ &= \sum_{\sigma',\sigma''}\iint \mathrm{d}^3r'\mathrm{d}^3r''\,\delta(\boldsymbol{r}-\boldsymbol{r}'')\,\delta(\boldsymbol{r}'-\boldsymbol{r}'')\,\delta_{\sigma'\sigma''}\widehat{\psi}^+_{\sigma'}(\boldsymbol{r}')\,\widehat{\psi}_{\sigma''}(\boldsymbol{r}'') = \\ &= \sum_{\sigma}\widehat{\psi}^+_{\sigma}(\boldsymbol{r})\widehat{\psi}_{\sigma}(\boldsymbol{r})\,.\end{aligned}$$

Solution 2.1.8

The Coulomb interaction H_{ee}:

$$H_{\mathrm{ee}} = \frac{1}{2}\sum_{\substack{ijkl\\ \sigma_1,\dots,\sigma_4}} v(i\sigma_1, j\sigma_2; k\sigma_3, l\sigma_4)\, a^+_{i\sigma_1}a^+_{j\sigma_2}a_{l\sigma_4}a_{k\sigma_3}\,.$$

Matrix element:

$$\begin{aligned}v(i\sigma_1, j\sigma_2; k\sigma_3, l\sigma_4) &= \frac{e^2}{4\pi\varepsilon_0}\left\langle (i\sigma_1)^{(1)}(j\sigma_2)^{(2)}\right| \frac{1}{\hat{\boldsymbol{r}}^{(1)}-\hat{\boldsymbol{r}}'^{(2)}}\left|(k\sigma_3)^{(1)}(l\sigma_4)^{(2)}\right\rangle = \\ &= \delta_{\sigma_1\sigma_3}\delta_{\sigma_2\sigma_4}v(ij,kl)\,,\end{aligned}$$

$$\begin{aligned}v(ij,kl) &= \frac{e^2}{4\pi\varepsilon_0}\iint \mathrm{d}^3r_1\mathrm{d}^3r_2\left\langle i^{(1)}j^{(2)}\right|\frac{1}{\hat{\boldsymbol{r}}^{(1)}-\hat{\boldsymbol{r}}'^{(2)}}\left|\boldsymbol{r}_1^{(1)}\boldsymbol{r}_2^{(2)}\right\rangle\left\langle \boldsymbol{r}_1^{(1)}\boldsymbol{r}_2^{(2)}\right| k^{(1)}l^{(2)}\rangle = \\ &= \frac{e^2}{4\pi\varepsilon_0}\iint \mathrm{d}^3r_1\mathrm{d}^3r_2\,\frac{1}{|\boldsymbol{r}_1-\boldsymbol{r}_2|}\left\langle i^{(1)}j^{(2)}\middle|\boldsymbol{r}_1^{(1)}\boldsymbol{r}_2^{(2)}\right\rangle\left\langle \boldsymbol{r}_1^{(1)}\boldsymbol{r}_2^{(2)}\middle| k^{(1)}l^{(2)}\right\rangle,\end{aligned}$$

$$\langle \boldsymbol{r}\mid i\rangle = \omega(\boldsymbol{r}-\boldsymbol{R}_i)\,:\quad \text{Wannier function}$$

$$\Longrightarrow\quad v(ij,kl) = \frac{e^2}{4\pi\varepsilon_0}\iint \mathrm{d}^3r_1\mathrm{d}^3r_2\,\frac{\omega^*(\boldsymbol{r}_1-\boldsymbol{R}_i)\,\omega^*(\boldsymbol{r}_2-\boldsymbol{R}_j)\,\omega(\boldsymbol{r}_1-\boldsymbol{R}_k)\,\omega(\boldsymbol{r}_2-\boldsymbol{R}_l)}{|\boldsymbol{r}_1-\boldsymbol{r}_2|}\,.$$

Hamiltonian:

$$H = \sum_{ij\sigma}T_{ij}a^+_{i\sigma}a_{j\sigma} + \frac{1}{2}\sum_{\substack{ijkl\\ \sigma\sigma'}} v(ij;kl)a^+_{i\sigma}a^+_{j\sigma'}a_{l\sigma'}a_{k\sigma}\,.$$

The jellium model:

$$\omega(\boldsymbol{r}-\boldsymbol{R}_i)=\frac{1}{\sqrt{VN}}\sum_{\boldsymbol{k}} \mathrm{e}^{\mathrm{i}\boldsymbol{k}\cdot(\boldsymbol{r}-\boldsymbol{R}_i)} .$$

As explained in detail in Sect. 2.1.2, an explicit calculation of the Coulomb matrix element requires the introduction of a *convergence-inducing* factor:

$$\begin{aligned}
v_\alpha(ij;kl) &= \frac{1}{V^2N^2}\frac{e^2}{4\pi\varepsilon_0}\iint \mathrm{d}^3r_1\mathrm{d}^3r_2 \sum_{\boldsymbol{k}_1,\dots,\boldsymbol{k}_4} \mathrm{e}^{-\alpha|\boldsymbol{r}_2-\boldsymbol{r}_1|} \cdot \\
&\quad\cdot \frac{\mathrm{e}^{-\mathrm{i}\boldsymbol{k}_1(\boldsymbol{r}_1-\boldsymbol{R}_i)}\mathrm{e}^{-\mathrm{i}\boldsymbol{k}_2(\boldsymbol{r}_2-\boldsymbol{R}_j)}\mathrm{e}^{\mathrm{i}\boldsymbol{k}_3(\boldsymbol{r}_1-\boldsymbol{R}_k)}\mathrm{e}^{\mathrm{i}\boldsymbol{k}_4(\boldsymbol{r}_2-\boldsymbol{R}_l)}}{|\boldsymbol{r}_1-\boldsymbol{r}_2|} = \\
&= \frac{1}{N^2}\sum_{\boldsymbol{k}_1,\dots,\boldsymbol{k}_4} \mathrm{e}^{\mathrm{i}(\boldsymbol{k}_1\boldsymbol{R}_i+\boldsymbol{k}_2\boldsymbol{R}_j-\boldsymbol{k}_3\boldsymbol{R}_k-\boldsymbol{k}_4\boldsymbol{R}_l)} \cdot \\
&\quad\cdot \frac{e^2}{4\pi\varepsilon_0}\frac{1}{V^2}\iint \mathrm{d}^3r_1\mathrm{d}^3r_2 \frac{\mathrm{e}^{-\mathrm{i}(\boldsymbol{k}_1-\boldsymbol{k}_3)\cdot\boldsymbol{r}_1}\mathrm{e}^{-\mathrm{i}(\boldsymbol{k}_2-\boldsymbol{k}_4)\cdot\boldsymbol{r}_2}}{|\boldsymbol{r}_1-\boldsymbol{r}_2|}\mathrm{e}^{-\alpha|\boldsymbol{r}_1-\boldsymbol{r}_2|} .
\end{aligned}$$

The integrals were already computed in (2.56) and (2.59):

$$\begin{aligned}
v_\alpha(ij;kl) &= \\
&= \frac{1}{N^2}\sum_{\boldsymbol{k}_1,\dots,\boldsymbol{k}_4} \mathrm{e}^{\mathrm{i}(\boldsymbol{k}_1\cdot\boldsymbol{R}_i+\boldsymbol{k}_2\boldsymbol{R}_j-\boldsymbol{k}_3\boldsymbol{R}_k-\boldsymbol{k}_4\cdot\boldsymbol{R}_l)}\delta_{\boldsymbol{k}_1-\boldsymbol{k}_3,\,\boldsymbol{k}_4-\boldsymbol{k}_2}\frac{e^2}{\varepsilon_0 V\left[(\boldsymbol{k}_1-\boldsymbol{k}_3)^2+\alpha^2\right]} .
\end{aligned}$$

Solution 2.1.9

1.

$$\widehat{\rho}(\boldsymbol{r})=\frac{1}{V}\sum_{\boldsymbol{q}}\widehat{\rho}_{\boldsymbol{q}}\,\mathrm{e}^{\mathrm{i}\boldsymbol{q}\cdot\boldsymbol{r}} .$$

It then follows that:

$$\begin{aligned}
G(\boldsymbol{r},t) &= \frac{1}{N}\frac{1}{V^2}\sum_{\boldsymbol{q},\boldsymbol{q}'}\left\langle\widehat{\rho}_{\boldsymbol{q}}\,\widehat{\rho}_{\boldsymbol{q}'}(t)\right\rangle\int \mathrm{d}^3r'\,\mathrm{e}^{\mathrm{i}\boldsymbol{q}\cdot(\boldsymbol{r}'-\boldsymbol{r})}\mathrm{e}^{\mathrm{i}\boldsymbol{q}'\cdot\boldsymbol{r}'} = \\
&= \frac{1}{NV}\sum_{\boldsymbol{q},\boldsymbol{q}'}\left\langle\widehat{\rho}_{\boldsymbol{q}}\,\widehat{\rho}_{\boldsymbol{q}'}(t)\right\rangle\mathrm{e}^{-\mathrm{i}\boldsymbol{q}\cdot\boldsymbol{r}}\delta_{-\boldsymbol{q},\boldsymbol{q}'} = \\
&= \frac{1}{NV}\sum_{\boldsymbol{q}}\left\langle\widehat{\rho}_{\boldsymbol{q}}\,\widehat{\rho}_{-\boldsymbol{q}}(t)\right\rangle\mathrm{e}^{-\mathrm{i}\boldsymbol{q}\cdot\boldsymbol{r}} .
\end{aligned}$$

This is the conditional probability of finding a particle at time t at the position $\boldsymbol{r}$, when at time $t = 0$, one was located at $\boldsymbol{r} = 0$.
An homogeneous system:

$$\left\langle \rho\left(\boldsymbol{r}' - \boldsymbol{r}, 0\right) \rho\left(\boldsymbol{r}', t\right)\right\rangle = \left\langle \rho(-\boldsymbol{r}, 0)\rho(0, t)\right\rangle = \left\langle \rho(0, 0)\rho(\boldsymbol{r}, t)\right\rangle$$

$$\Longrightarrow \quad G(\boldsymbol{r}, t) = \frac{V}{N}\left\langle \rho(0, 0)\rho(\boldsymbol{r}, t)\right\rangle .$$

2.

$$G(\boldsymbol{r}, 0) = \frac{1}{N}\int \mathrm{d}^3r' \sum_{i,j}\left\langle \delta\left(\boldsymbol{r}' - \boldsymbol{r} - \hat{\boldsymbol{r}}_i\right)\delta\left(\boldsymbol{r}' - \hat{\boldsymbol{r}}_j\right)\right\rangle = \frac{1}{N}\sum_{i,j}\left\langle \delta\left(\boldsymbol{r} + \hat{\boldsymbol{r}}_i - \hat{\boldsymbol{r}}_j\right)\right\rangle =$$

$$= \frac{1}{N}\sum_i \delta(\boldsymbol{r}) + \frac{1}{N}\sum_{i,j}^{i\neq j}\left\langle \delta\left(\boldsymbol{r} + \hat{\boldsymbol{r}}_i - \hat{\boldsymbol{r}}_j\right)\right\rangle =$$

$$= \delta(\boldsymbol{r}) + \frac{1}{N}\sum_{i,j}^{i\neq j}\left\langle \delta\left(\boldsymbol{r} + \hat{\boldsymbol{r}}_i(0) - \hat{\boldsymbol{r}}_j(0)\right)\right\rangle .$$

By comparison, we find:

$$ng(\boldsymbol{r}) = \frac{1}{N}\sum_{i,j}^{i\neq j}\left\langle \delta\left(\boldsymbol{r} + \hat{\boldsymbol{r}}_i(0) - \hat{\boldsymbol{r}}_j(0)\right)\right\rangle .$$

$g(\boldsymbol{r})$ is a measure of the probability of finding two particles at a mutual distance r at a particular time.

3. The dynamic structure factor:

$$S(\boldsymbol{q}, \omega) = \int \mathrm{d}^3r \int_{-\infty}^{+\infty} \mathrm{d}t\, G(\boldsymbol{r}, t)\mathrm{e}^{\mathrm{i}(\boldsymbol{q}\cdot\boldsymbol{r} - \omega t)} =$$

$$\stackrel{1.}{=} \frac{1}{NV}\sum_{\boldsymbol{q}'}\int \mathrm{d}^3r \int_{-\infty}^{+\infty} \mathrm{d}t\, \mathrm{e}^{\mathrm{i}(\boldsymbol{q} - \boldsymbol{q}')\cdot\boldsymbol{r}}\mathrm{e}^{-\mathrm{i}\omega t}\left\langle \widehat{\rho}_{\boldsymbol{q}'}\widehat{\rho}_{-\boldsymbol{q}'}(t)\right\rangle =$$

$$= \frac{1}{N}\sum_{\boldsymbol{q}'}\int_{-\infty}^{+\infty} \mathrm{d}t\, \mathrm{e}^{-\mathrm{i}\omega t}\delta_{\boldsymbol{q},\boldsymbol{q}'}\left\langle \widehat{\rho}_{\boldsymbol{q}'}\widehat{\rho}_{-\boldsymbol{q}'}(t)\right\rangle$$

$$\Longrightarrow \quad S(\boldsymbol{q}, \omega) = \frac{1}{N}\int_{-\infty}^{+\infty} \mathrm{d}t\, \mathrm{e}^{-\mathrm{i}\omega t}\left\langle \widehat{\rho}_{\boldsymbol{q}}\,\widehat{\rho}_{-\boldsymbol{q}}(t)\right\rangle .$$

With

$$\frac{1}{2\pi}\int\limits_{-\infty}^{+\infty} d\omega\, e^{-i\omega t} = \delta(t)\,,$$

it then finally follows that:

$$S(\boldsymbol{q}) = \frac{2\pi}{N}\left\langle \widehat{\rho}_{\boldsymbol{q}}\widehat{\rho}_{-\boldsymbol{q}}\right\rangle .$$

4. $T = 0 \Longrightarrow$ averaging over the ground state $|E_0\rangle$.

$$S(\boldsymbol{q},\omega) = \frac{1}{N}\int\limits_{-\infty}^{+\infty} dt\, e^{-i\omega t}\left\langle E_0\right|\widehat{\rho}_{\boldsymbol{q}}\,\widehat{\rho}_{-\boldsymbol{q}}(t)\left|E_0\right\rangle .$$

Time dependence:

$$\widehat{\rho}_{-\boldsymbol{q}}(t) = e^{\frac{i}{\hbar}Ht}\,\widehat{\rho}_{-\boldsymbol{q}}\,e^{-\frac{i}{\hbar}Ht}\,.$$

Completeness:

$$\mathbf{1} = \sum_n \left|E_n\right\rangle\left\langle E_n\right| .$$

This implies that:

$$\begin{aligned}
S(\boldsymbol{q},\omega) &= \frac{1}{N}\int\limits_{-\infty}^{+\infty} dt\, e^{-i\omega t}\sum_n \left\langle E_0\right|\widehat{\rho}_{\boldsymbol{q}}\left|E_n\right\rangle\left\langle E_n\right|\widehat{\rho}_{-\boldsymbol{q}}(t)\left|E_0\right\rangle = \\
&= \frac{1}{N}\sum_n \left\langle E_0\right|\widehat{\rho}_{\boldsymbol{q}}\left|E_n\right\rangle\left\langle E_n\right|\widehat{\rho}_{-\boldsymbol{q}}\left|E_0\right\rangle\int\limits_{-\infty}^{+\infty} dt\, e^{-i\omega t}\, e^{\frac{i}{\hbar}(E_n-E_0)t} = \\
&= \frac{2\pi}{N}\sum_n \left\langle E_0\right|\widehat{\rho}_{\boldsymbol{q}}\left|E_n\right\rangle\left\langle E_n\right|\widehat{\rho}_{-\boldsymbol{q}}\left|E_0\right\rangle\delta\left[\omega - \frac{1}{\hbar}(E_n - E_0)\right].
\end{aligned}$$

We further make use of $\widehat{\rho}_{-\boldsymbol{q}} = \widehat{\rho}_{\boldsymbol{q}}^{+}$:

$$S(\boldsymbol{q},\omega) = \frac{2\pi}{N}\sum_n \left|\left\langle E_n\right|\widehat{\rho}_{\boldsymbol{q}}^{+}\left|E_0\right\rangle\right|^2\delta\left[\omega - \frac{1}{\hbar}(E_n - E_0)\right].$$

Solution 2.1.10

1. The operator

$$\widehat{\rho}_{q}^{+} = \sum_{k\sigma} a_{k+q\sigma}^{+} a_{k\sigma}$$

creates particle-hole pairs. The ground state $|E_0\rangle$ corresponds to a *filled Fermi sphere*.

$$q = 0: \quad \left\langle E_n \,|\, \widehat{\rho}_q^+ \,|\, E_0 \right\rangle = N\delta_{n,0}\,,$$

$$q \neq 0: \quad \left\langle E_n \,|\, \widehat{\rho}_q^+ \,|\, E_0 \right\rangle = \begin{cases} 1\,, & \text{when } |E_n\rangle \text{ corresponds to a particle-hole excitation, which occurs in } \widehat{\rho}_q^+\,, \\ 0\,, & \text{otherwise.} \end{cases}$$

With the general result 4. from Ex. 2.1.9, it then follows that:

$$S(\boldsymbol{q}, \omega) \overset{(q\neq 0)}{=} 2\frac{2\pi}{N}\sum_{\boldsymbol{k}} \Theta\left(k_{\mathrm{F}} - k\right)\left[1 - \Theta\left(k_{\mathrm{F}} - |\boldsymbol{k}+\boldsymbol{q}|\right)\right] \cdot$$

$$\cdot\, \delta\left[\omega - \frac{1}{\hbar}\left(\varepsilon(\boldsymbol{k}+\boldsymbol{q}) - \varepsilon(\boldsymbol{k})\right)\right],$$

$$S(0, \omega) = 2\pi N\delta(\omega)\,.$$

The term $q = 0$ drops out in the jellium model. We therefore assume from now on that $q \neq 0$:

$$S(\boldsymbol{q}) = \int_{-\infty}^{+\infty} \mathrm{d}\omega\, S(\boldsymbol{q}, \omega) =$$

$$= \frac{4\pi}{N}\sum_{\boldsymbol{k}} \left[\Theta\left(k_{\mathrm{F}} - k\right) - \Theta\left(k_{\mathrm{F}} - k\right)\Theta\left(k_{\mathrm{F}} - |\boldsymbol{k}+\boldsymbol{q}|\right)\right] =$$

$$= 2\pi - \frac{4\pi}{N}\sum_{\boldsymbol{k}} \Theta\left(k_{\mathrm{F}} - k\right)\Theta\left(k_{\mathrm{F}} - |\boldsymbol{k}+\boldsymbol{q}|\right) =$$

$$= 2\pi\left[1 - \frac{2}{N}\frac{V}{(2\pi)^3}\int \mathrm{d}^3k\, \Theta\left(k_{\mathrm{F}} - k\right)\Theta\left(k_{\mathrm{F}} - |\boldsymbol{k}+\boldsymbol{q}|\right)\right].$$

The integral on the right-hand side was computed in Sect. 2.1.2. We adopt (2.95):

$$S(\boldsymbol{q}) = 2\pi\left[1 - \frac{2}{N}\frac{V}{8\pi^3}\frac{4\pi}{3}\Theta\left(2k_{\mathrm{F}} - q\right)\left(k_{\mathrm{F}}^3 - \frac{3}{4}qk_{\mathrm{F}}^2 + \frac{1}{16}q^3\right)\right].$$

With $k_F^3 = 3\pi^2 \frac{N}{V}$, we finally obtain:

$$S(q) = 2\pi \left[1 - \Theta\left(2k_F - q\right)\left(1 - \frac{3q}{4k_F} + \frac{q^3}{16k_F^3}\right)\right].$$

$S(q = 0) = 2\pi N.$

Fig. A.4

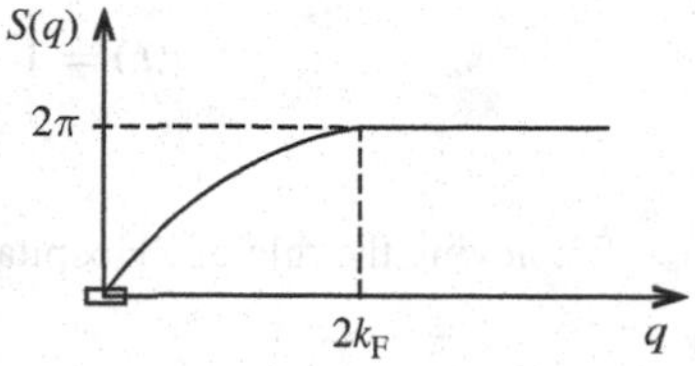

2.

$$\begin{aligned} G(\boldsymbol{r}, 0) &\overset{1.}{=} \frac{1}{NV} \sum_{\boldsymbol{q}} \langle \widehat{\rho}_{\boldsymbol{q}} \widehat{\rho}_{-\boldsymbol{q}} \rangle \mathrm{e}^{-\mathrm{i}\boldsymbol{q}\cdot\boldsymbol{r}} = \frac{1}{2\pi V} \sum_{\boldsymbol{q}} S(\boldsymbol{q}) \mathrm{e}^{-\mathrm{i}\boldsymbol{q}\cdot\boldsymbol{r}} = \\ &= \frac{2\pi N}{2\pi V} + \frac{1}{V} \sum_{\boldsymbol{q}} \mathrm{e}^{-\mathrm{i}\boldsymbol{q}\cdot\boldsymbol{r}} - \frac{2}{VN} \sum_{\boldsymbol{q}} \sum_{\boldsymbol{k}} \Theta\left(k_F - k\right) \Theta\left(k_F - |\boldsymbol{k} + \boldsymbol{q}|\right) \mathrm{e}^{-\mathrm{i}\boldsymbol{q}\cdot\boldsymbol{r}}. \end{aligned}$$

In the last two terms, we can add in the missing term $q = 0$, since it just cancels:

$$G(\boldsymbol{r}, 0) = n + \delta(\boldsymbol{r}) - \frac{2}{VN} \sum_{\boldsymbol{k}} \sum_{\boldsymbol{p}} \mathrm{e}^{-\mathrm{i}(\boldsymbol{p}-\boldsymbol{k})\cdot\boldsymbol{r}} \Theta\left(k_F - k\right) \Theta\left(k_F - p\right) \overset{!}{=} \delta(\boldsymbol{r}) + n g(\boldsymbol{r}).$$

From this, it follows that:

$$\begin{aligned} g(\boldsymbol{r}) &= 1 - \frac{2}{n^2} \frac{1}{(2\pi)^6} \iint \mathrm{d}^3p\, \mathrm{d}^3k\, \mathrm{e}^{-\mathrm{i}(\boldsymbol{p}-\boldsymbol{k})\cdot\boldsymbol{r}} \Theta\left(k_F - k\right) \Theta\left(k_F - p\right) = \\ &= 1 - \frac{2}{n^2} \left[\frac{1}{(2\pi)^3} \int \mathrm{d}^3p\, \mathrm{e}^{-\mathrm{i}\boldsymbol{p}\cdot\boldsymbol{r}} \Theta\left(k_F - p\right)\right]^2 = \\ &= 1 - \frac{2}{n^2} \left[\frac{1}{4\pi^2} \int_0^{k_F} \mathrm{d}p\, p^2 \frac{1}{-\mathrm{i}pr} \left(\mathrm{e}^{-\mathrm{i}pr} - \mathrm{e}^{\mathrm{i}pr}\right)\right]^2 = \\ &= 1 - \frac{2}{n^2} \left[\frac{1}{2\pi^2} \frac{1}{r} \int_0^{k_F} \mathrm{d}p\, p \sin pr\right]^2 = \end{aligned}$$

$$= 1 - \frac{1}{2\pi^4 n^2} \frac{1}{r^2} \left(\frac{\sin k_F r}{r^2} - \frac{k_F}{r} \cos k_F r \right)^2 =$$
$$= 1 - \frac{k_F^6}{2\pi^4 n^2} \left[\frac{\sin k_F r - (k_F r) \cos k_F r}{k_F^3 r^3} \right]^2 .$$

We thus have as final result:

$$g(r) = 1 - \frac{9}{2} \left[\frac{\sin k_F r - (k_F r) \cos k_F r}{k_F^3 r^3} \right]^2 .$$

Employing the rule of l'Hospital, one can show that:

$$\frac{\sin x - x \cos x}{x^3} \underset{x \to 0}{\longrightarrow} \frac{1}{3}$$
$$\Longrightarrow \quad g(r) \underset{r \to 0}{\longrightarrow} \frac{1}{2} \quad \textit{Fermi hole,}$$
$$g(r) \underset{r \to \infty}{\longrightarrow} 1 .$$

Fig. A.5

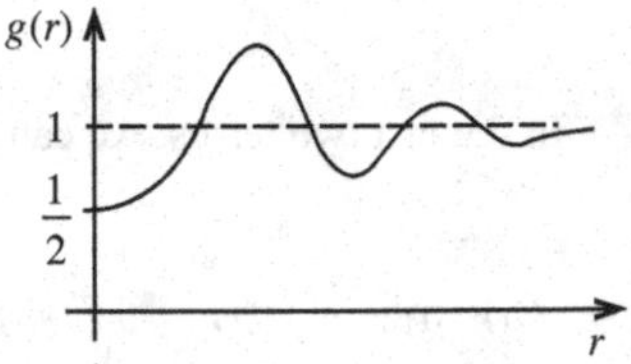

The *Fermi hole* results from the Pauli principle, which requires that two electrons with parallel spins cannot approach each other too closely. The value $g(r = 0) = 1/2$ is nevertheless not reasonable. The Sommerfeld model neglects the Coulomb interactions, so that two electrons with antiparallel spins could in principle approach each other arbitrarily closely.

Solution 2.1.11

$$\varepsilon(\boldsymbol{k}) = T_0 + \gamma_1 \sum_{\Delta} e^{i\boldsymbol{k} \cdot \boldsymbol{R}_\Delta} .$$

1. Body-centered cubic (b.c.c.)

   Number of nearest neighbours: $z_1 = 8$

$$\boldsymbol{R}_\Delta = \frac{a}{2}(\pm 1, \pm 1, \pm 1)$$

$$a: \quad \text{Lattice constant,}$$

$$\sum_\Delta \mathrm{e}^{\mathrm{i}\boldsymbol{k}\cdot\boldsymbol{R}_\Delta} = \left(\mathrm{e}^{\mathrm{i}k_x\frac{a}{2}} + \mathrm{e}^{-\mathrm{i}k_x\frac{a}{2}}\right)\left(\mathrm{e}^{\mathrm{i}k_y\frac{a}{2}} + \mathrm{e}^{-\mathrm{i}k_y\frac{a}{2}}\right)\left(\mathrm{e}^{\mathrm{i}k_z\frac{a}{2}} + \mathrm{e}^{-\mathrm{i}k_z\frac{a}{2}}\right)$$

$$\Longleftrightarrow \varepsilon_{\text{b.c.c.}}(\boldsymbol{k}) = T_0 + 8\gamma_1 \cos\left(\frac{1}{2}k_x a\right)\cos\left(\frac{1}{2}k_y a\right)\cos\left(\frac{1}{2}k_z a\right).$$

2. Face-centered cubic (f.c.c.)

   $z_1 = 12$

$$\boldsymbol{R}_\Delta = \frac{a}{2}(\pm 1, \pm 1, 0); \quad \frac{a}{2}(\pm 1, 0, \pm 1); \quad \frac{a}{2}(0, \pm 1, \pm 1),$$

$$\begin{aligned}\sum_\Delta \mathrm{e}^{\mathrm{i}\boldsymbol{k}\cdot\boldsymbol{R}_\Delta} &= \left(\mathrm{e}^{\mathrm{i}k_x\frac{a}{2}} + \mathrm{e}^{-\mathrm{i}k_x\frac{a}{2}}\right)\left(\mathrm{e}^{\mathrm{i}k_y\frac{a}{2}} + \mathrm{e}^{-\mathrm{i}k_y\frac{a}{2}}\right) + \\ &+ \left(\mathrm{e}^{\mathrm{i}k_x\frac{a}{2}} + \mathrm{e}^{-\mathrm{i}k_x\frac{a}{2}}\right)\left(\mathrm{e}^{\mathrm{i}k_z\frac{a}{2}} + \mathrm{e}^{-\mathrm{i}k_z\frac{a}{2}}\right) + \\ &+ \left(\mathrm{e}^{\mathrm{i}k_y\frac{a}{2}} + \mathrm{e}^{-\mathrm{i}k_y\frac{a}{2}}\right)\left(\mathrm{e}^{\mathrm{i}k_z\frac{a}{2}} + \mathrm{e}^{-\mathrm{i}k_z\frac{a}{2}}\right)\end{aligned}$$

$$\begin{aligned}\Longleftrightarrow \varepsilon_{\text{f.c.c.}}(\boldsymbol{k}) = T_0 + 4\gamma_1 \Bigg[&\cos\left(\frac{1}{2}k_x a\right)\cos\left(\frac{1}{2}k_y a\right) + \\ &+ \cos\left(\frac{1}{2}k_x a\right)\cos\left(\frac{1}{2}k_z a\right) + \cos\left(\frac{1}{2}k_y a\right)\cos\left(\frac{1}{2}k_z a\right)\Bigg].\end{aligned}$$

Solution 2.1.12

Tight-binding approach:

$$\psi_{n\boldsymbol{k}}(\boldsymbol{r}) = \frac{1}{\sqrt{N_i}}\sum_{j=1}^{N_i} \mathrm{e}^{\mathrm{i}\boldsymbol{k}\cdot\boldsymbol{R}_j}\varphi_n\left(\boldsymbol{r} - \boldsymbol{R}_j\right)$$

$$\Longrightarrow \quad \psi_{n\boldsymbol{k}}\left(\boldsymbol{r} + \boldsymbol{R}_i\right) = \frac{1}{\sqrt{N_i}}\sum_{j=1}^{N_i} \mathrm{e}^{\mathrm{i}\boldsymbol{k}\cdot\boldsymbol{R}_j}\varphi_n\left(\boldsymbol{r} + \boldsymbol{R}_i - \boldsymbol{R}_j\right)$$

$$(\text{Substitution: } \boldsymbol{R}_k = \boldsymbol{R}_j - \boldsymbol{R}_i)$$

$$\begin{aligned}\Longrightarrow \quad \psi_{n\boldsymbol{k}}\left(\boldsymbol{r} + \boldsymbol{R}_i\right) &= \mathrm{e}^{\mathrm{i}\boldsymbol{k}\cdot\boldsymbol{R}_i}\frac{1}{\sqrt{N_i}}\sum_{k=1}^{N_i} \mathrm{e}^{\mathrm{i}\boldsymbol{k}\cdot\boldsymbol{R}_k}\varphi_n\left(\boldsymbol{r} - \boldsymbol{R}_k\right) = \\ &= \mathrm{e}^{\mathrm{i}\boldsymbol{k}\cdot\boldsymbol{R}_i}\psi_{n\boldsymbol{k}}(\boldsymbol{r}).\end{aligned}$$

This is Bloch's theorem!

Section 2.2.3

Solution 2.2.1

1. Linear Bravais lattice with a two-atom basis:

Fig. A.6

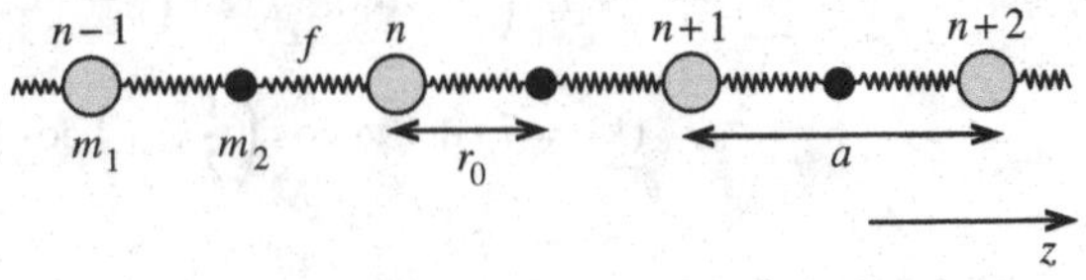

Primitive translations:

$$\boldsymbol{a} = a\boldsymbol{e}_z = 2r_0\boldsymbol{e}_z \, .$$

Basis:

$$\boldsymbol{R}_1 = 0, \quad \boldsymbol{R}_2 = r_0\boldsymbol{e}_z \, .$$

Lattice vectors:

$$R_1^n = na \, ; \quad R_2^m = \left(m + \frac{1}{2}\right)a \, ; \quad n, m \in \mathbf{Z} \, .$$

Primitive translations in the reciprocal lattice:

$$\boldsymbol{b} = b\boldsymbol{e}_z \, ; \quad b = \frac{2\pi}{a} \, .$$

First Brillouin zone:

$$-\frac{\pi}{a} \le q \le +\frac{\pi}{a} \, .$$

Reciprocal lattice vectors:

$$G^m = mb \, .$$

2. *Longitudinal waves*, i.e. the motion of the molecules along the chain consists of displacements parallel to the chain's direction.
   Force on the $(n, 1)$-th atom in the z-direction:

   from the right: $f\left(u_2^n - u_1^n\right)$; u : displacement from the equilibrium position

   from the left: $f\left(u_2^{n-1} - u_1^n\right)$

$\Longrightarrow$ Equation of motion for the $(n, 1)$-th atom:

$$m_1 \ddot{u}_1^n = f\left(u_2^n + u_2^{n-1} - 2u_1^n\right).$$

Analogously for the $(n, 2)$-th atom:

$$m_2 \ddot{u}_2^n = f\left(u_1^{n+1} + u_1^n - 2u_2^n\right).$$

3. This approach contains translational invariance with respect to the two-atom unit cell and takes into account the fact that the amplitudes can differ, owing to the different masses of the particles. Inserting into the above equations of motion yields the following system of equations:

$$m_1 \frac{c_1}{\sqrt{m_1}}(-\omega^2) = \frac{f c_2}{\sqrt{m_2}}\left(1 + \mathrm{e}^{-iqa}\right) - 2\frac{c_2}{\sqrt{m_1}} f,$$

$$m_2 \frac{c_2}{\sqrt{m_2}}(-\omega^2) = \frac{f c_1}{\sqrt{m_1}}\left(\mathrm{e}^{iqa} + 1\right) - 2\frac{c_2}{\sqrt{m_2}} f.$$

The secular equation of the homogeneous system of equations

$$0 = \begin{pmatrix} \frac{2f}{m_1} - \omega^2 & \frac{-f}{\sqrt{m_1 m_2}}\left(1 + \mathrm{e}^{-iqa}\right) \\ \frac{-f}{\sqrt{m_1 m_2}}\left(1 + \mathrm{e}^{iqa}\right) & \frac{2f}{m_2} - \omega^2 \end{pmatrix} \begin{pmatrix} c_1 \\ c_2 \end{pmatrix}$$

yields the eigenfrequencies (*branches of the dispersion relations*):

$$\omega_{1,2}^2(q) = f\left[\left(\frac{1}{m_1} + \frac{1}{m_2}\right) \pm \sqrt{\left(\frac{1}{m_1} + \frac{1}{m_2}\right)^2 - \frac{2}{m_1 m_2}(1 - \cos qa)}\,\right].$$

The two branches of the dispersion relations are periodic in q with a period of $\frac{2\pi}{a}$. For an arbitrary reciprocal-lattice vector $\boldsymbol{G}$,

$$G^m = m\boldsymbol{b} = m\frac{2\pi}{a}\boldsymbol{e}_z,$$

we clearly have:

$$\omega(q) = \omega(q + G^m).$$

All the physical information can thus be derived from only the first Brillouin zone,

$$-\frac{\pi}{a} \le q \le +\frac{\pi}{a}.$$

q's which lie outside it can be transformed into the first Brillouin zone by addition of a suitable reciprocal-lattice vector without affecting the dispersion relations.

4.

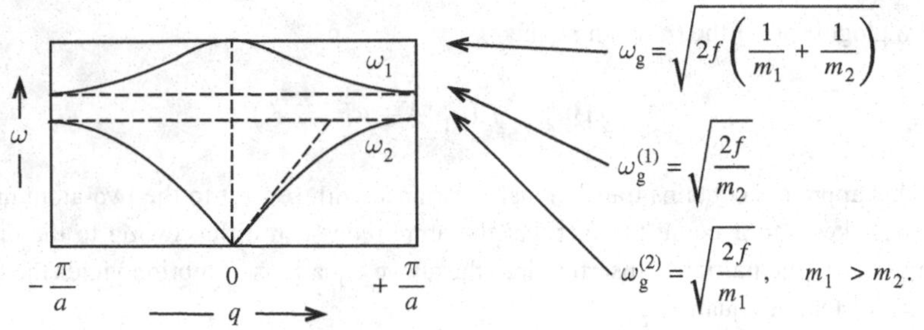

ω_1
ω_2
ω
$-\frac{\pi}{a}$
0
$+\frac{\pi}{a}$
q
$\omega_g = \sqrt{2f\left(\frac{1}{m_1} + \frac{1}{m_2}\right)}$
$\omega_g^{(1)} = \sqrt{\frac{2f}{m_2}}$
$\omega_g^{(2)} = \sqrt{\frac{2f}{m_1}}, \quad m_1 > m_2.$

Special cases:

a) $q = 0;\ \omega = \omega_2 \quad \Longrightarrow \quad \omega_2(q = 0) = 0$
From the system of homogeneous equations, we then find for the amplitudes:

$$\frac{c_1}{c_2} = \sqrt{\frac{m_1}{m_2}}\,.$$

Basis atoms oscillate *in phase*, but with different amplitudes.

b) $q \ll \frac{\pi}{a};\ \omega = \omega_2$

$$\omega_2^2 \approx f\left[\left(\frac{1}{m_1} + \frac{1}{m_2}\right) - \left(\frac{1}{m_1} + \frac{1}{m_2}\right)\sqrt{1 - \frac{m_1 m_2}{(m_1 + m_2)^2}(qa)^2}\,\right] \approx$$

$$\approx f\left[\left(\frac{1}{m_1} + \frac{1}{m_2}\right) - \left(\frac{1}{m_1} + \frac{1}{m_2}\right) + \frac{(qa)^2}{2\,(m_1 + m_2)}\right]$$

$$\Longrightarrow \quad \omega_2 \approx a\sqrt{\frac{f}{2\,(m_1 + m_2)}}\,q\,.$$

$q \ll \pi/a$ means that $\lambda \gg 2a$. For wavelengths in this range, the atomic structure of the solid is unimportant and a continuum theory can be applied as a good approximation. It yields for *sound waves* the relation:

$$\omega = v_s q \quad (v_s = \text{velocity of sound})\,.$$

Thus the lower dispersion branch ω_2 describes normal sound waves for long wavelengths (small values of q); it is therefore called the

acoustic branch.

c) $q = 0,\ \omega = \omega_1$

$$\Longrightarrow \quad \omega_1(q=0) = \omega_g = \sqrt{2f\left(\frac{1}{m_1}+\frac{1}{m_2}\right)}$$

limiting frequency of the spectrum.

We now obtain for the amplitudes:

$$\frac{c_1}{c_2} = -\sqrt{\frac{m_2}{m_1}}\,.$$

Basis atoms oscillate with different amplitudes and *opposite phases*. If the basis atoms have opposite electric charges (e.g. in a NaCl crystal), then an oscillating electric dipole moment results. It can interact with electromagnetic radiation and absorb or emit electromagnetic waves. Therefore, one refers to ω_1 as the

optical branch.

d) Zone boundary: $q = \pm\frac{\pi}{a}$

$$\omega_g^{(1)} = \omega_1\left(q=\pm\frac{\pi}{a}\right) = \sqrt{\frac{2f}{m_2}} \quad \text{(optical)}\,,$$

$$\omega_g^{(2)} = \omega_2\left(q=\pm\frac{\pi}{a}\right) = \sqrt{\frac{2f}{m_1}} \quad \text{(acoustic)}\,.$$

It follows from the system of homogeneous equations that:

$$\omega = \omega_g^{(1)} \quad \Longrightarrow \quad c_1 = 0: \quad \text{only } m_2 \text{ atoms oscillate,}$$

$$\omega = \omega_g^{(2)} \quad \Longrightarrow \quad c_2 = 0: \quad \text{only } m_1 \text{ atoms oscillate.}$$

A typical characteristic of the diatomic chain is the

$$\textit{frequency gap:} \quad \omega_g^{(2)} < \omega < \omega_g^{(1)}\,.$$

Solutions with real ω in the gap have complex wavenumbers q. The waves are then spatially damped.

Solution 2.2.2

Trial solution:

$$x_\alpha^n(t) = na + u_\alpha^n(t)\,,$$

$$u_\alpha^n(t) = \frac{c_\alpha}{\sqrt{m_\alpha}}\exp\left[\mathrm{i}\,(q_z na - \omega t)\right]\,.$$

Periodic boundary conditions:

$$u_\alpha^n(t) \stackrel{!}{=} u_\alpha^{n+N}(t)$$

$$\iff \quad e^{iNq_z a} \stackrel{!}{=} 1\,,$$

$$q_z = \bar{n}\frac{2\pi}{Na}\,; \quad \bar{n} = 0,\ \pm 1,\ \pm 2, \ldots, +\frac{N}{2}\,.$$

The term $-N/2$ is not counted, since q_z changes from $-N/2$ to $+N/2$ by just $2\pi/a$, i.e. by a reciprocal lattice vector.

$$\omega(q_z) = \omega(-q_z) = \omega(q)\,; \quad q = |q_z|$$

$$\Longrightarrow \quad D(\omega)\mathrm{d}\omega = D(q)\,\mathrm{d}q = 2D(q_z)\,\mathrm{d}q_z\,.$$

Every value q is associated twice with the frequency ω, but each wavenumber component $q_z = \pm q$ is associated with the frequency only once.

$$D(q_z) = \frac{1}{\frac{2\pi}{Na}} = \frac{Na}{2\pi}\,, \quad (D(q_z)\,:\ \text{number of } q_z \text{ per unit wavenumber})\,,$$

$$v_g = \frac{\mathrm{d}\omega}{\mathrm{d}q_z} \quad \Longrightarrow \quad D(\omega) = 2D(q_z)\frac{\mathrm{d}q_z}{\mathrm{d}\omega} = \frac{Na}{\pi}\frac{1}{v_g}\,.$$

When there are several dispersion branches, we then find all together:

$$D(\omega) = \frac{Na}{\pi}\sum_{s=1}^{3p}\frac{1}{v_g^{(s)}}\,.$$

Fig. A.7

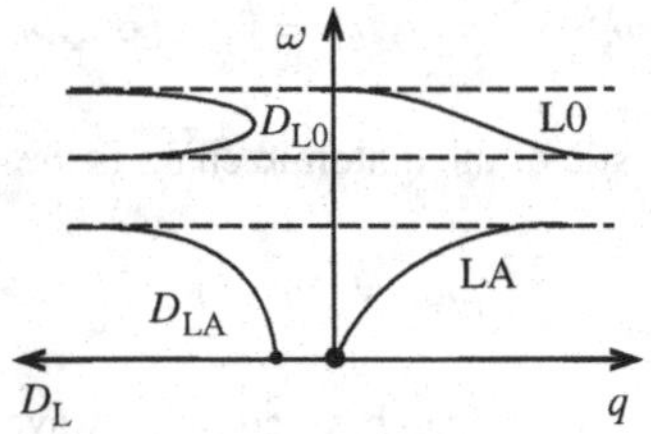

Solution 2.2.3

1. $\boldsymbol{a}_1, \boldsymbol{a}_2, \boldsymbol{a}_3$: primitive translations

$$V = (N_1\boldsymbol{a}_1)\cdot[(N_2\boldsymbol{a}_2)\times(N_3\boldsymbol{a}_3)] = NV_z$$

periodicity volume,

$$V_z = \boldsymbol{a}_1 \cdot [\boldsymbol{a}_2 \times \boldsymbol{a}_3]$$

unit cell,

$$N = N_1 N_2 N_3$$

number of primitive unit cells in the periodicity volume
= number of Bravais lattice points in the crystal.

Periodic boundary conditions: For the displacements from the equilibrium positions, we require:

$$u_{\mathrm{S},i}^{(m_1,m_2,m_3)} \stackrel{!}{=} u_{\mathrm{S},i}^{(m_1+N_1,m_2,m_3)} \stackrel{!}{=} u_{\mathrm{S},i}^{(m_1,m_2+N_2,m_3)} \stackrel{!}{=}$$

$$\stackrel{!}{=} u_{\mathrm{S},i}^{(m_1,m_2,m_3+N_3)}$$

$$\Longrightarrow \quad \boldsymbol{q} \cdot \boldsymbol{a}_i = \frac{2\pi}{N_i} n_i , \quad i = 1, 2, 3 ,$$

$$n_i = 0, \pm 1, \pm 2, \ldots , +\frac{N_i}{2} .$$

There are thus $N = N_1 N_2 N_3$ different wavenumbers $\boldsymbol{q}$:

$$\boldsymbol{q} = \sum_{j=1}^{3} \frac{n_j}{N_j} \boldsymbol{b}_j , \quad \boldsymbol{b}_j : \text{ primitive translations of the reciprocal lattice.}$$

2. *Grid volume*

$$\Delta^3 q = \frac{1}{N_1 N_2 N_3} \boldsymbol{b}_1 \cdot (\boldsymbol{b}_2 \times \boldsymbol{b}_3) \equiv \frac{V_z^*}{N} ,$$

$$\boldsymbol{b}_1 = \frac{2\pi}{V_z} (\boldsymbol{a}_2 \times \boldsymbol{a}_3) \quad \text{and cyclic permutations}$$

$$(\boldsymbol{a}_2 \times \boldsymbol{a}_3) \cdot (\boldsymbol{b}_2 \times \boldsymbol{b}_3) = (\boldsymbol{a}_2 \cdot \boldsymbol{b}_2)(\boldsymbol{a}_3 \cdot \boldsymbol{b}_3) - (\boldsymbol{a}_2 \cdot \boldsymbol{b}_3)(\boldsymbol{a}_3 \cdot \boldsymbol{b}_2) = (2\pi)^2$$

$$\Longrightarrow \quad \Delta^3 q = \frac{1}{N} \frac{(2\pi)^3}{V_z} ; \quad V_z^* = \frac{(2\pi)^3}{V_z} .$$

3.

$$D_r(\omega)\,\mathrm{d}\omega = \frac{1}{\Delta^3 q} \int\limits_{\substack{\text{shell} \\ (\omega_r,\, \omega_r + \mathrm{d}\omega)}} \mathrm{d}^3 q = \frac{V}{(2\pi)^3} \int\limits_{\substack{\text{shell} \\ (\omega_r,\, \omega_r + \mathrm{d}\omega)}} \mathrm{d}^3 q .$$

4.

$$\mathrm{d}f_\omega = \text{surface-area element } \omega = \text{const in } \boldsymbol{q}\text{-space},$$

$$\nabla_{\boldsymbol{q}}\omega = \text{vector perpendicular to the surface } \omega(\boldsymbol{q}) = \omega = \text{const}$$

$$\Longrightarrow \quad \mathrm{d}\omega = \left|\mathrm{d}\boldsymbol{q}\cdot\nabla_{\boldsymbol{q}}\omega\right| = \mathrm{d}q_\perp\,|\nabla_{\boldsymbol{q}}\omega| = v_\mathrm{g}\mathrm{d}q_\perp$$

$$\Longrightarrow \quad \text{volume element within the shell:} \quad \mathrm{d}^3q = \mathrm{d}f_\omega\mathrm{d}q_\perp = \frac{1}{v_\mathrm{g}}\mathrm{d}\omega\,\mathrm{d}f_\omega$$

$$\Longrightarrow \quad \text{density of states:} \quad D_r(\omega) = \frac{V}{(2\pi)^3}\int\limits_{\omega=\text{const}}\frac{\mathrm{d}f_\omega}{v_\mathrm{g}^{(r)}}\,.$$

5. Overall density of states:

$$D(\omega) = \sum_{r=1}^{3p} D_r(\omega)\,.$$

Solution 2.2.4

For the density of states, we make use of the expression obtained in part 4. of Ex. 2.2.3:

$$\text{group velocity:} \quad v_\mathrm{g}^{(r)} = \bar{v}_r\,,$$

$$\text{Bravais lattice:} \quad p = 1 \quad \Longrightarrow \quad r = 1, 2, 3\,.$$

There are one longitudinal acoustic and two (generally degenerate) transverse acoustic dispersion branches:

$$\int\limits_{\omega=\text{const}}\frac{\mathrm{d}f_\omega}{v_\mathrm{g}^{(r)}} \quad \Longrightarrow \quad \frac{1}{\bar{v}_r}\int\limits_{\omega=\text{const}}\mathrm{d}f_\omega = \frac{1}{\bar{v}_r}4\pi q^2(\omega) = \frac{4\pi\,\omega^2}{\bar{v}_r^3}\,.$$

Density of states:

$$D_r^\mathrm{D}(\omega) = \begin{cases} \dfrac{V}{2\pi^2\bar{v}_r^3}\omega^2 & \text{for} \quad 0 \le \omega \le \omega_r^\mathrm{D}\,, \\ 0 & \text{otherwise.} \end{cases}$$

Debye frequency:
Condition: the number of possible frequencies per dispersion branch = N,

$$N = \int\limits_0^{\omega_r^\mathrm{D}}\mathrm{d}\omega\,D_r^\mathrm{D}(\omega) = \frac{V}{6\pi^2\bar{v}_r^3}\left(\omega_r^\mathrm{D}\right)^3$$

$$\Longrightarrow \quad \omega_r^\mathrm{D} = \bar{v}_r\left(6\pi^2\frac{N}{V}\right)^{1/3}\,.$$

Solution 2.2.5

1. *Internal energy:*

$$U(T) = \langle H \rangle = \sum_{r=1}^{3p} \sum_{q}^{1.\,\mathrm{BZ}} \hbar\omega_r(\boldsymbol{q}) \left(\left\langle b_{qr}^{+} b_{qr} \right\rangle + \frac{1}{2} \right) ,$$

$$\left\langle b_{qr}^{+} b_{qr} \right\rangle = \left\{ \exp\left(\beta\hbar\omega_r(\boldsymbol{q})\right) - 1 \right\}^{-1} .$$

a) *High temperatures:* $k_\mathrm{B} T \gg \hbar\omega_r(\boldsymbol{q})$

$$\left\langle b_{qr}^{+} b_{qr} \right\rangle = \left\{ 1 + \frac{\hbar\omega_r(\boldsymbol{q})}{k_\mathrm{B}T} + \dots - 1 \right\}^{-1} \approx \frac{k_\mathrm{B}T}{\hbar\omega_r(\boldsymbol{q})}$$

$$\Longrightarrow \quad U(T) \approx \sum_{r,q} k_\mathrm{B}T \left(1 + \frac{1}{2} \frac{\hbar\omega_r(\boldsymbol{q})}{k_\mathrm{B}T} + \cdots \right) \approx 3pNk_\mathrm{B}T .$$

This is the well-known classical result. Each of the $3pN$ oscillators contributes on the average $k_\mathrm{B}T$ ($\frac{1}{2}k_\mathrm{B}T$ of kinetic energy and $\frac{1}{2}k_\mathrm{B}T$ of potential energy: equipartition theorem) to the internal energy! $\Longrightarrow$ Specific heat: $C_V \simeq 3pNk_\mathrm{B}$ (the "Law of Dulong and Petit").

b) *Low temperatures:* $k_\mathrm{B}T \ll \hbar\omega_r(\boldsymbol{q})$

The optical branches can be neglected, since

$$\left\langle b_{qr}^{+} b_{qr} \right\rangle^{\mathrm{opt.}} \approx 0 .$$

This does not apply to the three acoustic branches, since their energies go to zero for $q \to 0$.

In order to evaluate $U(T)$, we first convert the summation over $\boldsymbol{q}$ into an integral over ω. Justify the following representation of the density of states:

$$D_r(\omega) = \sum_{q} \delta\left(\omega - \omega_r(\boldsymbol{q})\right)$$

$$\Longrightarrow \quad U(T) - \underset{\substack{\uparrow \\ \text{Zero-point energy}}}{U(0)} = \sum_{r=1}^{3} \int_{-\infty}^{+\infty} \mathrm{d}\omega \, \frac{\hbar\omega D_r(\omega)}{\exp(\beta\hbar\omega) - 1} .$$

At low temperatures, we can use the Debye approximation from Ex. 2.2.4 for the acoustic branches:

$$U(T) - U(0) = \sum_{r=1}^{3} \frac{V}{2\pi^2 \bar{v}_r^3} \int_0^{\omega_r^\mathrm{D}} \mathrm{d}\omega \, \frac{\hbar\omega^3}{\mathrm{e}^{\beta\hbar\omega} - 1} = \sum_{r=1}^{3} \frac{3N}{\left(\omega_r^\mathrm{D}\right)^3} \int_0^{\omega_r^\mathrm{D}} \mathrm{d}\omega \, \frac{\hbar\omega^3}{\mathrm{e}^{\beta\hbar\omega} - 1} .$$

2. *Specific heat*

$$C_V = \left(\frac{\partial U}{\partial T}\right)_V = \sum_{r=1}^{3} \frac{3N}{\left(\omega_r^{\mathrm{D}}\right)^3} \frac{1}{k_{\mathrm{B}}T^2} \int\limits_0^{\omega_r^{\mathrm{D}}} \mathrm{d}\omega \, \frac{\hbar^2\omega^4 \mathrm{e}^{\beta\hbar\omega}}{\left(\mathrm{e}^{\beta\hbar\omega}-1\right)^2} \,,$$

$$x = \beta\hbar\omega \quad \Longrightarrow \quad \mathrm{d}\omega = \frac{k_{\mathrm{B}}T}{\hbar}\mathrm{d}x \,; \quad \hbar^2\omega^4 = \frac{x^4\,(k_{\mathrm{B}}T)^4}{\hbar^2} \,,$$

$$\Theta_{\mathrm{D}}^{(r)} = \frac{\hbar\omega_r^{\mathrm{D}}}{k_{\mathrm{B}}} \quad \text{(“Debye temperature”)}$$

$$\Longrightarrow \quad C_V \approx \sum_{r=1}^{3} 3Nk_{\mathrm{B}} \left(\frac{T}{\Theta_{\mathrm{D}}^{(r)}}\right)^3 \int\limits_0^{\Theta_{\mathrm{D}}^{(r)}/T} \mathrm{d}x \, \frac{x^4\mathrm{e}^x}{(\mathrm{e}^x-1)^2} \,.$$

Low temperatures:

$$T \ll \Theta_{\mathrm{D}}^{(r)} \quad \Longrightarrow \quad \int\limits_0^{\Theta_{\mathrm{D}}^{(r)}/T} \mathrm{d}x \, \frac{x^4\mathrm{e}^x}{(\mathrm{e}^x-1)^2} \approx \int\limits_0^{\infty} \mathrm{d}x \, \frac{x^4\mathrm{e}^x}{(\mathrm{e}^x-1)^2} = \frac{4}{15}\pi^4$$

$$\Longrightarrow \quad C_V = N\alpha T^3 \quad \text{(“Debye's } T^3 \text{ law”)} \,,$$

$$\alpha = \frac{4}{5}\pi^4 k_{\mathrm{B}} \sum_{r=1}^{3} \left(\Theta_{\mathrm{D}}^{(r)}\right)^{-3} \,.$$

Section 2.3.3

Solution 2.3.1

Creation and annihilation operators for Cooper pairs:

$$b_{\boldsymbol{k}}^{+} = a_{\boldsymbol{k}\uparrow}^{+} a_{-\boldsymbol{k}\downarrow}^{+} \,; \quad b_{\boldsymbol{k}} = a_{-\boldsymbol{k}\downarrow} a_{\boldsymbol{k}\uparrow} \,.$$

Here, we have used:

$a_{\boldsymbol{k}\sigma}, a_{\boldsymbol{k}\sigma}^{+}$: creation and annihilation operators for electrons in the Bloch representation!

Fundamental commutation relations:

a)

$$[b_k, b_{k'}]_- = \left[a_{-k\downarrow}a_{k\uparrow}, a_{-k'\downarrow}a_{k'\uparrow}\right]_- = 0\,,$$

$$\left[b_k^+, b_{k'}^+\right]_- = 0\,,$$ since the Fermion creation and annihilation operators mutually anticommute.

b)

$$\begin{aligned}\left[b_k, b_{k'}^+\right]_- &= \left[a_{-k\downarrow}a_{k\uparrow}, a_{k'\uparrow}^+ a_{-k'\downarrow}^+\right]_- = \\ &= \delta_{kk'} a_{-k\downarrow} a_{-k'\downarrow}^+ - \delta_{-k,-k'} a_{k'\uparrow}^+ a_{k\uparrow} \\ &= \delta_{kk'}\left(1 - \hat{n}_{-k\downarrow} - \hat{n}_{k\uparrow}\right)\,.\end{aligned}$$

Cooper pairs are thus, in spite of their spin 0, not *genuine* Bosons, since they fulfil only two of the three fundamental commutation relations.

c)

$$b_k^2 = \left(b_k^+\right)^2 = 0 \quad \text{as for Fermions,}$$

$$[b_k, b_{k'}]_+ = 2b_k b_{k'} \neq 0 \quad \text{for} \quad k \neq k'\,.$$

They are naturally also not genuine Fermions!

Solution 2.3.2

1. Model Hamiltonian:

$$H = \sum_{k\sigma} \varepsilon(k) a_{k\sigma}^+ a_{k\sigma} + \sum_{kq\sigma} V_k(q) a_{k+q\sigma}^+ a_{-k-q-\sigma}^+ a_{-k-\sigma} a_{k\sigma}\,.$$

Interactions occur only between the two *additional electrons*, which by construction have opposite spins and wavenumbers!

2. Cooper pair state:

$$\begin{aligned}|\psi\rangle &= \frac{1}{\sqrt{2}} \sum_{k\sigma} \alpha_\sigma(k) a_{k\sigma}^+ a_{-k-\sigma}^+ |\text{FS}\rangle = \\ &= \frac{1}{\sqrt{2}} \sum_{k'\sigma'} \alpha_{-\sigma'}\left(-k'\right) a_{-k'-\sigma'}^+ a_{k'\sigma'}^+ |\text{FS}\rangle = \\ &= -\frac{1}{\sqrt{2}} \sum_{k\sigma} \alpha_{-\sigma}(-k) a_{k\sigma}^+ a_{-k-\sigma}^+ |\text{FS}\rangle\end{aligned}$$

$$\Longrightarrow \quad \alpha_\sigma(k) = -\alpha_{-\sigma}(-k)\,.$$

3.

$$1 \stackrel{!}{=} \langle \psi \mid \psi \rangle = \frac{1}{2} \sum_{\substack{k\sigma \\ p\bar{\sigma}}} \alpha_\sigma^*(\boldsymbol{k}) \alpha_{\bar{\sigma}}(\boldsymbol{p}) \langle \mathrm{FS} | a_{-\boldsymbol{k}-\sigma} a_{\boldsymbol{k}\sigma} a^+_{\boldsymbol{p}\bar{\sigma}} a^+_{-\boldsymbol{p}-\bar{\sigma}} | \mathrm{FS} \rangle =$$

$$= \frac{1}{2} \sum_{\substack{k\sigma \\ p\bar{\sigma}}} \alpha_\sigma^*(\boldsymbol{k}) \alpha_{\bar{\sigma}}(\boldsymbol{p}) \Theta(k - k_\mathrm{F}) \Theta(p - k_\mathrm{F}) \cdot$$

$$\cdot \left\{ \delta_{\sigma\bar{\sigma}} \delta_{\boldsymbol{k}\boldsymbol{p}} \langle \mathrm{FS} | a_{-\boldsymbol{k}-\sigma} a^+_{-\boldsymbol{p}-\bar{\sigma}} | \mathrm{FS} \rangle - \right.$$

$$\left. - \delta_{\sigma - \bar{\sigma}} \delta_{\boldsymbol{k} - \boldsymbol{p}} \langle \mathrm{FS} | a_{-\boldsymbol{k}-\sigma} a^+_{\boldsymbol{p}\bar{\sigma}} | \mathrm{FS} \rangle \right\} =$$

$$= \frac{1}{2} \sum_{k\sigma} \alpha_\sigma^*(\boldsymbol{k}) \Theta(k - k_\mathrm{F}) \Theta(k - k_\mathrm{F}) \cdot$$

$$\cdot \langle \mathrm{FS} | (1 - \hat{n}_{-\boldsymbol{k}-\sigma}) | \mathrm{FS} \rangle (\alpha_\sigma(\boldsymbol{k}) - \alpha_{-\sigma}(-\boldsymbol{k})) =$$

$$\stackrel{2.}{=} \frac{1}{2} \sum_{k\sigma}^{k > k_\mathrm{F}} 2 \left| \alpha_\sigma(\boldsymbol{k}) \right|^2 \langle \mathrm{FS} \mid \mathrm{FS} \rangle = \sum_{k\sigma}^{k > k_\mathrm{F}} \left| \alpha_\sigma(\boldsymbol{k}) \right|^2 .$$

Solution 2.3.3

1.

$$2 \langle \psi \mid T \mid \psi \rangle =$$

$$= \sum_{\substack{kpq \\ \sigma\sigma'\sigma''}} \varepsilon(\boldsymbol{k}) \alpha^*_{\sigma'}(\boldsymbol{p}) \alpha_{\sigma''}(\boldsymbol{q}) \langle \mathrm{FS} | a_{-\boldsymbol{p}-\sigma'} a_{\boldsymbol{p}\sigma'} a^+_{\boldsymbol{k}\sigma} a_{\boldsymbol{k}\sigma} a^+_{\boldsymbol{q}\sigma''} a^+_{-\boldsymbol{q}-\sigma''} | \mathrm{FS} \rangle =$$

$$= \sum_{\substack{kpq \\ \sigma\sigma'\sigma''}} \varepsilon(\boldsymbol{k}) \alpha^*_{\sigma'}(\boldsymbol{p}) \alpha_{\sigma''}(\boldsymbol{q}) \Theta(p - k_\mathrm{F}) \Theta(q - k_\mathrm{F}) \cdot$$

$$\cdot \langle \mathrm{FS} | \left\{ \delta_{\sigma'\sigma} \delta_{\boldsymbol{p}\boldsymbol{k}} a_{-\boldsymbol{p}-\sigma'} a_{\boldsymbol{k}\sigma} a^+_{\boldsymbol{q}\sigma''} a^+_{-\boldsymbol{q}-\sigma''} + \right.$$

$$+ \delta_{\sigma'\sigma''} \delta_{\boldsymbol{p}\boldsymbol{q}} a_{-\boldsymbol{p}-\sigma'} a^+_{\boldsymbol{k}\sigma} a_{\boldsymbol{k}\sigma} a^+_{-\boldsymbol{q}-\sigma''} -$$

$$\left. - \delta_{\sigma'-\sigma''} \delta_{\boldsymbol{p}-\boldsymbol{q}} a_{-\boldsymbol{p}-\sigma'} a^+_{\boldsymbol{k}\sigma} a_{\boldsymbol{k}\sigma} a^+_{\boldsymbol{q}\sigma''} \right\} | \mathrm{FS} \rangle =$$

$$= \sum_{\substack{kq \\ \sigma\sigma''}} \varepsilon(\boldsymbol{k}) \alpha_{\sigma''}(\boldsymbol{q}) \Theta(q - k_\mathrm{F}) \cdot$$

$$\cdot \left\{ \alpha^*_\sigma(\boldsymbol{k}) \Theta(k - k_\mathrm{F}) \langle \mathrm{FS} | a_{-\boldsymbol{k}-\sigma} a_{\boldsymbol{k}\sigma} a^+_{\boldsymbol{q}\sigma''} a^+_{-\boldsymbol{q}-\sigma''} | \mathrm{FS} \rangle + \right.$$

$$+ \alpha^*_{\sigma''}(\boldsymbol{q}) \Theta(q - k_\mathrm{F}) \langle \mathrm{FS} | a_{-\boldsymbol{q}-\sigma''} a^+_{\boldsymbol{k}\sigma} a_{\boldsymbol{k}\sigma} a^+_{-\boldsymbol{q}-\sigma''} | \mathrm{FS} \rangle -$$

$$- \alpha^*_{-\sigma''}(-\boldsymbol{q})\Theta(q - k_F) \langle FS| a_{\boldsymbol{q}\sigma''} a^+_{\boldsymbol{k}\sigma} a_{\boldsymbol{k}\sigma} a^+_{\boldsymbol{q}\sigma''} |FS\rangle \Big\} =$$

$$= \sum_{\substack{\boldsymbol{kq}\\ \sigma\sigma''}} \varepsilon(\boldsymbol{k})\alpha^*_\sigma(\boldsymbol{k})\alpha_{\sigma''}(\boldsymbol{q})\Theta(q - k_F)\,\Theta(k - k_F)\,\cdot$$

$$\cdot \langle FS| \left(\delta_{\sigma\sigma''}\delta_{\boldsymbol{kq}}(1 - \hat{n}_{-\boldsymbol{k}-\sigma}) - \delta_{\sigma-\sigma''}\delta_{\boldsymbol{k}-\boldsymbol{q}}(1 - \hat{n}_{-\boldsymbol{k}-\sigma})\right) |FS\rangle +$$

$$+ \sum_{\substack{\boldsymbol{kq}\\ \sigma\sigma''}} \varepsilon(\boldsymbol{k})\alpha^*_{\sigma''}(\boldsymbol{q})\alpha_{\sigma''}(\boldsymbol{q})\,(\Theta(q - k_F))^2\,\cdot$$

$$\cdot \langle FS| \left(\delta_{\sigma-\sigma''}\delta_{-\boldsymbol{qk}}(1 - \hat{n}_{\boldsymbol{k}\sigma}) + \hat{n}_{\boldsymbol{k}\sigma} + \delta_{\sigma''\sigma}\delta_{\boldsymbol{qk}}(1 - \hat{n}_{\boldsymbol{k}\sigma}) + \hat{n}_{\boldsymbol{k}\sigma}\right) |FS\rangle$$

$$\Longrightarrow \quad \langle\psi \mid T \mid \psi\rangle = \frac{1}{2}\sum_{\boldsymbol{k}\sigma} \varepsilon(\boldsymbol{k})\alpha^*_\sigma(\boldsymbol{k})\,(\Theta(k - k_F))^2\,(\alpha_\sigma(\boldsymbol{k}) - \alpha_{-\sigma}(-\boldsymbol{k})) +$$

$$+ \frac{1}{2}\sum_{\boldsymbol{k}\sigma} \varepsilon(\boldsymbol{k})\,(\Theta(k - k_F))^2 \left(\left|\alpha_{-\sigma}(-\boldsymbol{k})\right|^2 + \left|\alpha_\sigma(\boldsymbol{k})\right|^2\right) +$$

$$+ \frac{1}{2} 2 \sum_{\boldsymbol{k}\sigma} \varepsilon(\boldsymbol{k})\Theta(k_F - k) \sum_{\boldsymbol{q}\sigma''}^{q > k_F} \left|\alpha_{\sigma''}(\boldsymbol{q})\right|^2 =$$

$$= \frac{1}{2}\sum_{\boldsymbol{k}\sigma}^{k > k_F} \varepsilon(\boldsymbol{k}) \left|\alpha_\sigma(\boldsymbol{k})\right|^2 (2 + 2) + \sum_{\boldsymbol{k}\sigma}^{k < k_F} \varepsilon(\boldsymbol{k}) =$$

$$= 2 \sum_{\boldsymbol{k}\sigma}^{k > k_F} \varepsilon(\boldsymbol{k}) \left|\alpha_\sigma(\boldsymbol{k})\right|^2 + 2 \sum_{\boldsymbol{k}}^{k < k_F} \varepsilon(\boldsymbol{k}) \quad \text{q. e. d.}$$

2.

$$2\,\langle\psi \mid V \mid \psi\rangle =$$

$$= \sum_{\substack{\boldsymbol{kq}\sigma\\ \boldsymbol{p}_1\sigma_1\boldsymbol{p}_2\sigma_2}} V_{\boldsymbol{k}}(\boldsymbol{q})\alpha^*_{\sigma_1}(\boldsymbol{p}_1)\alpha_{\sigma_2}(\boldsymbol{p}_2)\,\Theta(p_1 - k_F)\,\Theta(p_2 - k_F)\,\cdot$$

$$\cdot \langle FS| a_{-\boldsymbol{p}_1-\sigma_1} a_{\boldsymbol{p}_1\sigma_1} a^+_{\boldsymbol{k}+\boldsymbol{q}\sigma} a^+_{-\boldsymbol{k}-\boldsymbol{q}-\sigma} a_{-\boldsymbol{k}-\sigma} a_{\boldsymbol{k}\sigma}\,\cdot$$

$$\cdot\, a^+_{\boldsymbol{p}_2\sigma_2} a^+_{-\boldsymbol{p}_2-\sigma_2} |FS\rangle =$$

$$= \sum_{\substack{\boldsymbol{kq}\sigma\\ \boldsymbol{p}_1\sigma_1\boldsymbol{p}_2\sigma_2}} V_{\boldsymbol{k}}(\boldsymbol{q})\alpha^*_{\sigma_1}(\boldsymbol{p}_1)\alpha_{\sigma_2}(\boldsymbol{p}_2)\Theta(p_1 - k_F)\,\Theta(p_2 - k_F)\,\cdot$$

$$\cdot \langle FS| \Big\{ \delta_{\sigma_1\sigma}\delta_{\boldsymbol{p}_1,\boldsymbol{k}+\boldsymbol{q}} a_{-\boldsymbol{p}_1-\sigma_1} a^+_{-\boldsymbol{k}-\boldsymbol{q}-\sigma} a_{-\boldsymbol{k}-\sigma} a_{\boldsymbol{k}\sigma}\,\cdot$$

$$\cdot\, a^+_{\boldsymbol{p}_2\sigma_2} a^+_{-\boldsymbol{p}_2-\sigma_2} - \delta_{\sigma_1-\sigma}\delta_{\boldsymbol{p}_1,-\boldsymbol{k}-\boldsymbol{q}}\,\cdot$$

$$\cdot\, a_{-\boldsymbol{p}_1-\sigma_1} a^+_{\boldsymbol{k}+\boldsymbol{q}\sigma} a_{-\boldsymbol{k}-\sigma} a_{\boldsymbol{k}\sigma} a^+_{\boldsymbol{p}_2\sigma_2} a^+_{-\boldsymbol{p}_2-\sigma_2} +$$

$$+ \delta_{\sigma_1\sigma_2}\delta_{p_1p_2} a_{-p_1-\sigma_1} a^+_{k+q\sigma} a^+_{-q-k-\sigma} a_{-k-\sigma} a_{k\sigma} a^+_{-p_2-\sigma_2} -$$

$$- \delta_{\sigma_1-\sigma_2}\delta_{p_1,-p_2} a_{-p_1-\sigma_1} a^+_{k+q\sigma} a^+_{-k-q-\sigma} a_{-k-\sigma} a_{k\sigma} a^+_{p_2\sigma_2} \Big\} |\mathrm{FS}\rangle =$$

$$= \sum_{\substack{kq\sigma \\ p_2\sigma_2}} V_k(q) \alpha_{\sigma_2}(p_2) \Theta(p_2 - k_\mathrm{F}) \cdot$$

$$\cdot \langle \mathrm{FS}| \Big\{ \Big(\alpha^*_\sigma(k+q) \Theta(|k+q| - k_\mathrm{F}) \cdot$$

$$\cdot a_{-k-q-\sigma} a^+_{-k-q-\sigma} - \alpha^*_{-\sigma}(-k-q) \cdot$$

$$\cdot \Theta(|k+q| - k_\mathrm{F}) a_{k+q\sigma} a^+_{k+q\sigma} \Big) \cdot$$

$$\cdot a_{-k-\sigma} a_{k\sigma} a^+_{p_2\sigma_2} a^+_{-p_2-\sigma_2} +$$

$$+ \Big(\alpha^*_{\sigma_2}(p_2) \Theta(p_2 - k_\mathrm{F}) a_{-p_2-\sigma_2} a^+_{k+q\sigma} \cdot$$

$$\cdot a^+_{-q-k-\sigma} a_{-k-\sigma} a_{k\sigma} a^+_{-p_2-\sigma_2} -$$

$$- \alpha^*_{-\sigma_2}(-p_2) \Theta(p_2 - k_\mathrm{F}) a_{p_2\sigma_2} a^+_{k+q\sigma} \cdot$$

$$\cdot a^+_{-k-q-\sigma} a_{-k-\sigma} a_{k\sigma} a^+_{p_2\sigma_2} \Big) \Big\} |\mathrm{FS}\rangle =$$

$$= \sum_{\substack{kq\sigma \\ p_2\sigma_2}} V_k(q) \alpha_{\sigma_2}(p_2) \alpha^*_\sigma(k+q) \Theta(|k+q| - k_\mathrm{F}) \cdot$$

$$\cdot 2 \langle \mathrm{FS} | a_{-k-\sigma} a_{k\sigma} a^+_{p_2\sigma_2} a^+_{-p_2-\sigma_2} | \mathrm{FS} \rangle +$$

$$+ \sum_{\substack{kq\sigma \\ p_2\sigma_2}} V_k(q) |\alpha_{\sigma_2}(p_2)|^2 \Theta(p_2 - k_\mathrm{F}) \Big\{ \langle \mathrm{FS}| \Big(\delta_{-\sigma_2\sigma} \cdot$$

$$\cdot \delta_{-p_2,k+q} a^+_{-q-k-\sigma} a_{-k-\sigma} a_{k\sigma} a^+_{-p_2-\sigma_2} -$$

$$- \delta_{\sigma\sigma_2}\delta_{p_2,k+q} a^+_{k+q\sigma} a_{-k-\sigma} a_{k\sigma} a^+_{-p_2-\sigma_2} +$$

$$+ a^+_{k+q\sigma} a^+_{-k-q-\sigma} a_{-k-\sigma} a_{k\sigma} \left(1 - \hat{n}_{-p_2-\sigma_2}\right) +$$

$$+ \delta_{\sigma\sigma_2}\delta_{p_2,k+q} a^+_{-k-q-\sigma} a_{-k-\sigma} a_{k\sigma} a^+_{p_2\sigma_2} -$$

$$- \delta_{\sigma-\sigma_2}\delta_{p_2,-k-q} a^+_{k+q\sigma} a_{-k-\sigma} a_{k\sigma} a^+_{p_2\sigma_2} +$$

$$+ a^+_{k+q\sigma} a^+_{-k-q-\sigma} a_{-k-\sigma} a_{k\sigma} \left(1 - \hat{n}_{p_2\sigma_2}\right) \Big) \Big\} |\mathrm{FS}\rangle =$$

$$= \sum_{\substack{kq\sigma \\ p_2\sigma_2}} 2 V_k(q) \alpha_{\sigma_2}(p_2) \alpha^*_\sigma(k+q) \Theta(|k+q| - k_\mathrm{F}) \Theta(k - k_\mathrm{F}) \cdot$$

$$\cdot \langle \mathrm{FS}| \Big(\delta_{\sigma\sigma_2}\delta_{kp_2} (1 - \hat{n}_{k-\sigma}) - \delta_{\sigma-\sigma_2}\delta_{k-p_2} (1 - \hat{n}_{k-\sigma}) \Big) |\mathrm{FS}\rangle +$$

$$+ \sum_{kq\sigma} V_k(q) |\alpha_\sigma(k+q)|^2 \Theta(|k+q| - k_\mathrm{F}) \cdot$$

$$\cdot \langle \mathrm{FS}| \Big(a^+_{-k-q-\sigma} a_{-k-\sigma} a_{k\sigma} a^+_{k+q\sigma} - a^+_{k+q\sigma} a_{-k-\sigma} a_{k\sigma} a^+_{-k-q-\sigma} + \\ + a^+_{k+q\sigma} a^+_{-k-q-\sigma} a_{-k-\sigma} a_{k\sigma} + a^+_{-k-q-\sigma} a_{-k-\sigma} a_{k\sigma} a^+_{k+q\sigma} - \\ - a^+_{k+q\sigma} a_{-k-\sigma} a_{k\sigma} a^+_{-k-q-\sigma} + a^+_{k+q\sigma} a^+_{-k-q-\sigma} a_{-k-\sigma} a_{k\sigma} \Big) |\mathrm{FS}\rangle \,.$$

The second sum vanishes, since

$$\langle \mathrm{FS}| a^+_{k+q\sigma} = \langle \mathrm{FS}| a^+_{-k-q-\sigma} = 0\,, \quad \text{for} \quad |k+q| > k_\mathrm{F}\,.$$

In the first sum, we make use of:

$$\alpha_\sigma(k) = -\alpha_{-\sigma}(-k)\,,$$

as we have already done several times. We then find:

$$\langle \psi \mid V \mid \psi \rangle = 2 \sum_{kq\sigma}^{\substack{k>k_\mathrm{F} \\ |k+q|>k_\mathrm{F}}} V_k(q) \alpha^*_\sigma(k+q) \alpha_\sigma(k)\,. \quad \text{q. e. d.}$$

Solution 2.3.4

The energy of the model system in the *Cooper pair state* according to Ex. 2.3.3 is given by:

$$E = \langle \psi \mid H \mid \psi \rangle \\ = 2 \sum_{k,\sigma}^{k>k_\mathrm{F}} \varepsilon(k) \left|\alpha_\sigma(k)\right|^2 + 2 \sum_{k}^{k\le k_\mathrm{F}} \varepsilon(k) + 2 \sum_{k,q,\sigma}^{\substack{k>k_\mathrm{F} \\ |k+q|>k_\mathrm{F}}} V_k(q) \alpha^*_\sigma(k+q) \alpha_\sigma(k)\,.$$

1. For the determination of the $\alpha_\sigma(k)$, we minimise E, *coupling in* the boundary condition

$$\sum_{k,\sigma}^{k>k_\mathrm{F}} \left|\alpha_\sigma(k)\right|^2 = 1$$

with the aid of a Lagrange parameter λ:

$$\frac{\partial}{\partial \alpha^*_\sigma(k)} \left(E - \lambda \sum_{k',\sigma'}^{k'>k_\mathrm{F}} |\alpha_{\sigma'}(k')|^2 \right) = \\ \overset{(k>k_\mathrm{F})}{=} 2\varepsilon(k)\alpha_\sigma(k) + 2 \sum_{q}^{\substack{k>k_\mathrm{F} \\ |k+q|>k_\mathrm{F}}} V_{k-q}(q) \alpha_\sigma(k-q) - \lambda \alpha_\sigma(k) \overset{!}{=} 0\,.$$

Multiply by $\alpha_\sigma^*(\boldsymbol{k})$, then sum over all $\boldsymbol{k}$ and σ $(k > k_F)$:

$$\lambda = E - 2\sum_{\boldsymbol{k}}^{k<k_F} \varepsilon(\boldsymbol{k}) = \widehat{E}\,.$$

The Lagrange parameter thus corresponds to the additional energy due to the two electrons in the Cooper pair. With the simplification for the matrix element given in Ex. 2.3.2, we then find:

$$\left(2\varepsilon(\boldsymbol{k}) - \widehat{E}\right)\alpha_\sigma(\boldsymbol{k}) = 2V\sum_{\boldsymbol{k}'}^{k'>k_F} \alpha_\sigma\left(\boldsymbol{k}'\right) \equiv 2A_\sigma$$

$$\Longleftrightarrow\; A_\sigma = \sum_{\boldsymbol{k}} \frac{2VA_\sigma}{2\varepsilon(\boldsymbol{k}) - \widehat{E}}\,.$$

The summation over $\boldsymbol{k}$ naturally runs only over those wavevectors for which $V \neq 0$. We convert the sum into an integral:

$$1 = 2NV \int\limits_{\varepsilon_F}^{\varepsilon_F+\hbar\omega_D} \mathrm{d}x\, \frac{\rho_0(x)}{2x - \widehat{E}} \approx 2NV\rho_0(\varepsilon_F) \int\limits_{\varepsilon_F}^{\varepsilon_F+\hbar\omega_D} \frac{\mathrm{d}x}{2x - \widehat{E}} \approx$$

$$\approx 2NV\rho_0(\varepsilon_F)\frac{1}{2}\ln\frac{2\left(\varepsilon_F + \hbar\omega_D\right) - \widehat{E}}{2\varepsilon_F - \widehat{E}}$$

$$\Longrightarrow\quad \widehat{E} \approx 2\varepsilon_F - 2\hbar\omega_D \frac{\exp\left(-1/NV\rho_0(\varepsilon_F)\right)}{1 - \exp\left(-1/NV\rho_0(\varepsilon_F)\right)}\,.$$

For $V \neq 0$, the energy of the Cooper pair is thus less than the energy of two electrons at the Fermi edge which do not interact with each other. The Cooper pair therefore represents a bound state. The Fermi sphere is unstable with respect to the formation of Cooper pairs!

Solution 2.3.5

1.

$$1 \stackrel{!}{=} \langle \mathrm{BCS} \mid \mathrm{BCS}\rangle = \langle 0|\prod_{\boldsymbol{k}}\prod_{\boldsymbol{p}}\left(u_{\boldsymbol{k}} + v_{\boldsymbol{k}} b_{\boldsymbol{k}}\right)\left(u_{\boldsymbol{p}} + v_{\boldsymbol{p}} b_{\boldsymbol{p}}^{+}\right)|0\rangle\,.$$

All the operators commute for different wavenumbers. Due to $b_k\,|0\rangle = \langle 0|\,b_k^+ = 0$, we then have:

$$1 \overset{!}{=} \langle 0|\prod_k (u_k + v_k b_k)\left(u_k + v_k b_k^+\right)|0\rangle =$$

$$= \langle 0|\prod_k \left(u_k^2 + v_k\left(b_k + b_k^+\right)u_k + v_k^2 b_k b_k^+\right)|0\rangle =$$

$$= \langle 0|\prod_k \left(u_k^2 + v_k^2 b_k b_k^+\right)|0\rangle =$$

$$\overset{\text{Ex. 3.1}}{=} \langle 0|\prod_k \Big[u_k^2 + v_k^2\left(b_k^+ b_k + 1 - \hat{n}_{k\uparrow} - \hat{n}_{-k\downarrow}\right)\Big]|0\rangle$$

$$b_k^+ b_k\,|0\rangle = \hat{n}_{k\sigma}\,|0\rangle = 0$$

$$\Longrightarrow \quad 1 \overset{!}{=} \langle 0|\prod_k \left(u_k^2 + v_k^2\right)|0\rangle = \prod_k \left(u_k^2 + v_k^2\right).$$

All $\boldsymbol{k}$ terms are equivalent and mutually independent:

$$1 = u_k^2 + v_k^2 .$$

2.

$$\langle \mathrm{BCS} \mid b_k^+ b_k \mid \mathrm{BCS}\rangle =$$

$$= \langle 0|\prod_q\prod_p \left(u_q + v_q b_q\right) b_k^+ b_k \left(u_p + v_p b_p^+\right)|0\rangle =$$

$$= \langle 0|\left\{\prod_q^{\neq k}\prod_p^{\neq k}\left(u_q + v_q b_q\right)\left(u_p + v_p b_p^+\right)\right\}\left(u_k + v_k b_k\right) b_k^+ b_k \left(u_k + v_k b_k^+\right)|0\rangle =$$

$$= \langle 0|\left\{\prod_q^{\neq k}\prod_p^{\neq k}\left(u_q + v_q b_q\right)\left(u_p + v_p b_p^+\right)\right\} v_k^2 b_k b_k^+ b_k b_k^+\,|0\rangle .$$

As in part 1.: $b_k b_k^+\,|0\rangle = |0\rangle$

$$\Longrightarrow \quad \langle \mathrm{BCS} \mid b_k^+ b_k \mid \mathrm{BCS}\rangle = v_k^2\,\langle 0|\prod_q^{\neq k}\prod_p^{\neq k}\left(u_q + v_q b_q\right)\left(u_p + v_p b_p^+\right)|0\rangle =$$

$$\overset{1.}{=} v_k^2 \prod_p^{\neq k}\left(u_p^2 + v_p^2\right) \overset{1.}{=} v_k^2 .$$

v_k^2 is the probability that the Cooper pair $(\boldsymbol{k}\uparrow, -\boldsymbol{k}\downarrow)$ exists! The second expectation value is computed in a completely analogous manner:
$\boldsymbol{k} \neq \boldsymbol{p}$:

$$\langle \text{BCS} \mid b_k^+ b_k b_p^+ b_p \mid \text{BCS} \rangle = \langle 0| \left\{ \prod_q^{\neq k,p} \prod_{q'}^{\neq k,p} \left(u_q + v_q b_q\right) \left(u_{q'} + v_{q'} b_{q'}^+\right) \right\} \cdot$$

$$\cdot \left(u_k + v_k b_k\right) b_k^+ b_k \left(u_k + v_k b_k^+\right) \left(u_p + v_p b_p\right) b_p^+ b_p \left(u_p + v_p b_p^+\right) |0\rangle =$$

$$= \langle 0| \left\{ \prod_q^{\neq k,p} \prod_{q'}^{\neq k,p} \left(u_q + v_q b_q\right) \left(u_{q'} + v_{q'} b_{q'}^+\right) \right\} v_k^2 v_p^2 b_k b_p^+ b_k b_k^+ b_p b_p^+ b_p b_p^+ |0\rangle =$$

$$= v_k^2 v_p^2 \langle 0| \left\{ \prod_q^{\neq k,p} \prod_{q'}^{\neq k,p} \left(u_q + v_q b_q\right) \left(u_{q'} + v_{q'} b_{q'}^+\right) \right\} |0\rangle =$$

$$= v_k^2 v_p^2 \quad \left(= v_k^2 \quad \text{for} \quad \boldsymbol{k} = \boldsymbol{p}\right) .$$

$$\langle \text{BCS} \mid b_k^+ b_k \left(1 - b_p^+ b_p\right) \mid \text{BCS} \rangle = v_k^2 - v_k^2 v_p^2 = v_k^2 u_p^2 , \quad \text{when} \quad \boldsymbol{k} \neq \boldsymbol{p}$$

(= 0, when $\boldsymbol{k} = \boldsymbol{p}$). Here, u_p^2 is the probability that the Cooper pair $(\boldsymbol{p}\uparrow, -\boldsymbol{p}\downarrow)$ does *not* exist!
We still have to calculate:

$$\left\langle \text{BCS} \middle| b_p^+ b_k \middle| \text{BCS} \right\rangle =$$

$$= \langle 0| \left\{ \prod_q^{\neq k,p} \prod_{q'}^{\neq k,p} \left(u_q + v_q b_q\right) \left(u_{q'} + v_{q'} b_{q'}^+\right) \right\} \cdot$$

$$\cdot \left(u_p + v_p b_p\right) b_p^+ \left(u_p + v_p b_p^+\right) \left(u_k + v_k b_k\right) b_k \left(u_k + v_k b_k^+\right) |0\rangle =$$

$$= \langle 0| \left\{ \prod_q^{\neq k,p} \prod_{q'}^{\neq k,p} \left(u_q + v_q b_q\right) \left(u_{q'} + v_{q'} b_{q'}^+\right) \right\} u_p v_p b_p b_p^+ u_k v_k b_k b_k^+ |0\rangle =$$

$$= u_p v_p u_k v_k \langle 0| \left\{ \prod_q^{\neq k,p} \prod_{q'}^{\neq k,p} \left(u_q + v_q b_q\right) \left(u_{q'} + v_{q'} b_{q'}^+\right) \right\} |0\rangle =$$

$$\overset{\text{s.a.}}{=} u_p v_p u_k v_k .$$

Here, we have made use of results from Ex. 2.3.1:

$$\left(b_k^+\right)^2 = \left(b_k\right)^2 = 0 .$$

Solution 2.3.6

1. Phonon-induced electron-electron interaction (see Ex. 2.3.2):

$$H = \sum_{k\sigma} \varepsilon(\boldsymbol{k}) a^+_{k\sigma} a_{k\sigma} - V \sum_{\substack{kpq \\ \sigma,\sigma'}}^{q\neq 0} a^+_{k+q\sigma} a^+_{p-q\sigma'} a_{p\sigma'} a_{k\sigma} .$$

Variation is to be carried out with $|\mathrm{BCS}\rangle$. According to Ex. 2.3.5, this test state contains only Cooper pairs. Therefore, H can be reduced to

$$\overline{H} = \sum_{k\sigma} \varepsilon(\boldsymbol{k}) a^+_{k\sigma} a_{k\sigma} - V \sum_{p,k}^{p\neq k} b^+_p b_k .$$

All the other terms yield a contribution of zero when applied to $|\mathrm{BCS}\rangle$. Multiplying out the product in $|\mathrm{BCS}\rangle$, we find terms with a different number of creation operators b^+_k. It follows that: $|\mathrm{BCS}\rangle$ is not a state with a fixed number of particles! The side condition $N = \text{const}$ must therefore be *coupled in* by using a Lagrange parameter μ:

$$H_{\mathrm{BCS}} = \overline{H}(\varepsilon(\boldsymbol{k}) \to t(\boldsymbol{k})) \quad t(\boldsymbol{k}) = \varepsilon(\boldsymbol{k}) - \mu .$$

2. The expectation value for the potential energy was computed in part 2. of Ex. 2.3.5. All the operators commute as long as they belong to different wavenumbers $\boldsymbol{k}$: Therefore, $\langle \mathrm{BCS} \mid a^+_{k\sigma} a_{k\sigma} \mid \mathrm{BCS}\rangle$ can be sorted as follows:

$$\langle \mathrm{BCS} \mid a^+_{k\sigma} a_{k\sigma} \mid \mathrm{BCS}\rangle = \langle 0| \left\{ \prod_p^{\neq\pm k} \left(u_p + v_p b_p\right)\left(u_p + v_p b^+_p\right) \right\} \cdot$$

$$\cdot \left(u_k + v_k b_k\right)\left(u_{-k} + v_{-k} b_{-k}\right) a^+_{k\sigma} a_{k\sigma} \left(u_k + v_k b^+_k\right)\left(u_{-k} + v_{-k} b^+_{-k}\right) |0\rangle ,$$

$$\left[a^+_{k\sigma} a_{k\sigma}, b^+_k\right]_- = \left[a^+_{k\sigma} a_{k\sigma}, a^+_{k\uparrow} a^+_{-k\downarrow}\right]_- = \delta_{\sigma\uparrow} b^+_k ,$$

$$\left[a^+_{k\sigma} a_{k\sigma}, b^+_{-k}\right]_- = \delta_{\sigma\downarrow} b^+_{-k} ,$$

$$\left[a^+_{k\sigma} a_{k\sigma}, b^+_k b^+_{-k}\right]_- = \left(\delta_{\sigma\downarrow} + \delta_{\sigma\uparrow}\right) b^+_k b^+_{-k} = b^+_k b^+_{-k} .$$

We then find:

$$a^+_{k\sigma} a_{k\sigma} \left(u_k + v_k b^+_k\right)\left(u_{-k} + v_{-k} b^+_{-k}\right) |0\rangle =$$

$$= \left(v_k u_{-k} \delta_{\sigma\uparrow} b^+_k + u_k v_{-k} \delta_{\sigma\downarrow} b^+_{-k} + v_k v_{-k} b^+_k b^+_{-k}\right) |0\rangle .$$

And furthermore:

$$
\begin{aligned}
&\langle 0|\,(u_k+v_k b_k)(u_{-k}+v_{-k}b_{-k})\,a^+_{k\sigma}a_{k\sigma}\,\left(u_k+v_k b^+_k\right)\left(u_{-k}+v_{-k}b^+_{-k}\right)|0\rangle =\\
&=(v_k u_{-k})^2\,\delta_{\sigma\uparrow}\,\langle 0\mid b_k b^+_k\mid 0\rangle+(u_k v_{-k})^2\delta_{\sigma\downarrow}\,\langle 0\mid b_{-k}b^+_{-k}\mid 0\rangle+\\
&\quad+v^2_k v^2_{-k}\,\langle 0\mid b_k b_{-k} b^+_k b^+_{-k}\mid 0\rangle=\\
&=v^2_k\delta_{\sigma\uparrow}+v^2_{-k}\delta_{\sigma\downarrow}\,.
\end{aligned}
$$

The conclusions are as in the preceding exercise:

$$
\langle \mathrm{BCS}\mid a^+_{k\sigma}a_{k\sigma}\mid \mathrm{BCS}\rangle = v^2_k\delta_{\sigma\uparrow}+v^2_{-k}\delta_{\sigma\downarrow}\,.
$$

Owing to $t(-\boldsymbol{k})=t(\boldsymbol{k})$ and part 2. of Ex. 2.3.5, we finally obtain:

$$
E=2\sum_k t(\boldsymbol{k})v^2_k - V\sum_{k,p}^{k\neq p} v_k v_p u_k u_p\,.
$$

3. Minimum condition:

$$
\begin{aligned}
0 \overset{!}{=} \frac{\mathrm{d}E}{\mathrm{d}v_k} &= \left(\frac{\partial E}{\partial v_k}\right)_u+\left(\frac{\partial E}{\partial u_k}\right)_v\frac{\partial u_k}{\partial v_k}=\left(\frac{\partial E}{\partial v_k}\right)_u-\frac{v_k}{u_k}\left(\frac{\partial E}{\partial u_k}\right)_v=\\
&=4t(\boldsymbol{k})v_k-2V\sum_p^{\neq k} v_p u_p u_k+2V\frac{v_k}{u_k}\sum_p^{\neq k} v_k v_p u_p=\\
&=4t(\boldsymbol{k})v_k+2\Delta_k\left(\frac{v_k^2}{u_k}-u_k\right)\\
\Longleftrightarrow\quad 4t^2(\boldsymbol{k})v_k^2u_k^2&=\Delta_k^2\left(v_k^2-u_k^2\right)^2=\Delta_k^2\left(4v_k^4-4v_k^2+1\right)=\\
&=-\Delta_k^2 4v_k^2u_k^2+\Delta_k^2\,.
\end{aligned}
$$

This leads to:

$$
u_k v_k=\frac{1}{2}\frac{\Delta_k}{\sqrt{t^2(\boldsymbol{k})+\Delta_k^2}}\,.
$$

Inserting into the definition equation for Δ_k finally yields:

$$
\Delta_k=\frac{1}{2}V\sum_p^{\neq k}\frac{\Delta_p}{\sqrt{t^2(\boldsymbol{p})+\Delta_p^2}}\,.
$$

4.

$$u_k^2 v_k^2 = -v_k^4 + v_k^2 = -\left(v_k^2 - \frac{1}{2}\right)^2 + \frac{1}{4} = \frac{1}{4}\frac{\Delta_k^2}{t^2(\boldsymbol{k}) + \Delta_k^2}$$

$$\Longrightarrow \quad v_k^2 = \frac{1}{2}\left(1 + \frac{t(\boldsymbol{k})}{-\sqrt{t^2(\boldsymbol{k}) + \Delta_k^2}}\right), \qquad \text{negative root, since for } \Delta \to 0, \text{ no Cooper pairs exist.}$$

$$u_k^2 = \frac{1}{2}\left(1 + \frac{t(\boldsymbol{k})}{\sqrt{t^2(\boldsymbol{k}) + \Delta_k^2}}\right).$$

BCS-Ground-state energy:

$$E_0 = 2\sum_k t(\boldsymbol{k})\frac{1}{2}\left(1 - \frac{t(\boldsymbol{k})}{\sqrt{t^2(\boldsymbol{k}) + \Delta^2}}\right) - \sum_k v_k u_k \Delta_k$$

$$\Longrightarrow \quad E_0 = \sum_k \left\{ t(\boldsymbol{k}) - \frac{t^2(\boldsymbol{k}) + \frac{1}{2}\Delta_k^2}{\sqrt{t^2(\boldsymbol{k}) + \Delta_k^2}} \right\}.$$

Solution 2.3.7

$$e^S: \quad \text{unitary transformation} \quad \Longleftrightarrow \quad \left(e^S\right)^+ = \left(e^S\right)^{-1}$$

$$\Longleftrightarrow \quad S^+ = -S.$$

From (2.186):

$$S = \sum_{kq\sigma} T_q \left(x(\boldsymbol{k}, \boldsymbol{q}) b_q + y(\boldsymbol{k}, \boldsymbol{q}) b_{-q}^+\right) a_{k+q\sigma}^+ a_{k\sigma}$$

$$\Longrightarrow \quad S^+ = \sum_{kq\sigma} T_q^* \left(x^*(\boldsymbol{k}, \boldsymbol{q}) b_q^+ + y^*(\boldsymbol{k}, \boldsymbol{q}) b_{-q}\right) a_{k\sigma}^+ a_{k+q\sigma},$$

$$\boldsymbol{q} \to -\boldsymbol{q}; \quad \boldsymbol{k} \to \boldsymbol{k} + \boldsymbol{q}; \quad T_{-q}^* = T_q \quad (2.183),$$

$$\Longrightarrow \quad S^+ = \sum_{kq\sigma} T_q \left(x^*(\boldsymbol{k} + \boldsymbol{q}, -\boldsymbol{q}) b_{-q}^+ + y^*(\boldsymbol{k} + \boldsymbol{q}, -\boldsymbol{q}) b_q\right) a_{k+q\sigma}^+ a_{k\sigma}.$$

$S^+ = -S$ thus obviously holds only when

$$y^*(\boldsymbol{k} + \boldsymbol{q}, -\boldsymbol{q}) \stackrel{!}{=} -x(\boldsymbol{k}, \boldsymbol{q}),$$

$$x^*(\boldsymbol{k} + \boldsymbol{q}, -\boldsymbol{q}) \stackrel{!}{=} -y(\boldsymbol{k}, \boldsymbol{q})$$

is fulfilled. This is, from (2.190) and (2.191) and due to $\hbar\omega(-\boldsymbol{q}) = \hbar\omega(\boldsymbol{q})$, clearly the case!

Section 2.4.5

Solution 2.4.1

$$\left[S^{+}(\boldsymbol{k}_1), S^{-}(\boldsymbol{k}_2)\right]_{-} = \sum_{i,j} \mathrm{e}^{-\mathrm{i}(\boldsymbol{k}_1 \boldsymbol{R}_i + \boldsymbol{k}_2 \boldsymbol{R}_j)} \left[S_i^{+}, S_j^{-}\right]_{-} =$$

$$= \sum_{i,j} \mathrm{e}^{-\mathrm{i}(\boldsymbol{k}_1 \boldsymbol{R}_i + \boldsymbol{k}_2 \boldsymbol{R}_j)} 2\hbar \delta_{ij} S_i^z = 2\hbar \sum_i \mathrm{e}^{-\mathrm{i}(\boldsymbol{k}_1 + \boldsymbol{k}_2)\cdot \boldsymbol{R}_i} S_i^z =$$

$$= 2\hbar S^z(\boldsymbol{k}_1 + \boldsymbol{k}_2)\,.$$

$$\left[S^{z}(\boldsymbol{k}_1), S^{\pm}(\boldsymbol{k}_2)\right]_{-} = \sum_{i,j} \mathrm{e}^{-\mathrm{i}(\boldsymbol{k}_1 \cdot \boldsymbol{R}_i + \boldsymbol{k}_2 \cdot \boldsymbol{R}_j)} \left[S_i^{z}, S_j^{\pm}\right]_{-} =$$

$$= \pm\hbar \sum_{i,j} \mathrm{e}^{-\mathrm{i}(\boldsymbol{k}_1 \cdot \boldsymbol{R}_i + \boldsymbol{k}_2 \cdot \boldsymbol{R}_j)} \delta_{ij} S_i^{\pm} = \pm\hbar S^{\pm}(\boldsymbol{k}_1 + \boldsymbol{k}_2)\,.$$

Solution 2.4.2

$$\sum_{i,j} J_{ij} S_i^{\alpha} S_j^{\beta} =$$

$$= \frac{1}{N^3} \sum_{i,j} \sum_{\boldsymbol{kqp}} J(\boldsymbol{k}) S^{\alpha}(\boldsymbol{p}) S^{\beta}(\boldsymbol{q}) \mathrm{e}^{-\mathrm{i}\boldsymbol{k}\cdot(\boldsymbol{R}_i - \boldsymbol{R}_j)} \mathrm{e}^{\mathrm{i}\boldsymbol{p}\cdot \boldsymbol{R}_i} \mathrm{e}^{\mathrm{i}\boldsymbol{q}\cdot \boldsymbol{R}_j} =$$

$$= \frac{1}{N} \sum_{\boldsymbol{kqp}} J(\boldsymbol{k}) S^{\alpha}(\boldsymbol{p}) S^{\beta}(\boldsymbol{q}) \delta_{\boldsymbol{kp}} \delta_{\boldsymbol{k},-\boldsymbol{q}} = \frac{1}{N} \sum_{\boldsymbol{k}} J(\boldsymbol{k}) S^{\alpha}(\boldsymbol{k}) S^{\beta}(-\boldsymbol{k})\,,$$

$$\sum_i S_i^z = S^z(\boldsymbol{0})\,.$$

From this, it follows that:

$$H = -\sum_{i,j} J_{ij} \left(S_i^{+} S_j^{-} + S_i^z S_j^z\right) - g_J \frac{\mu_{\mathrm{B}}}{\hbar} B_0 \sum_i S_i^z =$$

$$= -\frac{1}{N} \sum_{\boldsymbol{k}} J(\boldsymbol{k}) \left(S^{+}(\boldsymbol{k}) S^{-}(-\boldsymbol{k}) + S^z(\boldsymbol{k}) S^z(-\boldsymbol{k})\right) - g_J \frac{\mu_{\mathrm{B}}}{\hbar} B_0 S^z(\boldsymbol{0})\,.$$

Solution 2.4.3

$$H = -\sum_{i,j} J_{ij} \left(S_i^{+} S_j^{-} + S_i^z S_j^z\right) - b \sum_i S_i^z =$$

$$\left(b = g_J \frac{\mu_{\mathrm{B}}}{\hbar} B_0\right)$$

$$= -\sum_{i,j} J_{ij} \Big(2S\hbar^2 \varphi(n_i)\, a_i a_j^+ \varphi(n_j) +$$

$$+ \hbar^2 (S - n_i)(S - n_j) \Big) - b \sum_i \hbar (S - n_i) ,$$

$$\sum_{i,j} J_{ij} = N J_0 ; \quad \sum_i J_{ij} = \sum_j J_{ij} = J_0 .$$

Ground-state energy:

$$E_0 = -N J_0 \hbar^2 S - N g_J \mu_B B_0 S = E_0(B_0) .$$

This implies that:

$$H = E_0(B_0) + 2S\hbar^2 J_0 \sum_i n_i - 2S\hbar^2 \sum_{i,j} J_{ij} \varphi(n_i)\, a_i a_j^+ \varphi(n_j) -$$

$$- \hbar^2 \sum_{i,j} J_{ij} n_i n_j \quad \text{q.e.d.}$$

Solution 2.4.4

$$\frac{M_0 - M_s(T)}{M_0} = \frac{1}{NS} \sum_q \frac{1}{\exp[\beta\hbar\omega(\boldsymbol{q})] - 1} .$$

First of all, we convert the sum into an integral:

$$\sum_q \frac{1}{\exp[\beta\hbar\omega(\boldsymbol{q})] - 1} = \frac{V}{(2\pi)^3} \int d^3q \left(e^{\beta\hbar\omega(\boldsymbol{q})} - 1 \right)^{-1} =$$

$$= \frac{V}{(2\pi)^3} \int d^3q \, \frac{e^{-\beta\hbar\omega(\boldsymbol{q})}}{1 - e^{-\beta\hbar\omega(\boldsymbol{q})}} = \frac{V}{(2\pi)^3} \sum_{n=1}^{\infty} \int d^3q \, e^{-n\beta\hbar\omega(\boldsymbol{q})} .$$

The integration is carried out over the first Brillouin zone. At low temperatures, β is very large, so that only the smallest magnon energies make a noticeable contribution. These are those with a small value of $|\boldsymbol{q}|$:

$$J_0 - J(\boldsymbol{q}) = \frac{1}{N} \sum_{i,j} J_{ij} \left(1 - e^{i\boldsymbol{q}\cdot\boldsymbol{R}_{ij}} \right) \approx \frac{1}{2N} \sum_{ij} J_{ij} \left(\boldsymbol{q} \cdot \boldsymbol{R}_{ij} \right)^2 \equiv \frac{D}{2S\hbar^2} q^2$$

$$\Longrightarrow \quad \hbar\omega(\boldsymbol{q}) \approx D q^2 .$$

For the same reason, we can replace the integration over the first Brillouin zone by an integral over the entire range of $\boldsymbol{q}$:

$$\frac{M_0 - M_s(T)}{M_0} = \frac{V}{2\pi^2 NS} \sum_{n=1}^{\infty} \int_0^{\infty} \mathrm{d}q\, q^2 \mathrm{e}^{-n\beta D q^2} =$$

$$= \frac{V}{2\pi^2 NS} \sum_{n=1}^{\infty} \frac{1}{2} (n\beta D)^{-3/2} \Gamma\left(\frac{3}{2}\right) .$$

Riemann's ζ function: $\zeta(m) = \sum_{n=1}^{\infty} \frac{1}{n^m}$.

From this we find for the low-temperature magnetisation:

$$\frac{M_0 - M_s(T)}{M_0} = C_{3/2} T^{3/2} \quad (T \overset{>}{\to} 0)$$

"Bloch's $T^{3/2}$ law",

$$C_{3/2} = \frac{V}{NS} \zeta\left(\frac{3}{2}\right) \left(\frac{k_\mathrm{B}}{4\pi D}\right)^{3/2} .$$

Solution 2.4.5

1. We verify the axioms of the scalar product:
   a) (A, B) is a complex number with

$$(A, B) = (B, A)^* ,$$

since

$$\frac{W_m - W_n}{E_n - E_m}$$

is a real number and

$$\big(\langle n \mid B^+ \mid m\rangle \langle m \mid A \mid n\rangle\big)^* = \langle n \mid A^+ \mid m\rangle \langle m \mid B \mid n\rangle .$$

   b) Linearity properties of the scalar product,

$$(A, \alpha_1 B_1 + \alpha_2 B_2) = \alpha_1 (A, B_1) + \alpha_2 (A, B_2) \quad \alpha_1, \alpha_2 \in \mathbf{C} ,$$

   follow immediately from those of the matrix element $\langle m \mid B \mid n\rangle$.

   c) $(A, A) \geq 0$, since

$$\frac{W_m - W_n}{E_n - E_m} \geq 0 \quad \Longrightarrow \quad (A, A) = \sum\nolimits_{n,m}' \left|\langle n \mid A^+ \mid m\rangle\right|^2 \frac{W_m - W_n}{E_n - E_m} \geq 0 .$$

   d) From $A = 0$, it naturally follows that $(A, A) = 0$. The converse, however, does not hold (see Ex. 2.4.6)! We are therefore dealing with a semidefinite scalar product!

2.

$$(A, B) = \sum_{n,m}{}' \langle n \mid A^+ \mid m\rangle \langle m \mid \left[C^+, H\right]_- \mid n\rangle \frac{W_m - W_n}{E_n - E_m} =$$

$$= \sum_{n,m} \langle n \mid A^+ \mid m\rangle \langle m \mid C^+ \mid n\rangle (W_m - W_n) .$$

Due to the factor on the right, the diagonal terms can now be counted. Using the completeness relation and the definition of W_n, we furthermore find:

$$(A, B) = -\sum_n W_n \langle n \mid A^+C^+ \mid n\rangle + \sum_m W_m \langle m \mid C^+A^+ \mid m\rangle =$$

$$= -\langle A^+C^+\rangle + \langle C^+A^+\rangle = \langle [C^+, A^+]_-\rangle ,$$

$$(B, B) = \left\langle \left[C^+, B^+\right]_- \right\rangle = \left\langle \left[C^+, [H, C]_-\right]_- \right\rangle .$$

For the third relation, we first carry out the following approximate estimate:

$$0 < \frac{W_m - W_n}{E_n - E_m} = \frac{1}{\mathrm{Tr}\, \mathrm{e}^{-\beta H}} \frac{\mathrm{e}^{-\beta E_m} + \mathrm{e}^{-\beta E_n}}{E_n - E_m} \frac{\mathrm{e}^{-\beta E_m} - \mathrm{e}^{-\beta E_n}}{\mathrm{e}^{-\beta E_m} + \mathrm{e}^{-\beta E_n}} =$$

$$= \frac{W_m + W_n}{E_n - E_m} \tanh\left[\frac{1}{2}\beta\,(E_n - E_m)\right] ,$$

$$\frac{\mathrm{d}}{\mathrm{d}x} \tanh x = \frac{1}{\cosh^2 x} < 1 \quad \text{for} \quad x \neq 0$$

$$\Longrightarrow \quad \frac{\tanh\left(\frac{1}{2}\beta\,(E_n - E_m)\right)}{E_n - E_m} = \frac{\tanh\left(\frac{1}{2}\beta\,|E_n - E_m|\right)}{|E_n - E_m|} \leq$$

$$\leq \frac{\frac{1}{2}\beta\,|E_n - E_m|}{|E_n - E_m|} = \frac{1}{2}\beta .$$

We thus have:

$$0 < \frac{W_m - W_n}{E_n - E_m} < \frac{1}{2}\beta\,(W_n + W_m) , \quad \text{when} \quad E_n \neq E_m .$$

And it then follows that:

$$(A, A) < \frac{1}{2}\beta \sum_{n,m}^{E_n \neq E_m} \langle n \mid A^+ \mid m\rangle \langle m \mid A \mid n\rangle (W_n + W_m) \leq$$

$$\leq \frac{1}{2}\beta \sum_{n,m} \langle n \mid A^+ \mid m\rangle \langle m \mid A \mid n\rangle (W_n + W_m) =$$

$$= \frac{1}{2}\beta \left(\langle A^+A\rangle + \langle AA^+\rangle\right) = \frac{1}{2}\beta \left\langle \left[A, A^+\right]_+ \right\rangle \quad \text{q. e. d.}$$

3. The scalar product here obeys *Schwarz's inequality:*

$$\left|(A, B)\right|^2 \leq (A, A)(B, B) .$$

According to 2., this means that

$$\left|\langle [C, A]_- \rangle\right|^2 \leq \frac{1}{2}\beta \left\langle [A, A^+]_+ \right\rangle \left\langle [C^+, [H, C]_-]_- \right\rangle$$

which proves the Bogoliubov inequality!

Solution 2.4.6

1.

$$(H, H) = \sum_{n,m}^{E_n \neq E_m} \langle n \mid H \mid m \rangle \langle m \mid H \mid n \rangle \frac{W_m - W_n}{E_n - E_m} =$$

$$= \sum_{n,m}^{E_n \neq E_m} E_n^2 \delta_{nm} \delta_{nm} \frac{W_m - W_n}{E_n - E_m} = 0 .$$

2.

$$\langle [C, A]_- \rangle = \left[\mathrm{Tr}\left(\mathrm{e}^{-\beta H}\right)\right]^{-1} \sum_n \mathrm{e}^{-\beta E_n} \langle n \mid CA - AC \mid n \rangle .$$

Because of $[C, H]_- = 0$, C and H have common eigenstates:

$$\langle [C, A]_- \rangle = \left[\mathrm{Tr}\left(\mathrm{e}^{-\beta H}\right)\right]^{-1} \sum_n \mathrm{e}^{-\beta E_n} c_n \langle n \mid A - A \mid n \rangle = 0 \quad \text{q. e. d.}$$

Solution 2.4.7

1a)

$$[C, A]_- = \left[S^+(\boldsymbol{k}), S^-(-\boldsymbol{k} - \boldsymbol{K})\right]_- = 2\hbar S^z(-\boldsymbol{K}) = 2\hbar \sum_i \mathrm{e}^{\mathrm{i}\boldsymbol{k} \cdot \boldsymbol{R}_i} S_i^z$$

$$\Longrightarrow \quad \left\langle [C, A]_- \right\rangle = 2\hbar \sum_i \mathrm{e}^{\mathrm{i}\boldsymbol{k} \cdot \boldsymbol{R}_i} \left\langle S_i^z \right\rangle = \frac{2\hbar N}{b} M(T, B_0) .$$

1b)

$$\sum_k \langle [A, A^+]_+ \rangle = \sum_k \langle [S^-(-k-K), S^+(k+K)]_+ \rangle =$$
$$= \sum_k \sum_{i,j} e^{i(k+k)\cdot(R_i - R_j)} \langle S_i^- S_j^+ + S_j^+ S_i^- \rangle =$$
$$= \sum_{i,j} e^{ik\cdot(R_i - R_j)} N \delta_{ij} \langle S_i^- S_i^+ + S_i^+ S_i^- \rangle =$$
$$= 2N \sum_i \langle (S_i^x)^2 + (S_i^y)^2 \rangle \le$$
$$\le 2N \sum_i \langle S_i^2 \rangle = 2\hbar^2 N^2 S(S+1)\,.$$

1c) Initially, we find:

$$R(k) \equiv \langle [[C, H]_-, C^+]_- \rangle = \sum_{m,n} e^{-ik\cdot(R_m - R_n)} \langle [[S_m^+, H]_-, S_n^-]_- \rangle .$$

We must therefore compute several commutators:

$$[S_m^+, H]_- =$$
$$= -\hbar \sum_i J_{im} \left\{ 2S_i^+ S_m^z - S_i^z S_m^+ - S_m^+ S_i^z \right\} + \hbar b B_0 S_m^+ e^{-ik\cdot R_m} =$$
$$= -2\hbar \sum_i J_{im} \left(S_i^+ S_m^z - S_i^z S_m^+ \right) + \hbar b B_0 S_m^+ e^{-ik\cdot R_m} ,$$

$$\left[[S_m^+, H]_-, S_n^- \right]_- =$$
$$= -2\hbar \sum_i J_{im} \left\{ S_i^+ [S_m^z, S_n^-]_- + [S_i^+, S_n^-]_- S_m^z - S_i^z [S_m^+, S_n^-]_- - \right.$$
$$\left. - [S_i^z, S_n^-]_- S_m^+ \right\} + \hbar b B_0 [S_m^+, S_n^-] e^{-ik\cdot R_m} =$$
$$= -2\hbar^2 \sum_i J_{im} \left\{ -\delta_{mn} S_i^+ S_n^- + 2\delta_{in} S_i^z S_m^z - 2\delta_{mn} S_i^z S_m^z + \delta_{in} S_i^- S_m^+ \right\} +$$
$$+ 2\hbar^2 b B_0 \delta_{mn} S_m^z e^{-ik\cdot R_m} =$$
$$= 2\hbar^2 \delta_{mn} \sum_i J_{im} \left(S_i^+ S_m^- + 2S_i^z S_m^z \right) - 2\hbar^2 J_{nm} \left(S_n^- S_m^+ + 2S_n^z S_m^z \right) +$$
$$+ 2\hbar^2 b B_0 \delta_{mn} S_m^z e^{-ik\cdot R_m} .$$

From this, we obtain the intermediate result:

$$R(\boldsymbol{k}) = 2\hbar^2 \sum_{m,n} J_{mn} \left(1 - \mathrm{e}^{-\mathrm{i}\boldsymbol{k}\cdot(\boldsymbol{R}_m - \boldsymbol{R}_n)}\right) \left\langle S_m^+ S_n^- + 2 S_m^z S_n^z \right\rangle +$$

$$+ 2\hbar^2 b B_0 \sum_m \mathrm{e}^{-\mathrm{i}\boldsymbol{k}\cdot\boldsymbol{R}_m} \left\langle S_m^z \right\rangle .$$

Here, we have repeatedly made use of $J_{ii} = 0$ and $J_{ij} = J_{ji}$.
Due to part 2. of Ex. 2.4.5, $R(\boldsymbol{k})$ cannot be negative. This naturally also holds for the corresponding expectation value $<R(\boldsymbol{k})>$, which is computed not with $C = S^+(\boldsymbol{k})$, but instead with $\widehat{C} = S^+(-\boldsymbol{k})$. It is thus clear that:

$$<R(\boldsymbol{k})> \le R(\boldsymbol{k}) + R(-\boldsymbol{k}) = +4\hbar^2 b B_0 \sum_m \mathrm{e}^{-\mathrm{i}\boldsymbol{k}\cdot\boldsymbol{R}_m} \left\langle S_m^z \right\rangle +$$

$$+ 4\hbar^2 \sum_{m,n} J_{mn} \Big[1 - \cos\left(\boldsymbol{k}\cdot(\boldsymbol{R}_m - \boldsymbol{R}_n)\right)\Big] \left\langle \boldsymbol{S}_m \cdot \boldsymbol{S}_n + S_m^z S_n^z \right\rangle .$$

To continue the estimate, we take the following form of the scalar product

$$(\boldsymbol{S}_m, \boldsymbol{S}_n) = \left\langle \boldsymbol{S}_m \cdot \boldsymbol{S}_n \right\rangle$$

and apply Schwarz's inequality:

$$\left|(\boldsymbol{S}_m, \boldsymbol{S}_n)\right|^2 \le (\boldsymbol{S}_m, \boldsymbol{S}_m) \cdot (\boldsymbol{S}_n, \boldsymbol{S}_n) .$$

Clearly, this implies that:

$$\left\langle \boldsymbol{S}_m \cdot \boldsymbol{S}_n \right\rangle^2 \le \hbar^4 [S(S+1)]^2 .$$

Furthermore, we also have:

$$\left\langle S_m^z S_n^z \right\rangle \le \hbar^2 S^2 .$$

With this, it then follows that:

$$R(\boldsymbol{k}) \le 4\hbar^2 N \left|B_0 M(T, B_0)\right| + 8\hbar^4 S(S+1) \sum_{m,n} J_{mn} \Big[1 - \cos\left(\boldsymbol{k}\cdot(\boldsymbol{R}_m - \boldsymbol{R}_n)\right)\Big] \le$$

$$\le 4\hbar^2 N \left|B_0 M(T, B_0)\right| + 8\hbar^4 S(S+1) \frac{1}{2} k^2 \underbrace{\sum_{m,n} J_{mn} \left|\boldsymbol{R}_m - \boldsymbol{R}_n\right|^2}_{NQ} .$$

We have thus shown that:

$$\left\langle \left[[C, H]_-, C^+ \right]_- \right\rangle \le 4N\hbar^2 \Big\{ \left|B_0 M(T, B_0)\right| + \hbar^2 k^2 Q S(S+1) \Big\} .$$

2a) As we know, $R(\boldsymbol{k}) \geq 0$. Therefore, we can write the Bogoliubov inequality as follows:

$$\frac{\beta}{2}\left\langle [A, A^+]_+ \right\rangle \geq \frac{\left| \langle [C, A]_- \rangle \right|^2}{\left\langle [[C, H]_-, C^+]_- \right\rangle} .$$

We sum this inequality over all $\boldsymbol{k}$ within the first Brillouin zone:

$$\beta S(S+1) \geq \frac{M^2}{\hbar^2 b^2} \frac{1}{N} \sum_{\boldsymbol{k}} \frac{1}{|B_0 M| + \hbar^2 k^2 Q S(S+1)} .$$

Taking the thermodynamic limit yields:

$$\frac{1}{N_d} \sum_{\boldsymbol{k}} \longrightarrow \frac{v_d}{(2\pi)^d} \int \mathrm{d}^d k ,$$

d: dimensionality of the system.
The d-dimensional *volume* V_d contains N_d spins ($v_d = V_d / N_d$). The integrand on the right-hand side of the inequality is positive. The inequality thus holds with certainty if we integrate not over the complete Brillouin zone, but instead over a sphere of radius k_0 which lies entirely within the zone:

$$S(S+1) \geq \frac{M^2 v_d \, \Omega_d}{\beta \hbar^2 b^2 (2\pi)^d} \int_0^{k_0} \mathrm{d}k \, \frac{k^{d-1}}{|B_0 M| + \hbar^2 k^2 Q S(S+1)} .$$

The angular integration has already been carried out and just gives the surface area of the unit sphere as Ω_d.

2b) $\boxed{d = 1}$

$$\int \frac{\mathrm{d}x}{a^2 x^2 + b^2} = \frac{1}{ab} \arctan \frac{ax}{b} + c$$

$$\Longrightarrow \quad S(S+1) \geq \frac{M^2 v_d}{2\pi \beta \hbar^2 b^2} \frac{\arctan\left(k_0 \sqrt{\frac{\hbar^2 Q S(S+1)}{|B_0 M|}} \right)}{\sqrt{\hbar^2 Q S(S+1) \, |B_0 M|}} .$$

We are interested in the behaviour at low fields:

$$\arctan\left(k_0 \sqrt{\frac{\hbar^2 Q S(S+1)}{|B_0 M|}} \right) \underset{B_0 \to 0}{\longrightarrow} \frac{\pi}{2} .$$

This implies that:

$$\left| M(T, B_0) \right| \underset{B_0 \to 0}{\overset{<}{\to}} \mathrm{const} \frac{B_0^{1/3}}{T^{2/3}}$$

and thus

$$M_s(T) = 0 \quad \text{for} \quad T \neq 0!$$

$\boxed{d = 2}$

$$\int \frac{\mathrm{d}x\, x}{a^2x^2 + b^2} = \frac{1}{2a^2} \ln c\left(a^2x^2 + b^2\right)$$

$$\Longrightarrow \quad S(S+1) \geq \frac{M^2 v_d}{2\pi\beta\hbar^2 b^2} \frac{\ln\left[\frac{\hbar^2 QS(S+1) + |B_0 M|}{|B_0 M|}\right]}{2\hbar^2 QS(S+1)} .$$

For low fields, we thus obtain:

$$\left|M(T, B_0)\right| \underset{B_0 \to 0}{\overset{<}{\to}} \text{const}_1 \frac{1}{\sqrt{T \ln\left(\frac{\text{const}_2 + |B_0 M|}{|B_0 M|}\right)}} .$$

This also results in

$$M_s(T) = 0 \qquad \text{for} \quad T \neq 0!$$

Section 3.1.6

Solution 3.1.1

The equation of motion for Heisenberg operators:

$$\mathrm{i}\hbar \frac{\mathrm{d}}{\mathrm{d}t} a_{\boldsymbol{k}\sigma}(t) = \left[a_{\boldsymbol{k}\sigma}, H_\mathrm{e}\right]_-(t) ,$$

$$\left[a_{\boldsymbol{k}\sigma}, H_\mathrm{e}\right]_- = \sum_{\boldsymbol{k}'\sigma'} \varepsilon\left(\boldsymbol{k}'\right)\left[a_{\boldsymbol{k}\sigma}, a^+_{\boldsymbol{k}'\sigma'} a_{\boldsymbol{k}'\sigma'}\right]_- =$$

$$= \sum_{\boldsymbol{k}'\sigma'} \varepsilon\left(\boldsymbol{k}'\right) \delta_{\boldsymbol{k}\boldsymbol{k}'} \delta_{\sigma\sigma'} a_{\boldsymbol{k}'\sigma'} = \varepsilon(\boldsymbol{k}) a_{\boldsymbol{k}\sigma}$$

$$\Longrightarrow \quad \mathrm{i}\hbar \frac{\mathrm{d}}{\mathrm{d}t} a_{\boldsymbol{k}\sigma}(t) = \varepsilon(\boldsymbol{k}) a_{\boldsymbol{k}\sigma}(t) ,$$

$$a_{\boldsymbol{k}\sigma}(t = 0) = a_{\boldsymbol{k}\sigma}$$

$$\Longrightarrow \quad a_{\boldsymbol{k}\sigma}(t) = a_{\boldsymbol{k}\sigma}\, \mathrm{e}^{-\frac{\mathrm{i}}{\hbar}\varepsilon(\boldsymbol{k})t} .$$

Analogously, one finds for the phonon gas:

$$i\hbar \frac{d}{dt} b_{qr}(t) = \left[b_{qr}, H_p\right]_- (t) = \hbar\omega_r(\boldsymbol{q}) b^+_{qr}(t)$$

$$\Longrightarrow \quad b_{qr}(t) = b_{qr}\, e^{-i\omega_r(q)t} \,.$$

An alternative derivation was used in (2.166)!

Solution 3.1.2

1.

$$f(\lambda) = e^{\lambda A} B e^{-\lambda A}\,; \quad A \neq A(\lambda)\,; \quad B \neq B(\lambda)$$

$$\Longrightarrow \quad \frac{d}{d\lambda} f(\lambda) = e^{\lambda A} [A, B]_- e^{-\lambda A}\,,$$

$$\frac{d^2}{d\lambda^2} f(\lambda) = e^{\lambda A} [A, [A, B]_-]_- e^{-\lambda A}\,,$$

$$\vdots$$

$$\frac{d^n}{d\lambda^n} f(\lambda) = e^{\lambda A} \Big[A, \big[A, \ldots [A, \underbrace{B\,]_- \ldots]_-\big]_-}_{n\text{-fold}} e^{-\lambda A}\,.$$

Taylor expansion around $\lambda = 0$:

$$f(\lambda) = B + \sum_{n=1}^{\infty} \frac{\lambda^n}{n!} \left[\frac{d^n}{d\lambda^n} f(\lambda)\right]_{\lambda=0} = B + \sum_{n=1}^{\infty} \frac{\lambda^n}{n!} \Big[A, \big[A, \ldots [A, \underbrace{B\,]_- \ldots]_-\big]_-}_{n-\text{fold}}\,.$$

The comparison yields:

$$\alpha_0 = B\,,$$

$$\alpha_n = \Big[A, [A, \ldots [A, B]_- \ldots]_-\Big]_- \frac{1}{n!}\,, \qquad n \geq 1\,.$$

2.

$$\alpha_n = 0 \quad \text{for} \quad n \geq 2\,,$$

$$\alpha_0 = B\,; \quad \alpha_1 = [A, B]_-$$

$$\Longrightarrow \quad f(\lambda) = B + \lambda [A, B]_-\,.$$

3.

$$g(\lambda) = \mathrm{e}^{\lambda A}\mathrm{e}^{\lambda B},$$

$$\frac{\mathrm{d}}{\mathrm{d}\lambda}g(\lambda) = \mathrm{e}^{\lambda A}(A+B)\mathrm{e}^{\lambda B} = \mathrm{e}^{\lambda A}(A+B)\mathrm{e}^{-\lambda A}g(\lambda) = (A + f(\lambda))\, g(\lambda).$$

Using part 2., we then obtain:

$$\frac{\mathrm{d}}{\mathrm{d}\lambda}g(\lambda) = (A + B + \lambda[A,B]_-)\, g(\lambda).$$

4. The preconditions give:

$$[(A+B), [A,B]_-]_- = 0.$$

The operator coefficient in the above differential equation thus behaves on integration just like a normal variable:

$$\frac{\mathrm{d}}{\mathrm{d}\lambda}g(\lambda) = (a_1 + \lambda a_2)\, g(\lambda),$$

$$g(0) = 1$$

$$\Longrightarrow \quad g(\lambda) = \mathrm{e}^{a_1\lambda + \frac{1}{2}a_2\lambda^2}$$

$$\Longrightarrow \quad g(\lambda = 1) = \mathrm{e}^{A}\mathrm{e}^{B} = \exp\left(A + B + \frac{1}{2}[A,B]_-\right) \quad \text{q. e. d.}$$

Solution 3.1.3

$$\rho\int_0^\beta \mathrm{d}\lambda\, \dot{A}(t - \mathrm{i}\lambda\hbar) = \rho\int_0^\beta \mathrm{d}\lambda \frac{\mathrm{i}}{\hbar}\frac{\mathrm{d}}{\mathrm{d}\lambda}A(t - \mathrm{i}\lambda\hbar) =$$

$$= \frac{\mathrm{i}}{\hbar}\rho\,[A(t - \mathrm{i}\hbar\beta) - A(t)] =$$

$$= \frac{\mathrm{i}}{\hbar}\rho\left[\mathrm{e}^{\frac{\mathrm{i}}{\hbar}(-\mathrm{i}\hbar\beta)\mathcal{H}}A(t)\mathrm{e}^{-\frac{\mathrm{i}}{\hbar}(-\mathrm{i}\hbar\beta)\mathcal{H}} - A(t)\right] =$$

$$= \frac{\mathrm{i}}{\hbar}\rho\left(\mathrm{e}^{\beta\mathcal{H}}A(t)\mathrm{e}^{-\beta\mathcal{H}} - A(t)\right) =$$

$$= \frac{\mathrm{i}}{\hbar}\left[\frac{\mathrm{e}^{-\beta\mathcal{H}}\mathrm{e}^{\beta\mathcal{H}}A(t)\mathrm{e}^{-\beta\mathcal{H}}}{\mathrm{Tr}\left(\mathrm{e}^{-\beta\mathcal{H}}\right)} - \rho A(t)\right] =$$

$$= \frac{\mathrm{i}}{\hbar}(A(t)\rho - \rho A(t)) = \frac{\mathrm{i}}{\hbar}[A(t), \rho]_- \quad \text{q. e. d.}$$

Solution 3.1.4

$$\left\langle \left[A(t), B\left(t'\right)\right]_{-}\right\rangle = \mathrm{Tr}\left\{\rho\left[A(t), B\left(t'\right)\right]_{-}\right\} = \mathrm{Tr}\left\{\rho A(t) B\left(t'\right) - \rho B\left(t'\right) A(t)\right\} =$$

$$= \mathrm{Tr}\left\{B\left(t'\right) \rho A(t) - \rho B\left(t'\right) A(t)\right\} = \mathrm{Tr}\left\{\left[B\left(t'\right), \rho\right]_{-} A(t)\right\}$$

(cyclic invariance of the trace!) .

Inserting the Kubo identity:

$$\left\langle\!\left\langle A(t); B\left(t'\right)\right\rangle\!\right\rangle^{\mathrm{ret}} = -\mathrm{i}\Theta\left(t-t'\right)\left\langle\left[A(t), B\left(t'\right)\right]_{-}\right\rangle =$$

$$= -\hbar\Theta\left(t-t'\right)\int_0^\beta \mathrm{d}\lambda\, \mathrm{Tr}\left\{\rho \dot{B}\left(t'-\mathrm{i}\lambda\hbar\right) A(t)\right\} =$$

$$= -\hbar\Theta\left(t-t'\right)\int_0^\beta \mathrm{d}\lambda \left\langle \dot{B}\left(t'-\mathrm{i}\lambda\hbar\right) A(t)\right\rangle \quad \text{q. e. d.}$$

Solution 3.1.5

In (3.84) we derived:

$$\sigma^{\beta\alpha}(E) = -\frac{1}{\hbar}\int_{-\infty}^{+\infty} \mathrm{d}t \left\langle\!\left\langle j^\beta(0); P^\alpha(-t)\right\rangle\!\right\rangle \mathrm{e}^{\frac{\mathrm{i}}{\hbar}(E+\mathrm{i}0^+)t}\,.$$

With the result from Ex. 3.1.4, it follows that:

$$\sigma^{\beta\alpha}(E) = \int_0^\infty \mathrm{d}t \int_0^\beta \mathrm{d}\lambda \left\langle \dot{P}^\alpha(-t-\mathrm{i}\lambda\hbar) j^\beta(0)\right\rangle \mathrm{e}^{\frac{\mathrm{i}}{\hbar}(E+\mathrm{i}0^+)t} =$$

$$\overset{3.3.79}{=} V \int_0^\infty \mathrm{d}t \int_0^\beta \mathrm{d}\lambda \left\langle j^\alpha(-t-\mathrm{i}\lambda\hbar) j^\beta(0)\right\rangle \mathrm{e}^{\frac{\mathrm{i}}{\hbar}(E+\mathrm{i}0^+)t}\,.$$

The correlation function depends only upon the time difference. Therefore, we also have:

$$\sigma^{\beta\alpha}(E) = V \int_0^\infty \mathrm{d}t \int_0^\beta \mathrm{d}\lambda \left\langle j^\alpha(0) j^\beta(t+\mathrm{i}\lambda\hbar)\right\rangle \mathrm{e}^{\frac{\mathrm{i}}{\hbar}(E+\mathrm{i}0^+)t} \quad \text{q. e. d.}$$

Solution 3.1.6

The dipole-moment operator (3.77)

$$P = \sum_{i=n}^{N} q_i \hat{\boldsymbol{r}}_i$$

is a single-particle operator. We consider identical particles:

$$q_i = q \quad \forall i\,.$$

1. In the Bloch representation:

$$\widehat{P} = q \sum_{\substack{k\sigma \\ k'\sigma'}} \langle \boldsymbol{k}\sigma \mid \hat{\boldsymbol{r}} \mid \boldsymbol{k}'\sigma' \rangle\, a^+_{\boldsymbol{k}\sigma} a_{\boldsymbol{k}'\sigma'}\,.$$

Matrix element:

$$\langle \boldsymbol{k}\sigma \mid \hat{\boldsymbol{r}} \mid \boldsymbol{k}'\sigma' \rangle = \int \mathrm{d}^3r\, \langle \boldsymbol{k}\sigma \mid \hat{\boldsymbol{r}} \mid \boldsymbol{r} \rangle \langle \boldsymbol{r} \mid \boldsymbol{k}'\sigma' \rangle = \delta_{\sigma\sigma'} \int \mathrm{d}^3r\, \langle \boldsymbol{k} \mid \boldsymbol{r} \rangle \boldsymbol{r} \langle \boldsymbol{r} \mid \boldsymbol{k}' \rangle =$$

$$= \delta_{\sigma\sigma'} \int \mathrm{d}^3r\, \psi^*_{\boldsymbol{k}\sigma}(\boldsymbol{r}) \boldsymbol{r} \psi_{\boldsymbol{k}'\sigma}(\boldsymbol{r})$$

$$\psi_{\boldsymbol{k}\sigma}(\boldsymbol{r}): \quad \text{Bloch function (2.20)}\,,$$

$$\langle \boldsymbol{k}\sigma \mid \hat{\boldsymbol{r}} \mid \boldsymbol{k}'\sigma' \rangle = \delta_{\sigma\sigma'} \boldsymbol{p}_{\boldsymbol{k}\boldsymbol{k}'\sigma}$$

$$\boldsymbol{p}_{\boldsymbol{k}\boldsymbol{k}'\sigma} \equiv \int \mathrm{d}^3r\, \psi^*_{\boldsymbol{k}\sigma}(\boldsymbol{r}) \boldsymbol{r} \psi_{\boldsymbol{k}'\sigma}(\boldsymbol{r})$$

$$\Longrightarrow \quad \widehat{P} = q \sum_{\boldsymbol{k}\boldsymbol{k}'\sigma} \boldsymbol{p}_{\boldsymbol{k}\boldsymbol{k}'\sigma} a^+_{\boldsymbol{k}\sigma} a_{\boldsymbol{k}'\sigma}\,.$$

2. In the Wannier representation:

$$\boldsymbol{p}_{ij\sigma} = \int \mathrm{d}^3r\, \omega^*_\sigma (\boldsymbol{r} - \boldsymbol{R}_i)\, \boldsymbol{r} \omega_\sigma \left(\boldsymbol{r} - \boldsymbol{R}_j\right)$$

$$\omega_\sigma (\boldsymbol{r} - \boldsymbol{R}_i): \quad \text{Wannier function (2.29)}$$

$$\Longrightarrow \quad \widehat{P} = q \sum_{ij\sigma} \boldsymbol{p}_{ij\sigma} a^+_{i\sigma} a_{j\sigma}\,.$$

Current-density operator:

$$\hat{j} = \frac{1}{V}\widehat{\dot{P}} = -\frac{\mathrm{i}}{\hbar V}\left[\widehat{\boldsymbol{P}}, H\right]_- .$$

1.

$$\hat{j} = -\frac{\mathrm{i}q}{\hbar V}\sum_{kk'\sigma} \boldsymbol{p}_{kk'\sigma}\left[a^+_{k\sigma}a_{k'\sigma}, H\right]_- .$$

2.

$$\hat{j} = -\frac{\mathrm{i}q}{\hbar V}\sum_{ij\sigma} \boldsymbol{p}_{ij\sigma}\left[a^+_{i\sigma}a_{j\sigma}, H\right]_- .$$

The conductivity tensor is found immediately by inserting into (3.85).

Solution 3.1.7

1. From Ex. 3.1.6:

$$\widehat{\boldsymbol{P}} \approx q\sum_{i,\sigma}\boldsymbol{R}_i n_{i\sigma} ,$$

$$\hat{j} \approx -\frac{\mathrm{i}q}{\hbar V}\sum_{i,\sigma}\boldsymbol{R}_i\left[n_{i\sigma}, H\right]_- .$$

2. $n_{i\sigma}$ commutes with all the occupation-number operators. Therefore, we have:

$$\begin{aligned}
\left[n_{i\sigma}, H\right]_- &= \sum_{l,m,\sigma'} T_{lm}\left[n_{i\sigma}, a^+_{l\sigma'}a_{m\sigma'}\right]_- = \\
&= \sum_{l,m} T_{lm}\left(\delta_{il}a^+_{i\sigma}a_{m\sigma} - \delta_{im}a^+_{l\sigma}a_{i\sigma}\right) = \\
&= \sum_{m}\left(T_{im}a^+_{i\sigma}a_{m\sigma} - T_{mi}a^+_{m\sigma}a_{i\sigma}\right) .
\end{aligned}$$

Current-density operator:

$$\hat{j} \approx -\frac{\mathrm{i}q}{\hbar V}\sum_{im\sigma}\boldsymbol{R}_i\left(T_{im}a^+_{i\sigma}a_{m\sigma} - T_{mi}a^+_{m\sigma}a_{i\sigma}\right)$$

$$\Longrightarrow \quad \hat{j} \approx -\frac{\mathrm{i}q}{\hbar V}\sum_{im\sigma}T_{im}\left(\boldsymbol{R}_i - \boldsymbol{R}_m\right)a^+_{i\sigma}a_{m\sigma} .$$

Conductivity tensor:

$$\sigma^{\alpha\beta}(E) = \mathrm{i}\hbar \frac{\frac{N}{V} q^2}{m\,(E + \mathrm{i}0^+)} - \frac{\mathrm{i}q^2}{\hbar^2 V\,(E + \mathrm{i}0^+)} \sum_{\substack{im\sigma \\ jn\sigma'}} T_{im} T_{jn} \cdot$$

$$\cdot \left(R_i^\alpha - R_m^\alpha\right)\left(R_j^\beta - R_n^\beta\right) \langle\langle a_{i\sigma}^+ a_{m\sigma} ; a_{j\sigma'}^+ a_{n\sigma'} \rangle\rangle_E^{\mathrm{ret}}$$

$$(\alpha, \beta = x, y, z)\,.$$

Section 3.2.6

Solution 3.2.1

$$\Theta\left(t - t'\right) = \int_{-\infty}^{t-t'} \mathrm{d}t''\, \delta\left(t''\right)$$

$$\Longrightarrow \quad \frac{\partial}{\partial t} \Theta\left(t - t'\right) = \frac{\mathrm{d}}{\mathrm{d}(t - t')} \Theta\left(t - t'\right) = \delta\left(t - t'\right)\,,$$

$$\frac{\partial}{\partial t'} \Theta\left(t - t'\right) = -\frac{\mathrm{d}}{\mathrm{d}(t - t')} \Theta\left(t - t'\right) = -\delta\left(t - t'\right)\,.$$

Solution 3.2.2

$$\begin{aligned} G_{AB}^{\mathrm{c}}\left(t, t'\right) &= -\mathrm{i}\langle T_\varepsilon \left(A(t) B\left(t'\right)\right)\rangle = \\ &= -\mathrm{i}\Theta\left(t - t'\right)\left\langle A(t) B\left(t'\right)\right\rangle - \mathrm{i}\varepsilon\Theta\left(t' - t\right)\left\langle B\left(t'\right) A(t)\right\rangle. \end{aligned}$$

From this, it follows that:

$$\begin{aligned} \mathrm{i}\hbar \frac{\partial}{\partial t} G_{AB}^{\mathrm{c}}\left(t, t'\right) &= +\hbar\delta\left(t - t'\right)\left\langle A(t) B\left(t'\right)\right\rangle - \mathrm{i}\Theta\left(t - t'\right)\left\langle [A, \mathcal{H}]_-(t) B\left(t'\right)\right\rangle - \\ &\quad - \hbar\varepsilon\delta\left(t - t'\right)\left\langle B\left(t'\right) A(t)\right\rangle - \mathrm{i}\varepsilon\Theta\left(t' - t\right)\left\langle B\left(t'\right) [A, \mathcal{H}]_-(t)\right\rangle = \\ &= \hbar\delta\left(t - t'\right)\left\langle \left[A(t), B\left(t'\right)\right]_{-\varepsilon}\right\rangle - \mathrm{i}\left\langle T_\varepsilon \left([A, \mathcal{H}](t) B\left(t'\right)\right)\right\rangle = \\ &= \hbar\delta\left(t - t'\right)\left\langle [A, B]_{-\varepsilon}\right\rangle + \langle\langle [A, \mathcal{H}]_-(t); B\left(t'\right)\rangle\rangle^{\mathrm{c}} \quad \text{q.e.d.} \end{aligned}$$

Solution 3.2.3

$$
\begin{aligned}
&\left\langle B(0)A(t+\mathrm{i}\beta)\right\rangle = \\
&= \frac{1}{\Xi}\,\mathrm{Tr}\left\{\mathrm{e}^{-\beta\mathcal{H}}B\,\mathrm{e}^{\frac{\mathrm{i}}{\hbar}\mathcal{H}(t+\mathrm{i}\hbar\beta)}A\,\mathrm{e}^{-\frac{\mathrm{i}}{\hbar}\mathcal{H}(t+\mathrm{i}\hbar\beta)}\right\} = \\
&= \frac{1}{\Xi}\,\mathrm{Tr}\left\{\mathrm{e}^{\beta\mathcal{H}}\mathrm{e}^{-\beta\mathcal{H}}B\,\mathrm{e}^{\frac{\mathrm{i}}{\hbar}\mathcal{H}t}\,\mathrm{e}^{-\beta\mathcal{H}}A\,\mathrm{e}^{-\frac{\mathrm{i}}{\hbar}\mathcal{H}t}\right\} = \\
&= \frac{1}{\Xi}\,\mathrm{Tr}\left\{\mathrm{e}^{-\beta\mathcal{H}}\mathrm{e}^{\frac{\mathrm{i}}{\hbar}\mathcal{H}t}A\,\mathrm{e}^{-\frac{\mathrm{i}}{\hbar}\mathcal{H}t}B\right\} = \left\langle A(t)B(0)\right\rangle .
\end{aligned}
$$

Here, we have made repeated use of the cyclic invariance of the trace.

Solution 3.2.4

1. $\boxed{t-t'>0}$

Fig. A.8

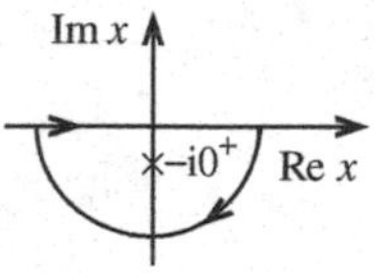

The integrand has a pole at $x = x_0 = -\mathrm{i}0^+$. Residue:

$$c_{-1} = \lim_{x\to x_0}(x-x_0)\frac{\mathrm{e}^{-\mathrm{i}x(t-t')}}{x+\mathrm{i}0^+} = \lim_{x\to x_0}\mathrm{e}^{-\mathrm{i}x(t-t')} = 1\,.$$

The semicircle is closed in the lower half-plane due to $t-t'>0$; then the exponential function ensures that the contribution on the semicircle vanishes. The pole is mathematically circumvented in a *negative* sense. It then follows that:

$$\Theta\left(t-t'\right) = \frac{\mathrm{i}}{2\pi}(-2\pi\mathrm{i})1 = 1\,.$$

2. $\boxed{t-t'<0}$

In order that no contribution result from the semicircle, it is now closed in the upper half-plane. It then follows that:

$$\Theta\left(t-t'\right) = 0\,,$$

since there is no pole within the region of integration.

Solution 3.2.5

$$f(\omega) = \int_{-\infty}^{+\infty} dt\, \bar{f}(t) e^{i\omega t} .$$

Suppose that the integral exists for real values of ω. Set:

$$\omega = \omega_1 + i\omega_2$$

$$\Longrightarrow \quad f(\omega) = \int_{-\infty}^{+\infty} dt\, \bar{f}(t) e^{i\omega_1 t} e^{-\omega_2 t} .$$

1. $\bar{f}(t) = 0$ for $t < 0$:

$$\Longrightarrow \quad f(\omega) = \int_{0}^{\infty} dt\, \bar{f}(t) e^{i\omega_1 t} e^{-\omega_2 t} .$$

   This converges for all $\omega_2 > 0$, and can thus be analytically continued in the upper half-plane!

2. $\bar{f}(t) = 0$ for $t > 0$:

$$\Longrightarrow \quad f(\omega) = \int_{-\infty}^{0} dt\, \bar{f}(t) e^{i\omega_1 t} e^{-\omega_2 t} .$$

   This converges for all $\omega_2 < 0$, and can thus be analytically continued in the lower half-plane.

Solution 3.2.6

It is expedient to first transform the conductivity tensor in Ex. 3.1.7 from the Bloch representation into a real-space representation. We have:

$$\sum_{k} (\nabla_k \varepsilon(\boldsymbol{k}))\, n_{k\sigma} =$$

$$= \frac{1}{N^2} \sum_{k} \sum_{i,j} \sum_{m,n} T_{ij} \left[-\mathrm{i} \left(\boldsymbol{R}_i - \boldsymbol{R}_j \right) \right] \mathrm{e}^{-\mathrm{i}\boldsymbol{k}\cdot(\boldsymbol{R}_i - \boldsymbol{R}_j)} \mathrm{e}^{\mathrm{i}\boldsymbol{k}\cdot(\boldsymbol{R}_m - \boldsymbol{R}_n)} a^+_{m\sigma} a_{n\sigma} =$$

$$= \frac{1}{N} \sum_{ij} \sum_{m,n} T_{ij} \left[-\mathrm{i} \left(\boldsymbol{R}_i - \boldsymbol{R}_j \right) \right] \delta_{n,\, m+j-i} a^+_{m\sigma} a_{n\sigma} =$$

$$= \frac{1}{N} \sum_{ijm} T_{ij} \left[-\mathrm{i} \left(\boldsymbol{R}_i - \boldsymbol{R}_j \right) \right] a^+_{m\sigma} a_{m+j-i\sigma} \,.$$

We insert this into the *interaction term* of the conductivity tensor, keeping in mind that because of translational symmetry,

$$\frac{1}{N} \sum_{m} \left\langle\!\left\langle a^+_{m\sigma} a_{m+j-i\sigma}; \dots \right\rangle\!\right\rangle^{\mathrm{ret}}_E = \left\langle\!\left\langle a^+_{i\sigma} a_{j\sigma}; \dots \right\rangle\!\right\rangle^{\mathrm{ret}}_E \,,$$

must hold. Then, from Ex. 3.1.7, we still have:

$$\sigma^{\alpha\beta}(E) = \mathrm{i}\hbar \frac{\frac{N}{V} e^2}{m\,(E + \mathrm{i}0^+)} + \frac{\mathrm{i}e^2}{\hbar^2 V\,(E + \mathrm{i}0^+)} \cdot$$

$$\cdot \sum_{\substack{k\sigma \\ k'\sigma'}} (\nabla_k \varepsilon(\boldsymbol{k})) \left(\nabla_{k'} \varepsilon\left(\boldsymbol{k}'\right) \right) \left\langle\!\left\langle n_{k\sigma}; n_{k'\sigma} \right\rangle\!\right\rangle^{\mathrm{ret}}_E \,.$$

For a system of non-interacting electrons:

$$\mathcal{H}_0 = \sum_{p\bar{\sigma}} \varepsilon(\boldsymbol{p}) a^+_{p\bar{\sigma}} a_{p\bar{\sigma}}$$

$$\Longrightarrow \quad [n_{k\sigma}, \mathcal{H}_0]_- = 0\,, \quad \left\langle [n_{k\sigma}, n_{k'\sigma'}]_- \right\rangle = 0\,.$$

With this, the equation of motion of the *higher-order* Green's function becomes trivial:

$$E \left\langle\!\left\langle n_{k\sigma}; n_{k'\sigma'} \right\rangle\!\right\rangle^{\mathrm{ret}}_E \equiv 0\,.$$

The interaction term thus vanishes, as expected:

$$\left(\sigma^{\alpha\beta}(E)\right)^{(0)} = \mathrm{i}\hbar \frac{\frac{N}{V} e^2}{m(E + \mathrm{i}0^+)}\,.$$

Solution 3.2.7

$$\left[G_{AB}^{\text{ret, adv}}(t,t')\right]^* = \left[\mp \mathrm{i}\Theta\left[\pm(t-t')\right]\langle[A(t),B(t')]_{-\varepsilon}\rangle\right]^* =$$
$$= \pm \mathrm{i}\Theta\left[\pm(t-t')\right]\langle[A(t),B(t')]_{-\varepsilon}\rangle^* = \pm \mathrm{i}\Theta\left[\pm(t-t')\right]\langle[A(t),B(t')]_{-\varepsilon}^{+}\rangle =$$
$$= \pm \mathrm{i}\Theta\left[\pm(t-t')\right]\langle B^+(t')\,A^+(t) - \varepsilon A^+(t)B^+(t')\rangle =$$
$$= \mp \mathrm{i}\varepsilon\Theta\left[\pm(t-t')\right]\langle[A^+(t),B^+(t')]_{-\varepsilon}\rangle =$$
$$= \varepsilon G_{A^+B^+}^{\text{ret, adv}} \quad \text{q. e. d.}$$

Solution 3.2.8

$$\int\limits_{-\infty}^{+\infty} \mathrm{d}E\,\left\{EG_{AB}^{\mathrm{c}}(E) - \hbar\langle[A,B]_{-\varepsilon}\rangle\right\} = \int\limits_{-\infty}^{+\infty} \mathrm{d}E\,\langle\langle [A,\mathcal{H}]_-;B\rangle\rangle_E^{\mathrm{c}} =$$
$$= \int\limits_{-\infty}^{+\infty} \mathrm{d}E \int\limits_{-\infty}^{+\infty} \mathrm{d}t\,\mathrm{e}^{\frac{\mathrm{i}}{\hbar}Et}\langle\langle [A,\mathcal{H}]_-(t);B(0)\rangle\rangle^{\mathrm{c}} =$$
$$= -\mathrm{i}\int\limits_{-\infty}^{+\infty} \mathrm{d}E\,\left\{\int\limits_{0}^{\infty} \mathrm{d}t\,\mathrm{e}^{\frac{\mathrm{i}}{\hbar}Et}\langle[A,\mathcal{H}]_-(t)B(0)\rangle + \right.$$
$$\left. +\varepsilon\int\limits_{-\infty}^{0} \mathrm{d}t\,\mathrm{e}^{\frac{\mathrm{i}}{\hbar}Et}\langle B(0)[A,\mathcal{H}]_-(t)\rangle\right\} =$$
$$= 2\pi\hbar^2\left\{\int\limits_{0}^{\infty}\mathrm{d}t\,\delta(t)\,\langle\dot{A}(t)B(0)\rangle + \varepsilon\int\limits_{-\infty}^{0}\mathrm{d}t\,\delta(t)\,\langle B(0)\dot{A}(t)\rangle\right\} =$$
$$= \pi\hbar^2\left\{\langle\dot{A}(0)B(0)\rangle + \varepsilon\,\langle B(0)\dot{A}(0)\rangle\right\} \quad \text{q. e. d.}$$

Solution 3.2.9

$$\mathcal{H} = \sum_{\boldsymbol{k}\sigma}\varepsilon(\boldsymbol{k})a_{\boldsymbol{k}\sigma}^{+}a_{\boldsymbol{k}\sigma} - \mu\widehat{N} = \sum_{\boldsymbol{k}\sigma}(\varepsilon(\boldsymbol{k})-\mu)\,a_{\boldsymbol{k}\sigma}^{+}a_{\boldsymbol{k}\sigma}\,.$$

One can readily calculate:

$$[a_{k\sigma}, \mathcal{H}] = \sum_{k'\sigma'} \left(\varepsilon\left(k'\right) - \mu\right) \left[a_{k\sigma}, a^+_{k'\sigma'} a_{k'\sigma'}\right]_- =$$

$$= \sum_{k'\sigma'} \left(\varepsilon\left(k'\right) - \mu\right) \delta_{kk'} \delta_{\sigma\sigma'} a_{k'\sigma'} = (\varepsilon(k) - \mu)\, a_{k\sigma} \,.$$

From this, it follows that:

$$\left[[a_{k\sigma}, \mathcal{H}]_- \mathcal{H}\right]_- = (\varepsilon(k) - \mu)\,[a_{k\sigma}, \mathcal{H}]_- = (\varepsilon(k) - \mu)^2\, a_{k\sigma} \,.$$

For the spectral moments, this implies that:

$$M^{(0)}_{k\sigma} = \left\langle \left[a_{k\sigma}, a^+_{k\sigma}\right]_+ \right\rangle = 1 \,,$$

$$M^{(1)}_{k\sigma} = \left\langle \left[[a_{k\sigma}, \mathcal{H}]_- \,, a^+_{k\sigma}\right]_+ \right\rangle =$$

$$= (\varepsilon(k) - \mu) \left\langle \left[a_{k\sigma}, a^+_{k\sigma}\right]_+ \right\rangle = (\varepsilon(k) - \mu) \,,$$

$$M^{(2)}_{k\sigma} = \left\langle \left[\left[[a_{k\sigma}, \mathcal{H}]_- \,, \mathcal{H}\right]_- \,, a^+_{k\sigma}\right]_+ \right\rangle =$$

$$= (\varepsilon(k) - \mu)^2 \left\langle \left[a_{k\sigma}, a^+_{k\sigma}\right]_+ \right\rangle = (\varepsilon(k) - \mu)^2 \,.$$

By complete induction, one then immediately obtains:

$$M^{(n)}_{k\sigma} = (\varepsilon(k) - \mu)^n \;; \quad n = 0, 1, 2, \dots \,.$$

The relation (3.166) with the spectral density,

$$M^{(n)}_{k\sigma} = \frac{1}{\hbar} \int\limits_{-\infty}^{+\infty} \mathrm{d}E\, E^n S_{k\sigma}(E) \,,$$

then leads to the solution:

$$S_{k\sigma}(E) = \hbar \delta\left(E - \varepsilon(k) + \mu\right) .$$

Solution 3.2.10

1.

$$\mathrm{Tr}(\rho) = \int\limits_{-\infty}^{+\infty} \mathrm{e}^{-\beta \frac{p^2}{2m}}\, \mathrm{d}p = \sqrt{2\pi m k_\mathrm{B} T}$$

2.

$$\begin{aligned}
\mathrm{Tr}(\rho)\,\langle H\rangle &= \mathrm{Tr}(\rho H)\\
&= \frac{1}{2m}\int\limits_{-\infty}^{+\infty} p^2\,\mathrm{e}^{-\beta\frac{p^2}{2m}}\,\mathrm{d}p\\
&= -\frac{\mathrm{d}}{\mathrm{d}\beta}\int\limits_{-\infty}^{+\infty}\mathrm{e}^{-\beta\frac{p^2}{2m}}\,\mathrm{d}p\\
&= -\frac{\mathrm{d}}{\mathrm{d}\beta}\sqrt{\frac{2\pi m}{\beta}} = \frac{1}{2}\sqrt{2\pi m}\,\beta^{-3/2}\\
\Longrightarrow\quad \langle H\rangle &= \frac{1}{2}\frac{\sqrt{2\pi m}}{\sqrt{2\pi m k_\mathrm{B}T}}(k_\mathrm{B}T)^{3/2} = \frac{1}{2}k_\mathrm{B}T
\end{aligned}$$

3.

$$EG_p^{(+)}(E) = \hbar\,\underbrace{\langle[p,p]_-\rangle}_{=0} + \langle\!\langle\underbrace{[p,H]_-}_{=0};p\rangle\!\rangle_E^{(+)} = 0$$

$$\Longrightarrow\quad G_p^{(+)}(E) \equiv 0\quad \text{for } E \neq 0$$

4.

$$\langle p^2\rangle = \frac{1}{\hbar}\int\limits_{-\infty}^{+\infty}\mathrm{d}E\,\frac{-\frac{1}{\pi}\,\mathrm{Im}\,G_p^{(+)}\left(E+\mathrm{i}0^+\right)}{\mathrm{e}^{\beta E}-1} + D = D$$

5.

$$EG_p^{(-)}(E) = \hbar\,\langle[p,p]_+\rangle + \langle\!\langle\underbrace{[p,H]_-}_{=0};p\rangle\!\rangle_E^{(-)} = 2\hbar\langle p^2\rangle$$

$$\begin{aligned}
&\Longrightarrow\quad \text{“combined” Green's function:}\quad G_p^{(-)}(E) = \frac{2\hbar\langle p^2\rangle}{E}\\
&\Longrightarrow\quad 2\hbar D = \lim_{E\to 0} EG_p^{(-)}(E) = 2\hbar\langle p^2\rangle\\
&\Longrightarrow\quad D = \langle p^2\rangle
\end{aligned}$$

The contradiction is removed, but no information is obtained from the spectral theorem.

6.

$$H' = \frac{p^2}{2m} + \frac{1}{2}m\omega^2x^2\quad(\omega\to 0)$$

$$[p, H']_- = \frac{1}{2} m\omega^2 [p, x^2]_- = \frac{1}{2} m\omega^2 (x[p, x]_- + [p, x]_- p) = -\mathrm{i}\hbar m\omega^2 x$$

$$[x, H']_- = \left[x, \frac{p^2}{2m} \right]_- = \frac{\mathrm{i}\hbar}{m} p$$

Chain of equations of motion:

$$EG_p^{(+)}(E) = 0 + \langle\langle [p, H']_-; p \rangle\rangle_E^{(+)} = -\mathrm{i}\hbar m\omega^2 \langle\langle x; p \rangle\rangle_E^{(+)}$$

$$E \langle\langle x; p \rangle\rangle_E^{(+)} = \mathrm{i}\hbar^2 + \langle\langle [x, H']_-; p \rangle\rangle = i\hbar^2 + \frac{\mathrm{i}\hbar}{m} G_p^{(+)}(E)$$

$$\Longrightarrow \quad E^2 G_p^{(+)}(E) = \hbar^3 m\omega^2 + \hbar^2\omega^2 G_p^{(+)}(E)$$

$$\Longrightarrow \quad G_p^{(+)}(E) = \frac{\hbar^3 m\omega^2}{E^2 - \hbar^2\omega^2} = \frac{1}{2} m\hbar^2\omega \left(\frac{1}{E - \hbar\omega} - \frac{1}{E + \hbar\omega} \right)$$

7. Anti-commutator Green's function:

$$EG_p^{(-)}(E) = 2\hbar \left\langle p^2 \right\rangle - \mathrm{i}\hbar m\omega^2 \langle\langle x; p \rangle\rangle_E^{(-)}$$
$$E \langle\langle x; p \rangle\rangle_E^{(-)} = \hbar \langle xp + px \rangle + \frac{\mathrm{i}\hbar}{m} G_p^{(-)}(E)$$
$$\Longrightarrow \quad E^2 G_p^{(-)}(E) = 2\hbar \left\langle p^2 \right\rangle E - i\hbar^2 m\omega^2 \langle xp + px \rangle + \hbar^2\omega^2 G_p^{(-)}(E)$$
$$\Longrightarrow \quad G_p^{(-)}(E) = \frac{2\hbar \left\langle p^2 \right\rangle E - \mathrm{i}\hbar^2 m\omega^2 \langle xp + px \rangle}{E^2 - \hbar^2\omega^2}$$

The poles naturally remain unchanged!

$$\Longrightarrow \quad 2\hbar D = \lim_{E \to 0} EG_p^{(-)}(E) = \frac{0}{-\hbar^2\omega^2} = 0$$
$$\Longrightarrow \quad D = 0$$

8.

$$\begin{aligned}
\langle H \rangle_\omega &= \frac{1}{2m} \left\langle p^2 \right\rangle_\omega \\
&= \frac{1}{2m\hbar} \int_{-\infty}^{+\infty} \mathrm{d}E \frac{-\frac{1}{\pi} \operatorname{Im} G_p^{(+)}(E + \mathrm{i}0^+)}{\mathrm{e}^{\beta E} - 1} \\
&= \frac{\hbar\omega}{4} \int_{-\infty}^{+\infty} \mathrm{d}E \frac{\delta(E - \hbar\omega) - \delta(E + \hbar\omega)}{\mathrm{e}^{\beta E} - 1} \\
&= \frac{\hbar\omega}{4} \left(\frac{1}{\mathrm{e}^{\beta\hbar\omega} - 1} - \frac{1}{\mathrm{e}^{-\beta\hbar\omega} - 1} \right)
\end{aligned}$$

9.

$$
\begin{aligned}
\lim_{\omega\to 0}\langle H\rangle_\omega &= \lim_{\omega\to 0}\frac{\hbar\omega}{4}\left(\frac{1}{e^{\beta\hbar\omega}-1}-\frac{1}{e^{-\beta\hbar\omega}-1}\right)\\
&= \lim_{\omega\to 0}\frac{\hbar\omega}{4}\left(\frac{1}{\beta\hbar\omega}-\frac{1}{-\beta\hbar\omega}\right)\\
\Longrightarrow\quad \lim_{\omega\to 0}\langle H\rangle_\omega &= \frac{1}{2}k_B T
\end{aligned}
$$

This agrees with the result in 2.!

Section 3.3.4

Solution 3.3.1

1. Phonons can be created and again annihilated in arbitrary numbers. In thermodynamic equilibrium, the particle number adjusts itself to the value for which the free energy F is minimised:

$$\frac{\partial F}{\partial N}\overset{!}{=}0\,.$$

The left-hand side, on the other hand, defines μ!

2. Equation of motion:

$$
\begin{aligned}
\left[b_{qr}, H\right]_- &= \sum_{q,r'}\hbar\omega_{r'}(q)\left[b_{qr}, b^+_{q'r'}b_{q'r'}\right]_- = \sum_{q',r'}\hbar\omega_{r'}(q')\left[b_{qr}, b^+_{q'r'}\right]_- b_{q'r'} =\\
&= \hbar\omega_r(q)b_{qr}\,.
\end{aligned}
$$

With this, it follows that:

$$
\begin{aligned}
\left[E-\hbar\omega_r(q)\right]G^\alpha_{qr}(E) &= \hbar\left\langle\left[b_{qr}, b^+_{qr}\right]_-\right\rangle = \hbar\\
\Longrightarrow\quad G^{\text{ret, adv}}_q(E) &= \frac{\hbar}{E-\hbar\omega_r(q)\pm i0^+}\,.
\end{aligned}
$$

3. Computed in Ex. 3.1.1:

$$b_{qr}(t) = b_{qr}\,\mathrm{e}^{-\mathrm{i}\omega_r(\boldsymbol{q})t}$$

$$\Longrightarrow \quad \left\langle \left[b_{qr}(t), b_{qr}^{+}(t')\right]_{-}\right\rangle = \mathrm{e}^{-\mathrm{i}\omega_r(\boldsymbol{q})(t-t')}\left\langle \left[b_{qr}, b_{qr}^{+}\right]_{-}\right\rangle$$

$$\Longrightarrow \quad G_{qr}^{\mathrm{ret}}\left(t, t'\right) = -\mathrm{i}\Theta\left(t - t'\right)\mathrm{e}^{-\mathrm{i}\omega_r(\boldsymbol{q})(t-t')},$$

$$G_{qr}^{\mathrm{adv}}\left(t, t'\right) = +\mathrm{i}\Theta\left(t' - t\right)\mathrm{e}^{-\mathrm{i}\omega_r(\boldsymbol{q})(t-t')}.$$

Check by means of Fourier transformation:

$$G_{qr}^{\mathrm{ret}}\left(t, t'\right) = \frac{1}{2\pi\hbar}\int\limits_{-\infty}^{+\infty} \mathrm{d}E\,\mathrm{e}^{-\frac{\mathrm{i}}{\hbar}E(t-t')}\frac{\hbar}{E - \hbar\omega_r(\boldsymbol{q}) + \mathrm{i}0^{+}} =$$

$$\overset{\overline{E}=E-\hbar\omega_r(\boldsymbol{q})}{=} \mathrm{e}^{-\mathrm{i}\omega_r(\boldsymbol{q})(t-t')}\frac{1}{2\pi}\int\limits_{-\infty}^{+\infty} \mathrm{d}\overline{E}\,\frac{\mathrm{e}^{-\frac{\mathrm{i}}{\hbar}\overline{E}(t-t')}}{\overline{E} + \mathrm{i}0^{+}} =$$

$$\overset{x=\overline{E}/\hbar}{=} \mathrm{e}^{-\mathrm{i}\omega_r(\boldsymbol{q})(t-t')}\frac{1}{2\pi}\int\limits_{-\infty}^{+\infty} \mathrm{d}x\,\frac{\mathrm{e}^{-\mathrm{i}x(t-t')}}{x + \mathrm{i}0^{+}} =$$

$$= -\mathrm{i}\Theta\left(t - t'\right)\mathrm{e}^{-\mathrm{i}\omega_r(\boldsymbol{q})(t-t')} \quad \text{(s. Ex. 3.2.4)}.$$

4. Spectral density:

$$S_{qr}(E) = -\frac{1}{\pi}\,\mathrm{Im}\,G_{qr}^{\mathrm{ret}}(E) = \hbar\delta(E - \hbar\omega_r(\boldsymbol{q})).$$

Mean occupation number, spectral theorem:

$$\langle m_{qr}\rangle = \langle b_{qr}^{+}b_{qr}\rangle = D_{qr}^{+}\left[\exp(\beta\hbar\omega_r(\boldsymbol{q})) - 1\right]^{-1},$$

D_{qr} from the *combined* anti-commutator Green's function. As a result of

$$\left\langle \left[b_{qr}, b_{qr}^{+}\right]_{+}\right\rangle = 1 + \langle m_{qr}\rangle,$$

we find for the latter:

$$G_{qr}^{(-)}(E) = \frac{\hbar\left(1 + \langle m_{qr}\rangle\right)}{E - \hbar\omega_r(\boldsymbol{q})},$$

$\omega_r(\boldsymbol{q}) = 0$ only for acoustic branches at $q = 0$:

$$q = 0 \Longleftrightarrow \lambda = \infty : \quad \begin{array}{l}\text{macroscopic translation of the whole crystal!}\\ \text{Uninteresting!}\end{array}$$

$q \neq 0$:

$$D_{qr} = \frac{1}{2\hbar} \lim_{E \to 0} E G_{qr}^{(0)}(E) = 0 \,.$$

We still have:

$$\langle m_{qr} \rangle = [\exp(\hbar\omega_r(\boldsymbol{q})) - 1]^{-1}$$

Bose-Einstein distribution function.

Internal energy:

$$U = \langle H \rangle = \sum_{qr} \hbar\omega_r(\boldsymbol{q}) \left(\langle m_{qr} \rangle + \frac{1}{2} \right) .$$

Solution 3.3.2

1. Equation of motion:

$$\begin{aligned}
&\left[a_{k\sigma}, H^*\right]_- = \\
&= \sum_{p\sigma'} t(\boldsymbol{p}) \left[a_{k\sigma}, a^+_{p\sigma'} a_{p\sigma'}\right]_- - \Delta \sum_{p} \left[a_{k\sigma}, a_{-p\downarrow} a_{p\uparrow} + a^+_{p\uparrow} a^+_{-p\downarrow}\right]_- = \\
&= \sum_{p\sigma'} t(\boldsymbol{p}) \delta_{\sigma\sigma'} \delta_{kp} a_{p\sigma'} - \Delta \sum_{p} \left(\delta_{kp} \delta_{\sigma\uparrow} a^+_{-p\downarrow} - \delta_{k-p} \delta_{\sigma\downarrow} a^+_{p\uparrow}\right) = \\
&= t(\boldsymbol{k}) a_{k\sigma} - \Delta \left(\delta_{\sigma\uparrow} - \delta_{\sigma\downarrow}\right) a^+_{-k-\sigma} ,
\end{aligned}$$

$$z_\sigma = \begin{cases} +1 & \text{for} \quad \sigma = \uparrow , \\ -1 & \text{for} \quad \sigma = \downarrow . \end{cases}$$

With this, the equation of motion becomes:

$$(E - t(\boldsymbol{k}))\, G_{k\sigma}(E) = \hbar - \Delta z_\sigma \langle\langle a^+_{-k-\sigma} ; a^+_{k\sigma} \rangle\rangle .$$

The Green's function on the right-hand side of the equation prevents direct solution. We formulate the corresponding equation of motion for it, also:

$$\begin{aligned}
&[a^+_{-k\,-\sigma}, H^*]_- = \\
&= -t(-k)a^+_{-k\,-\sigma} - \Delta \sum_p [a^+_{-k\,-\sigma}, a_{-p\downarrow}a_{p\uparrow}]_- = \\
&= -t(k)a^+_{-k\,-\sigma} - \Delta \sum_p \left(\delta_{kp}\delta_{-\sigma\downarrow}a_{p\uparrow} - \delta_{-kp}\delta_{-\sigma\uparrow}a_{-p\downarrow}\right) = \\
&= -t(k)a^+_{-k\,-\sigma} - \Delta z_\sigma a_{k\sigma}\,.
\end{aligned}$$

This yields the following equation of motion:

$$\begin{aligned}
&(E + t(k))\, \langle\!\langle a^+_{-k\,-\sigma}; a^+_{k\sigma} \rangle\!\rangle = -\Delta z_\sigma G_{k\sigma}(E) \\
\Longrightarrow \quad &\langle\!\langle a^+_{-k\,-\sigma}; a^+_{k\sigma} \rangle\!\rangle = -\frac{z_\sigma \Delta}{E + t(k)} G_{k\sigma}(E)\,.
\end{aligned}$$

This is to be inserted into the equation of motion for $G^{\mathrm{ret}}_{k\sigma}(E)$:

$$\left(E - t(k) - \frac{\Delta^2}{E + t(k)}\right) G_{k\sigma}(E) = \hbar\,.$$

Excitation energies:

$$E(k) = +\sqrt{t^2(k) + \Delta^2} \underset{t \to 0}{\longrightarrow} \Delta \quad \textit{Energy gap.}$$

Green's function:

$$G_{k\sigma}(E) = \frac{\hbar\,(E + t(k))}{E^2 - E^2(k)}\,.$$

Taking the boundary conditions into account:

$$G^{\mathrm{ret}}_{k\sigma}(E) = \frac{\hbar}{2E(k)} \left[\frac{t(k) + E(k)}{E - E(k) + \mathrm{i}0^+} - \frac{t(k) - E(k)}{E + E(k) + \mathrm{i}0^+}\right].$$

2. For Δ, we require the expectation value:

$$\langle a^+_{k\uparrow} a^+_{-k\downarrow} \rangle\,.$$

Its evaluation can be accomplished using the spectral theorem and the Green's function used in part 1.:

$$\langle\langle a^+_{-k\downarrow}; a^+_{k\uparrow} \rangle\rangle_E = \frac{-\Delta}{E + t(\boldsymbol{k})} G_{k\uparrow}(E) = \frac{-\hbar\Delta}{E^2 - E^2(\boldsymbol{k})} .$$

Taking the boundary conditions into account, we obtain for the corresponding retarded function:

$$\langle\langle a^+_{-k\downarrow}; a^+_{k\uparrow} \rangle\rangle^{\text{ret}}_E = \frac{\hbar\Delta}{2E(\boldsymbol{k})} \left(\frac{1}{E + E(\boldsymbol{k}) + \mathrm{i}0^+} - \frac{1}{E - E(\boldsymbol{k}) + \mathrm{i}0^+} \right) .$$

The corresponding spectral density:

$$S_{-k\downarrow; k\uparrow}(E) = \frac{\hbar\Delta}{2E(\boldsymbol{k})} \left[\delta\left(E + E(\boldsymbol{k})\right) - \delta\left(E - E(\boldsymbol{k})\right) \right] .$$

Spectral theorem:

$$\begin{aligned}
\langle a^+_{k\uparrow} a^+_{-k\downarrow} \rangle &= \frac{1}{\hbar} \int_{-\infty}^{+\infty} \mathrm{d}E \, \frac{S_{-k\downarrow; k\uparrow}(E)}{\exp(\beta E) + 1} = \\
&= \frac{\Delta}{2E(\boldsymbol{k})} \left(\frac{1}{\exp(-\beta E(\boldsymbol{k})) + 1} - \frac{1}{\exp(\beta E(\boldsymbol{k})) + 1} \right) = \\
&= \frac{\Delta}{2E(\boldsymbol{k})} \tanh\left(\frac{1}{2} \beta E(\boldsymbol{k}) \right) .
\end{aligned}$$

Finally, we obtain:

$$\Delta = \frac{1}{2} \Delta V \sum_{\boldsymbol{k}} \frac{\tanh\left(\frac{1}{2} \beta \sqrt{t^2(\boldsymbol{k}) + \Delta^2} \right)}{\sqrt{t^2(\boldsymbol{k}) + \Delta^2}} .$$

$\Delta = \Delta(T) \implies$ The energy gap is T-dependent.

Fig. A.9

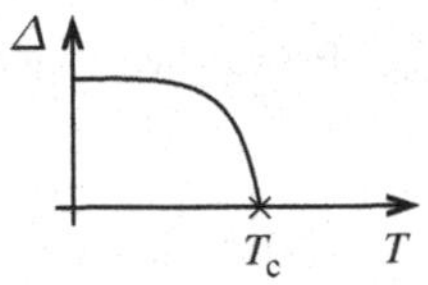

Special case:

$$T \to 0 \quad \implies \quad \tanh\left(\frac{1}{2} \beta \sqrt{t^2(\boldsymbol{k}) + \Delta^2} \right) \to 1$$

$\implies$ the same result as in Ex. 2.3.6 for $\Delta_{\boldsymbol{k}} \equiv \Delta$.

Solution 3.3.3

1. We prove the assertion using complete induction:
Initiation of induction $p = 1, 2$:

$$\left[a_{k\sigma}, H^*\right]_- = t(k)a_{k\sigma} - z_\sigma \Delta a^+_{-k-\sigma}$$

(s. Ex. 3.3.2),

$$\begin{aligned}\left[\left[a_{k\sigma}, H^*\right]_-, H^*\right]_- &= t(k)\left(t(k)a_{k\sigma} - z_\sigma \Delta a^+_{-k-\sigma}\right) - \\ &\quad - z_\sigma \Delta \left(-t(k)a^+_{-k-\sigma} - z_\sigma \Delta a_{k\sigma}\right) = \\ &= \left(t^2(k) + \Delta^2\right) a_{k\sigma} .\end{aligned}$$

Conclusion of induction $p \longrightarrow p+1$:

a) p even:

$$\underbrace{\left[\cdots\left[\left[a_{k\sigma}, H^*\right]_-, H^*\right]_-, \ldots, H^*\right]_-}_{(p+1)\text{-fold commutator}} =$$

$$= \left(t^2 + \Delta^2\right)^{p/2} \left[a_{k\sigma}, H^*\right]_- = \left(t^2 + \Delta^2\right)^{p/2} \left(t a_{k\sigma} - z_\sigma \Delta a^+_{-k-\sigma}\right) .$$

b) p odd:

$$\underbrace{\left[\cdots\left[\left[a_{k\sigma}, H^*\right]_-, H^*\right]_-, \ldots, H^*\right]_-}_{(p+1)\text{-fold commutator}} =$$

$$\begin{aligned}&= \left(t^2 + \Delta^2\right)^{(1/2)(p-1)} \left[t a_{k\sigma} - z_\sigma \Delta a^+_{-k-\sigma}, H^*\right]_- = \\ &= \left(t^2 + \Delta^2\right)^{(1/2)(p-1)} \left[t \left(t a_{k\sigma} - z_\sigma \Delta a^+_{-k-\sigma}\right) - z_\sigma \Delta \left(-t a^+_{-k-\sigma} - \Delta z_\sigma a_{k\sigma}\right)\right] = \\ &= \left(t^2 + \Delta^2\right)^{(1/2)(p+1)} a_{k\sigma} \quad \text{q. e. d.}\end{aligned}$$

For the spectral moments of the one-electron spectral density, we find immediately from this:
$n = 0, 1, 2, \ldots$

$$\begin{aligned}M^{(2n)}_{k\sigma} &= \left(t^2(k) + \Delta^2\right)^n , \\ M^{(2n+1)}_{k\sigma} &= \left(t^2(k) + \Delta^2\right)^n t(k) .\end{aligned}$$

2. We use:

$$M_{k\sigma}^{(n)} = \frac{1}{\hbar} \int_{-\infty}^{+\infty} \mathrm{d}E \, E^n \, S_{k\sigma}(E) \, .$$

Determining equations from the first four spectral moments:

$$\begin{aligned}
\alpha_{1\sigma} + \alpha_{2\sigma} &= \hbar \, , \\
\alpha_{1\sigma} E_{1\sigma} + \alpha_{2\sigma} E_{2\sigma} &= ht \, , \\
\alpha_{1\sigma} E_{1\sigma}^2 + \alpha_{2\sigma} E_{2\sigma}^2 &= \hbar \left(t^2 + \Delta^2\right) \, , \\
\alpha_{1\sigma} E_{1\sigma}^3 + \alpha_{2\sigma} E_{2\sigma}^3 &= \hbar \left(t^2 + \Delta^2\right) t \, .
\end{aligned}$$

This can be rearranged to:

$$\begin{aligned}
\alpha_{2\sigma} \left(E_{2\sigma} - E_{1\sigma}\right) &= \hbar \left(t - E_{1\sigma}\right) \, , \\
\alpha_{2\sigma} E_{2\sigma} \left(E_{2\sigma} - E_{1\sigma}\right) &= \hbar \left[t^2 + \Delta^2 - t E_{1\sigma}\right] \, , \\
\alpha_{2\sigma} E_{2\sigma}^2 \left(E_{2\sigma} - E_{1\sigma}\right) &= \hbar \left[\left(t^2 + \Delta^2\right)\left(t - E_{1\sigma}\right)\right] \, .
\end{aligned}$$

After division, it follows that:

$$E_{2\sigma}^2 = t^2 + \Delta^2 \quad \Longrightarrow \quad E_{2\sigma}(\boldsymbol{k}) = +\sqrt{t^2(\boldsymbol{k}) + \Delta^2} \equiv E(\boldsymbol{k}) \, .$$

This then leads to:

$$\begin{aligned}
& E(\boldsymbol{k}) = \frac{t^2 + \Delta^2 - t E_{1\sigma}}{t - E_{1\sigma}} = t + \frac{\Delta^2}{t - E_{1\sigma}} \\
\Longrightarrow \quad & \left(E(\boldsymbol{k}) - t(\boldsymbol{k})\right)^{-1} \Delta^2 = t(\boldsymbol{k}) - E_{1\sigma}(\boldsymbol{k}) \\
\Longrightarrow \quad & E_{1\sigma}(\boldsymbol{k}) = t(\boldsymbol{k}) - \frac{\Delta^2}{E(\boldsymbol{k}) - t(\boldsymbol{k})} = \frac{E(\boldsymbol{k}) t(\boldsymbol{k}) - E^2(k)}{E(\boldsymbol{k}) - t(\boldsymbol{k})} \\
\Longrightarrow \quad & E_{1\sigma}(\boldsymbol{k}) = -E(\boldsymbol{k}) = -E_{2\sigma}(\boldsymbol{k}) \, .
\end{aligned}$$

Spectral weights:

$$\alpha_{2\sigma}(\boldsymbol{k})2E(\boldsymbol{k}) = \hbar\,(t(\boldsymbol{k}) + E(\boldsymbol{k}))$$

$$\Longrightarrow \quad \alpha_{2\sigma}(\boldsymbol{k}) = \hbar\frac{t(\boldsymbol{k}) + E(\boldsymbol{k})}{2E(\boldsymbol{k})}\,,$$

$$\alpha_{1\sigma}(\boldsymbol{k}) = \hbar - \alpha_{2\sigma}(\boldsymbol{k}) = \hbar\frac{E(\boldsymbol{k}) - t(\boldsymbol{k})}{2E(\boldsymbol{k})}$$

$$\Longrightarrow \quad S_{\boldsymbol{k}\sigma}(E) = \hbar\left[\frac{E(\boldsymbol{k}) - t(\boldsymbol{k})}{2E(\boldsymbol{k})}\delta\,(E + E(\boldsymbol{k})) + \frac{E(\boldsymbol{k}) + t(\boldsymbol{k})}{2E(\boldsymbol{k})}\delta\,(E - E(\boldsymbol{k}))\right].$$

Solution 3.3.4

1. All the $H_{\boldsymbol{k}}$'s commute. We thus need consider only one fixed value $\boldsymbol{k}$. With the normalised vacuum state $|0\rangle$ and the fact that we are dealing with Fermions, only the following four states need be considered:

$$\begin{aligned}
|0,0\rangle &= |0\rangle\,;\\
|1,0\rangle &= a^+_{\boldsymbol{k}\uparrow}\,|0\rangle\,;\\
|0,1\rangle &= a^+_{-\boldsymbol{k}\downarrow}\,|0\rangle\,;\\
|1,1\rangle &= a^+_{\boldsymbol{k}\uparrow}a^+_{-\boldsymbol{k}\downarrow}\,|0\rangle\,.
\end{aligned}$$

The effect of $H_{\boldsymbol{k}}$ on these states can bc easily read off:

$$\begin{aligned}
H_{\boldsymbol{k}}\,|0,0\rangle &= -\Delta\,|1,1\rangle\,,\\
H_{\boldsymbol{k}}\,|1,0\rangle &= t(\boldsymbol{k})\,|1,0\rangle\,,\\
H_{\boldsymbol{k}}\,|0,1\rangle &= t(\boldsymbol{k})\,|0,1\rangle\,,\\
H_{\boldsymbol{k}}\,|1,1\rangle &= 2t(\boldsymbol{k})\,|1,1\rangle - \Delta\,|0,0\rangle\,.
\end{aligned}$$

This yields the following Hamiltonian matrix:

$$H_{\boldsymbol{k}} \equiv \begin{pmatrix} 0 & 0 & 0 & -\Delta\\ 0 & t(\boldsymbol{k}) & 0 & 0\\ 0 & 0 & t(\boldsymbol{k}) & 0\\ -\Delta & 0 & 0 & 2t(\boldsymbol{k}) \end{pmatrix}.$$

The eigenvalues are found from the requirement:

$$\det\left|H_k - E\mathbf{1}\right| \overset{!}{=} 0\,,$$

$$0 = (t-E)\det\begin{pmatrix} -E & 0 & -\Delta \\ 0 & t-E & 0 \\ -\Delta & 0 & 2t-E \end{pmatrix} =$$

$$= (t-E)\left[-E(t-E)(2t-E) - \Delta^2(t-E)\right]$$

$$\Longrightarrow \quad E_{1,2}(\boldsymbol{k}) = t(\boldsymbol{k})\,,$$

$$0 = -E(2t-E) - \Delta^2 \quad \Longleftrightarrow \quad \Delta^2 = E^2 - 2tE\,.$$

We thus find in summary the following energy eigenvalues:

$$E_0(\boldsymbol{k}) = t(\boldsymbol{k}) - \sqrt{t^2(\boldsymbol{k}) + \Delta^2} = t(\boldsymbol{k}) - E(\boldsymbol{k})\,,$$

$$E_1(\boldsymbol{k}) = E_2(\boldsymbol{k}) = t(\boldsymbol{k})\,,$$

$$E_3(\boldsymbol{k}) = t(\boldsymbol{k}) + \sqrt{t^2(\boldsymbol{k}) + \Delta^2} = t(\boldsymbol{k}) + E(\boldsymbol{k})\,.$$

2. *Ansatz*:

$$\left|E_0(\boldsymbol{k})\right\rangle = \alpha_0\,|0,0\rangle + \alpha_1\,|1,0\rangle + \alpha_2\,|0,1\rangle + \alpha_3\,|1,1\rangle\,,$$

$$(H_k - E_0(\boldsymbol{k})\mathbf{1})\,\left|E_0(\boldsymbol{k})\right\rangle = 0\,,$$

$$\begin{pmatrix} -E_0 & 0 & 0 & -\Delta \\ 0 & t-E_0 & 0 & 0 \\ 0 & 0 & t-E_0 & 0 \\ -\Delta & 0 & 0 & 2t-E_0 \end{pmatrix} \begin{pmatrix} \alpha_0 \\ \alpha_1 \\ \alpha_2 \\ \alpha_3 \end{pmatrix} = \begin{pmatrix} 0 \\ 0 \\ 0 \\ 0 \end{pmatrix}$$

$$\Longrightarrow \quad \alpha_1 = \alpha_2 = 0\,,$$

$$E_0\alpha_0 + \Delta\alpha_3 = 0\,; \quad \alpha_0^2 + \alpha_3^2 = 1$$

$$\Longrightarrow \quad \alpha_0^2 = \frac{\Delta^2}{E_0^2}\left(1 - \alpha_0^2\right) \quad \Longrightarrow \quad \alpha_0^2 = \frac{\Delta^2}{E_0^2 + \Delta^2}$$

$$\Longrightarrow \quad \alpha_0^2 = \frac{1}{2}\frac{\Delta^2}{t^2 + \Delta^2 - tE(\boldsymbol{k})} = \frac{1}{2}\frac{\Delta^2\left(t^2 + \Delta^2 + tE(\boldsymbol{k})\right)}{t^4 + \Delta^4 + 2t^2\Delta^2 - t^2\left(t^2 + \Delta^2\right)} =$$

$$= \frac{1}{2}\frac{t^2 + \Delta^2 + t\sqrt{t^2 + \Delta^2}}{\Delta^2 + t^2}$$

$$\Longrightarrow \quad \alpha_0^2 = \frac{1}{2}\left(1 + \frac{t(\boldsymbol{k})}{\sqrt{t^2 + \Delta^2}}\right) \equiv u_k^2 \quad \text{(s. Ex. 2.3.6)}\,.$$

This leads to:

$$\alpha_3^2 = \frac{1}{2}\left(1 - \frac{t(\boldsymbol{k})}{\sqrt{t^2 + \Delta^2}}\right) \equiv v_k^2 \quad \text{(s. Ex. 2.3.6)} .$$

The ground state is then given by:

$$\left| E_0(\boldsymbol{k}) \right\rangle = \left(u_k + v_k a^+_{\boldsymbol{k}\uparrow} a^+_{-\boldsymbol{k}\downarrow} \right) |0\rangle .$$

The two single-particle states are now found:

$$\left| E_1(\boldsymbol{k}) \right\rangle = a^+_{\boldsymbol{k}\uparrow} |0\rangle ,$$

$$\left| E_2(\boldsymbol{k}) \right\rangle = a^+_{-\boldsymbol{k}\downarrow} |0\rangle .$$

We finally still have to calculate $\left| E_3(\boldsymbol{k}) \right\rangle$:

$$\begin{pmatrix} -t(\boldsymbol{k}) - E(\boldsymbol{k}) & 0 & 0 & -\Delta \\ 0 & -E(\boldsymbol{k}) & 0 & 0 \\ 0 & 0 & -E(\boldsymbol{k}) & 0 \\ -\Delta & 0 & 0 & t(\boldsymbol{k}) - E(\boldsymbol{k}) \end{pmatrix} \begin{pmatrix} \gamma_0 \\ \gamma_1 \\ \gamma_2 \\ \gamma_3 \end{pmatrix} = 0$$

$$\Longrightarrow \quad \gamma_1 = 0 = \gamma_2 ,$$

$$(t + E)\gamma_0 + \Delta\gamma_3 = 0 ; \quad \gamma_0^2 + \gamma_3^2 = 1$$

$$\Longrightarrow \quad \gamma_0^2 = + \frac{\Delta^2}{(t + E)^2} \left(1 - \gamma_0^2\right)$$

$$\Longrightarrow \quad \gamma_0^2 = \frac{\Delta^2}{\Delta^2 + (t + E)^2} = \frac{\Delta^2}{2\Delta^2 + 2t^2 + 2tE} =$$

$$= \frac{1}{2} \frac{\Delta^2(\Delta^2 + t^2 - tE)}{\Delta^4 + t^4 + 2t^2\Delta^2 - t^2(t^2 + \Delta^2)} =$$

$$= \frac{1}{2}\left(1 - \frac{t}{\sqrt{t^2 + \Delta^2}}\right) = v_k^2$$

$$\Longrightarrow \quad \gamma_3^2 = u_k^2$$

$$\Longrightarrow \quad \left| E_3(\boldsymbol{k}) \right\rangle = \left(v_k - u_k a^+_{\boldsymbol{k}\uparrow} a^+_{-\boldsymbol{k}\downarrow} \right) |0\rangle ,$$

The minus sign ensures that $\langle E_0 \mid E_3 \rangle = 0$ holds!

3.

$$\Delta_{30} = 2\sqrt{t^2(\boldsymbol{k}) + \Delta^2}\,,$$

which is a *two*-particle excitation, does *not* appear as a pole of the one-electron Green's function!

$$\Delta_{32} = \Delta_{31} = \Delta_{20} = \Delta_{10} = \sqrt{t^2(\boldsymbol{k}) + \Delta^2} \equiv E(\boldsymbol{k})\,.$$

These *single*-particle excitations are identical to the poles of the Green's function in Ex. 3.3.3!

Solution 3.3.5

1.

$$\rho^+_{\boldsymbol{k}\uparrow}\left|E_0(\boldsymbol{k})\right\rangle = \left(u_{\boldsymbol{k}} a^+_{\boldsymbol{k}\uparrow} - v_{\boldsymbol{k}} a_{-\boldsymbol{k}\downarrow}\right)\left(u_{\boldsymbol{k}} + v_{\boldsymbol{k}} a^+_{\boldsymbol{k}\uparrow} a^+_{-\boldsymbol{k}\downarrow}\right)|0\rangle =$$

$$= \left(u_{\boldsymbol{k}}^2 + v_{\boldsymbol{k}}^2\right) a^+_{\boldsymbol{k}\uparrow}|0\rangle = \left|E_1(\boldsymbol{k})\right\rangle,$$

$$\rho^+_{-\boldsymbol{k}\downarrow}\left|E_0(\boldsymbol{k})\right\rangle = \left(u_{\boldsymbol{k}} a^+_{-\boldsymbol{k}\downarrow} + v_{\boldsymbol{k}} a_{\boldsymbol{k}\uparrow}\right)\left(u_{\boldsymbol{k}} + v_{\boldsymbol{k}} a^+_{\boldsymbol{k}\uparrow} a^+_{-\boldsymbol{k}\downarrow}\right)|0\rangle =$$

$$= \left(u_{\boldsymbol{k}}^2 + v_{\boldsymbol{k}}^2\right) a^+_{-\boldsymbol{k}\downarrow}|0\rangle = \left|E_2(\boldsymbol{k})\right\rangle,$$

$$\rho^+_{-\boldsymbol{k}\downarrow}\left|E_1(\boldsymbol{k})\right\rangle = \left(u_{\boldsymbol{k}} a^+_{-\boldsymbol{k}\downarrow} + v_{\boldsymbol{k}} a_{\boldsymbol{k}\uparrow}\right) a^+_{\boldsymbol{k}\uparrow}|0\rangle = \left(v_{\boldsymbol{k}} - u_{\boldsymbol{k}} a^+_{\boldsymbol{k}\uparrow} a^+_{-\boldsymbol{k}\downarrow}\right)|0\rangle =$$

$$= \left|E_3(\boldsymbol{k})\right\rangle,$$

$$\rho^+_{\boldsymbol{k}\uparrow}\left|E_2(\boldsymbol{k})\right\rangle = \left(u_{\boldsymbol{k}} a^+_{\boldsymbol{k}\uparrow} - v_{\boldsymbol{k}} a^+_{-\boldsymbol{k}\downarrow}\right) a^+_{-\boldsymbol{k}\downarrow}|0\rangle = -\left(v_{\boldsymbol{k}} - u_{\boldsymbol{k}} a^+_{\boldsymbol{k}\uparrow} a^+_{-\boldsymbol{k}\downarrow}\right)|0\rangle =$$

$$= -\left|E_3(\boldsymbol{k})\right\rangle.$$

2.

$$[\rho_{\boldsymbol{p}\uparrow}, \rho^+_{\boldsymbol{k}\uparrow}]_+ = \left[u_{\boldsymbol{p}} a_{\boldsymbol{p}\uparrow} - v_{\boldsymbol{p}} a^+_{-\boldsymbol{p}\downarrow}, u_{\boldsymbol{k}} a^+_{\boldsymbol{k}\uparrow} - v_{\boldsymbol{k}} a_{-\boldsymbol{k}\downarrow}\right]_+ =$$

$$= u_{\boldsymbol{p}} u_{\boldsymbol{k}}\left[a_{\boldsymbol{p}\uparrow}, a^+_{\boldsymbol{k}\uparrow}\right]_+ + v_{\boldsymbol{p}} v_{\boldsymbol{k}}\left[a^+_{-\boldsymbol{p}\downarrow}, a_{-\boldsymbol{k}\downarrow}\right]_+ =$$

$$= \left(u_{\boldsymbol{k}}^2 + v_{\boldsymbol{k}}^2\right)\delta_{\boldsymbol{p}\boldsymbol{k}} = \delta_{\boldsymbol{p}\boldsymbol{k}}\,,$$

$$\left[\rho_{\boldsymbol{p}\uparrow}, \rho_{\boldsymbol{k}\uparrow}\right]_+ = \left[u_{\boldsymbol{p}} a_{\boldsymbol{p}\uparrow} - v_{\boldsymbol{p}} a^+_{-\boldsymbol{p}\downarrow}, u_{\boldsymbol{k}} a_{\boldsymbol{k}\uparrow} - v_{\boldsymbol{k}} a^+_{-\boldsymbol{k}\downarrow}\right]_+ = 0\,.$$

3.

$$\begin{aligned}
\left[H^*, \rho^+_{k\uparrow}\right]_- &= \left[H_k, \rho^+_{k\uparrow}\right]_- = \\
&= t(\boldsymbol{k}) \left[a^+_{k\uparrow} a_{k\uparrow} + a^+_{-k\downarrow} a_{-k\downarrow}, u_k a^+_{k\uparrow} - v_k a_{-k\downarrow}\right]_- - \\
&\quad - \Delta \left[a^+_{k\uparrow} a^+_{-k\downarrow} + a_{-k\downarrow} a_{k\uparrow}, u_k a^+_{k\uparrow} - v_k a_{-k\downarrow}\right]_- = \\
&= t(\boldsymbol{k}) u_k \left[a^+_{k\uparrow} a_{k\uparrow}, a^+_{k\uparrow}\right]_- - t(\boldsymbol{k}) v_k \left[a^+_{-k\downarrow} a_{-k\downarrow}, a_{-k\downarrow}\right]_- + \\
&\quad + \Delta v_k \left[a^+_{k\uparrow} a^+_{-k\downarrow}, a_{-k\downarrow}\right] - \Delta u_k \left[a_{-k\downarrow} a_{k\uparrow}, a^+_{k\uparrow}\right]_- = \\
&= t(\boldsymbol{k}) \left(u_k a^+_{k\uparrow} + v_k a_{-k\downarrow}\right) + \Delta \left(v_k a^+_{k\uparrow} - u_k a_{-k\downarrow}\right),
\end{aligned}$$

$$\left(\frac{tu + \Delta v}{u}\right)^2 = \frac{t^2 u^2 + \Delta^2 v^2 + 2t\Delta uv}{u^2},$$

$$2tuv = 2t\frac{1}{2}\left(1 - \frac{t^2}{t^2 + \Delta^2}\right)^{1/2} = \frac{t\Delta}{\sqrt{t^2 + \Delta^2}} = \Delta\left(u^2 - v^2\right)$$

$$\Longrightarrow \quad \frac{tu + \Delta v}{u} = \sqrt{t^2 + \Delta^2}\,.$$

In an analogous manner, one shows that:

$$\frac{tv - \Delta u}{v} = -\sqrt{t^2 + \Delta^2}$$

$$\Longrightarrow \quad \left[H^*, \rho^+_{k\uparrow}\right]_- = E(\boldsymbol{k})\left\{u_k a^+_{k\uparrow} - v_k a_{-k\downarrow}\right\} = E(\boldsymbol{k})\rho^+_{k\uparrow}\,.$$

H^* describes a superconductor as a system of non-interacting *Bogolons*. These are the quasi-particles of superconductivity, created by ρ^+!

4.

$$\left[H^*, \rho_{k\uparrow}\right]_- = -E(\boldsymbol{k})\rho_{k\uparrow}\,,$$

$$\left\langle \left[\rho_{k\uparrow}, \rho^+_{k\uparrow}\right]_+ \right\rangle = 1$$

$$\Longrightarrow \quad \widehat{G}^{\mathrm{ret}}_{k\uparrow}(E) = \frac{\hbar}{E + E(\boldsymbol{k}) + \mathrm{i}0^+}\,.$$

Section 3.4.6

Solution 3.4.1
With

$$\mathcal{H}_0 = \sum_{k\sigma} (\varepsilon(\boldsymbol{k}) - \mu)\, a^+_{k\sigma} a_{k\sigma} \,,$$

we initially calculate:

$$\begin{aligned}
[a_{k\sigma}, \mathcal{H}_0]_- &= \sum_{k'\sigma'} \left(\varepsilon\left(\boldsymbol{k}'\right) - \mu\right) \left[a_{k\sigma}, a^+_{k'\sigma'} a_{k'\sigma'}\right]_- = \\
&= \sum_{k',\sigma'} \left(\varepsilon\left(\boldsymbol{k}'\right) - \mu\right) \delta_{kk'} \delta_{\sigma\sigma'} a_{k'\sigma'} = (\varepsilon(\boldsymbol{k}) - \mu)\, a_{k\sigma} \,.
\end{aligned}$$

The interaction term requires more effort:

$$\begin{aligned}
&[a_{k\sigma}, \mathcal{H} - \mathcal{H}_0]_- = \\
&= \frac{1}{2} \sum_{\substack{k'pq \\ \sigma''\sigma'}} v_{k'p}(\boldsymbol{q}) \left[a_{k\sigma}, a^+_{k'+q\sigma''} a^+_{p-q\sigma'} a_{p\sigma'} a_{k'\sigma''}\right]_- = \\
&= \frac{1}{2} \sum_{\substack{k',p,q \\ \sigma''\sigma'}} v_{k'p}(\boldsymbol{q}) \Big(\delta_{\sigma\sigma''} \delta_{k,k'+q} a^+_{p-q\sigma'} a_{p\sigma'} a_{k'\sigma''} - \\
&\quad - \delta_{\sigma\sigma'} \delta_{kp-q} a^+_{k'+q\sigma''} a_{p\sigma'} a_{k'\sigma''} \Big) = \\
&= \frac{1}{2} \sum_{pq\sigma'} v_{k-qp}(\boldsymbol{q}) a^+_{p-q\sigma'} a_{p\sigma'} a_{k-q\sigma} - \\
&\quad - \frac{1}{2} \sum_{k'q\sigma''} v_{k'k+q}(\boldsymbol{q}) a^+_{k'+q\sigma''} a_{k+q\sigma} a_{k'\sigma''} \,.
\end{aligned}$$

In the first term:

$$\boldsymbol{q} \to -\boldsymbol{q} \,; \quad v_{k+q,\,p}(-\boldsymbol{q}) = v_{p,\,k+q}(\boldsymbol{q}) \quad \text{(s. 3.299)} \,.$$

In the second term:

$$\boldsymbol{k}' \to \boldsymbol{p} \,; \quad \sigma'' \to \sigma' \,.$$

The two terms can then be combined:

$$[a_{k\sigma}, \mathcal{H} - \mathcal{H}_0]_- = \sum_{pq\sigma'} v_{p,\,k+q}(q) a^+_{p+q\sigma'} a_{p\sigma'} a_{k+q\sigma} \,.$$

Equation of motion:

$$(E - \varepsilon(k) + \mu)\, G^{\text{ret}}_{k\sigma}(E) = \hbar + \sum_{pq\sigma'} v_{p,\,k+q}(q) \langle\langle a^+_{p+q\sigma'} a_{p\sigma'} a_{k+q\sigma} ; a^+_{k\sigma} \rangle\rangle^{\text{ret}}_E \,.$$

Solution 3.4.2

$$\mathcal{H}_0 = \sum_{k\sigma} (\varepsilon(k) - \mu)\, a^+_{k\sigma} a_{k\sigma} \,.$$

With this, we readily calculate:

$$\begin{aligned}
[a_{k\sigma}, \mathcal{H}_0]_- &= (\varepsilon(k) - \mu)\, a_{k\sigma} \,, \\
\left[a^+_{k\sigma}, \mathcal{H}_0\right]_- &= -\,(\varepsilon(k) - \mu)\, a^+_{k\sigma} \,, \\
\left[a^+_{k\sigma} a_{k'\sigma'}, \mathcal{H}_0\right] &= \left[a^+_{k\sigma}, \mathcal{H}_0\right]_- a_{k'\sigma'} + a^+_{k\sigma} \left[a_{k'\sigma'}, \mathcal{H}_0\right]_- = \\
&= -\,(\varepsilon(k) - \mu)\, a^+_{k\sigma} a_{k'\sigma'} + \left(\varepsilon\left(k'\right) - \mu\right) a^+_{k\sigma} a_{k'\sigma'} = \\
&= \left(\varepsilon\left(k'\right) - \varepsilon(k)\right) a^+_{k\sigma} a_{k'\sigma'} \,.
\end{aligned}$$

$|\psi_0\rangle$ is an eigenstate of $\mathcal{H}_0$, since:

$$\begin{aligned}
\mathcal{H}_0 |\psi_0\rangle &= a^+_{k\sigma} a_{k'\sigma'} \mathcal{H}_0 |E_0\rangle - \left[a^+_{k\sigma} a_{k'\sigma'}, \mathcal{H}_0\right]_- |E_0\rangle = \\
&= \left(E_0 - \varepsilon\left(k'\right) + \varepsilon(k)\right) |\psi_0\rangle \,.
\end{aligned}$$

Time dependence:

$$\begin{aligned}
|\psi_0(t)\rangle &= a^+_{k\sigma}(t) a_{k'\sigma'}(t) |E_0\rangle = \mathrm{e}^{\frac{\mathrm{i}}{\hbar}\mathcal{H}_0 t} a^+_{k\sigma} a_{k'\sigma'} \mathrm{e}^{-\frac{\mathrm{i}}{\hbar}\mathcal{H}_0 t} |E_0\rangle = \\
&= \mathrm{e}^{-\frac{\mathrm{i}}{\hbar}E_0 t} \mathrm{e}^{\frac{\mathrm{i}}{\hbar}\mathcal{H}_0 t} |\psi_0\rangle = \mathrm{e}^{-\frac{\mathrm{i}}{\hbar}E_0 t} \mathrm{e}^{\frac{\mathrm{i}}{\hbar}(E_0 + \varepsilon(k') - \varepsilon(k))t} |\psi_0\rangle \\
\Longrightarrow \quad |\psi_0(t)\rangle &= \mathrm{e}^{-\frac{\mathrm{i}}{\hbar}(\varepsilon(k') - \varepsilon(k))t} |\psi_0\rangle \,.
\end{aligned}$$

With $\langle E_0 \mid E_0 \rangle = 1$, it now follows that:

$$\begin{aligned}
\left\langle \psi_0 \mid \psi_0 \right\rangle &= \left\langle E_0 \left| a^+_{k'\sigma'} a_{k\sigma} a^+_{k\sigma} a_{k'\sigma'} \right| E_0 \right\rangle = \\
&= \left\langle E_0 \left| a^+_{k'\sigma'} \left(1 - n_{k\sigma}\right) a_{k'\sigma'} \right| E_0 \right\rangle = \\
&= \left\langle E_0 \left| a^+_{k'\sigma'} a_{k'\sigma'} \right| E_0 \right\rangle = && (k > k_\mathrm{F}) \\
&= \left\langle E_0 \right| \left(1 - a_{k'\sigma'} a^+_{k'\sigma'}\right) \left| E_0 \right\rangle = \\
&= \left\langle E_0 \mid E_0 \right\rangle = && \left(k' < k_\mathrm{F}\right) \\
&= 1\,.
\end{aligned}$$

Finally, we obtain:

$$\left\langle \psi_0(t) \,\middle|\, \psi_0\left(t'\right) \right\rangle = \exp\left[-\frac{\mathrm{i}}{\hbar}\left(\varepsilon\left(\boldsymbol{k}'\right) - \varepsilon(\boldsymbol{k})\right)\left(t - t'\right)\right]$$

$$\Longrightarrow \quad \left|\left\langle \psi_0(t) \,\middle|\, \psi_0\left(t'\right) \right\rangle\right|^2 = 1 : \quad \textit{stationary state.}$$

Solution 3.4.3

$$G^{\mathrm{ret}}_{\boldsymbol{k}\sigma}(E) = \hbar\left(E - \varepsilon(\boldsymbol{k}) + \mu - \Sigma_\sigma(\boldsymbol{k}, E)\right)^{-1}$$

general representation.

1. The following must hold:

$$E - \varepsilon(\boldsymbol{k}) + \mu - \Sigma_\sigma(\boldsymbol{k}, E) \stackrel{!}{=} E - 2\varepsilon(\boldsymbol{k}) + \frac{E^2}{\varepsilon(\boldsymbol{k})} + \mathrm{i}\gamma\,|E|$$

$$\Longrightarrow \quad \Sigma_\sigma(\boldsymbol{k}, E) = R_\sigma(\boldsymbol{k}, E) + \mathrm{i} I_\sigma(\boldsymbol{k}, E) = \left(\varepsilon(\boldsymbol{k}) + \mu - \frac{e^2}{\varepsilon(\boldsymbol{k})}\right) - \mathrm{i}\gamma\,|E|$$

$$\Longrightarrow \quad R_\sigma(\boldsymbol{k}, E) = \varepsilon(\boldsymbol{k}) + \mu - \frac{e^2}{\varepsilon(\boldsymbol{k})}, \quad I_\sigma(\boldsymbol{k}, E) = -\gamma\,|E|\,.$$

2.

$$E_{i\sigma}(\mathbf{k}) \overset{!}{=} \varepsilon(\mathbf{k}) - \mu + R_\sigma\left(\mathbf{k}, E_{i\sigma}(\mathbf{k})\right) = 2\varepsilon(\mathbf{k}) - \frac{E_{i\sigma}^2(\mathbf{k})}{\varepsilon(\mathbf{k})}$$

$$\Longrightarrow \quad E_{i\sigma}^2(\mathbf{k}) + \varepsilon(\mathbf{k})E_{i\sigma}(\mathbf{k}) = 2\varepsilon^2(\mathbf{k}),$$

$$\left(E_{i\sigma}(\mathbf{k}) + \frac{1}{2}\varepsilon(\mathbf{k})\right)^2 = \frac{9}{4}\varepsilon^2(\mathbf{k}).$$

We obtain two quasi-particle energies:

$$E_{1\sigma}(\mathbf{k}) = -2\varepsilon(\mathbf{k}); \quad E_{2\sigma}(\mathbf{k}) = \varepsilon(\mathbf{k}).$$

Spectral weights (3.340):

$$\alpha_{i\sigma}(\mathbf{k}) = \left|1 - \frac{\partial}{\partial E}R_\sigma(\mathbf{k}, E)\right|^{-1}_{E = E_{i\sigma}} = \left|1 + 2\frac{E_{i\sigma}(\mathbf{k})}{\varepsilon(\mathbf{k})}\right|^{-1}$$

$$\Longrightarrow \quad \alpha_{1\sigma}(\mathbf{k}) = \alpha_{2\sigma}(\mathbf{k}) = \frac{1}{3}.$$

Lifetimes:

$$I_\sigma\left(\mathbf{k}, E_{1\sigma}(\mathbf{k})\right) = -2\gamma\left|\varepsilon(\mathbf{k})\right| = I_{1\sigma}(\mathbf{k}),$$

$$I_\sigma\left(\mathbf{k}, E_{2\sigma}(\mathbf{k})\right) = -\gamma\left|\varepsilon(\mathbf{k})\right| = I_{2\sigma}(\mathbf{k})$$

$$\Longrightarrow \quad \tau_{1\sigma}(\mathbf{k}) = \frac{3\hbar}{2\gamma\left|\varepsilon(\mathbf{k})\right|}; \quad \tau_{2\sigma}(\mathbf{k}) = \frac{3\hbar}{\gamma\left|\varepsilon(\mathbf{k})\right|}.$$

3. The quasi-particle concept is applicable, in the case that

$$\left|I_\sigma(\mathbf{k}, E)\right| \ll \left|\varepsilon(\mathbf{k}) - \mu + R_\sigma(\mathbf{k}, E)\right|$$

$$\Longleftrightarrow \quad \left|I_\sigma(\mathbf{k}, E_{i\sigma})\right| \ll \left|E_{i\sigma}(\mathbf{k})\right|$$

$$\Longleftrightarrow \quad \gamma\left|E_{i\sigma}(\mathbf{k})\right| \ll \left|E_{i\sigma}(\mathbf{k})\right|$$

$$\Longleftrightarrow \quad \gamma \ll 1.$$

4.

$$\left(\frac{\partial R_\sigma(\boldsymbol{k}, E)}{\partial E}\right)_{\varepsilon(\boldsymbol{k})} = -\frac{2E}{\varepsilon(\boldsymbol{k})},$$

$$\left(\frac{\partial R_\sigma(\boldsymbol{k}, E)}{\partial \varepsilon(\boldsymbol{k})}\right)_E = 1 + \frac{e^2}{\varepsilon^2(\boldsymbol{k})}$$

$$\Longrightarrow \quad m^*_{1\sigma}(\boldsymbol{k}) = m\frac{1-4}{1+5} = -\frac{1}{2}m,$$

$$m^*_{2\sigma}(\boldsymbol{k}) = m\frac{1+2}{1+2} = m.$$

Solution 3.4.4

The self-energy is real and independent of $\boldsymbol{k}$. Therefore, we find together with (3.362):

$$\rho_\sigma(E) = \rho_0\left[E - \Sigma_\sigma(E-\mu)\right] = \rho_0\left(E - a_\sigma\frac{E-b_\sigma}{E-c_\sigma}\right).$$

Lower band edge:

$$0 \overset{!}{=} E - a_\sigma\frac{E-b_\sigma}{E-c_\sigma}$$

$$\Longleftrightarrow \quad 0 = E^2 - (a_\sigma + c_\sigma)E + a_\sigma b_\sigma =$$

$$= \left[E - \frac{1}{2}(a_\sigma + c_\sigma)\right]^2 + a_\sigma b_\sigma - \frac{1}{4}(a_\sigma + c_\sigma)^2$$

$$\Longrightarrow \quad E^{(u)}_{1,2\sigma} = \frac{1}{2}\left[a_\sigma + c_\sigma \mp \sqrt{(a_\sigma + c_\sigma)^2 - 4a_\sigma b_\sigma}\right].$$

Upper band edge:

$$W \overset{!}{=} E - a_\sigma\frac{E-b_\sigma}{E-c_\sigma}$$

$$\Longleftrightarrow \quad -c_\sigma W = e^2 - (a_\sigma + c_\sigma + W)\,E + a_\sigma b_\sigma,$$

$$0 = \left[E - \frac{1}{2}(a_\sigma + c_\sigma + W)\right]^2 + (a_\sigma b_\sigma + c_\sigma W) - \frac{1}{4}(a_\sigma + c_\sigma + W)^2$$

$$\Longrightarrow \quad E^{(o)}_{1,2\sigma} = \frac{1}{2}\left[a_\sigma + c_\sigma + W \mp \sqrt{(a_\sigma + c_\sigma + W)^2 - 4\,(a_\sigma b_\sigma + c_\sigma W)}\right].$$

Quasi-particle density of states:

$$\rho_\sigma(E) = \begin{cases} 1/W\,, & \text{when} \quad E_{1\sigma}^{(u)} \le E \le E_{1\sigma}^{(o)}\,, \\ 1/W\,, & \text{when} \quad E_{2\sigma}^{(u)} \le E \le E_{2\sigma}^{(o)}\,, \\ 0\,, & \text{otherwise.} \end{cases}$$

Band splitting into two quasi-particle subbands!

Section 4.1.7

Solution 4.1.1

The Hubbard Hamiltonian in the Wannier representation:

$$H = \sum_{ij\sigma} T_{ij} a_{i\sigma}^+ a_{j\sigma} + \frac{1}{2} U \sum_{i,\sigma} n_{i\sigma} n_{i-\sigma}\,.$$

From (2.37), we find for the *hopping* integrals,

$$T_{ij} = \frac{1}{N} \sum_{\boldsymbol{k}} \varepsilon(\boldsymbol{k}) \mathrm{e}^{\mathrm{i}\boldsymbol{k}\cdot(\boldsymbol{R}_i - \boldsymbol{R}_j)}\,,$$

and for the creation and annihilation operators:

$$a_{i\sigma} = \frac{1}{\sqrt{N}} \sum_{\boldsymbol{k}} a_{\boldsymbol{k}\sigma} \mathrm{e}^{\mathrm{i}\boldsymbol{k}\cdot\boldsymbol{R}_i}\,.$$

We find from this for the single-particle contribution:

$$\begin{aligned}
&\sum_{ij\sigma} T_{ij} a_{i\sigma}^+ a_{j\sigma} = \\
&= \frac{1}{N^2} \sum_{\substack{\boldsymbol{k},\boldsymbol{p},\boldsymbol{q}\\ \sigma}} \varepsilon(\boldsymbol{k}) a_{\boldsymbol{p}\sigma}^+ a_{\boldsymbol{q}\sigma} \sum_{i,j} \mathrm{e}^{\mathrm{i}\boldsymbol{k}\cdot(\boldsymbol{R}_i - \boldsymbol{R}_j)} \mathrm{e}^{\mathrm{i}\boldsymbol{p}\cdot\boldsymbol{R}_i} \mathrm{e}^{-\mathrm{i}\boldsymbol{q}\cdot\boldsymbol{R}_j} = \\
&= \sum_{\substack{\boldsymbol{k},\boldsymbol{p},\boldsymbol{q}\\ \sigma}} \varepsilon(\boldsymbol{k}) a_{\boldsymbol{p}\sigma}^+ a_{\boldsymbol{q}\sigma} \delta_{\boldsymbol{k},-\boldsymbol{p}} \delta_{\boldsymbol{k},-\boldsymbol{q}} = \sum_{\boldsymbol{p},\sigma} \varepsilon(-\boldsymbol{p}) a_{\boldsymbol{p}\sigma}^+ a_{\boldsymbol{p}\sigma} = \\
&= \sum_{\boldsymbol{p}\sigma} \varepsilon(\boldsymbol{p}) a_{\boldsymbol{p}\sigma}^+ a_{\boldsymbol{p}\sigma}\,, \quad \text{da} \quad \varepsilon(-\boldsymbol{p}) = \varepsilon(\boldsymbol{p})\,.
\end{aligned}$$

For the interaction term, we require:

$$n_{i\sigma} = \frac{1}{N} \sum_{\boldsymbol{k}_1, \boldsymbol{k}_2} \mathrm{e}^{-\mathrm{i}(\boldsymbol{k}_1 - \boldsymbol{k}_2) \cdot \boldsymbol{R}_i} a^+_{\boldsymbol{k}_1\sigma} a_{\boldsymbol{k}_2\sigma} ,$$

$$\sum_{i,\sigma} n_{i\sigma} n_{i-\sigma} =$$

$$= \frac{1}{N^2} \sum_{i,\sigma} \sum_{\substack{\boldsymbol{k}_1, \boldsymbol{k}_2 \\ \boldsymbol{p}_1, \boldsymbol{p}_2}} a^+_{\boldsymbol{k}_1\sigma} a_{\boldsymbol{k}_2\sigma} a^+_{\boldsymbol{p}_1-\sigma} a_{\boldsymbol{p}_2-\sigma} \mathrm{e}^{-\mathrm{i}(\boldsymbol{k}_1 - \boldsymbol{k}_2 + \boldsymbol{p}_1 - \boldsymbol{p}_2) \cdot \boldsymbol{R}_i} =$$

$$= \frac{1}{N} \sum_{\substack{\boldsymbol{k}_1, \boldsymbol{k}_2, \boldsymbol{p}_1, \boldsymbol{p}_2 \\ \sigma}} \delta_{\boldsymbol{k}_1+\boldsymbol{p}_1, \boldsymbol{k}_2+\boldsymbol{p}_2} a^+_{\boldsymbol{k}_1\sigma} a_{\boldsymbol{k}_2\sigma} a^+_{\boldsymbol{p}_1-\sigma} a_{\boldsymbol{p}_2-\sigma} =$$

$$= \frac{1}{N} \sum_{\substack{\boldsymbol{k}_1, \boldsymbol{k}_2, \boldsymbol{p}_2 \\ \sigma}} a^+_{\boldsymbol{k}_1\sigma} a^+_{\boldsymbol{k}_2+\boldsymbol{p}_2-\boldsymbol{k}_1-\sigma} a_{\boldsymbol{p}_2-\sigma} a_{\boldsymbol{k}_2\sigma} =$$

$$= \frac{1}{N} \sum_{\boldsymbol{k}, \boldsymbol{p}, \boldsymbol{q}, \sigma} a^+_{\boldsymbol{k}+\boldsymbol{q}\sigma} a^+_{\boldsymbol{p}-\boldsymbol{q}-\sigma} a_{\boldsymbol{p}-\sigma} a_{\boldsymbol{k}\sigma} .$$

In the last step, we made the substitution: $\boldsymbol{k}_2 \to \boldsymbol{k}$, $\boldsymbol{p}_2 \to \boldsymbol{p}$, $\boldsymbol{k}_1 \to \boldsymbol{k} + \boldsymbol{q}$. Then the Hubbard Hamiltonian in the Bloch representation is given by:

$$H = \sum_{\boldsymbol{p}\sigma} \varepsilon(\boldsymbol{p}) a^+_{\boldsymbol{p}\sigma} a_{\boldsymbol{p}\sigma} + \frac{U}{2N} \sum_{\boldsymbol{kpq}\sigma} a^+_{\boldsymbol{k}+\boldsymbol{q}\sigma} a^+_{\boldsymbol{p}-\boldsymbol{q}-\sigma} a_{\boldsymbol{p}-\sigma} a_{\boldsymbol{k}\sigma} .$$

Comparison with (2.63):

Jellium		Hubbard	
$\dfrac{\hbar^2 k^2}{2m}$	$\longleftrightarrow$	$\varepsilon(\boldsymbol{k})$	(*Tight-binding* approximation)
$v_0(\boldsymbol{q}) = \dfrac{e^2}{V\varepsilon_0 q^2}$	$\longleftrightarrow$	$\dfrac{U}{N}\delta_{\sigma', -\sigma}$.	

Solution 4.1.2

First, consider $x \neq 0$:

$$\frac{1}{2} \lim_{\beta \to \infty} \frac{\beta}{1 + \cosh(\beta x)} = \lim_{\beta \to \infty} \beta \mathrm{e}^{-\beta |x|} = 0 .$$

For $x = 0$, this expression diverges. Furthermore, we have:

$$\int_{-\infty}^{+\infty} \mathrm{d}x \, \frac{1}{2} \lim_{\beta\to\infty} \frac{\beta}{1+\cosh(\beta x)} = \lim_{\beta\to\infty} \int_0^{\infty} \mathrm{d}x \, \frac{\beta}{1+\cosh(\beta x)} \,,$$

$$\int_0^{\infty} \mathrm{d}x \, \frac{\beta}{1+\cosh(\beta x)} = \int_0^{\infty} \mathrm{d}y \, \frac{1}{1+\cosh y} = \int_0^{\infty} \mathrm{d}y \, \frac{1}{2\cosh^2 \frac{y}{2}} =$$

$$= \int_0^{\infty} \mathrm{d}z \, \frac{1}{\cosh^2 z} = \tanh z \Big|_0^{\infty} = 1 - 0 = 1 \,.$$

The requirements for the δ-function are thus fulfilled!

Solution 4.1.3

1. Jellium model (2.63):

$$H = \sum_{k\sigma} \varepsilon_0(\boldsymbol{k}) a^+_{k\sigma} a_{k\sigma} + \frac{1}{2} \sum_{\substack{kpq\\ \sigma,\sigma'}}^{q\neq 0} v_0(\boldsymbol{q}) a^+_{k+q\sigma} a^+_{p-q\sigma'} a_{p\sigma'} a_{k\sigma} \,,$$

$$\varepsilon_0(\boldsymbol{k}) = \frac{\hbar^2 k^2}{2m} \,; \quad v_0(\boldsymbol{q}) = \frac{1}{V} \frac{e^2}{\varepsilon_0 q^2} \,.$$

Hartree-Fock approximation for the interaction term:

$$\begin{aligned}
a^+_{k+q\sigma} a^+_{p-q\sigma'} a_{p\sigma'} a_{k\sigma} &\overset{(q\neq 0)}{=} -\left(a^+_{k+q\sigma} a_{p\sigma'}\right)\left(a^+_{p-q\sigma'} a_{k\sigma}\right) \\
&\xrightarrow{\text{HFA}} -\left\langle a^+_{k+q\sigma} a_{p\sigma'}\right\rangle \left(a^+_{p-q\sigma'} a_{k\sigma}\right) - \left(a^+_{k+q\sigma} a_{p\sigma'}\right)\left\langle a^+_{p-q\sigma'} a_{k\sigma}\right\rangle + \\
&\quad + \left\langle a^+_{k+q\sigma} a_{p\sigma'}\right\rangle \left\langle a^+_{p-q\sigma'} a_{k\sigma}\right\rangle = \\
&= \delta_{p,k+q} \delta_{\sigma\sigma'} \Big(-\left\langle n_{k+q\sigma}\right\rangle n_{k\sigma} - n_{k+q\sigma} \left\langle n_{k\sigma}\right\rangle + \left\langle n_{k+q\sigma}\right\rangle \left\langle n_{k\sigma}\right\rangle \Big) \,.
\end{aligned}$$

Via the expectation values, we can make use of momentum and spin conservation in the last step. We furthermore define:

$$\langle \alpha_\sigma(\boldsymbol{k}) \rangle = \sum_q^{\neq 0} v_0(\boldsymbol{q}) \left\langle n_{k+q\sigma}\right\rangle = \sum_p^{\neq k} v_0(\boldsymbol{p}-\boldsymbol{k}) \left\langle n_{p\sigma}\right\rangle \,,$$

$$\langle \beta_\sigma \rangle = \frac{1}{2} \sum_{k,q,\sigma}^{q\neq 0} v_0(\boldsymbol{q}) \left\langle n_{k+q\sigma}\right\rangle \left\langle n_{k\sigma}\right\rangle \,.$$

We can then write the Hamiltonian for the jellium model as follows:

$$H_{\mathrm{HFA}} = \sum_{k\sigma} \Big\{ \varepsilon_0(\boldsymbol{k}) - \langle \alpha_\sigma(\boldsymbol{k}) \rangle \Big\} a^+_{\boldsymbol{k}\sigma} a_{\boldsymbol{k}\sigma} + \langle \beta_\sigma \rangle \,.$$

2. The equation of motion for the one-electron Green's function can readily be derived,

$$\left(E - \varepsilon_0(\boldsymbol{k}) + \mu + \langle \alpha_\sigma(\boldsymbol{k}) \rangle \right) G_{\boldsymbol{k}\sigma}(E) = \hbar \,,$$

and likewise solved:

$$G^{\mathrm{ret}}_{\boldsymbol{k}\sigma}(E) = \frac{\hbar}{E - \varepsilon_0(\boldsymbol{k}) + \mu + \langle \alpha_\sigma(\boldsymbol{k}) \rangle + \mathrm{i}0^+} \,.$$

From this, we can read off the spectral density directly:

$$S_{\boldsymbol{k}\sigma}(E) = \hbar \delta \left(E - \varepsilon_0(\boldsymbol{k}) + \mu + \langle \alpha_\sigma(\boldsymbol{k}) \rangle \right) .$$

3.

$$\langle n_{\boldsymbol{k}\sigma} \rangle = \left\langle a^+_{\boldsymbol{k}\sigma} a_{\boldsymbol{k}\sigma} \right\rangle = \frac{1}{\hbar} \int\limits_{-\infty}^{+\infty} \mathrm{d}E \, \frac{S_{\boldsymbol{k}\sigma}(E)}{\mathrm{e}^{\beta E} + 1} = f_-\left(E = \varepsilon_0(\boldsymbol{k}) - \langle \alpha_\sigma(\boldsymbol{k}) \rangle \right) .$$

The functional equation is implicit, since

$$\left\langle \alpha_\sigma(\boldsymbol{k}) \right\rangle = \sum_{\boldsymbol{p}}^{\neq \boldsymbol{k}} v_0(\boldsymbol{p} - \boldsymbol{k}) \left\langle n_{\boldsymbol{p}\sigma} \right\rangle .$$

4. According to (3.382), we have:

$$U(T) = \frac{1}{2\hbar} \sum_{\boldsymbol{k}\sigma} \int\limits_{-\infty}^{+\infty} \mathrm{d}E \, f_-(E) \left(E + \varepsilon_0(\boldsymbol{k}) \right) S_{\boldsymbol{k}\sigma}(E - \mu) .$$

This implies that:

$$U(T) = \frac{1}{2} \sum_{\boldsymbol{k}\sigma} \left(2\varepsilon_0(\boldsymbol{k}) - \langle \alpha_\sigma(\boldsymbol{k}) \rangle \right) \langle n_{\boldsymbol{k}\sigma} \rangle \,.$$

Obviously, then, we have: $U(T) = \langle H_{\mathrm{HFA}} \rangle$.

5. At $T = 0$, the averaging is performed over the ground state:

$$U(T = 0) = \sum_{\boldsymbol{k}\sigma} \varepsilon_0(\boldsymbol{k}) \, \langle n_{\boldsymbol{k}\sigma} \rangle_0 - \frac{1}{2} \sum_{\boldsymbol{k}\boldsymbol{q}\sigma}^{\boldsymbol{q} \neq 0} v_0(\boldsymbol{q}) \left\langle n_{\boldsymbol{k}+\boldsymbol{q}\sigma} \right\rangle_0 \langle n_{\boldsymbol{k}\sigma} \rangle_0 \,.$$

This is formally identical to the result from first-order perturbation theory (2.92). The difference lies in the different ground states which are used for the averaging. In (2.92), the ground state of the non-interacting system was used (*the filled Fermi sphere*).

Solution 4.1.4

Band limit:

$$G_{k\sigma}^{\text{ret}(0)}(E) = \frac{\hbar}{E - \varepsilon(\boldsymbol{k}) + \mu + \text{i}0^+} .$$

Atomic limit (4.11):

$$G_\sigma^{\text{ret}}(E) = \frac{\hbar\,(1 - \langle n_{-\sigma}\rangle)}{E - T_0 + \mu + \text{i}0^+} + \frac{\hbar\,\langle n_{-\sigma}\rangle}{E - T_0 - U + \mu + \text{i}0^+} .$$

1. Stoner approximation (4.23):

$$G_{k\sigma}^{\text{ret}}(E) = \frac{\hbar}{E - \varepsilon(\boldsymbol{k}) - U\,\langle n_{-\sigma}\rangle + \text{i}0^+} .$$

   The band limit is clearly applicable, but not the atomic limit!
2. Hubbard approximation (4.49), (4.50):

$$\text{Band limit } U \to 0 \quad \Longleftrightarrow \quad \Sigma_\sigma(E) \equiv 0,$$

   The Hubbard approximation is correct in this limit.

In the atomic limit ($\varepsilon(\boldsymbol{k}) \to T_0\ \forall \boldsymbol{k}$), from (4.50), the following holds:

$$\begin{aligned}
G_{k\sigma}(E) &= \frac{\hbar}{E - T_0 + \mu - \frac{U\langle n_{-\sigma}\rangle(E+\mu-T_0)}{E+\mu-U(1-\langle n_{-\sigma}\rangle)-T_0}} = \\
&= \frac{\hbar\,(E - T_0 + \mu - U\,(1 - \langle n_{-\sigma}\rangle))}{(E - T_0 + \mu)^2 - U\,(E - T_0 + \mu)} = \\
&= \frac{\hbar\,[(E - T_0 + \mu)\,\langle n_{-\sigma}\rangle + (E - T_0 - U + \mu)(1 - \langle n_{-\sigma}\rangle)]}{(E - T_0 + \mu)(E - T_0 - U + \mu)} = \\
&= \frac{\hbar\,\langle n_{-\sigma}\rangle}{E - T_0 - U + \mu} + \frac{\hbar\,(1 - \langle n_{-\sigma}\rangle)}{E - T_0 + \mu} .
\end{aligned}$$

This agrees with (4.11), if the boundary conditions for the retarded function are fulfilled by inserting $+\text{i}0^+$ into the denominator.

The Hubbard approximation is thus exact in *both* limiting cases.

Solution 4.1.5
The solution is found immediately from part 2. of Ex. 4.1.4:

$$\Sigma_\sigma(E) = U \langle n_{-\sigma}\rangle \frac{E+\mu-T_0}{E+\mu-T_0-U(1-\langle n_{-\sigma}\rangle)} .$$

The self-energy in the atomic limit is thus identical with that of the Hubbard solution (4.49)!

Solution 4.1.6

1.

$$\begin{aligned}
\left[S_i^+, S_j^-\right]_- &= \hbar^2\delta_{ij}\left[a_{i\uparrow}^+ a_{i\downarrow}, a_{i\downarrow}^+ a_{i\uparrow}\right]_- = \hbar^2\delta_{ij}\{n_{i\uparrow}-n_{i\downarrow}\} = 2\hbar\delta_{ij}S_i^z ,\\
\left[S_i^z, S_j^+\right]_- &= \hbar^2\frac{1}{2}\left[(n_{i\uparrow}-n_{i\downarrow}), a_{j\uparrow}^+ a_{j\downarrow}\right]_-\\
&= \frac{1}{2}\hbar^2\delta_{ij}\left\{\left[n_{i\uparrow}, a_{i\uparrow}^+ a_{i\downarrow}\right]_- - \left[n_{i\downarrow}, a_{i\uparrow}^+ a_{i\downarrow}\right]_-\right\} =\\
&= \frac{1}{2}\hbar^2\delta_{ij}\left\{a_{i\uparrow}^+ a_{i\downarrow} - \left(-a_{i\uparrow}^+ a_{i\downarrow}\right)\right\} = \hbar\delta_{ij}S_i^+ ,\\
\left[S_i^z, S_j^-\right] &= \frac{1}{2}\hbar^2\left\{\left[n_{i\uparrow}, a_{j\downarrow}^+ a_{j\uparrow}\right]_- - \left[n_{i\downarrow}, a_{j\downarrow}^+ a_{j\uparrow}\right]_-\right\} =\\
&= \frac{1}{2}\hbar^2\delta_{ij}\left\{-a_{i\downarrow}^+ a_{i\uparrow} - a_{i\downarrow}^+ a_{i\uparrow}\right\} = -\hbar\delta_{ij}S_i^- .
\end{aligned}$$

2. Quite generally, we have for spin operators:

$$\begin{aligned}
S_i^+ S_i^- &= \left(S_i^x + \mathrm{i}S_i^y\right)\left(S_i^x - \mathrm{i}S_i^y\right) = \left(S_i^x\right)^2 + \left(S_i^y\right)^2 + \mathrm{i}\left[S_i^y, S_i^x\right]_- =\\
&= \mathbf{S}_i\cdot\mathbf{S}_i - \left(S_i^z\right)^2 + \hbar S_i^z .
\end{aligned}$$

From this, it follows that:

$$\begin{aligned}
\frac{1}{\hbar}\mathbf{S}_i\cdot\mathbf{S}_i &= a_{i\uparrow}^+ a_{i\downarrow} a_{i\downarrow}^+ a_{i\uparrow} + \frac{1}{4}\left(n_{i\uparrow}-n_{i\downarrow}\right)^2 - \frac{1}{2}\left(n_{i\uparrow}-n_{i\downarrow}\right) =\\
&= n_{i\uparrow} - n_{i\uparrow}n_{i\downarrow} + \frac{1}{4}\left(n_{i\uparrow}^2 + n_{i\downarrow}^2 - 2n_{i\uparrow}n_{i\downarrow}\right) - \frac{1}{2}\left(n_{i\uparrow}-n_{i\downarrow}\right) =\\
&= \frac{3}{4}n_{i\uparrow} + \frac{3}{4}n_{i\downarrow} - \frac{3}{2}n_{i\uparrow}n_{i\downarrow} \qquad \left(n_{i\sigma}^2 = n_{i\sigma}\right)\\
&\Longrightarrow -\frac{2}{3\hbar^2}\sum_i \mathbf{S}_i\cdot\mathbf{S}_i = \sum_i n_{i\uparrow}n_{i\downarrow} - \frac{1}{2}\widehat{N} ,
\end{aligned}$$

$$\widehat{N} = \sum_{i,\sigma} n_{i\sigma} .$$

Field term:

$$\mu_B B_0 \sum_{i,\sigma} z_\sigma n_{i\sigma} \mathrm{e}^{-\mathrm{i}\boldsymbol{k}\cdot\boldsymbol{R}_i} = b\frac{\hbar}{2}\sum_{i,\sigma} z_\sigma n_{i\sigma}\mathrm{e}^{-\mathrm{i}\boldsymbol{k}\cdot\boldsymbol{R}_i} = b\sum_i S_i^z \mathrm{e}^{-\mathrm{i}\boldsymbol{k}\cdot\boldsymbol{R}_i}.$$

Here, we have made use of:

$$b = 2\frac{\mu_B}{\hbar}B_0\,; \quad z_\uparrow = +1\,; \quad z_\downarrow = -1\,.$$

The Hubbard Hamiltonian:

$$H = \sum_{ij\sigma} T_{ij} a_{i\sigma}^+ a_{j\sigma} - \frac{2U}{3\hbar^2}\sum_i \boldsymbol{S}_i\cdot\boldsymbol{S}_i + \frac{1}{2}U\widehat{N} - b\sum_i S_i^z \mathrm{e}^{-\mathrm{i}\boldsymbol{k}\cdot\boldsymbol{R}_i}.$$

3.

$$\left[S^z(\boldsymbol{k}), S^\pm(\boldsymbol{q})\right]_- = \sum_{i,j}\mathrm{e}^{-\mathrm{i}(\boldsymbol{k}\cdot\boldsymbol{R}_i + \boldsymbol{q}\cdot\boldsymbol{R}_j)}\left[S_i^z, S_j^\pm\right]_- =$$
$$= \pm\hbar\sum_i \mathrm{e}^{-\mathrm{i}(\boldsymbol{k}+\boldsymbol{q})\cdot\boldsymbol{R}_i} S_i^\pm = \pm\hbar S^\pm(\boldsymbol{k}+\boldsymbol{q})\,.$$

Analogously:

$$\left[S^+(\boldsymbol{k}), S^-(\boldsymbol{q})\right]_- = 2\hbar S^z(\boldsymbol{k}+\boldsymbol{q})\,.$$

Solution 4.1.7

1. For the spin-spin interaction, we can write:

$$\sum_i \boldsymbol{S}_i\cdot\boldsymbol{S}_i = \frac{1}{N^2}\sum_i\sum_{\boldsymbol{k},\boldsymbol{p}} \mathrm{e}^{\mathrm{i}(\boldsymbol{k}+\boldsymbol{p})\cdot\boldsymbol{R}_i}\boldsymbol{S}(\boldsymbol{k})\cdot\boldsymbol{S}(\boldsymbol{p}) =$$
$$= \frac{1}{N}\sum_{\boldsymbol{p},\boldsymbol{k}}\delta_{\boldsymbol{p},-\boldsymbol{k}}\boldsymbol{S}(\boldsymbol{k})\cdot\boldsymbol{S}(\boldsymbol{p}) = \frac{1}{N}\sum_{\boldsymbol{k}}\boldsymbol{S}(\boldsymbol{k})\cdot\boldsymbol{S}(-\boldsymbol{k})\,.$$

For the field term, we read off directly:

$$\sum_i S_i^z \mathrm{e}^{-\mathrm{i}\boldsymbol{k}\cdot\boldsymbol{R}_i} = S^z(\boldsymbol{K})\,.$$

We have already transformed the operator for the kinetic energy to the wavenumber representation:

$$\sum_{ij\sigma} T_{ij} a_{i\sigma}^+ a_{j\sigma} = \sum_{k\sigma} \varepsilon(\boldsymbol{k}) a_{k\sigma}^+ a_{k\sigma} .$$

With this, and with part 3. of Ex. 4.1.6, we have directly proved the assertion.

2.

$$\sum_{\boldsymbol{k}} \left[S^-(-\boldsymbol{k}-\boldsymbol{K}), S^+(\boldsymbol{k}+\boldsymbol{K})\right]_+ =$$

$$= \sum_{\boldsymbol{k}} \sum_{i,j} \left[S_i^-, S_j^+\right]_+ \mathrm{e}^{\mathrm{i}(\boldsymbol{k}+\boldsymbol{k}) \cdot \boldsymbol{R}_i} \mathrm{e}^{-\mathrm{i}(\boldsymbol{k}+\boldsymbol{k}) \cdot \boldsymbol{R}_j} =$$

$$= N \sum_{i,j} \delta_{ij} \left[S_i^-, S_j^+\right]_+ \mathrm{e}^{\mathrm{i}\boldsymbol{k} \cdot (\boldsymbol{R}_i - \boldsymbol{R}_j)} = N \sum_i \left(S_i^- S_i^+ + S_i^+ S_i^-\right) ,$$

$$S_i^- S_i^+ = \hbar^2 a_{i\downarrow}^+ a_{i\uparrow} a_{i\uparrow}^+ a_{i\downarrow} = \hbar^2 n_{i\downarrow} \left(1 - n_{i\uparrow}\right) ,$$

$$S_i^+ S_i^- = \hbar^2 a_{i\uparrow}^+ a_{i\downarrow} a_{i\downarrow}^+ a_{i\uparrow} = \hbar^2 n_{i\uparrow} \left(1 - n_{i\downarrow}\right) .$$

3.

$$\left[S^+(\boldsymbol{k}), \sum_{\boldsymbol{p}} \boldsymbol{S}(\boldsymbol{p}) \cdot \boldsymbol{S}(-\boldsymbol{p})\right]_- =$$

$$= \sum_{\boldsymbol{p}} \Big\{ \left[S^+(\boldsymbol{k}), S^z(\boldsymbol{p})\right]_- S^z(-\boldsymbol{p}) + S^z(\boldsymbol{p}) \left[S^+(\boldsymbol{k}), S^z(-\boldsymbol{p})\right]_- +$$

$$+ \frac{1}{2} \left[S^+(\boldsymbol{k}), S^+(\boldsymbol{p})\right]_- S^-(-\boldsymbol{p}) + \frac{1}{2} S^+(\boldsymbol{p}) \left[S^+(\boldsymbol{k}), S^-(-\boldsymbol{p})\right]_- +$$

$$+ \frac{1}{2} \left[S^+(\boldsymbol{k}), S^-(\boldsymbol{p})\right]_- S^+(-\boldsymbol{p}) + \frac{1}{2} S^-(\boldsymbol{p}) \left[S^+(\boldsymbol{k}), S^+(-\boldsymbol{p})\right]_- \Big\} =$$

$$= \sum_{\boldsymbol{p}} \{ -\hbar S^+(\boldsymbol{k}+\boldsymbol{p}) S^z(-\boldsymbol{p}) - \hbar S^z(\boldsymbol{p}) S^+(\boldsymbol{k}-\boldsymbol{p}) +$$

$$+ 0 + \underbrace{\hbar S^+(\boldsymbol{p}) S^z(\boldsymbol{k}-\boldsymbol{p})}_{\boldsymbol{p} \to \boldsymbol{p}+\boldsymbol{k}} + \underbrace{\hbar S^z(\boldsymbol{k}+\boldsymbol{p}) S^+(-\boldsymbol{p})}_{\boldsymbol{p} \to \boldsymbol{p}-\boldsymbol{k}} + 0 \} = 0 ,$$

$$\left[S^+(\boldsymbol{k}), \widehat{N}\right]_- = \hbar \sum_{\substack{i,j \\ \sigma}} \mathrm{e}^{-\mathrm{i}\boldsymbol{k} \cdot \boldsymbol{R}_i} \left[a_{i\uparrow}^+ a_{i\downarrow}, a_{j\sigma}^+ a_{j\sigma}\right]_- =$$

$$= \hbar \sum_{ij\sigma} \mathrm{e}^{-\mathrm{i}\boldsymbol{k} \cdot \boldsymbol{R}_i} \left(\delta_{ij} \delta_{\downarrow\sigma} a_{i\uparrow}^+ a_{j\sigma} - \delta_{ij} \delta_{\uparrow\sigma} a_{j\sigma}^+ a_{i\downarrow}\right) =$$

$$= \hbar \sum_i \mathrm{e}^{-\mathrm{i}\boldsymbol{k} \cdot \boldsymbol{R}_i} \left(a_{i\uparrow}^+ a_{i\downarrow} - a_{i\uparrow}^+ a_{i\downarrow}\right) = 0 .$$

4. First, we require:

$$\left[S^+(\boldsymbol{k}), \sum_{p\sigma} \varepsilon(\boldsymbol{p}) a^+_{p\sigma} a_{p\sigma}\right]_- =$$

$$= \sum_i \mathrm{e}^{-\mathrm{i}\boldsymbol{k}\cdot\boldsymbol{R}_i} \sum_{mn\sigma} T_{mn} \left[a^+_{i\uparrow} a_{i\downarrow}, a^+_{m\sigma} a_{n\sigma}\right]_- =$$

$$= \sum_i \mathrm{e}^{-\mathrm{i}\boldsymbol{k}\cdot\boldsymbol{R}_i} \sum_{mn\sigma} T_{mn} \left(\delta_{im}\delta_{\downarrow\sigma} a^+_{i\uparrow} a_{n\sigma} - \delta_{in}\delta_{\uparrow\sigma} a^+_{m\sigma} a_{i\downarrow}\right) =$$

$$= \sum_{m,n} T_{mn} \left(\mathrm{e}^{-\mathrm{i}\boldsymbol{k}\cdot\boldsymbol{R}_m} - \mathrm{e}^{-\mathrm{i}\boldsymbol{k}\cdot\boldsymbol{R}_n}\right) a^+_{m\uparrow} a_{n\downarrow} .$$

With

$$\left[S^+(\boldsymbol{k}), S^z(\boldsymbol{K})\right]_- = -\hbar S^+(\boldsymbol{k}+\boldsymbol{K}) ,$$

the assertion follows immediately!

5. The field term is simple:

$$\left[b\hbar S^+(\boldsymbol{k}+\boldsymbol{K}), S^-(-\boldsymbol{k})\right]_- = 2b\hbar^2 S^z(\boldsymbol{K}) .$$

Computation of the second term is somewhat more tedious:

$$\hbar \sum_{i,j} T_{ij} \left(\mathrm{e}^{-\mathrm{i}\boldsymbol{k}\cdot\boldsymbol{R}_i} - \mathrm{e}^{-\mathrm{i}\boldsymbol{k}\cdot\boldsymbol{R}_j}\right) \left[a^+_{i\uparrow} a_{j\downarrow}, S^-(-\boldsymbol{k})\right]_- =$$

$$= \hbar^2 \sum_{ijm} T_{ij} \left(\mathrm{e}^{-\mathrm{i}\boldsymbol{k}\cdot\boldsymbol{R}_i} - \mathrm{e}^{-\mathrm{i}\boldsymbol{k}\cdot\boldsymbol{R}_j}\right) \mathrm{e}^{\mathrm{i}\boldsymbol{k}\cdot\boldsymbol{R}_m} \underbrace{\left[a^+_{i\uparrow} a_{j\downarrow}, a^+_{m\downarrow} a_{m\uparrow}\right]_-}_{\delta_{jm} a^+_{i\uparrow} a_{m\uparrow} - \delta_{im} a^+_{m\downarrow} a_{j\downarrow}} =$$

$$= \hbar^2 \sum_{ij\sigma} T_{ij} \left(\mathrm{e}^{-\mathrm{i}z_\sigma \boldsymbol{k}(\boldsymbol{R}_i - \boldsymbol{R}_j)} - 1\right) a^+_{i\sigma} a_{j\sigma} .$$

This proves the assertion!

Solution 4.1.8

1. From part 2. of Ex. 4.1.7, we already know that:

$$\sum_{\boldsymbol{k}} \langle [A, A^+]_+ \rangle = N \sum_i \langle S_i^- S_i^+ + S_i^+ S_i^- \rangle =$$

$$= \hbar^2 N \sum_i \langle (n_{i\uparrow} - n_{i\downarrow})^2 \rangle \le \hbar^2 N \sum_i \langle n_i^2 \rangle \le 4\hbar^2 N^2 .$$

For the second inequality, we can use part 5. of Ex. 4.1.7:

$$0 \le \langle [[C, H]_-, C^+]_- \rangle \le \hbar^2 \sum_{ij\sigma} |T_{ij}| \left| e^{-iz_\sigma \boldsymbol{k} \cdot (\boldsymbol{R}_i - \boldsymbol{R}_j)} - 1 \right| \cdot$$

$$\cdot \left| \langle a^+_{i\sigma} a_{j\sigma} \rangle \right| + 2\,|b|\,\hbar^2 \left| \langle S^z(\boldsymbol{K}) \rangle \right| .$$

The first term on the right-hand side of the inequality can be further estimated:

$$\left| e^{-iz_\sigma \boldsymbol{k} \cdot (\boldsymbol{R}_i - \boldsymbol{R}_j)} - 1 \right| = \sqrt{\left[\cos\left(z_\sigma \boldsymbol{k} \cdot (\boldsymbol{R}_i - \boldsymbol{R}_j)\right) - 1\right]^2 + \sin^2\left(z_\sigma \boldsymbol{k} \cdot (\boldsymbol{R}_i - \boldsymbol{R}_j)\right)} \le$$

$$\le \left| \cos\left(\boldsymbol{k} \cdot (\boldsymbol{R}_i - \boldsymbol{R}_j)\right) - 1 \right| \le \frac{1}{2} k^2 \left(\boldsymbol{R}_i - \boldsymbol{R}_j\right)^2 ,$$

$$\langle a^+_{i\sigma} a_{j\sigma} \rangle = \frac{1}{N} \sum_{\boldsymbol{k}} e^{\boldsymbol{k}(\boldsymbol{R}_i - \boldsymbol{R}_j)} \langle n_{\boldsymbol{k}\sigma} \rangle$$

$$\Longrightarrow \quad \left| \langle a^+_{i\sigma} a_{j\sigma} \rangle \right| \le \frac{1}{N} \sum_{\boldsymbol{k}} \left| \langle n_{\boldsymbol{k}\sigma} \rangle \right| \le 1 .$$

From this, the assertion follows:

$$\left\langle [[C, H]_-, C^+]_- \right\rangle \le N \hbar^2 Q k^2 + 2\hbar^2\, |b|\, \left| \langle S^z(\boldsymbol{K}) \rangle \right| .$$

The third inequality follows immediately from the general commutation relations of the spin operators (Ex. 4.1.6):

$$\langle [C, A]_- \rangle = 2\hbar \langle S^z(-\boldsymbol{K}) \rangle .$$

2. The Bogoliubov inequality:

$$\frac{1}{2} \beta \langle [A, A^+]_+ \rangle \left\langle [[C, H]_-, C^+]_- \right\rangle \ge \left| \langle [C, A]_- \rangle \right|^2 .$$

According to part 2. of Ex. 2.4.5, $\left\langle [[C, H]_-, C^+]_- \right\rangle$ is not negative. Therefore, we also have:

$$\frac{1}{2} \beta \sum_{\boldsymbol{k}} \left\langle [A, A^+]_+ \right\rangle \ge \sum_{\boldsymbol{k}} \frac{\left| \langle [C, A]_- \rangle \right|^2}{\langle [[C, H]_-, C^+]_- \rangle} .$$

With

$$\left| \langle S^z(\boldsymbol{K}) \rangle \right| = \left| \langle S^z(-\boldsymbol{K}) \rangle \right| = \frac{N\hbar}{2\mu_B} \left| M(T, B_0) \right| ,$$

then, the assertion follows immediately by insertion of the results of part 1.

3. In Ex. 2.4.7, we found a corresponding inequality for the Heisenberg model:

$$\beta S(S+1) \geq \frac{M^2}{(g_j\mu_B)^2}\frac{1}{N}\sum_k \frac{1}{|B_0 M| + \hbar^2 k^2 Q S(S+1)} .$$

It is the same, apart from unimportant factors, as the inequality in part 2. The conclusions are the same, i.e. the Mermin-Wagner theorem holds also for the Hubbard model!

Solution 4.1.9

The Hubbard model in the *limiting case of an infinitely narrow band*:

$$H = T_0 \sum_{i,\sigma} n_{i\sigma} + \frac{1}{2} U \sum_{i,\sigma} n_{i\sigma} n_{i\,-\sigma} .$$

1. Making use of

$$[a_{i\sigma}, n_{i\sigma'}]_- = \delta_{\sigma\sigma'} a_{i\sigma} ,$$

we can readily compute the following commutators:

$$[a_{i\sigma}, H]_- = T_0 a_{i\sigma} + U a_{i\sigma} n_{i-\sigma} ,$$

$$[a_{i\sigma} n_{i\,-\sigma}, H]_- = (T_0 + U)\, a_{i\sigma} n_{i\,-\sigma} .$$

The proof is first carried out by complete induction:

$$\Big[\underbrace{\dots [[a_{i\sigma}, H]_-, H]_-, \dots, H}_{n\text{-fold commutator}}\Big]_- = T_0^n a_{i\sigma} + \left((T_0 + U)^n - T_0^n\right) a_{i\sigma} n_{i\,-\sigma} .$$

Initiation of induction $n = 1$: see above.
Conclusion of induction: $n \longrightarrow n + 1$:

$$\Big[\underbrace{\dots [[a_{i\sigma}, H]_-, H]_-, \dots, H}_{(n+1)\text{-fold commutator}}\Big]_- =$$

$$= T_0^n\, [a_{i\sigma}, H]_- + \left[(T_0 + U)^n - T_0^n\right] [a_{i\sigma} n_{i\,-\sigma}, H]_- =$$

$$= T_0^n (T_0 a_{i\sigma} + U a_{i\sigma} n_{i\,-\sigma}) + \left[(T_0 + U)^n - T_0^n\right] (T_0 + U)\, a_{i\sigma} n_{i\,-\sigma} =$$

$$= T_0^{n+1} a_{i\sigma} + a_{i\sigma} n_{i\,-\sigma} \left((T_0 + U)^{n+1} - T_0^{n+1}\right) \quad \text{q. e. d.}$$

With this, we find for the spectral moments:

$$M_{ii\sigma}^{(n)} = \left\langle \left[\underbrace{\left[\ldots [[a_{i\sigma}, H]_-, H]_-, \ldots, H \right]_-}_{n\text{-fold commutator}}, a_{i\sigma}^+ \right]_+ \right\rangle =$$

$$= T_0^n \left\langle \left[a_{i\sigma}, a_{i\sigma}^+ \right]_+ \right\rangle + \left[(T_0 + U)^n - T_0^n \right] \left\langle \left[a_{i\sigma} n_{i-\sigma}, a_{i\sigma}^+ \right]_+ \right\rangle =$$

$$= T_0^n + \left[(T_0 + U)^n - T_0^n \right] \langle n_{i-\sigma} \rangle \qquad \text{q. e. d.}$$

2.

$$D_{ii\sigma}^{(r)} = \begin{vmatrix} M_{ii\sigma}^{(0)} & \cdots & M_{ii\sigma}^{(r)} \\ \vdots & & \vdots \\ M_{ii\sigma}^{(r)} & \cdots & M_{ii\sigma}^{(2r)} \end{vmatrix} .$$

$\boxed{r = 1}$

$$D_{ii\sigma}^{(1)} = \begin{vmatrix} M_{ii\sigma}^{(0)} & M_{ii\sigma}^{(1)} \\ M_{ii\sigma}^{(1)} & M_{ii\sigma}^{(2)} \end{vmatrix} = M_{ii\sigma}^{(0)} M_{ii\sigma}^{(2)} - \left(M_{ii\sigma}^{(1)} \right)^2 =$$

$$= T_0^2 + \left[(T_0 + U)^2 - T_0^2 \right] \langle n_{i-\sigma} \rangle - (T_0 + U \langle n_{i-\sigma} \rangle)^2 =$$

$$= U^2 \langle n_{-\sigma} \rangle (1 - \langle n_{-\sigma} \rangle) \neq 0, \quad \text{when} \quad \langle n_{-\sigma} \rangle \neq 0, 1 .$$

For empty bands ($\langle n_{-\sigma} \rangle = 0$), fully-occupied bands ($\langle n_{-\sigma} \rangle = 1$), and completely polarised, half-filled bands ($\langle n_\sigma \rangle = 1$, $\langle n_{-\sigma} \rangle = 0$), the spectral density clearly consists of only *one* δ-function.

$\boxed{r = 2}$

$$D_{ii\sigma}^{(2)} = M_{ii\sigma}^{(0)} M_{ii\sigma}^{(2)} M_{ii\sigma}^{(4)} + 2 M_{ii\sigma}^{(1)} M_{ii\sigma}^{(2)} M_{ii\sigma}^{(3)} -$$

$$- \left(M_{ii\sigma}^{(2)} \right)^3 - M_{ii\sigma}^{(0)} \left(M_{ii\sigma}^{(3)} \right)^2 - \left(M_{ii\sigma}^{(1)} \right)^2 M_{ii\sigma}^{(4)} =$$

$$= M_{ii\sigma}^{(4)} \left[M_{ii\sigma}^{(2)} - \left(M_{ii\sigma}^{(1)} \right)^2 \right] + M_{ii\sigma}^{(2)} \left[M_{ii\sigma}^{(1)} M_{ii\sigma}^{(3)} - \left(M_{ii\sigma}^{(2)} \right)^2 \right] +$$

$$+ M_{ii\sigma}^{(3)} \left(M_{ii\sigma}^{(1)} M_{ii\sigma}^{(2)} - M_{ii\sigma}^{(3)} \right) =$$

$$= U^2 \langle n_{i-\sigma} \rangle (1 - \langle n_{i-\sigma} \rangle) \left[M_{ii\sigma}^{(4)} + T_0 (T_0 + U) M_{ii\sigma}^{(2)} - (U + 2T_0) M_{ii\sigma}^{(3)} \right] =$$

$$= 0 .$$

Thus, the spectral density is in general a two-pole function.

3.

$$S_{ii\sigma}(E) = \hbar\left[\alpha_{1\sigma}\,\delta\,(E - E_{1\sigma}) + \alpha_{2\sigma}\,\delta\,(E - E_{2\sigma})\right] .$$

The following must hold:

$$\frac{1}{\hbar}\int\limits_{-\infty}^{+\infty} \mathrm{d}E\; E^n\, S_{ii\sigma}(E) = M_{ii\sigma}^{(n)}$$

$$\Longleftrightarrow \quad \alpha_{1\sigma}\, E_{1\sigma}^n + \alpha_{2\sigma}\, E_{2\sigma}^n = T_0^n\,(1 - \langle n_{i\,-\sigma}\rangle) + (T_0 + U)^n\,\langle n_{i\,-\sigma}\rangle\ .$$

From this, we can read off directly:

$$E_{1\sigma} = E_{1\,-\sigma} = T_0\,; \qquad \alpha_{1\sigma} = 1 - \langle n_{i\,-\sigma}\rangle\ ,$$

$$E_{2\sigma} = E_{2\,-\sigma} = T_0 + U\,; \quad \alpha_{2\sigma} = \langle n_{i\,-\sigma}\rangle\ .$$

Solution 4.1.10

1. The spectral moments were calculated in Ex. 3.3.3:

$$M_{k\sigma}^{(2n)} = \left(t^2(\boldsymbol{k}) + \Delta^2\right)^n = (E(\boldsymbol{k}))^{2n}\ ,$$

$$M_{k\sigma}^{(2n+1)} = \left(t^2(\boldsymbol{k}) + \Delta^2\right)^n t(\boldsymbol{k}) = (E(\boldsymbol{k}))^{2n}\, t(\boldsymbol{k})\,.$$

2. Lonke determinant:

$$D_{k\sigma}^{(1)} = M_{k\sigma}^{(0)} M_{k\sigma}^{(2)} - \left(M_{k\sigma}^{(1)}\right)^2 = (E(\boldsymbol{k}))^2 - t^2(\boldsymbol{k}) = \Delta^2 \neq 0\,,$$

$$\begin{aligned} D_{k\sigma}^{(2)} &= M_{k\sigma}^{(0)} M_{k\sigma}^{(2)} M_{k\sigma}^{(4)} + 2M_{k\sigma}^{(1)} M_{k\sigma}^{(2)} M_{k\sigma}^{(3)} - \\ &\quad - \left(M_{k\sigma}^{(2)}\right)^3 - M_{k\sigma}^{(0)}\left(M_{k\sigma}^{(3)}\right)^2 - \left(M_{k\sigma}^{(1)}\right)^2 M_{k\sigma}^{(4)} = \\ &= (E(\boldsymbol{k}))^6 + 2t^2(\boldsymbol{k})\,(E(\boldsymbol{k}))^4 - (E(\boldsymbol{k}))^6 - \\ &\quad - t^2(\boldsymbol{k})\,(E(\boldsymbol{k}))^4 - t^2(\boldsymbol{k})\,(E(\boldsymbol{k}))^4 = 0 \quad \text{q.e.d.} \end{aligned}$$

Section 4.2.4

Solution 4.2.1

$$
\begin{aligned}
\left[a^+_{k\sigma} a_{k+q\sigma}, \widehat{N}\right]_- &= \sum_{p,\sigma'} \left[a^+_{k\sigma} a_{k+q\sigma}, a_{p\sigma'} a_{p\sigma'}\right] = \\
&= \sum_{p,\sigma'} \left\{\delta_{\sigma\sigma'} \delta_{p,\,k+q} a^+_{k\sigma} a_{p\sigma'} - \delta_{\sigma\sigma'} \delta_{kp} a_{p\sigma'} a_{k+q\sigma}\right\} = \\
&= a^+_{k\sigma} a_{k+q\sigma} - a^+_{k\sigma} a_{k+q\sigma} = 0\,.
\end{aligned}
$$

Solution 4.2.2

1.

$$
\rho_\sigma(E) = \rho_0(E) \qquad\qquad \rho_\sigma(E) = \rho_0\left(E + z_\sigma \mu_B B_0\right)\,,
$$

$$
B_0 = \mu_0 H\,.
$$

The magnetisation:

$$
\begin{aligned}
N_\sigma &= N \int_{-\infty}^{+\infty} \mathrm{d}E\ \underbrace{f_-(E)}_{\text{Fermi function}} \rho_\sigma(E) = \\
&= N \int_{-z_\sigma \mu_B B_0}^{\infty} \mathrm{d}E\ f_-(E) \rho_0\left(E + z_\sigma \mu_B B_0\right) = \\
&= N \int_0^\infty \mathrm{d}\overline{E}\ f_-\left(\overline{E} - z_\sigma \mu_B B_0\right) \rho_0(\overline{E})\,,
\end{aligned}
$$

$\mu_B B_0 = \mathrm{E}{-4} \ldots \mathrm{E}{-3}\mathrm{eV}$ in the usual fields.

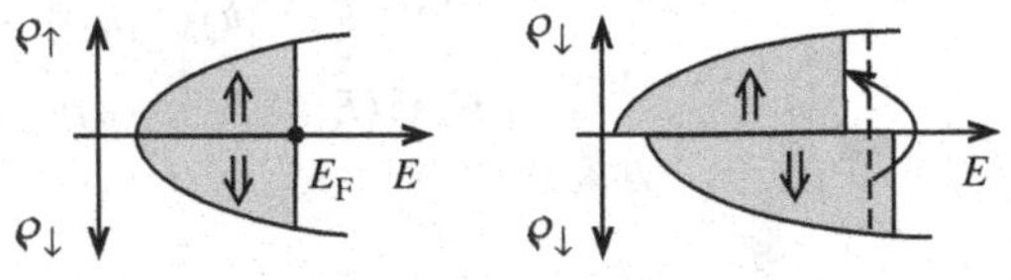

Fig. A.10

The Taylor expansion for the Fermi function can therefore be terminated after the linear term:

$$N_\sigma \approx N \int\limits_0^\infty \mathrm{d}\overline{E} \left(f_-(\overline{E}) - z_\sigma \mu_\mathrm{B} B_0 \frac{\partial f_-}{\partial \overline{E}} \right) \rho_0(\overline{E}),$$

$$\frac{\partial f_-}{\partial \overline{E}} \approx -\delta\left(\overline{E} - E_\mathrm{F}\right)$$

$$\Longrightarrow \quad M \approx 2\mu_\mathrm{B}^2 \mu_0 N \rho_0 (E_\mathrm{F}) H .$$

The Pauli susceptibility:

$$\chi_{\text{Pauli}} \approx 2\mu_\mathrm{B}^2 \mu_0 N \rho_0 (E_\mathrm{F}) .$$

2.

$$\begin{aligned}
&\chi_0(\boldsymbol{q}, E = 0) = \\
&= \frac{2V\hbar}{(2\pi)^3} \int \mathrm{d}^3k \, \frac{\Theta\left(k_\mathrm{F} - |\boldsymbol{k} + \boldsymbol{q}|\right) - \Theta\left(k_\mathrm{F} - k\right)}{\frac{\hbar^2}{2m}\left[(\boldsymbol{k} + \boldsymbol{q})^2 - k^2\right]} = \\
&= \frac{4mV}{\hbar\,(2\pi)^3} \int \mathrm{d}^3k \, \Theta\left(k_\mathrm{F} - k\right) \left\{ \left(2\boldsymbol{k}\cdot\boldsymbol{q} - q^2\right)^{-1} - \left(2\boldsymbol{k}\cdot\boldsymbol{q} + q^2\right)^{-1} \right\} = \\
&= -\frac{mV}{\hbar\pi^2} \int\limits_0^{k_\mathrm{F}} \mathrm{d}k\, k^2 \int\limits_{-1}^{+1} \mathrm{d}x \left\{ \frac{1}{q^2 + 2kqx} + \frac{1}{q^2 - 2kqx} \right\} = \\
&= -\frac{mV}{2\hbar\pi^2 q} \int\limits_0^{k_\mathrm{F}} \mathrm{d}k\, k \left\{ \ln\left(\frac{q^2 + 2kq}{q^2 - 2kq} \right) - \ln\left(\frac{q^2 - 2kq}{q^2 + 2kq} \right) \right\} = \\
&= -\frac{mV}{\hbar\pi^2 q} \int\limits_0^{k_\mathrm{F}} \mathrm{d}k\, k \ln\left| \frac{q^2 + 2kq}{q^2 - 2kq} \right| = \frac{-mV}{4\hbar\pi^2 q^3} \int\limits_0^{2k_\mathrm{F}q} \mathrm{d}x\, x \ln\left| \frac{q^2 + x}{q^2 - x} \right| .
\end{aligned}$$

We use the integral formulation (4.158):

$$\begin{aligned}
&\chi_0(\boldsymbol{q}, E = 0) = \\
&= \frac{-mV}{4\hbar\pi^2 q^3} \left\{ \frac{1}{2}\left(x^2 - q^4\right) \ln\left(q^2 + x\right) - \frac{1}{2}\left(x^2 - q^4\right) \ln\left(q^2 - x\right) - \frac{1}{2}\left(\frac{x^2}{2} - q^2 x \right) + \right. \\
&\quad \left. + \frac{1}{2}\left(\frac{x^2}{2 + q^2 x} \right) \right\}_0^{2k_\mathrm{F}q} = \\
&= \frac{-mVk_\mathrm{F}}{\hbar\pi^2} \left\{ \frac{1}{2} + \frac{k_\mathrm{F}}{2q}\left(1 - \frac{q^2}{4k_\mathrm{F}^2} \right) ln \left| \frac{2k_\mathrm{F} + 1}{2k_\mathrm{F} - q} \right| \right\} .
\end{aligned}$$

In the brackets, we can identify the function defined in (4.160):

$$g\left(n = \frac{q}{2k_{\mathrm{F}}}\right) .$$

With the density of states of the non-interacting electron gas, which was introduced in part 4. of Ex. 2.1.4 (the *Sommerfeld model*),

$$\rho_0(E) = \frac{V}{4\pi^2 N}\left(\frac{2m}{\hbar^2}\right)^{3/2}\sqrt{E}\,\Theta(E) ,$$

we can reformulate the prefactor somewhat:

$$\rho_0\left(E_{\mathrm{F}}\right) = \frac{V}{4\pi^2 N}\left(\frac{2m}{\hbar^2}\right)^{3/2}\sqrt{\frac{\hbar^2}{2m}k_{\mathrm{F}}} = \frac{mV}{2N\pi^2\hbar^2}k_{\mathrm{F}}$$

$$\Longrightarrow \quad \chi_0(\boldsymbol{q}, E = 0) = -2N\hbar\rho_0\left(E_{\mathrm{F}}\right) g\left(q/2k_{\mathrm{F}}\right) =$$

$$= -\frac{\hbar}{\mu_0\mu_{\mathrm{B}}^2}\chi_{\mathrm{Pauli}}\, g\left(q/2k_{\mathrm{F}}\right) .$$

Solution 4.2.3

1. We need to calculate:

$$\chi_{\boldsymbol{q}}^{zz}(E) = -\frac{\mu_0\mu_{\mathrm{B}}^2}{V\hbar}\sum_{\sigma,\sigma'}(2\delta_{\sigma\sigma'} - 1)\frac{1}{N}\sum_{\boldsymbol{k},\boldsymbol{k}'}\langle\langle a_{\boldsymbol{k}\sigma}^+ a_{\boldsymbol{k}+\boldsymbol{q}\sigma}; a_{\boldsymbol{k}'\sigma'}^+ a_{\boldsymbol{k}'-\boldsymbol{q}\sigma'}\rangle\rangle .$$

It is reasonable to start from the following Green's function:

$$\widehat{\chi}_{\boldsymbol{kq}}^{\sigma\sigma'}(E) = \sum_{\boldsymbol{k}'}\langle\langle a_{\boldsymbol{k}\sigma}^+ a_{\boldsymbol{k}+\boldsymbol{q}\sigma}; a_{\boldsymbol{k}'\sigma'}^+ a_{\boldsymbol{k}'-\boldsymbol{q}\sigma'}\rangle\rangle .$$

Hubbard model:

$$\mathcal{H} = \mathcal{H}_0 + H_1 ,$$

$$\mathcal{H}_0 = \sum_{\boldsymbol{k}\sigma}(\varepsilon(\boldsymbol{k}) - \mu)\, a_{\boldsymbol{k}\sigma}^+ a_{\boldsymbol{k}\sigma} ,$$

$$H_1 = \frac{U}{N}\sum_{\boldsymbol{kpq}} a_{\boldsymbol{k}\uparrow}^+ a_{\boldsymbol{k}-\boldsymbol{q}\uparrow} a_{\boldsymbol{p}\downarrow}^+ a_{\boldsymbol{p}+\boldsymbol{q}\downarrow} .$$

For the equation of motion of the function $\chi_{\boldsymbol{kq}}^{\sigma\sigma'}(E)$, we require:

a) *Inhomogeneity*

$$\left\langle\left[a^+_{k\sigma}a_{k+q\sigma},\, a^+_{k'\sigma'}a_{k'-q\sigma'}\right]_-\right\rangle =$$

$$= \delta_{\sigma\sigma'}\delta_{k',\,k+q}\left\langle a^+_{k\sigma}a_{k'-q\sigma'}\right\rangle - \delta_{\sigma\sigma'}\delta_{k,\,k'-q}\left\langle a^+_{k'\sigma'}a_{k+q\sigma}\right\rangle$$

$$= \delta_{\sigma\sigma'}\delta_{k',\,k+q}\left(\langle n_{k\sigma}\rangle - \langle n_{k+q\sigma}\rangle\right) .$$

b)

$$\left[a^+_{k\sigma}a_{k+q\sigma},\, \mathcal{H}_0\right]_- =$$

$$= \sum_{k'\sigma'}\left(\varepsilon\left(\boldsymbol{k}'\right) - \mu\right)\left[a^+_{k\sigma}a_{k+q\sigma},\, a^+_{k'\sigma'}a_{k'\sigma'}\right]_- =$$

$$= \sum_{k',\sigma'}\left(\varepsilon\left(\boldsymbol{k}'\right) - \mu\right)\left(\delta_{\sigma\sigma'}\delta_{k',\,k+q}a^+_{k\sigma}a_{k'\sigma'} - \delta_{\sigma\sigma'}\delta_{k',\,k}a^+_{k'\sigma'}a_{k+q\sigma}\right) =$$

$$= \left(\varepsilon(\boldsymbol{k}+\boldsymbol{q}) - \varepsilon(\boldsymbol{k})\right)a^+_{k\sigma}a_{k+q\sigma} .$$

c)

$$\left[a^+_{k\sigma}a_{k+q\sigma},\, H_1\right]_- =$$

$$= \frac{U}{N}\sum_{k',\,p,\,q'}\left[a^+_{k\sigma}a_{k+q\sigma},\, a^+_{k'\uparrow}a_{k'-q'\uparrow}a^+_{p\downarrow}a_{p+q'\downarrow}\right]_- =$$

$$= \frac{U}{N}\sum_{k',\,p,\,q'}\Big\{\delta_{\sigma\uparrow}\delta_{k',\,k+q}a^+_{k\sigma}a_{k'-q'\uparrow}a^+_{p\downarrow}a_{p+q'\downarrow} -$$

$$- \delta_{\sigma\uparrow}\delta_{k,\,k'-q'}a^+_{k'\uparrow}a_{k+q\sigma}a^+_{p\downarrow}a_{p+q'\downarrow} +$$

$$+ \delta_{\sigma\downarrow}\delta_{k+q,\,p}a^+_{k'\uparrow}a_{k'-q'\uparrow}a^+_{k\sigma}a_{p+q'\downarrow} -$$

$$- \delta_{\sigma\downarrow}\delta_{k,\,p+q'}a^+_{k'\uparrow}a_{k'-q'\uparrow}a^+_{p\downarrow}a_{k+q\sigma}\Big\} =$$

$$= \frac{U}{N}\delta_{\sigma\uparrow}\sum_{p,\,q'}\left\{a^+_{k\uparrow}a_{k+q-q'\uparrow}a^+_{p\downarrow}a_{p+q'\downarrow} - a^+_{k+q'\uparrow}a_{k+q\uparrow}a^+_{p\downarrow}a_{p+q'\downarrow}\right\} +$$

$$+ \frac{U}{N}\delta_{\sigma\downarrow}\sum_{k',\,q'}\left\{a^+_{k'\uparrow}a_{k'-q'\uparrow}a^+_{k\downarrow}a_{k+q+q'\downarrow} - a^+_{k'\uparrow}a_{k'-q'\uparrow}a^+_{k-q'\downarrow}a_{k+q\downarrow}\right\} .$$

The interaction term H_1 thus leads to the following *higher-order* Green's functions:

$$\begin{aligned}
&\langle\langle [a_{k\sigma}^{+} a_{k+q\sigma}, H_1]_{-} ; \dots \rangle\rangle = \\
&= \frac{U}{N} \sum_{p,q'} \Bigg[\delta_{\sigma\uparrow} \Big[\langle\langle a_{k\uparrow}^{+} a_{k+q-q'\uparrow} a_{p\downarrow}^{+} a_{p+q'\downarrow}; \dots \rangle\rangle - \\
&- \langle\langle a_{k+q'\uparrow}^{+} a_{k+q\uparrow} a_{p\downarrow}^{+} a_{p+q'\downarrow}; \dots \rangle\rangle \Big] + \\
&+ \delta_{\sigma\downarrow} \Big[\langle\langle a_{p\uparrow}^{+} a_{p-q'\uparrow} a_{k\downarrow}^{+} a_{k+q+q'\downarrow}; \dots \rangle\rangle - \\
&- \langle\langle a_{p\uparrow}^{+} a_{p-q'\uparrow} a_{k-q'\downarrow}^{+} a_{k+q\downarrow}; \dots \rangle\rangle \Big] \Bigg] .
\end{aligned}$$

The *higher order* Green's functions are decoupled using the RPA method from Sect. 4.2.2, whereby special attention must be paid to the conservation of spin and momentum:

$$\begin{aligned}
&\langle\langle [a_{k\sigma}^{+} a_{k+q\sigma}, H_1]_{-} ; \dots \rangle\rangle \xrightarrow{\text{RPA}} \\
&\xrightarrow{\text{RPA}} \frac{U}{N} \sum_{p,q'} \Bigg[\delta_{\sigma\uparrow} \Big[\delta_{qq'} \langle n_{k\uparrow} \rangle \langle\langle a_{p\downarrow}^{+} a_{p+q'\downarrow}; \dots \rangle\rangle + \\
&+ \delta_{q',0} \langle n_{p\downarrow} \rangle \langle\langle a_{k\uparrow}^{+} a_{k+q-q'\uparrow}; \dots \rangle\rangle - \delta_{qq'} \langle n_{k+q\uparrow} \rangle \langle\langle a_{p\downarrow}^{+} a_{p+q'\downarrow}; \dots \rangle\rangle - \\
&- \delta_{q',0} \langle n_{p\downarrow} \rangle \langle\langle a_{k+q'\uparrow}^{+} a_{k+q\uparrow}; \dots \rangle\rangle \Big] + \delta_{\sigma\downarrow} \Big[\delta_{q'0} \langle n_{p\uparrow} \rangle \langle\langle a_{k\downarrow}^{+} a_{k+q+q'\downarrow}; \dots \rangle\rangle + \\
&+ \delta_{q,-q'} \langle n_{k\downarrow} \rangle \langle\langle a_{p\uparrow}^{+} a_{p-q'\uparrow}; \dots \rangle\rangle - \delta_{q',0} \langle n_{p\uparrow} \rangle \langle\langle a_{k-q'\downarrow}^{+} a_{k+q\downarrow}; \dots \rangle\rangle - \\
&- \delta_{-q',q} \langle n_{k+q\downarrow} \rangle \langle\langle a_{p\uparrow}^{+} a_{p-q'\uparrow}; \dots \rangle\rangle \Big] \Bigg] = \\
&= \frac{U}{N} \sum_{p} \left(\langle n_{k\sigma} \rangle - \langle n_{k+q\sigma} \rangle \right) \langle\langle a_{p-\sigma}^{+} a_{p+q-\sigma}; \dots \rangle\rangle .
\end{aligned}$$

Equation of motion:

$$\begin{aligned}
&[E - (\varepsilon(\boldsymbol{k}+\boldsymbol{q}) - \varepsilon(\boldsymbol{k}))] \chi_{kq}^{\sigma\sigma'}(E) = \\
&= \hbar \delta_{\sigma\sigma'} \left(\langle n_{k\sigma} \rangle - \langle n_{k+q\sigma} \rangle \right) + \frac{U}{N} \sum_{p} \left(\langle n_{k\sigma} \rangle - \langle n_{k+q\sigma} \rangle \right) \chi_{pq}^{-\sigma\sigma'}(E) .
\end{aligned}$$

In the sense of the RPA, the expectation values can be considered to be those of the non-interacting system. They are thus independent of spin ($\langle n_{k\sigma}\rangle^{(0)} = \langle n_{k-\sigma}\rangle^{(0)}$).

$$\chi_q^{zz}(E) = -\frac{\mu_0\mu_B^2}{V\hbar}\frac{1}{N}\sum_{k\sigma}\left(\chi_{kq}^{\sigma\sigma}(E) - \chi_{kq}^{-\sigma\sigma}(E)\right) =$$

$$\stackrel{(4.134)}{=} -\frac{\mu_0\mu_B^2}{V\hbar}\left[\frac{1}{N}\chi_0(q,E) + \frac{U}{N^2}\frac{1}{\hbar}\chi_0(q,E)\sum_p \chi_{pq}^{-\sigma\sigma}(E) - \right.$$

$$\left. - \frac{U}{N^2}\frac{1}{\hbar}\chi_0(q,E)\sum_p \chi_{pq}^{-\sigma-\sigma}(E)\right]$$

$$\Longrightarrow \quad \chi_q^{zz}(E)\left[1 + \frac{U}{N}\frac{1}{2\hbar}\chi_0(q,E)\right] = -\frac{\mu_0\mu_B^2}{V\hbar N}\chi_0(q,E),$$

$$\chi_q^{zz}(E) = -\frac{\mu_0\mu_B^2}{V\hbar N}\frac{\chi_0(q,E)}{1 + \frac{U}{2N\hbar}\chi_0(q,E)}.$$

2. Making use of the result of Ex. 4.2.2 for χ_0, we obtain:

$$\left[\lim_{(q,E)\to 0}\chi_q^{zz}(E)\right]^{-1} = V\frac{1 - U\rho_0(E_F)}{2\mu_0\mu_B^2\rho_0(E_F)}.$$

According to 3.71, the zero of this expression yields a criterion for the occurrence of ferromagnetism:

$$1 \stackrel{!}{=} U\rho_0(E_F).$$

This is the well-known Stoner criterion (4.38).

Solution 4.2.4

1. Transformation to wavenumbers, making use of translational symmetry:

$$D_{ij}(E) = \frac{1}{N}\sum_q D_q(E)e^{iq\cdot(R_i - R_j)},$$

$$D_q(E) = \frac{1}{N}\sum_{k,p} D_{kp}(q,E),$$

$$D_{kp}(q,E) = \langle\langle a_{k-\sigma}a_{q-k\sigma}; a^+_{q-p\sigma}a^+_{p-\sigma}\rangle\rangle_E^{\text{ret}}.$$

Setting up the equation of motion:

$$\left[a_{k-\sigma}a_{q-k\sigma}, a^+_{q-p\sigma}a^+_{p-\sigma}\right]_- = \delta_{kp}a_{k-\sigma}a^+_{p-\sigma} - \delta_{kp}a^+_{q-p\sigma}a_{q-k\sigma} =$$

$$= \delta_{kp}\left(1 - n_{k-\sigma} - n_{q-k\sigma}\right) ,$$

$$\left[a_{k-\sigma}a_{q-k\sigma}, \mathcal{H}_s\right]_- = \sum_{k',\sigma'}\left(E_{\sigma'}(k') - \mu\right)\left[a_{k-\sigma}a_{q-k\sigma}, a^+_{k'\sigma'}a_{k'\sigma'}\right]_- =$$

$$= \sum_{k',\sigma'}\left(E_{\sigma'}(k') - \mu\right)\left(\delta_{k',q-k}\delta_{\sigma\sigma'}a_{k-\sigma}a_{k'\sigma'} - \delta_{k',k}\delta_{\sigma'-\sigma}a_{q-k\sigma}a_{k'\sigma'}\right) =$$

$$= \left(E_\sigma(q-k) + E_{-\sigma}(k) - 2\mu\right)a_{k-\sigma}a_{q-k\sigma}$$

$$\Longrightarrow \quad \left[E + 2\mu - \left(E_\sigma(q-k) + E_{-\sigma}(k)\right)\right]D_{kp}(q,E) =$$

$$= \hbar\delta_{kp}\left(1 - \langle n_{k-\sigma}\rangle - \langle n_{q-k\sigma}\rangle\right) .$$

Solution with a *suitable* boundary condition:

$$D_{kp}(q,E) = \delta_{kp}\frac{\hbar\left(1 - \langle n_{k-\sigma}\rangle - \langle n_{q-k\sigma}\rangle\right)}{E + 2\mu - \left(E_\sigma(q-k) + E_{-\sigma}(k)\right) + \mathrm{i}0^+} .$$

We require:

$$S^{(2)}_{ii}(E - 2\mu) =$$

$$= \frac{1}{N^2}\sum_{kpq}\left(-\frac{1}{\pi}\,\mathrm{Im}\,D_{kp}(q, E - 2\mu)\right)$$

$$= \frac{1}{N^2}\hbar\sum_{kq}\left(1 - \langle n_{k-\sigma}\rangle - \langle n_{q-k\sigma}\rangle\right)\delta\Big(E - \left(E_\sigma(q-k) + E_{-\sigma}(k)\right)\Big) .$$

In the Stoner model, we have for the one-electron spectral density given in Eq. (4.27):

$$S^{(S)}_{k\sigma}(E) = \hbar\delta\left(E - \varepsilon(k) - U\langle n_{-\sigma}\rangle + \mu\right) = \hbar\delta\left(E - E_\sigma(k) + \mu\right) .$$

With the spectral theorem, one therefore finds:

$$\langle n_\sigma\rangle = f_-\left(E_\sigma(k)\right) .$$

For the two-particle spectral density, we can further write:

$$S_{ii}^{(2)}(E-2\mu) = \frac{\hbar}{N^2}\sum_{kp}\Big[1 - f_-(E_{-\sigma}(\boldsymbol{k})) - f_-(E_\sigma(\boldsymbol{p}))\Big]\cdot$$

$$\cdot\,\delta\Big[E - (E_\sigma(\boldsymbol{p}) + E_{-\sigma}(\boldsymbol{k}))\Big] =$$

$$= \hbar\int \mathrm{d}x\,\frac{1}{N}\sum_{k}[1 - f_-(E_{-\sigma}(\boldsymbol{k})) - f_-(x)]\cdot$$

$$\cdot\,\delta(E - E_{-\sigma}(\boldsymbol{k}) - x)\,\frac{1}{N}\sum_{p}\delta(x - E_\sigma(\boldsymbol{p})) =$$

$$= \hbar\int \mathrm{d}x\,\rho_\sigma^{(S)}(x)[1 - f_-(E-x) - f_-(x)]\cdot$$

$$\cdot\,\frac{1}{N}\sum_{k}\delta(E - E_{-\sigma}(\boldsymbol{k}) - x) =$$

$$= \hbar\int \mathrm{d}x\,\rho_\sigma^{(S)}(x)\rho_{-\sigma}^{(S)}(E-x)[1 - f_-(E-x) - f_-(x)]\,.$$

Here, for the Stoner quasi-particle density of states, we have:

$$\rho_\sigma^{(S)}(E) = \frac{1}{N\hbar}\sum_{k} S_{k\sigma}^{(S)}(E-\mu) = \frac{1}{N}\sum_{k}\delta(E - E_\sigma(\boldsymbol{k})) = \rho_0(E - U\langle n_{-\sigma}\rangle)\,.$$

2. The width of the spectrum is determined by the densities of states:

$$\begin{aligned}
&E_{-\sigma}^{\min}(\boldsymbol{k}) \le E - x \le E_{-\sigma}^{\max}(\boldsymbol{k})\\
\Longrightarrow\quad &E_{\max} = E_{-\sigma}^{\max}(\boldsymbol{k}) + x_{\max} = E_{-\sigma}^{\max}(\boldsymbol{k}) + E_\sigma^{\max}(\boldsymbol{k})\,,\\
&E_{\min} = E_{-\sigma}^{\min}(\boldsymbol{k}) + x_{\min} = E_{-\sigma}^{\min}(\boldsymbol{k}) + E_\sigma^{\min}(\boldsymbol{k})\,,\\
\text{Width} = E_{\max} - E_{\min} = &\left(E_{-\sigma}^{\max}(\boldsymbol{k}) - E_{-\sigma}^{\min}(\boldsymbol{k})\right) + \left(E_\sigma^{\max}(\boldsymbol{k}) - E_\sigma^{\min}(\boldsymbol{k})\right) =\\
&= W_{-\sigma} + W_\sigma\,.
\end{aligned}$$

In the Stoner model, $W_\sigma = W_{-\sigma} = W \quad\Longrightarrow\quad$ width of the spectrum: $2W$.

Solution 4.2.5

The two-particle spectral density:

$$S_{ii}^{(2)}(E) = \int_{-\infty}^{+\infty}\mathrm{d}\left(t-t'\right)\,\mathrm{e}^{\frac{\mathrm{i}}{\hbar}E(t-t')}\frac{1}{2\pi}\left\langle\left[(a_{i-\sigma}a_{i\sigma})(t),\left(a_{i\sigma}^{+}a_{i-\sigma}^{+}\right)(t')\right]_-\right\rangle\,.$$

We calculate the two expectation values separately; Ξ = grand canonical partition function:

$$\Xi \left\langle \left(a_{i\sigma}^{+} a_{i-\sigma}^{+}\right)(t')\,(a_{i-\sigma} a_{i\sigma})(t)\right\rangle =$$

$$= \mathrm{Tr}\left\{ \mathrm{e}^{-\beta\mathcal{H}} \left(a_{i\sigma}^{+} a_{i-\sigma}^{+}\right)(t')\,(a_{i-\sigma} a_{i\sigma})(t)\right\} =$$

$$= \sum_N \sum_n \mathrm{e}^{-\beta E_m(N)} \left\langle E_n(N)\right| \left(a_{i\sigma}^{+} a_{i-\sigma}^{+}\right)(t')\,(a_{i-\sigma} a_{i\sigma})(t) \left| E_n(N)\right\rangle =$$

$$= \sum_{N,\,N'} \sum_{n,\,m} \mathrm{e}^{-\beta E_n(N)} \left\langle E_n(N)\right| a_{i\sigma}^{+} a_{i-\sigma}^{+} \left| E_m(N')\right\rangle \cdot$$

$$\cdot \left\langle E_m(N')\right| a_{i-\sigma} a_{i\sigma} \left| E_n(N)\right\rangle \cdot$$

$$\cdot \exp\left(\frac{\mathrm{i}}{\hbar}\left(E_n(N) - E_m(N')\right)(t - t')\right) =$$

$$= \sum_N \sum_{n,\,m} \mathrm{e}^{-\beta E_n(N)} \left\langle E_n(N)\right| a_{i\sigma}^{+} a_{i-\sigma}^{+} \left| E_m(N-2)\right\rangle \cdot$$

$$\cdot \left\langle E_m(N-2)\right| a_{i-\sigma} a_{i\sigma} \left| E_n(N)\right\rangle \cdot$$

$$\cdot \exp\left(-\frac{\mathrm{i}}{\hbar}\left(E_n(N) - E_m(N-2)\right)(t - t')\right).$$

In complete analogy, we find for the second term:

$$\Xi \left\langle (a_{i-\sigma} a_{i\sigma})(t)\left(a_{i\sigma}^{+} a_{i-\sigma}^{+}\right)(t')\right\rangle =$$

$$= \sum_{N'} \sum_{n,\,m} \mathrm{e}^{-\beta E_n(N)} \mathrm{e}^{-\beta(E_m(N-2) - E_n(N))} \cdot$$

$$\cdot \left\langle E_n(N)\right| a_{i\sigma}^{+} a_{i-\sigma}^{+} \left| E_m(N-2)\right\rangle \left\langle E_m(N-2)\right| a_{i-\sigma} a_{i\sigma} \left| E_n(N)\right\rangle \cdot$$

$$\cdot \exp\left(-\frac{\mathrm{i}}{\hbar}\left(E_n(N) - E_m(N-2)\right)(t - t')\right).$$

Thus for the spectral density, using

$$E_n(N) \approx E_n - \mu N \quad (N \gg 1),$$

we find the following *spectral representation:*

$$S_{ii}^{(2)}(E) = \frac{\hbar}{\Xi} \sum_N \sum_{n,\,m} \mathrm{e}^{-\beta E_n(N)} \left\langle E_n(N)\right| a_{i\sigma}^{+} a_{i-\sigma}^{+} \left| E_m(N-2)\right\rangle \cdot$$

$$\cdot \left\langle E_m(N-2)\right| a_{i-\sigma} a_{i\sigma} \left| E_n(N)\right\rangle \left(\mathrm{e}^{\beta E} - 1\right) \delta\left[E - (E_n - E_m - 2\mu)\right].$$

With it, we calculate:

$$\int_{-\infty}^{+\infty} \mathrm{d}E\, I_{\mathrm{AES}}(E-2\mu) = \frac{1}{\hbar}\int_{-\infty}^{+\infty} \mathrm{d}E\, \frac{S_{ii}^{(2)}(E-2\mu)}{\mathrm{e}^{\beta(E-2\mu)}-1} =$$

$$= \frac{1}{\Xi}\sum_N \sum_{N'} \sum_{n,m} \mathrm{e}^{-\beta E_n(N)} \left\langle E_n(N) \middle| a_{i\sigma}^+ a_{i-\sigma}^+ \middle| E_m(N') \right\rangle \cdot$$

$$\cdot \left\langle E_m(N') \middle| a_{i-\sigma} a_{i\sigma} \middle| E_n(N) \right\rangle =$$

$$= \frac{1}{\Xi}\sum_N \sum_n \mathrm{e}^{-\beta E_n(N)} \left\langle E_n(N) \middle| a_{i\sigma}^+ a_{i-\sigma}^+ a_{i-\sigma} a_{i\sigma} \middle| E_n(N) \right\rangle =$$

$$= \left\langle a_{i\sigma}^+ a_{i-\sigma}^+ a_{i-\sigma} a_{i\sigma} \right\rangle = \langle n_{i\sigma} n_{i-\sigma} \rangle = \langle n_\sigma n_{-\sigma} \rangle \quad \text{q. e. d.}$$

Analogously, one finds:

$$\int_{-\infty}^{+\infty} \mathrm{d}E\, I_{\mathrm{APS}}(E-2\mu) = \frac{1}{\Xi}\sum_N \sum_{n,m} \mathrm{e}^{-\beta E_n(N)} \mathrm{e}^{\beta(E_n - E_m - 2\mu)} \,.$$

$$\cdot \left\langle E_n(N) \middle| a_{i\sigma}^+ a_{i-\sigma}^+ \middle| E_m(N-2) \right\rangle \left\langle E_m(N-2) \middle| a_{i-\sigma} a_{i\sigma} \middle| E_n(N) \right\rangle =$$

$$= \frac{1}{\Xi}\sum_N \sum_{N'} \sum_{n,m} \mathrm{e}^{-\beta E_m(N')} \left\langle E_n(N) \middle| a_{i\sigma}^+ a_{i-\sigma}^+ \middle| E_m(N') \right\rangle \cdot$$

$$\cdot \left\langle E_m(N') \middle| a_{i-\sigma} a_{i\sigma} \middle| E_n(N) \right\rangle =$$

$$= \frac{1}{\Xi}\sum_{N'} \sum_m \mathrm{e}^{-\beta E_m(N')} \left\langle E_m(N') \middle| a_{i-\sigma} a_{i\sigma} a_{i\sigma}^+ a_{i-\sigma}^+ \middle| E_m(N') \right\rangle =$$

$$= \left\langle a_{i-\sigma} a_{i\sigma} a_{i\sigma}^+ a_{i-\sigma}^+ \right\rangle = \langle (1 - n_{i-\sigma})(1 - n_{i\sigma}) \rangle = 1 - n + \langle n_{-\sigma} n_\sigma \rangle \quad \text{q. e. d.}$$

For both of these intermediate results, we have made use of the completeness relation,

$$\sum_N \sum_n \left| E_n(N) \right\rangle \left\langle E_n(N) \right| = \mathbf{1} \,.$$

Furthermore, we were able to use

$$\left\langle E_n(N) \middle| a_{i\sigma}^+ a_{i-\sigma}^+ \middle| E_m(N') \right\rangle \sim \delta_{N',\, N-2}$$

repeatedly.

Solution 4.2.6

We compute the retarded Green's function

$$D^{\text{ret}}_{mn;\,jj}(E) = \langle\langle a_{m\sigma} a_{n-\sigma}; a^+_{j-\sigma} a^+_{j\sigma} \rangle\rangle^{\text{ret}}_E$$

with the aid of its equation of motion. Due to the assumed empty band, we set $\mu \to -\infty$, i.e.

$$\frac{e^{\beta(E-2\mu)}}{e^{\beta(E-2\mu)}-1} \longrightarrow 1\,; \quad \frac{1}{e^{\beta(E-2\mu)}-1} \longrightarrow 0$$

$$\Longrightarrow \quad I_{\text{AES}} \equiv 0\,; \quad I_{\text{APS}}(E-2\mu) \longrightarrow -\frac{1}{\hbar\pi}\,\text{Im}\, D^{\text{ret}}_{ii;\,ii}(E-2\mu)\,.$$

The μ-dependence on the right is now only formal. The chemical potential μ no longer occurs explicitly in $D^{\text{ret}}_{ii;\,ii}(E-2\mu)$, so that we can already set it to zero for simplicity in the Hamiltonian:

$$I^{(n=0)}_{\text{APS}}(E) = -\frac{1}{\hbar\pi}\,\text{Im}\, D^{\text{ret}}_{ii;\,ii}(E)\,.$$

We require the commutator:

$$\begin{aligned}
&[a_{m\sigma} a_{n-\sigma}, H]_- = \\
&= \sum_{ij\sigma'} T_{ij}\left[a_{m\sigma} a_{n-\sigma}, a^+_{i\sigma'} a_{j\sigma'}\right]_- + \frac{1}{2} U \sum_{i\sigma'} \left[a_{m\sigma} a_{n-\sigma}, n_{i\sigma'} n_{i-\sigma'}\right]_- = \\
&= \sum_j \left(T_{nj} a_{m\sigma} a_{j-\sigma} - T_{mj} a_{n-\sigma} a_{j\sigma}\right) + \frac{1}{2} U \Big[a_{m\sigma}\left(a_{n-\sigma} n_{n\sigma} + n_{n\sigma} a_{n-\sigma}\right) + \\
&\quad + \left(a_{m\sigma} n_{m-\sigma} + n_{m-\sigma} a_{m\sigma}\right) a_{n-\sigma}\Big] = \\
&= \sum_j \left(T_{nj} a_{m\sigma} a_{j-\sigma} - T_{mj} a_{n-\sigma} a_{j\sigma}\right) + U\left(a_{m\sigma} a_{n-\sigma} n_{n\sigma} + n_{m-\sigma} a_{m\sigma} a_{n-\sigma}\right).
\end{aligned}$$

This yields the still-exact equation of motion:

$$\begin{aligned}
&(E - U\delta_{mn})\, D^{\text{ret}}_{mn;\,jj}(E) = \\
&= \hbar\Big(\delta_{nj}\left\langle a_{m\sigma} a^+_{j\sigma}\right\rangle - \delta_{mj}\left\langle a^+_{j-\sigma} a_{n-\sigma}\right\rangle\Big) + \\
&\quad + \sum_l \left(T_{nl} D^{\text{ret}}_{ml,\,jj}(E) + T_{ml} D^{\text{ret}}_{ln,\,jj}(E)\right) + \\
&\quad + U(1-\delta_{mm})\, \langle\langle a_{m\sigma}\left(n_{n\sigma} + n_{m-\sigma}\right) a_{n-\sigma}; a^+_{j-\sigma} a^+_{j\sigma} \rangle\rangle^{\text{ret}}_E\,.
\end{aligned}$$

We can make use of the assumed empty energy band ($n = 0$):

$$\left\langle a_{m\sigma} a^+_{j\sigma} \right\rangle = \delta_{mj} - \left\langle a^+_{j\sigma} a_{m\sigma} \right\rangle \longrightarrow \delta_{mj} \,,$$
$$\left\langle a^+_{j-\sigma} a_{n-\sigma} \right\rangle \longrightarrow 0 \,,$$
$$\langle\langle a_{m\sigma} \left(n_{n\sigma} + n_{m-\sigma}\right) a_{n-\sigma j} a^+_{j-\sigma} ; a^+_{j\sigma} \rangle\rangle^{\text{ret}}_E \underset{m \neq n}{\longrightarrow} 0 \,.$$

The last relation is to be verified directly via the definition of the Green's function. Now, only the greatly simplified equation of motion remains:

$$(E - U\delta_{mn}) \, D^{\text{ret}}_{mn;\,jj}(E) = \hbar \delta_{nj} \delta_{mj} + \sum_l \left(T_{nl} D^{\text{ret}}_{ml;\,jj}(E) + T_{ml} D^{\text{ret}}_{ln;\,jj}(E) \right).$$

It can be solved via Fourier transformation:

$$D^{\text{ret}}_{\boldsymbol{kp};\,jj}(E) = \frac{1}{N} \sum_{m,n} \mathrm{e}^{-\mathrm{i}(\boldsymbol{k} \cdot \boldsymbol{R}_m + \boldsymbol{p} \cdot \boldsymbol{R}_n)} D^{\text{ret}}_{mn;\,jj}(E) \,.$$

In detail, one then finds:

$$\sum_l T_{nl} D^{\text{ret}}_{ml;\,jj}(E) = \frac{1}{N} \sum_{\boldsymbol{k},\boldsymbol{p}} \mathrm{e}^{\mathrm{i}(\boldsymbol{p} \cdot \boldsymbol{R}_n + \boldsymbol{k} \cdot \boldsymbol{R}_m)} \varepsilon(\boldsymbol{p}) D^{\text{ret}}_{\boldsymbol{kp};\,jj}(E) \,,$$

$$\sum_l T_{ml} D^{\text{ret}}_{ln;\,jj}(E) = \frac{1}{N} \sum_{\boldsymbol{k},\boldsymbol{p}} \mathrm{e}^{\mathrm{i}(\boldsymbol{p} \cdot \boldsymbol{R}_n + \boldsymbol{k} \cdot \boldsymbol{R}_m)} \varepsilon(\boldsymbol{k}) D^{\text{ret}}_{\boldsymbol{kp};\,jj}(E) \,,$$

$$\delta_{mn} D^{\text{ret}}_{mn;\,jj}(E) = \frac{1}{N^2} \sum_{\boldsymbol{q}} \sum_{\boldsymbol{k},\boldsymbol{p}} \mathrm{e}^{\mathrm{i}(\boldsymbol{p} \cdot \boldsymbol{R}_n + \boldsymbol{k} \cdot \boldsymbol{R}_m)} D^{\text{ret}}_{\boldsymbol{k}-\boldsymbol{q},\,\boldsymbol{p}+\boldsymbol{q};\,jj}(E) \,,$$

$$\delta_{mj} \delta_{nj} = \delta_{mj} \delta_{mn} = \frac{1}{N^2} \sum_{\boldsymbol{p},\boldsymbol{k}} \mathrm{e}^{\mathrm{i}(\boldsymbol{p} \cdot \boldsymbol{R}_n + \boldsymbol{k} \cdot \boldsymbol{R}_m)} \mathrm{e}^{-\mathrm{i}(\boldsymbol{p}+\boldsymbol{k}) \cdot \boldsymbol{R}_j} \,.$$

This yields the following Fourier-transformed equation of motion:

$$[E - \varepsilon(\boldsymbol{k}) - \varepsilon(\boldsymbol{p})] \, D^{\text{ret}}_{\boldsymbol{kp};\,jj}(E) = \frac{\hbar}{N} \mathrm{e}^{-\mathrm{i}(\boldsymbol{p}+\boldsymbol{k}) \cdot \boldsymbol{R}_j} + \frac{U}{N} \sum_{\boldsymbol{q}} D^{\text{ret}}_{\boldsymbol{k}-\boldsymbol{q},\,\boldsymbol{p}+\boldsymbol{q};\,jj}(E) \,.$$

The following change of variables now appears expedient:

$$\boldsymbol{\rho} = \boldsymbol{k} + \boldsymbol{p}\,; \quad \bar{\boldsymbol{\rho}} = \frac{1}{2}(\boldsymbol{k} - \boldsymbol{p})$$

$$\Longrightarrow \quad \left[E - \varepsilon\left(\frac{1}{2}\boldsymbol{\rho} + \bar{\boldsymbol{\rho}}\right) - \varepsilon\left(\frac{1}{2}\boldsymbol{\rho} - \bar{\boldsymbol{\rho}}\right)\right] D^{\mathrm{ret}}_{\frac{1}{2}\boldsymbol{\rho}+\bar{\boldsymbol{\rho}},\,\frac{1}{2}\boldsymbol{\rho}-\bar{\boldsymbol{\rho}};\,jj}(E) =$$

$$= \frac{\hbar}{N}\mathrm{e}^{-\mathrm{i}\boldsymbol{\rho}\cdot\boldsymbol{R}_j} + \frac{U}{N}\sum_{\boldsymbol{q}} D^{\mathrm{ret}}_{\frac{1}{2}\boldsymbol{\rho}+\bar{\boldsymbol{q}},\,\frac{1}{2}\boldsymbol{\rho}-\bar{\boldsymbol{q}};\,jj}(E)\,.$$

We require:

$$D^{\mathrm{ret}}_{ii;\,ii}(E) = \frac{1}{N}\sum_{\boldsymbol{k},\,\boldsymbol{p}} \mathrm{e}^{\mathrm{i}(\boldsymbol{k}+\boldsymbol{p})\cdot\boldsymbol{R}_i} D^{\mathrm{ret}}_{\boldsymbol{k}\boldsymbol{p};\,ii}(E) = \frac{1}{N}\sum_{\boldsymbol{\rho},\,\bar{\boldsymbol{\rho}}} \mathrm{e}^{\mathrm{i}\boldsymbol{\rho}\cdot\boldsymbol{R}_i} D^{\mathrm{ret}}_{\frac{1}{2}\boldsymbol{\rho}+\bar{\boldsymbol{\rho}},\,\frac{1}{2}\boldsymbol{\rho}-\bar{\boldsymbol{\rho}};\,ii}(E)\,.$$

Initially, the equation of motion can be condensed after summation over $\bar{\boldsymbol{\rho}}$, making use of the conventional definition of $\Lambda^{(0)}_{\boldsymbol{k}}(E)$, yielding:

$$\frac{1}{N}\sum_{\bar{\boldsymbol{\rho}}} D^{\mathrm{ret}}_{\frac{1}{2}\boldsymbol{\rho}+\bar{\boldsymbol{\rho}},\,\frac{1}{2}\boldsymbol{\rho}-\bar{\boldsymbol{\rho}};\,ii}(E) = \frac{\hbar}{N}\mathrm{e}^{-\mathrm{i}\boldsymbol{\rho}\cdot\boldsymbol{R}_i}\,\frac{\Lambda^{(0)}_{\boldsymbol{\rho}}(E)}{1 - U\Lambda^{(0)}_{\boldsymbol{\rho}}(E)}\,.$$

From this, we obtain the assertion:

$$I^{(n=0)}_{\mathrm{APS}}(E) = -\frac{1}{\pi}\,\mathrm{Im}\,\frac{1}{N}\sum_{\boldsymbol{k}}\frac{\Lambda^{(0)}_{\boldsymbol{k}}(E)}{1 - U\Lambda^{(0)}_{\boldsymbol{k}}(E)}\,.$$

For small values of U, this expression can be simplified to:

$$I^{(n=0)}_{\mathrm{APS}}(E) \approx \frac{1}{N^2}\sum_{\boldsymbol{k}}\sum_{\boldsymbol{q}}\delta\left[E - \varepsilon(\boldsymbol{k}) - \varepsilon(\boldsymbol{k}-\boldsymbol{q})\right] =$$

$$= \int_{-\infty}^{+\infty}\mathrm{d}x\,\rho_0(x)\frac{1}{N}\sum_{\boldsymbol{k}}\delta\left(E - \varepsilon(\boldsymbol{k}) - x\right) = \int_{-\infty}^{+\infty}\mathrm{d}x\,\rho_0(x)\rho_0(E-x)\,.$$

This is the *self-convolution* of the Bloch density of states:

$$\rho_0(x) = \frac{1}{N}\sum_{\boldsymbol{p}}\delta\left(x - \varepsilon(\boldsymbol{p})\right)\,.$$

Solution 4.2.7

Precisely speaking, we should choose the range $\mu \to +\infty$ for the chemical potential in this case. This implies that:

$$\begin{aligned} I_{\text{AES}}(E-2\mu) &\longrightarrow +\frac{1}{\hbar\pi}\,\text{Im}\,D^{\text{ret}}_{ii;\,ii}(E-2\mu)\,, \\ I_{\text{APS}}(E-2\mu) &\longrightarrow 0\,. \end{aligned}$$

In the exact and generally valid equation of motion for $D^{\text{ret}}_{mn;\,jj}(E)$ in the solution of Ex. 4.2.6, we can now make the following simplifications due to $n=2$:

$$\begin{aligned} \left\langle a_{m\sigma}a^{+}_{j\sigma}\right\rangle &\longrightarrow 0\,, \\ \left\langle a^{+}_{j-\sigma}a_{n-\sigma}\right\rangle &\longrightarrow \delta_{nj}\,, \\ \left\langle\!\left\langle a_{m\sigma}n_{n\sigma}a_{n-\sigma};a^{+}_{j-\sigma}a^{+}_{j\sigma}\right\rangle\!\right\rangle^{\text{ret}}_{E} &\longrightarrow D^{\text{ret}}_{mn;\,jj}(E)\,, \\ \left\langle\!\left\langle a_{m\sigma}n_{m-\sigma}a_{n-\sigma};a^{+}_{j-\sigma}a^{+}_{j\sigma}\right\rangle\!\right\rangle^{\text{ret}}_{E} &\longrightarrow (1-\delta_{mn})\,D^{\text{ret}}_{mn;\,jj}(E)\,. \end{aligned}$$

This then leads to the simplified equation of motion:

$$\begin{aligned} &[E+2\mu-U\,(2-\delta_{nm})]\,D^{\text{ret}}_{mn;\,jj}(E)= \\ &=-\hbar\delta_{mj}\delta_{nj}+\sum_{l}\left(T_{nl}\,D^{\text{ret}}_{ml;\,jj}(E)+T_{ml}\,D^{\text{ret}}_{ln;\,jj}(E)\right)\,. \end{aligned}$$

This is very similar to the corresponding equation of motion for $n=0$. We have only to replace E by $E+2\mu-2U$ and U by $-U$ We can thus adopt that result directly:

$$\begin{aligned} I^{(n=2)}_{\text{AES}}(E-2\mu) &= +\frac{1}{\pi}\,\text{Im}\,\frac{1}{N}\sum_{k}\frac{\Lambda^{(2)}_{k}(E)}{1+U\Lambda^{(2)}_{k}(E)}\,, \\ \Lambda^{(2)}_{k}(E) &= \frac{1}{N}\sum_{p}\frac{1}{E-2U-\varepsilon(\boldsymbol{k})-\varepsilon(\boldsymbol{k}-\boldsymbol{p})+\text{i}0^{+}}\,. \end{aligned}$$

Section 4.4.3

Solution 4.4.1

We have according to (4.292):

$$\sigma\equiv\frac{\langle S^{z}\rangle}{\hbar S}=\frac{1}{1+2\varphi}=1-2\varphi+(2\varphi)^{2}-\cdots$$

At low temperatures, we can limit ourselves to the first terms of the expansion:

$$\varphi = \frac{1}{N}\sum_{\boldsymbol{q}} \frac{1}{\exp(\beta E(\boldsymbol{q})) - 1}\,.$$

From (4.288), for the quasi-particle energies, we have:

$$E(\boldsymbol{q}) = 2\hbar\left\langle S^z\right\rangle(J_0 - J(\boldsymbol{q})) \quad (B_0 = 0^+)\,.$$

We are interested in the *spontaneous* magnetisation. There is thus no external magnetic field present.

In the thermodynamic limit, we can convert the wavenumber summation into an integration:

$$\begin{aligned}
\varphi &= \frac{V}{N(2\pi)^3}\int \mathrm{d}^3q\,\mathrm{e}^{-\beta E(\boldsymbol{q})}\frac{1}{1-\mathrm{e}^{-\beta E(\boldsymbol{q})}} = \\
&= \frac{V}{N(2\pi)^3}\int \mathrm{d}^3q\,\mathrm{e}^{-\beta E(\boldsymbol{q})}\sum_{n=0}^{\infty}\mathrm{e}^{-n\beta E(\boldsymbol{q})} = \\
&= \frac{V}{N(2\pi)^3}\sum_{n=1}^{\infty}\int \mathrm{d}^3q\,\mathrm{e}^{-n\beta 2\hbar\langle S^z\rangle(J_0 - J(\boldsymbol{q}))}\,.
\end{aligned}$$

At low temperatures, ($\beta \to \infty$), the integrand is practically zero except at small values of $|\boldsymbol{q}|$. We may therefore make the following approximation:

$$\varphi \approx \frac{V}{N4\pi^2}\sum_{n=1}^{\infty}\int_0^{\infty}\mathrm{d}q\,q^2\mathrm{e}^{-n\beta\sigma DQ^2} = \frac{V}{8\pi^2 N}\sum_{n=1}^{\infty}(n\beta\sigma D)^{-3/2}\,\Gamma\left(\frac{3}{2}\right),$$

$$\Gamma\left(\frac{3}{2}\right) = \frac{1}{2}\sqrt{\pi}\,,$$

$$\zeta\left(\frac{3}{2}\right) = \sum_{n=1}^{\infty}\frac{1}{n^{3/2}} \approx 2.612 \quad \text{(Riemann's } \zeta \text{ function)}\,.$$

Finally, this means that:

$$\varphi \approx \frac{V}{N}\left(\frac{k_{\mathrm{B}}T}{4\pi\sigma D}\right)^{3/2}\zeta\left(\frac{3}{2}\right).$$

In the neighbourhood of ferromagnetic saturation, $\varphi \ll 1$:

$$1 - \frac{\left\langle S^z\right\rangle}{\hbar S} \equiv 1 - \sigma \approx 2\varphi \sim T^{3/2} \quad \text{q. e. d.}$$

Solution 4.4.2

From the operator identity (4.307),

$$\prod_{m_s=-1}^{+1} \left(S_i^z - \hbar m_s\right) = \left(S_i^z + \hbar\right) S_i^z \left(S_i^z - \hbar\right) ,$$

it follows for $S = 1$ that:

$$\left(S_i^z\right)^3 = \hbar^2 S_i^z .$$

The system of eqs. (4.311) is now to be evaluated for $n = 0, 1$:

$\boxed{n = 0}$

$$2\hbar^2 - \hbar \left\langle S^z \right\rangle - \left\langle (S^z)^2 \right\rangle = 2\hbar \left\langle S^z \right\rangle \varphi(1) .$$

$\boxed{n = 1}$

$$2\hbar^2 \left\langle S^z \right\rangle - \hbar \left\langle (S^z)^2 \right\rangle - \left\langle (S^z)^3 \right\rangle = \left(3\hbar \left\langle (S^z)^2 \right\rangle - \hbar^2 \left\langle S^z \right\rangle - 2\hbar^3 \right)\varphi(1) .$$

The following relation from the $n = 0$ equation,

$$\left\langle (S^z)^2 \right\rangle = 2\hbar^2 - \hbar \left\langle S^z \right\rangle (1 + 2\varphi(1)) ,$$

is inserted into the $n = 1$ equation:

$$2\hbar^2 \left\langle S^z \right\rangle - 2\hbar^3 + \hbar^2 \left\langle S^z \right\rangle (1 + 2\varphi(1)) - \hbar^2 \left\langle S^z \right\rangle =$$

$$= \left[6\hbar^3 - 3\hbar^2 \left\langle S^z \right\rangle (1 + 2\varphi(1)) - \hbar^2 \left\langle S^z \right\rangle - 2\hbar^3 \right]\varphi(1) .$$

Solving for $\langle S^z \rangle$, this yields the assertion:

$$\left\langle S^z \right\rangle = \hbar \frac{1 + 2\varphi(1)}{1 + 3\varphi(1) + 3\varphi^2(1)} .$$

Since $\langle S^z \rangle$ is also contained in $\varphi(1)$, this is an *implicit* functional equation for $\langle S^z \rangle$. Furthermore, one finds by substitution that:

$$\left\langle (S^z)^2 \right\rangle_{S=1} = \hbar^2 \frac{1 + 2\varphi(1) + 2\varphi^2(1)}{1 + 3\varphi(1) + 3\varphi^2(1)} .$$

Solution 4.4.3

1. Proof through complete induction:
   $n = 1$:

$$\left[S_i^-, S_i^z\right]_- = \hbar S_i^- .$$

$n \longrightarrow n+1$:

$$\left[\left(S_i^-\right)^{n+1}, S_i^z\right]_- = \left(S_i^-\right)^n \left[S_i^-, S_i^z\right]_- + \left[\left(S_i^-\right)^n, S_i^z\right]_- S_i^- =$$
$$= \hbar \left(S_i^-\right)^{n+1} + n\hbar \left(S_i^-\right)^n S_i^- = (n+1)\hbar \left(S_i^-\right)^{n+1} .$$

2. Proof using the partial result from 1.:

$$\left[\left(S_i^-\right)^n, \left(S_i^z\right)^2\right]_- = \left[\left(S_i^-\right)^n, S_i^z\right]_- S_i^z + S_i^z \left[\left(S_i^-\right)^n, S_i^z\right]_- = n\hbar \left(\left(S_i^-\right)^n S_i^z + S_i^z \left(S_i^-\right)^n\right) =$$
$$= n\hbar \left(n\hbar \left(S_i^-\right)^n + 2S_i^z \left(S_i^-\right)^n\right) = n^2\hbar^2 \left(S_i^-\right)^n + 2n\hbar S_i^z \left(S_i^-\right)^n .$$

3. Proof through complete induction:
   $n = 1$:

$$\left[S_i^+, S_i^-\right]_- = 2\hbar S_i^z .$$

$n \longrightarrow n+1$:

$$\left[S_i^+, \left(S_i^-\right)^{n+1}\right]_- = S_i^- \left[S_i^+, \left(S_i^-\right)^n\right]_- + \left[S_i^+, S_i^-\right]_- \left(S_i^-\right)^n =$$
$$= S_i^- \left[2n\hbar S_i^z + \hbar^2 n\,(n-1)\right] \left(S_i^-\right)^{n-1} + 2\hbar S_i^z \left(S_i^-\right)^n =$$
$$= \hbar^2 n\,(n-1) \left(S_i^-\right)^n + 2n\hbar \left(\hbar S_i^- + S_i^z S_i^-\right) \left(S_i^-\right)^{n-1} + 2\hbar S_i^z \left(S_i^-\right)^n =$$
$$= \hbar^2 n\,(n+1) \left(S_i^-\right)^n + 2\hbar(n+1) S_i^z \left(S_i^-\right)^n \quad \text{q. e. d.}$$

Solution 4.4.4

$$
\begin{aligned}
&\left(S_i^-\right)^n \left(S_i^+\right)^n = \\
&= \left(S_i^-\right)^{n-1}\left[\hbar^2 S\,(S+1) - \hbar S_i^z - \left(S_i^z\right)^2\right]\left(S_i^+\right)^{n-1} = \\
&= \left\{\hbar^2 S\,(S{+}1) - \hbar S_i^z - \left(S_i^z\right)^2\right\}\left(S_i^-\right)^{n-1}\left(S_i^+\right)^{n+1} - \hbar\left[\left(S_i^-\right)^{n-1}, S_i^z\right]_-\left(S_i^+\right)^{n-1} - \\
&\quad - \left[\left(S_i^-\right)^{n-1}, \left(S_i^z\right)^2\right]_-\left(S_i^+\right)^{n-1} = \\
&= \left\{\hbar^2 S\,(S+1) - \hbar S_i^z - \left(S_i^z\right)^2\right\}\left(S_i^-\right)^{n-1}\left(S_i^+\right)^{n+1} - \\
&\quad - \hbar\left\{(n-1)\hbar\left(S_i^-\right)^{n-1}\right\}\left(S_i^+\right)^{n-1} - \\
&\quad - \left\{(n-1)^2\hbar^2 + 2(n-1)\hbar S_i^z\right\}\left(S_i^-\right)^{n-1}\left(S_i^+\right)^{n-1} = \\
&= \left\{\hbar^2 S\,(S+1) - n\,(n-1)\hbar^2 - (2n-1)\hbar S_i^z - \left(S_i^z\right)^2\right\}\left(S_i^-\right)^{n-1}\left(S_i^+\right)^{n-1} = \\
&= \prod_{p=1}^{n}\left\{\hbar^2 S\,(S+1) - (n-p)(n-p+1)\hbar^2 - \right. \\
&\quad \left. - (2n-2p+1)\hbar S_i^z - \left(S_i^z\right)^2\right\} \quad \text{q. e. d.}
\end{aligned}
$$

Solution 4.4.5

The *active* operator for the equation of motion S_i^+ to the left of the semicolon is the same as in(4.281). The Tyablikow approximation for (4.287) therefore leads to a completely analogous solution:

$$
\begin{aligned}
G_{\boldsymbol{q}}^{(n)}(E) &= \left\langle\left[S_i^+, \left(S_i^-\right)^{n+1}\left(S_i^+\right)^n\right]_-\right\rangle \frac{1}{E - E(\boldsymbol{q}) + \mathrm{i}0^+}\,, \\
E(\boldsymbol{q}) &= 2\hbar\left\langle S^z\right\rangle\left(J_0 - J(\boldsymbol{q})\right)\,.
\end{aligned}
$$

The spectral theorem then yields:

$$
\begin{aligned}
\left\langle\left(S_i^-\right)^{n+1}\left(S_i^+\right)^{n+1}\right\rangle &= \left\langle\left[S_i^+, \left(S_i^-\right)^{n+1}\left(S_i^+\right)^n\right]_-\right\rangle\varphi(S)\,, \\
\varphi\,(S) &= \frac{1}{N}\sum_{\boldsymbol{q}}\left(\mathrm{e}^{\beta E(\boldsymbol{q})} - 1\right)^{-1}\,.
\end{aligned}
$$

Here, we now insert the partial results from the two preceding exercises:

$n = 0$:

$$\hbar^2 S(S+1) - \hbar \langle S^z \rangle - \left\langle (S^z)^2 \right\rangle = 2\hbar \langle S^z \rangle \varphi(S) .$$

$n \geq 1$:

$$\left\langle \prod_{p=1}^{n+1} \left\{ \hbar^2 S(S+1) - (n+1-p)(n+2-p)\hbar^2 - (2n-2p+3)\hbar S^z - (S^z)^2 \right\} \right\rangle =$$

$$= \varphi(S) \left\langle \left[\hbar^2 n(n+1) + 2\hbar(n+1)S^z \right] \prod_{p=1}^{n} \left\{ \hbar^2 S(S+1) - (n-p)(n+1-p)\hbar^2 - \right. \right.$$

$$\left. \left. - (2n-2p+1)\hbar S^z - (S^z)^2 \right\} \right\rangle .$$

Evaluation for $\mathbf{S} = 1$:
Due to $2S - 1 = 1$, we need the equations for $n = 0$ and $n = 1$:
$n = 0$:

$$2\hbar^2 - \hbar \langle S^z \rangle - \left\langle (S^z)^2 \right\rangle = 2\hbar \langle S^z \rangle \varphi(1) .$$

$n = 1$:

$$\left\langle (S^z)^4 + 4\hbar\,(S^z)^3 + \hbar^2 (S^z)^2 - 6\hbar^3 S^z \right\rangle =$$

$$= \varphi(1) \left\langle 4\hbar^4 + 6\hbar^3 S^z - 6\hbar^2 (S^z)^2 - 4\hbar\,(S^z)^3 \right\rangle .$$

Furthermore, from (4.307), we still have:

$$(S^z)^3 = \hbar^2 S^z \quad \Longleftrightarrow \quad (S^z)^4 = \hbar^2 (S^z)^2 .$$

Then the $n = 1$ equation becomes:

$$2\hbar^2 \left\langle (S^z)^2 \right\rangle - 2\hbar^3 \langle S^z \rangle = \varphi(1) \left\{ 4\hbar^4 + 2\hbar^3 \langle S^z \rangle - 6\hbar^2 \left\langle (S^z)^2 \right\rangle \right\} .$$

The $n = 0$ equation yields:

$$\left\langle (S^z)^2 \right\rangle = 2\hbar^2 - \hbar \langle S^z \rangle (1 + 2\varphi(1)) .$$

This is inserted:

$$4\hbar^4 - 4\hbar^3 \langle S^z \rangle (1 + \varphi(1)) = \varphi(1) \left\{ -8\hbar^4 + 2\hbar^3 \langle S^z \rangle (4 + 6\varphi(1)) \right\}$$

$$\Longrightarrow \quad 4\hbar^4 (1 + 2\varphi(1)) = 4\hbar^3 \langle S^z \rangle \left(1 + 3\varphi(1) + 3\varphi^2(1) \right) .$$

From this, the relation known from Ex. 4.4.2 follows:

$$\left\langle S^z \right\rangle_{S=1} = \hbar \frac{1 + 2\varphi(1)}{1 + 3\varphi(1) + 3\varphi^2(1)} \quad \text{q. e. d.}$$

Section 4.5.5

Solution 4.5.1

For the equation of motion, we require a series of commutators:

$$\left[S_i^z, H_f\right]_- = -\sum_{m,n} J_{mn}\left[S_i^z, S_m^+ S_n^-\right]_- = \hbar \sum_m J_{im}\left(S_m^+ S_i^- - S_i^+ S_m^-\right),$$

$$\left[S_i^z, H_{s-f}\right]_- = -\frac{1}{2} g\hbar \sum_{m,\sigma}\left[S_i^z, S_m^\sigma\right]_- a_{m-\sigma}^+ a_{m\sigma} = -\frac{1}{2} g\hbar^2 \sum_\sigma z_\sigma S_i^\sigma a_{i-\sigma}^+ a_{i\sigma}.$$

We then find all together:

$$\left[S_i^z, H\right]_- = \hbar \sum_m J_{im}\left(S_m^+ S_i^- - S_i^+ S_m^-\right) - \frac{1}{2} g\hbar^2 \sum_\sigma z_\sigma S_i^\sigma a_{i-\sigma}^+ a_{i\sigma}.$$

We combine this with (4.395):

$$\begin{aligned}
\left[S_i^z a_{k\sigma}, H\right]_- &= S_i^z\left[a_{k\sigma}, H\right]_- + \left[S_i^z, H\right]_- a_{k\sigma} = \\
&= \sum_m T_{km} S_i^z a_{m\sigma} + U S_i^z n_{k-\sigma} a_{k\sigma} - \\
&\quad - \frac{1}{2} g\hbar z_\sigma S_i^z S_k^z a_{k\sigma} - \frac{1}{2} g\hbar S_i^z S_k^{-\sigma} a_{k-\sigma} + \\
&\quad + \hbar \sum_m J_{im}\left(S_m^+ S_i^- - S_i^+ S_m^-\right) a_{k\sigma} - \\
&\quad - \frac{1}{2} g\hbar^2 \left(S_i^+ a_{i\downarrow}^+ a_{i\uparrow} - S_i^- a_{i\uparrow}^+ a_{i\downarrow}\right) a_{k\sigma}.
\end{aligned}$$

We define several new Green's functions:

$$
\begin{aligned}
D^{(1)}_{ik,\,j\sigma}(E) &= \langle\!\langle S_i^z n_{k-\sigma} a_{k\sigma} ; a^+_{j\sigma} \rangle\!\rangle_E ,\\
D^{(2)}_{ik,\,j\sigma}(E) &= \langle\!\langle S_i^z S_k^z a_{k\sigma} ; a^+_{j\sigma} \rangle\!\rangle_E ,\\
D^{(3)}_{ik,\,j\sigma}(E) &= \langle\!\langle S_i^z S_k^{-\sigma} a_{k-\sigma} ; a^+_{j\sigma} \rangle\!\rangle_E ,\\
H_{imk,\,j\sigma}(E) &= \langle\!\langle \left(S_m^+ S_i^- - S_i^+ S_m^-\right) a_{k\sigma} ; a^+_{j\sigma} \rangle\!\rangle_E ,\\
L_{ik,\,j\sigma}(E) &= \left\langle\!\!\left\langle \left(S_i^+ a^+_{i\downarrow} a_{i\uparrow} - S_i^- a^+_{i\uparrow} a_{i\downarrow}\right) a_{k\sigma} ; a^+_{j\sigma} \right\rangle\!\!\right\rangle_E .
\end{aligned}
$$

With these definitions, the already rather complicated, complete equation of motion becomes:

$$
\begin{aligned}
&\sum_m \left(E\delta_{km} - T_{km}\right) D_{im,\,j\sigma}(E) =\\
&= \hbar\delta_{kj} \langle S^z \rangle + U D^{(1)}_{ik,\,j\sigma}(E) - \frac{1}{2} g\hbar z_\sigma D^{(2)}_{ik,\,j\sigma}(E) -\\
&\quad - \frac{1}{2} g\hbar D^{(3)}_{ik,\,j\sigma}(E) + \hbar \sum_m J_{im} H_{imk,\,j\sigma}(E) - \frac{1}{2} g\hbar^2 L_{ik,\,j\sigma}(E) .
\end{aligned}
$$

Solution 4.5.2

We require once again several commutators for the equation of motion:

$$
\begin{aligned}
\left[n_{i-\sigma}, H\right]_- &= \left[n_{i-\sigma}, H_s\right]_- + \left[n_{i-\sigma}, H_{s-f}\right]_- ,\\
\left[n_{i-\sigma}, H_s\right]_- &= \sum_{\substack{m,n\\ \sigma'}} T_{mn} \left[n_{i-\sigma}, a^+_{m\sigma'} a_{n\sigma'}\right]_- =\\
&= \sum_{\substack{m,n\\ \sigma'}} T_{mn} \left\{\delta_{im}\delta_{\sigma'-\sigma} a^+_{i-\sigma} a_{n\sigma'} - \delta_{in}\delta_{\sigma'-\sigma} a^+_{m\sigma'} a_{i-\sigma}\right\} =\\
&= \sum_m T_{im} \left(a^+_{i-\sigma} a_{m-\sigma} - a^+_{m-\sigma} a_{i-\sigma}\right) ,\\
\left[n_{i-\sigma}, H_{s-f}\right]_- &= -\frac{1}{2} g\hbar \sum_{m,\sigma'} S_m^{\sigma'} \left[n_{i-\sigma}, a^+_{m-\sigma'} a_{m\sigma'}\right]_- =\\
&= -\frac{1}{2} g\hbar \sum_{m,\sigma'} S_m^{\sigma'} \delta_{im} \left(\delta_{\sigma\sigma'} a^+_{i-\sigma} a_{m\sigma'} - \delta_{-\sigma\sigma'} a^+_{m-\sigma'} a_{i-\sigma}\right) =\\
&= -\frac{1}{2} g\hbar \left(S_i^\sigma a^+_{i-\sigma} a_{i\sigma} - S_i^{-\sigma} a^+_{i\sigma} a_{i-\sigma}\right) .
\end{aligned}
$$

All together, with (4.395) this gives:

$$[n_{i\,-\sigma} a_{k\sigma}, H]_- =$$
$$= \sum_m T_{km} n_{i\,-\sigma} a_{m\sigma} + \sum_m T_{im} \left(a^+_{i\,-\sigma} a_{m\,-\sigma} - a^+_{m\,-\sigma} a_{i\,-\sigma}\right) a_{k\sigma} -$$
$$- \frac{1}{2} g\hbar z_\sigma S^z_k n_{i\,-\sigma} a_{k\sigma} - \frac{1}{2} g\hbar S^{-\sigma}_k n_{i\,-\sigma} a_{k\,-\sigma} -$$
$$- \frac{1}{2} g\hbar \left(S^\sigma_i a^+_{i\,-\sigma} a_{i\sigma} - S^{-\sigma}_i a^+_{i\sigma} a_{i\,-\sigma}\right) a_{k\sigma} .$$

We define several new Green's functions:

$$K_{imk,\,j\sigma}(E) = \langle\!\langle \left(a^+_{i\,-\sigma} a_{m\,-\sigma} - a^+_{m\,-\sigma} a_{i\,-\sigma}\right) a_{k\sigma} ; a^+_{j\sigma} \rangle\!\rangle_E ,$$
$$P^{(1)}_{ik,\,j\sigma}(E) = \langle\!\langle S^z_k n_{i\,-\sigma} a_{k\sigma} ; a^+_{j\sigma} \rangle\!\rangle_E ,$$
$$P^{(2)}_{ik,\,j\sigma}(E) = \langle\!\langle S^{-\sigma}_k n_{i\,-\sigma} a_{k\,-\sigma} ; a^+_{j\sigma} \rangle\!\rangle_E ,$$
$$P^{(3)}_{ik,\,j\sigma}(E) = \langle\!\langle \left(S^\sigma_i a^+_{i\,-\sigma} a_{i\sigma} - S^{-\sigma}_i a^+_{i\sigma} a_{i\,-\sigma}\right) a_{k\sigma} ; a^+_{j\sigma} \rangle\!\rangle_E .$$

With this, the complete equation of motion is given by:

$$\sum_m (E\delta_{km} - T_{km})\, P_{im,\,j\sigma}(E) =$$
$$= \hbar\delta_{kj} \langle n_{-\sigma} \rangle + \sum_m T_{im} K_{imk,\,j\sigma}(E) -$$
$$- \frac{1}{2} g\hbar z_\sigma P^{(1)}_{ik,\,j\sigma}(E) - \frac{1}{2} g\hbar \left(P^{(2)}_{ik,\,j\sigma}(E) + P^{(3)}_{ik,\,j\sigma}(E)\right) .$$

Solution 4.5.3

For the equation of motion, we require the following commutators:

$$[S^\sigma_i, H]_- = [S^\sigma_i, H_{\mathrm{f}}]_- + [S^\sigma_i, H_{\mathrm{s-f}}]_- ,$$
$$[S^\sigma_i, H_{\mathrm{f}}]_- = - \sum_{m,n} J_{mn} \left([S^\sigma_i, S^+_m S^-_n]_- + [S^\sigma_i, S^z_m S^z_n]_-\right) =$$
$$= - \sum_{m,n} J_{mn} \Big[\delta_{\sigma\downarrow}(-2\hbar S^z_i \delta_{im}) S^-_n +$$
$$+ S^+_m \delta_{\sigma\uparrow} \left(2\hbar S^z_i \delta_{in}\right) + S^z_m \left(-z_\sigma \hbar S^\sigma_i \delta_{in}\right) + \left(-z_\sigma \hbar S^\sigma_i \delta_{im}\right) S^z_n \Big] =$$
$$= 2\hbar z_\sigma \sum_m J_{im} \left(S^z_m S^\sigma_i - S^\sigma_m S^z_i\right) .$$

In the last step, we made use of $J_{ii} = 0$:

$$\left[S_i^\sigma, H_{\rm sf}\right]_- = -\frac{1}{2} g\hbar \sum_{m,\sigma'} \left(z_{\sigma'} \left[S_i^\sigma, S_m^z\right]_- n_{m\sigma'} + \left[S_i^\sigma, S_m^{\sigma'}\right]_- a_{m-\sigma'}^+ a_{m\sigma'} \right) =$$

$$= +\frac{1}{2} g\hbar^2 S_i^\sigma \left(n_{i\sigma} - n_{i-\sigma}\right) - g\hbar^2 z_\sigma S_i^z a_{i\sigma}^+ a_{i-\sigma} \, .$$

This is now combined with the commutator (4.395):

$$\begin{aligned}
\left[S_i^{-\sigma} a_{k-\sigma}, H\right]_- &= S_i^{-\sigma} \left[a_{k-\sigma}, H\right]_- + \left[S_i^{-\sigma}, H\right]_- a_{k-\sigma} = \\
&= \sum_m T_{km} S_i^{-\sigma} a_{m-\sigma} + U S_i^{-\sigma} n_{k\sigma} a_{k-\sigma} + \\
&+ \frac{1}{2} g\hbar z_\sigma S_i^{-\sigma} S_k^z a_{k-\sigma} - \frac{1}{2} g\hbar S_i^{-\sigma} S_k^\sigma a_{k\sigma} - \\
&- \frac{1}{2} g\hbar^2 S_i^{-\sigma} \left(n_{i\sigma} - n_{i-\sigma}\right) a_{k-\sigma} + \\
&+ g\hbar^2 z_\sigma S_i^z a_{i-\sigma}^+ a_{i\sigma} a_{k-\sigma} - \\
&- 2\hbar z_\sigma \sum_m J_{im} \left(S_m^z S_i^{-\sigma} - S_m^{-\sigma} S_i^z\right) a_{k-\sigma} \, .
\end{aligned}$$

We define the following *higher-order* Green's functions:

$$\begin{aligned}
F_{ik,\,j\sigma}^{(1)}(E) &= \langle\!\langle S_i^{-\sigma} S_k^z a_{k-\sigma}; a_{j\sigma}^+ \rangle\!\rangle_E \, , \\
F_{ik,\,j\sigma}^{(2)}(E) &= \langle\!\langle S_i^{-\sigma} S_k^\sigma a_{k\sigma}; a_{j\sigma}^+ \rangle\!\rangle_E \, , \\
F_{ik,\,j\sigma}^{(3)}(E) &= \langle\!\langle S_i^{-\sigma} (n_{i\sigma} - n_{i-\sigma}) a_{k-\sigma}; a_{j\sigma}^+ \rangle\!\rangle_E \, , \\
F_{ik,\,j\sigma}^{(4)}(E) &= \langle\!\langle S_i^{-\sigma} n_{k\sigma} a_{k-\sigma}; a_{j\sigma}^+ \rangle\!\rangle_E \, , \\
R_{ik,\,j\sigma}(E) &= \langle\!\langle S_i^z a_{i-\sigma}^+ a_{i\sigma} a_{k-\sigma}; a_{j\sigma}^+ \rangle\!\rangle_E \, , \\
Q_{imk,\,j\sigma}(E) &= \langle\!\langle \left(S_i^{-\sigma} S_m^z - S_m^{-\sigma} S_i^z\right) a_{k-\sigma}; a_{j\sigma}^+ \rangle\!\rangle_E \, .
\end{aligned}$$

The equation of motion:

$$\begin{aligned}
&\sum_m \left(E\delta_{km} - T_{km}\right) F_{im,\,j\sigma}(E) = \\
&= U F_{ik,\,j\sigma}^{(4)}(E) + \frac{1}{2} g\hbar \left(z_\sigma F_{ik,\,j\sigma}^{(1)}(E) - F_{ik,\,j\sigma}^{(2)}(E) \right) - \\
&- \frac{1}{2} g\hbar^2 \left(F_{ik,\,j\sigma}^{(3)}(E) - 2 z_\sigma R_{ik,\,j\sigma}(E) \right) - 2\hbar z_\sigma \sum_m J_{im} Q_{imk,\,j\sigma}(E) \, .
\end{aligned}$$

Solution 4.5.4

1. The exact equation of motion for the one-electron Green's function (i.e., See 4.395):

$$\sum_m (E\delta_{im} - T_{im})\, G_{mj\sigma}(E) =$$

$$= \hbar\delta_{ij} + U\, P_{ii,\, j\sigma}(E) - \frac{1}{2} g\hbar \left(z_\sigma D_{ii,\, j\sigma}(E) + F_{ii,\, j\sigma}(E)\right).$$

For the special case of ($n = 2$, $T = 0$), we can use the following:

$$D_{ii,\, j\sigma}(E) \equiv \langle\langle S_i^z a_{i\sigma}; a_{j\sigma}^+ \rangle\rangle_E \xrightarrow[(n=2,\, T=0)]{} \hbar S G_{ij\sigma}(E),$$

$$P_{ii,\, j\sigma}(E) \equiv \langle\langle n_{i-\sigma} a_{i\sigma}; a_{j\sigma}^+ \rangle\rangle_E \xrightarrow[(n=2,\, T=0)]{} G_{ij\sigma}(E).$$

We obtain the still exact, but – owing to ($n = 2$, $T = 0$) already greatly simplified – equation of motion:

$$\sum_m \left[\left(E - U + \frac{1}{2} g\hbar^2 S z_\sigma\right)\delta_{im} - T_{im}\right] G_{mj\sigma}(E) = \hbar\delta_{ij} - \frac{1}{2} g\hbar F_{ii,\, j\sigma}(E).$$

For the spin-flip function, we furthermore have:

$$F_{ii,\, j\downarrow}(E) \equiv \langle\langle S_i^+ a_{i\uparrow}; a_{j\downarrow}^+ \rangle\rangle \xrightarrow[(n=2,\, T=0)]{} 0.$$

This can best be seen from the time-dependent function:

$$F_{ii,\, j\downarrow}(t, t') = -\mathrm{i}\Theta(t - t')\, \langle E_0| \Big[\underbrace{\langle (S_i^+ a_{i\uparrow})(t) a_{j\downarrow}^+(t') \rangle}_{=0,\ \text{due to } n=2} + \underbrace{\langle a_{j\downarrow}^+(t') (S_i^+ a_{i\uparrow})(t) \rangle}_{=0,\ \text{due to } T=0} \Big] |E_0\rangle.$$

The remaining equation of motion can be readily solved by Fourier transformation:

$$G_{k\downarrow}^{(n=2,\, T=0)}(E) = \hbar \left[E - \varepsilon(k) - U - \frac{1}{2} g\hbar^2 S + \mathrm{i}0^+\right]^{-2}.$$

2. For $\sigma = \uparrow$ electrons, the spin-flip function is non-vanishing. Its equation of motion was calculated in Ex. 4.5.3:

$$\sum_m (E\delta_{km} - T_{km})\, F_{im,\, j\uparrow}(E) = U F_{ik,\, j\uparrow}^{(4)}(E) + \frac{1}{2} g\hbar \left(F_{ik,\, j\uparrow}^{(1)}(E) - F_{ik,\, j\uparrow}^{(2)}(E)\right) -$$

$$- \frac{1}{2} g\hbar^2 \left(F_{ik,\, j\uparrow}^{(3)}(E) - 2R_{ik,\, j\uparrow}(E)\right) -$$

$$- 2\hbar \sum_m J_{im}\, Q_{imk,\, j\uparrow}(E).$$

The *higher-order* Green's functions can be simplified to some extent due to the condition ($n = 2, T = 0$):

$$F^{(4)}_{ik,\,j\uparrow}(E) \equiv \langle\!\langle S_i^- n_{k\uparrow} a_{k\downarrow}; a^+_{j\uparrow} \rangle\!\rangle_E \xrightarrow[(n=2,\,T=0)]{} F_{ik,\,j\uparrow}(E)\,,$$

$$F^{(1)}_{ik,\,j\uparrow}(E) \equiv \langle\!\langle S_i^- S_k^z a_{k\downarrow}; a^+_{j\uparrow} \rangle\!\rangle_E \xrightarrow[(n=2,\,T=0)]{} \hbar S F_{ik,\,j\uparrow}(E)\,,$$

$$F^{(2)}_{ik,\,j\uparrow}(E) = \langle\!\langle S_i^- S_k^+ a_{k\uparrow}; a^+_{j\downarrow} \rangle\!\rangle_E \xrightarrow[(n=2,\,T=0)]{} 0\,,$$

$$F^{(3)}_{ik,\,j\uparrow}(E) \equiv \langle\!\langle S_i^- (n_{i\uparrow} - n_{i\downarrow}) a_{k\downarrow}; a^+_{j\uparrow} \rangle\!\rangle_E \xrightarrow[(n=2,\,T=0)]{} +\,\delta_{ik} F_{ik,\,j\uparrow}(E)\,,$$

$$R_{ik,\,j\uparrow}(E) \equiv \langle\!\langle S_i^z a^+_{i\downarrow} a_{i\uparrow} a_{k\downarrow}; a^+_{j\uparrow} \rangle\!\rangle_E \xrightarrow[(n=2,\,T=0)]{} -\,\delta_{ik} \hbar S G_{ij\uparrow}(E)\,,$$

$$Q_{imk,\,j\uparrow}(E) \equiv \langle\!\langle \left(S_i^- S_m^z - S_m^- S_i^z\right) a_{k\downarrow}; a^+_{j\uparrow} \rangle\!\rangle_E$$

$$\xrightarrow[(n=2,\,T=0)]{} \hbar S \left(F_{ikj,\,\uparrow}(E) - F_{mk,\,j\uparrow}(E)\right)\,.$$

This yields the greatly simplified equation of motion:

$$\left[E - U - \frac{1}{2} g\hbar^2 \left(S - \delta_{ik}\right)\right] F_{ik,\,j\uparrow}(E) =$$

$$= \sum_m T_{km} F_{im,\,j\uparrow}(E) - g\hbar^3 S \delta_{ik} G_{ij\uparrow}(E) + 2\hbar^2 S \sum_m J_{im} \left(F_{ik,\,j\uparrow}(E) - F_{mk,\,j\uparrow}(E)\right)\,.$$

The equation for the single-particle Green's function is also a part of this system:

$$\sum_m \left[\left(E - U + \frac{1}{2} g\hbar^2 S\right) \delta_{im} - T_{im}\right] G_{mj\uparrow}(E) = \hbar \delta_{ij} - \frac{1}{2} g\hbar F_{ii,\,j\uparrow}(E)\,.$$

For the solution of this system of equations, we apply the Fourier transformation defined in (4.412) and (4.413), which leads us in a manner quite analogous to (4.414) and (4.417) to the following equations:

$$\left(E - U + \frac{1}{2} g\hbar^2 S - \varepsilon(\boldsymbol{k})\right) G^{(2,0)}_{\boldsymbol{k}\uparrow}(E) = \hbar - \frac{1}{2} g\hbar \frac{1}{\sqrt{N}} \sum_{\boldsymbol{q}} F^{(2,0)}_{\boldsymbol{kq}\uparrow}(E)\,,$$

$$\left[E - U - \frac{1}{2} g\hbar^2 S - \varepsilon(\boldsymbol{k} - \boldsymbol{q}) + \hbar\omega(\boldsymbol{q})\right] F^{(2,0)}_{\boldsymbol{kq}\uparrow}(E) =$$

$$= -\frac{1}{2} g\hbar^2 \frac{1}{N} \sum_{\bar{\boldsymbol{q}}} F^{(2,0)}_{\boldsymbol{k}\bar{\boldsymbol{q}}\uparrow}(E) - g\hbar^3 S \frac{1}{\sqrt{N}} G^{(2,0)}_{\boldsymbol{k}\uparrow}(E)\,.$$

The spin-wave energies are defined as in (2.232). We abbreviate:

$$B^{(2)}_{\boldsymbol{k}}(E) = \frac{1}{N} \sum_{\boldsymbol{q}} \left[E - U - \frac{1}{2} g\hbar^2 S - \varepsilon(\boldsymbol{k} - \boldsymbol{q}) + \hbar\omega(\boldsymbol{q})\right]^{-1}\,.$$

We then find:

$$\frac{1}{\sqrt{N}}\sum_{q} F^{(2,0)}_{kq\uparrow}(E) = \frac{-g\hbar^3 S B^{(2)}_{k}(E)}{1+\frac{1}{2}g\hbar^2 B^{(2)}_{k}(E)} G^{(2,0)}_{k\uparrow}(E)\,.$$

This yields the equation of motion for the one-electron Green's function:

$$\left\{E - U + \frac{1}{2}g\hbar^2 S - \varepsilon(k) - \frac{\frac{1}{2}g^2\hbar^4 S B^{(2)}_{k}(E)}{1+\frac{1}{2}g\hbar^2 B^{(2)}_{k}(E)}\right\} G^{(2,0)}_{k\uparrow}(E) = \hbar\,.$$

From it, we finally obtain the self-energy:

$$\Sigma^{(2,0)}_{k\uparrow}(E) = U - \frac{1}{2}g\hbar^2 S\left(1 - \frac{g\hbar^2 B^{(2)}_{k}(E)}{1+\frac{1}{2}g\hbar^2 B^{(2)}_{k}(E)}\right).$$

With this, we have solved the problem; compare the result with (4.419). Further evaluation can be carried out as described in Sect. 4.5.4.

Solution 4.5.5

The Hartree-Fock approximation:

$$\begin{aligned} D_{ii,\,j\sigma}(E) &\longrightarrow \langle S^z\rangle\, G_{ij\sigma}(E)\,,\\ P_{ii,\,j\sigma}(E) &\longrightarrow \langle n_{-\sigma}\rangle\, G_{ij\sigma}(E)\,,\\ F_{ii,\,j\sigma}(E) &\longrightarrow 0\,. \end{aligned}$$

This simplified equation of motion,

$$\sum_{m}\left[\left(E - U\langle n_{-\sigma}\rangle + \frac{1}{2}g\hbar z_\sigma \langle S^z\rangle\right)\delta_{im} - T_{im}\right] G^{(\mathrm{HFA})}_{mj\sigma}(E) = \hbar\delta_{ij}\,,$$

can be readily solved through Fourier transformation:

$$G^{\mathrm{HFA}}_{k\sigma}(E) = \frac{\hbar}{E - \varepsilon(k) - U\langle n_{-\sigma}\rangle + \frac{1}{2}g\hbar z_\sigma\langle S^z\rangle + \mathrm{i}0^+}\,.$$

"Band limit" ($U = g = 0$) :	true,
Atomic limit ($\varepsilon(k) = T_0\ \forall k$) :	false,
($n = 0, T = 0$) :	true for $\sigma = \uparrow$, false for $\sigma = \downarrow$,
($n = 2, T = 0$) :	true for $\sigma = \downarrow$, false for $\sigma = \uparrow$.

The principal disadvantage of the Hartree-Fock approximation no doubt lies in its complete suppression of the spin-flip processes!

Section 5.1.4

Solution 5.1.1

$$[P_0, \mathcal{H}_0]_- = |\eta\rangle \langle \eta \mid \mathcal{H}_0 - \mathcal{H}_0 \mid \eta\rangle \langle \eta| = (\eta - \eta) |\eta\rangle \langle \eta| = 0 ,$$

since $\mathcal{H}_0$ is Hermitian,

$$[Q_0, \mathcal{H}_0]_- = [\mathbf{1} - P_0, \mathcal{H}_0]_- = -[P_0, \mathcal{H}_0]_- = 0 .$$

Solution 5.1.2

$$\begin{aligned}
\frac{d}{d\lambda} E_0(\lambda) &= \frac{d}{d\lambda} \langle E_0(\lambda) | H(\lambda) | E_0(\lambda) \rangle = \\
&= \langle E_0(\lambda) | v | E_0(\lambda) \rangle + \left\langle \frac{d}{d\lambda} E_0(\lambda) \middle| H(\lambda) \middle| E_0(\lambda) \right\rangle + \left\langle E_0(\lambda) \middle| H(\lambda) \middle| \frac{d}{d\lambda} E_0(\lambda) \right\rangle = \\
&= \langle E_0(\lambda) | v | E_0(\lambda) \rangle + E_0(\lambda) \left\langle \frac{d}{d\lambda} E_0(\lambda) \middle| E_0(\lambda) \right\rangle + E_0(\lambda) \left\langle E_0(\lambda) \middle| \frac{d}{d\lambda} E_0(\lambda) \right\rangle = \\
&= \langle E_0(\lambda) | v | E_0(\lambda) \rangle + E_0(\lambda) \frac{d}{d\lambda} \langle E_0(\lambda) | E_0(\lambda) \rangle = \\
&= \langle E_0(\lambda) | v | E_0(\lambda) \rangle .
\end{aligned}$$

With $\eta_0 = E_0(0)$, we then find:

$$\Delta E_0 = E_0 - \eta_0 = \int_0^\lambda d\lambda' \, \langle E_0(\lambda') | v | E_0(\lambda') \rangle .$$

Solution 5.1.3

1. Clearly, we have:

$$H_0 = \sum_{k\sigma} (H_{k\sigma})_0 ,$$

where in the basis

$$\left|\psi_{k\sigma}^{\alpha}\right\rangle = a_{k\sigma\alpha}^{+}\,|0\rangle\,; \quad \alpha = \mathrm{A,B}$$

the following holds:

$$(H_{k\sigma})_0 \equiv \begin{pmatrix} \varepsilon(\boldsymbol{k}) & t(\boldsymbol{k}) \\ t^*(\boldsymbol{k}) & \varepsilon(\boldsymbol{k}) \end{pmatrix},$$

$$\det\left[\eta - \left(H_{k\sigma}^{\alpha\beta}\right)_0\right] \stackrel{!}{=} 0$$

$$\Longrightarrow \quad \eta_{\pm}^{(0)}(\boldsymbol{k}) = \varepsilon(\boldsymbol{k}) \pm \left|t(\boldsymbol{k})\right|\,.$$

Eigenstates:

$$\begin{pmatrix} \mp\left|t(\boldsymbol{k})\right| & t(\boldsymbol{k}) \\ t^*(\boldsymbol{k}) & \mp\left|t(\boldsymbol{k})\right| \end{pmatrix} \begin{pmatrix} C_\mathrm{A} \\ C_\mathrm{B} \end{pmatrix} = 0\,,$$

$$C_\mathrm{A}^{\pm} = \pm\gamma C_\mathrm{B}\,; \quad \gamma = \frac{t(\boldsymbol{k})}{\left|t(\boldsymbol{k})\right|}\,.$$

Normalisation:

$$\left|\eta_{\pm}^{(0)}(\boldsymbol{k})\right\rangle = \frac{1}{\sqrt{2}}\left(a_{k\sigma\mathrm{A}}^{+} \pm \gamma a_{k\sigma\mathrm{B}}^{+}\right)|0\rangle\,.$$

Because of $(-\sigma, \mathrm{B}) \Longleftrightarrow (\sigma, \mathrm{A})$, the right-hand side is not really spin dependent!

2. First-order energy correction:

$$\left\langle\eta_{\pm}^{(0)}(\boldsymbol{k})\right|H_1\left|\eta_{\pm}^{(0)}(\boldsymbol{k})\right\rangle = \frac{1}{2}\left(-\frac{1}{2}g\left\langle S^z\right\rangle\right)\langle 0|\left(a_{k\sigma\mathrm{A}} \pm \gamma^* a_{k\sigma\mathrm{B}}\right)\sum_{\sigma'} z_{\sigma'}\,\cdot$$

$$\cdot\left(a_{k\sigma'\mathrm{A}}^{+}a_{k\sigma'\mathrm{A}} - a_{k\sigma'\mathrm{B}}^{+}a_{k\sigma'\mathrm{B}}\right)\left(a_{k\sigma\mathrm{A}}^{+} \pm \gamma a_{k\sigma\mathrm{B}}^{+}\right)|0\rangle =$$

$$= -\frac{1}{4}gz_\sigma\left\langle S^z\right\rangle\langle 0|\left(a_{k\sigma\mathrm{A}} \pm \gamma^* a_{k\sigma\mathrm{B}}\right)\left(a_{k\sigma\mathrm{A}}^{+} \mp \gamma a_{k\sigma\mathrm{B}}^{+}\right)|0\rangle =$$

$$= -\frac{1}{4}gz_\sigma\left\langle S^z\right\rangle\langle 0|\left(\mathbf{1} - |\gamma|^2\mathbf{1}\right)|0\rangle = 0$$

$$\Longrightarrow \quad \eta_{\pm}^{(1)}(\boldsymbol{k}) \equiv 0\,.$$

Second-order energy correction:

$$\begin{aligned}
&\left\langle \eta_-^{(0)}(\boldsymbol{k}) \middle| H_1 \middle| \eta_+^{(0)}(\boldsymbol{k}) \right\rangle = \\
&= -\frac{1}{4} g z_\sigma \left\langle S^z \right\rangle \langle 0| \left(a_{\boldsymbol{k}\sigma\mathrm{A}} - \gamma^* a_{\boldsymbol{k}\sigma\mathrm{B}}\right)\left(a^+_{\boldsymbol{k}\sigma\mathrm{A}} - \gamma a^+_{\boldsymbol{k}\sigma\mathrm{B}}\right) |0\rangle = \\
&= -\frac{1}{4} g z_\sigma \left\langle S^z \right\rangle \langle 0| \left(\mathbf{1} + |\gamma|^2 \mathbf{1}\right) |0\rangle = -\frac{1}{2} g z_\sigma \left\langle S^z \right\rangle
\end{aligned}$$

$$\Longrightarrow \quad \eta_\pm^{(2)}(\boldsymbol{k}) = \frac{\left|\left\langle \eta_\mp^{(0)}(\boldsymbol{k}) \middle| H_1 \middle| \eta_\pm^{(0)}(\boldsymbol{k}) \right\rangle\right|^2}{\eta_\pm^{(0)}(\boldsymbol{k}) - \eta_\mp^{(0)}(\boldsymbol{k})} = \pm \frac{1}{8} g^2 \frac{\langle S^z \rangle^2}{|t(\boldsymbol{k})|^2} .$$

Up to second order, Schrödinger perturbation theory thus yields:

$$\eta_\pm^{(S)}(\boldsymbol{k}) = \varepsilon(\boldsymbol{k}) \pm |t(\boldsymbol{k})| \pm \frac{1}{8} g^2 \frac{\langle S^z \rangle^2}{|t(\boldsymbol{k})|^2} + O\left(g^3\right) .$$

There are problems at the zone boundary, since $t(\boldsymbol{k})$ vanishes there.

3. The first-order energy correction of Brillouin-Wigner is the same as that of Schrödinger:

$$\eta_\pm^{(1)}(\boldsymbol{k}) \equiv 0 .$$

In second order, we have:

$$\begin{aligned}
\eta_\pm^{(2)}(\boldsymbol{k}) &= \frac{\left|\left\langle \eta_\mp^{(0)}(\boldsymbol{k}) \middle| H_1 \middle| \eta_\pm^{(0)}(\boldsymbol{k}) \right\rangle\right|^2}{\eta_\pm(\boldsymbol{k}) - \eta_\mp^{(0)}(\boldsymbol{k})} \\
\Longrightarrow \quad \eta_\pm^{(\mathrm{BW})}(\boldsymbol{k}) &= \eta_\pm^{(0)}(\boldsymbol{k}) + \frac{1}{4} g^2 \left\langle S^z \right\rangle^2 \frac{1}{\eta_\pm^{(\mathrm{BW})}(\boldsymbol{k}) - \eta_\mp^{(0)}(\boldsymbol{k})} \\
\Longrightarrow \quad & \left(\eta_\pm^{(\mathrm{BW})}(\boldsymbol{k})\right)^2 - \eta_\pm^{(\mathrm{BW})}(\boldsymbol{k}) \left(\eta_\pm^{(0)}(\boldsymbol{k}) + \eta_\mp^{(0)}(\boldsymbol{k})\right) = \\
&= \frac{1}{4} g^2 \left\langle S^z \right\rangle^2 - \eta_\pm^{(0)}(\boldsymbol{k}) \eta_\mp^{(0)}(\boldsymbol{k}) \\
\Longrightarrow \quad & \left(\eta_\pm^{(\mathrm{BW})}(\boldsymbol{k}) - \varepsilon(\boldsymbol{k})\right)^2 = \frac{1}{4} g^2 \left\langle S^z \right\rangle^2 + |t(\boldsymbol{k})|^2 , \\
\eta_\pm^{(\mathrm{BW})}(\boldsymbol{k}) &= \varepsilon(\boldsymbol{k}) \pm \sqrt{\frac{1}{4} g^2 \langle S^z \rangle^2 + |t(\boldsymbol{k})|^2} .
\end{aligned}$$

There is now no problem at the zone boundary; a splitting of $\left| g \langle S^z \rangle \right|$ appears there (*Slater gap*).

4. Exact eigenenergies:

$$H = \sum_{k\sigma} H_{k\sigma}\,,$$

$$H_{k\sigma} = \begin{pmatrix} \varepsilon(\boldsymbol{k}) - \frac{1}{2} g z_\sigma \langle S^z\rangle & t(\boldsymbol{k}) \\ t^*(\boldsymbol{k}) & \varepsilon(\boldsymbol{k}) + \frac{1}{2} g z_\sigma \langle S^z\rangle \end{pmatrix},$$

$$\det(E - H_{k\sigma}) \stackrel{!}{=} 0$$

$$\Longrightarrow \quad (E - \varepsilon(\boldsymbol{k}))^2 - \frac{1}{4} g^2 \langle S^z\rangle^2 = \left|t(\boldsymbol{k})\right|^2$$

$$\Longrightarrow \quad E_\pm(\boldsymbol{k}) = \varepsilon(\boldsymbol{k}) \pm \sqrt{\frac{1}{4} g^2 \langle S^z\rangle^2 + \left|t(\boldsymbol{k})\right|^2}\,.$$

Brillouin-Wigner perturbation theory is thus exact already to second order, whilst Schrödinger perturbation theory gives only the first term in the expansion of the root!

Section 5.2.3

Solution 5.2.1

1. We employ Wick's theorem:

$$T_\varepsilon\left\{a_{k\sigma}(t_1)\, a^+_{l\sigma'}(t_2)\, a_{m\sigma}(t_3)\, a^+_{n\sigma'}(t_3)\right\} =$$

$$= N\left\{a_{k\sigma}(t_1)\, a^+_{l\sigma'}(t_2)\, a_{m\sigma}(t_3)\, a^+_{n\sigma'}(t_3)\right\} +$$

$$+ \underbrace{a_{k\sigma}(t_1)\, a^+_{l\sigma'}(t_2)} N\left\{a_{m\sigma}(t_3)\, a^+_{n\sigma'}(t_3)\right\} +$$

$$+ \underbrace{a_{m\sigma}(t_3)\, a^+_{n\sigma'}(t_3)} N\left\{a_{k\sigma}(t_1)\, a^+_{l\sigma'}(t_2)\right\} +$$

$$+ \underbrace{a_{k\sigma}(t_1)\, a^+_{n\sigma'}(t_3)} N\left\{a^+_{l\sigma'}(t_2)\, a_{m\sigma}(t_3)\right\} +$$

$$+ \underbrace{a^+_{l\sigma'}(t_2)\, a_{m\sigma}(t_3)} N\left\{a_{k\sigma}(t_1)\, a^+_{n\sigma'}(t_3)\right\} +$$

$$+ \underbrace{a_{k\sigma}(t_1)\, a^+_{l\sigma'}(t_2)}\underbrace{a_{m\sigma}(t_3)\, a^+_{n\sigma'}(t_3)} +$$

$$+ \underbrace{a_{k\sigma}(t_1)\, a^+_{n\sigma'}(t_3)}\underbrace{a^+_{l\sigma'}(t_2)\, a_{m\sigma}(t_3)}\,.$$

Only the contractions between creation and annihilation operators can be non-vanishing!

2. The expectation value of a normal product in the ground state $|\eta_0\rangle$ is always zero:

$$\langle\eta_0|\,T_\varepsilon\left\{a_{k\sigma}(t_1)\,a^+_{l\sigma'}(t_2)\,a_{m\sigma}(t_3)\,a^+_{n\sigma'}(t_3)\right\}|\eta_0\rangle =$$

$$= \underbrace{a_{k\sigma}(t_1)\,a^+_{l\sigma'}}(t_2)\underbrace{a_{m\sigma}(t_3)\,a^+_{n\sigma'}}(t_3) + \underbrace{a_{k\sigma}(t_1)\,a^+_{n\sigma'}}(t_3)\underbrace{a^+_{l\sigma'}(t_2)\,a_{m\sigma}}(t_3) =$$

$$= -\delta_{kl}\delta_{mn}\delta_{\sigma\sigma'}G^{0,c}_{k\sigma}(t_1-t_3)\,G^{0,c}_{m\sigma}\left(0^-\right) + \delta_{kn}\delta_{lm}\delta_{\sigma\sigma'}G^{0,c}_{k\sigma}(t_1-t_3)\,G^{0,c}_{m\sigma}(t_3-t_2) =$$

$$= \delta_{\sigma\sigma'}\left[\delta_{kn}\delta_{lm}G^{0,c}_{k\sigma}(t_1-t_3) - \mathrm{i}\delta_{kl}\delta_{mn}G^{0,c}_{k\sigma}(t_1-t_3)\,\langle n_{m\sigma}\rangle^{(0)}\right].$$

Solution 5.2.2

We adopt the solution of the previous exercise:

$$\langle\eta_0|\,T_\varepsilon\left\{a_{k\sigma}(t_1)\,a^+_{k\sigma}(t_2)\,a_{k\sigma}(t_3)\,a^+_{k\sigma}(t_3)\right\}|\eta_0\rangle =$$
$$= \mathrm{i}G^{0,c}_{k\sigma}(t_1-t_3)\left[-\mathrm{i}G^{0,c}_{k\sigma}(t_3-t_2) - \langle n_{k\sigma}\rangle^{(0)}\right].$$

1. $t_1 > t_2 > t_3$:

$$\langle\eta_0|\,T_\varepsilon\left\{a_{k\sigma}(t_1)\,a^+_{k\sigma}(t_2)\,a_{k\sigma}(t_3)\,a^+_{k\sigma}(t_3)\right\}|\eta_0\rangle =$$
$$= \mathrm{e}^{-\frac{\mathrm{i}}{\hbar}(\varepsilon(k)-\mu)(t_1-t_3)}\left(1-\langle n_{k\sigma}\rangle^{(0)}\right)\langle n_{k\sigma}\rangle^{(0)}\left(\mathrm{e}^{-\frac{\mathrm{i}}{\hbar}(\varepsilon(k)-\mu)(t_3-t_2)}-1\right) = 0\,.$$

Check by direct computation:

$$\langle\eta_0|\,T_\varepsilon\left\{a_{k\sigma}(t_1)\,a^+_{k\sigma}(t_2)\,a_{k\sigma}(t_3)\,a^+_{k\sigma}(t_3)\right\}|\eta_0\rangle =$$
$$= -\langle\eta_0|\underbrace{a_{k\sigma}}_{\overleftarrow{=0\ \text{for}\ k\,\le\,k_\mathrm{F}}}(t_1)\,a^+_{k\sigma}(t_2)\underbrace{n_{k\sigma}}_{\overrightarrow{=0\ \text{for}\ k\,>\,k_\mathrm{F}}}(t_3)|\eta_0\rangle = 0\,.$$

2. $t_1 > t_3 > t_2$:

$$\begin{aligned}
&\left\langle \eta_0 \middle| T_\varepsilon \left\{a_{k\sigma}(t_1)\, a^+_{k\sigma}(t_2)\, a_{k\sigma}(t_3)\, a^+_{k\sigma}(t_3)\right\} \middle| \eta_0 \right\rangle = \\
&= \mathrm{e}^{-\frac{i}{\hbar}(\varepsilon(\boldsymbol{k})-\mu)(t_1-t_3)} \left(1 - \langle n_{k\sigma}\rangle^{(0)}\right) \cdot \\
&\quad \cdot \left[-\left(1-\langle n_{k\sigma}\rangle^{(0)}\right) \mathrm{e}^{-\frac{i}{\hbar}(\varepsilon(\boldsymbol{k})-\mu)(t_3-t_2)} - \langle n_{k\sigma}\rangle^{(0)}\right] = \\
&= -\left(1-\langle n_{k\sigma}\rangle^{(0)}\right) \mathrm{e}^{-\frac{i}{\hbar}(\varepsilon(\boldsymbol{k})-\mu)(t_1-t_2)} = \\
&= \begin{cases} 0 & \text{for } \; k \le k_\mathrm{F}\,, \\ -\exp\left[-\frac{i}{\hbar}(\varepsilon(\boldsymbol{k})-\mu)(t_1-t_2)\right] & \text{for } \; k > k_\mathrm{F}\,. \end{cases}
\end{aligned}$$

Check through direct computation:

$$\begin{aligned}
&\left\langle \eta_0 \middle| T_\varepsilon \left\{a_{k\sigma}(t_1)\, a^+_{k\sigma}(t_2)\, a_{k\sigma}(t_3)\, a^+_{k\sigma}(t_3)\right\} \middle| \eta_0 \right\rangle = \\
&= -\left\langle \eta_0 \middle| a_{k\sigma}(t_1)\, n_{k\sigma}(t_3)\, a^+_{k\sigma}(t_2) \middle| \eta_0 \right\rangle = \\
&= -\mathrm{e}^{-\frac{i}{\hbar}(\varepsilon(\boldsymbol{k})-\mu)(t_1-t_2)} \left\langle \eta_0 \middle| a_{k\sigma} n_{k\sigma} a^+_{k\sigma} \middle| \eta_0 \right\rangle = \\
&= -\left(1-\langle n_{k\sigma}\rangle^{(0)}\right) \mathrm{e}^{-\frac{i}{\hbar}(\varepsilon(\boldsymbol{k})-\mu)(t_1-t_2)}\,.
\end{aligned}$$

Section 5.3.4

Solution 5.3.1

For the vacuum amplitude, according to (5.92) we have from first-order perturbation theory:

$$\left\langle \eta_0 \middle| U_\alpha^{(1)}(t,t') \middle| \eta_0 \right\rangle = \overline{U}_1 \left(\frac{i}{2\hbar} \int_{t'}^{t} \mathrm{d}t_1\, \mathrm{e}^{-\alpha|t_1|}\right).$$

The integral over time can be easily computed:

$$\overline{U}_1 \equiv \sum_{kl} \langle n_k\rangle \langle n_l\rangle \left[v(kl;lk) - v(kl;kl)\right]\,.$$

1. Hubbard model

$$k \equiv (\boldsymbol{k}, \sigma_k), \ldots$$

From Ex. 4.1.1, we find for the interaction term:

$$V = \frac{1}{2}\sum_{klmn} v_{\mathrm{H}}(kl;nm) a_k^+ a_l^+ a_m a_n \,,$$

$$v_{\mathrm{H}}(kl;nm) \equiv \frac{U}{N}\delta_{\boldsymbol{k}+\boldsymbol{l},\,\boldsymbol{m}+\boldsymbol{n}}\delta_{\sigma_k\sigma_n}\delta_{\sigma_l\sigma_m}\delta_{\sigma_k\,-\sigma_l}\,.$$

One can see immediately that:

$$v_{\mathrm{H}}(kl;lk) = \frac{U}{N}\delta_{\boldsymbol{k}+\boldsymbol{l},\,\boldsymbol{k}+\boldsymbol{l}}\delta_{\sigma_k\sigma_l}\delta_{\sigma_l\sigma_k}\delta_{\sigma_k\,-\sigma_l} = 0\,,$$

$$v_{\mathrm{H}}(kl;kl) = \frac{U}{N}\delta_{\boldsymbol{k}+\boldsymbol{l},\,\boldsymbol{l}+\boldsymbol{k}}\delta_{\sigma_k\sigma_k}\delta_{\sigma_l\sigma_l}\delta_{\sigma_k\,-\sigma_l} = \frac{U}{N}\delta_{\sigma_k\,-\sigma_l}\,.$$

We thus have:

$$\overline{U}_1 = -\frac{U}{N}\sum_{\boldsymbol{k}\boldsymbol{l}\sigma}\langle n_{\boldsymbol{k}\sigma}\rangle\,\langle n_{\boldsymbol{l}\,-\sigma}\rangle = -\frac{U}{N}N_\sigma N_{-\sigma}\,,$$

$N_\sigma = \sum_{\boldsymbol{k}}\langle n_{\boldsymbol{k}\sigma}\rangle$ is the number of electrons with spin σ.

2. Jellium model

$$v_j(kl;nm) = v(\boldsymbol{k}-\boldsymbol{n})(1-\delta_{\boldsymbol{k}\boldsymbol{n}})\,\delta_{\boldsymbol{k}+\boldsymbol{l},\,\boldsymbol{m}+\boldsymbol{n}}\delta_{\sigma_k\sigma_n}\delta_{\sigma_m\sigma_l}\,.$$

For the special cases required here, this means that:

$$v_j(kl;lk) = v(\boldsymbol{k}-\boldsymbol{l})(1-\delta_{\boldsymbol{k}\boldsymbol{l}})\delta_{\sigma_k\sigma_l}\,,$$

$$v_j(kl;kl) = v(\boldsymbol{0})(1-\delta_{\boldsymbol{k}\boldsymbol{k}}) = 0\,.$$

Bubbles make no contribution!
We then find:

$$\overline{U}_1 = \sum_{\boldsymbol{k}\boldsymbol{l}\sigma} v(\boldsymbol{k}-\boldsymbol{l})(1-\delta_{\boldsymbol{k}\boldsymbol{l}})\,\langle n_{\boldsymbol{k}\sigma}\rangle\,\langle n_{\boldsymbol{l}\sigma}\rangle\,.$$

This term was explicitly evaluated in Sect. 2.1.2 (see (2.92)).

Solution 5.3.2

1. Contribution of the diagram according to the rules in Sect. 5.3.1:

$$(D) = \frac{1}{2!}\left(-\frac{\mathrm{i}}{2\hbar}\right)^2 \int_{t'}^{t} \cdots \int \mathrm{d}t_1\, \mathrm{d}t_1'\, \mathrm{d}t_2\, \mathrm{d}t_2'\, \delta\left(t_1 - t_1'\right) \delta\left(t_2 - t_2'\right) \mathrm{e}^{-\alpha(|t_1|+|t_2|)} \cdot$$

$$\cdot \sum_{\substack{k_1 l_1 m_1 n_1 \\ k_2 l_2 m_2 n_2}} v\left(k_1 l_1; n_1 m_1\right) v\left(k_2 l_2; n_2 m_2\right) (-1)^2 \cdot$$

$$\cdot \left[\mathrm{i}G_{l_1}^{0,\mathrm{c}}\left(t_2' - t_1'\right) \delta_{l_1 m_2}\right] \left[\mathrm{i}G_{n_1}^{0,\mathrm{c}}\left(t_1 - t_2'\right) \delta_{n_1 l_2}\right] \cdot$$

$$\cdot \left(-\left\langle n_{k_1}\right\rangle \delta_{k_1 m_1}\right) \left(-\left\langle n_{k_2}\right\rangle \delta_{k_2 n_2}\right) =$$

$$= \frac{1}{8\hbar^2} \iint_{t'}^{t} \mathrm{d}t_1\, \mathrm{d}t_2\, \mathrm{e}^{-\alpha(|t_1|+|t_2|)} \cdot$$

$$\cdot \sum_{k_1, l_1, n_1, k_2} v\left(k_1 l_1; n_1 k_1\right) v\left(k_2 n_1; k_2 l_1\right) \cdot$$

$$\cdot\, G_{l_1}^{0,\mathrm{c}}\left(t_2 - t_1\right) G_{n_1}^{0,\mathrm{c}}\left(t_1 - t_2\right) \left\langle n_{k_1}\right\rangle \left\langle n_{k_2}\right\rangle .$$

2. Hubbard model:

$$v_{\mathrm{H}}\left(k_1 l_1; n_1 k_1\right) = \frac{U}{N} \delta_{l_1, n_1} \delta_{\sigma_{k_1} \sigma_{n_1}} \delta_{\sigma_{l_1} \sigma_{k_1}} \delta_{\sigma_{k_1}\, -\sigma_{l_1}} = 0$$

$$\Longrightarrow \quad (D) = 0 .$$

3. Jellium model:

$$v_j\left(k_2 n_1; k_2 l_1\right) = v(\mathbf{0}) \left(1 - \delta_{k_2 k_2}\right) \delta_{n_1 l_1} \delta_{\sigma_{l_1} \sigma_{n_1}} = 0$$

$$\Longrightarrow \quad (D) = 0 .$$

Solution 5.3.3

In the following, the indices correspond to the diagram notation as in Sect. 5.3.1:

$$h(\Theta_1) = 8 \longrightarrow A(\Theta_1) = 1\,,$$

$$h(\Theta_2) = 4 \longrightarrow A(\Theta_2) = 1\,,$$
same contributions from the diagrams (2), (8),

$$h(\Theta_3) = 2 \longrightarrow A(\Theta_3) = 4\,,$$
same contributions from the diagrams (3), (6), (15), (22),

$$h(\Theta_4) = 1 \longrightarrow A(\Theta_4) = 8\,,$$
same contributions from the diagrams (4), (5), (9), (12), (13), (16), (20), (21),

$$h(\Theta_7) = 8 \longrightarrow A(\Theta_7) = 1\,,$$

$$h(\Theta_{10}) = 2 \longrightarrow A(\Theta_{10}) = 4\,,$$
same contributions from the diagrams (10), (11), (14), (19),

$$h(\Theta_{17}) = 4 \longrightarrow A(\Theta_{17}) = 2\,,$$
same contributions from the diagrams (17), (24),

$$h(\Theta_{18}) = 4 \longrightarrow A(\Theta_{18}) = 2\,,$$
same contributions from the diagrams (18), (23).

Section 5.4.3

Solution 5.4.1

For the electron-electron interaction, we find in the Hubbard model (cf. exercise 5.3.1):

$$v_{\mathrm{H}}(kl; nm) = \frac{U}{N} \delta_{\boldsymbol{k}+\boldsymbol{l},\,\boldsymbol{n}+\boldsymbol{m}} \delta_{\sigma_k \sigma_n} \delta_{\sigma_l \sigma_m} \delta_{\sigma_k\,-\sigma_l}\,.$$

The following two diagrams contribute in first order to the self-energy:

Fig. A.11

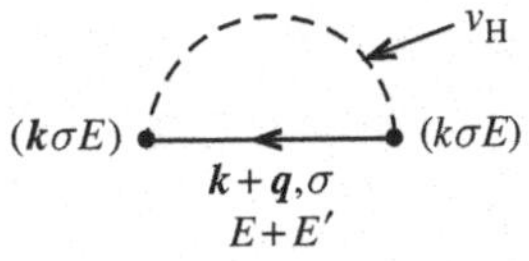

1.

$$k=(\boldsymbol{k}+\boldsymbol{q},\sigma)\,,\quad l=(\boldsymbol{k},\sigma)\,,\quad m=(\boldsymbol{k}+\boldsymbol{q},\sigma)\,,\quad n=(\boldsymbol{k},\sigma)$$

$$\Longrightarrow\quad v_{\mathrm{H}}(kl;nm)=0\quad\text{due to}\quad \delta_{\sigma_k-\sigma_l}=0\,.$$

Fig. A.12

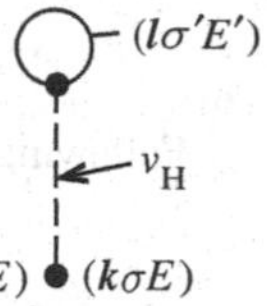

2.

$$k=(\boldsymbol{k},\sigma)\,,\quad l=\left(\boldsymbol{l},\sigma'\right)\,,\quad m=\left(\boldsymbol{l},\sigma'\right)\,,\quad n=(\boldsymbol{k},\sigma)\,,$$

$$v_{\mathrm{H}}(kl;nm)=\frac{U}{N}\delta_{\sigma-\sigma'}$$

$\Longrightarrow$ Contribution to the self-energy:

$$-\frac{\mathrm{i}}{\hbar}\Sigma^{(1)}_{\boldsymbol{k}\sigma}(E)=-\frac{\mathrm{i}}{\hbar}(-1)\frac{1}{2\pi\hbar}\frac{U}{N}\sum_{\boldsymbol{l}}\int \mathrm{d}E'\left(\mathrm{i}G^{0,\mathrm{c}}_{\boldsymbol{l}-\sigma}\left(E'\right)\right)$$

$$\Longrightarrow\quad \Sigma^{(1)}_{\boldsymbol{k}\sigma}(E)=-\frac{U}{N}\sum_{\boldsymbol{l}}\left(\mathrm{i}G^{0,\mathrm{c}}_{\boldsymbol{l}-\sigma}(0^-)\right)=\frac{U}{N}\sum_{\boldsymbol{l}}\langle n_{\boldsymbol{l}-\sigma}\rangle^{(0)}=U\,\langle n_{-\sigma}\rangle^{(0)}\,.$$

This yields the following causal single-particle Green's function:

$$G^{\mathrm{c}}_{\boldsymbol{k}\sigma}(E)=\frac{\hbar}{E-\left(\varepsilon(\boldsymbol{k})+U\,\langle n_{-\sigma}\rangle^{(0)}-\varepsilon_{\mathrm{F}}\right)\pm\mathrm{i}0^+}\,.$$

It is essentially identical to that of the $T=0$-Stoner model (4.23), and thus corresponds to the Hartree-Fock approximation of the equation of motion method. However, here $\langle n_{-\sigma}\rangle$ is the expectation value of the number operator for the *non*-interacting system. The same holds for the chemical potential, $\mu(T=0)=\varepsilon_{\mathrm{F}}$.

Solution 5.4.2

1. The annotation of the diagram is given by conservation of momentum and energy at the vertex, conservation of spin at the vertex point, and

$$v_{\mathrm{H}}(kl;nm)\sim\delta_{\sigma_k-\sigma_l}\,.$$

Fig. A.13

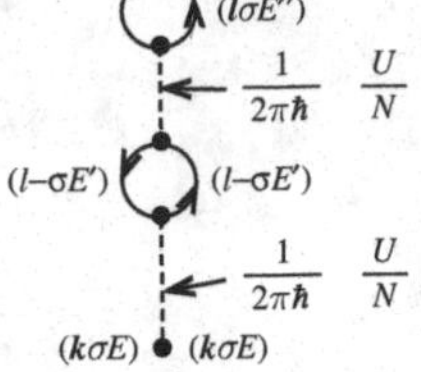

Following the diagram rules from Sect. 5.4.1, we still have to evaluate the following:

$$-\frac{\mathrm{i}}{\hbar}\Sigma_{k\sigma}^{(2,a)}(E) = \iint \mathrm{d}E'\mathrm{d}E'' \sum_{l,\bar{l}}(-1)^2\left(-\frac{\mathrm{i}}{\hbar}\right)^2\left(\frac{1}{2\pi\hbar}\frac{U}{N}\right)^2 \cdot$$
$$\cdot\left(\mathrm{i}G_{l-\sigma}^{0,\,c}(E')\right)^2\left(\mathrm{i}G_{\bar{l}\sigma}^{0,\,c}(E'')\right)$$
$$\Longrightarrow \quad \Sigma_{k\sigma}^{(2,a)}(E) = U^2\langle n_\sigma\rangle^{(0)}\left(-\frac{\mathrm{i}}{\hbar}\right)\frac{1}{N}\sum_l \frac{1}{2\pi\hbar}\int \mathrm{d}E'\left(\mathrm{i}G_{l-\sigma}^{0,\,c}(E')\right)^2 .$$

2. The annotation of the diagram is motivated as above.

Fig. A.14

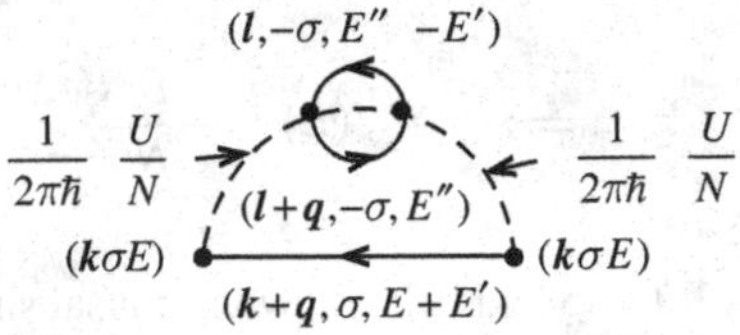

$$\Longrightarrow \quad \Sigma_{k\sigma}^{(2,b)}(E) = U^2\frac{\mathrm{i}}{\hbar}\frac{1}{N^2}\sum_{lq}\frac{1}{(2\pi\hbar)^2}\iint \mathrm{d}E'\mathrm{d}E''\left(\mathrm{i}G_{k+q\sigma}^{0,\,c}(E+E')\right)\cdot$$
$$\cdot\left(\mathrm{i}G_{l+q-\sigma}^{0,\,c}(E'')\right)\left(\mathrm{i}G_{l-\sigma}^{0,\,c}(E''-E')\right).$$

All of the other second-order diagrams are zero due to

$$v_{\mathrm{H}}(kl;nm) \sim \delta_{\sigma_k-\sigma_l} .$$

Solution 5.4.3

First-order perturbation theory:

$$\mathrm{i}G^{\mathrm{c}}_{k\sigma}(E) \approx \mathrm{i}G^{0,\mathrm{c}}_{k\sigma}(E) - \frac{\mathrm{i}}{\hbar} U \langle n_{-\sigma}\rangle^{(0)} \left(\mathrm{i}G^{0,\mathrm{c}}_{k\sigma}(E)\right)^2$$

$$\Longrightarrow \quad G^{\mathrm{c}}_{k\sigma}(E) \approx G^{0,\mathrm{c}}_{k\sigma}(E)\left[1 + \frac{1}{\hbar} U \langle n_{-\sigma}\rangle^{(0)} G^{0,\mathrm{c}}_{k\sigma}(E)\right].$$

Dyson equation (Ex. 5.4.1):

$$G^{\mathrm{c}}_{k\sigma}(E) \approx G^{0,\mathrm{c}}_{k\sigma}(E)\left[1 + \frac{1}{\hbar} U \langle n_{-\sigma}\rangle^{(0)} G^{\mathrm{c}}_{k\sigma}(E)\right].$$

First-order perturbation theory thus corresponds to the first term in the expansion of the infinite partial series, which is mediated by the Dyson equation.

Solution 5.4.4

Fig. A.15

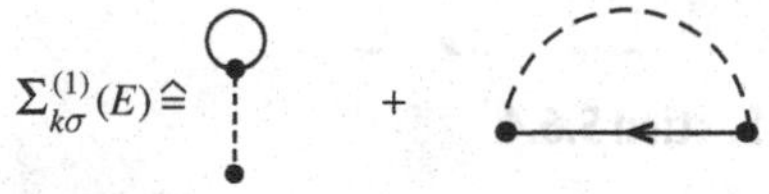

Via the Dyson equation, this gives the following diagrams for the one-electron Green's function up to second order:

Fig. A.16

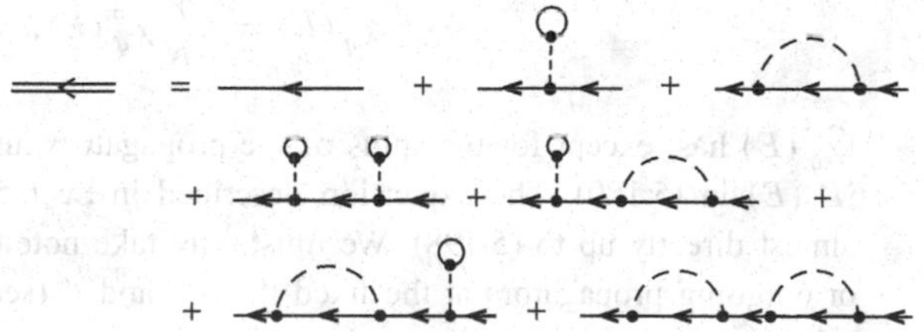

Solution 5.4.5

$$-\frac{\mathrm{i}}{\hbar}\widehat{\Sigma}^{(1)}_{k\sigma}(E) =$$

$$= -\frac{\mathrm{i}}{\hbar}\frac{1}{2\pi\hbar}\sum_{l,\sigma'}\int \mathrm{d}E'\left[-v(\boldsymbol{kl};\boldsymbol{kl})\left(\mathrm{i}G^{\mathrm{c}}_{l\sigma'}(E')\right) + v(\boldsymbol{lk};\boldsymbol{kl})\delta_{\sigma'\sigma}\left(\mathrm{i}G^{\mathrm{c}}_{l\sigma}(E')\right)\right].$$

We have:

$$\frac{1}{2\pi\hbar}\int dE'\, G^{c}_{l\sigma'}(E') = -i\left\langle T_{\varepsilon}\left[a_{l\sigma'}(t)c^{+}_{l\sigma'}(t+0^{+})\right]\right\rangle = +i\,\langle n_{l\sigma'}\rangle\ .$$

Here, we have made use of the *equal-time convention*:

$$\widehat{\Sigma}^{(1)}_{k\sigma}(E) = \sum_{l,\sigma'}\left[v(kl;kl) - v(lk;kl)\delta_{\sigma'\sigma}\right]\langle n_{l\sigma'}\rangle\ .$$

The difference compared to the solution of Ex. 5.4.4 consists *merely* in the fact that the expectation value of the occupation-number operator is now to be taken for the interacting system, not for the *free* system.

The *renormalisation* leads to a whole series of *new* diagrams, such as e.g.

Fig. A.17

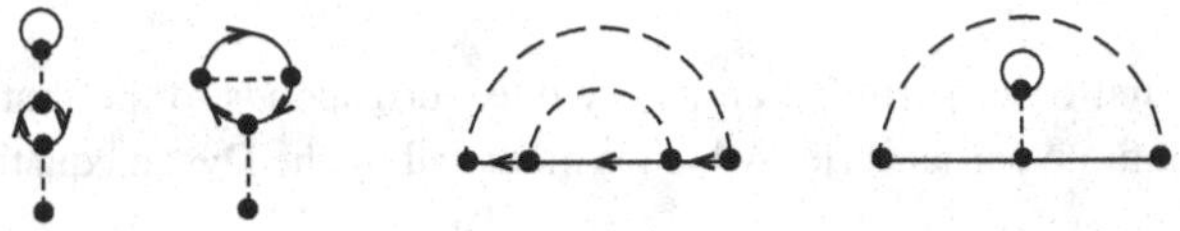

Section 5.6.4

Solution 5.6.1

1. We write:

$$\chi^{\pm}_{q}(E) = -\frac{\gamma}{N}\widehat{\chi}^{\pm}_{q}(E)\,.$$

$i\widehat{\chi}^{\pm}_{q}(E)$ has, except for the spins of the propagators involved, the same structure as $iD_q(E)$ in (5.180). The expansion described in Sect. 5.6 can therefore be adopted almost directly up to (5.198). We must only take note of the fact that the incoming or outgoing propagators at the fixed times t and t' (see e.g. (5.182)) have different spins.

Fig. A.18

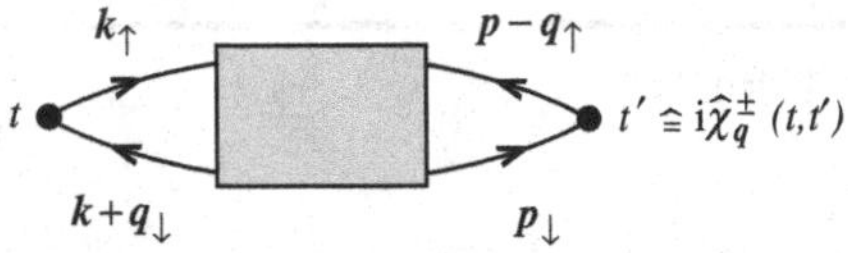

Following Fourier transformation to the energy domain, the second term in the Dyson equation which is analogous to (5.184) vanishes, since owing to conservation of spin at the vertex point at the endpoints of the above diagrams, *no* interaction line can be attached.

$$\Longrightarrow \quad \mathrm{i}\widehat{\chi}_{q}^{\pm}(E) \quad \triangleq$$

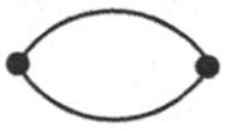

$\mathrm{i}\hbar\Lambda_{q\uparrow\downarrow}(E)$

2. The vertex function in the ladder approximation:

Fig. A.19

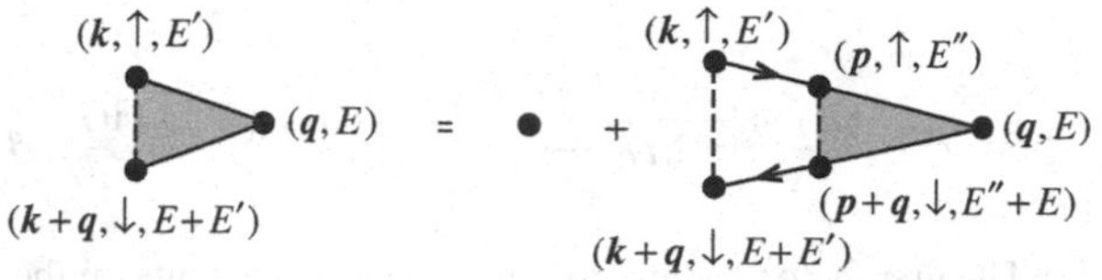

$$\Gamma_L^{\uparrow\downarrow}\left(qE;kE'\right) =$$

$$= 1 + \frac{1}{2\pi\hbar}\frac{U}{N}\left(-\frac{\mathrm{i}}{\hbar}\right)\sum_{p}\int \mathrm{d}E''\left(\mathrm{i}G_{p\uparrow}^{0,\mathrm{c}}\left(E''\right)\right)\left(\mathrm{i}G_{p+q\downarrow}^{0,\mathrm{c}}\left(E''+E\right)\right)\Gamma_L^{\uparrow\downarrow}\left(qE;pE''\right) .$$

Since in the Hubbard model, the interaction matrix element is a constant, the right-hand side is independent of (k, E'). This means that

$$\Gamma_L^{\uparrow\downarrow}\left(qE;kE'\right) \equiv \Gamma_L^{\uparrow\downarrow}(qE)$$

and therefore:

$$\Gamma_L^{\uparrow\downarrow}(qE) =$$

$$= 1 + \Gamma_L^{\uparrow\downarrow}(qE)\left\{\frac{\mathrm{i}}{\hbar}\frac{U}{N}\left(-\frac{1}{2\pi\hbar}\right)\sum_{p}\int \mathrm{d}E''\left(\mathrm{i}G_{p\uparrow}^{0,\mathrm{c}}\left(E''\right)\right)\left(\mathrm{i}G_{p+q\downarrow}^{0,\mathrm{c}}\left(E''+E\right)\right)\right\} =$$

$$= 1 + \Gamma_L^{\uparrow\downarrow}(qE)\left\{\frac{\mathrm{i}}{\hbar}\frac{U}{N}\left(\mathrm{i}\hbar\Lambda_{q\uparrow\downarrow}^{(0)}(E)\right)\right\} .$$

The ladder approximation for the Hubbard model can thus be summed exactly:

$$\Gamma_L^{\uparrow\downarrow}(qE) = \frac{1}{1 + \frac{U}{N}\Lambda_{q\uparrow\downarrow}^{(0)}(E)} .$$

3. It holds exactly that:

Fig. A.20

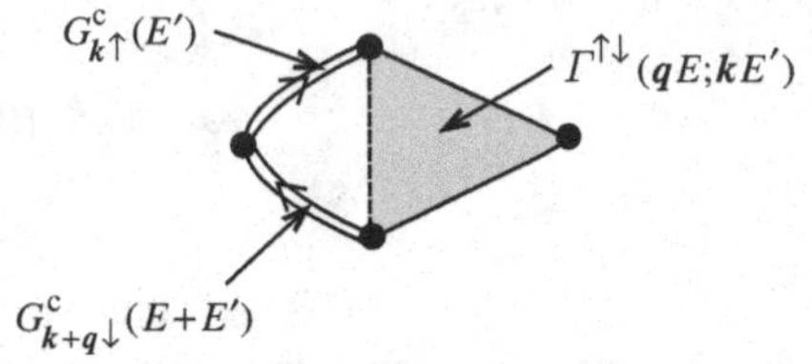

4.

$$\mathrm{i}\widehat{\chi}^{\pm}_{q}(E) \approx -\frac{1}{2\pi\hbar}\sum_{k}\int \mathrm{d}E' \left(\mathrm{i}G^{0,\mathrm{c}}_{k\uparrow}(E')\right)\left(\mathrm{i}G^{(0,\mathrm{c})}_{k+q\downarrow}(E+E')\right)\Gamma^{\uparrow\downarrow}_{L}(q,E).$$

The first factor comes from the outer attachments on the left!

$$\mathrm{i}\widehat{\chi}^{\pm}_{q}(E) \approx \mathrm{i}\hbar\Lambda^{(0)}_{q\uparrow\downarrow}(E)\Gamma^{\uparrow\downarrow}_{L}(q,E).$$

For the susceptibility, we then obtain:

$$\chi^{\pm}_{q}(E) = -\gamma\frac{\frac{\hbar}{N}\Lambda^{(0)}_{q\uparrow\downarrow}(E)}{1+\frac{U}{N}\Lambda^{(0)}_{q\uparrow\downarrow}(E)}.$$

Except for the factor $\left(-\frac{\gamma}{N}\right)$, $\Lambda^{(0)}_{q\uparrow\downarrow}$ is identical to the *free* susceptibility. The above result thus agrees with (4.183)! $\Lambda^{(0)}_{q\uparrow\downarrow}(E)$ was computed in ((5.192)).

Solution 5.6.2

 $: \; -\frac{\mathrm{i}}{\hbar}T_{k\sigma}(E)$

All the other symbols have the same meanings as in the text:
T-matrix equation:

Fig. A.21

$$\mathrm{i}G^{\mathrm{c}}_{k\sigma}(E) = \mathrm{i}G^{0,\mathrm{c}}_{k\sigma}(E) + \mathrm{i}G^{0,\mathrm{c}}_{k\sigma}(E)\left(-\frac{\mathrm{i}}{\hbar}T_{k\sigma}(E)\right)\mathrm{i}G^{0,\mathrm{c}}_{k\sigma}(E),$$

$$G^{\mathrm{c}}_{k\sigma}(E) = G^{0,\mathrm{c}}_{k\sigma}(E) + \frac{1}{\hbar}G^{0,\mathrm{c}}_{k\sigma}(E)T_{k\sigma}(E)G^{0,\mathrm{c}}_{k\sigma}(E).$$

Comparison with the Dyson equation:

Fig. A.22

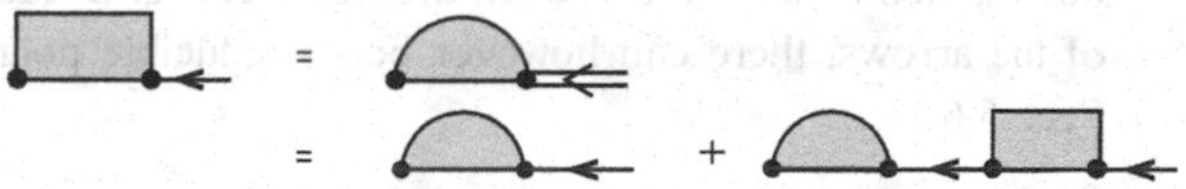

This implies that:

Fig. A.23

$$-\frac{\mathrm{i}}{\hbar}T_{k\sigma}(E) = -\frac{\mathrm{i}}{\hbar}\Sigma_{k\sigma}(E) + \left(-\frac{\mathrm{i}}{\hbar}\Sigma_{k\sigma}(E)\right)\mathrm{i}G^{0,\,\mathrm{c}}_{k\sigma}(E)\left(-\frac{\mathrm{i}}{\hbar}T_{k\sigma}(E)\right)$$

$$\Longrightarrow \quad T_{k\sigma}(E) = \frac{\Sigma_{k\sigma}(E)}{1 - \frac{1}{\hbar}G^{0,\,\mathrm{c}}_{k\sigma}(E)\Sigma_{k\sigma}(E)}\,.$$

Solution 5.6.3

The following two-particle spectral density is to be computed:

$$S^{(2)}_{ii\sigma}(E - 2\mu) = -\frac{1}{\pi}\frac{1}{N}\sum_{q}\mathrm{Im}\,\widehat{D}_q(E - 2\mu)\,.$$

Here, we have:

$$\mathrm{i}\widehat{D}_{q\sigma}(E) = \int\limits_{-\infty}^{+\infty}\mathrm{d}\left(t - t'\right)\,\mathrm{e}^{\frac{\mathrm{i}}{\hbar}E(t-t')}\,.$$

$$\cdot\sum_{\substack{kp\\ \sigma}}\left\langle E_0\right|T_\varepsilon\left\{a_{k\,-\sigma}(t)a_{q\,-\,k\sigma}(t)a^+_{q\,-\,p\sigma}\left(t'\right)a^+_{p\,-\sigma}\left(t'\right)\right\}\left|E_0\right\rangle\,.$$

1. The general diagram has the form:

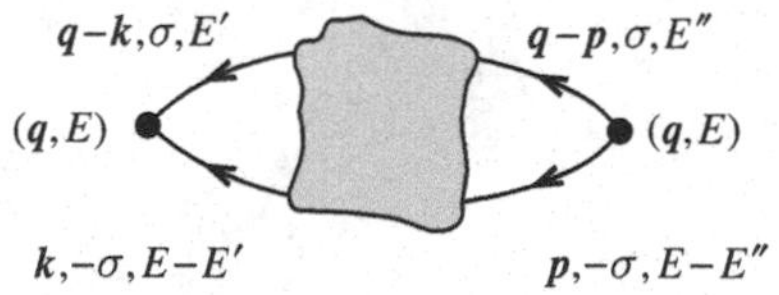

Fig. A.24

Except for the annotation and the directions of the arrows, we have the same diagram types as in the density correlation $D_q(E)$ in Sect. 5.6. The diagram rules correspond for the most part to those following (5.183) in Sect. 5.6.1. We merely have to index the outer attachments (Rule 4) as in the figure above. Because of the particular directions of the arrows, there can however be *no* reducible polarisation parts in the sense of Sect. 5.6.1.

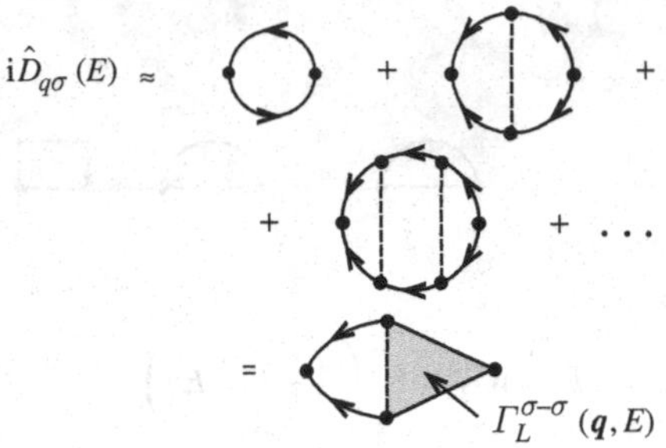

Fig. A.25

2.

$$i\hbar\widehat{\Lambda}^{(0)}_{q\sigma} = -\frac{1}{2\pi\hbar}\sum_{k}\int dE' \left(iG^{0,c}_{q-k\sigma}(E')\right)\left(iG^{0,c}_{k-\sigma}(E-E')\right).$$

As in Ex. 5.6.1, we find:

$$\Gamma_L^{\sigma-\sigma}(q,E) = \frac{1}{1+\frac{U}{N}\widehat{\Lambda}^{(0)}_{q\sigma}(E)}.$$

This yields:

$$\widehat{D}_{q\sigma}(E) = \hbar\frac{\widehat{\Lambda}^{(0)}_{q\sigma}(E)}{1+\frac{U}{N}\widehat{\Lambda}^{(0)}_{q\sigma}(E)}.$$

$\Lambda^{(0)}_{q\sigma}(E)$ is computed in complete analogy to (5.192).

3. Replace the *free* propagators in $\widehat{\Lambda}^{(0)}_{q\sigma}(E)$ by the *full* propagators!

Index

Page numbers in boldface indicate a definition, the formulation of a theorem, or an important statement. Italic page numbers refer to an exercise.